Current Topics in Behavioral Neurosciences

Volume 76

Current Topics in Behavioral Neurosciences provides critical and comprehensive discussions of the most significant areas of behavioral neuroscience research, written by leading international authorities. Each volume in the series represents the most informative and contemporary account of its subject available, making it an unrivalled reference source. Each volume will be made available in both print and electronic form.

With the development of new methodologies for brain imaging, genetic and genomic analyses, molecular engineering of mutant animals, novel routes for drug delivery, and sophisticated cross-species behavioral assessments, it is now possible to study behavior relevant to psychiatric and neurological diseases and disorders on the physiological level. The Behavioral Neurosciences series focuses on translational medicine and cutting-edge technologies. Preclinical and clinical trials for the development of new diagnostics and therapeutics as well as prevention efforts are covered whenever possible. Special attention is also drawn on epigenetical aspects, especially in psychiatric disorders.

CTBN series is indexed in PubMed and Scopus.

Ziva D. Cooper • Sachin Patel
Stephen P. H. Alexander
Editors

Behavioral Impact of Cannabinoids

Springer

Editors
Ziva D. Cooper
UCLA Center for Cannabis
and Cannabinoids, Jane & Terry
Semel Institute for Neuroscience
and Human Behavior, Department
of Psychiatry and Biobehavioral
Sciences, Department of Anesthesiology
and Perioperative Medicine
David Geffen School of Medicine
University of California
Los Angeles, CA, USA

Sachin Patel
Psychiatry and Behavioral Sciences
Northwestern University
Chicago, IL, USA

Stephen P. H. Alexander
School Of Life Sciences
University of Nottingham Medical School
Nottingham, UK

ISSN 1866-3370 ISSN 1866-3389 (electronic)
Current Topics in Behavioral Neurosciences
ISBN 978-3-032-09132-1 ISBN 978-3-032-09133-8 (eBook)
https://doi.org/10.1007/978-3-032-09133-8

This Springer imprint is published by the registered company Springer Nature Switzerland AG
The registered company address is: Gewerbestrasse 11, 6330 Cham, Switzerland

Preface

Although the *Cannabis* plant has been cultured by humans for thousands of years and consumed for both medicinal and non-medicinal intentions, there is a complexity about the plant and the actions of individual chemical constituents that is still being unravelled.

In this volume, we have engaged prominent authors to provide an overview of our current understanding of the Behavioural Neurobiology of the Endocannabinoid System. This edition highlights the progress in the topic area since the first volume in this series was published in 2009.

The organization of the chapters can be divided into two parts. In the first four chapters, there is coverage of the chemistry, pharmacology and molecular biology of the endocannabinoid system. The first is a reflection of the plant-derived cannabinoids and their molecular targets (Alexander), while the second chapter identifies the best characterized targets of the cannabinoids, the cannabinoid receptors (Ehrlich et al.). The third chapter reflects on the chemical probes available to explore the endocannabinoid system (Hanske et al.), and the fourth chapter describes the impact of genetic manipulation of the endocannabinoid system (Monory et al.).

The second group of chapters focuses on the neurobiology and neuropathology of the endocannabinoid system. The fifth chapter looks at the impact of ageing and neurodegeneration on the endocannabinoid system and vice versa (Fernandez-Ruiz et al.). The sixth and seventh chapters review the treatment of *Cannabis* use disorder (Bedillion et al.) and broader substance use disorders (Lujuan et al.). Subsequent chapters investigate the role of the endocannabinoid system in nausea and vomiting (Rock and Parker), pain (Harris et al.) and particular neurological disorders (Szejko et al.).

One of the features of the last 15 years, since the first volume in this series, is the improved quality of research in humans, not just in preclinical models, so many of these chapters are directed towards evidence in that area. This work is increasingly relevant with global increases in cannabis availability and use with widespread legalization. It is clear, however, that the impact of the endocannabinoid system on human behavioural neuroscience is still developing and we anticipate the advent of better targeted therapeutic agents exploiting this overlap in the future.

Los Angeles, CA, USA — Ziva D. Cooper
Chicago, IL, USA — Sachin Patel
Nottingham, UK — Stephen P. H. Alexander

Contents

The Pharmacological Profile of Plant-Derived Cannabinoids In Vitro 1
Stephen P. H. Alexander

Canonical Cannabinoid Receptors 37
Beth Ehrlich, Monica Patel, Xiaoxi Zheng, and Michelle Glass

Chemical Probes for Investigating the Endocannabinoid System 69
Annaleah Hanske, Marc Nazaré, and Uwe Grether

Genetic Tools in Rodents to Study Cannabinoid Functions 145
Krisztina Monory, Inigo Ruiz de Azua, and Beat Lutz

Ageing, Neurodegeneration and the Endocannabinoid System 209
Javier Fernández-Ruiz, Onintza Sagredo, María Gómez-Ruiz, and Eva de Lago

Pharmacotherapy for Cannabis Use Disorder: Preclinical and Clinical Models 249
Margaret F. Bedillion, Catherine F. Moore, Elise M. Weerts, Caroline A. Arout, Hannah M. Harris, and Margaret Haney

Cannabinoid-based Pharmacology for the Management of Substance Use Disorders 297
M. Á. Luján, Y. Kim, L. Y. Zhang, and J. F. Cheer

The Role of Cannabinoids and the Endocannabinoid System in the Treatment and Regulation of Nausea and Vomiting 339
Erin M. Rock and Linda A. Parker

Delta-9-tetrahydrocannabinol and Cannabidiol for Pain: Preclinical and Clinical Models 389
H. M. Harris, C. F. Moore, B. W. Jenkins, M. F. Bedillion, E. M. Weerts, and C. A. Arout

The Use of Cannabis-Based Medicine in Selected Neurological Disorders 433
Natalia Szejko, Kamila Saramak, and Kirsten R. Müller-Vahl

Correction to: Genetic Tools in Rodents to Study Cannabinoid Functions 455
Krisztina Monory, Inigo Ruiz de Azua, and Beat Lutz

The Pharmacological Profile of Plant-Derived Cannabinoids In Vitro

Stephen P. H. Alexander

Contents

1 The Genus *Cannabis* and Its Diversity 2
  1.1 A Brief History 2
  1.2 Biosynthetic Pathways of the Major Phytocannabinoids 3
  1.3 The Major Phytocannabinoids 5
2 Phytocannabinoids and the Endocannabinoid System 7
  2.1 Phytocannabinoid Interaction with Cannabinoid Receptors In Vitro 8
  2.2 Phytocannabinoid Interaction with Endocannabinoid-Related Enzymes In Vitro 8
3 Phytocannabinoid Effects at Non-canonical Targets of Endocannabinoids In Vitro 12
  3.1 Transient Receptor Potential Ion Channels 12
  3.2 Nuclear Hormone Receptors 15
4 Phytocannabinoid Effects at Other Mammalian Targets In Vitro 16
  4.1 Non-cannabinoid G Protein-Coupled Receptors 16
  4.2 Other Ligand-Gated Ion Channels 18
  4.3 Voltage-Gated Ion Channels 19
  4.4 Transporters 19
  4.5 Other Enzymes 22
  4.6 Catalytic Receptors 23
5 ADME Profile of the Major Phytocannabinoids 23
6 Conclusions 26
References 27

Abstract With over 100 apparently unique metabolites in the *Cannabis* plant, there are a plethora of potential molecular targets. The molecular actions of Δ^9-tetrahydrocannabinol are the best-investigated, which have centred mostly on a relatively high-potency, low-efficacy action at the CB_1 cannabinoid receptor. By contrast, cannabidiol has a multitude of lower potency molecular targets identified in vitro, but with limited clarity about their relevance for the effects of cannabidiol in

S. P. H. Alexander (✉)
Division of Pharmacology, Physiology and Neuroscience, School of Life Sciences, University of Nottingham Medical School, Nottingham, UK
e-mail: Steve.Alexander@nottingham.ac.uk

Curr Topics Behav Neurosci (2026) 76: 1–36
https://doi.org/10.1007/7854_2025_605
Published Online: 10 October 2025

humans. This review highlights the incomplete nature of studies of the major phytocannabinoids, particularly the acid phytocannabinoids.

Keywords Cannabidiol · Cannabinoids · Cannabis · Tetrahydrocannabinol

Abbreviations

ADME	Absorption distribution metabolism and excretion
CBC	Cannabichromene
CBCA	Cannabichromenic acid
CBD	Cannabidiol
CBDA	Cannabidiolic acid
CBDV	Cannabidivarin
CBDVA	Cannabidivarinic acid
CBG	Cannabigerol
CBGA	Cannabigerolic acid
CBN	Cannabinol
DAGL	Diacylglycerol lipase
FAAH	Fatty acid amide hydrolase
MAGL	Monoacylglycerol lipase
NAAA	*N*-Acylethanolamine acid amidase
NAPE-PLD	*N*-arachidonoylphosphatidylethanolamine-phospholipase D
THC	Δ^9-tetrahydrocannabinol
THCA	Δ^9-tetrahydrocannabinolic acid
THCV	Δ^9-tetrahydrocannabivarin
THCVA	Δ^9-tetrahydrocannabivarinic acid

1 The Genus *Cannabis* and Its Diversity

1.1 A Brief History

Plants belonging to the *Cannabis* genus have been associated with humans for millennia, with evidence for cultivation since 10,000 BCE in China (Russo 2007). Documentation of medicinal uses of *Cannabis* effectively mirrors written history with papyrus and sacred texts from Ancient Egypt and India. A large part of the beginnings of awareness of medicinal *Cannabis* in the West is often credited to William O'Shaughnessy who worked as an engineer in British-occupied India and wrote a book in the 1840s entitled the 'Bengal dispensatory and pharmacopoeia', which prompted some celebrity upon his return to the UK (O'Shaughnessy 1841). While medicinal *Cannabis* was adopted in many locations in the West in the nineteenth century, usage diminished in the early twentieth century as alternative medicines that were more controllable and/or capable of mass production took over market share. In many parts of Europe, North America, South America, Asia and Australasia, the last decade has seen a resurgence in access to medicinal *Cannabis*, which has prompted an expanded interest in both the plant as a 'natural remedy' and what has been referred to as a pharmacological treasure trove of individual component chemicals (Mechoulam 2005).

The chemical definition of *Cannabis* constituents in the mid-twentieth century allowed identification of the most investigated phytocannabinoids cannabidiol, CBD and Δ^9-tetrahydrocannabinol (THC) after the most stable of the phytocannabinoids, cannabinol (CBN), was isolated much earlier (Cahn 1932). The structure of CBD was claimed first following its isolation from Minnesota wild hemp (Adams et al. 1940); however, the structure was later corrected (Mechoulam and Shvo 1963). The structure of THC was identified following purification from contraband hashish (Gaoni and Mechoulam 1964a); Raphi Mechoulam suggested that, at the time, he thought he was isolating a steroid-like structure. At the same period, what is often described as a 'minor phytocannabinoid', cannabigerol, CBG, was also identified (Gaoni and Mechoulam 1964b). Over the next 10 years, cannabichromene, CBC (Gaoni and Mechoulam 1966a), cannabidivarin, CBDV (Vollner et al. 1969) and Δ^9-tetrahydrocannabivarin (THCV) (Merkus 1971) were all isolated from the *Cannabis* plant.

1.2 Biosynthetic Pathways of the Major Phytocannabinoids

The *Cannabis* genus expresses a seemingly unique enzyme series, which is able to generate apparently unique metabolites, more than 100, the majority of which are resorcinols. The enzyme pathways responsible for phytocannabinoid synthesis are illustrated in Fig. 1. Geranyl diphosphate is a common intermediate in the production of farnesyl diphosphate and geranylgeranyl diphosphate, which themselves are precursors for terpenes in plants. Naturally occurring terpenes, of which there are thousands identified, are generally insoluble in water, but are often volatile and provide aromas characteristic of herbs and other plants. Examples include limonene, myrcene, β-caryophyllene, linalool and pinene, which themselves are found in the *Cannabis* plant (Russo 2011) and are reported to be biologically active.

The first committing step in phytocannabinoid synthesis is conducted by a poorly characterized enzyme identified as geranylphosphate:olivetolate geranyltransferase (Got, Enzyme 1 in Fig. 1) (Fellermeier and Zenk 1998). Alongside geranyl diphosphate, a co-substrate is olivetolic acid, which the enzyme utilizes to displace the terminal inorganic pyrophosphate. In Fig. 1, the alkyl substituent denoted as 'R' for olivetolic acid is a pentyl sidechain. A lower abundant metabolite with a propyl sidechain can substitute for olivetolic acid in this reaction, divarinic acid. The organic product of combining geranyl diphosphate and olivetolic acid is cannabigerolic acid (CBGA), while cannabigerovarinic acid (CBGVA) results from enzymatic combination of geranyldiphosphate and divarinic acid (Fig. 1).

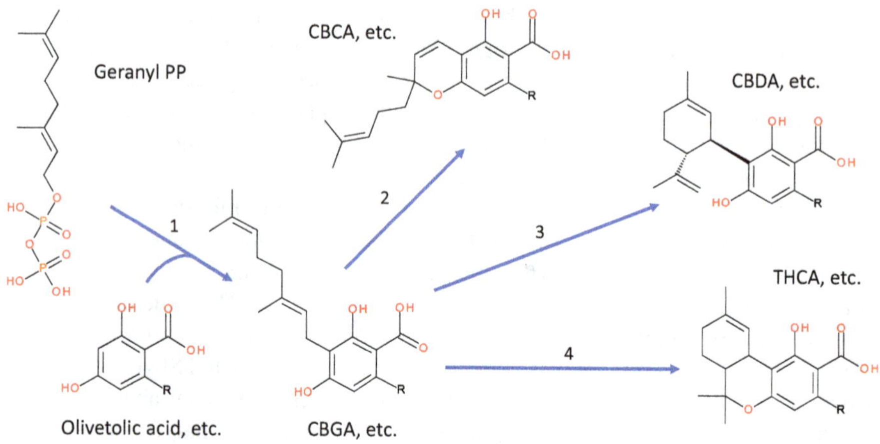

Fig. 1 Unique enzymatic pathways in the *Cannabis* plant. Enzyme 1 is geranylphosphate: olivetolate geranyltransferase; Enzyme 2 is cannabichromenic acid synthase. Enzyme 3 is cannabidiolic acid synthase. The last enzyme (Enzyme 4) is tetrahydrocannabinolic acid synthase

Thus, dependent on the co-substrate available, at least two distinct families of phytocannabinoids are generated in the *Cannabis* plant derived from CBGA and CBGVA, where the pentyl analogues are regarded as generally the most abundant (Fig. 1). Further variations in these sidechains have been described as low abundance metabolites, including butyl (Linciano et al. 2020a) and hexyl (Linciano et al. 2020b) analogues. Other minor metabolites in the *Cannabis* plant which may be products of biosynthesis or ageing of the preparation (see below) include two four-ring structures, cannabicyclol (Crombie et al. 1968) and cannabicitran (Wood et al. 2023), which may not be exclusive to *Cannabis* (Iwata and Kitanaka 2011). Other metabolites, such as cannabielsohn (Yamamoto et al. 1988) and cannabimovone (Iannotti et al. 2020), illustrate the complexity of secondary metabolite production in *Cannabis*.

Enzyme 2 in Fig. 1 is cannabichromenic acid synthase (Cbcas), which is capable of producing cannabichromenic acid (CBCA) and cannabichromevarinic acid (CBCVA). Enzyme 3 in Fig. 1 is cannabidiolic acid synthase (Cbdas), which is able to produce cannabidiolic acid (CBDA) and cannabidivarinic acid (CBDVA). The last enzyme (Enzyme 4 in Fig. 1) is tetrahydrocannabinolic acid synthase (Thcas), which can produce Δ^9-tetrahydrocannabinolic acid (THCA) and Δ^9-tetrahydrocannabivarinic acid (THCVA).

While the *Cannabis* genus is often identified as having three species, *C. sativa*, *C. indica* and *C. ruderalis*, the precise mapping of metabolites and genetic profiling to these three species has proved contentious (Lapierre et al. 2023). This may well be the product of the extended human history of living with *Cannabis* plants. For distinguishing medicinal/investigational uses of *Cannabis* preparations, there is often a pragmatic division of sources into three groups on the basis of THC:CBD ratios of abundance. These three groups (THC > CBD, THC=CBD and

THC < CBD) were analysed in over 80 plants of 21 *Cannabis* 'strains' from the Canadian market for multiple secondary metabolites, including flavonoids, sterols and terpenoids (Jin et al. 2021b). The study suggested a good separation between THC > CBD and THC < CBD 'strains' but also concluded that further analysis in the context of different growing environments was warranted. The same group conducted a genetic analysis, whole genome sequencing, of 23 Canadian *Cannabis* strains, which identified that more than one-third of the influential single nucleotide polymorphisms were found on the same chromosome as the genes encoding *Cbdas* and *Thcas* (Jin et al. 2021a). The five clusters of variants split into three THC > CBD groups, one THC=CBD and one THC < CBD group. Given that the USA has a legal definition to profile *Cannabis* strains as 'hemp' or marijuana' based on THC content, there does, therefore, appear to be good genetic correlates to separate them (Cheng and Houston 2024).

1.3 The Major Phytocannabinoids

From a chemistry perspective, the phytocannabinoids depicted in Figs. 1 and 2 have features in common and other elements that distinguish them. CBGA, CBCA, CBDA and THCA are all meroterpenoid, also known as terpenophenolics. Meroterpenoids are also found in the plant most closely related to *Cannabis*; humulene is an example from the hops plant (*Humulus* sp.), also in the family *Cannabaceae*. Humulene is structurally quite distinct from the phytocannabinoids, more similar to β-caryophyllene (see the discussion of the terpenoids above) as it is an 11-carbon-ring hydrocarbon. CBCA (and CBC, Fig. 2) has two rings fused together, while in CBDA (and CBD, Fig. 2), the two rings are separated. In

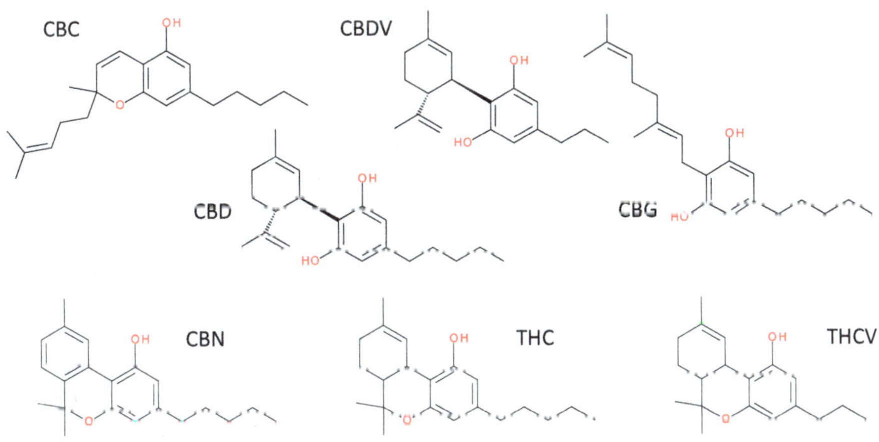

Fig. 2 Structures of the major neutral phytocannabinoids. CBC cannabichromene, CBD cannabidiol, CBDV cannabidivarin, CBG cannabigerol, CBN cannabinol, THC Δ^9-tetrahydrocannabinol, THCV Δ^9-tetrahydrocannabivarin

Table 1 Physicochemical properties of the major phytocannabinoids

Phytocannabinoid acids	CBGA	CBCA	CBDA	THCA	
Molecular formula	$C_{22}H_{32}O_4$	$C_{22}H_{30}O_4$	$C_{22}H_{30}O_4$	$C_{22}H_{30}O_4$	
Molecular weight g/mol	360.5	358.5	358.5	358.5	
XLogP3-AA	7.5	6.9	6.6	7.0	
Hydrogen bond donor count	3	2	3	2	
Hydrogen bond acceptor count	4	4	4	4	
Rotatable bond count	10	8	7	5	
Neutral phytocannabinoids	CBG	CBC	CBD	THC	CBN
Molecular formula	$C_{21}H_{32}O_2$	$C_{21}H_{30}O_2$	$C_{21}H_{30}O_2$	$C_{21}H_{30}O_2$	$C_{21}H_{26}O_2$
Molecular weight, g/mol	316.5	314.5	314.5	314.5	310.4
XLogP3-AA	7.4	6.9	6.5	7.0	6.1
Hydrogen bond donor count	2	1	2	1	1
Hydrogen bond acceptor count	2	2	2	2	2
Rotatable bond count	9	7	6	4	4

Data are from PubChem

THCA (and THC, Fig. 2), there are three rings fused together, which is indicative of limited flexibility in the molecule. This is in contrast to CBD where the separated rings allow for rotation around the linking bond (4 vs 6 rotatable bonds, Table 1). This flexibility has been used to rationalise the broad, 'non-specific' nature of CBD in comparison to THC (see Sect. 3).

The three enzymes that metabolise CBGA as a common substrate (identified as 2, 3 and 4 in Fig. 1) are all 62 kDa members of an oxygen-dependent flavin adenine dinucleotide-linked oxidoreductase family. The balance between their activities will determine the metabolite content of different strains/chemovars of *Cannabis*.

One feature to note about these natural phytocannabinoids is that they are acids. Storage and drying/heating allow decarboxylation of most of these acids, so the most frequently consumed versions are neutral: CBGA to CBG (cannabigerol, Fig. 2); CBCA gives rise to CBC (cannabichromene, Fig. 2); CBCVA to CBCV (cannabichromevarin); CBDA to CBD (cannabidiol, Fig. 2); CBDVA to CBDV (cannabidivarin, Fig. 2); THCA to THC (Δ^9-tetrahydrocannabinol, Fig. 2) and THCVA produces THCV (Δ^9-tetrahydrocannabivarin, Fig. 2).

Table 1 describes some of the physiochemical attributes of the major acid and neutral phytocannabinoids. It illustrates the identical molecular formulae and molecular weights of three acids (CBCA, CBDA and THCA) as a group, and three neutral phytocannabinoids (CBC, CBD and THC) as a second group. This commonality of physicochemical properties underlines the difficulties associated with the isolation of individual phytocannabinoids from the complex mixtures of plant preparations. A reservation to be noted when interpreting data from apparently 'purified' phytocannabinoids is that the purity of the phytocannabinoid acids may be a particular problem, with one study showing 2% contamination of a THCA sample with THC (McPartland et al. 2017), which may be abundant enough to distort the apparent pharmacological activity.

In drug discovery, a common consideration is whether drug candidates conform to Lipinski's rule of five, which were compared to the oral bioavailability of licensed drugs (Lipinski et al. 2001). These phytocannabinoids (Table 1) comply with Lipinski's rules, with the exception of hydrophobicity, which is approximately 7.0 for all of the compounds – in an octanol:water partition, ten million molecules would be found in the organic phase for every molecule in the aqueous phase. This limited water solubility suggests the likelihood for binding to protein partners in biological matrices.

A consequence of the inevitable variation in genetic background, growth conditions, harvesting times, preparation handling, etc., means that the abundance of phytocannabinoids can vary hugely between different preparations. This has implications for the interpretation of 'real world' experiences of *Cannabis* consumption as to whether therapeutic (or unwanted) effects are consistent and predictable. However, in recent years, there are numerous commercial suppliers where quality control of *Cannabis* preparations is more evident. At the same time, it should be noted that nabiximols, a licensed product approved in many countries for therapeutic use, is composed of 40% CBD, 40% THC and 20% 'other' unspecified *Cannabis*-derived compounds, including minor cannabinoids, terpenoids, and sterols (Russo and Guy 2006).

'Excessive' heating/oxidation of plant sources can generate a further metabolite of THC, CBN (cannabinol, Fig. 2), which has three fused rings, two of which are fully unsaturated, implying a very limited structural flexibility. A structural analogue of Δ^9-THC, Δ^8-THC is also found at minor abundance in the *Cannabis* plant (Hively et al. 1966). It may also be chemically synthesised from CBD in the presence of acid (Gaoni and Mechoulam 1966b; Watanabe et al. 2007). In many jurisdictions, there is uncertainty about the legal status of Δ^8-THC and so it has become more prevalent in the wider community in the last decade (Harlow et al. 2024).

Given the energy costs to the plant of synthesizing this broad range of secondary metabolites, there is an expectation that some evolutionary advantage accrues from this bounty. CBC, CBD, CBG, CBN and THC all show antibacterial effects in vitro (Appendino et al. 2008), while crude plant extracts exhibit antiviral (Sea et al. 2023), antibacterial and antifungal (Serventi et al. 2023) properties. The projection, therefore, is that many of the compounds described are of protective value for the plant.

2 Phytocannabinoids and the Endocannabinoid System

Following the identification of stereoselective binding sites for cannabinoids in the brain (Devane et al. 1988), cannabinoid receptors were cloned in 1990 from rat brain (Matsuda et al. 1990) and human brain (Gerard et al. 1990). Within 3 years, a second cannabinoid receptor was cloned from peripheral immune tissues (Munro et al. 1993). These were named CB_1 and CB_2 cannabinoid receptors, respectively (Howlett et al. 2002). Further information about these cannabinoid receptors can be found in an accompanying review (Ehrlich et al. 2025).

At around the same time as these receptors were recognised, two arachidonic acid derivatives, anandamide (Devane et al. 1992) and 2-arachidonoylglycerol (Sugiura et al. 1995), were identified as endogenous agonists for these two receptors. These two endocannabinoids have parallel, independent pathways for synthesis and hydrolysis. Further detail on the endocannabinoid system and probes for examining its functional roles can be found in an adjacent publication (Hanske et al. 2025).

2.1 Phytocannabinoid Interaction with Cannabinoid Receptors In Vitro

THC, the most (in)famous phytocannabinoid, is a partial agonist at the human recombinant CB_1 cannabinoid receptor (Gerard et al. 1991), with lower intrinsic activity at recombinant CB_2 cannabinoid receptor (Bayewitch et al. 1996). Tables 2 and 3 identify that of the major phytocannabinoids, consistently the most potent agonist at CB_1 and CB_2 receptors is THC.

Of the other phytocannabinoids, CBD and CBDA both appear to function as either negative allosteric modulators or antagonists of the CB_1 cannabinoid receptor (Table 2). THCV has a complicated profile with high affinity displacement of binding to CB_1 and CB_2 receptors, but antagonist actions at CB_1 receptors and pathway-dependent partial agonist action at CB_2 receptors (Tables 2 and 3).

2.2 Phytocannabinoid Interaction with Endocannabinoid-Related Enzymes In Vitro

Both cannabinoid receptors and the enzymes that synthesize and hydrolyse the endocannabinoids would be anticipated to share a common pharmacophore, so it would be a cogent hypothesis that the endocannabinoid-metabolizing enzymes might be sensitive to the phytocannabinoids. Further information about probes targeting these enzymes can be found in an accompanying review (Hanske et al. 2025).

N-Arachidonoylphosphatidylethanolamine-phospholipase D (NAPE-PLD) is able to generate the endocannabinoid anandamide, and other *N*-acylethanolamines (Sugiura et al. 1996), with selective inhibitors described, such as LEI-401 (Mock et al. 2020). However, there don't appear to be reports of the effects of phytocannabinoids on NAPE-PLD activity.

The other family of endocannabinoids, 2-arachidonoylglycerol and its analogues, are generated by diacylglycerol lipases (DAGL) α and β (Bisogno et al. 2003), for which selective inhibitors have also been described (Baggelaar et al. 2015). In the few instances where phytocannabinoids have been investigated as potential inhibitors of DAGLα, none of the compounds had an effect (Table 4).

Table 2 Phytocannabinoid action at CB_1 cannabinoid receptors in vitro

Neutral ligand	Binding affinity (pK_i value)	Agonist potency (pEC_{50} value)	Acid ligand	Binding affinity (pK_i value)	Agonist potency (pEC_{50} value)
CBC	5.8 (Rosenthaler et al. 2014) 8.0 (Zagzoog et al. 2020)	<4.5 (Udoh et al. 2019) 6.7[a] (Zagzoog et al. 2020)	CBCA	<10% displacement at 10 μM (Santillo et al. 2024)	
CBD	6.8 (Husni et al. 2014) 5.8 (Navarro et al. 2020) 6.7 (Zagzoog et al. 2020)	5.8 (Husni et al. 2014) <5.0 (Zagzoog et al. 2020) NAM[c] activity at 0.5–1 μM (Laprairie et al. 2015)	CBDA	6.0[d] (Bolognini et al. 2013) <5.5 (McPartland et al. 2017) 6.2 (Navarro et al. 2020) <5.0 (Zagzoog et al. 2020)	5.7[e] (Bolognini et al. 2013) 7.5 (Zagzoog et al. 2020)
CBDV	6.3 (Husni et al. 2014) 4.8 (Navarro et al. 2020) <5.0 (Zagzoog et al. 2020)	<5.0 (Husni et al. 2014) <5.0 (Zagzoog et al. 2020)	CBDVA		
CBG	5.5 (Husni et al. 2014) 5.5 (Navarro et al. 2020) 5.9 (Zagzoog et al. 2020)	<5.0 (Husni et al. 2014) 6.9[a] (Zagzoog et al. 2020)	CBGA	5.3 (Husni et al. 2014) 4.9 (Navarro et al. 2020)	6.7 (Husni et al. 2014)
CBGV	5.5 (Navarro et al. 2020)		CBGVA		
CBN	5.9 (Felder et al. 1995) 7.1 (Husni et al. 2014)	6.9[b] (Rhee et al. 1997) 6.5 (Husni et al. 2014)			
THC	7.3 (Felder et al. 1995) 7.7 (Husni et al. 2014) 7.4 (Zagzoog et al. 2020)	7.8[a] (Felder et al. 1995) 8.0[a,b] (Rhee et al. 1997) 6.6 (Husni et al. 2014) 6.6 (Zagzoog et al. 2020)	THCA	5.9 (Husni et al. 2014) 7.2 (Rosenthaler et al. 2014) 5.5 (McPartland et al. 2017) 6.6 (Palomares et al. 2020) 8.9 (Zagzoog et al. 2020)	<5.0 (Husni et al. 2014) 5.4 (Palomares et al. 2020) <5.0 (Zagzoog et al. 2020)
THCV	7.1[d] (Thomas et al. 2005) 7.7 (Husni et al. 2014) 7.7 (Zagzoog et al. 2020)	7.0[d,e] (Thomas et al. 2005) <5.0 (Husni et al. 2014) 6.6[a] (Zagzoog et al. 2020)	THCVA		

[a]Partial agonist activity. [b]Rat receptor. [c]Negative allosteric modulator. [d]Mouse receptor. [e]Antagonist activity. Unless otherwise indicated, the proteins are from human

Table 3 Phytocannabinoid action at CB_2 cannabinoid receptors in vitro

Neutral ligand	Binding affinity (pK_i value)	Agonist potency (pEC_{50} value)	Acid ligand	Binding affinity (pK_i value)	Agonist potency (pEC_{50} value)
CBC	5.9 (Rosenthaler et al. 2014) 7.6 (Zagzoog et al. 2020)	8.1[a] (Zagzoog et al. 2020) 5.8[a] (Udoh et al. 2025)	CBCA	<10% displacement at 10 μM (Santillo et al. 2024)	
CBD	5.3 (Husni et al. 2014) 5.8 (Navarro et al. 2020) 6.6 (Zagzoog et al. 2020)	7.0 (Husni et al. 2014) <5.0 (Zagzoog et al. 2020)	CBDA	<5.0 (McPartland et al. 2017) 6.1 (Navarro et al. 2020) 7.9 (Zagzoog et al. 2020)	6.9 (Zagzoog et al. 2020)
CBDV	5.4 (Husni et al. 2014) 4.8 (Navarro et al. 2020) 6.9 (Zagzoog et al. 2020)	8.5 (Husni et al. 2014) 8.3[a] (Zagzoog et al. 2020)	CBDVA		
CBG	5.5 (Husni et al. 2014) 5.5 (Navarro et al. 2020) 6.3 (Zagzoog et al. 2020)	6.9[a] (Zagzoog et al. 2020)	CBGA	<5.0 (Husni et al. 2014) 4.8 (Navarro et al. 2020)	6.9 (Husni et al. 2014)
CBGV	5.5 (Navarro et al. 2020)		CBGVA		
CBN	6.5(Felder et al. 1995) 7.1 (Husni et al. 2014)	6.6 (Rhee et al. 1997) 6.5 (Husni et al. 2014)			
THC	7.1 (Felder et al. 1995) 7.4 (Husni et al. 2014) 7.5 (Zagzoog et al. 2020)	7.4[a] (Felder et al. 1995) 6.5 (Husni et al. 2014)	THCA	6.5 (Rosenthaler et al. 2014) 5.8 (Husni et al. 2014) 4.9 (McPartland et al. 2017) 6.3 (Palomares et al. 2020)	<5.0 (Husni et al. 2014) 6.0[b] (Palomares et al. 2020) 8.9[a] (Zagzoog et al. 2020)
THCV	7.2 (Thomas et al. 2005) 7.0 (Husni et al. 2014) 7.3 (Zagzoog et al. 2020)	8.0[c] (Thomas et al. 2005) 7.4[a] (Bolognini et al. 2010) <5.0 (Husni et al. 2014) 6.6[a] (Zagzoog et al. 2020)	THCVA		

[a]Partial agonist activity. [b]Inverse agonist activity. [c]Antagonist activity. Unless otherwise indicated, proteins are from human

Table 4 Phytocannabinoid action at endocannabinoid-related enzymes in vitro

Ligand	Target	Inhibitor potency (pIC_{50} value)	Acid ligand	Inhibitor potency (pIC_{50} value)
CBC	rFAAH	<4.0 (De Petrocellis et al. 2011)	CBCA	
	DAGLα	<4.0 (De Petrocellis et al. 2011)		
	MAGL	4.3 (De Petrocellis et al. 2011)		
	NAAA	<4.0 (De Petrocellis et al. 2011)		
CBD	mFAAH	4.7 (Bisogno et al. 2001)	CBDA	
	rFAAH	4.8 (De Petrocellis et al. 2011)		
	DAGLα	<4.0 (De Petrocellis et al. 2011)		4.7 (De Petrocellis et al. 2011)
	MAGL	<4.0 (De Petrocellis et al. 2011)		<4.3 (De Petrocellis et al. 2011)
	NAAA	<4.0 (De Petrocellis et al. 2011)		4.6 (De Petrocellis et al. 2011)
CBDV	rFAAH	<4.0 (De Petrocellis et al. 2011)	CBDVA	
	DAGLα	4.8 (De Petrocellis et al. 2011)		<4.3 (De Petrocellis et al. 2011)
	MAGL	<4.0 (De Petrocellis et al. 2011)		<4.3 (De Petrocellis et al. 2011)
	NAAA	4.1 (De Petrocellis et al. 2011)		
CBG	rFAAH	<4.0 (De Petrocellis et al. 2011)	CBGA	
	DAGLα	<4.0 (De Petrocellis et al. 2011)		
	MAGL	4.0 (De Petrocellis et al. 2011)		
	NAAA	<4.0 (De Petrocellis et al. 2011)		<4.3 (De Petrocellis et al. 2011)
CBN	rFAAH	<4.3 (De Petrocellis et al. 2011)		
	DAGLα	<4.3 (De Petrocellis et al. 2011)		
	MAGL	<4.3 (De Petrocellis et al. 2011)		
	NAAA	<4.3 (De Petrocellis et al. 2011)		
THC	rFAAH		THCA	<4.3 (De Petrocellis et al. 2011)
	DAGLα			4.6 (De Petrocellis et al. 2011)

(continued)

Table 4 (continued)

Ligand	Target	Inhibitor potency (pIC_{50} value)	Acid ligand	Inhibitor potency (pIC_{50} value)
	MAGL			4.3 (De Petrocellis et al. 2011)
	NAAA			<4.3 (De Petrocellis et al. 2011)
THCV	rFAAH	<4.0 (De Petrocellis et al. 2011)	THCVA	<4.0 (De Petrocellis et al. 2011)
	DAGLα	<4.3 (De Petrocellis et al. 2011)		4.5 (De Petrocellis et al. 2011)
	MAGL	<4.3 (De Petrocellis et al. 2011)		
	NAAA	<4.0 (De Petrocellis et al. 2011)		<4.0 (De Petrocellis et al. 2011)

Unless otherwise indicated (*m* mouse, *r* rat), the proteins are from human

Anandamide signalling may be terminated by two enzymes, one associated with intracellular membranes (FAAH, fatty acid amide hydrolase) and the other with lysosomes (NAAA, *N*-acylethanolamine acid amidase). FAAH was originally cloned from rat tissues as an amidase (Cravatt et al. 1996) and has multiple selective inhibitors described, including JNJ1661010 (Keith et al. 2008). NAAA appears to prefer saturated *N*-acylethanolamines as substrates (Ueda et al. 1999) and can be inhibited selectively by ARN726 (Ribeiro et al. 2015).

Monoacylglycerol lipase (MAGL) is the predominant enzyme responsible for hydrolysis of 2-acylglycerols (Dinh et al. 2002), where there are a number of selective inhibitors available, including MJN110 (Niphakis et al. 2013).

While there are only a limited number of investigations of the effects of phytocannabinoids as inhibitors of these enzymes (Table 4), none have identified a potency consistent with in vivo concentrations (see Sect. 5). It seems very unlikely that these enzymes mediate any of the acute actions of phytocannabinoids. On the other hand, the increase in anandamide and other *N*-acylethanolamines in the brains of mice treated with CBD in vivo was identified to be dependent on NAPE-PLD expression (Leishman et al. 2018). This suggests a complex mechanism of action in vivo which has yet to be drawn out in vitro.

3 Phytocannabinoid Effects at Non-canonical Targets of Endocannabinoids In Vitro

3.1 Transient Receptor Potential Ion Channels

Both anandamide (Smart et al. 2000) and 2-arachidonoylglycerol (Zygmunt et al. 2013) are agonists at TRPV1 ion channels, which are polymodal integrators of pain processing in primary afferent neurones (Caterina et al. 1997). The TRPV1 ion channel is representative of 28 ion channels, divided into six families (Wu et al. 2010).

Table 5 emphasizes the incomplete nature of studies of phytocannabinoid action at TRP ion channels. The most potently activated TRP ion channels are the TRPA1 and TRPM8, characterized as cold-sensing receptors. These channels are activated

Table 5 Phytocannabinoid action at TRP ion channels in vitro

Ligand	Target	Activity	Potency/affinity	Acid ligand	Potency/affinity
CBC	rTRPA1	Activator	7.0 (De Petrocellis et al. 2011)	CBCA	
	rTRPM8	Inhibitor	4.4 (De Petrocellis et al. 2011)		
	TRPV1	Activator	4.6 (De Petrocellis et al. 2011)		
	rTRPV2	Activator	<5.0 (De Petrocellis et al. 2011)		
	rTRPV3	Activator	5.7 (De Petrocellis et al. 2012)		
	rTRPV4	Activator	6.2 (De Petrocellis et al. 2008)		
CBD	rTRPA1	Activator	6.3 (Chianese et al. 2020) 7.0 (De Petrocellis et al. 2011)	CBDA	5.3 (De Petrocellis et al. 2011)
	rTRPM8	Inhibitor	7.2 (De Petrocellis et al. 2011) 5.6 (Chianese et al. 2020)		5.3 (De Petrocellis et al. 2011)
	TRPV1	Activator	5.5 (Bisogno et al. 2001) 6.0 (De Petrocellis et al. 2011) 5.4 (Chianese et al. 2020)		4.7 (De Petrocellis et al. 2011)
	rTRPV2	Activator	5.4 (Qin et al. 2008) 5.9 (De Petrocellis et al. 2011) 5.9 (Chianese et al. 2020)		<5.0 (De Petrocellis et al. 2011)
	rTRPV3	Activator	5.4 (De Petrocellis et al. 2012) 6.3 (Chianese et al. 2020)		<5.0 (De Petrocellis et al. 2012)
	rTRPV4	Activator	6.1 (De Petrocellis et al. 2008)		5.1 (De Petrocellis et al. 2008)
CBDV	rTRPA1	Activator	6.4 (De Petrocellis et al. 2011)		
	rTRPM8	Inhibitor	6.0 (De Petrocellis et al. 2011)		
	TRPV1	Activator	5.4 (De Petrocellis et al. 2011)		
	rTRPV2	Activator	5.1 (De Petrocellis et al. 2011)		
	rTRPV3	Activator	5.8 (De Petrocellis et al. 2012)		
	rTRPV4	Activator	6.0 (De Petrocellis et al. 2012)		
CBG	rTRPA1	Activator	6.2 (De Petrocellis et al. 2011)	CBGA	5.1 (De Petrocellis et al. 2011)
	rTRPM8	Inhibitor	6.8 (De Petrocellis et al. 2011)		5.9 (De Petrocellis et al. 2011)
	TRPV1	Activator	5.9 (De Petrocellis et al. 2011)		4.7 (De Petrocellis et al. 2011)
		Inhibitor			4.7 (Anderson et al. 2021)
	rTRPV2	Activator	5.8 (De Petrocellis et al. 2011)		<5.0 (De Petrocellis et al. 2011)

(continued)

Table 5 (continued)

Ligand	Target	Activity	Potency/affinity	Acid ligand	Potency/affinity
	rTRPV3	Activator	6.0 (De Petrocellis et al. 2012)		4.9 (De Petrocellis et al. 2012)
	rTRPV4	Activator	5.3 (De Petrocellis et al. 2012)		4.5 (De Petrocellis et al. 2012)
CBGV	rTRPA1	Activator	5.8 (De Petrocellis et al. 2011)		
	rTRPM8	Inhibitor	5.8 (De Petrocellis et al. 2011)		
	TRPV1	Activator	5.7 (De Petrocellis et al. 2011)		
	rTRPV2	Activator	5.9 (De Petrocellis et al. 2011)		
	rTRPV3	Activator	5.6 (De Petrocellis et al. 2012)		
	rTRPV4	Activator	4.7 (De Petrocellis et al. 2012)		
CBN	rTRPA1	Activator	6.7 (De Petrocellis et al. 2011)		
	rTRPM8	Inhibitor	6.7 (De Petrocellis et al. 2011)		
	TRPV1	Activator	5.2 (De Petrocellis et al. 2011)		
	rTRPV2	Activator	4.1 (Qin et al. 2008) 4.7 (De Petrocellis et al. 2011)		
	rTRPV3	Activator	5.3 (De Petrocellis et al. 2012)		
	rTRPV4	Activator	4.8 (De Petrocellis et al. 2012)		
THC	rTRPA1	Activator	4.9 (Jordt et al. 2004) 6.6 (De Petrocellis et al. 2008)	THCA	5.6 (De Petrocellis et al. 2011)
	rTRPM8	Inhibitor	6.8 (De Petrocellis et al. 2008)		6.8 (De Petrocellis et al. 2011)
	TRPV1	Inactive	<5.0 (De Petrocellis et al. 2011)		<5.0 (De Petrocellis et al. 2011)
	rTRPV2	Activator	4.8 (Qin et al. 2008) 6.5 (De Petrocellis et al. 2011)		4.7 (De Petrocellis et al. 2011)
	rTRPV3	Activator	5.0 (De Petrocellis et al. 2012)		<5.0 (De Petrocellis et al. 2012)
	rTRPV4	Activator	5.1 (De Petrocellis et al. 2012)		5.5 (De Petrocellis et al. 2012)
THCV	rTRPA1	Activator	5.8 (De Petrocellis et al. 2011)	THCVA	4.8 (De Petrocellis et al. 2011)
	rTRPM8	Inhibitor	6.1 (De Petrocellis et al. 2011)		5.9 (De Petrocellis et al. 2011)
	TRPV1	Activator	5.8 (De Petrocellis et al. 2011)		4.6 (De Petrocellis et al. 2011)
	rTRPV2	Activator	5.4 (De Petrocellis et al. 2011)		<5.0 (De Petrocellis et al. 2011)
	rTRPV3	Activator	5.4 (De Petrocellis et al. 2012)		4.3 (De Petrocellis et al. 2012)
	rTRPV4	Activator	5.2 (De Petrocellis et al. 2012)		
	drTRPV5	Inhibitor	5.3 (Janssens et al. 2018)		
	drTRPV6	Inhibitor	5.0 (Janssens et al. 2018)		

Unless otherwise indicated (*r* rat, *dr Danio rerio* zebrafish, the proteins are from human)

and inhibited, respectively, by sub-micromolar concentrations of CBD, CBDV, CBG, CBN and THC although the reports of THC activation are more variable (Table 5). While studies are not comprehensive, there is an apparent reduction in activity for the acids compared to the neutral phytocannabinoids, with the exception of THCA which shows nanomolar inhibitor potency at TRPM8.

3.2 Nuclear Hormone Receptors

Peroxisome proliferator-activated receptors (PPARs) are general lipid sensors with multiple endogenous agonists, which tend to be fatty acids and their derivatives (Michalik et al. 2006). Given this broad specificity, the agonist activity of anandamide (Rockwell and Kaminski 2004) and 2-arachidonoylglycerol (Kozak et al. 2002) as well as the oxidative metabolites is logical, with the benefit of hindsight. Table 6 identifies that a complete molecular investigation of phytocannabinoid interaction

Table 6 Phytocannabinoid action at nuclear hormone receptors in vitro

Neutral ligand	Target	Activity	Potency/affinity (pEC_{50}/pK_i value)	Acid ligand	Potency/affinity (pEC_{50}/pK_i value)
CBC	PPARγ	Binding potency	4.6 (Granja et al. 2012)	CBCA	
CBD	PPARα	Agonist potency		CBDA	Active at 10 μM[a] (D'Aniello et al. 2019)
	PPARβ	Agonist potency			Active at 5 μM (Hirao-Suzuki et al. 2022)
	PPARγ	Binding potency/ affinity	4.7 (Granja et al. 2012) 4.9 (Nadal et al. 2017)		5.1 (Nadal et al. 2017)
CBDV				CBDVA	
CBG	PPARα	Agonist potency	Active at 10 μM[a] (D'Aniello et al. 2019)	CBGA	Active at 10 μM[a] (D'Aniello et al. 2019)
	PPARγ	Agonist potency	Active at 10 μM[a] (D'Aniello et al. 2019)		
	PPARγ	Binding potency/ affinity	4.9 [a] (Granja et al. 2012) 4.9 (Nadal et al. 2017)		6.3 (Nadal et al. 2017)
CBGV				CBGVA	
CBN					
THC	PPARα	Agonist potency	Takeda et al. (2014)	THCA	
	PPARγ	Agonist potency	O'Sullivan et al. (2005)		Active at 1 μM (Nadal et al. 2017)
	PPARγ	Binding potency/ affinity	4.7 (Granja et al. 2012) 5.2 (Nadal et al. 2017)		6.6 (Nadal et al. 2017)
THCV				THCVA	

[a]Mouse receptor. Unless otherwise indicated, the proteins are from human

with these receptors is lacking. However, it is apparent that many of the major phytocannabinoids are able to active PPARs, albeit at higher concentrations. While there are relatively few head-to-head investigations of the phenomenon, there is an interesting trend for the phytocannabinoid acids to be more potent than the neutral phytocannabinoids (Table 6). Given that PPARs are typically associated with anti-inflammatory profiles, targeted in the clinic to treat dyslipidemias and type II diabetes, there is the potential for future development of alternative therapeutic options.

4 Phytocannabinoid Effects at Other Mammalian Targets In Vitro

4.1 Non-cannabinoid G Protein-Coupled Receptors

The investigation of non-cannabinoid GPCR targets for phytocannabinoids has been quite limited (Table 7) and most of the observations reported are of low potency and have yet to be confirmed. A general observation would be to link some orphan GPCR (GPR3, GPR6, GPR18 and GPR55) with CBD and THC in particular, but there is clearly a need for replication of these findings (Table 7).

While it may simply be a consequence of the limited number of investigations accumulated so far, there is an overlap with monoamine receptors (Table 7). One

Table 7 Phytocannabinoid action at non-CB G protein-coupled receptors in vitro

Neutral Ligand	Target	Activity	Potency/affinity (pEC_{50}/pK_i value)	Acid ligand	Potency/affinity (pEC_{50}/pK_i value)
CBC	GPR6	Inverse agonist potency	~5.5 (Laun et al. 2018)	CBCA	
	H1	Binding potency			5.2 (Santillo et al. 2024)
CBD	$5HT_{1A}$	Binding potency	~5.0 (Russo et al. 2005)	CBDA	
	GPR3	Inverse agonist potency	6.0 (Laun and Song 2017)		
	GPR6	Inverse agonist potency	6.7 (Laun and Song 2017) 7.1 (Laun et al. 2018)		
	GPR18	Agonist potency	4.3[a] (McHugh et al. 2012)		
	GPR52	Inverse agonist potency	5.6 (Stott et al. 2021)		
	GPR55	Antagonist potency	6.3–6.5 (Ryberg et al. 2007)		
	DOP	Binding potency	5.0 (Kathmann et al. 2006)		

(continued)

Table 7 (continued)

Neutral Ligand	Target	Activity	Potency/affinity (pEC_{50}/pK_i value)	Acid ligand	Potency/affinity (pEC_{50}/pK_i value)
	MOP	NAM activity	(Kathmann et al. 2006) 4.8 (Bosquez-Berger et al. 2023)		
CBDV	GPR6	Inverse agonist potency	~5.5 (Laun et al. 2018)	CBDVA	
	GPR55	Antagonist potency			5.0 (Udoh et al. 2022)
	MOP	Binding potency	5.2 (Santillo et al. 2024)		
CBG	α_2-AR	Agonist potency	9.7[b] (Cascio et al. 2010)	CBGA	
	GPR6	Inverse agonist potency	~5.5 (Laun et al. 2018)		
	GPR55	Antagonist potency			5.3 (Anderson et al. 2021)
	$5HT_{1A}$	Antagonist affinity	7.3[b] (Cascio et al. 2010)		
	$5HT_{2A}$	Binding potency	5.7 (Santillo et al. 2024)		
	MOP	Binding potency	5.9 (Santillo et al. 2024)		
CBGV	GPR55	Antagonist potency		CBGVA	5.1 (Udoh et al. 2022)
CBN	$5HT_{2A}$	Binding potency	5.3 (Santillo et al. 2024)		
	GPR6	Inverse agonist potency	~5.5 (Laun et al. 2018)		
THC	$5HT_{1A}$	Binding potency	<4.5 (Russo et al. 2005)	THCA	
	GPR6	Inverse agonist potency	~5.5 (Laun et al. 2018)		
	GPR18	Agonist potency	6.0 (McHugh et al. 2012) 5.5 (Mahardhika et al. 2024)		
	GPR55	Agonist potency	8.1 (Ryberg et al. 2007) <5.0 (Mahardhika et al. 2024)		
THCV	$5HT_{1A}$	PAM[c] activity	5.7 (Cascio et al. 2015)	THCVA	

[a]Partial agonist activity. [b]Mouse receptors. [c]Negative allosteric modulator. [d]Positive allosteric modulator. Unless otherwise indicated, activities are at human proteins

remarkable observation is the very high potency of CBG in vitro as an agonist at mouse α_2-adrenoceptors while antagonising mouse $5HT_{1A}$ receptors with good affinity (Cascio et al. 2010). Clearly, this observation is in need of repeating with human receptors.

4.2 Other Ligand-Gated Ion Channels

Table 8 summarises data on the effects of phytocannabinoids on ligand-gated ion channels other than the TRP channels. Notably, the preponderance of evidence,

Table 8 Phytocannabinoid action at ligand-gated ion channels in vitro

Ligand	Target	Activity	Potency (pEC_{50} value)	Acid ligand	Potency (pEC_{50} value)
CBC	P2X4	Binding affinity	<4.0 (Puopolo et al. 2023)	CBCA	
CBD	$5HT_{3A}$	NAM potency	6.2 (Yang et al. 2010a) 6.5 (Xiong et al. 2011b)	CBDA	
	Glycine	PAM potency	4.7–4.9 (Ahrens et al. 2009)		
	$GABA_A$	PAM potency	4.6–6.0 (Bakas et al. 2017)		
	nACh	Inhibitor potency	5.0 (Mahgoub et al. 2013) 6.3 (Chrestia et al. 2022)		
	P2X4	Binding affinity	6.0 (Puopolo et al. 2023)		<4.0 (Puopolo et al. 2023)
CBDV	P2X4	Binding affinity	5.0 (Puopolo et al. 2023)	CBDVA	
CBG	$GABA_A$	PAM potency		CBGA	6.0 (Anderson et al. 2021)
	P2X4	Binding affinity	<4.0 (Puopolo et al. 2023)		
CBGV	P2X4	Binding affinity	5.5 (Puopolo et al. 2023)	CBGVA	5.0 (Puopolo et al. 2023)
CBN	nACh	Inhibitor potency	Inactive at 10 μM (Mahgoub et al. 2013)		
THC	$5HT_{3A}$	NAM potency	7.4 (Barann et al. 2002) 5.9 (Yang et al. 2010b) 6.9 (Xiong et al. 2011b)		
	Glycine	PAM potency	7.1 (Hejazi et al. 2006)	THCA	
	$GABA_A$		Inactive at 300 nM (Hejazi et al. 2006)		
	nACh	Inhibitor potency	Inactive at 10 μM (Chrestia et al. 2022) Inactive at 10 μM (Mahgoub et al. 2013)		

(continued)

Table 8 (continued)

Ligand	Target	Activity	Potency (pEC_{50} value)	Acid ligand	Potency (pEC_{50} value)
	P2X4	Binding affinity			4.0 (Puopolo et al. 2023)
THCV	P2X4	Binding affinity	4.8 (Puopolo et al. 2023)	THCVA	

Unless otherwise indicated (prefix *m* mouse, *r* rat), the proteins are from human

sparse as it is, covers four out of the five cys-loop receptors, but not the zinc-activated channel (Schmiedhofer et al. 2022). It is apparent that both CBD and THC have sub-micromolar potency as negative allosteric modulators at $5HT_3$ receptors, which might contribute to their antiemetic properties (Grimison et al. 2020). The potency of THC as a positive allosteric modulator at glycine receptors has been suggested to contribute to the analgesic actions of THC in vivo (Xiong et al. 2011a).

There is still a paucity of data on these actions of the phytocannabinoids, suggesting that further work to confirm (or not) these observations is needed.

4.3 *Voltage-Gated Ion Channels*

Of the phytocannabinoids, CBD has been the most investigated for its potential effects on voltage-gated ion channels (Table 9). While many of the effects in vitro described suggest low potency action, the combination of inhibition of voltage-gated sodium channels with activation of voltage-gated potassium channels would be expected to reduce excessive neuronal activity, consistent with the use of CBD in the treatment of epilepsy.

As with many other molecular targets, it appears that investigation of the acid phytocannabinoids is lagging behind that of the neutral compounds.

4.4 *Transporters*

In terms of phytocannabinoid interactions with transporters, most of the studies reported are low potency (above micromolar concentrations, Table 10). However, the equilibrative nucleoside transporter (ENT1, SLC29A1) exhibits sub-micromolar inhibition by CBD and THC. In the clinic, ENT1 is a high potency target of dipyridamole as an antiplatelet and anti-anginal therapy, as inhibition of the transporter increases local levels of adenosine (and other nucleosides) to act through cell surface adenosine receptors.

Table 9 Phytocannabinoid action at voltage-gated ion channels in vitro

Ligand	Target	Activity	Potency/affinity	Acid ligand	Potency/affinity
CBC				CBCA	
CBD	$Na_v1.1$	Inhibitor potency	4.5 (Hill et al. 2014) 5.7 (Ghovanloo et al. 2018) 4.7 (Milligan et al. 2022)	CBDA	
	$Na_v1.2$	Inhibitor potency	4.5 (Hill et al. 2014) 5.6 (Ghovanloo et al. 2018) Active at 10^{-6} M (Mason and Cummins 2020) 4.7 (Milligan et al. 2022)		
	$Na_v1.3$	Inhibitor potency	4.5 (Hill et al. 2014) 5.5 (Ghovanloo et al. 2018)		
	$Na_v1.4$	Inhibitor potency	5.4 (Huang et al. 2021) 5.7 (Ghovanloo et al. 2018)		
	$Na_v1.5$	Inhibitor potency	5.1 (Le Marois et al. 2020) 5.4 (Ghovanloo et al. 2018)		
	$Na_v1.6$	Inhibitor potency	Active at 10^{-6} M (Patel et al. 2016) 5.5 (Ghovanloo et al. 2018) 4.8 (Milligan et al. 2022)		
	$Na_v1.7$	Inhibitor potency	5.5 (Ghovanloo et al. 2018) 4.9 (Milligan et al. 2022)		
	$Na_v1.8$	Inhibitor potency	5.7 (Ghovanloo et al. 2025)		
	$Ca_v1.2$	Inhibitor potency	5.0 (Le Marois et al. 2020)		
	$Ca_v2.2$	Inhibitor potency	Active at 10^{-6} M (Mendis et al. 2019)		
	$K_v4.3$	Inhibitor potency	5.1 (Le Marois et al. 2020)		
	$K_v7.1$	Inhibitor potency	5.6 (Le Marois et al. 2020)		
	$K_v7.2$	Enhancing potency	5.5 (Pokl et al. 2023)		
	$K_v7.3$	Enhancing potency	4.7 (Pokl et al. 2023)		
	$K_v7.2/7.3$	Enhancing potency	6.7 (Zhang et al. 2022)		
	$K_v11.1$	Inhibitor potency	4.8 (Le Marois et al. 2020)		
	$K_{ir}2.1$	Inactive	<4.3 (Le Marois et al. 2020)		
CBDV	$Ca_v3.1$	Inhibitor potency		CBDVA	5.7 (Udoh et al. 2022)
	$Ca_v3.2$	Inhibitor potency			5.0 (Udoh et al. 2022)

(continued)

Table 9 (continued)

Ligand	Target	Activity	Potency/affinity	Acid ligand	Potency/affinity
CBG	$Na_v1.1$	Inhibitor potency	4.1 (Hill et al. 2014)	CBGA	
	$Na_v1.2$	Inhibitor potency	4.1 (Hill et al. 2014)		
	$Na_v1.3$	Inhibitor potency	4.4 (Hill et al. 2014)		
	$Na_v1.8$	Inhibitor potency	5.3 (Ghovanloo et al. 2025)		
CBGV	$Ca_v3.1$	Inhibitor potency		CBGVA	5.2 (Udoh et al. 2022)
	$Ca_v3.2$	Inhibitor potency			5.0 (Udoh et al. 2022)
CBN	$Na_v1.8$	Inhibitor potency	4.8 (Ghovanloo et al. 2025)		
THC	$Na_v1.2$	Inhibitor potency	5.6 (Ghovanloo et al. 2018)	THCA	
	$Ca_v3.1$	Inhibitor potency	6.0 (Mirlohi et al. 2022)		5.6 (Mirlohi et al. 2022)
THCV				THCVA	

Unless otherwise indicated, the proteins are from human

Table 10 Phytocannabinoid action at transporters in vitro

Ligand	Target	Activity	Potency/affinity	Acid ligand	Potency/affinity
CBC				CBCA	
CBD	ENT1	Inhibitor affinity	6.6 (Carrier et al. 2006) 5.5 (Pandolfo et al. 2011)	CBDA	
CBDV	DAT	Binding potency	5.5 (Santillo et al. 2024)	CBDVA	
CBG	DAT	Binding potency	6.0 (Santillo et al. 2024)	CBGA	5.4 (Santillo et al. 2024)
	NET	Binding potency	5.6 (Santillo et al. 2024)		
	SERT	Binding potency	5.8 (Santillo et al. 2024)		
CBN					
THC	ENT1	Inhibitor affinity	6.3 (Carrier et al. 2006)	THCA	5.3 (Zhu et al. 2006)
THCV				THCVA	

Unless otherwise indicated, the proteins are from human

4.5 Other Enzymes

The potencies of any phytocannabinoids investigated as inhibitors of lactate dehydrogenase and cholinesterases are poor (Table 11). When the inflammation-associated eicosanoid-metabolizing enzymes 5-lipoxygenase and microsomal

Table 11 Phytocannabinoid action at other enzymes in vitro

Ligand	Target	Activity	Potency/affinity	Acid ligand	Potency/affinity
CBC	5-Lipoxygenase	Inhibitory potency	5.8 (Martin et al. 2021; Peltner et al. 2023)	CBCA	
	LDHA	Inhibitor affinity	5.1 (Martin et al. 2021)		
	mPGES-1	Inhibitory potency	5.2 (Peltner et al. 2023)		
CBD	eeAChE	Inhibitor affinity		CBDA	5.1 (Vitale et al. 2025)
	eBuChE	Inhibitor affinity			5.2 (Vitale et al. 2025)
	5-Lipoxygenase	Inhibitory potency	5.4 (Peltner et al. 2023)		
	mPGES-1	Inhibitory potency	<5.0 (Peltner et al. 2023)		
CBDV	5-Lipoxygenase	Inhibitory potency	5.3 (Peltner et al. 2023)	CBDVA	
	mPGES-1	Inhibitory potency	<5.0 (Peltner et al. 2023)		
CBG	eeAChE	Inhibitor affinity		CBGA	5.0 (Vitale et al. 2025)
	eBuChE	Inhibitor affinity			4.6 (Vitale et al. 2025)
	5-Lipoxygenase	Inhibitory potency	6.2 (Peltner et al. 2023)		
	mPGES-1	Inhibitory potency	5.6 (Peltner et al. 2023)		
CBN	5-Lipoxygenase	Inhibitory potency	5.4 (Peltner et al. 2023)		
	mPGES-1	Inhibitory potency	<5.0 (Peltner et al. 2023)		
THC	5-Lipoxygenase	Inhibitory potency	5.8 (Peltner et al. 2023)	THCA	
	mPGES-1	Inhibitory potency	<5.0 (Peltner et al. 2023)		
	LDHA	Inhibitor affinity			5.1 (Martin et al. 2021)
THCV				THCVA	

Unless otherwise indicated, the proteins are from human

prostaglandin synthase-1 were investigated for effects of the neutral phytocannabinoids, only CBG had a potency below micromolar concentrations (Table 11). As elsewhere in this analysis, Table 11 highlights how much more is to be studied in screening for molecular targets of the phytocannabinoids.

4.6 Catalytic Receptors

Very few studies have been conducted to test for the impact of cannabinoids on catalytic receptor function. One, very recent study, described high potency inhibition of HER2 tyrosine kinase activity by CBD (7.4), CBG (7.6) and CBN (7.1) (Lamtha et al. 2025).

5 ADME Profile of the Major Phytocannabinoids

Following smoked administration of THC-containing *Cannabis* cigarettes under controlled conditions in humans, a three-compartment model was fitted to plasma concentrations (Sempio et al. 2020). A steady-state volume of distribution was calculated of over 3,000 L, indicative of accumulation in peripheral tissues. Smoke inhalation is associated with relatively rapid peaks of plasma THC concentrations – usually within 5–10 min, with a broad range of bioavailability of 10–35% (Grotenhermen 2003). Peak plasma concentration of THC following smoking a cigarette with 34 mg THC reached 150 µg/L – about 0.5 µM (Huestis et al. 1992).

Oral administration of THC in volunteers is associated with a slower rise to peak plasma concentrations and with large inter-individual variation (1–4 h) (Grotenhermen 2003). Similarly, the licensed pure version of CBD administered orally in sesame oil only had a bioavailability of about 6%, with a peak plasma concentration at 4–5 h (Taylor et al. 2018). Again, a large volume of distribution was observed (20,000–43,000 L). An interesting observation was that the plasma concentrations of CBD were enhanced in the fed state, implying improved absorption from the gut in the presence of a more hydrophobic environment. Thus, for 1,500 mg administered, the peak plasma concentration increased from ~300 ng/mL to 1,600 ng/mL (about 1 µM to 5 µM).

In vitro investigations of the key drug-metabolizing enzymes, carboxylesterase and cytochrome P450s, suggest potencies above the anticipated in vivo concentrations of the phytocannabinoids (Table 12). Exceptions would be particular cytochrome P450 (CYP1) activities, where sub-micromolar potencies/affinities of CBD, CBDV and CBN could alter CYP450-mediated metabolism of other drugs in vivo. There appear to be no studies of the acid phytocannabinoid regulation of these enzymes.

Table 12 Phytocannabinoid action at drug-metabolizing enzymes in vitro

Ligand	Target	Activity	Potency/affinity	Acid ligand	Potency/ affinity
CBC	CYP2J2	Substrate affinity	4.1 (Arnold et al. 2018)	CBCA	
CBD	Carboxylesterase 1	Inhibitor affinity	6.0 (Qian et al. 2019)	CBDA	
	CYP1A1	Inhibitor potency	6.8 (Yamaori et al. 2010)		
	CYP1A1	Inhibitor affinity	6.8 (Yamaori et al. 2013)		
	CYP1A2	Inhibitor potency	5.6 (Yamaori et al. 2010)		
	CYP1B1	Inhibitor potency	5.4 (Yamaori et al. 2010)		
	CYP2C9	Inhibitor potency	5.6 (Yamaori et al. 2012)		
	CYP2D6	Inhibitor potency	5.6–5.9 (Yamaori et al. 2011)		
	CYP2J2	Substrate affinity	4.2 (Arnold et al. 2018)		
	CYP3A4	Inhibitor potency	4.9 (Yamaori et al. 2011)		
	CYP3A5	Inhibitor potency	5.8 (Yamaori et al. 2011)		
	CYP3A7	Inhibitor potency	4.6 (Yamaori et al. 2011)		
	UGT2B7	Inhibitor potency	5.0–5.3 (Coates et al. 2024)		
CBDV	CYP1A1	Inhibitor affinity	6.3 (Yamaori et al. 2013)	CBDVA	
	CYP2D6	Inhibitor potency	5.0 (Yamaori et al. 2011)		
CBG	CYP2J2	Substrate affinity	5.1 (Arnold et al. 2018)	CBGA	
CBN	Carboxylesterase 1	Inhibitor affinity	6.6 (Qian et al. 2019)		
	CYP1A1	Inhibitor potency	6.3 (Yamaori et al. 2010)		
	CYP1A2	Inhibitor potency	7.1 (Yamaori et al. 2010)		
	CYP1B1	Inhibitor potency	6.8 (Yamaori et al. 2010)		
	CYP2C9	Inhibitor potency	5.5 (Yamaori et al. 2012)		
	CYP2J2	Substrate affinity	4.1 (Arnold et al. 2018)		

(continued)

Table 12 (continued)

Ligand	Target	Activity	Potency/affinity	Acid ligand	Potency/ affinity
	CYP3A4	Inhibitor potency	<4.3 (Yamaori et al. 2011)		
	CYP3A5	Inhibitor potency	<4.3 (Yamaori et al. 2011)		
	CYP3A7	Inhibitor potency	4.6 (Yamaori et al. 2011)		
THC	Carboxylesterase 1	Inhibitor affinity	6.3 (Qian et al. 2019)	THCA	
	CYP1A1	Inhibitor potency	5.3 (Yamaori et al. 2010)		
	CYP1A1	Inhibitor affinity	5.5 (Yamaori et al. 2013)		
	CYP1A2	Inhibitor potency	5.1 (Yamaori et al. 2010)		
	CYP1B1	Inhibitor potency	5.6 (Yamaori et al. 2010)		
	CYP2C9	Inhibitor potency	5.5 (Yamaori et al. 2012)		
	CYP2D6	Inhibitor potency	4.7 (Yamaori et al. 2011)		
	CYP2J2	Substrate affinity	4.9 (Arnold et al. 2018)		
	CYP3A4	Inhibitor potency	<4.3 (Yamaori et al. 2011)		
	CYP3A5	Inhibitor potency	4.4 (Yamaori et al. 2011)		
	CYP3A7	Inhibitor potency	4.5 (Yamaori et al. 2011)		
	UGT2B7	Inhibitor potency	4.7–5.1 (Coates et al. 2024)		
THCV				THCVA	

Unless otherwise indicated, the proteins are from human

Drug disposition is regulated by both drug-metabolizing enzymes and particular transporters, notably of the ATP-binding cassette family (Table 13). Most of the reported interactions of cannabinoids with drug transporters are of low potency. However, there is indirect evidence for CBD and THC inhibiting ABCB11 and ABCG2 in the low micromolar range (Anderson et al. 2022).

The other groups of transporters that would be associated with drug disposition would be members of the SLCO and SLC22 family. Although there is very little evidence in the literature, it suggests that CBD and THC do not interact with SLC22A1, SLC22A2, SLC22A6, SLC22A8, SLCOB1, SLCOB3, SLC47A1 and SLC47A2 (Anderson et al. 2022).

Table 13 Phytocannabinoid action at drug transporters in vitro

Ligand	Target	Activity	Potency/affinity	Acid ligand	Potency/affinity
CBC	ABCG2	Inhibitor potency	No effect at 10 μM (Etchart et al. 2022)	CBCA	No effect at 10 μM (Etchart et al. 2022)
CBD	ABCB1	Substrate affinity	4.7 (Zhu et al. 2006)	CBDA	
	ABCB1	Inhibitor potency	4.4 (Zhu et al. 2006)		
	ABCB1	ATPase inhibition potency	5.3 (Holland et al. 2008)		
	ABCG2	Inhibitor potency	5.1 (Holland et al. 2007)		
CBDV				CBDVA	
CBG				CBGA	
CBN	ABCB1	Substrate affinity	5.4 (Zhu et al. 2006)		
	ABCG2	Inhibitor potency	5.3 (Holland et al. 2007)		
THC	ABCB1	Substrate affinity	4.3 (Zhu et al. 2006)	THCA	5.3 (Zhu et al. 2006)
	ABCB1	ATPase inhibition potency	4.8 (Holland et al. 2008)		
	ABCG2	Inhibitor potency	5.4 (Holland et al. 2007)		
THCV				THCVA	

6 Conclusions

Given that plasma levels of the major phytocannabinoids are typically micromolar levels or below, it could logically be asserted that the majority of in vitro phytocannabinoid interactions with the distinct molecular targets described in the tables are irrelevant in terms of any in vivo actions of these agents. However, the peripheral accumulation of cannabinoids, which probably includes adipose tissue as well as other organs, means that the local concentrations may exceed plasma concentrations. Furthermore, the hydrophobicity of these metabolites means that they are not freely soluble in either extracellular fluids or cytoplasm. They would be expected to be bound to extracellular proteins, such as albumin, as well as intracellular carriers, such as fatty acid-binding proteins (Huang et al. 2018).

Overall, it is apparent that despite the long history of humans with the *Cannabis* plant, there is still a lot more to be learnt about the molecular targets of individual components of the plant.

References

Adams R, Hunt M, Clark JH (1940) Structure of cannabidiol, a product isolated from the marihuana extract of Minnesota wild hemp. I. J Am Chem Soc 62:196–200

Ahrens J, Demir R, Leuwer M, de la Roche J, Krampfl K, Foadi N et al (2009) The nonpsychotropic cannabinoid cannabidiol modulates and directly activates alpha-1 and alpha-1-Beta glycine receptor function. Pharmacology 83:217–222. https://doi.org/10.1159/000201556

Anderson LL, Heblinski M, Absalom NL, Hawkins NA, Bowen MT, Benson MJ et al (2021) Cannabigerolic acid, a major biosynthetic precursor molecule in cannabis, exhibits divergent effects on seizures in mouse models of epilepsy. Br J Pharmacol 178:4826–4841. https://doi.org/10.1111/bph.15661

Anderson LL, Etchart MG, MacNair L, Land MH, Mosesova IA, Bonn-Miller MO et al (2022) In vitro screening of three commercial cannabis-based products on ATP-binding cassette and solute-carrier transporter function. Cannabis Cannabinoid Res 7:304–317. https://doi.org/10.1089/can.2020.0053

Appendino G, Gibbons S, Giana A, Pagani A, Grassi G, Stavri M et al (2008) Antibacterial cannabinoids from Cannabis sativa: a structure-activity study. J Nat Prod 71:1427–1430. https://doi.org/10.1021/np8002673

Arnold WR, Weigle AT, Das A (2018) Cross-talk of cannabinoid and endocannabinoid metabolism is mediated via human cardiac CYP2J2. J Inorg Biochem 184:88–99. https://doi.org/10.1016/j.jinorgbio.2018.03.016

Baggelaar MP, Chameau PJ, Kantae V, Hummel J, Hsu KL, Janssen F et al (2015) Highly selective, reversible inhibitor identified by comparative chemoproteomics modulates diacylglycerol lipase activity in neurons. J Am Chem Soc 137:8851–8857. https://doi.org/10.1021/jacs.5b04883

Bakas T, van Nieuwenhuijzen PS, Devenish SO, McGregor IS, Arnold JC, Chebib M (2017) The direct actions of cannabidiol and 2-arachidonoyl glycerol at GABA(A) receptors. Pharmacol Res 119:358–370. https://doi.org/10.1016/j.phrs.2017.02.022

Barann M, Molderings G, Bruss M, Bonisch H, Urban BW, Gothert M (2002) Direct inhibition by cannabinoids of human 5-HT3A receptors: probable involvement of an allosteric modulatory site. Br J Pharmacol 137:589–596. https://doi.org/10.1038/sj.bjp.0704829

Bayewitch M, Rhee MH, Avidor-Reiss T, Breuer A, Mechoulam R, Vogel Z (1996) (−)-Delta9-tetrahydrocannabinol antagonizes the peripheral cannabinoid receptor-mediated inhibition of adenylyl cyclase. J Biol Chem 271:9902–9905. https://doi.org/10.1074/jbc.271.17.9902

Bisogno T, Hanus L, De Petrocellis L, Tchilibon S, Ponde DE, Brandi I et al (2001) Molecular targets for cannabidiol and its synthetic analogues: effect on vanilloid VR1 receptors and on the cellular uptake and enzymatic hydrolysis of anandamide. Br J Pharmacol 134:845–852. https://doi.org/10.1038/sj.bjp.0704327

Bisogno T, Howell F, Williams G, Minassi A, Cascio MG, Ligresti A et al (2003) Cloning of the first sn1-DAG lipases points to the spatial and temporal regulation of endocannabinoid signaling in the brain. J Cell Biol 163:463–468. https://doi.org/10.1083/jcb.200305129

Bolognini D, Costa B, Maione S, Comelli F, Marini P, Di Marzo V et al (2010) The plant cannabinoid Delta9-tetrahydrocannabivarin can decrease signs of inflammation and inflammatory pain in mice. Br J Pharmacol 160:677–687. https://doi.org/10.1111/j.1476-5381.2010.00756.x

Bolognini D, Rock EM, Cluny NL, Cascio MG, Limebeer CL, Duncan M et al (2013) Cannabidiolic acid prevents vomiting in *Suncus murinus* and nausea-induced behaviour in rats by enhancing 5-HT_{1A} receptor activation. Br J Pharmacol 168:1456–1470. https://doi.org/10.1111/bph.12043

Bosquez-Berger T, Gudorf JA, Kuntz CP, Desmond JA, Schlebach JP, VanNieuwenhze MS et al (2023) Structure-activity relationship study of cannabidiol-based analogs as negative allosteric modulators of the mu-opioid receptor. J Med Chem 66:9466–9494. https://doi.org/10.1021/acs.jmedchem.3c00061

Cahn RS (1932) Cannabis indica resin. Part III. The constitution of cannabinol. J Chem Soc:1342–1353
Carrier EJ, Auchampach JA, Hillard CJ (2006) Inhibition of an equilibrative nucleoside transporter by cannabidiol: a mechanism of cannabinoid immunosuppression. Proc Natl Acad Sci USA 103:7895–7900. https://doi.org/10.1073/pnas.0511232103
Cascio MG, Gauson LA, Stevenson LA, Ross RA, Pertwee RG (2010) Evidence that the plant cannabinoid cannabigerol is a highly potent alpha2-adrenoceptor agonist and moderately potent 5HT1A receptor antagonist. Br J Pharmacol 159:129–141. https://doi.org/10.1111/j.1476-5381.2009.00515.x
Cascio MG, Zamberletti E, Marini P, Parolaro D, Pertwee RG (2015) The phytocannabinoid, Delta (9)-tetrahydrocannabivarin, can act through 5-HT(1)A receptors to produce antipsychotic effects. Br J Pharmacol 172:1305–1318. https://doi.org/10.1111/bph.13000
Caterina MJ, Schumacher MA, Tominaga M, Rosen TA, Levine JD, Julius D (1997) The capsaicin receptor: a heat-activated ion channel in the pain pathway. Nature 389:816–824. https://doi.org/10.1038/39807
Cheng YC, Houston R (2024) The development of a next-generation sequencing panel targeting cannabinoid synthase genes to distinguish between marijuana and hemp. Electrophoresis 45: 948–957. https://doi.org/10.1002/elps.202300233
Chianese G, Lopatriello A, Schiano-Moriello A, Caprioglio D, Mattoteia D, Benetti E et al (2020) Cannabitwinol, a dimeric Phytocannabinoid from Hemp, Cannabis sativa L., is a selective thermo-TRP modulator. J Nat Prod 83:2727–2736. https://doi.org/10.1021/acs.jnatprod.0c00668
Chrestia JF, Esandi MDC, Bouzat C (2022) Cannabidiol as a modulator of alpha7 nicotinic receptors. Cell Mol Life Sci 79:564. https://doi.org/10.1007/s00018-022-04600-y
Coates S, Bardhi K, Lazarus P (2024) Cannabinoid-induced inhibition of morphine glucuronidation and the potential for in vivo drug-drug interactions. Pharmaceutics 16. https://doi.org/10.3390/pharmaceutics16030418
Cravatt BF, Giang DK, Mayfield SP, Boger DL, Lerner RA, Gilula NB (1996) Molecular characterization of an enzyme that degrades neuromodulatory fatty-acid amides. Nature 384:83–87. https://doi.org/10.1038/384083a0
Crombie L, Ponsford R, Shani A, Yagnitinsky B, Mechoulam R (1968) Hashish components. Photochemical production of cannabicyclol from cannabichromene. Tetrahedron Lett:5771–5772. https://doi.org/10.1016/s0040-4039(00)76346-5
D'Aniello E, Fellous T, Iannotti FA, Gentile A, Allara M, Balestrieri F et al (2019) Identification and characterization of phytocannabinoids as novel dual PPARalpha/gamma agonists by a computational and in vitro experimental approach. Biochim Biophys Acta Gen Subj 1863: 586–597. https://doi.org/10.1016/j.bbagen.2019.01.002
De Petrocellis L, Vellani V, Schiano-Moriello A, Marini P, Magherini PC, Orlando P et al (2008) Plant-derived cannabinoids modulate the activity of transient receptor potential channels of ankyrin type-1 and melastatin type-8. J Pharmacol Exp Ther 325:1007–1015. https://doi.org/10.1124/jpet.107.134809
De Petrocellis L, Ligresti A, Moriello AS, Allara M, Bisogno T, Petrosino S et al (2011) Effects of cannabinoids and cannabinoid-enriched *Cannabis* extracts on TRP channels and endocannabinoid metabolic enzymes. Br J Pharmacol 163:1479–1494. https://doi.org/10.1111/j.1476-5381.2010.01166.x
De Petrocellis L, Orlando P, Moriello AS, Aviello G, Stott C, Izzo AA et al (2012) Cannabinoid actions at TRPV channels: effects on TRPV3 and TRPV4 and their potential relevance to gastrointestinal inflammation. Acta Physiol (Oxf) 204:255–266. https://doi.org/10.1111/j.1748-1716.2011.02338.x
Devane WA, Dysarz FA 3rd, Johnson MR, Melvin LS, Howlett AC (1988) Determination and characterization of a cannabinoid receptor in rat brain. Mol Pharmacol 34:605–613

Devane WA, Hanus L, Breuer A, Pertwee RG, Stevenson LA, Griffin G et al (1992) Isolation and structure of a brain constituent that binds to the cannabinoid receptor. Science 258:1946–1949. https://doi.org/10.1126/science.1470919

Dinh TP, Carpenter D, Leslie FM, Freund TF, Katona I, Sensi SL et al (2002) Brain monoglyceride lipase participating in endocannabinoid inactivation. Proc Natl Acad Sci USA 99:10819–10824. https://doi.org/10.1073/pnas.152334899

Ehrlich B, Patel M, Zheng X, Glass M (2025) Canonical Cannabinoid Receptors. Curr Top Behav Neurosci. https://doi.org/10.1007/7854_2024_556

Etchart MG, Anderson LL, Ametovski A, Jones PM, George AM, Banister SD et al (2022) In vitro evaluation of the interaction of the cannabis constituents cannabichromene and cannabichromenic acid with ABCG2 and ABCB1 transporters. Eur J Pharmacol 922:174836. https://doi.org/10.1016/j.ejphar.2022.174836

Felder CC, Joyce KE, Briley EM, Mansouri J, Mackie K, Blond O et al (1995) Comparison of the pharmacology and signal transduction of the human cannabinoid CB1 and CB2 receptors. Mol Pharmacol 48:443–450

Fellermeier M, Zenk MH (1998) Prenylation of olivetolate by a hemp transferase yields cannabigerolic acid, the precursor of tetrahydrocannabinol. FEBS Lett 427:283–285. https://doi.org/10.1016/s0014-5793(98)00450-5

Gaoni Y, Mechoulam R (1964a) Isolation, structure and partial synthesis of active constituent of hashish. J Am Chem Soc 86:1646–1647

Gaoni Y, Mechoulam R (1964b) The structure and synthesis of cannabigerol, a new hashish constituent. P Chem Soc London 1:82–82

Gaoni Y, Mechoulam R (1966a) Cannabichromene, a new active principle in hashish. J Chem Soc D: Chem Commun:20–21

Gaoni Y, Mechoulam R (1966b) Concerning isomerization of Delta1- to Delta1(6)-tetrahydrocannabinol. J Am Chem Soc 88:5673

Gerard C, Mollereau C, Vassart G, Parmentier M (1990) Nucleotide sequence of a human cannabinoid receptor cDNA. Nucleic Acids Res 18:7142. https://doi.org/10.1093/nar/18.23.7142

Gerard CM, Mollereau C, Vassart G, Parmentier M (1991) Molecular cloning of a human cannabinoid receptor which is also expressed in testis. Biochem J 279(Pt 1):129–134. https://doi.org/10.1042/bj2790129

Ghovanloo MR, Shuart NG, Mezeyova J, Dean RA, Ruben PC, Goodchild SJ (2018) Inhibitory effects of cannabidiol on voltage-dependent sodium currents. J Biol Chem 293:16546–16558. https://doi.org/10.1074/jbc.RA118.004929

Ghovanloo MR, Tyagi S, Zhao P, Waxman SG (2025) Nav1.8, an analgesic target for nonpsychotomimetic phytocannabinoids. Proc Natl Acad Sci USA 122:e2416886122. https://doi.org/10.1073/pnas.2416886122

Granja AG, Carrillo-Salinas F, Pagani A, Gomez-Canas M, Negri R, Navarrete C et al (2012) A cannabigerol quinone alleviates neuroinflammation in a chronic model of multiple sclerosis. J Neuroimmune Pharmacol 7:1002–1016. https://doi.org/10.1007/s11481-012-9399-3

Grimison P, Mersiades A, Kirby A, Lintzeris N, Morton R, Haber P et al (2020) Oral THC:CBD cannabis extract for refractory chemotherapy-induced nausea and vomiting: a randomised, placebo-controlled, phase II crossover trial. Ann Oncol 31:1553–1560. https://doi.org/10.1016/j.annonc.2020.07.020

Grotenhermen F (2003) Pharmacokinetics and pharmacodynamics of cannabinoids. Clin Pharmacokinet 42:327–360. https://doi.org/10.2165/00003088-200342040-00003

Hanske A, Nazare M, Grether U (2025) Chemical probes for investigating the endocannabinoid system. Curr Top Behav Neurosci. https://doi.org/10.1007/7854_2024_563

Harlow AF, Miech RA, Leventhal AM (2024) Adolescent Delta8 THC and marijuana use in the US. JAMA 331:861–865. https://doi.org/10.1001/jama.2024.0865

Hejazi N, Zhou C, Oz M, Sun H, Ye JH, Zhang L (2006) Delta9-tetrahydrocannabinol and endogenous cannabinoid anandamide directly potentiate the function of glycine receptors. Mol Pharmacol 69:991–997. https://doi.org/10.1124/mol.105.019174

Hill AJ, Jones NA, Smith I, Hill CL, Williams CM, Stephens GJ et al (2014) Voltage-gated sodium (Na_V) channel blockade by plant cannabinoids does not confer anticonvulsant effects *per se*. Neurosci Lett 566:269–274. https://doi.org/10.1016/j.neulet.2014.03.013

Hirao-Suzuki M, Takayuki K, Takiguchi M, Peters JM, Takeda S (2022) Cannabidiolic acid activates the expression of the PPARbeta/delta target genes in MDA-MB-231 cells. Arch Biochem Biophys 731:109428. https://doi.org/10.1016/j.abb.2022.109428

Hively RL, Mosher WA, Hoffmann FW (1966) Isolation of trans-delta-tetrahydrocannabinol from marijuana. J Am Chem Soc 88:1832–1833. https://doi.org/10.1021/ja00960a056

Holland ML, Lau DT, Allen JD, Arnold JC (2007) The multidrug transporter ABCG2 (BCRP) is inhibited by plant-derived cannabinoids. Br J Pharmacol 152:815–824. https://doi.org/10.1038/sj.bjp.0707467

Holland ML, Allen JD, Arnold JC (2008) Interaction of plant cannabinoids with the multidrug transporter ABCC1 (MRP1). Eur J Pharmacol 591:128–131. https://doi.org/10.1016/j.ejphar.2008.06.079

Howlett AC, Barth F, Bonner TI, Cabral G, Casellas P, Devane WA et al (2002) International Union of Pharmacology. XXVII. Classification of cannabinoid receptors. Pharmacol Rev 54:161–202. https://doi.org/10.1124/pr.54.2.161

Huang H, McIntosh AL, Martin GG, Dangott LJ, Kier AB, Schroeder F (2018) Structural and functional interaction of Delta(9)-tetrahydrocannabinol with liver fatty acid binding protein (FABP1). Biochemistry 57:6027–6042. https://doi.org/10.1021/acs.biochem.8b00744

Huang CW, Lin PC, Chen JL, Lee MJ (2021) Cannabidiol selectively binds to the voltage-gated sodium channel Na(v)1.4 in its slow-inactivated state and inhibits sodium current. Biomedicine 9. https://doi.org/10.3390/biomedicines9091141

Huestis MA, Henningfield JE, Cone EJ (1992) Blood cannabinoids. I. Absorption of THC and formation of 11-OH-THC and THCCOOH during and after smoking marijuana. J Anal Toxicol 16:276–282. https://doi.org/10.1093/jat/16.5.276

Husni AS, McCurdy CR, Radwan MM, Ahmed SA, Slade D, Ross SA et al (2014) Evaluation of phytocannabinoids from high potency Cannabis sativa using in vitro bioassays to determine structure-activity relationships for cannabinoid receptor 1 and cannabinoid receptor 2. Med Chem Res 23:4295–4300. https://doi.org/10.1007/s00044-014-0972-6

Iannotti FA, De Maio F, Panza E, Appendino G, Taglialatela-Scafati O, De Petrocellis L et al (2020) Identification and characterization of Cannabimovone, a cannabinoid from Cannabis sativa, as a novel PPARgamma agonist via a combined computational and functional study. Molecules 25. https://doi.org/10.3390/molecules25051119

Iwata N, Kitanaka S (2011) New cannabinoid-like chromane and chromene derivatives from *Rhododendron anthopogonoides*. Chem Pharm Bull (Tokyo) 59:1409–1412. https://doi.org/10.1248/cpb.59.1409

Janssens A, Silvestri C, Martella A, Vanoevelen JM, Di Marzo V, Voets T (2018) Delta(9)-tetrahydrocannabivarin impairs epithelial calcium transport through inhibition of TRPV5 and TRPV6. Pharmacol Res 136:83–89. https://doi.org/10.1016/j.phrs.2018.08.021

Jin D, Henry P, Shan J, Chen J (2021a) Classification of cannabis strains in the Canadian market with discriminant analysis of principal components using genome-wide single nucleotide polymorphisms. PLoS One 16:e0253387. https://doi.org/10.1371/journal.pone.0253387

Jin D, Henry P, Shan J, Chen J (2021b) Identification of chemotypic markers in three chemotype categories of cannabis using secondary metabolites profiled in inflorescences, leaves, stem bark, and roots. Front Plant Sci 12:699530. https://doi.org/10.3389/fpls.2021.699530

Jordt SE, Bautista DM, Chuang HH, McKemy DD, Zygmunt PM, Hogestatt ED et al (2004) Mustard oils and cannabinoids excite sensory nerve fibres through the TRP channel ANKTM1. Nature 427:260–265. https://doi.org/10.1038/nature02282

Kathmann M, Flau K, Redmer A, Trankle C, Schlicker E (2006) Cannabidiol is an allosteric modulator at mu- and delta-opioid receptors. Naunyn Schmiedeberg's Arch Pharmacol 372: 354–361. https://doi.org/10.1007/s00210-006-0033-x

Keith JM, Apodaca R, Xiao W, Seierstad M, Pattabiraman K, Wu J et al (2008) Thiadiazolopiperazinyl ureas as inhibitors of fatty acid amide hydrolase. Bioorg Med Chem Lett 18:4838–4843. https://doi.org/10.1016/j.bmcl.2008.07.081

Kozak KR, Gupta RA, Moody JS, Ji C, Boeglin WE, DuBois RN et al (2002) 15-Lipoxygenase metabolism of 2-arachidonylglycerol. Generation of a peroxisome proliferator-activated receptor alpha agonist. J Biol Chem 277:23278–23286. https://doi.org/10.1074/jbc.M201084200

Lamtha T, Jongkon N, Lertvanithphol T, Horprathum M, Seetaha S, Choowongkomon K (2025) Cannabinoids as promising inhibitors of HER2-tyrosine kinase: a novel strategy for targeting HER2-positive ovarian cancer. ACS Omega 10:6191–6200. https://doi.org/10.1021/acsomega.4c11108

Lapierre E, Monthony AS, Torkamaneh D (2023) Genomics-based taxonomy to clarify cannabis classification. Genome 66:202–211. https://doi.org/10.1139/gen-2023-0005

Laprairie RB, Bagher AM, Kelly ME, Denovan-Wright EM (2015) Cannabidiol is a negative allosteric modulator of the cannabinoid CB_1 receptor. Br J Pharmacol 172:4790–4805. https://doi.org/10.1111/bph.13250

Laun AS, Song ZH (2017) GPR3 and GPR6, novel molecular targets for cannabidiol. Biochem Biophys Res Commun 490:17–21. https://doi.org/10.1016/j.bbrc.2017.05.165

Laun AS, Shrader SH, Song ZH (2018) Novel inverse agonists for the orphan G protein coupled receptor 6. Heliyon 4:e00933. https://doi.org/10.1016/j.heliyon.2018.e00933

Le Marois M, Ballet V, Sanson C, Maizieres MA, Carriot T, Chantoiseau C et al (2020) Cannabidiol inhibits multiple cardiac ion channels and shortens ventricular action potential duration in vitro. Eur J Pharmacol 886:173542. https://doi.org/10.1016/j.ejphar.2020.173542

Leishman E, Manchanda M, Thelen R, Miller S, Mackie K, Bradshaw HB (2018) Cannabidiol's upregulation of N-acyl ethanolamines in the central nervous system requires N-acyl phosphatidyl ethanolamine-specific phospholipase D. Cannabis Cannabinoid Res 3:228–241. https://doi.org/10.1089/can.2018.0031

Linciano P, Citti C, Luongo L, Belardo C, Maione S, Vandelli MA et al (2020a) Isolation of a high-affinity cannabinoid for the human CB1 receptor from a medicinal Cannabis sativa variety: Delta(9)-tetrahydrocannabutol, the butyl homologue of Delta(9)-tetrahydrocannabinol. J Nat Prod 83:88–98. https://doi.org/10.1021/acs.jnatprod.9b00876

Linciano P, Citti C, Russo F, Tolomeo F, Lagana A, Capriotti AL et al (2020b) Identification of a new cannabidiol n-hexyl homolog in a medicinal cannabis variety with an antinociceptive activity in mice: cannabidihexol. Sci Rep 10:22019. https://doi.org/10.1038/s41598-020-79042-2

Lipinski CA, Lombardo F, Dominy BW, Feeney PJ (2001) Experimental and computational approaches to estimate solubility and permeability in drug discovery and development settings. Adv Drug Deliv Rev 46:3–26. https://doi.org/10.1016/s0169-409x(00)00129-0

Mahardhika AB, Zaluski M, Schoeder CT, Boshta NM, Schabikowski J, Perri F et al (2024) Potent, selective agonists for the cannabinoid like orphan G protein-coupled receptor GPR18: a promising drug target for cancer and immunity. J Med Chem 67:9896–9926. https://doi.org/10.1021/acs.jmedchem.3c02423

Mahgoub M, Keun-Hang SY, Sydorenko V, Ashoor A, Kabbani N, Al Kury L et al (2013) Effects of cannabidiol on the function of alpha7-nicotinic acetylcholine receptors. Eur J Pharmacol 720: 310–319. https://doi.org/10.1016/j.ejphar.2013.10.011

Martin LJ, Cairns EA, Heblinski M, Fletcher C, Krycer JR, Arnold JC et al (2021) Cannabichromene and Delta(9)-tetrahydrocannabinolic acid identified as lactate dehydrogenase-a inhibitors by in silico and in vitro screening. J Nat Prod 84:1469–1477. https://doi.org/10.1021/acs.jnatprod.0c01281

Mason ER, Cummins TR (2020) Differential inhibition of human Nav1.2 resurgent and persistent sodium currents by cannabidiol and GS967. Int J Mol Sci 21. https://doi.org/10.3390/ijms21072454

Matsuda LA, Lolait SJ, Brownstein MJ, Young AC, Bonner TI (1990) Structure of a cannabinoid receptor and functional expression of the cloned cDNA. Nature 346:561–564. https://doi.org/10.1038/346561a0

McHugh D, Page J, Dunn E, Bradshaw HB (2012) Delta(9) -tetrahydrocannabinol and N-arachidonyl glycine are full agonists at GPR18 receptors and induce migration in human endometrial HEC-1B cells. Br J Pharmacol 165:2414–2424. https://doi.org/10.1111/j.1476-5381.2011.01497.x

McPartland JM, MacDonald C, Young M, Grant PS, Furkert DP, Glass M (2017) Affinity and efficacy studies of tetrahydrocannabinolic acid A at cannabinoid receptor types one and two. Cannabis Cannabinoid Res 2:87–95. https://doi.org/10.1089/can.2016.0032

Mechoulam R (2005) Plant cannabinoids: a neglected pharmacological treasure trove. Br J Pharmacol 146:913–915. https://doi.org/10.1038/sj.bjp.0706415

Mechoulam R, Shvo Y (1963) Hashish. I. The structure of cannabidiol. Tetrahedron 19:2073–2078. https://doi.org/10.1016/0040-4020(63)85022-x

Mendis GDC, Berecki G, Morrisroe E, Pachernegg S, Li M, Varney M et al (2019) Discovering the pharmacodynamics of conolidine and cannabidiol using a cultured neuronal network based workflow. Sci Rep 9:121. https://doi.org/10.1038/s41598-018-37138-w

Merkus FW (1971) Cannabivarin and tetrahydrocannabivarin, two new constituents of hashish. Nature 232:579–580. https://doi.org/10.1038/232579a0

Michalik L, Auwerx J, Berger JP, Chatterjee VK, Glass CK, Gonzalez FJ et al (2006) International Union of Pharmacology. LXI Peroxisome proliferator-activated receptors. Pharmacol Rev 58: 726–741. https://doi.org/10.1124/pr.58.4.5

Milligan CJ, Anderson LL, Bowen MT, Banister SD, McGregor IS, Arnold JC et al (2022) A nutraceutical product, extracted from Cannabis sativa, modulates voltage-gated sodium channel function. J Cannabis Res 4:30. https://doi.org/10.1186/s42238-022-00136-x

Mirlohi S, Bladen C, Santiago M, Connor M (2022) Modulation of recombinant human T-type calcium channels by Delta(9)-tetrahydrocannabinolic acid in vitro. Cannabis Cannabinoid Res 7:34–45. https://doi.org/10.1089/can.2020.0134

Mock ED, Mustafa M, Gunduz-Cinar O, Cinar R, Petrie GN, Kantae V et al (2020) Discovery of a NAPE-PLD inhibitor that modulates emotional behavior in mice. Nat Chem Biol 16:667–675. https://doi.org/10.1038/s41589-020-0528-7

Munro S, Thomas KL, Abu-Shaar M (1993) Molecular characterization of a peripheral receptor for cannabinoids. Nature 365:61–65. https://doi.org/10.1038/365061a0

Nadal X, Del Rio C, Casano S, Palomares B, Ferreiro-Vera C, Navarrete C et al (2017) Tetrahydrocannabinolic acid is a potent PPARgamma agonist with neuroprotective activity. Br J Pharmacol 174:4263–4276. https://doi.org/10.1111/bph.14019

Navarro G, Varani K, Lillo A, Vincenzi F, Rivas-Santisteban R, Raich I et al (2020) Pharmacological data of cannabidiol- and cannabigerol-type phytocannabinoids acting on cannabinoid CB(1), CB(2) and CB(1)/CB(2) heteromer receptors. Pharmacol Res 159:104940. https://doi.org/10.1016/j.phrs.2020.104940

Niphakis MJ, Cognetta AB 3rd, Chang JW, Buczynski MW, Parsons LH, Byrne F et al (2013) Evaluation of NHS carbamates as a potent and selective class of endocannabinoid hydrolase inhibitors. ACS Chem Neurosci 4:1322–1332. https://doi.org/10.1021/cn400116z

O'Shaughnessy WB (1841) The Bengal dispensatory and pharmacopoeia. vol. Valerianeae to Convolvulacea. Calcutta, India

O'Sullivan SE, Tarling EJ, Bennett AJ, Kendall DA, Randall MD (2005) Novel time-dependent vascular actions of Delta9-tetrahydrocannabinol mediated by peroxisome proliferator-activated receptor gamma. Biochem Biophys Res Commun 337:824–831. https://doi.org/10.1016/j.bbrc.2005.09.121

Palomares B, Garrido-Rodriguez M, Gonzalo-Consuegra C, Gomez-Canas M, Saen-Oon S, Soliva R et al (2020) Delta(9) -tetrahydrocannabinolic acid alleviates collagen-induced arthritis: role of PPARgamma and CB(1) receptors. Br J Pharmacol 177:4034–4054. https://doi.org/10.1111/bph.15155

Pandolfo P, Silveirinha V, dos Santos-Rodrigues A, Venance L, Ledent C, Takahashi RN et al (2011) Cannabinoids inhibit the synaptic uptake of adenosine and dopamine in the rat and mouse striatum. Eur J Pharmacol 655:38–45. https://doi.org/10.1016/j.ejphar.2011.01.013

Patel RR, Barbosa C, Brustovetsky T, Brustovetsky N, Cummins TR (2016) Aberrant epilepsy-associated mutant Nav1.6 sodium channel activity can be targeted with cannabidiol. Brain 139:2164–2181. https://doi.org/10.1093/brain/aww129

Peltner LK, Gluthmann L, Borner F, Pace S, Hoffstetter RK, Kretzer C et al (2023) Cannabidiol acts as molecular switch in innate immune cells to promote the biosynthesis of inflammation-resolving lipid mediators. Cell Chem Biol 30(1508–1524):e1507. https://doi.org/10.1016/j.chembiol.2023.08.001

Pokl M, Sridhar A, Frampton DJA, Linhart VA, Delemotte L, Liin SI (2023) Subtype-specific modulation of human K(V) 7 channels by the anticonvulsant cannabidiol through a lipid-exposed pore-domain site. Br J Pharmacol 180:2956–2972. https://doi.org/10.1111/bph.16183

Puopolo T, Cai A, Liu C, Ma H, Seeram NP (2023) Investigating cannabinoids as P2X purinoreceptor 4 ligands by using surface plasmon resonance and computational docking. Heliyon 9:e21265. https://doi.org/10.1016/j.heliyon.2023.e21265

Qian Y, Wang X, Markowitz JS (2019) In vitro inhibition of carboxylesterase 1 by major cannabinoids and selected metabolites. Drug Metab Dispos 47:465–472. https://doi.org/10.1124/dmd.118.086074

Qin N, Neeper MP, Liu Y, Hutchinson TL, Lubin ML, Flores CM (2008) TRPV2 is activated by cannabidiol and mediates CGRP release in cultured rat dorsal root ganglion neurons. J Neurosci 28:6231–6238. https://doi.org/10.1523/JNEUROSCI.0504-08.2008

Rhee MH, Vogel Z, Barg J, Bayewitch M, Levy R, Hanus L et al (1997) Cannabinol derivatives: binding to cannabinoid receptors and inhibition of adenylylcyclase. J Med Chem 40:3228–3233. https://doi.org/10.1021/jm970126f

Ribeiro A, Pontis S, Mengatto L, Armirotti A, Chiurchiu V, Capurro V et al (2015) A potent systemically active N-acylethanolamine acid amidase inhibitor that suppresses inflammation and human macrophage activation. ACS Chem Biol 10:1838–1846. https://doi.org/10.1021/acschembio.5b00114

Rockwell CE, Kaminski NE (2004) A cyclooxygenase metabolite of anandamide causes inhibition of interleukin-2 secretion in murine splenocytes. J Pharmacol Exp Ther 311:683–690. https://doi.org/10.1124/jpet.104.065524

Rosenthaler S, Pohn B, Kolmanz C, Huu CN, Krewenka C, Huber A et al (2014) Differences in receptor binding affinity of several phytocannabinoids do not explain their effects on neural cell cultures. Neurotoxicol Teratol 46:49–56. https://doi.org/10.1016/j.ntt.2014.09.003

Russo EB (2007) History of cannabis and its preparations in saga, science, and sobriquet. Chem Biodivers 4:1614–1648. https://doi.org/10.1002/cbdv.200790144

Russo EB (2011) Taming THC: potential cannabis synergy and phytocannabinoid-terpenoid entourage effects. Br J Pharmacol 163:1344–1364. https://doi.org/10.1111/j.1476-5381.2011.01238.x

Russo E, Guy GW (2006) A tale of two cannabinoids: the therapeutic rationale for combining tetrahydrocannabinol and cannabidiol. Med Hypotheses 66:234–246. https://doi.org/10.1016/j.mehy.2005.08.026

Russo EB, Burnett A, Hall B, Parker KK (2005) Agonistic properties of cannabidiol at 5-HT1a receptors. Neurochem Res 30:1037–1043. https://doi.org/10.1007/s11064-005-6978-1

Ryberg E, Larsson N, Sjogren S, Hjorth S, Hermansson NO, Leonova J et al (2007) The orphan receptor GPR55 is a novel cannabinoid receptor. Br J Pharmacol 152:1092–1101. https://doi.org/10.1038/sj.bjp.0707460

Santillo MF, Glenn IS, Paris L, Sprando RL (2024) In vitro biological target screening and colloidal aggregation of minor cannabinoids. J Nat Prod 87:2881–2886. https://doi.org/10.1021/acs.jnatprod.4c00608

Schmiedhofer P, Vogel FD, Koniuszewski F, Ernst M (2022) Cys-loop receptors on cannabinoids: all high? Front Physiol 13:1044575. https://doi.org/10.3389/fphys.2022.1044575

Sea YL, Gee YJ, Lal SK, Choo WS (2023) Cannabis as antivirals. J Appl Microbiol 134. https://doi.org/10.1093/jambio/lxac036

Sempio C, Huestis MA, Mikulich-Gilbertson SK, Klawitter J, Christians U, Henthorn TK (2020) Population pharmacokinetic modeling of plasma Delta9-tetrahydrocannabinol and an active and inactive metabolite following controlled smoked cannabis administration. Br J Clin Pharmacol 86:611–619. https://doi.org/10.1111/bcp.14170

Serventi L, Flores GA, Cusumano G, Barbaro D, Tirillini B, Venanzoni R et al (2023) Comparative investigation of antimicrobial and antioxidant effects of the extracts from the inflorescences and leaves of the Cannabis sativa L cv strawberry. Antioxidants (Basel) 12. https://doi.org/10.3390/antiox12020219

Smart D, Gunthorpe MJ, Jerman JC, Nasir S, Gray J, Muir AI et al (2000) The endogenous lipid anandamide is a full agonist at the human vanilloid receptor (hVR1). Br J Pharmacol 129:227–230. https://doi.org/10.1038/sj.bjp.0703050

Stott LA, Brighton CA, Brown J, Mould R, Bennett KA, Newman R et al (2021) Characterisation of inverse agonism of the orphan-G protein-coupled receptor GPR52 by cannabinoid ligands Cannabidiol and O-1918. Heliyon 7:e07201. https://doi.org/10.1016/j.heliyon.2021.e07201

Sugiura T, Kondo S, Sukagawa A, Nakane S, Shinoda A, Itoh K et al (1995) 2-Arachidonoylglycerol: a possible endogenous cannabinoid receptor ligand in brain. Biochem Biophys Res Commun 215:89–97. https://doi.org/10.1006/bbrc.1995.2437

Sugiura T, Kondo S, Sukagawa A, Tonegawa T, Nakane S, Yamashita A et al (1996) Transacylase-mediated and phosphodiesterase-mediated synthesis of *N*-arachidonoylethanolamine, an endogenous cannabinoid-receptor ligand, in rat brain microsomes. Comparison with synthesis from free arachidonic acid and ethanolamine. Eur J Biochem 240:53–62. https://doi.org/10.1111/j.1432-1033.1996.0053h.x

Takeda S, Ikeda E, Su S, Harada M, Okazaki H, Yoshioka Y et al (2014) Delta(9)-THC modulation of fatty acid 2-hydroxylase (FA2H) gene expression: possible involvement of induced levels of PPARalpha in MDA-MB-231 breast cancer cells. Toxicology 326:18–24. https://doi.org/10.1016/j.tox.2014.09.011

Taylor L, Gidal B, Blakey G, Tayo B, Morrison G (2018) A phase I, randomized, double-blind, placebo-controlled, single ascending dose, multiple dose, and food effect trial of the safety, tolerability and pharmacokinetics of highly purified Cannabidiol in healthy subjects. CNS Drugs 32:1053–1067. https://doi.org/10.1007/s40263-018-0578-5

Thomas A, Stevenson LA, Wease KN, Price MR, Baillie G, Ross RA et al (2005) Evidence that the plant cannabinoid Delta9-tetrahydrocannabivarin is a cannabinoid CB1 and CB2 receptor antagonist. Br J Pharmacol 146:917–926. https://doi.org/10.1038/sj.bjp.0706414

Udoh M, Santiago M, Devenish S, McGregor IS, Connor M (2019) Cannabichromene is a cannabinoid CB(2) receptor agonist. Br J Pharmacol 176:4537–4547. https://doi.org/10.1111/bph.14815

Udoh M, Bladen C, Heblinski M, Luo JL, Janve VS, Anderson LL et al (2022) The anticonvulsant phytocannabinoids CBGVA and CBDVA inhibit recombinant T-type channels. Front Pharmacol 13:1048259. https://doi.org/10.3389/fphar.2022.1048259

Udoh M, Santiago M, Haneef S, Rodger A, Marlowe CK, Barr PJ et al (2025) Activation of CB2 receptors by (−)-Cannabichromene but Not (+)-Cannabichromene. Cannabis Cannabinoid Res 10:e94–e101. https://doi.org/10.1089/can.2023.0212

Ueda N, Yamanaka K, Terasawa Y, Yamamoto S (1999) An acid amidase hydrolyzing anandamide as an endogenous ligand for cannabinoid receptors. FEBS Lett 454:267–270. https://doi.org/10.1016/s0014-5793(99)00820-0

Vitale RM, Morace AM, D'Errico A, Ricciardi F, Fusco A, Boccella S et al (2025) Identification of Cannabidiolic and Cannabigerolic acids as MTDL AChE, BuChE, and BACE-1 inhibitors against Alzheimer's disease by in silico, in vitro, and in vivo studies. Phytother Res 39:233–245. https://doi.org/10.1002/ptr.8369

Vollner L, Bieniek D, Korte F (1969) Hashish. XX. Cannabidivarin, a new hashish constituent. Tetrahedron Lett 10:145–147. https://doi.org/10.1016/s0040-4039(01)87494-3

Watanabe K, Itokawa Y, Yamaori S, Funahashi T, Kimura T, Kaji T et al (2007) Conversion of cannabidiol to Δ9-tetrahydrocannabinol and related cannabinoids in artificial gastric juice, and their pharmacological effects in mice. Forensic Toxicol 25:16–21. https://doi.org/10.1007/s11419-007-0021-y

Wood JS, Gordon WH, Morgan JB, Williamson RT (2023) Cannabicitran: its unexpected racemic nature and potential origins. Chirality 35:540–548. https://doi.org/10.1002/chir.23571

Wu LJ, Sweet TB, Clapham DE (2010) International Union of Basic and Clinical Pharmacology. LXXVI. Current progress in the mammalian TRP ion channel family. Pharmacol Rev 62:381–404. https://doi.org/10.1124/pr.110.002725

Xiong W, Cheng K, Cui T, Godlewski G, Rice KC, Xu Y et al (2011a) Cannabinoid potentiation of glycine receptors contributes to cannabis-induced analgesia. Nat Chem Biol 7:296–303. https://doi.org/10.1038/nchembio.552

Xiong W, Koo BN, Morton R, Zhang L (2011b) Psychotropic and nonpsychotropic cannabis derivatives inhibit human 5-HT(3A) receptors through a receptor desensitization-dependent mechanism. Neuroscience 184:28–37. https://doi.org/10.1016/j.neuroscience.2011.03.066

Yamamoto I, Gohda H, Narimatsu S, Yoshimura H (1988) Identification of cannabielsoin, a new metabolite of cannabidiol formed by Guinea-pig hepatic microsomal enzymes, and its pharmacological activity in mice. J Pharmacobiodyn 11:833–838. https://doi.org/10.1248/bpb1978.11.833

Yamaori S, Kushihara M, Yamamoto I, Watanabe K (2010) Characterization of major phytocannabinoids, cannabidiol and cannabinol, as isoform-selective and potent inhibitors of human CYP1 enzymes. Biochem Pharmacol 79:1691–1698. https://doi.org/10.1016/j.bcp.2010.01.028

Yamaori S, Okamoto Y, Yamamoto I, Watanabe K (2011) Cannabidiol, a major phytocannabinoid, as a potent atypical inhibitor for CYP2D6. Drug Metab Dispos 39:2049–2056. https://doi.org/10.1124/dmd.111.041384

Yamaori S, Koeda K, Kushihara M, Hada Y, Yamamoto I, Watanabe K (2012) Comparison in the *in vitro* inhibitory effects of major phytocannabinoids and polycyclic aromatic hydrocarbons contained in marijuana smoke on cytochrome P450 2C9 activity. Drug Metab Pharmacokinet 27:294–300. https://doi.org/10.2133/dmpk.dmpk-11-rg-107

Yamaori S, Okushima Y, Masuda K, Kushihara M, Katsu T, Narimatsu S et al (2013) Structural requirements for potent direct inhibition of human cytochrome P450 1A1 by cannabidiol: role of pentylresorcinol moiety. Biol Pharm Bull 36:1197–1203. https://doi.org/10.1248/bpb.b13-00183

Yang KH, Galadari S, Isaev D, Petroianu G, Shippenberg TS, Oz M (2010a) The nonpsychoactive cannabinoid cannabidiol inhibits 5-hydroxytryptamine3A receptor-mediated currents in Xenopus laevis oocytes. J Pharmacol Exp Ther 333:547–554. https://doi.org/10.1124/jpet.109.162594

Yang KH, Isaev D, Morales M, Petroianu G, Galadari S, Oz M (2010b) The effect of Delta9-tetrahydrocannabinol on 5-HT3 receptors depends on the current density. Neuroscience 171:40–49. https://doi.org/10.1016/j.neuroscience.2010.08.044

Zagzoog A, Mohamed KA, Kim HJJ, Kim ED, Frank CS, Black T et al (2020) In vitro and in vivo pharmacological activity of minor cannabinoids isolated from *Cannabis sativa*. Sci Rep 10:20405. https://doi.org/10.1038/s41598-020-77175-y

Zhang HB, Heckman L, Niday Z, Jo S, Fujita A, Shim J et al (2022) Cannabidiol activates neuronal Kv7 channels. elife 11. https://doi.org/10.7554/eLife.73246

Zhu HJ, Wang JS, Markowitz JS, Donovan JL, Gibson BB, Gefroh HA et al (2006) Characterization of P-glycoprotein inhibition by major cannabinoids from marijuana. J Pharmacol Exp Ther 317:850–857. https://doi.org/10.1124/jpet.105.098541

Zygmunt PM, Ermund A, Movahed P, Andersson DA, Simonsen C, Jonsson BA et al (2013) Monoacylglycerols activate TRPV1 – a link between phospholipase C and TRPV1. PLoS One 8:e81618. https://doi.org/10.1371/journal.pone.0081618

Canonical Cannabinoid Receptors

Beth Ehrlich, Monica Patel, Xiaoxi Zheng, and Michelle Glass

Contents

1 G Protein-Coupled Receptors ... 39
2 The Cannabinoid Receptors ... 40
3 Receptor Distribution and Function ... 42
3.1 CB_1 ... 42
3.2 CB_2 ... 44
4 CB_1-Mediated Intracellular Signalling ... 45
4.1 Inhibition of Adenylate Cyclase ... 45
4.2 Activation of Mitogen-Activated Protein Kinases ... 48
4.3 Activation of PI3K/Akt and Transactivation of Receptor Tyrosine Kinases (RTKs) 51
4.4 Modulation of Ion Channels ... 51
5 CB_1 Regulation ... 52
5.1 G Protein-Coupled Receptor Kinases (GRKs) ... 52
5.2 β-Arrestins ... 53
6 Conclusions ... 54
References ... 55

Abstract This chapter will review the basic pharmacology of the canonical cannabinoid receptors. The endocannabinoid system is a complex signalling network involved in a wide range of physiological processes, including pain modulation, appetite regulation, and synaptic plasticity. The canonical cannabinoid receptors, CB_1 and CB_2, are central in orchestrating this system. CB_1 is highly enriched in the central nervous system (CNS), where it plays a crucial role in modulating neurotransmitter release and synaptic plasticity. In contrast, CB_2 is predominantly expressed in peripheral tissues and immune cells, participating in anti inflammatory

B. Ehrlich and M. Patel
Department of Pharmacology and Toxicology, University of Otago, Dunedin, New Zealand

X. Zheng and M. Glass (✉)
Department of Pharmacology and Toxicology, University of Otago, Dunedin, New Zealand

Institute of Environmental Science and Research Limited (ESR) Kenepuru Science Centre, Porirua, New Zealand
e-mail: michelle.glass@otago.ac.nz

Curr Topics Behav Neurosci (2026) 76: 37–68
https://doi.org/10.1007/7854_2024_556
Published Online: 07 March 2025

processes. Here, we focus on cannabinoid receptor distribution, intracellular signalling, and receptor regulation. We describe the intracellular signalling pathways activated by CB_1, including the modulation of ion channels, second messengers, and protein kinases. Overall, this chapter provides an overview of the canonical cannabinoid receptors and their role in the regulation of neuronal signalling and plasticity, highlighting the molecular and cellular mechanisms underlying their effects in the CNS.

Keywords 2-Arachidonyl glycerol · Anandamide · cAMP · Cannabinoid · CB_1 · CB_2 · Endocannabinoid · ERK · G protein · Protein kinase · Retrograde neurotransmitter · Signalling

Abbreviations

AP2	Adapter protein 2
AEA	*N*-arachidonoylethanolamide (anandamide)
ATP	Adenosine triphosphate
AC	Adenylate cyclase
BDNF	Brain-derived neurotrophic factor
BRET	Bioluminescence resonance energy transfer
eCB	Endocannabinoid
ECS	Endocannabinoid system
EGFP	Enhanced green fluorescent protein
Epac	Exchange protein directly activated by cAMP
ERK	Extracellular signal-regulated kinase
cAMP	Cyclic adenosine monophosphate
CBD	Cannabidiol
CB_1	Cannabinoid CB_1 receptor
CB_2	Cannabinoid CB_2 receptor
CHO	Chinese hamster ovary
CNS	Central nervous system
D_2	Dopamine D_2 receptor
GABA	ɣ-Aminobutyric acid
GDP	Guanosine diphosphate
GEF	Guanine nucleotide exchange factor
GIRK	G protein-coupled inwardly rectifying K^+ channel
GPCR	G protein-coupled receptor
GRK	G protein-coupled receptor kinase
GTP	Guanosine triphosphate
HEK	Human embryonic kidney
IEG	Immediate early gene
JNK	c-Jun N-terminal kinase
MAPK	Mitogen-activated protein kinase

MAPKK	MAPK kinase
MAPKKK	MAPKK kinase
MEK	MAPK-ERK kinase
mGluR	Metabotropic glutamate receptor
$mtCB_1$	Mitochondrial CB_1
PCR	Polymerase chain reaction
PI3K	Phosphatidylinositol-3-kinase
PLC	Phospholipase C
PKA	Protein kinase A
PTX	Pertussis toxin
RTK	Receptor tyrosine kinase
S1P	Sphingosine 1-phosphate receptor
SCRA	Synthetic cannabinoid receptor agonist
SNP	Single nucleotide polymorphism
THC	Δ^9-Tetrahydrocannabinol
TrkB	Tropomyosin receptor kinase B
VGCC	Voltage-gated Ca^{2+} channel.
2-AG	2-Arachidonylglycerol

1 G Protein-Coupled Receptors

The interaction between receptors and ligands is central to a plethora of biological events and, therefore, underlies the activity of many drugs. G protein-coupled receptors (GPCRs) are a family of integral membrane proteins and are responsible for the therapeutic effects of approximately one-third of clinically approved agents (Hauser et al. 2017). GPCRs transduce external stimuli into specific intracellular signals that regulate many physiological processes, including vision, olfaction, cognition, and pain. Consequently, GPCRs are attractive drug targets with over 800 members subdivided into distinct classes based on homology and physiology; class A (rhodopsin), class B (secretin), class C (metabotropic glutamate), and class F (frizzled) (Schiöth and Fredriksson 2005). Accordingly, the endogenous ligands for GPCRs are chemically diverse, consisting of carbohydrates, growth factors, ions, lipids, hormones, neurotransmitters, photons, and peptides. Despite this, the overall structural architecture and mechanisms of activation are remarkably conserved across families, often allowing for the translation of novel pharmacological insights from one receptor to another. All GPCRs consist of an extracellular N-terminus, connected to seven transmembrane α-helices joined by alternating intra- and extra-cellular loops, ending with an intracellular C-terminus (Hill 2006).

In their active state, GPCRs engage with and stimulate heterotrimeric G proteins, setting off a chain of molecular events within the cell that leads to various physiological responses. G proteins are comprised of three distinct subunits designated α, β and γ, in order of decreasing mass. While the α subunit is known for its ability to bind and hydrolyse GTP, the β and γ subunits are tightly associated and form an obligate dimer. The subunits are products of sixteen, five and twelve genes, respectively, offering a vast array of potential combinations and functions (Downes and Gautam 1999). The Gα subunits are categorised into four families based on sequence homology and the specific signalling cascade that they initiate: the $G\alpha_s$ and $G\alpha_{i/o}$ families directly affect the activity of adenylate cyclase (AC) in opposing manners, with $G\alpha_s$ stimulating and $G\alpha_{i/o}$ inhibiting its function; the $G\alpha_{q/11}$ family is responsible for triggering intracellular Ca^{2+} release; while $G\alpha_{12/13}$ are involved in cytoskeletal rearrangement. For a detailed review of G protein signalling, regulation, and function, see Syrovatkina et al. (2016).

The active GPCR causes a conformational change in the Gα subunit that reduces its affinity for GDP, facilitating the exchange of GDP for GTP. One ligand-receptor complex may activate multiple G proteins. The exchange of GDP for GTP prompts the dissociation of Gα from Gβγ. Each subunit then activates various primary effectors; it was long believed that Gα was responsible for initiating the majority of signalling events, however, the role of Gβγ continues to expand (as reviewed by Khan et al. (2013)). The primary effectors downstream of Gα (or, Gβγ) may also modulate second messengers, triggering a cascade of signalling events that amplify the signal generated by a small number of receptors, thereby broadening the cellular response.

2 The Cannabinoid Receptors

The discovery of the Endocannabinoid System (ECS) was initially prompted by an inquisition into the pharmacological effects of cannabis. Extracts of *Cannabis sativa* have been exploited recreationally and medicinally for centuries (Crocq 2020). However, the chemical structure of its primary psychoactive component, Δ^9-tetrahydrocannabinol (THC) was not elucidated until 1964 (Gaoni and Mechoulam 1964). The physiological effects of cannabinoids were initially ascribed to their high lipophilicity, where intercalation into cell membranes was proposed to disrupt the binding or uptake of hormones and neurotransmitters (Hillard et al. 1985; Howlett and Fleming 1984). The synthesis and characterisation of synthetic analogues of THC suggested the existence of a cannabinoid-responsive $G\alpha_{i/o}$-linked GPCR (Devane et al. 1988; Howlett et al. 1986), which was subsequently cloned (Matsuda et al. 1990) and termed the cannabinoid CB_1 receptor (CB_1). Degenerate PCR strategies were then employed to clone a second cannabinoid GPCR enriched in immune cells (Munro et al. 1993) – the cannabinoid CB_2 receptor (CB_2). CB_1 and CB_2 share 44% amino acid homology overall, with divergences in tissue distribution and function.

Following the discovery of the cannabinoid receptors, several putative endocannabinoids (eCBs) have been uncovered. In 1992, Devane et al. (1992) reported the isolation of the first eCB. This ligand was identified in porcine brain preparations and was found to compete with radiolabelled [^{3}H]HU-243 in a centrifugal radioligand binding assay using rat synaptosomal membranes. The structure, determined by mass spectrometry and nuclear magnetic resonance spectroscopy, resembled *N*-arachidonoylethanolamide (AEA), a derivative of arachidonic acid. It was named anandamide, after a Sanskrit word (*ananda*) that means bliss, happiness, or pleasure. This name reflects the role of the endocannabinoid system (ECS) in cannabis-related intoxication. AEA acts with partial efficacy at both cannabinoid receptors but is more efficacious at CB_1 (Bayewitch et al. 1995; Howlett et al. 2002; Pertwee 1999; Soethoudt et al. 2017).

A few years later, a second eCB, 2-arachidonoylglycerol (2-AG), was identified in canine intestinal tissue (Mechoulam et al. 1995; Sugiura et al. 1995). Since then, several other putative eCBs have been identified (Bradshaw and Walker 2005) including *N*-dihomo-γ-linolenoylethanolamide and *N*-docosatetraenoylethanolamide (Hanus et al. 1993), 2-arachidonylglycerol ether (noladin ether) (Hanuš et al. 2001), O-arachidonoylethanolamine (virodhamine, O-AEA) (Porter et al. 2002), *N*-arachidonoyldopamine (NADA) (Bisogno et al. 2000), *N*-oleoyldopamine (OLDA) (Chu et al. 2003), and oleamide (Leggett et al. 2004). 2-AG is considered the dominant eCB in the brain and acts as a full agonist at both receptors (Ben-Shabat et al. 1998; Mechoulam et al. 1995). Both 2-AG and AEA exhibit sub-micromolar potency at CB_1 and CB_2 (Soethoudt et al. 2017), and are classed as eicosanoids; lipophilic molecules derived from fatty acids. Although several eCBs have been identified, AEA and 2-AG remain the most investigated and are considered the primary endogenous ligands of the cannabinoid receptor family.

The CB_1 receptor is encoded by the CNR1 gene, which is 26.1 kb in length and located on chromosome 6q14-q15 in humans. The full-length variant is 472 amino acids long. In addition to the full-length variant, two hCB_1 splice variants have been identified in humans: hCB_{1a} (Shire et al. 1995) and hCB_{1b} (Ryberg et al. 2005). The evidence for the expression of each variant across different tissues and cell types is conflicting, however, all agree that the full-length variant is the most abundantly expressed variant in the brain (González-Mariscal et al. 2016; Ryberg et al. 2005; Xiao et al. 2008). The CB_2 receptor is encoded by CNR2, which purportedly transcribes two isoforms in humans. The predominant isoform is expressed in immune cells, particularly the spleen (CB_{2B}), and the alternative isoform is transcribed in the testes and reward regions of the brain (CB_{2A}) (Liu et al. 2009).

Both cannabinoid receptors belong to class A, the rhodopsin-like, GPCR family. Typically, the (water-soluble) ligands of Class A GPCRs bind to the extracellular domains of their receptors. However, lipid receptors, such as CB_1 and CB_2, require binding sites that are accessible to hydrophobic ligands. The ligand binding site for CB_1 is therefore located deep within the protein structure and is believed to be accessed via a "lipid portal" from within the lipid bi-layer between transmembrane helix (TMH) 1 and TMH7 (Krishna Kumar et al. 2019; Shao et al. 2016). This mode of entry is typical for other Class A lipid receptors, such as the sphingosine 1-phosphate receptor (S1P) (Hanson et al. 2012).

In addition to the two primary cannabinoid receptors, the ECS consists of eCBs, and the proteins responsible for their transport, synthesis and degradation. The eCBs, AEA and 2-AG, are synthesised on demand in response to neuronal activity and act as retrograde messengers to supress neurotransmitter release. Through these receptors, the ECS regulates a wide range of neurological processes, including the perception of pain, central regulation of appetite and energy expenditure, neuroprotection, neurotransmission, synaptic plasticity, and the cellular basis of learning and memory. This chapter focuses on the canonical cannabinoid receptors, with an emphasis on CB_1, the predominant neurological cannabinoid receptor.

3 Receptor Distribution and Function

CB_1 is the predominant cannabinoid receptor in the central nervous system (CNS), regulating neurotransmitter release, synaptic plasticity, and neuronal excitability (reviewed by Kano et al. (2009)). CB_2 is mainly expressed in peripheral tissues and immune cells, modulating inflammation, cell proliferation and apoptosis (Simard et al. 2022). In this section, we describe the localisation of CB_1 and CB_2, and their roles in various physiological processes.

3.1 *CB_1*

CB_1 is one of the most prevalent GPCRs in the CNS (Glass et al. 1997; Herkenham et al. 1991; Tsou et al. 1998), found in both excitatory and inhibitory neurons. Quantitative receptor autoradiography in rat and human brain tissue revealed the highest densities of [^{3}H]CP55940 binding in cortical brain regions, the basal ganglia, hippocampus, and the molecular layer of the cerebellum – areas crucial for cognition, learning, memory, and motor and sensory functions (Glass et al. 1997; Herkenham et al. 1991). Although CP55940 is a non-selective cannabinoid ligand, the distribution of CB_1 has been substantiated by in situ hybridisation (Mailleux and Vanderhaeghen 1992; Westlake et al. 1994) and immunohistochemical studies (Farquhar-Smith et al. 2000; Mailleux and Vanderhaeghen 1992; Tsou et al. 1998). Over the years, the localisation of CB_1 has been further verified through disease and lesion model studies, as well as high-resolution electron microscopy. Its distribution aligns well with the effects of cannabinoid receptor activation.

CB_1 expression is highly polarised to axons and pre-synaptic sites, while the enzymes responsible for eCB synthesis are located post-synaptically in the dendritic spine (Katona et al. 1999, 2001, 2006; Mátyás et al. 2008). These findings contributed to the discovery that 2-AG signals backwards across the synapse, as a retrograde neurotransmitter (Fig. 1). The retrograde action of 2-AG is essential for regulating the strength and duration of synaptic transmission (Kreitzer and Regehr 2001; Ohno-Shosaku et al. 2001; Wilson and Nicoll 2001). It provides a feedback mechanism that helps to prevent excessive neural activity, making it an important component of synaptic plasticity. In addition to retrograde signalling, 2-AG released from pyramidal neurons activates somatodendritic CB_1 at its own membrane; a form

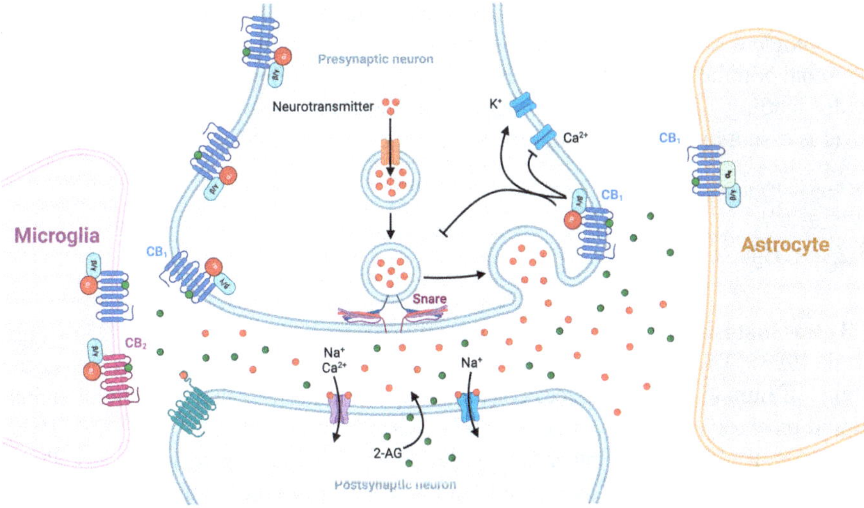

Fig. 1 Canonical cannabinoid receptors at the synapse. The presynaptic neuron releases neurotransmitters, such as glutamate, into the synaptic cleft, which depolarises the post-synaptic neuron. This depolarisation triggers the release of endocannabinoids, such as 2-AG. 2-AG then traverses the synapse, in a retrograde manner, activating nearby cannabinoid receptors (CB_1, shown in blue; CB_2, shown in pink) on the pre-synaptic nerve terminal, and surrounding glial cells. Activation of CB_1 leads to dissociation of the G protein, $G\alpha_{i/o}$, subsequently stimulating G protein-coupled inwardly rectifying K^+ (GIRK) channels and inhibiting voltage-gated Ca^{2+} channels (VGCCs). This modulation reduces the probability of further neurotransmitter release. This figure was created using Biorender.com

of autocrine signalling to control intrinsic neuronal excitability (Marinelli et al. 2009).

CB_1 is widely expressed across most cell types in the CNS. They are particularly abundant in GABAergic neurons, which display significantly higher densities of CB_1 compared to glutamatergic neurons (Marsicano and Lutz 1999). In addition to projection neurons and interneurons, CB_1 receptors are present in glial cells, including astrocytes, where they contribute to the modulation of neuronal signalling. Recent studies also provide evidence of CB_1 receptors in microglia and oligodendrocytes, further highlighting their extensive distribution throughout the CNS (Marinelli et al. 2023; Stella 2010).

While CB_1 exhibits punctate labelling along the entire length of axons, it is also found in dense intracellular pools, with smaller subpopulations present in the membranes of soma and dendrites (Fletcher-Jones et al. 2020). Subcellular populations of CB_1 have been discovered in mitochondria (Bénard et al. 2012; Djeungoue-Petga and Hebert-Chatelain 2017; Hebert-Chatelain et al. 2014; Jong et al. 2018; Melser et al. 2017) and other intracellular organelles. These subcellular receptors are purported to initiate very different signalling cascades compared to their extracellular counterparts (Soria-Gomez et al. 2021), and have been reviewed in detail elsewhere (Djeungoue-Petga and Hebert-Chatelain 2017).

The widespread expression of CB_1 throughout the CNS highlights its diverse physiological repertoire. Given the heterogeneous quantity of CB_1 within each neuronal population and the breadth of physiological actions it commands, a detailed understanding of cannabinoid signalling at the regional, cellular, and subcellular level is essential for developing novel pharmacological interventions.

3.2 CB_2

CB_2 was first isolated in HL60 human promyelocytic leukemic cells in 1993 (Munro et al. 1993). The receptor was later identified in human leukocytes but not in the brains of humans, rats, or mice, consolidating the notion of the "peripheral cannabinoid receptor" (Bouaboula et al. 1993; Galiègue et al. 1995; McCoy et al. 1999; Schatz et al. 1997). The receptor's predominant localisation in immune cells is, therefore, unsurprisingly associated with various immunomodulatory functions, particularly immune suppression and anti-inflammation (Galiègue et al. 1995; Simard et al. 2022; Turcotte et al. 2016). However, CB_2 expression is not solely restricted to leukocytes, it is also moderately expressed in other peripheral tissues, such as bone, gastrointestinal tract, lung, kidney, spleen, tonsils, and testes (Atwood and Mackie 2010; Bouaboula et al. 1993; Galiègue et al. 1995; Jenkin et al. 2011; Vidlarova et al. 2022; Whyte et al. 2012; Wright et al. 2008).

The initial characterisation as the peripheral cannabinoid receptor has faced scrutiny as growing evidence indicates the presence of functional CB_2 in the CNS. Notably, CB_2 has been described in activated microglia during neuroinflammation and other neurological disorders, such as Alzheimer's disease, Huntington's disease, and multiple sclerosis (Benito et al. 2003; Bouchard et al. 2012; Carlisle et al. 2002; Concannon et al. 2016; Duffy et al. 2021; Maresz et al. 2005; Palazuelos et al. 2009; Yiangou et al. 2006). More recently, using mice with CB_2-dependent enhanced green fluorescent protein (EGFP) expression, López et al. (2018) reported CB_2 expression in the microglial cells of mice co-expressing Alzheimer's disease mutations, while CB_2 was undetectable in the CNS of their wild-type counterparts. This conflicted with an earlier transgenic GFP mouse model, which demonstrated CB_2 expression in the microglia of healthy brains (Schmöle et al. 2015). However, GFP expression may have occurred independently of CB_2 expression, leaving the presence of CB_2 in healthy brains unconfirmed (Jordan and Xi 2019).

CB_2 has also been detected in sensory, enteric, and central neurons, albeit with controversy, particularly surrounding the specificity of CB_2 antibodies employed in early studies (Ashton et al. 2014; Baek et al. 2013; Grabon et al. 2023; Marchalant et al. 2014; Zhang et al. 2019). Several studies have demonstrated its presence in dopaminergic (Humburg et al. 2021; Ma et al. 2019; Zhang et al. 2014), GABAergic (Morgan et al. 2009; Sadanandan et al. 2020), and glutamatergic neurons (Sánchez-Zavaleta et al. 2018; Stempel et al. 2016; Stumpf et al. 2018; Zhang et al. 2021), modulating both excitatory and inhibitory synaptic transmission via pre- or postsynaptic receptors (Andó et al. 2012; Brusco et al. 2008; Foster et al. 2016; Kim and Li

2015; Tang et al. 2015; Zhang et al. 2017; Zheng et al. 2015). Expression in these neurons may be associated with addiction, cognition, mood, nociception, and neuroinflammation (Grabon et al. 2023; Jordan and Xi 2019). The molecular mechanisms and consequences of CB_2 activation in the brain remain comparatively undefined, especially regarding neuropsychiatry. However, its low physiological expression and sparse distribution in the CNS has rendered the receptor an attractive therapeutic target for inflammatory disorders, due to a lack of associated psychoactivity.

4 CB_1-Mediated Intracellular Signalling

4.1 Inhibition of Adenylate Cyclase

CB_1 predominantly interacts with $G\alpha_{i/o}$ proteins; a finding that has been extensively proven using various experimental approaches, such as [^{35}S]-GTPγS binding and investigations into the sensitivity of CB_1-mediated responses to pertussis toxin (PTX). In response to CB_1 activation, $G\alpha_{i/o}$ proteins (in addition to $G\beta\gamma$) decrease the production of cyclic adenosine monophosphate (cAMP) by inhibiting adenylate cyclase (AC) (Fig. 2). This response has been observed following CB_1 stimulation in brain tissue and neuronal cells expressing endogenous CB_1, as well as transformed cell lines expressing recombinant CB_1 (Felder et al. 1995; Howlett 1985; Howlett et al. 1986, 2002; Matsuda et al. 1990).

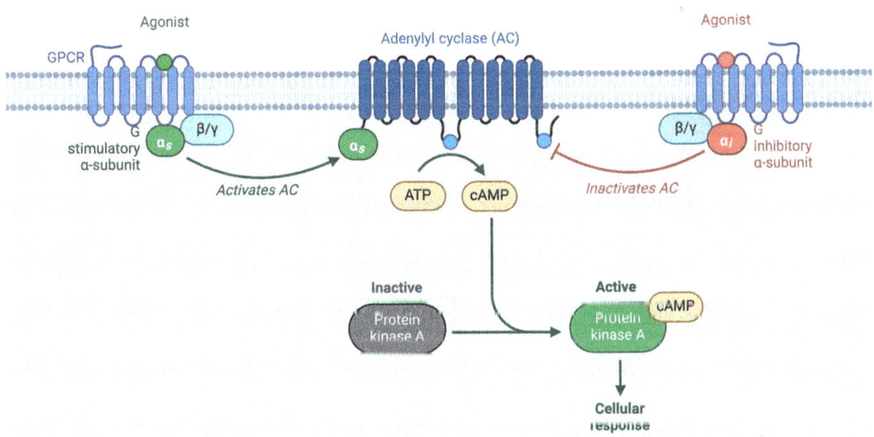

Fig. 2 G protein-mediated regulation of adenylate cyclase. G protein-coupled receptor activation of $G\alpha_s$ increases the activity of adenylate cyclase, enhancing cAMP production. This, in turn, fosters the activation of protein kinase A. In opposition, G protein-coupled receptor activation of $G\alpha_{i/o}$ instigates adenylate cyclase inhibition, attenuating cAMP production, which, in turn, diminishes protein kinase A activity. CB_1 canonically couples to the inhibitory pathway. This figure was created using Biorender.com

AC is an enzyme with two hydrophobic domains (each with six transmembrane α-helices), and two cytoplasmic domains where the catalytic activity takes place (Krupinski et al. 1989); its role is to catalyse the conversion of adenosine triphosphate (ATP) into cAMP. There are many isoforms of AC, each with distinct sensitivities to regulatory proteins ($G\alpha_s$, $G\alpha_{i/o}$, Gβɣ) and divergences in tissue distribution (see Hanoune and Defer (2001) for review). In addition to its other targets (reviewed by Gancedo (2013)), cAMP binds to the regulatory subunits of protein kinase A (PKA, also known as cAMP-dependent protein kinase), causing a conformational change that releases PKA's catalytic subunits, enabling it to phosphorylate various target substrates.

PKA is a serine/threonine protein kinase that phosphorylates proteins containing an Arg-Arg/Lys-X-Ser/Thr motif; this phosphorylation can either enhance or reduce the activity of its substrates. PKA has a broad spectrum of targets which differ between cell types. The cAMP/PKA signalling pathway plays a critical role in various biological processes, such as synaptic plasticity, neurotransmitter release, proliferation, neuronal differentiation, and neurogenesis.

Although CB_1 canonically couples to members of the $G\alpha_{i/o}$ family, it has also been reported to couple "promiscuously" to members of the other three families of Gα; namely, the $G\alpha_s$, $G\alpha_{q/11}$ and $G\alpha_{12/13}$ families. This was first observed in primary rat striatal neurons endogenously expressing both the dopamine D_2 receptor and CB_1; agonists of either receptor alone produced cAMP inhibition, as expected, whereas stimulation of both receptors simultaneously led to a concentration-dependent reversal of this inhibition (Glass and Felder 1997). Specific inactivation of $G\alpha_{i/o}$ by PTX also allows cAMP levels to accumulate following CB_1 stimulation in multiple cell types, with or without D_2 expression or stimulation (Bonhaus et al. 1998; Finlay et al. 2017; Glass and Felder 1997; Scotter et al. 2010). Further investigation into CB_1/D_2 receptor dimers in HEK293 cells ultimately revealed that CB_1 was capable of 'switching' to $G\alpha_s$-like signalling when maximally effective concentrations of both CB_1 and D_2 agonists were used simultaneously (Kearn et al. 2005).

These early studies suggested that CB_1 might couple to alternative G proteins when its preferred $G\alpha_{i/o}$ subtypes are unavailable, due to inactivation by PTX or sequestration by D_2 or other GPCRs. Direct evidence from Eldeeb et al. (2016) using a $[^{35}S]$-GTPɣS scintillation proximity assay showed that CB_1 activates $G\alpha_s$ to a much lesser extent than $G\alpha_{i/o}$, and suggests that these conditions reduce the ability of $G\alpha_{i/o}$ to inhibit AC, leading to an overall nett increase in cAMP.

The culminating evidence would then suggest that if receptor numbers were multiplied to a point where the CB_1:$G\alpha_{i/o}$ ratio was insufficient, CB_1-$G\alpha_s$ coupling would occur, enhancing cAMP accumulation following CB_1 stimulation. This hypothesis was supported in a study by Finlay et al. (2017), who investigated CB_1-mediated $G\alpha_s$ signalling in HEK293 cells at various levels of receptor expression. The stimulation of cAMP production was able to overcome the inhibition when receptor expression was elevated; this effect could be abrogated by increasing the expression of $G\alpha_{i1}$ or by reducing functional receptor number with an irreversible antagonist. It was suggested that this effect may be due to a more efficacious response between $G\alpha_s$ and AC in comparison to $G\alpha_{i/o}$. A computational model has since been used to model this hypothesis, confirming its feasibility and indicating that CB_1-G protein pre-coupling, was necessary for the $G\alpha_s$-switch to occur (Yang et al. 2023).

Evidence of an interaction between CB_1 and $G\alpha_q$ has also been reported in the literature (Diez-Alarcia et al. 2016; Lauckner et al. 2005). A crystal structure of CB_1-coupled to a mini$G\alpha_{s/q}$ protein was published very recently (Wang et al. 2024). To date, the most compelling evidence for CB_1-$G\alpha_q$ coupling arises from astrocytes, as these cells communicate through Ca^{2+} mobilisation (Navarrete and Araque 2008), see (Covelo et al. 2021) for review.

CB_1-G protein coupling occurs in a ligand-dependent manner; a type of ligand bias known as G protein subtype selectivity (Ibsen et al. 2017). Several in-depth reviews on bias/functional selectivity have been published, both cannabinoid-related (Ibsen et al. 2017; Manning et al. 2021; Patel et al. 2021; Wouters et al. 2019) and more general (Finlay et al. 2020; Jelinek et al. 2021; Kenakin 2004; Urban et al. 2007; Wootten et al. 2018). Ligand-dependent G protein subtype selectivity at CB_1 was identified early on using the end-point style GTPγS assay. This assay utilises a non-hydrolysable form of GTP to compare the level of G protein activation between ligands. It is often coupled with immunoprecipitation using antibodies specific to each Gα isoform in order to discriminate between subtypes for a more revealing analysis of G protein activation. Using a G protein reconstitution assay in SF9 cell membranes, Glass and Northup (1999) showed that cannabinoid ligands, WIN55,212-2 and anandamide, preferentially induce $[^{35}S]$GTPγS binding of $G\alpha_i$ over $G\alpha_o$. Likewise, using N18TG2 neuroblastoma cell membranes, Mukhopadhyay and Howlett (2005) quantified the $G\alpha_i$ subtypes and Gβγ proteins that co-immunoprecipitated with CB_1 following stimulation with cannabinoid ligands, WIN55-212,2, DALN, and (R)-methanandamide. They found that each ligand evoked a differential pattern of G protein dissociation; WIN55,212-2 produced partial dissociation of all three $G\alpha_i$ subtypes, whereas DALN produced concentration-dependent dissociation of just $G\alpha_{i2}$ and Ga_{i3}. Furthermore, (R)-methanandamide only disrupted CB_1 receptor-$G\alpha_{i3}$ complexes, strongly indicative of ligand-specific receptor conformations. In contrast, Diez-Alarcia et al. (2016) found that THC, WIN55,212-2, and ACEA all induced differential coupling to $G\alpha_{i/o}$ subtypes, and that none of these agonists activated $G\alpha_{i2}$ binding in mouse brain cortex using a GTPγS immunoprecipitation approach. However, since the expression of $G\alpha_{i2}$ was never confirmed, this may be due to an absence of this subtype in the tissue tested, rather than the inability of CB_1-$G\alpha_{i2}$ coupling to occur. Indeed, recent evidence indicates that several synthetic cannabinoid receptor agonists induced $G\alpha_{i2}$ activation at CB_1 in transiently transfected in HEK293 cells (Ryalls et al. 2024).

More modern kinetic assays utilising bioluminescence resonance energy transfer (BRET) biosensors do not show any evidence of G protein subtype selectivity for a range of cannabinoid ligands at CB_1 and CB_2 (Patel et al. 2023; Ryalls et al. 2024). These studies employ HEK293 cells and each of the $G\alpha_{i/o}$ BRET constructs (Olsen et al. 2020). However, a similar BRET-based G protein dissociation assay was used to compare the G protein subtype selectivity between HEK293 cells and primary neurons. It was found that in primary neurons, cannabinoid ligands produced a more robust response with $G\alpha_i$ subtypes than they did at $G\alpha_{oA/oB}$, while the opposite was true in HEK293 cells (Xu et al. 2024). Although, it remains possible that the distinct profile of endogenous G protein expression for each cell line in this latter study may be influencing the results, due to differing competition between exogenous and endogenous $G\alpha$ subtypes.

4.2 Activation of Mitogen-Activated Protein Kinases

Protein kinases regulate many cellular processes by phosphorylating specific amino acids on target proteins, thereby modulating their activity, localisation, and interaction with other molecules. This is essential for signal transduction and the regulation of various cellular pathways. The mitogen-activated protein kinase (MAPK) signalling cascade consists of three protein families, MAPKKKs, MAPKKs, and MAPKs, the former phosphorylating the latter in each case. The MAPKs in the bottom tier include extracellular signal-regulated kinases (ERKs), c-Jun N-terminal kinases (JNKs), and p38 MAPKs.

CB_1 activation causes enhanced ERK1/2 phosphorylation (Bouaboula et al. 1995) (Fig. 3). ERK phosphorylation coincides with multiple signalling cascades and there are many mechanisms by which ERK phosphorylation can occur (Fig. 3). ERK1/2 are regulated by the specific MAPKKK, Raf and MAPKK, MAPK-ERK kinase (MEK) (Kyosseva 2004). PKA phosphorylates Raf-1 (C-Raf) (Häfner et al. 1994; Wu et al. 1993), inhibiting its activation of MEK, which has only one known target, ERK. Thus, CB_1-mediated inhibition of AC-cAMP-PKA leads to an increase in ERK phosphorylation, as there is less PKA-mediated inhibition of Raf-1. Raf-1 is then free to activate the subsequent members of the MAPK cascade.

While the suppression of the cAMP/PKA pathway is thought to be a principal route for CB_1-induced ERK activation in cells of neural origin (Davis et al. 2003), ERK activity can also be heightened through interactions with β-arrestin, phosphatidylinositol-3-kinase (PI3K) and through CB_1-mediated transactivation of multiple receptor tyrosine kinases (RTKs) (Dalton et al. 2020; Dalton and Howlett 2012), which can facilitate ERK phosphorylation independent of cAMP/PKA signalling (Daigle et al. 2008a). These mechanisms underscore the complexity of ERK regulation and its pivotal role in signal transduction.

ERK activation is also highly dependent on cell type, as cAMP levels display opposing effects in regulating the MAPK cascade; stimulating proliferation and neuronal differentiation in some cell lines, while inhibiting these outcomes in others (Stork and Schmitt 2002). This dual mechanism of cAMP-mediated ERK regulation can be explained by another member of the Raf family of MAPKKKs, B-Raf, which is highly expressed in neurons and shows very selective patterns of distribution

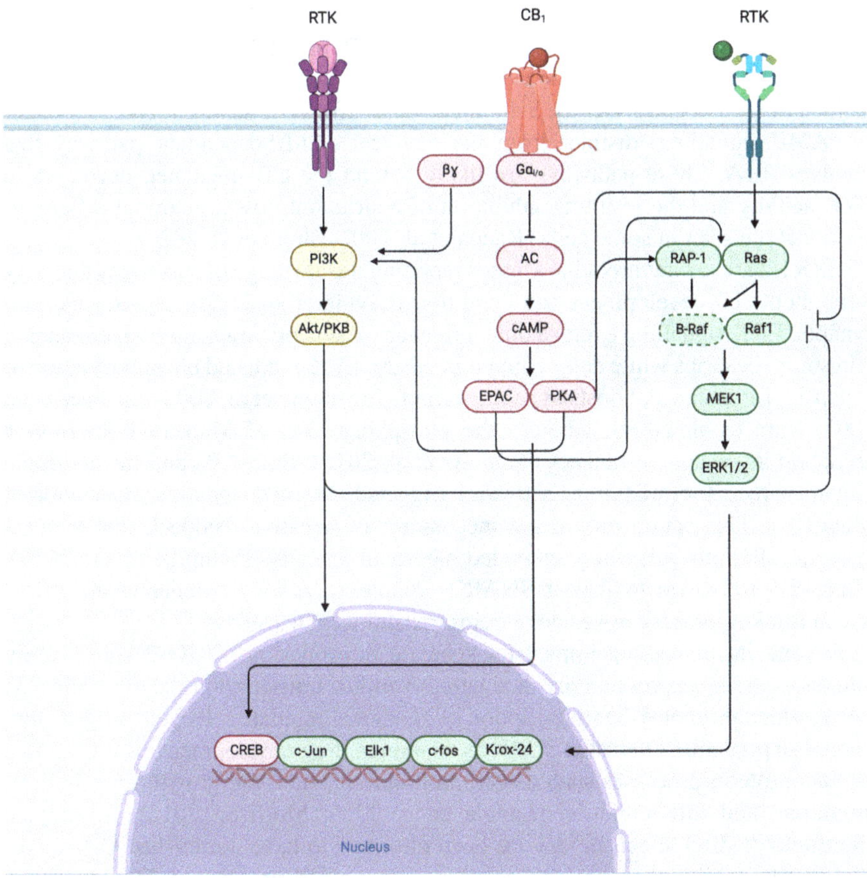

Fig. 3 Signal transduction and crosstalk between pathways downstream of CB_1. $G\alpha_{i/o}$ inhibits the activity of adenylate cyclase, reducing cAMP production. This diminishes active protein kinase A (PKA) and exchange proteins directly activated by cAMP (Epac) within the cell. Typically, this boosts the effects of extracellular signal-regulated kinases (ERK), as the inhibition of Raf-1 by PKA is reduced. However, in cells that express B-Raf, ERK activity may be decreased as activation of small GTPases, such as RAP-1 and Ras, is reduced. Due to significant overlap across these signalling pathways, the resulting responses in different models can vary unpredictably. Gβɣ and PKA can stimulate phosphatidylinositol-3-kinase (PI3K) leading to increased protein kinase B (PKB)/Akt activity. Notable crosstalk also exists between CB_1 and various receptor tyrosine kinases (RTKs). This figure was created using Biorender.com

(Barnier et al. 1995). B-Raf has been shown to switch the effect of cAMP/PKA on ERK from inhibitory to stimulatory (Vossler et al. 1997). In B-Raf expressing cells, ERK phosphorylation is *stimulated* by cAMP, and possibly PKA, through the activation of the GTPase Rap-1. This mechanism mirrors the pathway taken by growth factors, which initiate the MAPK cascade through the GTPase Ras. Another possible mechanism of cAMP-mediated ERK activation is via the PKA-independent regulation of Epac (exchange protein directly activated by cAMP). Epac is a guanine nucleotide exchange factor (GEFs) that directly activates Ras-like small GTPases (like Rap-1) by facilitating their exchange of GDP for GTP, subsequently stimulating ERK phosphorylation. For a review of Epac and its many effectors, which include PI3K/Akt and JNK, see Schmidt et al. (2013). Epac serves as a key mediator of cAMP signalling, distinct from the classical cAMP-dependent pathway that involves PKA. These pathways are likely behind the CB_1-mediated decreases in ERK activity and the resulting inhibition of neurite outgrowth neuronal differentiation that are seen in some cells (Rueda et al. 2002; Shum et al. 2020).

ERK1/2 are serine/threonine kinases targeting a wide range of substrates that vary throughout cell development and condition (Lavoie et al. 2020). In neurons, presynaptic ERK signalling exerts a dual influence on synaptic transmission, enhancing it in some scenarios while dampening it in others. ERK-mediated phosphorylation of synapsin, for instance, boosts synaptic output (Jovanovic et al. 2000; Kushner et al. 2005; Vara et al. 2009), whereas the phosphorylation of Munc18-1 leads to a reduction in synaptic efficiency (Schmitz et al. 2016). Both CB_1 and metabotropic glutamate receptor ($mGluR_{2/3}$) activation triggers ERK-mediated phosphorylation of Munc18-1. This action provides a mechanism of negative feedback that tempers synaptic strength following extended neuronal activity (Schmitz et al. 2016). Munc18-1 is known to inhibit SNARE complex assembly (Meijer et al. 2018), which is a key process in vesicular neurotransmitter release.

Beyond its immediate impact on synaptic transmission, ERK1/2 significantly influences gene expression by activating various transcription factors, such as Elk-1, which promotes the transcription of genes that regulate cell cycle progression. There is a particular interest in pERK as a target in drug development, as it is known to play a role in processes such as cell proliferation, survival, growth, metabolism, migration, and differentiation (Lavoie et al. 2020; Morrison 2012). Notably, a Huntington's disease cell model has been observed to have diminished CB_1-mediated pERK, while other signalling pathways remained unchanged, alluding to its potential as a therapeutic target (Laprairie et al. 2016).

In addition to ERK1/2, other MAPK signalling pathways may be activated by CB_1, with conflicting reports of JNK and p38 activation, likely due to different cell lines exhibiting different responses (Derkinderen et al. 2001; Graham et al. 2006; Rueda et al. 2000). While ERKs are triggered by growth factors through tyrosine kinase receptors and tumour-inducing phorbol esters, the JNK and p38 MAPK pathways are predominantly activated by stress-related signals. The first of these studies identified heightened JNK phosphorylation following CB_1 stimulation by THC, CP55940, and HU-210 in transfected CHO cells and multiple glial cell types, but not in primary neurons (Rueda et al. 2000). A build-up of p38 phosphorylation was also seen in transfected CHO cells following incubation with THC.

A subsequent study reported p38 activation in hippocampal slices following CB_1 stimulation with 2-AG, AEA, THC and synthetic cannabinoids WIN55,212-2 and CP55940, but there was no evidence of JNK activation (Derkinderen et al. 2001). However, as rodent hippocampal slices were used, this could be mediated by glial or neuronal CB_1. Lastly, neither p38 nor JNK phosphorylation was increased in response to CB_1 stimulation by HU-210 in Neuro-2a cells (Graham et al. 2006).

4.3 *Activation of PI3K/Akt and Transactivation of Receptor Tyrosine Kinases (RTKs)*

The PI3K/Akt (Protein Kinase B) signalling pathway is another signalling pathway, distinct from the MAPKs, that may be influenced by CB_1 (Blázquez et al. 2015; Gómez del Pulgar et al. 2000, 2002). Activation of PI3K is not required for cannabinoid-induced ERK activation in hippocampal slices (Derkinderen et al. 2003). A recent study used western blotting to determine that THC, but not 2-AG, activates Akt in human iPSC-derived glutamatergic cortical neurons (Shum et al. 2020). However, this effect was only seen at high agonist concentrations, so its physiological relevance is not clear.

The ERK and PI3K/Akt pathways are believed to bestow protective effects and encourage cell survival by increasing the expression of immediate early genes (IEGs), such as Zif268 (also known as Egr1, NGFI-A, and Krox-24), c-Fos, and brain-derived neurotrophic factor (BDNF) (Graham et al. 2006; Marsicano et al. 2003). Although BDNF is not an IEG in the strictest sense, BDNF expression is rapidly upregulated in response to neuronal activity, including CB_1 activation (Blázquez et al. 2015), promoting neuronal growth, survival, and plasticity. Like CB_1, BDNF is also involved in the suppression of neurotransmission and assists in protection against excitotoxicity. Recent studies show that BDNF activates its canonical receptor, tropomyosin receptor kinase B (TrkB), at the postsynaptic membrane, triggering 2-AG synthesis and the subsequent suppression of neurotransmitter release by retrograde activation of pre-synaptic CB_1 receptors (Wu et al. 2020). CB_1 has also been reported to activate TrkB in the absence of BDNF, leading to TrkB-mediated chemotaxis (Berghuis et al. 2005).

4.4 *Modulation of Ion Channels*

CB_1 modulates several types of voltage-gated Ca^{2+} channels (VGCCs), including N-type ("neural"), L-type ("long-lasting"), P-type ("Purkinje neuron"), and Q-type Ca^{2+} channels (Mackie et al. 1995; Twitchell et al. 1997). Activation of CB_1 typically suppresses these channels through a $G\alpha_{i/o}$-dependent mechanism, which diminishes calcium influx, thereby reducing neuronal excitability and neurotransmitter release. Notably, the influence of CB_1 on L-type channels has shown inconsistent effects across studies, suggesting a modulation pattern that may vary by neuronal type and experimental conditions. Further research is needed to elucidate the precise mechanisms through which CB_1 affects different VGCC subtypes and their implications for diseases, such as epilepsy and cardiovascular disorders.

CB_1 also regulates potassium (K^+) conductance by activating G protein-coupled inwardly rectifying K^+ (GIRK) channels (Mackie et al. 1995) and fast-inactivating I_A potassium currents. This activity hyperpolarises the cell membrane, decreasing neuronal excitability.

The inhibition of N-type VGCCs (NTCCs) is one of the more well-characterised effects of CB_1 activation (Harding et al. 2023; Ho et al. 2000; Mackie and Hille 1992; Twitchell et al. 1997). CB_1-mediated inhibition of NTCCs was discovered in 1992 using murine NG108-15 neuroblastoma cells, which revealed a mechanism for the analgesic and antiepileptic effects of CB_1 agonists (Caulfield and Brown 1992; Mackie and Hille 1992). The $\beta\gamma$ subunit of G proteins binds directly to the NTCC pore-forming subunit $\alpha 1$, inhibiting channel activity (De Waard et al. 2005), and playing a role in pain perception, seizure susceptibility and immune function.

The effects of CB_1 on LTCCs remain less clear, with studies reporting varied outcomes from increased Ca^{2+} influx (Drysdale et al. 2006; Rubovitch et al. 2002) to inhibition (Caulfield and Brown 1992; Endoh 2006; Hoddah et al. 2009; Straiker et al. 1999), or no effect (Ho et al. 2000; Mackie and Hille 1992). This variability highlights the need for further research to understand the context-dependent regulation of LTCCs by CB_1. Similarly, recent findings on voltage-gated sodium channels (VGSCs) show that AEA-mediated inhibition, typically thought to be direct (Kim et al. 2005; Nicholson et al. 2003), is CB_1 and G protein-dependent in the soma of neocortical neurons (Steiger et al. 2023). The application of a $G\alpha_{q/11}$ inhibitor, YM-254890, reversed a fraction of the effect, however the specificity of the inhibitor was brought into question. The authors suggested that the inhibition may be mediated by intracellular CB_1 as intracellular administration of CB_1 antagonists significantly enhanced their effectiveness, however further exploration is needed to confirm this hypothesis.

5 CB_1 Regulation

The regulation of CB_1, as for most GPCRs, involves a complex interplay of proteins, such as GPCR kinases (GRKs) and arrestins, which control receptor activation, desensitisation, and internalisation (reviewed in Fletcher-Jones et al. (2020).

5.1 G Protein-Coupled Receptor Kinases (GRKs)

GRKs phosphorylate the serine and threonine residues within the third intracellular domain and/or the C-terminus of many GPCRs (Krupnick and Benovic 1998; Nobles et al. 2011). GRK-mediated phosphorylation initiates the recruitment of arrestins to the intracellular C-terminal of the receptor, occluding the G protein binding site and thus, blocking subsequent G protein signalling (Seyedabadi et al. 2021; Staus et al. 2020). This phenomenon is known as desensitisation.

GRK2/3 are thought to be the canonical regulators of Class A GPCRs. For CB_1, the first reports of GRK/β-arrestin-mediated desensitisation demonstrated that co-expression of GRK3 and β-arrestin 2 were required for rapid desensitisation of CB_1-mediated GIRK channel activity in Xenopus oocytes (Jin et al. 1999). Shortly thereafter, presynaptic expression of dominant-negative mutant forms of GRK2 and β-arrestin was reported to weaken CB_1 desensitisation in response to prolonged agonist treatment in primary hippocampal neurons (Kouznetsova et al. 2002). CB_1 can be phosphorylated by all GRK isoforms (GRK1-6), but GRK2/3 are thought to mediate most of the phosphorylation required for β-arrestin recruitment (Manning et al. 2024). It is sometimes overlooked that this process is not always G protein independent, indeed GRK2/3 are activated by the Gβɣ-complex (Eichmann et al. 2003; Manning et al. 2024; Pitcher et al. 1992).

Most of the evidence for GRK-mediated phosphorylation of CB_1 comes from site-directed mutagenesis studies. Mutation of a cluster of six serine/threonine residues within the extreme distal C-terminus of CB_1 diminishes receptor internalisation and β-arrestin recruitment, consequently enhancing G protein signalling (Daigle et al. 2008b; Hsieh et al. 1999; Jin et al. 1999). Additionally, mutation of two serine residues more proximal, but still within the C-terminus, was initially reported to be critical for CB_1 desensitisation but not for β-arrestin recruitment (Daigle et al. 2008a; Jin et al. 1999). Recent BRET-based studies have revised this view, establishing that both proximal and distal serine/threonine clusters are required for arrestin recruitment to CB_1 (Manning et al. 2022). Moreover, Manning et al. (2022) also reported identical phosphorylation dependence for both β arrestin 1 and 2. Mutation of the more proximal serine residues also leads to slower β-arrestin translocation kinetics, perhaps explaining the discrepancy with earlier reports of unchanged β-arrestin recruitment to this mutated form of CB_1 using endpoint immunocytochemistry assays.

5.2 *β-Arrestins*

β-arrestins have three roles: (i) they are key mediators of receptor desensitisation, controlling the temporal tone of receptor signalling; (ii) they recruit endocytic machinery, facilitating receptor internalisation, and (iii) they provide a scaffold for numerous intracellular signalling cascades. Arrestins have two main structural domains: an N-terminal domain and a C-terminal domain, along with a polar core between them. The N-terminal domain of arrestin interacts with the phosphorylated C-terminal tail of the GPCR. This interaction is facilitated by the recognition of the phosphorylated residues and is crucial for arrestin to bind to the receptor with high affinity. The current evidence suggests that CB_1 more strongly recruits β-arrestin 2 than it does β-arrestin 1 (Finlay et al. 2019; Ibsen et al. 2019; Manning et al. 2023; Ryalls et al. 2024). This aligns with an early study that identified two classes of GPCRs with regard to their interaction with the different isoforms of arrestin; some that recruit β-arrestin 2 more strongly than β-arrestin 1, and those that recruit both β-arrestins equally and also have the ability to recruit the visual arrestin, arrestin 1 (Oakley et al. 2000). The two isoforms of β-arrestin share a high degree of sequence homology (78%), with most differences found in the C-terminus. So it is interesting that CB_1 (and some other GPCRs) couple to one isoform more strongly than the other.

Arrestin recruits endocytic proteins, such as clathrin and adapter protein AP2, to interact with the receptor, facilitating receptor internalisation into clathrin-coated vesicles in a dynamin-dependent manner (DeWire et al. 2007; Goodman et al. 1996; Hsieh et al. 1999; Jin et al. 1999; Krupnick et al. 1997). Many in vitro studies show that CB_1 rapidly internalises following agonist stimulation (Daigle et al. 2008b; Grimsey et al. 2010; Rozenfeld and Devi 2008; Zhu et al. 2019). Class A GPCRs typically display a transient, low-affinity interaction with β-arrestin that is not maintained upon internalisation. This has been investigated for CB_1, where Gyombolai et al. (2013) used immunocytochemistry to show that CB_1 recruits β-arrestin 2-GFP at the plasma membrane but there was no evidence of β-arrestin 2-GFP in endosomal vesicles following stimulation, in contrast to the AT_1 angiotensin receptor. Once endocytosed, GPCRs are destined for either degradation or for recycling back to the plasma membrane. Current evidence suggests that CB_1 undergoes both degradation (Grimsey et al. 2010) and recycling (Hsieh et al. 1999; Leterrier et al. 2004). The fate of endocytosed CB_1 is reliant on both exposure and concentration of agonist. Acute agonist exposure and low concentrations appear to lead to recycling of the receptor, whereas increased exposure and concentrations push the receptor towards degradation. CB_1 is also tonically internalised in a manner that can be reversed by inverse agonists (Leterrier et al. 2004). This constitutive internalisation may be responsible for the polarised nature of CB_1 expression in neurons (Leterrier et al. 2006; McDonald et al. 2007), as the constant cycling between intracellular vesicles and the plasma membrane occurs at different rates in axons and dendrites (Fletcher-Jones et al. 2019, 2020).

Additional signalling roles, beyond receptor desensitisation and internalisation, have been reported for β-arrestins. They serve as molecular scaffolds to provide the GPCR with which they are bound, easy access to specific signalling partners, such as the MAPK cascade and the non-receptor tyrosine kinase, c-Src (reviewed by DeWire et al. (2007)). Evidence supporting β-arrestin signalling at CB_1 includes the activation of JNK3 by β-arrestin 2 (Henderson-Redmond et al. 2020) and the initiation of a delayed wave of ERK phosphorylation (Nogueras-Ortiz and Yudowski 2016). Both JNK3 and β-arrestin 2 are highly expressed in the brain. The scaffolding function of β-arrestin holds the signalling proteins at the receptor, thereby reducing the effects that these kinases produce in the nucleus, like the activation of certain transcription factors. Evidence suggests that this form of signalling may be involved in cytoskeletal rearrangement and chemotaxis.

6 Conclusions

This chapter has detailed the fundamental roles and intricate mechanisms of canonical cannabinoid receptor signalling within the ECS. These receptors are pivotal in a myriad of physiological processes central to human health, including pain regulation, immune response, and neural plasticity. CB_1, predominantly located in the CNS, modulates neurotransmission and synaptic architecture, thereby influencing memory, learning, and neuroprotection. CB_2, while classically associated with peripheral tissues, also plays a role in neuroinflammation and neurodegeneration.

Despite significant advances in our understanding, numerous challenges and opportunities for future research remain. One such area is the exploration of the function of CB_2 in the CNS, a relatively uncharted domain. Additionally, the pharmacological landscape of cannabinoid receptors invites further refinement. Specifically, the development of biased ligands and allosteric modulators could harness the therapeutic potential of these receptors while minimising adverse effects. Ongoing research to elucidate the complex signalling pathways and interactions at play highlights the potential of cannabinoids in clinical applications. Ultimately, a deeper grasp of cannabinoid receptor biology not only advances our understanding of fundamental neuroscience but also opens the door to innovative and more effective therapeutic modalities.

References

Andó RD, Bíró J, Csölle C, Ledent C, Sperlágh B (2012) The inhibitory action of exo- and endocannabinoids on [3H]GABA release are mediated by both CB1 and CB2 receptors in the mouse hippocampus. https://doi.org/10.1016/J.NEUINT.2011.11.012

Ashton JC, Zheng Y, Darlington CL, Baek J-H, Smith PF (2014) Cannabinoid CB2 receptor immunolabelling in the healthy brain – still a live possibility. Naunyn Schmiedebergs Arch Pharmacol. https://doi.org/10.1007/S00210-013-0948-Y

Atwood BK, Mackie K (2010) CB2: a cannabinoid receptor with an identity crisis. Br J Pharmacol. https://doi.org/10.1111/J.1476-5381.2010.00729.X

Baek J-H, Darlington CL, Smith PF, Ashton JC (2013) Antibody testing for brain immunohistochemistry: brain immunolabeling for the cannabinoid CB2 receptor. J Neurosci Methods. https://doi.org/10.1016/J.JNEUMETH.2013.03.021

Barnier JV, Papin C, Eychène A, Lecoq O, Calothy G (1995) The mouse B-raf gene encodes multiple protein isoforms with tissue-specific expression. J Biol Chem 270(40):23381–23389. https://doi.org/10.1074/jbc.270.40.23381

Bayewitch M, Avidor-Reiss T, Levy R, Barg J, Mechoulam R, Vogel Z (1995) The peripheral cannabinoid receptor: adenylate cyclase inhibition and G protein coupling. FEBS Lett 375(1–2): 143–147. https://doi.org/10.1016/0014-5793(95)01207-u

Ben-Shabat S, Fride E, Sheskin T, Tamiri T, Rhee MH, Vogel Z, Bisogno T, De Petrocellis L, Di Marzo V, Mechoulam R (1998) An entourage effect: inactive endogenous fatty acid glycerol esters enhance 2-arachidonoyl-glycerol cannabinoid activity. Eur J Pharmacol 353(1):23–31. https://doi.org/10.1016/s0014-2999(98)00392-6

Bénard G, Massa F, Puente N, Lourenço J, Bellocchio L, Soria-Gómez E, Matias I, Delamarre A, Metna-Laurent M, Cannich A, Hebert-Chatelain E, Mulle C, Ortega-Gutiérrez S, Martín-Fontecha M, Klugmann M, Guggenhuber S, Lutz B, Gertsch J, Chaouloff F et al (2012) Mitochondrial CB_1 receptors regulate neuronal energy metabolism. Nat Neurosci 15(4): 558–564. https://doi.org/10.1038/nn.3053

Benito C, Núñez E, Tolón R, Carrier EJ, Rábano A, Hillard C, Romero J (2003) Cannabinoid CB2 receptors and fatty acid amide hydrolase are selectively overexpressed in neuritic plaque-associated glia in Alzheimer's disease brains. J Neurosci. https://doi.org/10.1523/JNEUROSCI.23-35-11136.2003

Berghuis P, Dobszay MB, Wang X, Spano S, Ledda F, Sousa KM, Schulte G, Ernfors P, Mackie K, Paratcha G, Hurd YL, Harkany T (2005) Endocannabinoids regulate interneuron migration and morphogenesis by transactivating the TrkB receptor. Proc Natl Acad Sci 102(52):19115–19120. https://doi.org/10.1073/pnas.0509494102

Bisogno T, Melck D, Myu B, Gretskaya N, Bezuglov V, De Petrocellis L, Di Marzo V (2000) N-acyl-dopamines: novel synthetic CB(1) cannabinoid-receptor ligands and inhibitors of anandamide inactivation with cannabimimetic activity in vitro and in vivo. Biochem J. https://doi.org/10.1042/BJ3510817

Blázquez C, Chiarlone A, Bellocchio L, Resel E, Pruunsild P, García-Rincón D, Sendtner M, Timmusk T, Lutz B, Galve-Roperh I, Guzmán M (2015) The CB1 cannabinoid receptor signals striatal neuroprotection via a PI3K/Akt/mTORC1/BDNF pathway. Cell Death Differ 22(10): 1618–1629. https://doi.org/10.1038/cdd.2015.11

Bonhaus DW, Chang LK, Kwan J, Martin GR (1998) Dual activation and inhibition of adenylyl cyclase by cannabinoid receptor agonists: evidence for agonist-specific trafficking of intracellular responses. J Pharmacol Exp Ther 287(3):884–888. https://jpet.aspetjournals.org/content/287/3/884.long

Bouaboula M, Poinot-Chazel C, Bourrié B, Canat X, Calandra B, Rinaldi-Carmona M, Le Fur G, Casellas P (1995) Activation of mitogen-activated protein kinases by stimulation of the central cannabinoid receptor CB1. Biochem J 312(2):637–641. https://doi.org/10.1042/bj3120637

Bouaboula M, Rinaldi M, Carayon P, Carillon C, Delpech B, Shire D, Le Fur G, Casellas P (1993) Cannabinoid-receptor expression in human leukocytes. Eur J Biochem. https://doi.org/10.1111/J.1432-1033.1993.TB17910.X

Bouchard J, Truong J, Bouchard K, Dunkelberger DL, Desrayaud S, Moussaoui S, Tabrizi S, Stella N, Muchowski P (2012) Cannabinoid receptor 2 signaling in peripheral immune cells modulates disease onset and severity in mouse models of Huntington's disease. J Neurosci. https://doi.org/10.1523/JNEUROSCI.4008-12.2012

Bradshaw HB, Walker JM (2005) The expanding field of cannabimimetic and related lipid mediators. Br J Pharmacol 144(4):459–465. https://doi.org/10.1038/sj.bjp.0706093

Brusco A, Tagliaferro P, Saez T, Onaivi E (2008) Postsynaptic localization of CB2 cannabinoid receptors in the rat hippocampus. Synapse. https://doi.org/10.1002/SYN.20569

Carlisle SJ, Marciano-Cabral F, Staab A, Ludwick C, Cabral GA (2002) Differential expression of the CB2 cannabinoid receptor by rodent macrophages and macrophage-like cells in relation to cell activation. Int Immunopharmacol. https://doi.org/10.1016/S1567-5769(01)00147-3

Caulfield MP, Brown DA (1992) Cannabinoid receptor agonists inhibit Ca current in NG108-15 neuroblastoma cells via a pertussis toxin-sensitive mechanism. Br J Pharmacol 106(2):231–232. https://doi.org/10.1111/j.1476-5381.1992.tb14321.x

Chu CJ, Huang SM, De Petrocellis L, Bisogno T, Ewing SA, Miller JD, Zipkin R, Daddario N, Appendino G, Di Marzo V, Walker J (2003) N-Oleoyldopamine, a novel endogenous capsaicin-like lipid that produces hyperalgesia*. J Biol Chem. https://doi.org/10.1074/JBC.M211231200

Concannon RM, Okine BN, Finn DP, Dowd E (2016) Upregulation of the cannabinoid CB2 receptor in environmental and viral inflammation-driven rat models of Parkinson's disease. Exp Neurol. https://doi.org/10.1016/J.EXPNEUROL.2016.06.014

Covelo A, Eraso-Pichot A, Fernández-Moncada I, Serrat R, Marsicano G (2021) CB1R-dependent regulation of astrocyte physiology and astrocyte-neuron interactions. Neuropharmacology 195: 108678. https://doi.org/10.1016/j.neuropharm.2021.108678

Crocq M-A (2020) History of cannabis and the endocannabinoid system. Dialogues Clin Neurosci 22(3):223–228. https://doi.org/10.31887/dcns.2020.22.3/mcrocq

Daigle TL, Kearn CS, Mackie K (2008a) Rapid CB1 cannabinoid receptor desensitization defines the time course of ERK1/2 MAP kinase signaling. Neuropharmacology 54(1):36–44. https://doi.org/10.1016/j.neuropharm.2007.06.005

Daigle TL, Kwok ML, Mackie K (2008b) Regulation of CB1cannabinoid receptor internalization by a promiscuous phosphorylation-dependent mechanism. J Neurochem 106(1):70–82. https://doi.org/10.1111/j.1471-4159.2008.05336.x

Dalton GD, Carney ST, Marshburn JD, Norford DC, Howlett AC (2020) CB1 cannabinoid receptors stimulate Gβγ-GRK2-mediated FAK phosphorylation at tyrosine 925 to regulate ERK activation involving neuronal focal adhesions. Front Cell Neurosci. https://doi.org/10.3389/FNCEL.2020.00176

Dalton GD, Howlett AC (2012) Cannabinoid CB1 receptors transactivate multiple receptor tyrosine kinases and regulate serine/threonine kinases to activate ERK in neuronal cells. Br J Pharmacol 165(8):2497–2511. https://doi.org/10.1111/j.1476-5381.2011.01455.x

Davis MI, Ronesi J, Lovinger DM (2003) A predominant role for inhibition of the adenylate cyclase/protein kinase A pathway in ERK activation by cannabinoid receptor 1 in N1E-115 neuroblastoma cells. J Biol Chem 278(49):48973–48980. https://doi.org/10.1074/jbc.M305697200

De Waard M, Hering J, Weiss N, Feltz A (2005) How do G proteins directly control neuronal Ca2+ channel function? Trends Pharmacol Sci 26(8):427–436. https://doi.org/10.1016/j.tips.2005.06.008

Derkinderen P, Ledent C, Parmentier M, Girault JA (2001) Cannabinoids activate p38 mitogen-activated protein kinases through CB1 receptors in hippocampus. J Neurochem 77(3):957–960. https://doi.org/10.1046/j.1471-4159.2001.00333.x

Derkinderen P, Valjent E, Toutant M, Corvol J-C, Enslen H, Ledent C, Trzaskos J, Caboche J, Girault J-A (2003) Regulation of extracellular signal-regulated kinase by cannabinoids in hippocampus. J Neurosci 23(6):2371–2382. https://doi.org/10.1523/jneurosci.23-06-02371.2003

Devane WA, Dysarz FA 3rd, Johnson MR, Melvin LS, Howlett AC (1988) Determination and characterization of a cannabinoid receptor in rat brain. Mol Pharmacol 34(5):605–613

Devane WA, Hanus L, Breuer A, Pertwee RG, Stevenson LA, Griffin G, Gibson D, Mandelbaum A, Etinger A, Mechoulam R (1992) Isolation and structure of a brain constituent that binds to the cannabinoid receptor. Science 258(5090):1946–1949. https://doi.org/10.1126/science.1470919

DeWire SM, Ahn S, Lefkowitz RJ, Shenoy SK (2007) Beta-arrestins and cell signaling. Annu Rev Physiol 69(1):483–510. https://doi.org/10.1146/annurev.physiol.69.022405.154749

Diez-Alarcia R, Ibarra-Lecue I, Lopez-Cardona ÁP, Meana J, Gutierrez-Adán A, Callado LF, Agirregoitia E, Urigüen L (2016) Biased Agonism of three different cannabinoid receptor agonists in mouse brain cortex [original research]. Front Pharmacol 7. https://doi.org/10.3389/fphar.2016.00415

Djeungoue-Petga M-A, Hebert-Chatelain E (2017) Linking mitochondria and synaptic transmission: the CB1 receptor. Bioessays 39(12):1700126. https://doi.org/10.1002/bies.201700126

Downes GB, Gautam N (1999) The G protein subunit gene families. Genomics 62(3):544–552. https://doi.org/10.1006/geno.1999.5992

Drysdale AJ, Ryan D, Pertwee RG, Platt B (2006) Cannabidiol-induced intracellular Ca2+ elevations in hippocampal cells. Neuropharmacology 50(5):621–631. https://doi.org/10.1016/j.neuropharm.2005.11.008

Duffy SS, Hayes JP, Fiore NT, Moalem-Taylor G (2021) The cannabinoid system and microglia in health and disease. Neuropharmacology. https://doi.org/10.1016/J.NEUROPHARM.2021.108555

Eichmann T, Lorenz K, Hoffmann M, Brockmann J, Krasel C, Lohse MJ, Quitterer U (2003) The amino-terminal domain of G-protein-coupled receptor kinase 2 is a regulatory Gβγ binding site. J Biol Chem 278(10):8052–8057. https://doi.org/10.1074/jbc.m204795200

Eldeeb K, Leone-Kabler S, Howlett AC (2016) CB1 cannabinoid receptor-mediated increases in cyclic AMP accumulation are correlated with reduced Gi/o function. J Basic Clin Physiol Pharmacol 27(3):311–322. https://doi.org/10.1515/jbcpp-2015-0096

Endoh T (2006) Pharmacological characterization of inhibitory effects of postsynaptic opioid and cannabinoid receptors on calcium currents in neonatal rat nucleus tractus solitarius. Br J Pharmacol 147(4):391–401. https://doi.org/10.1038/sj.bjp.0706623

Farquhar-Smith WP, Egertová M, Bradbury EJ, McMahon SB, Rice AS, Elphick MR (2000) Cannabinoid CB(1) receptor expression in rat spinal cord. Mol Cell Neurosci 15(6):510–521. https://doi.org/10.1006/mcne.2000.0844

Felder CC, Joyce KE, Briley EM, Mansouri J, Mackie K, Blond O, Lai Y, Ma AL, Mitchell RL (1995) Comparison of the pharmacology and signal transduction of the human cannabinoid CB1 and CB2 receptors. Mol Pharmacol 48(3):443–450

Finlay DB, Cawston EE, Grimsey NL, Hunter MR, Korde A, Vemuri VK, Makriyannis A, Glass M (2017) Gαs signalling of the CB1 receptor and the influence of receptor number. Br J Pharmacol 174(15):2545–2562. https://doi.org/10.1111/bph.13866

Finlay DB, Duffull SB, Glass M (2020) 100 years of modelling ligand-receptor binding and response: a focus on GPCRs. Br J Pharmacol 177(7):1472–1484. https://doi.org/10.1111/bph.14988

Finlay DB, Manning JJ, Ibsen MS, Macdonald CE, Patel M, Javitch JA, Banister SD, Glass M (2019) Do toxic synthetic cannabinoid receptor agonists have signature in vitro activity profiles? A case study of AMB-FUBINACA. ACS Chem Nerosci 10(10):4350–4360. https://doi.org/10.1021/acschemneuro.9b00429

Fletcher-Jones A, Hildick KL, Evans AJ, Nakamura Y, Henley JM, Wilkinson KA (2020) Protein interactors and trafficking pathways that regulate the cannabinoid type 1 receptor (CB1R). Front Mol Neurosci 13. https://doi.org/10.3389/fnmol.2020.00108

Fletcher-Jones A, Hildick KL, Evans AJ, Nakamura Y, Wilkinson KA, Henley JM (2019) The C-terminal helix 9 motif in rat cannabinoid receptor type 1 regulates axonal trafficking and surface expression. Elife 8. https://doi.org/10.7554/eLife.44252

Foster DJ, Wilson JM, Remke DH, Mahmood MS, Uddin MJ, Wess J, Patel S, Marnett LJ, Niswender CM, Jones CK, Xiang Z, Lindsley CW, Rook JM, Conn PJ (2016) Antipsychotic-like effects of M4 positive allosteric modulators are mediated by CB2 receptor-dependent inhibition of dopamine release. Neuron. https://doi.org/10.1016/J.NEURON.2016.08.017

Galiègue S, Mary S, Marchand J, Dussossoy D, Carrière D, Carayon P, Bouaboula M, Shire D, Le Fur G, Casellas P (1995) Expression of central and peripheral cannabinoid receptors in human immune tissues and leukocyte subpopulations. Eur J Biochem. https://doi.org/10.1111/J.1432-1033.1995.TB20780.X

Gancedo JM (2013) Biological roles of cAMP: variations on a theme in the different kingdoms of life. Biol Rev 88(3):645–668. https://doi.org/10.1111/brv.12020

Gaoni Y, Mechoulam R (1964) Isolation, structure, and partial synthesis of an active constituent of hashish. J Am Chem Soc 86(8):1646–1647. https://doi.org/10.1021/ja01062a046

Glass M, Faull RLM, Dragunow M (1997) Cannabinoid receptors in the human brain: a detailed anatomical and quantitative autoradiographic study in the fetal, neonatal and adult human brain. Neuroscience 77(2):299–318. https://doi.org/10.1016/S0306-4522(96)00428-9

Glass M, Felder CC (1997) Concurrent stimulation of cannabinoid CB1 and dopamine D2 receptors augments cAMP accumulation in striatal neurons: evidence for a Gs-linkage to the CB1 receptor. J Neurosci 17(14):5327–5333. https://doi.org/10.1523/jneurosci.17-14-05327.1997

Glass M, Northup JK (1999) Agonist selective regulation of G proteins by cannabinoid CB1and CB2Receptors. Mol Pharmacol 56(6):1362–1369. https://doi.org/10.1124/mol.56.6.1362

Gómez del Pulgar T, De Ceballos MAL, Guzmán M, Velasco G (2002) Cannabinoids protect astrocytes from ceramide-induced apoptosis through the phosphatidylinositol 3-kinase/protein kinase B pathway. J Biol Chem 277(39):36527–36533. https://doi.org/10.1074/jbc.m205797200

Gómez del Pulgar T, Velasco G, Guzmán M (2000) The CB1 cannabinoid receptor is coupled to the activation of protein kinase B/Akt. Biochem J 347(Pt 2):369–373. https://doi.org/10.1042/0264-6021:3470369

González-Mariscal I, Krzysik-Walker SM, Doyle ME, Liu QR, Cimbro R, Santa-Cruz Calvo S, Ghosh S, Cieśla Ł, Moaddel R, Carlson OD, Witek RP, O'Connell JF, Egan JM (2016) Human CB1 receptor isoforms, present in hepatocytes and β-cells, are involved in regulating metabolism. Sci Rep 6:33302. https://doi.org/10.1038/srep33302

Goodman OB, Krupnick JG, Santini F, Gurevich VV, Penn RB, Gagnon AW, Keen JH, Benovic JL (1996) β-Arrestin acts as a clathrin adaptor in endocytosis of the β2-adrenergic receptor. Nature 383(6599):447–450. https://doi.org/10.1038/383447a0

Grabon W, Bodennec J, Rheims S, Belmeguenai A, Bezin L (2023) Update on the controversial identity of cells expressing cnr2 gene in the nervous system. CNS Neurosci Ther. https://doi.org/10.1111/CNS.13977

Graham ES, Ball N, Scotter EL, Narayan P, Dragunow M, Glass M (2006) Induction of Krox-24 by endogenous cannabinoid type 1 receptors in Neuro2A cells is mediated by the MEK-ERK MAPK pathway and is suppressed by the phosphatidylinositol 3-kinase pathway. J Biol Chem 281(39):29085–29095. https://doi.org/10.1074/jbc.M602516200

Grimsey NL, Graham ES, Dragunow M, Glass M (2010) Cannabinoid receptor 1 trafficking and the role of the intracellular pool: implications for therapeutics. Biochem Pharmacol 80(7): 1050–1062. https://doi.org/10.1016/j.bcp.2010.06.007

Gyombolai P, Boros E, Hunyady L, Turu G (2013) Differential β-arrestin2 requirements for constitutive and agonist-induced internalization of the CB1 cannabinoid receptor. Mol Cell Endocrinol. https://doi.org/10.1016/J.MCE.2013.03.013

Häfner S, Adler HS, Mischak H, Janosch P, Heidecker G, Wolfman A, Pippig S, Lohse M, Ueffing M, Kolch W (1994) Mechanism of inhibition of Raf-1 by protein kinase A. Mol Cell Biol 14(10):6696–6703. https://doi.org/10.1128/mcb.14.10.6696-6703.1994

Hanoune J, Defer N (2001) Regulation and role of adenylyl cyclase isoforms. Annu Rev Pharmacol Toxicol 41:145–174. https://doi.org/10.1146/annurev.pharmtox.41.1.145

Hanson MA, Roth CB, Jo E, Griffith MT, Scott FL, Reinhart G, Desale H, Clemons B, Cahalan SM, Schuerer SC, Sanna MG, Han GW, Kuhn P, Rosen H, Stevens RC (2012) Crystal structure of a lipid G protein–coupled receptor. Science 335(6070):851–855. https://doi.org/10.1126/science.1215904

Hanuš L, Abu-Lafi S, Fride E, Breuer A, Vogel Z, Shalev DE, Kustanovich I, Mechoulam R (2001) 2-Arachidonyl glyceryl ether, an endogenous agonist of the cannabinoid CB_1 receptor. Proc Natl Acad Sci 98(7):3662–3665. https://doi.org/10.1073/pnas.061029898

Hanus L, Gopher A, Almog S, Mechoulam R (1993) Two new unsaturated fatty acid ethanolamides in brain that bind to the cannabinoid receptor. J Med Chem 36(20):3032–3034. https://doi.org/10.1021/jm00072a026

Harding EK, Souza IA, Gandini M, Gadotti VM, Ali MY, Huang S, Antunes FTT, Trang T, Zamponi G (2023) Differential regulation of Cav3.2 and Cav2.2 calcium channels by CB1 receptors and cannabidiol. Br J Pharmacol. https://doi.org/10.1111/BPH.16035

Hauser AS, Attwood MM, Rask-Andersen M, Schiöth HB, Gloriam DE (2017) Trends in GPCR drug discovery: new agents, targets and indications. Nat Rev Drug Discov 16(12):829–842. https://doi.org/10.1038/nrd.2017.178

Hebert-Chatelain E, Reguero L, Puente N, Lutz B, Chaouloff F, Rossignol R, Piazza PV, Benard G, Grandes P, Marsicano G (2014) Cannabinoid control of brain bioenergetics: exploring the subcellular localization of the CB1 receptor. Mol Metab 3(4):495–504. https://doi.org/10.1016/j.molmet.2014.03.007

Henderson-Redmond AN, Nealon CM, Davis BJ, Yuill MB, Sepulveda DE, Blanton HL, Piscura MK, Zee ML, Haskins CP, Marcus DJ, Mackie K, Guindon J, Morgan DJ (2020) c-Jun N terminal kinase signaling pathways mediate cannabinoid tolerance in an agonist-specific manner. Neuropharmacology 164:107847–107847. https://doi.org/10.1016/j.neuropharm.2019.107847

Herkenham M, Lynn AB, Johnson MR, Melvin LS, de Costa BR, Rice KC (1991) Characterization and localization of cannabinoid receptors in rat brain: a quantitative in vitro autoradiographic study. J Neurosci 11(2):563–583. https://doi.org/10.1523/JNEUROSCI.11-02-00563.1991

Hill SJ (2006) G-protein-coupled receptors: past, present and future. Br J Pharmacol 147(S1):S27–S37

Hillard CJ, Harris RA, Bloom AS (1985) Effects of the cannabinoids on physical properties of brain membranes and phospholipid vesicles: fluorescence studies. J Pharmacol Exp Ther 232(3): 579–588. https://jpet.aspetjournals.org/content/232/3/579.long

Ho BY, Stadnicka A, Prather PL, Buckley AR, Current LL, Bosnjak ZJ, Kwok W-M (2000) Cannabinoid CB1 receptor-mediated inhibition of prolactin release and signaling mechanisms in GH4C1 cells*. Endocrinology 141(5):1675–1685. https://doi.org/10.1210/endo.141.5.7454

Hoddah H, Marcantoni A, Comunanza V, Carabelli V, Carbone E (2009) L-type channel inhibition by CB1 cannabinoid receptors is mediated by PTX-sensitive G proteins and cAMP/PKA in GT1-7 hypothalamic neurons. Cell Calcium 46(5–6):303–312. https://doi.org/10.1016/j.ceca.2009.08.007

Howlett AC (1985) Cannabinoid inhibition of adenylate cyclase. Biochemistry of the response in neuroblastoma cell membranes. Mol Pharmacol 27(4):429–436. https://www.ncbi.nlm.nih.gov/pubmed/2984538

Howlett AC, Barth F, Bonner TI, Cabral G, Casellas P, Devane WA, Felder CC, Herkenham M, Mackie K, Martin BR, Mechoulam R, Pertwee RG (2002) International Union of Pharmacology. XXVII. Classification of cannabinoid receptors. Pharmacol Rev 54(2):161–202. https://doi.org/10.1124/pr.54.2.161

Howlett AC, Fleming RM (1984) Cannabinoid inhibition of adenylate cyclase. Pharmacology of the response in neuroblastoma cell membranes. Mol Pharmacol 26(3):532–538. https://www.ncbi.nlm.nih.gov/pubmed/6092901

Howlett AC, Qualy JM, Khachatrian LL (1986) Involvement of Gi in the inhibition of adenylate cyclase by cannabimimetic drugs. Mol Pharmacol 29(3):307–313

Hsieh C, Brown S, Derleth C, Mackie K (1999) Internalization and recycling of the CB1 cannabinoid receptor. J Neurochem 73(2):493–501. https://doi.org/10.1046/j.1471-4159.1999.0730493.x

Humburg BA, Jordan C, Zhang H-Y, Shen H, Han X, Bi G-H, Hempel BJ, Galaj E, Baumann M, Xi Z (2021) Optogenetic brain-stimulation reward: a new procedure to re-evaluate the rewarding versus aversive effects of cannabinoids in dopamine transporter-Cre mice. Addict Biol. https://doi.org/10.1111/ADB.13005

Ibsen MS, Connor M, Glass M (2017) Cannabinoid CB1 and CB2 receptor signaling and bias. Cannabis Cannabinoid Res 2(1):48–60. https://doi.org/10.1089/can.2016.0037

Ibsen MS, Finlay DB, Patel M, Javitch JA, Glass M, Grimsey NL (2019) Cannabinoid CB1 and CB2 receptor-mediated arrestin translocation: species, subtype, and agonist-dependence. Front Pharmacol 10:350–350. https://doi.org/10.3389/fphar.2019.00350

Jelinek V, Mösslein N, Bünemann M (2021) Structures in G proteins important for subtype selective receptor binding and subsequent activation. Commun Biol 4(1). https://doi.org/10.1038/s42003-021-02143-9

Jenkin K, McAinch A, Grinfeld E, Hryciw D (2011) Role for cannabinoid receptors in human proximal tubular hypertrophy. Cell Physiol Biochem. https://doi.org/10.1159/000323997

Jin W, Brown S, Roche JP, Hsieh C, Celver JP, Kovoor A, Chavkin C, Mackie K (1999) Distinct domains of the CB1 cannabinoid receptor mediate desensitization and internalization. J Neurosci 19(10):3773–3780. https://doi.org/10.1523/jneurosci.19-10-03773.1999

Jong Y-JI, Harmon SK, O'Malley KL (2018) Intracellular GPCRs play key roles in synaptic plasticity. ACS Chem Nerosci 9(9):2162–2172. https://doi.org/10.1021/acschemneuro.7b00516

Jordan CJ, Xi Z (2019) Progress in brain cannabinoid CB2 receptor research: from genes to behavior. Neurosci Biobehav Rev. https://doi.org/10.1016/J.NEUBIOREV.2018.12.026

Jovanovic JN, Czernik AJ, Fienberg AA, Greengard P, Sihra TS (2000) Synapsins as mediators of BDNF-enhanced neurotransmitter release. Nat Neurosci 3(4):323–329. https://doi.org/10.1038/73888

Kano M, Ohno-Shosaku T, Hashimotodani Y, Uchigashima M, Watanabe M (2009) Endocannabinoid-mediated control of synaptic transmission. Physiol Rev 89(1):309–380. https://doi.org/10.1152/physrev.00019.2008

Katona I, Rancz EA, Acsády L, Ledent C, Mackie K, Hájos N, Freund TF (2001) Distribution of CB1 cannabinoid receptors in the amygdala and their role in the control of GABAergic transmission. J Neurosci 21(23):9506–9518. https://doi.org/10.1523/jneurosci.21-23-09506.2001

Katona I, Sperlágh B, Sík A, Käfalvi A, Vizi ES, Mackie K, Freund TF (1999) Presynaptically located CB1 cannabinoid receptors regulate GABA release from axon terminals of specific hippocampal interneurons. J Neurosci 19(11):4544–4558. https://doi.org/10.1523/jneurosci.19-11-04544.1999

Katona I, Urbán GM, Wallace M, Ledent C, Jung KM, Piomelli D, Mackie K, Freund TF (2006) Molecular composition of the endocannabinoid system at glutamatergic synapses. J Neurosci 26(21):5628–5637. https://doi.org/10.1523/jneurosci.0309-06.2006

Kearn CS, Blake-Palmer K, Daniel E, Mackie K, Glass M (2005) Concurrent stimulation of cannabinoid CB1 and dopamine D2 receptors enhances heterodimer formation: a mechanism for receptor cross-talk? Mol Pharmacol 67(5):1697–1704. https://doi.org/10.1124/mol.104.006882

Kenakin T (2004) Principles: receptor theory in pharmacology. Trends Pharmacol Sci 25(4): 186–192. https://doi.org/10.1016/j.tips.2004.02.012

Khan SM, Sleno R, Gora S, Zylbergold P, Laverdure J-P, Labbé J-C, Miller GJ, Hébert TE (2013) The expanding roles of Gβγ subunits in G protein coupled receptor signaling and drug action. Pharmacol Rev 65(2):545–577. https://doi.org/10.1124/pr.111.005603

Kim HI, Kim TH, Shin YK, Lee CS, Park M, Song J-H (2005) Anandamide suppression of Na+ currents in rat dorsal root ganglion neurons. Brain Res. https://doi.org/10.1016/J.BRAINRES.2005.09.004

Kim J, Li Y (2015) Chronic activation of CB2 cannabinoid receptors in the hippocampus increases excitatory synaptic transmission. J Physiol. https://doi.org/10.1113/JPHYSIOL.2014.286633

Kouznetsova M, Kelley BG, Shen M, Thayer SA (2002) Desensitization of cannabinoid-mediated presynaptic inhibition of neurotransmission between rat hippocampal neurons in culture. Mol Pharmacol. https://doi.org/10.1124/MOL.61.3.477

Kreitzer AC, Regehr WG (2001) Retrograde inhibition of presynaptic calcium influx by endogenous cannabinoids at excitatory synapses onto Purkinje cells. Neuron 29(3):717–727. https://doi.org/10.1016/s0896-6273(01)00246-x

Krishna Kumar K, Shalev-Benami M, Robertson MJ, Hu H, Banister SD, Hollingsworth SA, Latorraca NR, Kato HE, Hilger D, Maeda S, Weis WI, Farrens DL, Dror RO, Malhotra SV, Kobilka BK, Skiniotis G (2019) Structure of a signaling cannabinoid receptor 1-G protein complex. Cell 176(3):448–458.e412. https://doi.org/10.1016/j.cell.2018.11.040

Krupinski J, Coussen F, Bakalyar HA, Tang WJ, Feinstein PG, Orth K, Slaughter C, Reed RR, Gilman AG (1989) Adenylyl cyclase amino acid sequence: possible channel- or transporter-like structure. Science 244(4912):1558–1564. https://doi.org/10.1126/science.2472670

Krupnick JG, Benovic JL (1998) The role of receptor kinases and arrestins in G protein-coupled receptor regulation. Annu Rev Pharmacol Toxicol 38(1):289–319. https://doi.org/10.1146/annurev.pharmtox.38.1.289

Krupnick JG, Goodman OB Jr, Keen JH, Benovic JL (1997) Arrestin/clathrin interaction. Localization of the clathrin binding domain of nonvisual arrestins to the carboxy terminus. J Biol Chem 272(23):15011–15016. https://doi.org/10.1074/jbc.272.23.15011

Kushner SA, Elgersma Y, Murphy GG, Jaarsma D, Van Woerden GM, Hojjati MR, Cui Y, Leboutillier JC, Marrone DF, Choi ES, De Zeeuw CI, Petit TL, Pozzo-Miller L, Silva AJ (2005) Modulation of presynaptic plasticity and learning by the H-ras/extracellular signal-regulated kinase/Synapsin I signaling pathway. J Neurosci 25(42):9721–9734. https://doi.org/10.1523/jneurosci.2836-05.2005

Kyosseva SV (2004) Mitogen-activated protein kinase signaling. In: International review of neurobiology, vol 59. Academic Press, pp 201–220. https://doi.org/10.1016/S0074-7742(04)59008-6

Laprairie RB, Bagher AM, Kelly MEM, Denovan-Wright EM (2016) Biased type 1 cannabinoid receptor signaling influences neuronal viability in a cell culture model of Huntington disease. Mol Pharmacol 89(3):364–375. https://doi.org/10.1124/mol.115.101980

Lauckner JE, Hille B, Mackie K (2005) The cannabinoid agonist WIN55,212-2 increases intracellular calcium via CB1 receptor coupling to Gq/11 G proteins. Proc Natl Acad Sci 102(52): 19144–19149. https://doi.org/10.1073/pnas.0509588102

Lavoie H, Gagnon J, Therrien M (2020) ERK signalling: a master regulator of cell behaviour, life and fate. Nat Rev Mol Cell Biol 21(10):607–632. https://doi.org/10.1038/s41580-020-0255-7

Leggett JD, Aspley S, Beckett SRG, D'Antona AM, Kendall DA (2004) Oleamide is a selective endogenous agonist of rat and human CB1 cannabinoid receptors. Br J Pharmacol. https://doi.org/10.1038/SJ.BJP.0705607

Leterrier C, Bonnard D, Carrel D, Rossier J, Lenkei Z (2004) Constitutive endocytic cycle of the CB1 cannabinoid receptor. J Biol Chem 279(34):36013–36021. https://doi.org/10.1074/jbc.M403990200

Leterrier C, Lainé J, Darmon M, Boudin H, Rossier J, Lenkei Z (2006) Constitutive activation drives compartment-selective endocytosis and axonal targeting of type 1 cannabinoid receptors. J Neurosci 26(12):3141–3153. https://doi.org/10.1523/jneurosci.5437-05.2006

Liu QR, Pan C, Hishimoto A, Li CY, Xi Z, Llorente-Berzal Á, Viveros M, Ishiguro H, Arinami T, Onaivi E, Uhl G (2009) Species differences in cannabinoid receptor 2 (CNR2 gene): identification of novel human and rodent CB2 isoforms, differential tissue expression and regulation by cannabinoid receptor ligands. Genes Brain Behav. https://doi.org/10.1111/J.1601-183X.2009.00498.X

López A, Aparicio N, Pazos MR, Grande MT, Barreda-Manso MA, Benito-Cuesta I, Vázquez C, Amores M, Ruiz-Pérez G, García-García E, Beatka M, Tolón RM, Dittel BN, Hillard CJ, Romero J (2018) Cannabinoid CB2 receptors in the mouse brain: relevance for Alzheimer's disease. J Neuroinflammation. https://doi.org/10.1186/S12974-018-1174-9

Ma Z, Gao F, Larsen B, Gao M, Luo Z, Chen D, Ma X, Qiu S, Zhou Y, Xie J, Xi Z, Wu J (2019) Mechanisms of cannabinoid CB2 receptor-mediated reduction of dopamine neuronal excitability in mouse ventral tegmental area. Soc Sci Res Netw. https://doi.org/10.2139/SSRN.3314420

Mackie K, Hille B (1992) Cannabinoids inhibit N-type calcium channels in neuroblastoma-glioma cells. Proc Natl Acad Sci 89(9):3825–3829. https://doi.org/10.1073/pnas.89.9.3825

Mackie K, Lai Y, Westenbroek R, Mitchell R (1995) Cannabinoids activate an inwardly rectifying potassium conductance and inhibit Q-type calcium currents in AtT20 cells transfected with rat brain cannabinoid receptor. J Neurosci 15(10):6552–6561. https://doi.org/10.1523/jneurosci.15-10-06552.1995

Mailleux P, Vanderhaeghen JJ (1992) Distribution of neuronal cannabinoid receptor in the adult rat brain: a comparative receptor binding radioautography and in situ hybridization histochemistry. Neuroscience 48(3):655–668. https://doi.org/10.1016/0306-4522(92)90409-u

Manning JJ, Finlay DB, Glass M (2024) GPCR kinase subtype requirements for arrestin-2 and -3 translocation to the type-1 cannabinoid receptor and the consequences on G protein signalling. Biochem Pharmacol:116190. https://doi.org/10.1016/j.bcp.2024.116190

Manning JJ, Green HM, Glass M, Finlay DB (2021) Pharmacological selection of cannabinoid receptor effectors: Signalling, allosteric modulation and bias. Neuropharmacology 193:108611. https://doi.org/10.1016/j.neuropharm.2021.108611

Manning JJ, Rawcliffe G, Finlay DB, Glass M (2022) Cannabinoid 1 (CB1) receptor arrestin subtype-selectivity and phosphorylation dependence. Br J Pharmacol. https://doi.org/10.1111/BPH.15973

Manning JJ, Rawcliffe G, Finlay DB, Glass M (2023) Cannabinoid 1 (CB(1)) receptor arrestin subtype-selectivity and phosphorylation dependence. Br J Pharmacol 180(3):369–382. https://doi.org/10.1111/bph.15973

Marchalant Y, Brownjohn P, Bonnet A, Kleffmann T, Ashton J (2014) Validating antibodies to the cannabinoid CB2 receptor. J Histochem Cytochem 62:395. https://doi.org/10.1369/0022155414530995

Maresz K, Carrier EJ, Ponomarev E, Hillard C, Dittel B (2005) Modulation of the cannabinoid CB2 receptor in microglial cells in response to inflammatory stimuli. J Neurochem. https://doi.org/10.1111/J.1471-4159.2005.03380.X

Marinelli S, Marrone MC, Di Domenico M, Marinelli S (2023) Endocannabinoid signaling in microglia. Glia 71(1):71–90. https://doi.org/10.1002/glia.24281

Marinelli S, Pacioni S, Cannich A, Marsicano G, Bacci A (2009) Self-modulation of neocortical pyramidal neurons by endocannabinoids. Nat Neurosci 12(12):1488–1490. https://doi.org/10.1038/nn.2430

Marsicano G, Goodenough S, Monory K, Hermann H, Eder M, Cannich A, Azad SC, Cascio MG, Gutiérrez SO, van der Stelt M, López-Rodriguez ML, Casanova E, Schütz G, Zieglgänsberger W, Di Marzo V, Behl C, Lutz B (2003) CB1 cannabinoid receptors and on-demand defense against excitotoxicity. Science 302(5642):84–88. https://doi.org/10.1126/science.1088208

Marsicano G, Lutz B (1999) Expression of the cannabinoid receptor CB1 in distinct neuronal subpopulations in the adult mouse forebrain. Eur J Neurosci 11(12):4213–4225. https://doi.org/10.1046/j.1460-9568.1999.00847.x

Matsuda LA, Lolait SJ, Brownstein MJ, Young AC, Bonner TI (1990) Structure of a cannabinoid receptor and functional expression of the cloned cDNA. Nature 346(6284):561–564. https://doi.org/10.1038/346561a0

Mátyás F, Urbán GM, Watanabe M, Mackie K, Zimmer A, Freund TF, Katona I (2008) Identification of the sites of 2-arachidonoylglycerol synthesis and action imply retrograde endocannabinoid signaling at both GABAergic and glutamatergic synapses in the ventral tegmental area. Neuropharmacology 54(1):95–107. https://doi.org/10.1016/j.neuropharm.2007.05.028

McCoy K, Matveyeva M, Carlisle S, Cabral G (1999) Cannabinoid inhibition of the processing of intact lysozyme by macrophages: evidence for CB2 receptor participation. J Pharmacol Exp Ther

McDonald NA, Henstridge CM, Connolly CN, Irving AJ (2007) An essential role for constitutive endocytosis, but not activity, in the axonal targeting of the CB1 cannabinoid receptor. Mol Pharmacol 71(4):976–984. https://doi.org/10.1124/mol.106.029348

Mechoulam R, Ben-Shabat S, Hanus L, Ligumsky M, Kaminski NE, Schatz AR, Gopher A, Almog S, Martin BR, Compton DR, Pertwee RG, Griffin G, Bayewitch M, Barg J, Vogel Z (1995) Identification of an endogenous 2-monoglyceride, present in canine gut, that binds to cannabinoid receptors. Biochem Pharmacol 50(1):83–90. https://doi.org/10.1016/0006-2952(95)00109-D

Meijer M, Dörr B, Lammertse HCA, Blithikioti C, van Weering JRT, Toonen RF, Söllner TH, Verhage M (2018) Tyrosine phosphorylation of Munc18-1 inhibits synaptic transmission by preventing SNARE assembly. EMBO J. https://doi.org/10.15252/EMBJ.201796484

Melser S, Pagano Zottola AC, Serrat R, Puente N, Grandes P, Marsicano G, Hebert-Chatelain E (2017) Functional analysis of mitochondrial CB1 cannabinoid receptors (mtCB1) in the brain. Methods Enzymol 593:143–174. https://doi.org/10.1016/bs.mie.2017.06.023

Morgan NH, Stanford IM, Woodhall GL (2009) Functional CB2 type cannabinoid receptors at CNS synapses. Neuropharmacology. https://doi.org/10.1016/J.NEUROPHARM.2009.07.017

Morrison DK (2012) MAP kinase pathways. Cold Spring Harb Perspect Biol 4(11):a011254–a011254. https://doi.org/10.1101/cshperspect.a011254

Mukhopadhyay S, Howlett AC (2005) Chemically distinct ligands promote differential CB1 cannabinoid receptor-Gi protein interactions. Mol Pharmacol 67(6):2016–2024. https://doi.org/10.1124/mol.104.003558

Munro S, Thomas KL, Abu-Shaar M (1993) Molecular characterization of a peripheral receptor for cannabinoids. Nature 365(6441):61–65. https://doi.org/10.1038/365061a0

Navarrete M, Araque A (2008) Endocannabinoids mediate neuron-astrocyte communication. Neuron 57(6):883–893. https://doi.org/10.1016/j.neuron.2008.01.029

Nicholson RA, Liao C, Zheng J, David LS, Coyne L, Errington AC, Singh G, Lees G (2003) Sodium channel inhibition by anandamide and synthetic cannabimimetics in brain. Brain Res 978(1):194–204. https://doi.org/10.1016/S0006-8993(03)02808-7

Nobles KN, Xiao K, Ahn S, Shukla AK, Lam CM, Rajagopal S, Strachan RT, Huang T-Y, Bressler EA, Hara MR, Shenoy SK, Gygi SP, Lefkowitz RJ (2011) Distinct phosphorylation sites on the β(2)-adrenergic receptor establish a barcode that encodes differential functions of β-arrestin. Sci Signal 4(185):ra51. https://doi.org/10.1126/scisignal.2001707

Nogueras-Ortiz CJ, Yudowski G (2016) The multiple waves of cannabinoid 1 receptor signaling. Mol Pharmacol. https://doi.org/10.1124/MOL.116.104539

Oakley RH, Laporte SA, Holt JA, Caron MG, Barak LS (2000) Differential affinities of visual Arrestin, βArrestin1, and βArrestin2 for G protein-coupled receptors delineate two major classes of receptors. J Biol Chem 275(22):17201–17210. https://doi.org/10.1074/jbc.m910348199

Ohno-Shosaku T, Maejima T, Kano M (2001) Endogenous cannabinoids mediate retrograde signals from depolarized postsynaptic neurons to presynaptic terminals. Neuron 29(3):729–738. https://doi.org/10.1016/s0896-6273(01)00247-1

Olsen RHJ, DiBerto JF, English JG, Glaudin AM, Krumm BE, Slocum ST, Che T, Gavin AC, McCorvy JD, Roth BL, Strachan RT (2020) TRUPATH, an open-source biosensor platform for interrogating the GPCR transducerome. Nat Chem Biol 16(8):841–849. https://doi.org/10.1038/s41589-020-0535-8

Palazuelos J, Aguado T, Pazos M, Julien B, Carrasco C, Resel E, Sagredo O, Benito C, Romero J, Azcoitia Í, Fernández-Ruiz J, Guzmán M, Galve-Roperh I (2009) Microglial CB2 cannabinoid receptors are neuroprotective in Huntington's disease excitotoxicity. Brain. https://doi.org/10.1093/BRAIN/AWP239

Patel M, Finlay DB, Glass M (2021) Biased agonism at the cannabinoid receptors – evidence from synthetic cannabinoid receptor agonists. Cell Signal 78:109865. https://doi.org/10.1016/j.cellsig.2020.109865

Patel M, Grimsey NL, Banister SD, Finlay DB, Glass M (2023) Evaluating signaling bias for synthetic cannabinoid receptor agonists at the cannabinoid CB2 receptor. Pharmacol Res Perspect 11(6). https://doi.org/10.1002/prp2.1157

Pertwee RG (1999) Pharmacology of cannabinoid receptor ligands. Curr Med Chem 6(8):635–664

Pitcher JA, Inglese J, Higgins JB, Arriza JL, Casey PJ, Kim C, Benovic JL, Kwatra MM, Caron MG, Lefkowitz RJ (1992) Role of beta gamma subunits of G proteins in targeting the beta-adrenergic receptor kinase to membrane-bound receptors. Science 257(5074):1264–1267. https://doi.org/10.1126/science.1325672

Porter AC, Sauer JM, Knierman MD, Becker GW, Berna MJ, Bao J, Nomikos GG, Carter P, Bymaster FP, Leese AB, Felder CC (2002) Characterization of a novel endocannabinoid, virodhamine, with antagonist activity at the CB1 receptor. J Pharmacol Exp Ther 301(3):1020–1024. https://doi.org/10.1124/jpet.301.3.1020

Rozenfeld R, Devi LA (2008) Regulation of CB1 cannabinoid receptor trafficking by the adaptor protein AP-3. FASEB J 22(7):2311–2322. https://doi.org/10.1096/fj.07-102731

Rubovitch V, Gafni M, Sarne Y (2002) The cannabinoid agonist DALN positively modulates L-type voltage-dependent calcium-channels in N18TG2 neuroblastoma cells. https://doi.org/10.1016/S0169-328X(02)00174-2

Rueda D, Galve-Roperh I, Haro A, Guzmán M (2000) The CB(1) cannabinoid receptor is coupled to the activation of c-Jun N-terminal kinase. Mol Pharmacol 58(4):814–820. https://doi.org/10.1124/mol.58.4.814

Rueda D, Navarro B, Martinez-Serrano A, Guzman M, Galve-Roperh I (2002) The endocannabinoid anandamide inhibits neuronal progenitor cell differentiation through attenuation of the Rap1/B-Raf/ERK pathway. J Biol Chem 277(48):46645–46650. https://doi.org/10.1074/jbc.M206590200

Ryalls B, Patel M, Sparkes E, Banister SD, Finlay DB, Glass M (2024) Investigating selectivity and bias for G protein subtypes and beta-arrestins by synthetic cannabinoid receptor agonists at the cannabinoid CB(1) receptor. Biochem Pharmacol 222:116052. https://doi.org/10.1016/j.bcp.2024.116052

Ryberg E, Vu HK, Larsson N, Groblewski T, Hjorth S, Elebring T, Sjögren S, Greasley PJ (2005) Identification and characterisation of a novel splice variant of the human CB1 receptor. FEBS Lett 579(1):259–264. https://doi.org/10.1016/j.febslet.2004.11.085

Sadanandan SM, Kreko-Pierce T, Khatri SN, Pugh JR (2020) Cannabinoid type 2 receptors inhibit GABAA receptor-mediated currents in cerebellar Purkinje cells of juvenile mice. PLoS One. https://doi.org/10.1371/JOURNAL.PONE.0233020

Sánchez-Zavaleta R, Cortés H, Avalos-Fuentes JA, García U, Vila J, Erlij D, Florán B (2018) Presynaptic cannabinoid CB2 receptors modulate [3 H]-glutamate release at subthalamo-nigral terminals of the rat. Synapse. https://doi.org/10.1002/SYN.22061

Schatz AR, Lee M, Condie R, Pulaski J, Kaminski N (1997) Cannabinoid receptors CB1 and CB2: a characterization of expression and adenylate cyclase modulation within the immune system. Toxicol Appl Pharmacol. https://doi.org/10.1006/TAAP.1996.8034

Schiöth HB, Fredriksson R (2005) The GRAFS classification system of G-protein coupled receptors in comparative perspective. Gen Comp Endocrinol 142(1–2):94–101. https://doi.org/10.1016/j.ygcen.2004.12.018

Schmidt M, Dekker FJ, Maarsingh H (2013) Exchange protein directly activated by cAMP (epac): a multidomain cAMP mediator in the regulation of diverse biological functions. Pharmacol Rev 65(2):670–709. https://doi.org/10.1124/pr.110.003707

Schmitz SK, King C, Kortleven C, Huson V, Kroon T, Kevenaar JT, Schut D, Saarloos I, Hoetjes JP, de Wit H, Stiedl O, Spijker S, Li KW, Mansvelder HD, Smit AB, Cornelisse LN, Verhage M, Toonen RF (2016) Presynaptic inhibition upon CB1 or mGlu2/3 receptor activation requires ERK/MAPK phosphorylation of Munc18-1. EMBO J 35(11):1236–1250. https://doi.org/10.15252/embj.201592244

Schmöle A-C, Lundt R, Gennequin B, Schrage H, Beins E, Krämer A, Zimmer TS, Limmer A, Zimmer A, Otte D (2015) Expression analysis of CB2-GFP BAC transgenic mice. PLoS One. https://doi.org/10.1371/JOURNAL.PONE.0138986

Scotter E, Goodfellow C, Graham E, Dragunow M, Glass M (2010) Neuroprotective potential of CB1 receptor agonists in an *in vitro* model of Huntington's disease. Br J Pharmacol 160(3):747–761. https://doi.org/10.1111/j.1476-5381.2010.00773.x

Seyedabadi M, Gharghabi M, Gurevich EV, Gurevich VV (2021) Receptor-arrestin interactions: the GPCR perspective. Biomolecules 11(2):218. https://doi.org/10.3390/biom11020218

Shao Z, Yin J, Chapman K, Grzemska M, Clark L, Wang J, Rosenbaum DM (2016) High-resolution crystal structure of the human CB1 cannabinoid receptor. Nature 540(7634):602–606. https://doi.org/10.1038/nature20613

Shire D, Carillon C, Kaghad M, Calandra B, Rinaldi-Carmona M, Le Fur G, Caput D, Ferrara P (1995) An amino-terminal variant of the central cannabinoid receptor resulting from alternative splicing. J Biol Chem 270(8):3726–3731. https://doi.org/10.1074/jbc.270.8.3726

Shum C, Dutan L, Annuario E, Warre-Cornish K, Taylor SE, Taylor RD, Andreae LC, Buckley NJ, Price J, Bhattacharyya S, Srivastava DP (2020) Δ9-tetrahydrocannabinol and 2-AG decreases neurite outgrowth and differentially affects ERK1/2 and Akt signaling in hiPSC-derived cortical neurons. Mol Cell Neurosci 103:103463. https://doi.org/10.1016/j.mcn.2019.103463

Simard M, Rakotoarivelo V, Di Marzo V, Flamand N (2022) Expression and functions of the CB2 receptor in human leukocytes. Front Pharmacol. https://doi.org/10.3389/FPHAR.2022.826400

Soethoudt M, Grether U, Fingerle J, Grim TW, Fezza F, De Petrocellis L, Ullmer C, Rothenhäusler B, Perret C, Van Gils N, Finlay D, Macdonald C, Chicca A, Gens MD, Stuart J, De Vries H, Mastrangelo N, Xia L, Alachouzos G et al (2017) Cannabinoid CB2 receptor ligand profiling reveals biased signalling and off-target activity. Nat Commun 8(1):13958. https://doi.org/10.1038/ncomms13958

Soria-Gomez E, Pagano Zottola AC, Mariani Y, Desprez T, Barresi M, Bonilla-Del Río I, Muguruza C, Le Bon-Jego M, Julio-Kalajzić F, Flynn R, Terral G, Fernández Moncada I, Robin LM, Oliveira Da Cruz JF, Corinti S, Amer YO, Goncalves J, Varilh M, Cannich A et al

(2021) Subcellular specificity of cannabinoid effects in striatonigral circuits. Neuron 109(9): 1513–1526.e1511. https://doi.org/10.1016/j.neuron.2021.03.007

Staus DP, Hu H, Robertson MJ, Kleinhenz ALW, Wingler LM, Capel WD, Latorraca NR, Lefkowitz RJ, Skiniotis G (2020) Structure of the M2 muscarinic receptor–β-arrestin complex in a lipid nanodisc. Nature 579(7798):297–302. https://doi.org/10.1038/s41586-020-1954-0

Steiger LJ, Tsintsadze T, Mattheisen GB, Smith SM (2023) Somatic and terminal CB1 receptors are differentially coupled to voltage-gated sodium channels in neocortical neurons. Cell Rep 42(3). https://doi.org/10.1016/j.celrep.2023.112247

Stella N (2010) Cannabinoid and cannabinoid-like receptors in microglia, astrocytes, and astrocytomas. Glia 58(9):1017–1030. https://doi.org/10.1002/glia.20983

Stempel AV, Stumpf A, Zhang HY, Özdogan T, Pannasch U, Theis AK, Otte DM, Wojtalla A, Racz I, Ponomarenko A, Xi Z, Zimmer A, Schmitz D (2016) Cannabinoid type 2 receptors mediate a cell type-specific plasticity in the hippocampus. Neuron. https://doi.org/10.1016/J.NEURON.2016.03.034

Stork PJS, Schmitt JM (2002) Crosstalk between cAMP and MAP kinase signaling in the regulation of cell proliferation. Trends Cell Biol 12(6):258–266. https://doi.org/10.1016/s0962-8924(02)02294-8

Straiker A, Stella N, Piomelli D, Mackie K, Karten HJ, Maguire G (1999) Cannabinoid CB1 receptors and ligands in vertebrate retina: localization and function of an endogenous signaling system. Proc Natl Acad Sci 96(25):14565–14570. https://doi.org/10.1073/pnas.96.25.14565

Stumpf A, Parthier D, Sammons RP, Stempel AV, Breustedt J, Rost BR, Schmitz D (2018) Cannabinoid type 2 receptors mediate a cell type-specific self-inhibition in cortical neurons. Neuropharmacology. https://doi.org/10.1016/J.NEUROPHARM.2018.07.020

Sugiura T, Kondo S, Sukagawa A, Nakane S, Shinoda A, Itoh K, Yamashita A, Waku K (1995) 2-Arachidonoylglycerol: a possible endogenous cannabinoid receptor ligand in brain. Biochem Biophys Res Commun 215(1):89–97. https://doi.org/10.1006/bbrc.1995.2437

Syrovatkina V, Alegre KO, Dey R, Huang XY (2016) Regulation, signaling, and physiological functions of G-proteins. J Mol Biol 428(19):3850–3868. https://doi.org/10.1016/j.jmb.2016.08.002

Tang J, Tao Y, Tan L, Yang L, Niu Y, Chen Q, Yang Y, Feng H, Chen Z, Zhu G (2015) Cannabinoid receptor 2 attenuates microglial accumulation and brain injury following germinal matrix hemorrhage via ERK dephosphorylation in vivo and in vitro. Neuropharmacology. https://doi.org/10.1016/J.NEUROPHARM.2015.04.028

Tsou K, Brown S, Sañudo-Peña MC, Mackie K, Walker JM (1998) Immunohistochemical distribution of cannabinoid CB1 receptors in the rat central nervous system. Neuroscience 83(2): 393–411. https://doi.org/10.1016/s0306-4522(97)00436-3

Turcotte C, Blanchet MR, Laviolette M, Flamand N (2016) The CB2 receptor and its role as a regulator of inflammation. Cell Mol Life Sci. https://doi.org/10.1007/S00018-016-2300-4

Twitchell W, Brown S, Mackie K (1997) Cannabinoids inhibit N- and P/Q-type calcium channels in cultured rat hippocampal neurons. J Neurophysiol 78(1):43–50. https://doi.org/10.1152/jn.1997.78.1.43

Urban JD, Clarke WP, von Zastrow M, Nichols DE, Kobilka B, Weinstein H, Javitch JA, Roth BL, Christopoulos A, Sexton PM, Miller KJ, Spedding M, Mailman RB (2007) Functional selectivity and classical concepts of quantitative pharmacology. J Pharmacol Exp Ther 320(1):1–13. https://doi.org/10.1124/jpet.106.104463

Vara H, Onofri F, Benfenati F, Sassoè-Pognetto M, Giustetto M (2009) ERK activation in axonal varicosities modulates presynaptic plasticity in the CA3 region of the hippocampus through synapsin I. Proc Natl Acad Sci 106(24):9872–9877. https://doi.org/10.1073/pnas.0900077106

Vidlarova M, Berta E, Prasil P, Prokopova A, Gurska S, Khoylou M, Rehulkova A, Kourilova P, Chudacek J, Szkorupa M, Klein J, Skarda J, Srovnal J, Hajduch M, Vidlarova M, Berta E, Prasil P, Prokopova A, Gurska S et al (2022) Cannabinoid receptor 2 expression in early-stage non-small cell lung cancers identifies patients with good prognosis and longer survival. https://doi.org/10.21037/TLCR-22-247

Vossler MR, Yao H, York RD, Pan M-G, Rim CS, Stork PJS (1997) cAMP activates MAP kinase and Elk-1 through a B-Raf- and Rap1-dependent pathway. Cell 89(1):73–82. https://doi.org/10.1016/s0092-8674(00)80184-1

Wang T, Tang W, Zhao Z, Zhao R, Lv Z, Guo X, Gu Q, Liu B, Lv H, Chen J, Zhang K, Li F, Wang J (2024) Fenofibrate recognition and Gq protein coupling mechanisms of the human cannabinoid receptor CB1. Adv Sci. https://doi.org/10.1002/advs.202306311

Westlake TM, Howlett AC, Bonner TI, Matsuda LA, Herkenham M (1994) Cannabinoid receptor binding and messenger RNA expression in human brain: an in vitro receptor autoradiography and in situ hybridization histochemistry study of normal aged and Alzheimer's brains. Neuroscience 63(3):637–652. https://doi.org/10.1016/0306-4522(94)90511-8

Whyte L, Ford L, Ridge SM, Cameron G, Rogers M, Ross R (2012) Cannabinoids and bone: endocannabinoids modulate human osteoclast function in vitro. Br J Pharmacol. https://doi.org/10.1111/J.1476-5381.2011.01519.X

Wilson RI, Nicoll RA (2001) Endogenous cannabinoids mediate retrograde signalling at hippocampal synapses. Nature 410(6828):588–592. https://doi.org/10.1038/35069076

Wootten D, Christopoulos A, Marti-Solano M, Babu MM, Sexton PM (2018) Mechanisms of signalling and biased agonism in G protein-coupled receptors. Nat Rev Mol Cell Biol 19(10): 638–653. https://doi.org/10.1038/s41580-018-0049-3

Wouters E, Walraed J, Banister SD, Stove CP (2019) Insights into biased signaling at cannabinoid receptors: synthetic cannabinoid receptor agonists. Biochem Pharmacol 169:113623. https://doi.org/10.1016/j.bcp.2019.08.025

Wright KL, Duncan M, Sharkey KA (2008) Cannabinoid CB2 receptors in the gastrointestinal tract : a regulatory system in states of inflammation. Br J Pharmacol. https://doi.org/10.1038/SJ.BJP.0707486

Wu J, Dent P, Jelinek T, Wolfman A, Weber MJ, Sturgill TW (1993) Inhibition of the EGF-activated MAP kinase signaling pathway by adenosine 3′,5′-monophosphate. Science 262(5136):1065–1069. https://doi.org/10.1126/science.7694366

Wu Y, Liu Q, Guo B, Ye F, Ge J, Xue L (2020) BDNF activates postsynaptic TrkB receptors to induce endocannabinoid release and inhibit presynaptic calcium influx at a calyx-type synapse. J Neurosci 40(42):8070–8087. https://doi.org/10.1523/jneurosci.2838-19.2020

Xiao JC, Jewell JP, Lin LS, Hagmann WK, Fong TM, Shen CP (2008) Similar in vitro pharmacology of human cannabinoid CB1 receptor variants expressed in CHO cells. Brain Res 1238: 36–43. https://doi.org/10.1016/j.brainres.2008.08.027

Xu C, Zhou Y, Liu Y, Lin L, Liu P, Wang X, Xu Z, Pin J-P, Rondard P, Liu J (2024) Specific pharmacological and Gi/o protein responses of some native GPCRs in neurons. Nat Commun 15(1). https://doi.org/10.1038/s41467-024-46177-z

Yang L, Finlay DB, Green HM, Zhu X, Glass M, Duffull S (2023) A quantitative pharmacology model for cannabinoid CB1 receptor mediated by Gi/Gs protein competition. Br J Pharmacol 181(8):1324–1340. https://doi.org/10.1111/bph.16293

Yiangou Y, Facer P, Durrenberger PF, Chessell IP, Naylor A, Bountra C, Banati RR, Anand P (2006) COX-2, CB2 and P2X7-immunoreactivities are increased in activated microglial cells/macrophages of multiple sclerosis and amyotrophic lateral sclerosis spinal cord. BMC Neurol. https://doi.org/10.1186/1471-2377-6-12

Zhang, H.-Y., Gao, M., Shen, H., Bi, G.-h., Yang, H.-J., Liu, Q.-R., Wu, J., Gardner, E., Bonci, A., & Xi, Z. (2017). Expression of functional cannabinoid CB2 receptor in VTA dopamine neurons in rats. Addict Biol https://doi.org/10.1111/ADB.12367

Zhang H-Y, Shen H, Gao M, Ma Z, Hempel B, Bi G-H, Gardner EL, Wu J, Xi Z (2021) Cannabinoid CB2 receptors are expressed in glutamate neurons in the red nucleus and functionally modulate motor behavior in mice. Neuropharmacology. https://doi.org/10.1016/J.NEUROPHARM.2021.108538

Zhang HY, Gao M, Liu Q-R, Bi GH, Li X, Yang HJ, Gardner EL, Wu J, Xi Z (2014) Cannabinoid CB2 receptors modulate midbrain dopamine neuronal activity and dopamine-related behavior in mice. Proc Natl Acad Sci U S A. https://doi.org/10.1073/PNAS.1413210111

Zhang HY, Shen H, Jordan CJ, Liu Q-R, Gardner EL, Bonci A, Xi Z (2019) CB 2 receptor antibody signal specificity: correlations with the use of partial CB 2-knockout mice and anti-rat CB 2 receptor antibodies. Acta Pharmacol Sin. https://doi.org/10.1038/S41401-018-0037-3
Zheng L, Wu X, Dong X, Ding X, Song C (2015) Effects of chronic alcohol exposure on the modulation of ischemia-induced glutamate release via cannabinoid receptors in the dorsal hippocampus. Alcohol Clin Exp Res. https://doi.org/10.1111/ACER.12845
Zhu X, Finlay DB, Glass M, Duffull SB (2019) Model-free and kinetic modelling approaches for characterising non-equilibrium pharmacological pathway activity: internalisation of cannabinoid CB1 receptors. Br J Pharmacol 176(14):2593–2607. https://doi.org/10.1111/bph.14684

Chemical Probes for Investigating the Endocannabinoid System

Annaleah Hanske, Marc Nazaré, and Uwe Grether

Contents

1 The Endocannabinoid System 73
2 Pharmacological Tools and Chemical Probes 76
3 The Cannabinoid Receptors 78
3.1 Ligands Targeting CB_1R 80
3.2 Ligands Targeting CB_2R 84
4 Metabolic Enzymes of AEA 1.2 92
4.1 NAPE-PLD Inhibitors 93
4.2 FAAH Inhibitors 94
4.3 NAAA Inhibitors 100
5 Metabolic Enzymes of 2-AG 1.1 102
5.1 DAGL Inhibitors 103
5.2 MAGL Inhibitors 106
6 Transmembrane, Intracellular, and Extracellular Transporters 112
6.1 Selective eCB Reuptake Inhibitors 112
6.2 FABP Inhibitors 114
7 Additional Targets Within the "Endocannabinoidome" 115
8 Conclusion 117
References 119

Abstract *Cannabis sativa* has been used therapeutically since early civilizations, with key cannabinoids Δ^9-tetrahydrocannabinol (THC) **3.1** and cannabidiol characterized in the 1960s, leading to the discovery of cannabinoid receptors type 1 (CB_1R) and type 2 (CB_2R) and the endocannabinoid system (ECS) in the 1990s. The ECS, involving endogenous ligands like 2-arachidonoylglycerol (2-AG) **1.1**, anandamide (*N*-arachidonoylethanolamine (AEA)) **1.2,** and various proteins, regulates vital

A. Hanske and M. Nazaré
Leibniz-Forschungsinstitut für Molekulare Pharmakologie FMP, Berlin, Germany
e-mail: hanske@fmp-berlin.de; nazare@fmp-berlin.de

U. Grether (✉)
Pharma Research and Early Development (pRED), Roche Innovation Center Basel, F. Hoffmann-La Roche Ltd, Basel, Switzerland
e-mail: uwe.grether@roche.com

Curr Topics Behav Neurosci (2026) 76: 69–144
https://doi.org/10.1007/7854_2024_563
Published Online: 03 January 2025

processes such as sleep, appetite, and memory, and holds significant therapeutic potential, especially for neurological disorders. Small molecule-derived pharmacological tools, or chemical probes, target key components of the ECS and are crucial for target validation, mechanistic studies, pathway elucidation, phenotypic screening, and drug discovery. These probes selectively interact with specific proteins or pathways, enabling researchers to modulate target activity and observe biological effects. When they carry an additional reporter group, they are referred to as labeled chemical probes. Developed through medicinal chemistry, structural biology, and high-throughput screening, effective chemical probes must be selective, potent, and depending on their purpose meet additional criteria such as cell permeability and metabolic stability.

This chapter describes high-quality labeled and unlabeled chemical probes targeting ECS constituents that have been successfully applied for various research purposes. CB_1R and CB_2R, class A G protein-coupled receptors, are activated by 2-AG **1.1,** AEA **1.2**, and THC **3.1**, with numerous ligands developed for these receptors. Imaging techniques like single-photon emission computed tomography, positron emission tomography, and fluorescently labeled CB_1R and CB_2R probes have enhanced CB receptor studies. CB_2R activation generally results in immunosuppressive effects, limiting tissue injury. AEA **1.2** is mainly degraded by fatty acid amide hydrolase (FAAH) or *N*-acylethanolamine acid amidase (NAAA) into ethanolamine and arachidonic acid (AA) **1.3**. FAAH inhibitors increase endogenous fatty acid amides, providing analgesic effects without adverse effects. NAAA inhibitors reduce inflammation and pain in animal models. Diacylglycerol lipase (DAGL) is essential for 2-AG **1.1** biosynthesis, while monoacylglycerol lipase (MAGL) degrades 2-AG **1.1** into AA **1.3**, thus regulating cannabinoid signaling. Multiple inhibitors targeting FAAH and MAGL have been generated, though NAAA and DAGL probe development lags behind. Similarly, advancements in inhibitors targeting endocannabinoid (eCB) cellular uptake or trafficking proteins like fatty acid-binding proteins have been slower. The endocannabinoidome (eCBome) includes the ECS and related molecules and receptors, offering therapeutic opportunities from non-THC cannabinoids and eCBome mediators. Ongoing research aims to refine chemical tools for ECS and eCBome study, addressing unmet medical needs in central nervous system disorders and beyond.

Keywords 2-Arachidonoylglycerol (2-AG) · Cannabinoid receptor type-1 (CB_1R) · Cannabinoid receptor type-2 (CB_2R) · Chemical probes · Diacylglycerol lipase (DAGL) · Endocannabinoid system (ECS) · Fatty acid amide hydrolase (FAAH) · Fatty acid-binding protein (FABP) · Monoacylglycerol lipase (MAGL) · *N*-Acylethanolamine acid amidase (NAAA) · *N*-Acylphosphatidylethanolamine-specific phospholipase D (NAPE-PLD) · *N*-Arachidonoylethanolamine (AEA) · Reverse-design

Abbreviations

2-AG	2-Arachidonoylglycerol
AA	Arachidonic acid

ABHD	α/β-Hydrolase domain
ABP	Activity-based probes
ABPP	Activity-based protein/proteome profiling
ACE	Angiotensin-converting enzyme
ADMET	Absorption, distribution, metabolism, excretion, and toxicology
AD	Alzheimer's disease
AM	Allosteric modulator
BALB	Bagg Albino
BBB	Blood–brain barrier
BODIPY	Boron-dipyrromethene
cAMP	Cyclic adenosine monophosphate
CBD	Cannabidiol
CB_1R	Cannabinoid receptor type 1
CB_2R	Cannabinoid receptor type-2
CC	Click-chemistry
CFA	Complete Freund's adjuvant
CIPN	Chemotherapy-induced peripheral neuropathy
CNS	Central nervous system
CoA	Coenzyme A
COX-2	Cyclooxygenase-2
cPLA2ε	Cytosolic phospholipase A2ε
CuAAC	Copper(I)-catalyzed azide-alkyne cycloaddition
CYP450	Cytochrome P450
$Cy5\text{-}N_3$	Sulfo-cyanine5-azide ($Cy5\text{-}N_3$)
DAG	Diacylglycerol
DAGL	Diacylglycerol lipase
dSTORM	Direct stochastic optical reconstruction microscopy
EAE	Experimental autoimmune encephalomyelitis
eCB	Endocannabinoid
eCBome	Endocannabinoidome
ECS	Endocannabinoid system
ED_{50}	Median effective dose
EIT	Intracellular eCB transporters
EIU	Endotoxin-induced uveitis
EMT	Endocannabinoid membrane transporter
ERK	Extracellular signal-regulated kinase
FAAH	Fatty acid amide hydrolase
FAAs	Fatty acid amides
FABP	Fatty acid-binding protein
FP-Rh	Fluorophosphonate rhodamine
FTD	Frontotemporal dementia
GIRK	G protein-coupled inwardly rectifying potassium channel
GIT	Gastrointestinal tract
GPCR	G protein-coupled receptor
GTP	Guanosine-5′-triphosphate
h	Human

hCB_2R	Human cannabinoid receptor type-2
HETE	Hydroxyeicosatetraenoic acid
HFD	High-fat diet
HFIP	Hexafluoroisopropanol
HPA	Hypothalamus–pituitary–adrenal
HTS	High-throughput screening
IEDDA	Inverse electron demand Diels–Alder reaction
iNAT	Ca^{2+}-independent *N*-acyltransferase
i.p.	Intraperitoneal
i.v.	Intravenous
KO	Knockout
LAPS	Ligand-assisted protein structure
LDC	Ligand-directed covalent
LOX	Lipoxygenase
LPS	Lipopolysaccharide
m	Mouse
MAGL	Monoacylglycerol lipase
MAPK	Mitogen-activated protein kinases
mCB_2R	Mouse cannabinoid receptor type 2
MIA	Monosodium iodoacetate
MS	Multiple sclerosis
MT	Meniscus tear
NAAA	*N*-acylethanolamine acid amidase
NAM	Negative allosteric modulator
NAPE	*N*-acylphosphatidylethanolamine
NAPE-PLD	*N*-acylphosphatidylethanolamine-specific phospholipase D
NArPE	*N*-arachidonoylphosphatidylethanolamine
NASA	*N*-acyl-*N*-alkyl sulfonamide
NAT	*N*-acyltransferase
NF-kB	Nuclear factor "kappa-light-chain-enhancer" of activated B-cells
N-SP	*N*-sulfonyl pyridone
OEA	Oleoylethanolamide
O-NBD	*O*-nitrobenzoxadiazole
PAM	Positive allosteric modulator
pCBs	Phytocannabinoids
PD	Pharmacodynamic
PED6	(*N*-{[6-(2,4-dinitrophenyl) amino]hexanoyl}-2-[4, 4-difluoro-5,7-dimethyl-4-bora-3a,-4a-diaza- s-indacene-3-pentanoyl}-1-hexadecanoyl-sn-glycero-3-phospho ethanolamine
PEA	*N*-palmitoylethanolamine
PET	Positron emission tomography
PG	Prostaglandin
PK	Pharmacokinetic
PLAAT	Phospholipase A and acyltransferase
PLC	Phospholipase C
p.o.	Oral

PPAR	Peroxisome proliferator-activated receptor
PTSD	Post-traumatic stress disorder
r	Rat
RXR	Retinoid X receptor
SAR	Structure–activity relationship
SDS-PAGE	Sodium dodecyl sulfate–polyacrylamide gel electrophoresis
SERI	Selective eCB reuptake inhibitor
Sn	Stereospecific numbering
SPECT	Single-photon emission computed tomography.
STAT3	Signal transducer and activator of transcription 3
$t_{1/2}$	Half-life
TAMRA	Tetramethyl-6-carboxyrhodamine
Th17	T helper 17 cells
THC	Δ^9-Tetrahydrocannabinol
TR-FRET	Time-resolved fluorescence resonance energy transfer
TRP	Transient receptor potential
TRPV	Transient receptor potential vanilloid
TU	Triazole ureas
WT	Wild-type

1 The Endocannabinoid System

Cannabis sativa is a plant that contains psychotropic compounds and was used in early civilizations for therapeutic purposes (Stephen P. H. Alexander 2016). Δ^9-Tetrahydrocannabinol (THC) **3.1** and cannabidiol (CBD) are natural products derived from *Cannabis* plants (Raphael Mechoulam and Gaoni 1965; R Mechoulam and Shvo 1963). These cannabinoids were characterized in the 1960s and were instrumental in understanding the plant's effects. Insights into the mechanism of action of THC **3.1**, the psychotropic component of *Cannabis*, led to the identification of cannabinoid receptors type 1 and type 2 (CB_1R and CB_2R, respectively) in the 1990s (Matsuda et al. 1990; Munro et al. 1993). This discovery catalyzed the identification of endogenous ligands for these receptors, known as endocannabinoids (eCBs) (William A. Devane et al. 1992; Raphael Mechoulam et al. 1995; Sugiura et al. 1995; V Di Marzo and Fontana 1995).

Subsequently, it became apparent that, besides the CBRs and eCBs, additional receptors, enzymes, and transporters form a complex lipid signaling network ubiquitously expressed in mammalian tissues, termed the endocannabinoid system (ECS) (Maccarrone et al. 2023; Cristino et al. 2020; Maccarrone 2020). The ECS plays a crucial role in re-establishing homeostasis after pathological insults and regulates several vital bodily functions, including sleep, body temperature, appetite, learning, memory, and emotional processing. Thus, it holds great therapeutic potential for multiple pathologies in both the periphery (Maccarrone et al. 2015) and the central nervous system (Iannotti et al. 2016; Marc P Baggelaar et al. 2018). Alterations in the ECS are found in experimental models and patients with most neurological diseases (Jung et al. 2012; Altamura et al. 2015; Di Iorio et al. 2013), and genetic manipulation of the ECS in mouse models alters susceptibility to neurodegenerative disorders (Bilsland et al. 2006; Mulder et al. 2011; D'Addario et al. 2012; Aymerich et al. 2018; Celorrio et al. 2016). These findings suggest that targeting components of the ECS is a viable therapeutic strategy, motivating efforts to translate these insights into treatments (Vincenzo Di Marzo 2018; Maccarrone et al. 2023).

The most active and well-studied eCBs are 2-arachidonoylglycerol (2-AG) **1.1** (Raphael Mechoulam et al. 1995; Sugiura et al. 1995) and anandamide (*N*-arachidonoylethanolamine, AEA) **1.2** (William A. Devane et al. 1992) (Fig. 1). These eCBs or their derivatives modulate various receptors localized on the plasma membrane and within the nucleus. Key eCB-binding receptors include the G protein-coupled receptors (GPCRs) CB_1R and CB_2R. Additionally, the orphan GPCRs GPR55, GPR119, and GPR18 can also bind cannabinoid-like ligands (Roger Guy Pertwee et al. 2010; Godlewski et al. 2009; Pingwei Zhao and Abood 2013; Shore and Reggio 2015; Morales and Reggio 2017; Alhouayek et al. 2018; Morales et al. 2020; Im 2021).

Further receptors that interact with eCBs include the ionotropic transient receptor potential vanilloid (TRPV) channels 1, 2, 3, and 4, TRP cation channel A1, and melastatin 8. These are six-transmembrane-spanning receptors located on the plasma membrane with intracellular binding sites (Kawahara et al. 2011; Benito et al. 2012). Additionally, nuclear peroxisome proliferator-activated receptors (PPARs) α, β (δ) and γ, are transcription factors that regulate gene expression (Guzman et al. 2004; Rockwell et al. 2006; Paterniti et al. 2013; Maccarrone 2020; Figuerola-Asencio et al. 2023). The receptor-mediated activities of eCBs are spatiotemporally tightly controlled, and dynamic homeostatic balance between synthesis and degradation regulates their cellular concentration via various biosynthetic and hydrolytic enzymes (Cristino et al. 2020; Maccarrone 2020).

Redundancy and promiscuity are hallmarks of eCB biosynthesis and degradation. Membrane phospholipid precursors can produce AEA **1.2** via multiple pathways. The classical route for AEA **1.2** release from phosphatidylethanolamine and phosphatidylcholine precursors involves *N*-acyltransferase (NAT), either Ca^{2+}-dependent or independent (iNAT), and *N*-acylphosphatidylethanolamine-specific phospholipase D (NAPE-PLD). Multiple pathways also exist for AEA **1.2** degradation. Cleavage into ethanolamine and arachidonic acid (AA) **1.3**, thus terminating its biological activity, is primarily catalyzed by fatty acid amide hydrolase-1 (FAAH-1), but also by the less widespread FAAH-2 or the lysosomal enzyme *N*-acylethanolamine acid amidase (NAAA) (Daniele Piomelli et al. 2020). AEA **1.2** can be transformed by oxygenation of the AA moiety, catalyzed by lipoxygenase 5, 12, and 15 isozymes (5-, 12-, 15-LOX), cyclooxygenase-2 (COX-2), or cytochrome P450 (CYP450) (Rouzer and Marnett 2011; Fezza et al. 2014; Simard et al. 2022). Remarkably, COX-2-generated prostamides and other oxidative derivatives of AEA **1.2** have their own biological activities (van der Stelt et al. 2002; Simard et al. 2022).

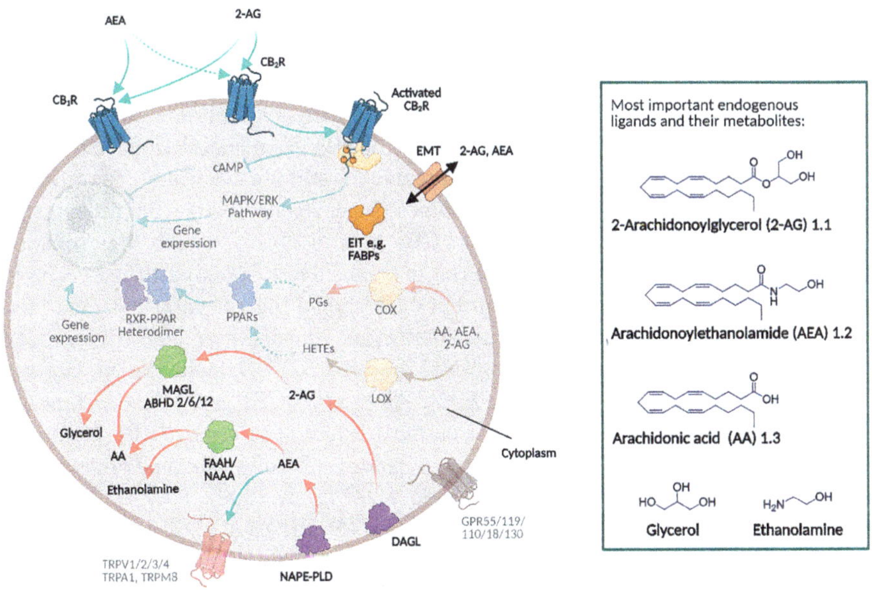

Fig. 1 Schematic illustration of the essential elements of the endocannabinoid system (ECS). 2-AG **1.1** and AEA **1.2** extracellularly activate G protein-coupled receptors (GPCRs) such as cannabinoid receptor type 1 (CB_1R) and type 2 (CB_2R). Intracellularly, they interact with transient receptor potential (TRP) channels and, directly or indirectly after metabolism, with nuclear hormone receptors such as peroxisome proliferator-activated receptors (PPARs) in the nucleus. 2-AG **1.1** is released from membrane lipids through the action of diacylglycerol lipase (DAGL) and can be hydrolyzed to glycerol and arachidonic acid (AA) **1.3** by cytosolic monoacylglycerol lipase (MAGL). AEA **1.2** is primarily synthesized by *N*-acylphosphatidylethanolamine-specific phospholipase D (NAPE-PLD) and degraded mainly by fatty acid amide hydrolase (FAAH) and to a smaller extent by *N*-acylethanolamine acid amidase (NAAA) to form AA **1.3** and ethanolamine. A specific endocannabinoid membrane transporter (EMT) facilitates the transmembrane exchange of eCBs. Intracellular traffic is facilitated by the intracellular eCB transporters (EIT) such as fatty acid-binding proteins (FABPs). Oxidative metabolism of eCBs involves the addition of oxygen to the fatty acid moiety by cyclooxygenase-2 (COX-2) to generate prostaglandins (PGs) and by lipoxygenases (LOXs) to form hydroxyeicosatetraenoic acids (HETEs). ABHD = α/β-hydrolase domain; cAMP = cyclic adenosine monophosphate; ERK = extracellular signal-regulated kinase; GPR = G protein-coupled receptor; MAPK = mitogen-activated protein kinases; RXR = retinoid X receptor

The classical biosynthetic route to 2-AG **1.1** involves Ca^{2+}- and glutathione-dependent DAG lipases (DAGL) α and β, releasing arachidonic acid-containing diacylglycerol (Tiziana Bisogno et al. 2003). 2-AG **1.1** is primarily cleaved to glycerol and AA **1.3** by monoacylglycerol lipase (MAGL). Alternative pathways involve α/β hydrolase domain (ABHD) proteins 2, 6, and 12, carboxylesterases 1 and 2, and palmitoyl-protein thioesterase 1 (Marc P. Baggelaar et al. 2018; Maccarrone 2020). Furthermore, 2-AG **1.1** can be oxygenated by COX-2, 12-, and 15-LOX (Rouzer and Marnett 2011; Fezza et al. 2014; Simard et al. 2022), leading to oxidative derivatives like prostaglandin- or thromboxane-glyceryl esters with their own biological activities (Marc P. Baggelaar et al. 2018; Simard et al. 2022).

The stringent metabolic control of eCB tone is further modulated by distinct transporters that facilitate the movement of eCBs across the plasma membrane, as well as intracellularly and extracellularly. eCBs can be released from membrane precursors "on demand" when the cell receives a stimulus, but they can also be stored in cytosolic organelles like adiposomes (Maccarrone 2020). The mechanisms underlying the membrane transport of eCBs have been extensively investigated. Two prevailing models exist: eCBs are transported either by passive diffusion (Fasia et al. 2003) or through a membrane carrier (Vincenzo Di Marzo et al. 1994; Beltramo et al. 1997). The mechanisms of transmembrane transport of eCBs have been comprehensively reviewed (Christopher J Fowler 2013; Nicolussi and Gertsch 2015; Kaczocha and Haj-Dahmane 2022). As lipids, eCBs cannot travel through the aqueous cytosol without a suitable carrier (Maccarrone et al. 2010). Cytosolic eCB transporters identified in recent years include structurally unrelated proteins like heat shock protein 70 and albumin (Sergio Oddi et al. 2009), fatty acid-binding proteins (FABPs) 1, 5, and 7 (Kaczocha et al. 2009; Elmes et al. 2019), and sterol carrier protein 2 (Hillard et al. 2017).

The understanding of the endocannabinoid system is complicated by the promiscuity of mediators, the overlap with other pathways, and alternative metabolic processes. Modulation of its components affects a wider endocannabinoid-related network known as the endocannabinoidome (eCBome) (Cristino et al. 2020). The main elements of the eCBome are endocannabinoid-like mediators, such as other long-chain *N*-acyl-amides, with their own downstream targets and metabolic enzymes, which are often shared with the ECS and form together a complex signaling system.

2 Pharmacological Tools and Chemical Probes

Pharmacological tools are indispensable in drug discovery and biomedical research for several reasons. These tools, which include small molecules, peptides, antibodies, and other chemical entities, interact with specific biological targets such as receptors, enzymes, ion channels, or transporters. Pharmacological tools are essential for characterizing drug targets, identifying and validating the biological targets of drugs (Arrowsmith et al. 2015), understanding the detailed mechanisms by which drugs exert their effects, and mapping out biochemical and signaling pathways (Bradner et al. 2010; Bunnage et al. 2013). These insights can lead to the development of new therapeutic agents and the discovery of biomarkers for disease states or drug responses.

Often highly specific and potent, pharmacological tools allow researchers to precisely dissect complex biological systems. They are critical in both basic research and the drug development process. This chapter focuses on the most essential group of pharmacological tools: small molecule-derived agents, also known as chemical probes, which target key components of the endocannabinoid system (ECS) most relevant to neurological disorders.

A chemical probe is a small molecule designed to selectively interact with a specific protein or biological pathway, enabling the study of its function within a biological system. These probes address fundamental biological questions, linking chemistry, biology, and drug discovery, and influencing all phases of research and development.

Chemical probes bind to their targets with high specificity and affinity, allowing researchers to modulate the function of the target protein and observe the resulting biological effects. This provides valuable insights into the role of proteins in various cellular processes and disease states.

The use of chemical probes is crucial in verifying target expression and addressing mechanistic questions during the target validation process. These probes play a significant role in setting up customized in vitro pharmacology assays during the hit identification and lead optimization phases (Schenone et al. 2013). They also verify target engagement, identify biomarkers, and support preclinical studies and human trials. Although their history spans over four decades, the development of chemical probes has largely been guided by empirical methods (Atlas and Levitzki 1977).

For a chemical probe to produce efficient and conclusive results, it must meet certain criteria (Guberman et al. 2022; Workman and Collins 2010; Blagg and Workman 2017). It should selectively bind to the target protein without significant affinity to other proteins, ensuring high selectivity. The probe must have a high affinity for the target to ensure effective modulation at low concentrations, indicating high potency. Depending on its purpose, it should be capable of entering cells to reach intracellular targets, thereby demonstrating cell permeability. Additionally, the probe should be stable under physiological conditions to ensure consistent activity during experiments.

Developing chemical probes involves a combination of medicinal chemistry, structural biology, and high-throughput screening techniques. These probes are invaluable in both, basic research and the early stages of drug development, and they have been instrumental in unraveling the ECS.

Although high-quality probes are designed to bind their target protein specifically, they often lack specialized reporters that enable direct spatiotemporal readouts across diverse validation settings. This unexploited potential can be leveraged by further functionalization with reporter groups that are often not drug-like—such as fluorescent dyes, click handles, reactive units, that is, warheads, photocages, protein tags, or proximity-inducing degradation tags (Prevet and Collins 2019; Muir et al. 2022; Auberson et al. 2023). These labeling modifications extend the utility of small-molecule probes, enhancing their ability to validate biological hypotheses, including those related to the eCBS. However, the design and introduction of such labeled probes without compromising target affinity, selectivity, and physicochemical properties pose a considerable challenge, as the introduction of medium to large-sized, often non-drug-like reporter entities may risk altering the probe's overall properties detrimentally. In this regard, the reverse design concept has proven to be a viable

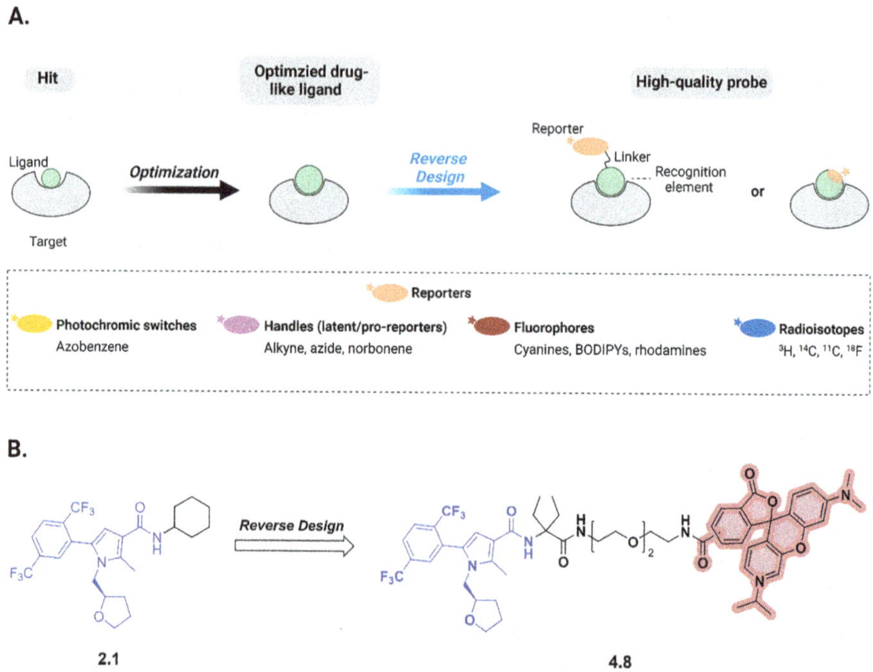

Fig. 2 (**a**) Reverse-design approach to labeled chemical probes and illustrative examples of reporter units. (**b**) Representative reverse design example for the generation of fluorescently labeled CB_1 inverse agonist **4.8**, starting from optimized drug candidate **2.1**

approach for the rapid conversion of drugs into high-quality labeled probes. This strategy leverages the synthetically optimized properties of drugs that have already undergone multiple chemical optimization cycles (Fig. 2a) (Kindermann et al. 2010; Gazzi et al. 2022; Mach et al. 2024). For the design of potent and selective labeled probes, this prior optimization work by medicinal chemists, resulting in favorable pharmacokinetic properties, and the extensive structure–activity relationship (SAR) data serve as invaluable insights into potential labeling sites. As an illustrative recent example, fluorescently labeled CB_1 ligand **4.8** (Mach et al. 2024) was generated from highly potent and selective CB_1 inverse agonist **2.1** through identification of a suitable exit vector for linker and subsequent TAMRA dye attachment (Fig. 2b).

3 The Cannabinoid Receptors

CB_1R and CB_2R belong to the class A (rhodopsin-like) GPCR family and share a high degree of homology, with 44% overall sequence identity and 68% homology in the ligand-binding domain (Matsuda et al. 1990; Munro et al. 1993; A. C. Howlett 2002). 2-AG **1.1** ($pK_i CB_1R/CB_2R$ 7.15/6.94 (Soethoudt et al. 2017)), AEA **1.2**

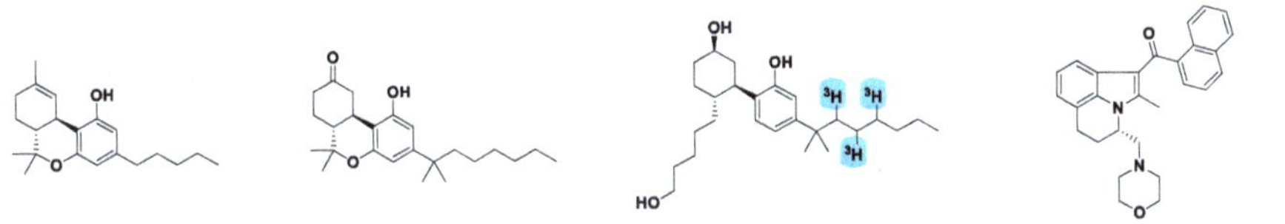

Δ⁹-THC 3.1
*K*ᵢCB₁R/CB₂R [nM] 41/36[a]
Partial agonist
Brain penetrant ✓
Oral bioavailability ✓
Launched ✓

Nabilone 3.2
CB$_{2/1}$$K_i$ = 1.84 nM / 2.19 nM[a]
Partial agonist
Brain penetrant ✓
Oral bioavailability ✓
Launched ✓

[^{3}H]CP-55,940 3.3
CB$_{2/1}$pK_i = 8.44 nM / 9.26 nM[a]
Full agonist
Active across species ✓

WIN 55212-2 3.4
CB$_{2/1}$pK_i = 8.57 nM / 8.72 nM[a]
Full agonist
Active across species ✓

Fig. 3 Chemical structures, in vitro pharmacology data, key characteristics, and highest development phase of dual CB_1R/CB_2R agonists. (**a**) Binding affinity on human cannabinoid receptors (Han et al. 2013). Radioisotopes are highlighted in blue

(p$K_i$$CB_1R/CB_2R$ 7.04/6.91 (Soethoudt et al. 2017)), and THC **3.1** are dual effectors at both CB_1R and CB_2R (Fig. 3). Remarkably, no natural ligands have been identified that exhibit exclusive selectivity for either of the two CBRs. Selectivity appears to be regulated by varying receptor expression levels in different cell types and tissues. CB_1R is predominantly expressed in the CNS but is also found in many peripheral tissues and organs to a lesser degree (A. C. Howlett 2002; Pál Pacher et al. 2006). In contrast, CB_2R is highly but not exclusively abundant in immune tissues such as the spleen, lymph nodes, and bone marrow, where it is found in immune cells, including macrophages, T and B cells, natural killer cells, monocytes, and polymorphonuclear neutrophils (Bouaboula et al. 1993; Galiegue et al. 1995; Atwood and Mackie 2010).

Due to the enormous therapeutic potential of the CBRs, thousands of ligands with varying property profiles targeting CB_1R and/or CB_2R have been generated. These can be grouped into endogenous cannabinoids and related fatty acid derivatives, phytocannabinoids, and synthetic cannabinoids. THC **3.1** is a dual CB_1R and CB_2R partial agonist exhibiting multiple therapeutically interesting physiological properties involving both receptor types, including anti-inflammatory, immunosuppressive, and analgesic effects. THC **3.1** was the first cannabinoid agonist approved as a medication by the US Food and Drug Administration (FDA) under the generic name dronabinol (marinol), although its use was restricted due to CNS-mediated psychotropic side effects. The determination of the absolute stereochemistry of THC **3.1** in 1967 (Raphael Mechoulam and Gaoni 1965) was key for the generation of many additional nonselective cannabinoid agonists (Razdan 1986; Robson 2001; Roger G Pertwee 2012; Roger G. Pertwee 2008b), enabling the discovery of nabilone (LY-109514) **3.2**, which was launched to ameliorate chemotherapy-induced nausea and (Adis Insight Web Page, *Drug report on nabilone*, https://adisinsight.springer.com/drugs/800025856, accessed November 2024a). Clinical trials for the treatment of Parkinson's disease and pain are in advanced stages.

Intending to develop cannabinoid, nonopioid analgesics, Pfizer produced a series of "nonclassical" THC analogs (M. R. Johnson et al. 1981; Allyn C. Howlett et al. 1990; L S Melvin et al. 1993, 1995), resulting in the development of the dual CB_1R/

CB_2R full agonist CP-55,940 **3.3**. It outperformed THC **3.1** concerning CB_1R/CB_2R-binding affinity and analgesic activity (William A Devane et al. 1988; A C Howlett et al. 1988; Showalter et al. 1996) and became an important research tool for identifying CBR cellular and systems physiology. Tritiated CP-55,940 **3.3** was critically involved in the deorphanization of both CB_1R (Matsuda et al. 1990) and CB_2R (Munro et al. 1993) and has been broadly applied to quantify the SARs of many novel ligands developed for the investigation of cannabinoid receptors. Sterling-Winthrop discovered that structural modifications of the nonsteroidal anti-inflammatory agent pravadoline resulted in greater antinociceptive activity with diminished potential to block prostaglandin production (Bell et al. 1991), leading to the design of the dual CB_1R/CB_2R full agonist WIN 55212-2 **3.4**. Aminoalkylindole **3.4** has greatly contributed to the understanding of cannabinoid receptor physiology and pharmacology, and its tritiated analog is a standard reference and broadly applied CB_1R/CB_2R radioligand (D'Ambra et al. 1992; Eissenstat et al. 1995). CB_1R was identified as the specific brain receptor for THC by the Howlett group in 1988 (William A Devane et al. 1988). It is critical in regulating mood, appetite, pain perception, learning and memory, and motor control (Marsicano and Lutz 2006; Kano et al. 2009; De Laurentiis et al. 2014). CB_1R has been recognized as a target for pharmacotherapeutic development based on extensive preclinical data (for reviews, see Mackie 2008; R. G. Pertwee 2008a, 2012; Tsang and Giudice 2016; Lu and Anderson 2017; Amin and Ali 2019; Lesley D Schurman et al. 2020; Wilkerson et al. 2021; Roger G. Pertwee 2008b). However, the development of CB_1R agonists and antagonists has faced challenges related to target-mediated psychotropic side effects. Despite these challenges, targeting CB_1R for unmet therapeutic needs has continued to evolve based on preclinical investigations (Maccarrone et al. 2023).

3.1 Ligands Targeting CB_1R

Several experimental studies have shown that THC **3.1** and 2-AG **1.1** stimulate appetite and food intake in animal models (Milano et al. 1988; Kirkham et al. 2002). CB_1R knockout mice exhibit markedly lower daily food intake and body weight, even when fed a high-fat diet (HFD) (Ravinet Trillou et al. 2004). This led to the search for CB_1R inverse agonists for controlling appetite, food intake, and ultimately body weight, which commenced in the late 1990s (Roger Guy Pertwee 2006). Target-based drug discovery has progressed and delivered many different series of CB_1-receptor antagonists. The most advanced CB_1R antagonist, rimonabant (SR141716A) **4.1** (Van Gaal et al. 2005) (Fig. 4), received marketing authorization in the European Union as an antiobesity medication. Further selective CB_1R agonists, including taranabant (Merck), otenabant (Pfizer), ibipinabant (Solvay), and surinabant (Sanofi), progressed into clinics. In 2008, rimonabant **4.1** was withdrawn from the market due to an increased incidence of anxiety, depression, and suicidality (Sam et al. 2011). These psychiatric side effects appear to preclude the use of CB_1R

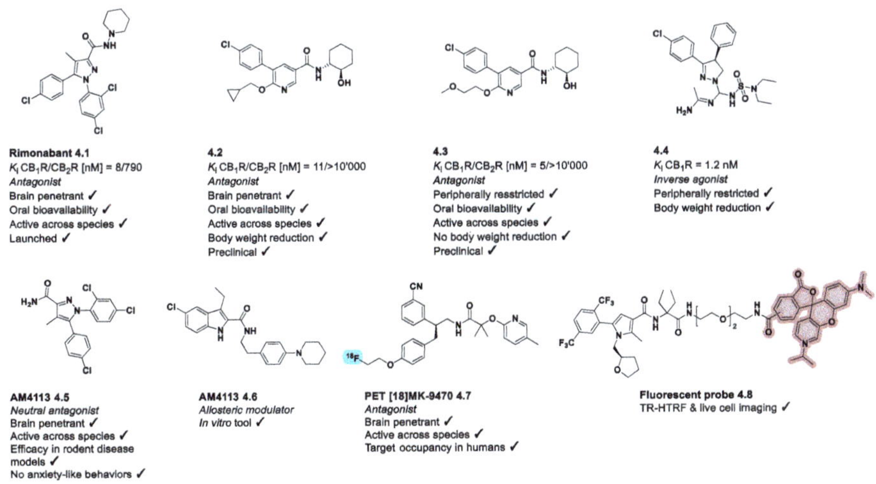

Fig. 4 Chemical structures, in vitro pharmacology data, key characteristics, and highest development phase of CB$_1$R ligands. Radioisotopes are highlighted in blue, and fluorophore reporter groups in pink

receptor antagonists for non-life-threatening diseases or chronic treatment. Thus, the development of further brain-penetrant CB$_1$R antagonists was discontinued by other pharmaceutical companies. However, it has become apparent that the CB$_1$ receptor has peripheral functions, suggesting that peripherally active and selective agents may have therapeutic potential with reduced psychiatric side effects.

Roche identified 6-alkoxy-5-aryl-3-pyridinecarboxamides (cf. compound **4.2**) as potent CB$_1$ receptor antagonists with high selectivity over CB$_2$ receptors (Stephan Röver et al. 2013). The series was optimized to reduce lipophilicity, resulting in peripherally active molecules with minimal central effects. This optimization led to the development of a highly potent and selective CB$_1$R inverse agonist, pyridine **4.3**, which exhibited markedly reduced brain exposure compared to rimonabant. Despite high peripheral exposure, the ligand had no significant effect on body weight and only a marginal impact on body fat reduction in high-fat-fed rats. These data are consistent with the reported effects in tissue-specific CB$_1$R knockout (KO) mice and suggest that the metabolic benefits of CB$_1$R antagonists are primarily centrally mediated. Other groups have reported efficacy with small-molecule CB$_1$R antagonists/inverse agonists, for example, dihydropyrazoline **4.4**, that exhibit minimal brain exposure in animal models of obesity and metabolic syndrome, as well as in alcoholic and nonalcoholic liver steatosis, liver fibrosis, and renal diseases, as recently reviewed by the Kunos group (Resat Cinar and Kunos 2020). The primary methods used to determine brain permeability are pharmacokinetic (PK) studies (Yue-Mei Zhang et al. 2018; Iyer et al. 2022), while specific engagement of brain CB$_1$R is assessed using positron emission tomography tracers (Tam et al. 2012; Chun-Ping Chang et al. 2019), as well as antagonism of the tetrad effects induced by CB$_1$R agonists (Fulp et al. 2013; Amato et al. 2019). Given the challenge of completely excluding small molecules from the brain, peripherally restricted therapeutic CB_1R antibodies might unequivocally determine whether central or peripheral CB_1R antagonism/inverse agonism is responsible for improving obesity and metabolic syndrome. Two candidates that have been in clinical development for renal diseases and diabetic complications are nimacimab (Bird Rock Bio) and GFB-024 (Goldfinch Bio) (Maccarrone et al. 2023).

CB_1R is a constitutively active receptor that exists in equilibrium between active and inactive states, even in the absence of ligands, resulting in increased basal activity (Roger G. Pertwee 2005; Fong 2014), which may be important for cellular homeostasis. While inverse agonists reduce the basal activity of receptors, neutral antagonists do not significantly affect it (Bond and IJzerman 2006; Kelly S. Sink et al. 2008; Kenakin and Kenakin 2024). Additionally, the ECS exhibits an endogenously active tone controlled by the cellular production of eCBs (Allyn C. Howlett et al. 2011). Therefore, CB_1R neutral antagonists can compete with endogenous cannabinoid ligands without affecting the basal activity of the receptor. It was hypothesized that CB_1R neutral antagonists could produce the therapeutic benefits of CB_1R antagonism without the unwanted CNS and gastrointestinal side effects.

The most extensively studied CB_1R neutral antagonist AM4113 **4.5**, exhibited therapeutic efficacy together with a better tolerability profile. In animal models of obesity, AM4113 **4.5** was shown to reduce food intake and weight gain, as well as to suppress food-reinforced operant responding and feeding (Chambers et al. 2007; Kelly S. Sink et al. 2008; Gueye et al. 2016). In addiction-related models, AM4113 **4.5** was effective in suppressing alcohol consumption, reducing drug-seeking behavior for nicotine and THC **3.1**, and inhibiting the self-administration of heroin (Xiang-Hu He et al. 2019; Gueye et al. 2016; Schindler et al. 2016; Balla et al. 2018). Moreover, AM4113 **4.5** did not induce anxiety-like behaviors in the elevated plus maze and electrical brain stimulation reward paradigm, unlike the CB_1R antagonist/inverse agonist AM251 (K.S. Sink et al. 2010; Gueye et al. 2016; Xiang-Hu He et al. 2019).

The development of biased CB_1R agonists (Laprairie et al. 2016; Ibsen et al. 2017; Leo and Abood 2021; Manning et al. 2021), which are ligands designed to alter the conformation of the CB_1R to preferentially interact with either the Gi/o family or facilitate receptor phosphorylation via G-protein receptor kinases to enhance interaction with β-arrestins 1 or 2 (Priestley et al. 2017; Al-Zoubi et al. 2019), is likely to play an increasingly significant role in drug discovery (Ahn et al. 2012; Khajehali et al. 2015; Saleh et al. 2018; Garai et al. 2021). This selectivity targets the signal transduction pathways involved in beneficial effects while minimizing those responsible for unwanted side effects. However, the functional consequences of CB_1R-biased signaling remain poorly understood since all known orthosteric agonists lack signaling specificity. Additionally, no highly CB_1R-selective orthosteric agonist has been reported to date. Therefore, the development of biased allosteric ligands may be a viable strategy to dissociate the activation of G-proteins, β-arrestin1, or β-arrestin2 and refine CB_1R-targeted therapeutics. Org27569 **4.6** (Dalzell et al. 1981), initially developed by Organon, was among the first allosteric modulators for the CB_1 receptor. Recently, it was co-crystallized with the CP-55,940-bound CB_1 receptor, leading to the characterization of an allosteric site for the receptor (Shao et al. 2019). Furthermore, ZCZ011 increased the potency and reduced tolerance development in the anti-nociceptive activity of CB_1R agonists (Ignatowska-Jankowska et al. 2015), and GAT211 synergized with FAAH- or MAGL-inhibitor-mediated eCB accumulation to attenuate inflammatory and neuropathic pain (Slivicki et al. 2018, 2020). Preclinical studies of CB_1R allosteric modulators have been reviewed recently (Khurana et al. 2017; Hryhorowicz et al. 2019; Manning et al. 2021; Leo and Abood 2021). While allosteric CB_1R ligands have been successfully used to selectively trigger or inhibit specific signal transducers, suggesting that allosteric-binding sites hold promise for the development of strongly biased CB_1R ligands, additional ligands with improved absorption, distribution, metabolism, excretion, and toxicology (ADMET) profiles are required to further explore this potential.

Over the past decade, substantial efforts have been made to develop CB_1R single-photon emission computed tomography (SPECT) and positron emission tomography (PET) ligands to enable in vivo imaging of this receptor system (Hou et al. 2021). These ligands are valuable tools for investigating CB_1R neurobiology. Successful in vivo ligands typically exhibit high affinity, moderate lipophilicity, high specific binding, and effective brain penetration (Honer et al. 2014). Several tracers have transitioned into clinical research studies, including [^{11}C]OMAR (Wong et al. 2008, 2010) and [^{18}F]MK-9470 **4.7** (Burns et al. 2007; Terry et al. 2010). These tracers have demonstrated utility in preclinical and clinical imaging studies for noninvasive CB_1R biology and pharmacology investigations and for measuring receptor occupancy of potential therapeutic CB_1R inverse agonists. They have provided promising results in mental disorders such as schizophrenia and addiction.

Specifically decorated pyrrole carboxamides have been identified as selective CB_1R antagonists and have served as seed structures for the generation of fluorescently labeled CB_1R probes based on a reverse-design approach (Mach et al. 2024). A modular design concept, featuring a diethyl glycine-based building block as the centerpiece and supported by computational docking studies, facilitated the creation of pyrrole-based CB_1R fluorescent probes, such as the tetramethyl-6-carboxyrhodamine (TAMRA) derivative **4.8**. These probes have been successfully applied in time-resolved fluorescence resonance energy transfer (TR-FRET) target-engagement studies and CB_1R live cell imaging studies, offering significant potential for deepening the understanding of mechanistic aspects of CB_1R localization, trafficking, and activation, which are essential for elucidating the receptor's function and role in pathological conditions.

The design of next-generation orthosteric and allosteric ligands will benefit from recent structural insights into CB_1R. Studies using ligand-assisted protein structure (LAPS) methodologies have characterized the action sites at CB_1R and CB_2R (Janero et al. 2017). A detailed CB_1R crystal structure in its inactive conformation was reported in 2016 using the antagonist AM6538 (Hua et al. 2016) and taranabant (Shao et al. 2016), enabling the docking of several CB_1R antagonist analogs. Subsequent studies of agonist-bound CB_1R (Hua et al. 2017, 2020; Krishna

Kumar et al. 2019) revealed significant conformational changes upon activation, elucidating how antagonists fit into the receptor.

3.2 *Ligands Targeting CB_2R*

CB_2R was identified in 1993 (Munro et al. 1993). It is predominantly found in immune tissues such as the spleen, lymph nodes, and bone marrow, where it is present in various immune cells, including macrophages, T and B cells, natural killer cells, monocytes, mast cells, and polymorphonuclear neutrophils (Pál Pacher et al. 2006; Bouaboula et al. 1993; Galiègue et al. 1995). Additionally, CB_2R is detected in other peripheral tissues like the liver (Ariane Mallat et al. 2013; Atwood and Mackie 2010), kidney (Mukhopadhyay et al. 2010b; Mukhopadhyay et al. 2010a), spleen (Galiègue et al. 1995), gastrointestinal tract (Atwood et al. 2012), and bone (Atwood et al. 2012; Ofek et al. 2006), mainly in immune-derived cells such as Kupffer cells, osteoblasts, and osteoclasts. CB_2R mRNA is also observed in microglia, especially during neuroinflammation, where microglia activation leads to increased CB_2R expression (Cabral et al. 2008). Outside of immune cells, CB_2R expression is generally low, making its detection challenging (Cabral et al. 2008; Marchalant et al. 2014).

CB_2R is expressed both on the cell surface and intracellularly (Kleyer et al. 2012; Brailoiu et al. 2014; Castaneda et al. 2017). Inflammation and tissue injury can cause rapid increases in local eCB levels, likely due to contributions from both infiltrating immune cells and damaged parenchymal cells. This often results in elevated CB_2R expression, primarily from the infiltration and activation of immune cells. Such changes in eCB levels and CB_2R expression have been documented in numerous human diseases (P. Pacher and Mechoulam 2011). These alterations modulate CB_2R-mediated signaling responses in immune and other cells.

At the molecular level, CB_2R couples with Gαi/o protein, inhibiting cAMP production via adenylate cyclase, which stimulates downstream signaling pathways. Internalization and ERK phosphorylation are also significant signaling mechanisms. The signaling profile can vary among orthosteric binders, indicating biased signaling (Soethoudt et al. 2017). This variability leads to distinct cellular responses, potentially enabling the development of tailored CB_2R therapeutics (Oyagawa et al. 2018).

Generally, CB_2R activation in immune or immune-derived cells results in immunosuppressive effects, such as inhibited proliferation, induced apoptosis, and suppressed cytokine and chemokine production and migration of stimulated immune cells (Kishimoto et al. 2005; Ziring et al. 2006; Börner et al. 2009; Cencioni et al. 2010; Correa et al. 2011). Additionally, T regulatory cells are recruited, reducing autoimmune inflammatory responses (Hegde et al. 2008). Overall, these functions help prevent, attenuate, and repair damage, thereby limiting tissue injury in various pathological conditions, including cardiovascular (Pál Pacher et al. 2008; Pál Pacher and Gao 2008), gastrointestinal/inflammatory bowel disease (Karen Wright et al.

2005; K. L. Wright et al. 2008; Izzo and Camilleri 2009; Izzo and Sharkey 2010), liver (A. Mallat and Lotersztajn 2008; Pál Pacher and Gao 2008; Horvath et al. 2012; Matyas et al. 2020), kidney (Zoja et al. 2016; Mukhopadhyay et al. 2010a, b; Trojnar et al. 2020), lung (Pál Pacher et al. 2006), neurodegenerative (Centonze et al. 2007; Fernández-Ruiz et al. 2007; Fernández-Ruiz 2009), systemic sclerosis and scleroderma (García-Martín et al. 2019; Xingwu Jiang et al. 2020; Bei-Er Jiang et al. 2021; Spiera et al. 2023), diabetes and its complications (retinopathy, nephropathy, and cardiomyopathy) (Barutta et al. 2010, 2011; El-Remessy et al. 2011; Rajesh et al. 2012, 2022; Jourdan et al. 2014; Gruden et al. 2016; Spyridakos et al. 2022), and pain (Guindon and Hohmann 2008; Anand et al. 2009), particularly those associated with sterile inflammatory responses. In many forms of tissue injury or inflammation, CB_2R activation is generally protective. However, in certain disease states or stages, the immunosuppressive effects of CB_2R activation may exacerbate tissue damage context-dependently (P. Pacher and Mechoulam 2011). In such cases, CB_2R inverse agonists and antagonists might offer therapeutic benefits.

The significant therapeutic potential of CB_2R modulation has been recognized by both academic and industry institutions, leading to the discovery of numerous ligands and chemical probes targeting CB_2R (Ilan et al. 2013; Thakur et al. 2009; Riether 2012; Manera et al. 2015; Morales et al. 2016; Aghazadeh Tabrizi et al. 2016; Cooper et al. 2017; Brennecke et al. 2021; Guba et al. 2021; Kosar et al. 2024a). The first manuscripts and patents appeared in 1991 (Bell et al. 1991) and 1996 (Rinaldi et al. 1996), respectively. Since then, over 1,132 basic patent applications have been filed up to this article's submission date. The reported structures encompass a wide, structurally diverse chemical space, including endogenous cannabinoids and related fatty acid derivatives, marijuana-derived cannabinoids, and synthetic cannabinoids.

Recently, there has been a notable shift in CB_2R ligand development, focusing on agonists with high selectivity over CB_1R. This shift aligns with the growing interest in inflammation or tissue injury indications. Concurrently, more than 20 CB_2R agonists have been evaluated in human studies (Brennecke et al. 2021).

The list of notable CB_2R tool compounds includes HU-910 **5.1** (Horváth et al. 2012), a nonclassical cannabinoid in which the pyranyl ring from THC **3.1** has been replaced (Fig. 5). HU-910 **5.1** demonstrated a favorable overall profile, including high binding affinity and functional selectivity against CB_1R as well as toward a representative set of other off-targets. Consequently, it has been recommended as a selective CB_2R agonist for studying the role of CB_2R in biological and disease processes (Soethoudt et al. 2017). Compared to the commonly used HU-308 (Hanuš et al. 1999) and JWH133 (Roger G Pertwee 1999), HU-910 **5.1** exhibits a superior pharmacokinetic profile, which translated into sustained beneficial effects in liver post-ischemia paradigms (Horváth et al. 2012). Additionally, HU-910 **5.1** (10 mg/kg p.o. for 11 weeks) significantly reduced albuminuria and renal injury in mice with type 2 diabetic nephropathy, comparable to the clinically approved angiotensin-converting enzyme (ACE) inhibitor lisinopril, suggesting that CB_2R agonism could be a potential treatment for type 2 diabetic nephropathy (Zoja et al. 2016). In an in vivo mouse model of endotoxin-induced uveitis (EIU), HU-910 **5.1** exhibited

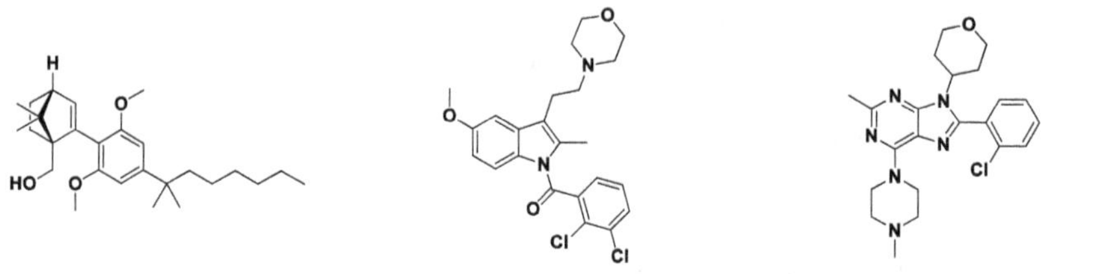

HU-910 5.1
Balanced agonist
Brain penetrant ✓
Oral bioavailability ✓
Activity across species ✓
In vivo efficacy ✓

GW-405,833 5.2
$CB_{2/1}$ K_i = 3.8 nM / 4800 nM
Ratio CB_1/CB_2 = 1263

LY-2828360 5.3
$CB_{2/1}$ EC_{50} = 20 nM / > 100 000 nM
Ratio CB_1/CB_2 = > 5000
Agonist
Brain penetrant ✓
Bioavailability ✓
Activity across species ✓
In vivo efficacy ✓

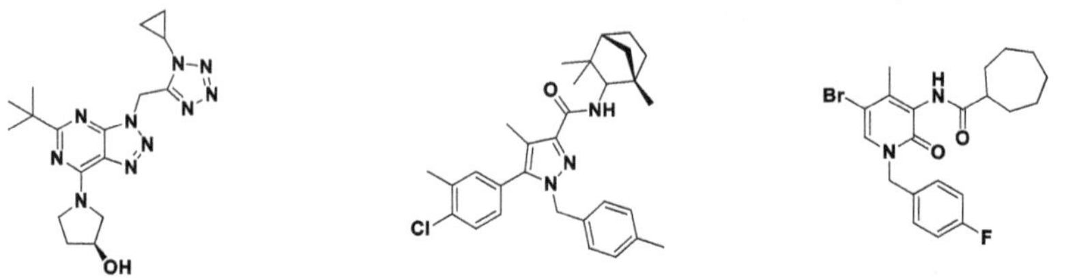

RO6871304 5.4
cAMP EC_{50} [nM] (% effect)
CB_2R / CB_1R 0.7 /103) / > 10'000
Agonist
Bioavailability ✓
Activity across species ✓
In vivo efficacy ✓
Preclinical ✓

SR144528 5.5
$CB_{2/1}$ pK_i = 7.88 / 5.77
CB_2 = selectivity 129
Antagonist
Brain penetrant ✓
Oral bioavailability ✓
Activity across species ✓

Compound C2 5.6
Positive allosteric modulator
Brain penetrant ✓
Oral bioavailability ✓
In vivo efficacy ✓

Fig. 5 Chemical structures, in vitro pharmacology data, key characteristics, and highest development phase of CB_2R ligands

beneficial immunomodulating actions in the eye by decreasing leukocyte adhesion through the inhibition of resident ocular immune cells (Porter et al. 2019). Importantly, HU-910's selective CB_2R activation shows promise across multiple neuroscience indications in preclinical models, including neuroprotection (Horváth et al. 2012), schizophrenia (Cortez et al. 2022), and pain (Horváth et al. 2012), without producing detectable catalepsy, antinociception, or hypothermia at doses up to 100 mg/kg, indicating a lack of CB_1R-mediated side effects (Soethoudt et al. 2017). Notably, both HU-308 and HU-910 **5.1** acted as well-balanced agonists across all five signal transduction pathways on human CB_2R (hCB_2R) (Soethoudt et al. 2017). However, both ligands exhibited differences in signaling preference between human and mouse CB_2R. For mouse CB_2R (mCB_2R), a significant bias toward G-protein activation compared to β-arrestin recruitment and cAMP signaling has been reported (Soethoudt et al. 2017). These interspecies differences in signaling preference must be considered when translating preclinical models to clinical settings for testing novel drug candidates.

The aminoalkylindole-derived CB_2R partial agonist GW-405,833 **5.2** is highly selective against CB_1R in a recombinant-binding assay, with a selectivity ratio greater than 1,200-fold (Valenzano et al. 2005). However, the receptor selectivity in rats was lower, with a rat CB_1R/CB_2R K_i ratio of 78. GW-405,833 **5.2** demonstrated dose-dependent antihyperalgesic effects in rodent pain models, including sciatic nerve ligation (ED_{50} = 0.1 mg/kg, intraperitoneal (i.p.)), skin incision (ED_{50} = 2.58 mg/kg, i.p.), and complete Freund's adjuvant (CFA)-induced inflammatory pain (ED_{50} = 0.17 mg/kg, i.p.). Studies conducted in CB_2R knockout mice confirmed that the observed in vivo pharmacological effects were "on target" and independent of opioid receptor activation (Valenzano et al. 2005; Clayton et al. 2002).

Imidazopyrimidine LY-2828360 **5.3**, developed by Eli Lilly for the oral treatment of osteoarthritic knee pain (ClinicalTrials.gov identifier: NCT01319929) (Lilly 2010), is a purine-derived ligand that exhibits high selectivity over CB_1R, with a CB_1R/CB_2R EC_{50} guanosine5′-thiotriphosphate (GTPγS) ratio greater than 5,000 (Hollinshead et al. 2013). LY-2828360 **5.3** has similar binding affinities for human and rat CB_2R (human K_i = 37 nM, rat K_i = 40 nM), is capable of central nervous system penetration, and advanced to phase 2 clinical trials before being terminated in January 2012, despite an acceptable side effect profile in humans (Hollinshead et al. 2013). Preclinical studies demonstrated that LY-2828360 **5.3** significantly reduced pain at doses of 0.3 and 1 mg/kg p.o. in the monosodium iodoacetate (MIA)-induced knee injury model and the meniscal tear (MT) model of osteoarthritis (M. P. Johnson et al. 2012). Additionally, it suppressed neuropathic pain with sustained efficacy and attenuated morphine dependence and tolerance (Lin et al. 2018). LY-2828360 **5.3** also inhibited paw flinches, thermal hyperalgesia, and tactile allodynia in rodent models (M. P. Johnson et al. 2012).

Researchers from Roche have reported the generation of highly selective and potent bicyclic (het)aryl-derived CB_2R agonists, which were developed from a high-throughput screening (HTS) hit (Nettekoven et al. 2016). This hit-to-lead optimization led to the identification of triazolopyrimidine RO6871304 **5.4**, which demonstrated equal potency toward human and mouse CB_2R isoforms in the sub-nanomolar range (human CB_2R EC_{50} cAMP = 1 nM, mouse CB_2R EC_{50} cAMP = 0.5 nM). The compound exhibited very high selectivity in binding versus the CB_1 receptor (>1,190-fold). RO6871304 **5.4** showed high average free plasma concentrations following both intravenous (i.v.) and oral (p.o.) administration, translating into efficacy in a kidney ischemia-reperfusion model and protective properties in a renal fibrosis model. Additionally, RO6871304 **5.4** effectively reduced leukocyte-neutrophil activity in an in vivo mouse model of EIU after topical administration (Porter et al. 2019). In vitro, RO6871304 **5.4** decreased neutrophil migration in wild-type (WT) neutrophils but not in neutrophils from CB_2R−/− mice, and attenuated the adhesion of adoptively transferred leukocytes in EIU. Furthermore, a brain-penetrant close analog, RO6866945, demonstrated efficacy against TDP-43-induced neuropathology in frontotemporal dementia (FTD) models (Gonzalo-Consuegra et al. 2024; Ouali Alami et al. 2018). Furthermore, RO6866945 was used in mouse models of Alzheimer's disease (AD) to study the distribution of

CB_2 receptors and their roles in microglial function demonstrating that CB_2Rs are expressed in dystrophic neurite-associated microglia, and their modulation affects the number and activity of microglial cells (de Martin Esteban et al. 2022). In addition, the activation of CB_2R with RO6866945 or genetic deletion of CB_2R impacts microglial phagocytic activity and amyloid metabolism. CB_2R receptor activation increased insoluble amyloid levels, while CB_2R receptor deletion decreased these levels, suggesting a role in amyloid plaque dynamics. Another brain-penetrant analog of RO6871304, known as vicasinabin, was investigated in phase 2 clinical trials for the treatment of diabetic retinopathy (ClinicalTrials.gov identifier: NCT04265261) (Grether et al. 2024).

SR144528 **5.5** is one of the earliest identified selective CB_2R inverse agonists (Rinaldi-Carmona et al. 1998; Portier et al. 1999). It demonstrates high binding selectivity against human CB_1R (CB_2R selectivity ratio = 129) and mouse CB_1R (mouse CB_1R $pK_i = 6.92$, mouse CB_2R $pK_i = 10.7$, CB_2R selectivity ratio (10^(pK_i CB_2R − pK_i CB_1R) = 6.026)) (Soethoudt et al. 2017). SR144528 **5.5** has been extensively used as a tool to antagonize the pharmacological effects induced by CB_2R agonists both, in vitro and in vivo (Nackley et al. 2003). Notably, SR144528 **5.5** exhibits specificity across different signal transduction pathways. While it effectively antagonizes CB_2R-mediated modulation of cAMP signaling, it is less potent in other pathways such as GIRK and pERK (Soethoudt et al. 2017). Another frequently used CB_2R antagonist in both in vitro and in vivo studies is AM630 (Roger Pertwee et al. 1995).

Regarding allosteric modulators (AMs), despite their great potential, the chemical matter identified for CB_2R remains limited compared to CB_1R, and the exploration of AMs by exogenous synthetic ligands for CB_2R is still largely uncharted territory. Petrucci et al. (2017) identified pepcan-12, an endogenous RVD-hemopressin peptide, as a highly potent endogenous-positive allosteric modulator (PAM) of the CB_2 receptor. Pepcan-12 enhances the binding of orthosteric ligands and exhibits positive cooperativity, potentiating CP55,940 **3.3**- and 2-AG **1.1**-induced hCB_2R signaling, in addition to its negative allosteric modulator (NAM) properties for CB_1R. Recently, Gado et al. reported a series of the first synthetic PAMs for CB_2R, including compound C2 **5.6**, which displayed potent in vitro activity, illustrating the potential of CB_2R AMs for future therapeutics. PAM **5.6** significantly enhanced the ability of CP55,940 **3.3** to stimulate [^{35}S]GTPγ binding to CB_2R at 100 nM, as well as for 2-AG **1.1**. Additionally, at the same concentration, the dissociation of [^{3}H]CP55,940 **3.3** was completely prevented for at least 2 h. In an experimental mouse model of neuropathic pain, significant antinociceptive activity was observed in vivo in a dose-dependent manner upon oral administration (20 mg/kg). This activity was demonstrated to be mediated by CB_2R and could be potentiated with the coadministration of a MAGL inhibitor (Gado et al. 2019). Furthermore, CBD-DMH (Cordova et al. 1980) was reported to negatively modulate the G protein activation of reference agonists CP55,940 **3.3**, AEA **1.2**, and 2-AG **1.1**, but not the agonist-induced β-arrestin-2 recruitment. However, CBD-DMH also displayed competitive binding to CB_2R and partial agonism on G protein activation, inhibition of cAMP production, and β-arrestin-2 recruitment, suggesting a dual-binding mode of

CBD-DMH (Bouma et al. 2023). In light of recent advancements in structural biology, including the elucidation of the active and inactive conformations of CB_2R receptors (Li et al. 2019; Xing et al. 2020; Hua et al. 2020), progress in identifying and characterizing allosteric binding sites may be expected soon. These advancements have provided deep insights into the multiple protein–ligand interactions, underlying receptor dynamics, and activation processes. These results and conclusions will tremendously influence the field, allowing for the rationalization of the structure–function relationships of the described ligands and greatly influencing the design of future ligands.

Given the potential of CB_2R activation for treating (neuro)inflammatory diseases, several research groups have focused on understanding CB_2R expression and its molecular and cellular mechanisms. This has been achieved by developing labeled chemical probes for translational imaging, addressing the specificity issues associated with existing anti-CB_2R antibodies (Marchalant et al. 2014). Such translational imaging, including the generation of therapeutically effective occupancy levels, is crucial for designing clinical trials.

Recently, tetrasubstituted pyrazole CB_2R full agonists [^{3}H]-RNB-61 **6.1** and [^{14}C]-RNB-61, which exhibit high CB_2R potency (K_i 0.13–1.81 nM, depending on species) and excellent selectivity over CB_1R, have been successfully used to detect and quantify CB_2R in cells and tissues (Fig. 6) (Chicca et al. 2024). Additionally, the tritiated pyridine-derived CB_2R inverse agonist [^{3}H]RO6957022 **6.2**, which shows high binding selectivity against CB_1R, has been developed and applied for studying drug–target-binding kinetics (Martella et al. 2017). Its [^{11}C]-labeled analog [^{11}C]RSR-056 serves as a PET imaging agent, demonstrating specific binding to CB_2R-positive spleen tissue in rats and mice both in vitro and in vivo (Slavik et al. 2015). [^{11}C]RSR-056 was the starting point for the discovery of the fluorine-18-labeled CB_2R PET tracer [^{18}F]RoSMA-18-d$_6$ **6.3** (Haider et al. 2019, 2020). This CB_2R PET tracer exhibits sub-nanomolar affinity (K_i for $CB_2R = 0.8$ nM) and a remarkable selectivity factor of >12,000 over CB_1R. [^{18}F]RoSMA-18-d$_6$ **6.3** demonstrated exceptional CB_2R attributes through in vitro autoradiography, ex vivo biodistribution, and PET across species, and has progressed into clinical trials to monitor neuroinflammatory processes in amyotrophic lateral sclerosis (ALS) patients (ClinicalTrials.gov identifier: NCT05880563).

Based on LEI101, a selective CB_2R partial agonist, Soethoudt et al. incorporated two functionalities into this probe: a photoactivatable group and a clickable alkyne (Soethoudt et al. 2018). LEI121 **6.4** was designed to first covalently cross-link to CB_2R and subsequently allow flexible fluorophore conjugation. The probe selectively binds to CB_2R with a K_i of 63 nM and shows no affinity for CB_1R ($K_i > 10$ μM) in a radioligand assay, behaving as an inverse agonist. LEI121 **6.4** has been effectively used for target engagement studies in affinity-based protein profiling in CB_2R-overexpressing and endogenous CB_2R-expressing live human cells, utilizing covalent SDS-PAGE visualization, mass spectrometry-based proteomics, and flow cytometry.

Furthermore, the CB_2R-selective HU-308-biotin **6.5** has been effectively used for visualizing endogenous CB_2R through the conjugation of streptavidin-

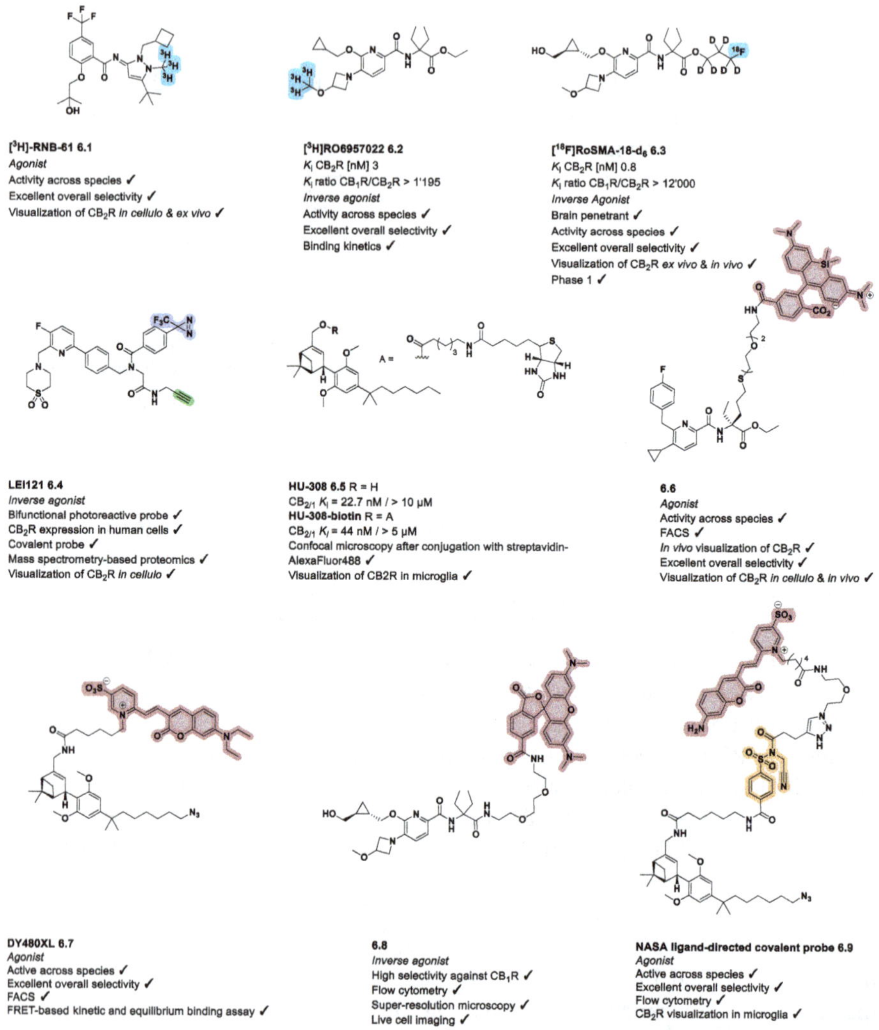

Fig. 6 Chemical structures, in vitro pharmacology data, and key characteristics of labeled CB_2R probes. Radioisotopes are highlighted in blue, affinity handles designed for click-chemistry in green, photoreactive groups in purple, fluorophore reporter groups in pink, and NASA warheads in yellow

AlexaFluor488, enabling confocal microscopy in primary cultured microglia (Martín-Couce et al. 2012). In microglia, **6.5** successfully reflected changes in CB_2R expression between resting and activated (pro-inflammatory) states induced by lipopolysaccharide (LPS) exposure.

The Nazaré and Carreira groups utilized a reverse design concept to create reversible fluorescent CB_2R probes. The key to success was the structure-based design of core linker tag constructs, which positioned the fluorophore toward the extracellular space, thereby avoiding the loss of binding affinity encountered in earlier attempts where the reporter dye was incorporated within the binding cavity. The silicon-rhodamine-labeled CB_2R agonist **6.6** is cell-permeable, allowing for labeling both extra- and intracellular CB_2R pools (Gazzi et al. 2022). It maintains interspecies affinity and selectivity for both mouse and human CB_2R and has been used for CB_2R detection in endogenously expressing living cells and zebrafish larvae.

The DY480XL-labeled CB_2R agonist **6.7**, based on an HU-308 probe platform (Westphal et al. 2020), is another versatile tool for studying CB_2R. It has been successfully applied in setting up FRET-based CB_2R kinetic and equilibrium binding assays (Sarott et al. 2020). The rational design enabled the conversion of functionality, allowing for the creation of fluorescently labeled CB_2R inverse agonists, thus expanding the toolbox for a comprehensive understanding of CB_2 receptor pharmacology (Kosar et al. 2024b). A recently reported reverse design-based matched fluorescent probe pair, including the TAMRA-labeled CB_2R inverse agonist **6.8**, has been developed to selectively address and visualize both the active and inactive states of CB_2R (Wasinka-Kalwa et al. 2024). These chemical tools successfully visualized CB_2R pools at the cell surface and intracellular membranes in living cells using super-resolution microscopy.

Additionally, a sophisticated tool for visualizing and studying CB_2R while maintaining its ability to bind other ligands at the orthosteric site was developed by using ligand-directed covalent (LDC) labeling concepts (Kosar et al. 2023). This method involves covalent labeling of a peripheral CB_2R lysine residue using fluorescent *O*-nitrobenzoxadiazole (*O*-NBD), *N*-sulfonyl pyridone (N-SP), and *N*-acyl-*N*-alkyl sulfonamide (NASA) (e.g., probe **6.9**). These LDC-functionalized probes have been applied to specifically visualize CB_2R in conventional and imaging flow cytometry, as well as in confocal fluorescence microscopy using overexpressing and endogenously expressing microglial live cells.

These and other recently developed high-quality labeled CB_2R fluorescent and PET probes have significantly advanced our understanding of CB_2R expression and function. However, many questions remain regarding the receptor's mechanism of action and its clinical implications. The phenomenon of biased signaling induced by different ligands, resulting from various ligand-dependent CB_2R conformations, is not yet fully understood but is highly relevant due to its impact on distinct pharmacological responses (Soethoudt et al. 2017; Morales et al. 2018). Additionally, understanding binding kinetics, including association rates and ligand residence time, may provide crucial insights and rationale for in vivo efficacy and could even influence signaling bias (Morales et al. 2018). The association equilibria of CB_2R receptor homo- and heterodimers might translate into different functional properties, offering another potential mode of therapeutic intervention. Lastly, improving the understanding of the translatability from promising preclinical in vivo efficacy studies to humans is essential. A feedback loop and back-translation of clinical data and knowledge from patients or human systems to preclinical research will greatly enhance the chances of success.

4 Metabolic Enzymes of AEA 1.2

AEA **1.2** is synthesized on demand from the membrane phospholipid precursor *N*-arachidonoylphosphatidylethanolamine (NArPE) through a two-step enzymatic process (Maccarrone et al. 2023) involving the *N*-acyl transferase NAT and iNAT followed by the action of *N*-acylphosphatidylethanolamine-specific phospholipase D (NAPE-PLD). The initial step involves the formation of NArPE via the *N*-acylation of phosphatidylethanolamine, mediated by Ca^{2+}-dependent or independent *N*-acyl transferase (NAT and iNAT). The acyl donor in this reaction is another phospholipid molecule, such as phosphatidylcholine, rather than acyl-CoA. The presence of NArPE in mammalian tissues and the *N*-acyl-transferase activity responsible for its production was first reported in the late 1990s (Cadas et al. 1997) and later identified molecularly as cytosolic phospholipase A2ε (cPLA2ε) (Ogura et al. 2016). Members of the phospholipase A and acyltransferase (PLAAT) family (Jin et al. 2007; Uyama et al. 2012) were identified as Ca^{2+}-dependent and -independent NAT, respectively. cPLA2ε, part of the cPLA2 family, uses a serine residue as its catalytic nucleophile. For *N*-acylation of phosphatidylethanolamine, cPLA2ε selectively abstracts an acyl chain from the sn-1 position of the glycerol backbone of glycerophospholipid, which is rich in saturated and monounsaturated fatty acids rather than polyunsaturated fatty acids like AA **1.3**. Consequently, NArPE and AEA **1.2** constitute a small fraction of cells' *N*-acylphosphatidylethanolamine (NAPE) and fatty acid ethanolamides. Studies on cPLA2ε-deficient mice have highlighted the enzyme's central role in the accumulation of NAPEs and *N*-acylethanolamines in models of imiquimod-induced psoriasis (Liang et al. 2022) and brain ischemia (Rahman et al. 2022). The NAPE-forming activity of cPLA2ε in the skin is suggested to be protective against inflammation, such as psoriasis, by producing anti-inflammatory *N*-acylethanolamines. Conversely, PLAAT enzymes, which have a cysteine residue as the catalytic nucleophile (Uyama et al. 2017), form a small protein family. Among the five human members (1–5), PLAAT1, 2, and 5 exhibit relatively high NAT activity compared to their coexisting PLA1/A2 activity (Uyama et al. 2012). The NAT activity of PLAATs is easily detectable in cells expressing recombinant PLAAT without any cellular stimulus, suggesting their role in maintaining basal levels of NAPEs and *N*-acylethanolamines in unstimulated cells. However, their contribution to NAPE formation in vivo remains unclear.

NAPE-PLD catalyzes the second step of AEA **1.2** formation. This enzyme releases AEA **1.2** and other fatty acid ethanolamides from their corresponding NAPEs through a PLD-type hydrolytic reaction (Okamoto et al. 2004). NAPE-PLD, a member of the metallo-β-lactamase superfamily, shows no sequence similarity to classical PLDs that convert phosphatidylcholine to phosphatidic acid. Multiple conserved aspartic acid and histidine residues are essential for its catalytic activity, with metal analysis indicating the presence of Zn^{2+} coordinated by these residues (Wang and Casero 2006). The crystal structure of human NAPE-PLD revealed homodimer formation adapted to associate with phospholipids and a binuclear Zn^{2+} center at the active site (Magotti et al. 2015). Purified recombinant

NAPE-PLD selectively hydrolyzes NAPE among various other phospholipids (Wang and Casero 2006). The enzyme does not distinguish between *N*-acyl species in NAPE, explaining why the composition of naturally occurring fatty acid ethanolamides mirrors the *N*-acyl composition of NAPEs. Recently, NAPE-PLD's role in energy metabolism has garnered attention. A common NAPE-PLD haplotype was reported to be protective against obesity (Wangensteen et al. 2011). Conditional knockout of adipocyte, intestinal, or hepatic NAPE-PLD tended to induce obesity (Geurts et al. 2015; Everard et al. 2019; Lefort et al. 2020).

Multiple pathways for the degradation of AEA **1.2** exist. It can be hydrolyzed into ethanolamine and AA **1.3**, thereby terminating its biological action. This hydrolysis is primarily catalyzed by FAAH-1, but can also be facilitated by the less common FAAH-2 or the lysosomal enzyme NAAA (Daniele Piomelli et al. 2020). Alternatively, AEA **1.2** can undergo biotransformation through oxygenation of the AA moiety, a process catalyzed by lipoxygenase isozymes 5, 12, and 15 (5-, 12-, 15-LOX), COX-2, or CYP450, as reviewed by Rouzer and Marnett (2011), Fezza et al. (2014), and Simard et al. (2022). Notably, COX-2-generated prostamides and other oxidative derivatives of AEA **1.2** each possess their own biological activities (Simard et al. 2022; van der Stelt et al. 2002). Although their pathophysiological roles are not yet fully understood, they appear to include neuroprotective effects on the brain (Veldhuis et al. 2003). In addition to the classical pathways for the release of AEA **1.2** from phosphatidylethanolamine and phosphatidylcholine precursors, several enzymes catalyze alternative biosynthetic routes. These enzymes include soluble phospholipase A2, α/β hydrolase domain protein 4, phospholipase C, lysophospholipase D, protein tyrosine phosphatase non-receptor type 22, SH2 domain-containing polyinositol-5-phosphatase 1, and various members of the glycerophosphodiesterase family. These parallel pathways contribute to the complex regulation of AEA **1.2** biosynthesis (Maccarrone et al. 2023).

4.1 NAPE-PLD Inhibitors

NAPE-PLD catalyzes the hydrolytic release of AEA **1.2** and other fatty acid ethanolamides from NAPEs. Recent studies have highlighted NAPE-PLD's role in energy metabolism and its association with obesity (Wangensteen et al. 2011). Efforts aimed at selectively inhibiting NAPE-PLD have not been ongoing for long; however, LEI-401 **7.1** was identified as the first brain-active inhibitor of NAPE-PLD (Fig. 7) (Mock et al. 2020). This compound significantly decreased a wide range of NAEs, including AEA **1.2**, in neuronal cells through a NAPE-PLD-dependent mechanism. In WT mice treated with LEI-401 **7.1**, levels of AEA **1.2** and other NAEs were reduced in the brain, whereas no such reduction was observed in NAPE-PLD KO mice. LEI-401 **7.1** also activated the hypothalamus–pituitary–adrenal (HPA) axis and impaired the extinction of aversive memories in mice, effects similar to those seen with cannabinoid CB_1R antagonism. A FAAH inhibitor prevented these behavioral changes. Overall, LEI-401 **7.1** represents a pioneering

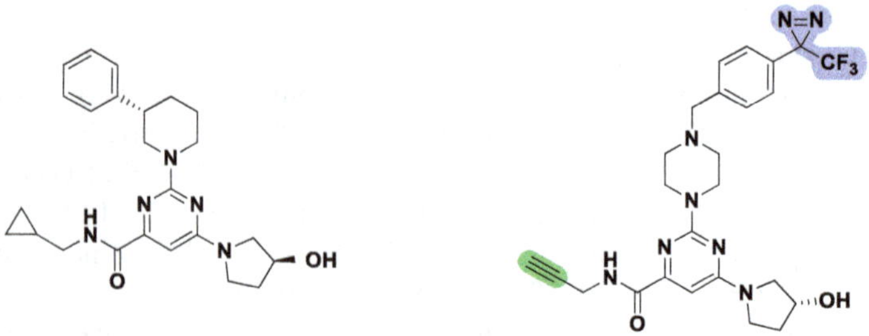

LEI-401 7.1
Brain penetrant ✓
Active across species ✓
In vivo efficacy ✓

7.2
Bifunctional photoreactive probe ✓
Covalent probe ✓
Visualization of NAPE-PLD *in cellulo* ✓

Fig. 7 Chemical structures and key characteristics of NAPE-PLD ligands. Affinity handles designed for click-chemistry are highlighted in green and photoreactive groups in purple

brain-active NAPE-PLD inhibitor that inhibits NAE biosynthesis in the brains of freely moving mice, suggesting a potential endogenous role for this lipid family in regulating emotional behavior.

The target engagement and selectivity of LEI-401 **7.1** in cellular systems were validated using a photoaffinity labeling probe. To explore the structure–activity relationships of LEI-401 **7.1**, a photoactivatable alkyne probe, **7.2**, was synthesized. This involved replacing the 3-phenyl-piperidine with a trifluoromethyl-diazirine-benzyl-piperazine as the photoreactive group. Additionally, the methylcyclopropyl group was substituted with a propargyl group to act as a ligation handle, facilitating the introduction of reporter groups via copper(I)-catalyzed azide-alkyne cycloaddition (CuAAC, or "click") chemistry for studying target engagement. Probe **7.2** inhibits NAPE-PLD with a K_i of 0.18 μM in the biochemical PED6 assay. Visualization of NAPE-PLD in HEK293T cells transfected with human NAPE-PLD was achieved after in situ treatment with compound **7.2**, followed by irradiation at 350 nm, cell lysis, and labeling by clicking of a Cy5-N_3 fluorophore, which was then detected through in-gel fluorescence scanning. The development of the photoaffinity probe **7.2** enabled the detection of NAPE-PLD in living cells and facilitated the visualization of target engagement by LEI-401 **7.1** with NAPE-PLD.

4.2 *FAAH Inhibitors*

FAAH is an intracellular, membrane-bound enzyme that plays a significant role in the signal termination of a class of bioactive lipids called fatty acid amides in the CNS and peripheral tissues. Its primary substrate is the endogenous cannabinoid ligand AEA **1.2,** which is synthesized on demand and subsequently hydrolyzed by FAAH to AA **1.3** and ethanolamine (Keith et al. 2015). FAAH acts as a serine hydrolase using an atypical catalytic triad (Ser-Ser-Lys) to efficiently hydrolase both amides and esters (McKinney and Cravatt 2005). In vivo, inactivation of FAAH by genetic knockout or pharmacological inhibition increases endogenous levels of fatty acid amides and induces a range of analgesic effects in various animal models of inflammation and chronic pain. Importantly, these phenotypes occur without the adverse effects commonly associated with direct cannabinoid receptor agonists, suggesting that FAAH may serve as a promising therapeutic target to treat inflammatory pain and related conditions (Douglas S. Johnson et al. 2011). Owing to its highly promising therapeutic potential, extensive efforts have been dedicated to develop novel ligands in academia and industry.

To date, several classes of irreversible and reversible FAAH inhibitors have been reported (Seierstad and Breitenbucher 2008; Tripathi 2020; Fazio et al. 2020). The first generation of irreversible FAAH inhibitors was predominantly inspired by substrates and natural products. Various approaches led to the development of a series of AEA-like arachidonoyl-amides, such as arachidonoyl serotonin (T. Bisogno et al. 1998). Although these compounds were poor drug candidates due to their intrinsically high reactivity, lipophilicity, and low selectivity versus other serine hydrolases, they significantly advanced the understanding of FAAH biochemistry and pharmacology. Consequently, developing potent and selective small-molecule FAAH inhibitors remains crucial in validating the enzyme as a potential therapeutic target (Maccarrone 2020).

In the pursuit of novel and druggable FAAH ligands, a multidisciplinary team led by Tarzia, Mor, and Piomelli developed URB597 **8.1**, the first potent ($IC_{50} = 4.6$ nM rFAAH), selective, and in vivo active carbamate-based irreversible FAAH inhibitor (Mor et al. 2004) (Fig. 8). URB597 **8.1** demonstrated robust anxiolytic and antidepressant properties through indirect CB_1R activation by the accumulated anandamide. URB597 **8.1** remains one of the most extensively studied FAAH inhibitor tools, significantly contributing to the pharmacological advancement of FAAH inhibitors.

Besides carbamate-based inhibitors, irreversible ligands derived from urea structures have also been developed. In line with carbamate compounds, urea-based derivatives also catalyze carbamoylation of the catalytic Ser241 residue within the binding pocket of FAAH, crucially differentiated by the structure and reactivity of the warhead (Keith et al. 2015). In 2011, Pfizer disclosed the characterization of a new category of benzylidene piperidine urea inhibitors, among them PF-04457845 **8.2**, a promising, potent ($IC_{50} = 7.2$ nM hFAAH), selective, and orally bioavailable FAAH irreversible inhibitor (Douglas S. Johnson et al. 2011). PF-04457845 **8.2** demonstrated significant in vivo efficacy and exhibited notable effectiveness in the CFA model for inflammatory pain in rats. Given its encouraging preclinical outcomes and favorable PK profile, PF 04457845 **8.2** entered human clinical trials for the treatment of chronic pain and other CNS disorders (ClinicalTrials.gov identifier: NCT00836082/NCT03386487).

In the same year, Pfizer reported the chemical optimization of a spirocyclic urea inhibitor series bearing five-membered heterocycles as warheads (Meyers et al. 2011). Within this series, PF-04862853 **8.3** was selected due to its remarkable potency, selectivity, pharmacokinetic properties, and in vivo efficacy. Preclinical

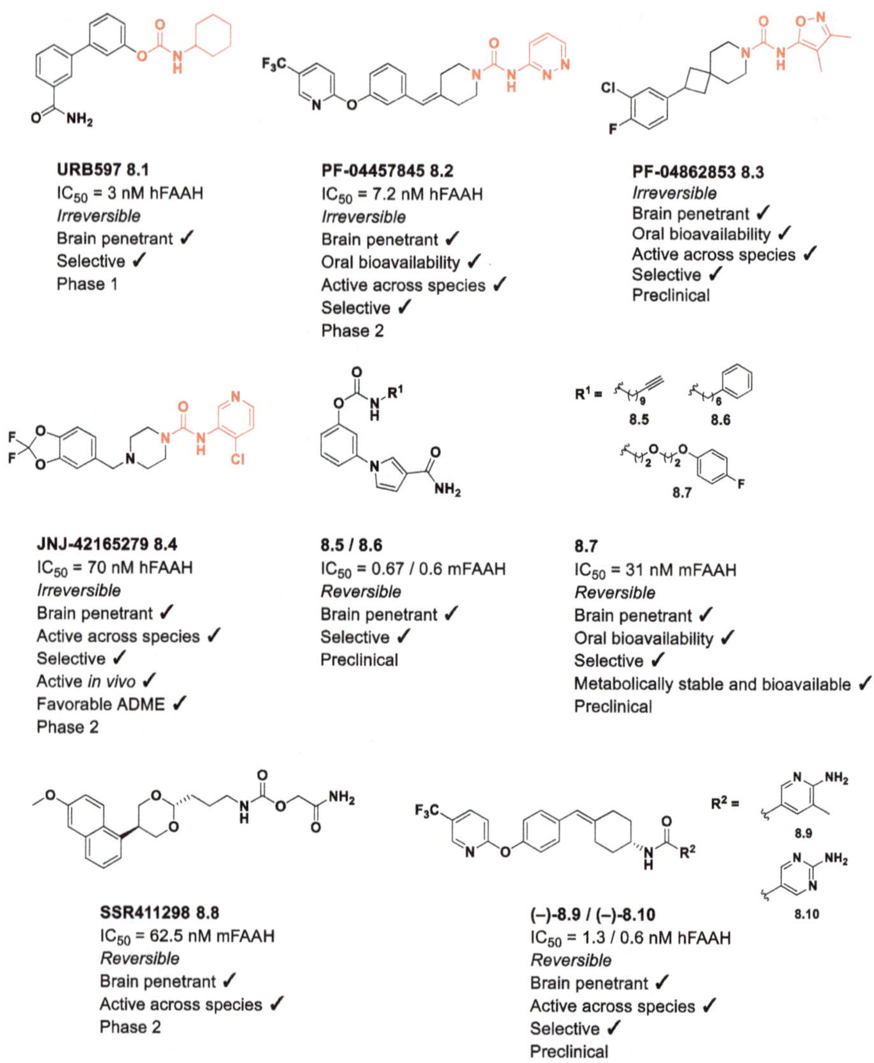

Fig. 8 Chemical structures, IC_{50} values, key properties, and highest development phase of representative covalent FAAH inhibitors. Warheads of irreversible inhibitors are highlighted in red

studies in rats and dogs demonstrated, for example, complete inhibition of FAAH activity in leukocytes and brain samples, as well as favorable pharmacokinetic profiles with high bioavailability and intermediate half-life, resulting in the selection of PF-04862853 **8.3** as a clinical candidate.

In 2015, Johnson & Johnson reported a novel potent inhibitor using an aryl piperazinyl urea moiety as a covalent warhead (IC_{50} = 70 nM hFAAH) (Keith et al. 2015). JNJ-42165279 **8.4** exhibits high selectivity, favorable physical and pharmacodynamic properties, as well as excellent ADME characteristics. In rats,

JNJ-42165279 **8.4** effectively inhibited FAAH both in the brain and periphery, resulting in significantly increased AEA **1.2** levels. These preclinical results paved the way for JNJ-42165279 **8.4** to enter clinical trials for various neurological disorders: a study on depressive disorders was concluded in February 2019 (ClinicalTrials.gov identifier: NCT02498392).

In January 2016, a first-in-human trial was initiated for the FAAH inhibitor BIA 10-2474 **8.5** (Van Esbroeck et al. 2017). Initially regarded as a promising CNS-active inhibitor due to its high potency and prolonged action (Hayes et al. 2020; Kiss et al. 2018), it later garnered negative attention following severe neurological side effects and the death of a participant in a phase 1 trial in the highest dosing cohort (Kerbrat et al. 2016). Subsequent in-depth investigations revealed that BIA 10-2474 exhibited significantly higher off-target activity against human serine hydrolases than other clinical FAAH inhibitors, such as PF-04457845 **8.2**. Although the cause of the clinical neurotoxicity remains unclear, it has been postulated that these off-target interactions, in conjunction with BIA 10-2474's irreversible mode of action, may have played a crucial role in the observed adverse effects (Van Esbroeck et al. 2017; Griebel et al. 2018). Numerous ongoing clinical trials were suspended as a precautionary response to this incident, which has slowed down the advancement of irreversible FAAH inhibitors. Nevertheless, since other FAAH inhibitors already tested in clinical trials (e.g., URB597 **8.1**, PF-04457845 **8.2**, SSR411298 **8.8**, and JNJ-42165279 **8.4**) did not elicit any adverse neurological findings, FAAH inhibition is still considered a safe therapeutic strategy, and led to the assumption that the severe adverse effects observed with BIA 10-2474 were not target related (Maccarrone et al. 2023).

The fatal incident involving BIA 10-2474 has highlighted the potential risk of irreversible inhibitors. Their ability to permanently bind to and inactivate not only their intended target but also off-target proteins can lead to severe toxic effects that are difficult to predict. This is particularly concerning because once bound, these effects persist, increasing the risk of toxicity. In contrast, reversible inhibitors exhibit a temporary binding, which reduces the risk of prolonged off-target activity, making them a more controlled and potentially safer therapeutic option (Griebel et al. 2018; Ming Jiang et al. 2023; Van Egmond et al. 2021). Within this context, Butini et al. developed a series of reversible phenyl-heterocyclic-derived derivatives (Butini et al. 2012). Two analogs, **8.5**, and **8.6**, emerged as particularly potent ($IC_{50} = 0.67$ nM and 0.60 nM mFAAH) and selective FAAH inhibitors, featuring a reversible mechanism of action. **8.5** and **8.6** demonstrated antinociceptive effects in vivo and subsequent optimization trials yielded compound **8.7** with even better pharmacokinetic properties and metabolic stability.

In 2018, Griebel et al. at Sanofi identified SSR411298 **8.8** as a potent ($IC_{50} = 62.5$ nM mFAAH) and selective reversible FAAH inhibitor, demonstrating the promising therapeutic potential for anxiety and depression disorders (Griebel et al. 2018). In mice, this compound induced elevated levels of FAAH substrate [AEA, oleoylethanolamide (OEA), *N*-palmitoylethanolamine (PEA)] in the hippocampus after oral administration of a dose of 0.3 mg/kg without mimicking the interoceptive state or causing typical behavioral side effects such as memory deficit and motor impairment associated with directly acting cannabinoid ligands. In rodent models of anxiety, SSR411298 **8.8** exhibited pronounced anxiolytic and antidepressant-like effects. Taken together, these findings suggest SSR411298 **8.8** as a promising candidate for the treatment of post-traumatic stress disorder (PTSD). First clinical studies regarding depressive disorder have already been conducted (ClinicalTrials.gov identifier: NCT00822744).

Building on the structural framework of the benchmarking irreversible FAAH inhibitor PF-04457845 **8.2**, Bhuniya and colleagues embarked on the development of a novel series of reversible nicotinamide-based lead compounds (Bhuniya et al. 2019). Their investigation focussed on modifying the urea moiety while maintaining the core structure of PF-04457845 **8.2** to keep the well-established selectivity profile of the chemotype. From extensive SAR studies, two optimized compounds, **(−)-8.9**, and **(−)-8.10**, evolved for extended efficacy assessment in a rat model of chemotherapy-induced peripheral neuropathy (CIPN), considering their nanomolar FAAH potency, high selectivity, metabolic stability, and the ability to penetrate the brain. Both compounds demonstrated dose-dependent antihyperalgesic effects, warranting further exploration to validate their clinical potential.

The significant role of labeled chemical probes in drug development has spurred considerable efforts toward the discovery and implementation of effective PET tracers in recent years (Varlow et al. 2020). Significant advancements in the field were achieved by directing radiotracer development toward analogs of the potent and selective FAAH inhibitor URB597 **8.1**. Utilizing ^{11}C-carbonylation via [^{11}C]CO_2 fixation led to the successful synthesis of several URB597-derived radiotracers circumventing the formation of brain-penetrant radiometabolites. Among these, [^{11}C]CURB **9.1** stands out as a relevant clinically applied imaging probe, particularly employed in the investigation of FAAH binding in the brains of chronic *Cannabis* users (Boileau et al. 2016). These studies unveiled a significant reduction in FAAH binding within the brains of chronic cannabis users during early withdrawal, underscoring the pivotal role of FAAH in neurobiological alterations associated with chronic cannabis exposure. Such insights hold promise in guiding the development of novel FAAH and endocannabinoid-targeted strategies for mitigating cannabis abuse-related adverse effects.While radiotracers like [^{11}C]CURB **9.1** hold considerable promise, their use is limited by the short half-life of the radionuclide carbon-11 ($t_{1/2}$ = 20.4 min), necessitating on-site cyclotron availability for ^{11}C production (Fig. 9) (Shoup et al. 2015). In contrast, fluorine-18 ($t_{1/2}$ = 109.7 min) boasts a significantly longer half-life, which allows for prolonged imaging protocols and facilitates transport to remote facilities. Shoup et al. described a novel fluorine-18-labeled FAAH inhibitor, [^{18}F]FCHC **9.2**, designed for investigating addictive and neurological disorders. Through ex vivo biodistribution and PET imaging studies in rats, [^{18}F]FCHC **9.2** demonstrated remarkable sensitivity as a radiotracer for imaging FAAH in the central nervous system, thereby paving the way for further investigations in larger animal models.

A valuable approach for investigating enzymes and their activity in their native cellular environment is the use of covalent click probes, which can be used in two-step labeling strategies. First, the selective click probe forms a covalent bond

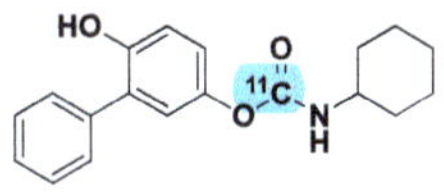

[^{11}C]CURB 9.1
PET tracer
Brain penetrant ✓
Active across species ✓
Selective ✓
Visualization of FAAH *in vivo* ✓

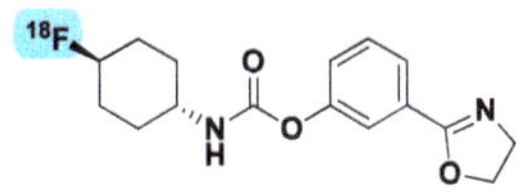

[^{18}F]FCHC 9.2
PET tracer
Brain penetrant ✓
Visualization of FAAH *in vivo* ✓

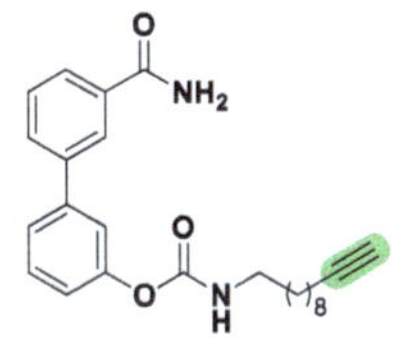

JP-104 9.3
Covalent click-probe
Brain penetrant ✓
Active *in vivo / in cellulo* ✓

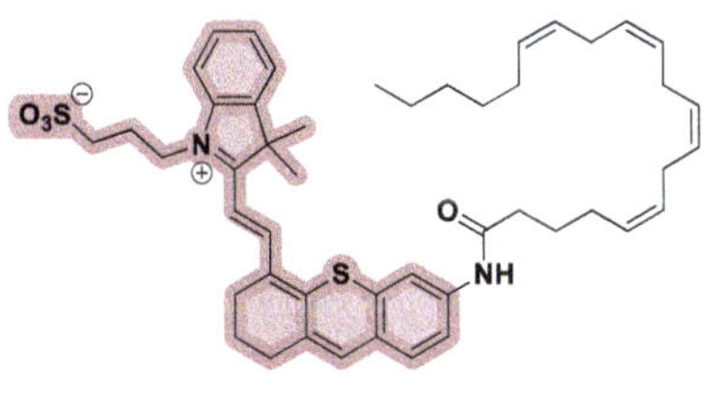

PA-HD-FAAH 9.4
Photoacoustic probe
Active *in vivo* ✓
Selective ✓
Visualization of FAAH *in vivo* ✓

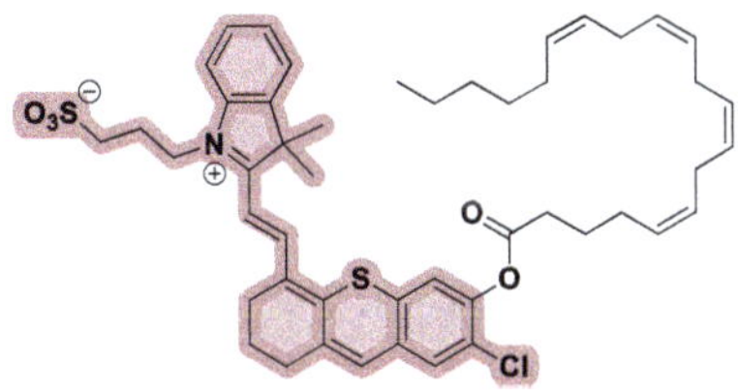

PA-HD-MAGL 9.5
Photoacoustic probe
Active *in vivo* ✓
Selective ✓
Visualization of MAGL *in vivo* ✓

Fig. 9 Chemical structures and key properties of selected labeled FAAH probes. Radioisotopes are highlighted in blue, affinity handles designed for click-chemistry in green, and fluorophore reporter groups in pink

with the enzyme, carrying a functional group (e.g., an alkyne or azide moiety). In a second step, this functional group enables the attachment of a reporter moiety (e.g., a fluorophore or a biotin tag) via a simple Cu-catalyzed click reaction. This two-step functionalization strategy offers the advantage of flexible probe modification, allowing specific adaptation to different experiments. Moreover, since click reactions are often bioorthogonal, this approach facilitates real-time investigation of intracellular targets within living cells or tissues (Parker and Pratt 2020; Speers and Cravatt 2004). Building on this approach, Alexander and Cravatt introduced two novel and potent FAAH inhibitors (Jessica P. Alexander and Cravatt 2005). The respective URB597 carbamate analogs were initially conceptualized to validate the hypothesized irreversible binding mechanism to FAAH. However, by converting one agent into an active site-directed "click-chemistry" (CC) probe (JP-104 **9.3**), simultaneously the development of an FAAH probe tailored for in vivo functional analysis of the enzyme activity and inhibition was achieved. Leveraging the CC probe enabled the exploration of target selectivity among FAAH-directed carbamates in vivo, thereby offering a comprehensive understanding of the "hot spots" of off-target activity within the proteome.

Based on activatable arachidonic acid ester substrates, Lucero et al. designed a probe for the selective and precise visualization of FAAH and MAGL enzyme activities in vivo (Lucero et al. 2022). Their study unveiled two activity-based photoacoustic probes, designated as PA-HD-FAAH **9.4** and PA-HD-MAGL **9.5**, engineered for the specific detection of FAAH and MAGL activities in vivo. The technique of photoacoustic imaging enabled the detection of enzyme activities with high resolution at depths of up to 10 cm, which is a crucial capability for imaging deep organ systems such as the gastrointestinal tract (GIT). Based on this, the probes were employed to assess FAAH and MAGL activity in a mouse model of obesity. The results revealed a significant increase in the activity of both enzymes in the gastrointestinal tract, shedding light on the involvement of endocannabinoid metabolism in obesity-associated inflammation.

4.3 NAAA Inhibitors

NAAA is a lysosomal hydrolase that shares 33% amino acid identity with acid ceramidase, an enzyme responsible for hydrolyzing ceramide into sphingosine and fatty acid (Tsuboi et al. 2007a; Ueda et al. 2010). Like other members of the *N*-terminal nucleophile hydrolase family (Linhorst and Lübke 2022), NAAA is initially synthesized as an inactive precursor. It undergoes post-translational autoproteolytic cleavage to form a heterodimer composed of α and β subunits (Zhao et al. 2007). This maturation process occurs in vitro only under acidic conditions, indicating that it likely takes place in endosomes or lysosomes following the enzyme's transport from the endoplasmic reticulum through the Golgi apparatus. The *N*-terminal cysteine residue of the β-subunit (Cys-126 in human NAAA, Cys-131 in rodents) acts as the catalytic nucleophile and is essential for autoproteolytic cleavage. Crystal structures of NAAA have shown that autoproteolysis exposes the active site, enabling catalysis (Gorelik et al. 2018).

NAAA hydrolyzes various fatty acid ethanolamides in vitro, with the highest reactivity observed for *N*-palmitoylethanolamine (Ghidini et al. 2021). The enzyme is highly expressed in macrophages (Tsuboi et al. 2007b) and other immune cells (Ribeiro et al. 2015), suggesting a role in regulating fatty acid ethanolamide levels at sites of inflammation. In a mouse model of dermatitis induced by 2,4-dinitrofluorobenzene, NAAA-deficient mice exhibited elevated levels of *N*-palmitoylethanolamine, but not *N*-oleoylethanolamine, in ear tissue compared to wild-type controls, along with a significant reduction in the inflammatory response (Sasso et al. 2018). Additionally, NAAA deficiency in mice led to increased levels of PEA and AEA **1.2** in the bone marrow and macrophages, as well as elevated AEA **1.2** levels in the lungs (Xie et al. 2022). The anti-inflammatory effects of PEA are believed to be mediated through the activation of the nuclear receptor PPARα (Lo Verme et al. 2005). When activated by PEA, PPARα forms a dimer with the retinoid X receptor. This complex then binds to specific DNA elements, resulting in

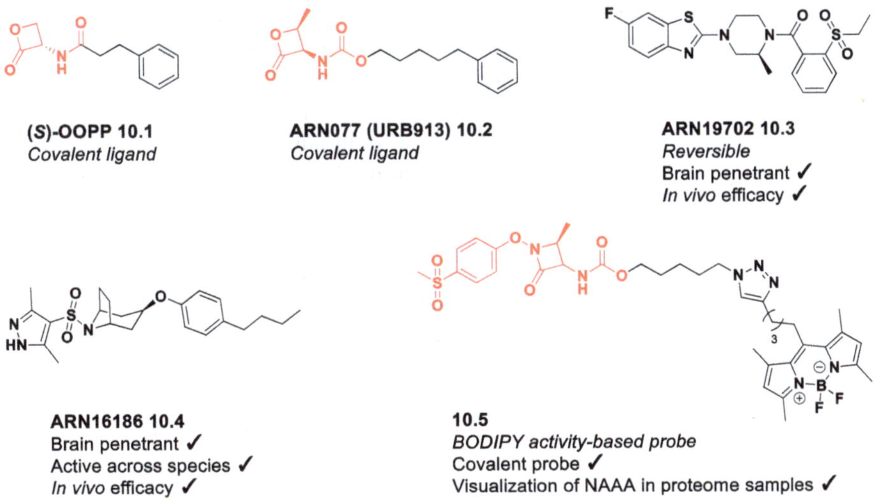

Fig. 10 Chemical structures and key characteristics of NAAA ligands. Covalent warheads are highlighted in red

the trans-repression of pro-inflammatory transcription factors (Pontis et al. 2016; Rankin and Fowler 2020).

The search for potent, selective, and systemically active NAAA inhibitors began in 2009 with the identification of the β-lactone derivative *N*-[(3*S*)-2-oxo-3-oxetanyl]-3-phenylpropanamide ((*S*)-OOPP) **10.1**, which inhibits rat NAAA with submicromolar potency ($IC_{50} = 420$ nM) via a noncompetitive and partially reversible mechanism (Fig. 10) (Solorzano et al. 2009). Despite its rapid hydrolytic deactivation due to β-lactone ring opening, (*S*)-OOPP **10.1** is selective for NAAA over other lipid amidases and exhibits stereospecific inhibitory effects, increasing PEA levels in macrophages and blocking inflammation in mice (Solorzano et al. 2009). These findings highlighted NAAA as a druggable target for inflammation treatment and prompted the development of more stable inhibitors, such as ARN077 (URB913) **10.2**, a β-lactone derivative with improved properties including chemical stability and greater NAAA inhibitory potency ($IC_{50} = 50$ nM), compared to (*S*)-OOPP **10.1**. ARN077 **10.2** is selective for NAAA and effectively reduces inflammation and pain-related responses in animal models (Sasso et al. 2013, 2018; Ponzano et al. 2013). Despite these advances, the β-lactone ring's low stability continues to pose challenges for systemic use. This prompted the development of new NAAA inhibitors such as β-lactam, isothiocyanate, azetidine-nitrile, and benzothiazole derivatives (Ribeiro et al. 2015; Alhouayek et al. 2015; Malamas et al. 2020; Migliore et al. 2016). Reversible inhibitor ARN19702 **10.3** exhibited an IC_{50} of 0.23 ± 0.04 μM for human NAAA (hNAAA), crossed the blood–brain barrier (BBB), and increased brain levels of PEA and OEA (Migliore et al. 2016; Fotio et al. 2024). It delayed disease onset and reduced symptom intensity in the experimental autoimmune encephalomyelitis (EAE) model, suggesting that regulating NAAA in vivo may be beneficial for multiple sclerosis (MS). Further improved pharmacological tool ARN16186 **10.4** (IC_{50} value of 23 ± 4 nM for hNAAA) confirmed efficacy in the EAE mouse model and was efficacious in slowing disease progression and preserving locomotor activity (Sgroi et al. 2021). Additionally, inhibiting NAAA decreased the infiltration of immune cells into the spinal cord and regulated the overactivation of the NF-kB and STAT3 transcription factors, resulting in a reduced expansion of Th17 cells during the disease progression.

Moreover, NAAA-reactive activity-based probes (ABP) have been developed as research tools for discovering new inhibitors and for detailed characterization of NAAA in its native cellular environment (Petracca et al. 2017). The boron-dipyrromethene activity-based probe (BODIPY-ABP) *N-O*-substituted threonine β-lactam **10.5** can be used to identify novel classes of NAAA inhibitors. A norbornene-ABP analog can be employed to label NAAA in situ, offering the advantage of selectively detecting the active subunit of the enzyme in its natural cellular context. Furthermore, the copper(I)-free conditions utilized in the inverse electron demand Diels–Alder reaction (IEDDA)-mediated ABPP protocol make the probe a valuable tool for investigating the cellular and subcellular localization of NAAA in living cells and tissues.

5 Metabolic Enzymes of 2-AG 1.1

The endocannabinoid 2-AG **1.1** is produced through two main pathways: the metabolic pathway and the signaling pathway. The metabolic pathway involves the hydrolysis of sn-2 arachidonoyl-containing triglycerides by hormone-sensitive lipase, carboxyl esterases, or other lipases to form sn-2 arachidonoyl DAGs (Marc P. Baggelaar et al. 2018). The signaling pathway involves the conversion of phosphatidylinositol-4,5-bisphosphate by phospholipase C (PLC) β in the CNS or PLCγ2 in immune cells, activated by Ca^{2+} ions and Gq protein-coupled receptor activation (Marc P. Baggelaar et al. 2018). Both pathways produce diglycerides that activate protein kinase C and serve as precursors for 2-AG **1.1**, which is generated by DAGLα and DAGLβ (Tiziana Bisogno et al. 2003). These enzymes were discovered by Doherty's group in 2003, and studies with genetically modified animals lacking *Dagla* and *Daglb* genes confirmed their essential role in 2-AG **1.1** production in the brain (Gao et al. 2010; Tanimura et al. 2010). DAGLs also terminate protein kinase C signaling by hydrolyzing DAGs, connecting lipid and kinase signaling. Termination of 2-AG **1.1** signaling at CB_1R or CB_2R occurs through hydrolysis by MAGL, resulting in the production of AA **1.3** and glycerol (Dinh et al. 2002). MAGL is the main enzyme responsible for 2-AG **1.1** inactivation in the brain, while α/β-hydrolase domain proteins 6 and 12 may play roles in specific cell types (Marrs et al. 2010; Blankman et al. 2007). 2-AG **1.1** is a major source of AA **1.3**, a precursor for pro-inflammatory lipids like prostaglandins, making MAGL a central node in endocannabinoid and eicosanoid signaling (Nomura et al. 2010). The therapeutic potential of modulators of 2-AG **1.1** metabolism is discussed in further sections, with an extensive review available by Punt et al. (2022).

5.1 DAGL Inhibitors

Mammalian DAGL is the key enzyme responsible for the biosynthesis of 2-AG **1.1**, which is together with AEA **1.2** the most important endocannabinoid activating the CB_1 and CB_2 receptors. In 2003, Bisogno et al. discovered two different isoforms, DAGLα and DAGLβ (Tiziana Bisogno et al. 2003). Both DAGL isoforms are intracellular multidomain integral membrane proteins showing extensive homology in their encoded gene products but differ in size (DAGLα = 120 kDa and DAGLβ = 70 kDa) (Marc P. Baggelaar et al. 2015). DAGLs belong to the serine hydrolase family that utilizes the same catalytic Ser-His-Asp triad to process the sn-1 ester of 1-acyl-2-arachidonoylglycerides, generating 2-AG **1.1** on demand. DAGLα is primarily responsible for 2-AG **1.1** production in the CNS with significant expression in brain regions, including the cortex, hippocampus, and cerebellum, whereas DAGLβ is the main enzyme for 2-AG **1.1** production in immune cells (Maccarrone et al. 2023). Numerous classes of DAGL inhibitors have been documented (Tiziana Bisogno et al. 2003; Appiah et al. 2013; Janssen et al. 2014; Tiziana Bisogno et al. 2013). However, these inhibitors typically originate from natural substrates and incorporate reactive chemical warheads, which compromises their selectivity against other serine hydrolases. Still, methodological approaches to selectively measure endogenous DAGLs activity are crucial for a detailed investigation of the cellular role of these biosynthetic enzymes (Van Rooden et al. 2018).

DO34 and DH376 represent prototypical covalent dual DAGL inhibitors inhibiting both isoforms and were extensively used for investigating the role of these enzymes in physiological pathways (Ogasawara et al. 2016). In genetic models, these inhibitors were employed to elucidate that 2-AG **1.1** distinctly is synthesized on demand by DAGL (Ogasawara et al. 2016; Marc P. Baggelaar et al. 2015). Moreover, DAGL inhibitors have significantly enlarged our understanding of 2-AG's involvement in various processes, such as food intake (Deng et al. 2017), neuroinflammation (Ogasawara et al. 2016), anxiety, stress (Bluett et al. 2017), and regulation of learning and memory (Lesley D. Schurman et al. 2019). Although these inhibitors have significantly improved our understanding of DAGL's role within the endocannabinoid system, it is essential to emphasize limitations due to their restricted selectivity as they do not exclusively inhibit DAGL but also interfere with other serine hydrolases like ABHD6.

The most selective DAGL inhibitor to date was described by Baggelaar et al. (2015). Through comparative and competitive activity-based proteome profiling, the α-keto heterocycle LEI-105 **11.1** was discovered, which is a potent and covalent-reversible dual inhibitor of DAGLα and DAGLβ, exerting high selectivity toward other endocannabinoid-metabolizing enzymes such as ABHD6, ABHD12, MAGL, and FAAH (Fig. 11). This was further demonstrated in Neuro2A cells, a mouse

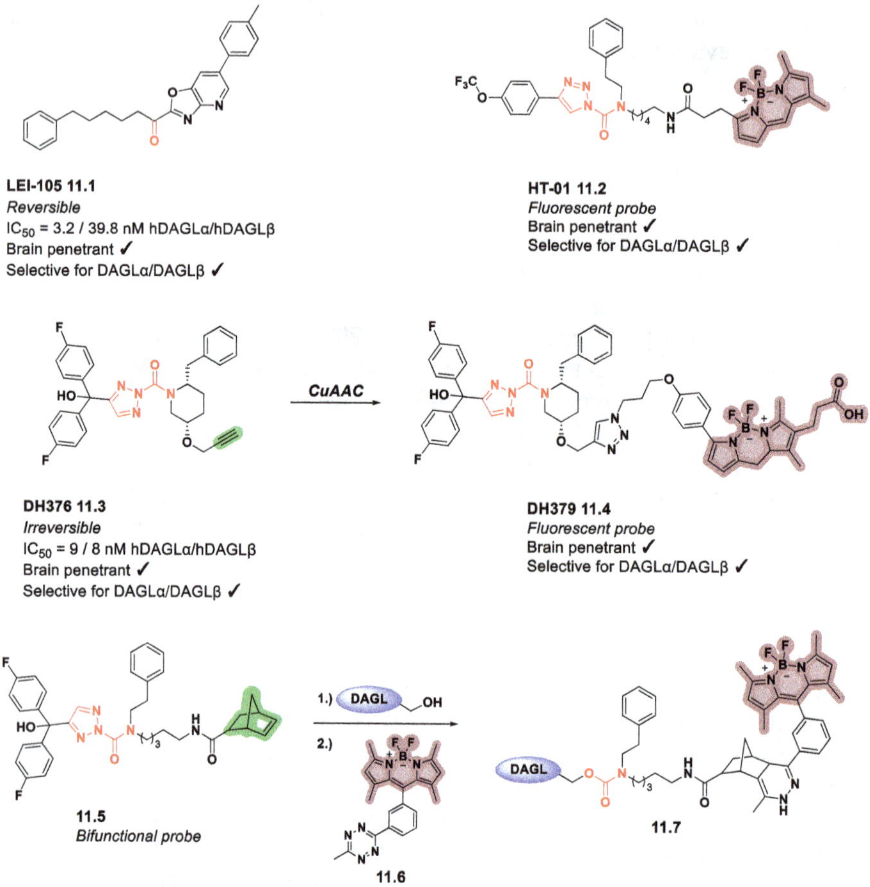

Fig. 11 Chemical structures of potent preclinical DAGL covalent inhibitors and labeled DAGL probes as well as a schematic representation of the copper-catalyzed azide-alkyne cycloaddition to generate probe DH379 **11.4** (middle) and the inverse electron-demand Diels–Alder reaction starting from the two-step activity-based DAGL probe **11.5**. Affinity handles are highlighted in green, fluorophore reporter groups in pink, and warheads in red

neuroblastoma cell line, where LEI-105 **11.1** selectively reduced 2-AG **1.1** but not AEA **1.2** levels in a concentration-dependent manner. A key advantage of reversible inhibitors compared to their irreversible congeners is the ability to allow more precise control over the extent of target inhibition. This is crucial for physiological modulation and homeostatic maintenance of endocannabinoid signaling in the CNS, where complete CB_1R blockade is associated with severe side effects. Therefore, as shown by studies using LEI-105 **11.1** a reversible mode of action is critical for the development of clinical candidates to provide future treatment options for diseases, such as obesity, related metabolic disorders, and neuroinflammation, all conditions linked to excessive 2-AG **1.1** signaling.

A DAGLβ-selective inhibitor was reported by Cravatt et al. investigating the role of DAGL in macrophages and inflammatory processes (Ku-Lung Hsu et al. 2012). By screening a library of thirty 1,2,3-triazole ureas (1,2,3-TUs), a chemotype well-suited for serine hydrolase inhibitor development, two peripherally restricted covalent DAGLβ inhibitors, KT109 and KT172, were identified, showing high selectivity over DAGLα in vivo. Despite extensive medicinal chemistry optimization efforts, off-target activity on ABHD6 persisted, complicating further investigations. This problem was addressed by generating a structurally related ABHD6-selective control probe, which was used in further biological studies investigating the inhibitory activity of KT109 and KT172 against DAGLβ. To overcome challenges arising from low expression levels of DAGLβ in neuronal cells and precisely measure its activity with high sensitivity, a fluorescent activity-based probe was synthesized, based on the 1,2,3-TU derived DAGLβ-inhibitors. Modification of the piperidine ring enabled the conjugation of a BODIPY fluorophore, leading to the development of HT-01 **11.2**, which effectively labeled both DAGLα and DAGLβ. In addition to demonstrating significantly higher affinity toward DAGLα and DAGLβ, HT-01 **11.2** exhibited a much more restricted labeling profile in proteomes from mouse brains, primary neurons, and Neuro2A neuroblastoma cells compared to other broad-spectrum ABPs like fluorophosphonate rhodamine (FP-Rh). This enabled comprehensive characterization of the inhibitors' activity and selectivity using competitive ABPP. Experiments in Neuro2A cells and in vivo studies in mice revealed that DAGLβ inhibition significantly reduces 2-AG **1.1** production, thereby modulating inflammatory responses in macrophages. Further in vivo studies in mice using HT-01 **11.2** investigated the effects of DAGLβ inhibition in peritoneal macrophages on lipid metabolism.

By further modification of the 1,2,3-TU-based scaffold, Ogasawara et al. successfully developed selective and CNS-active inhibitors targeting DAGLs (Ogasawara et al. 2016). In addition to the previously mentioned DO34, DH376 **11.3** emerged as a highly potent and selective probe compound, exhibiting inhibition of DAGLα ($pIC_{50} = 8.3$ hMAGL) and DAGLβ ($pIC_{50} = 8.6$ mMAGL). The chiral propargyl ether at C4 position of the piperidine ring served as an attachment handle for the introduction of a variety of reporter groups using CuAAC chemistry. By leveraging this handle, the tailored BODIPY-labeled analog DH379 **11.4** was constructed as a fluorescent probe. DH379 **11.4** detected both recombinant DAGLα and DAGLβ within the proteome of mouse brain membranes. However, the incorporation of large reporter groups may detrimentally impact various properties such as affinity, selectivity, stability, cell permeability, protein binding, oral availability, brain penetration, and toxicity (Van Rooden et al. 2018). To avoid such challenges arising from large reporter groups, bioorthogonal two-step labeling strategies are preferred, in which the attachment of the reporter group occurs after the covalent binding of the target of interest. Merging structural elements of DAGL inhibitor DH376 **11.3** with features of fluorescent DAGL probe HT-01 **11.2** led to the construction of the bifunctional probe **11.5**. Here, the electrophilic 1,2,3-triazole-urea moiety ensures covalent and irreversible binding to the serine 472 in the active center of DAGL, while the norbornene moiety serves as a bioorthogonal conjugation handle for the introduction of reporters via IEDDA reactions, providing a rapid and metal-free alternative to CuAAC. In this context, **11.5** was successfully employed for live cell labeling in U2OS cells overexpressing DAGLα, following in situ treatment with BODIPY-tetrazine **11.6**. This resulted in the formation of the Diels–Alder adduct **11.7**.

5.2 MAGL Inhibitors

MAGL is a cytosolic, membrane-associated lipase belonging to the serine hydrolases family. This key catabolic enzyme in the endocannabinoid system regulates the level of the most abundant brain endocannabinoid 2-AG **1.1**, thus significantly influencing retrograde cannabinoid signaling pathways (Ming Jiang et al. 2023). 2-AG **1.1** acts as an endogenous agonist for the cannabinoid receptors CB_1R and CB_2R, making MAGL essential for intracellular signal termination by deactivating the cannabinoid receptors through the hydrolysis of 2-AG **1.1**. The hydrolysis of 2-AG **1.1** leads to the release in AA **1.3**, a precursor for the biosynthesis of inflammatory and pain-mediating secondary messengers such as prostaglandins (Cisar et al. 2018; Serhan et al. 2008). Therefore, MAGL acts as a pivotal connecting switch of the endocannabinoid and eicosanoid signaling pathway, impacting a variety of physiological and pathological processes, including nociception, learning, addiction, appetite regulation, pain perception, and emotional regulation (Nomura et al. 2010, 2011; Deng and Li 2020; Gil-Ordóñez et al. 2018). Furthermore, MAGL is highly elevated in various cancer cell lines from multiple tissues, underscoring the major role of AA-level regulation in tumorigenesis and cancer proliferation (Qin and Ruan 2014). This highlights the importance of MAGL as a therapeutic target within the endocannabinoid system and has led to significant research efforts to explore its pharmacology over the past few decades.

Several highly potent and selective inhibitors have been developed to modulate MAGL and thus the 2-AG **1.1** level. Acute pharmacological blockade of MAGL results in an increase in 2-AG **1.1** levels and thus in the indirect activation of the cannabinoid receptors while limiting the CB_1R-mediated adverse side effects observed by exogenous cannabinoid ligands such as psychotropic effects resulting in psychological and cognitive difficulties and addictive potential (Bononi et al. 2021). The drug discovery efforts have led to the identification of several classes of compounds that can be categorized as irreversible and reversible inhibitors, depending on their mode of action. The numerous inhibitors developed as well as the current state of the art have been summarized recently (Zanfirescu et al. 2021; Granchi et al. 2017; Bononi et al. 2021; Deng and Li 2020). Irreversible inhibitors are characterized by an activated carbamate or urea group that interacts covalently with the catalytic serine (Ser122) of MAGL and is attached to various leaving groups (compounds **12.1-3**, Fig. 12) (Karlsson et al. 1997).

In 2009, Long et al. identified JZL184 **12.1**, the first selective, in vivo active irreversible MAGL inhibitor, which is considered one of the important

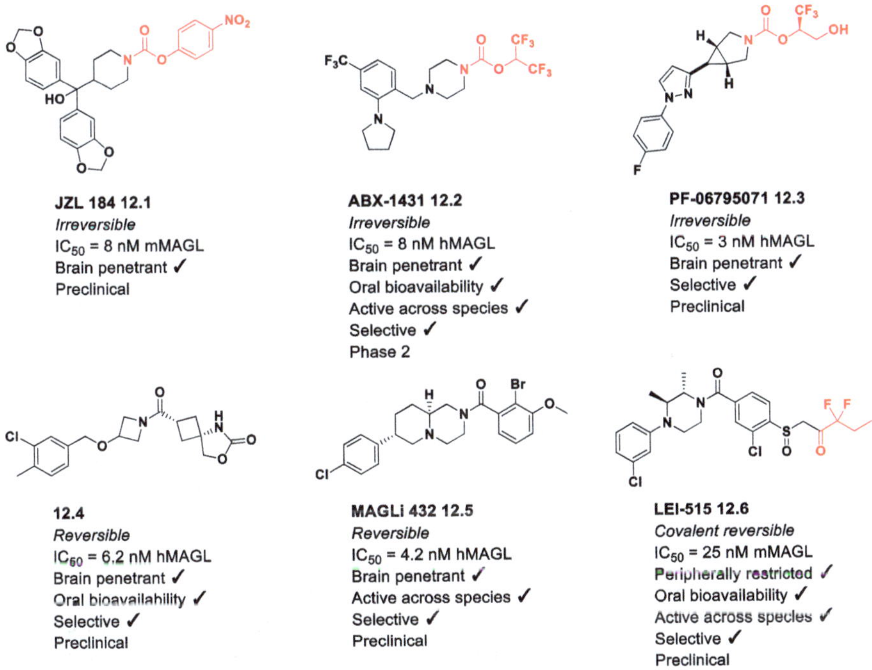

Fig. 12 Chemical structures, IC_{50} values, key properties, and highest development phase of representative in vivo active MAGL inhibitors. Top: irreversible inhibitors with warheads highlighted in red; bottom: reversible inhibitors

breakthroughs for MAGL inhibitor development (Long et al. 2009). JZL184 **12.1** showed high potency ($IC_{50} = 8$ nM mMAGL) and demonstrated beneficial effects in relevant pharmacological rodent models of pain, inflammation (Ghosh et al. 2013), anxiety (Nawata et al. 2019), neurodegeneration (Choi et al. 2018), and cancer (Ma et al. 2016). However, studies have indicated that chronic and complete inhibition by JZL184 **12.1** leads to CNS-mediated side effects (Schlosburg et al. 2010; Ming Jiang et al. 2023). Additionally, JZL184 **12.1** cross-reacts with several off-target proteins, including ABHD6, esterase 1, and triacylglycerol hydrolase (Deng and Li 2020). To address these drawbacks, JZL184 **12.1** was subjected to further structural modifications resulting in new carbamate inhibitors. These second-generation ligands KML29 (Jae Won Chang et al. 2012), JJKK-048 (Aaltonen et al. 2013), JW651 (Jae Won Chang et al. 2013), and MJN110 (Niphakis et al. 2013) exhibited high potencies, and in particular improved cross-species activities as well as enhanced selectivity over other hydrolases in vitro and in vivo.

The pioneering and elegant investigations of the Cravatt lab have greatly advanced the field of MAGL inhibitors and led to the foundation of Abide Therapeutics, which is now part of Lundbeck. In 2018, Cisar et al. introduced ABX-1431 **12.2** (Cisar et al. 2018), a highly potent ($IC_{50} = 8$ nM hMAGL) and selective irreversible MAGL inhibitor with favorable CNS penetration, interspecies activity, and oral bioavailability. ABX-1431 **12.2** demonstrated notable MAGL inhibition in human PC3 lysates, a human prostate cancer cell line, and rodent brain homogenates with no significant activity against a panel of 95 common off-targets. Due to its favorable characteristics, ABX-1431 **12.2** was selected as the lead compound for clinical evaluation and is still the most advanced experimental clinical drug to date. It has been examined in various phase 2 clinical trials involving patients with focal epilepsy and Tourette syndrome, among other indications. However, adverse CNS effects have been observed during these trials (Müller-Vahl et al. 2021; Müller-Vahl et al. 2022). Nonetheless, ongoing clinical phase 2 trials investigate its potential in alternative indications, including post-traumatic stress disorder and multiple sclerosis-related spasticity (ClinicalTrials.gov identifier: NCT04597450/ NCT04990219) (Ming Jiang et al. 2023).

Researchers at Pfizer (McAllister et al. 2018) reported an aryl pyrazole-derived lead series focusing on improving the physicochemical properties and reactivity of the warhead thereby optimizing the in vivo performance of the MAGL inhibitors. These efforts culminated in the identification of PF-06795071 **12.3**, a highly selective covalent MAGL inhibitor featuring a novel chiral [3.1.0] core structure and a highly optimized trifluoromethyl glycol carbamate as the reactive warhead, which significantly improved physicochemical properties and selectivity. The discovery of this glycol-leaving group represented a major advancement, providing a new crucial modality for the development of novel covalent MAGL inhibitors (Van Egmond et al. 2021). PF-06795071 **12.3** shows high potency ($IC_{50} = 3$ nM) toward human MAGL, excellent CNS distribution in mouse brains, and a clean in vitro safety profile. Moreover, in vivo studies demonstrated that PF-06795071 **12.3** is able to substantially increase 2-AG **1.1** levels in the brains of mice and effectively reduce inflammatory markers induced by an inflammatory challenge.

In addition to the observed adverse side effects within the CNS associated with chronic blockade of MAGL by irreversible inhibitors (Schlosburg et al. 2010), the persistent inhibition of the target uncouples the pharmacodynamic response from the pharmacokinetic properties. As a result, precise dosing control becomes more difficult for irreversible inhibitors, requiring a broad therapeutic window to mitigate adverse side effects due to mechanism-based off-target activity. Therefore, the development of highly selective reversible MAGL inhibitors may have the potential to circumvent these drawbacks as they allow for better-adapted dose regimens and have an improved safety profile (Ming Jiang et al. 2023; Van Egmond et al. 2021).

The development of reversible MAGL inhibitors has been significantly advanced by researchers at Takeda Pharmaceutical. Ikeda et al. published a series of new spiro-scaffolds (Ikeda et al. 2021). Among these, **12.4** emerged as a promising MAGL inhibitor, demonstrating potent in vitro activity ($IC_{50} = 6.2$ nM hMAGL) and good selectivity toward all closely related serine hydrolases. In vivo, profiling in mice showed favorable pharmacokinetics with good oral absorption and blood–brain barrier penetration.

Hoffman La Roche recently reported about MAGLi 432 **12.5**, a novel noncovalent MAGL inhibitor exhibiting high selectivity and potency in lysates derived from human and mouse brain tissues (Kemble et al. 2022). The research by Kemble et al. elucidated crucial insights into the species-dependent MAGL expression in mouse and human cells of the neurovascular unit in vitro, thereby providing a basis for future cell-specific studies of MAGLi 432 **12.5** to modulate MAGL activity in neurological disorders, including diseases associated with blood–brain barrier dysfunction.

Peripherally restricted inhibitors are an alternative approach mitigating undesired CNS effects while preserving the therapeutic effects of MAGL inhibitors for peripheral indications (Maccarrone et al. 2023). Recently, LEI-515 **12.6** was developed as a novel peripherally restricted and potent MAGL inhibitor with favorable pharmacokinetic properties based on an extensive optimization of a high-throughput screening hit (Ming Jiang et al. 2023). Interestingly, analysis of the co-crystal structure showed that LEI-515 **12.6** inhibits MAGL by a less explored covalent reversible mechanism, thereby forming a reversible enzyme-inhibitor hemiketal adduct between the electrophilic carbonyl moiety of the ligand and the OH-backbone of catalytic Ser122. Jiang et al. demonstrated that LEI-515 **12.6** selectively elevates 2-AG **1.1** levels in peripheral tissues without affecting the 2-AG **1.1** levels in the brain. Furthermore, LEI-515 **12.6** exhibits the capacity to ameliorate tissue injury, inflammation, and neuropathic pain without inducing adverse effects associated with direct CB_1R activation in the brain such as the development of tolerance, or physical dependence.

Labeled probes can provide critical evidence of target engagement, biological activity, and the expression or distribution of targets, confirming on-target drug effects (Willmann et al. 2008). Furthermore, such probes facilitate the development of tailored in vitro pharmacology assays to identify hits and optimize lead structures. Their application also supports preclinical studies and human trials, making them a crucial element of modern drug discovery (Guberman et al. 2022).

Pioneering research by the Cravatt lab on active-site-directed probes for ECS enzymes, particularly serine hydrolases, led to the introduction of the first dual MAGL/ABHD6 activity-based imaging probe in 2013 (Jae Won Chang et al. 2013). Leveraging the high specificity of hexafluoroisopropanol (HFIP) carbamates for MAGL and ABHD6, the fluorescent HFIP carbamate probe JW912 **13.1** was developed by conjugating the reactive group with a BODIPY fluorophore (Fig. 13). In both mouse and human cells, JW912 **13.1** exhibited selective labeling and inhibition of the enzymes, thus enabling visualization of the subcellular distribution of MAGL and ABHD6, for example, in prostate cancer cells. The lack of selectivity of JW912 **13.1** between MAGL and ABHD6 was addressed by pre-blocking with competitive selective inhibitors. In the same work, the reverse design concept was applied to introduce the dual MAGL/ABHD6 activity-based probe JW651yne **13.6**. By modifying the MAGL-selective irreversible ligand JW651 **13.5**, an alkyne moiety was incorporated, which can act as a latent click-handle for two-step labeling.

Similarly derived from JW651 **13.5** is the recently introduced valuable fluorescent so-called PharmacoSTORM probe LEI-463 **13.7.** Utilizing super-resolution fluorescent microscopy, in particular direct stochastic optical reconstruction microscopy (dSTORM), these fluorescently labeled PharmacoSTORM probes enable the visualization and quantification of drug–target interactions at the nanoscale in living

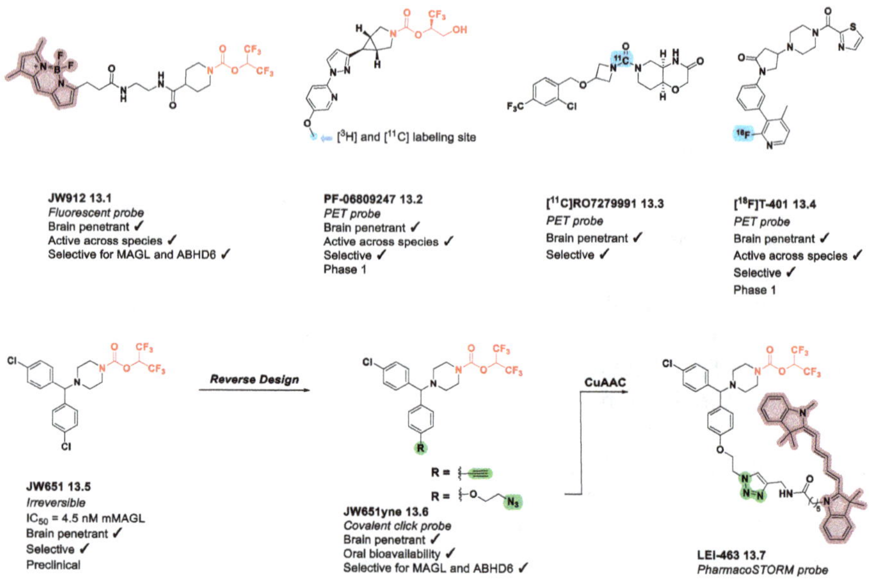

Fig. 13 Chemical structures of selected labeled probes applied for in vitro and in vivo visualization of MAGL. Radioisotopes are highlighted in blue, affinity handles designed for click-chemistry in green, fluorophore reporter groups in pink, and warheads in red

cells, achieving exceptional precision (Prokop et al. 2021). Here, a Cy5-conjugated covalent MAGL inhibitor was constructed by combining the azide-containing derivative **13.6** with the fluorescent reporter through CuAAC. The selectivity of the probe was confirmed by competitive ligand-binding assays with an unlabeled MAGL inhibitor, demonstrating LEI-463s **13.7** suitability for quantitative nanoscale imaging of MAGL in PharmacoSTORM experiments.

Due to minimal or nonexistent structural modification within PET labeling, this technique allows for unperturbed affinity, selectivity, and exceptional sensitivity. Consequently, PET radiotracers are optimal tools for evaluating target engagement of MAGL inhibitors in vivo, studying drug–target interactions, and validating new MAGL inhibitors in clinical trials (Gil-Ordóñez et al. 2018). Therefore, significant efforts have been made recently to develop new radioligands for PET imaging.

The extensive discovery program on MAGL inhibitors at Pfizer also included the development of covalent [^{11}C]-PET and [^{3}H]-labeled probes derived from their most advanced preclinical candidates (Lei Zhang et al. 2019). For this purpose, the irreversible PF-06795071 **12.3** inhibitor was modified based on previously reported SAR data by exchange of the fluorine substituent to a methoxy group. The subsequent selection of candidates' analogs was further supported by in silico-calculated PET-specific properties, resulting in the potent and selective MAGL inhibitor PF-06809247 **13.2**. This radioligand was successfully utilized for cross-species measurement of MAGL brain expression, in vivo binding assessment in rats, and PET imaging in nonhuman primates, underscoring its potential as a promising radioligand for clinical MAGL PET imaging. Preliminary investigations of the binding kinetics and test–retest characteristics have been conducted in healthy adult participants volunteers conforming to the reliability of the pharmacokinetics and biodistribution (ClinicalTrials.gov identifier: NCT03100136).

In a complementary approach, Roche investigated a range of morpholine-3-one derivatives for [^{11}C]-labeling, aiming to craft novel PET tracers with a reversible binding mode of action (Yingfang He et al. 2022). The candidates were meticulously chosen based on their high affinity toward the target as well as their physicochemical properties, which indicate high brain permeation, making them favorable for neuro-imaging. The radiotracers were evaluated via in vitro autoradiography and demonstrated remarkable specificity and selectivity. The most promising compound [^{11}C]RO7279991 **13.3** enabled successful in vivo visualization of MAGL in the mouse brain. Based on these studies, [^{11}C]RO7279991 **13.3** could serve as a lead compound for further structural optimization, promising enhanced noninvasive PET imaging of MAGL in vivo.

Further reversible PET probes have been developed by Takeda Pharmaceuticals. Based on a piperazinyl pyrrolidine-2-one derivative, Hattori et al. introduced the PET imaging agent [^{18}F]T-401 **13.4**, which demonstrates high selectivity and optimized permeability across the blood–brain barrier (Hattori et al. 2019). In vivo, PET imaging of wild-type and MAGL-deficient mice revealed specific uptake in regions with elevated MAGL expression. Moreover, initial in vivo studies have been conducted in rhesus monkeys, where [^{18}F]T-401 **13.4** was employed to visualize MAGL availability and measure the effective inhibition by the tool compound JW642 (Hattori et al. 2021). The promising outcomes of these studies underscore the potential of [^{18}F]T-401 **13.4** for PET imaging of MAGL of healthy and diseased human brains, as well as for the in vivo assessment of interactions between MAGL and potential therapeutic agents. In this context, [^{18}F]T-401 **13.4** was used in a clinical phase 1 multipart PET imaging study in healthy adult subjects to assess its biodistribution and radiation dosimetry and evaluate the availability of MAGL and FAAH in the central nervous system after oral administration of the dual CC-97489 inhibitor (ClinicalTrials.gov identifier: NCT05065541).

Overall, since the identification of MAGL as the key enzyme responsible for the degradation of 2-AG **1.1**, extensive research has revealed its critical role in 2-AG-mediated signaling as well as the therapeutic potential of MAGL inhibitors. The ligands discussed here represent only a fraction of the compounds explored and investigated for modulating MAGL activity. This wide array of inhibitors and imaging probes, differing in mechanism, distribution, and functionality, provides the flexibility to select appropriate ligands for different experiments, specific targets, or assays. The available diversity of different probes significantly enhances the ability to investigate and validate the role of MAGL across various biological and therapeutic contexts.

6 Transmembrane, Intracellular, and Extracellular Transporters

While substantial translational efforts have focused on developing ECS modulators, particularly inhibitors of eCB degradation, particularly on FAAH and MAGL blockers (Blankman and Cravatt 2013; Van Egmond et al. 2021; C. J. Fowler 2021), progress on inhibitors targeting eCB cellular uptake or trafficking has been slower. The development of these transport inhibitors is complicated by the promiscuity of the extra- and intracellular eCB-binding proteins and the absence of an unambiguously clearly identified target responsible for plasma membrane transport. Despite these challenges, the pharmaceutical industry is developing RT126, a FABP inhibitor that competes with eCB binding for FABP4 and FABP5, and SYT510, a selective eCB reuptake inhibitor (SERI) that increases extracellular eCB levels in the brain by targeting the putative eCB membrane transporter (Artelo Biosciences Web Page, Artelo Biosciences Reports Positive Pre-Clinical Results with its Novel Inhibitor to Fatty Acid Binding Protein 5, https://ir.artelobio.com/news-events/press-releases/detail/90/artelo-biosciences-reports-positive-pre-clinical-results, accessed November 2022).

Unlike active cellular transport mechanisms that require energy, lipophilic eCBs move between membranes and across aqueous barriers through interactions with binding proteins and traverse the plasma membrane via energy-independent processes. The facilitated diffusion of eCBs is influenced by their interactions with both extracellular and intracellular binding proteins, as well as metabolic enzymes, making their measurement challenging and necessitating specialized assays (Sergio Oddi et al. 2010; Christopher J Fowler 2013; Rau et al. 2016; Reynoso-Moreno et al. 2023). Differentiating between FAAH and AEA **1.2** uptake inhibitors has been difficult due to their intrinsic coupling; however, selective and potent inhibitors of eCB cellular uptake have recently been developed (Chicca et al. 2017). The discovery of intracellular carrier proteins, mainly FABPs, and lipid droplets as potential sequestration domains for AEA **1.2** has provided new insights into AEA **1.2** transport (Kaczocha et al. 2009; S. Oddi et al. 2008, 2009). These eCB transport inhibitors exhibit CB_1R/CB_2R-dependent indirect cannabimimetic effects, such as analgesia, anti-inflammatory, and anxiolytic effects, representing a new class of pharmacological inhibitors that indirectly activate the ECS.

6.1 *Selective eCB Reuptake Inhibitors*

The *N*-isobutylamide guineensine **14.1** from *Piper* species is a highly selective and potent nanomolar inhibitor of AEA **1.2** cellular uptake (Fig. 14) (Nicolussi et al. 2014b). Guineensine **14.1** showed negligible inhibition of FAAH activity ($IC_{50} > 50$ μM) and other ECS targets including FABP5 ($K_i > 100$ μM), and induced dose-dependent cannabimimetic effects in BALB/c mice, such as catalepsy,

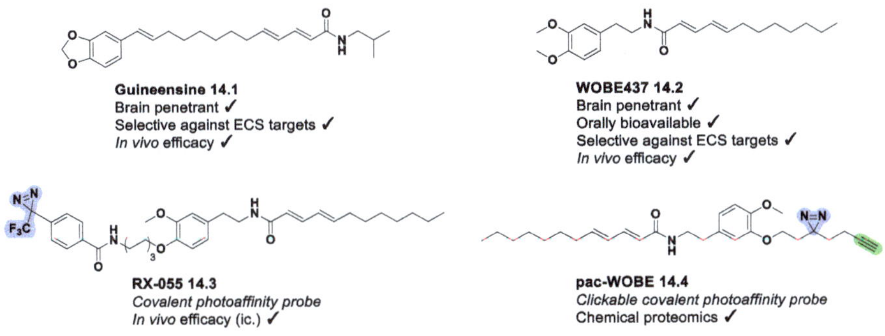

Fig. 14 Chemical structures and key characteristics of eCB reuptake Inhibitors. Affinity handles designed for click-chemistry are highlighted in green and photoreactive groups in purple

hypothermia, reduced locomotion, and analgesia, which were blocked by the CB_1R antagonist rimonabant **4.1** (Reynoso-Moreno et al. 2017). This suggests guineensine **14.1** targets eCB cellular reuptake in vivo. An efficient total synthesis of guineensine **14.1** has been published, aiding research in this area (Bartholomäus et al. 2019).

N-Benzyl-(9*Z*,12*Z*)-octadecadieneamide (macamide 7) demonstrated a nanomolar IC_{50} value for AEA **1.2** uptake inhibition but also inhibits FAAH at low micromolar concentrations (Hajdu et al. 2014). An analog of farinosone-C (BSL-34) was identified as a more selective inhibitor of AEA **1.2** uptake (IC_{50} = 232 nM) over FAAH inhibition (IC_{50} > 10 μM), showing structural similarity to OMDM-2 (Burch et al. 2014). Research on *N*-alkyl-2,4-dodecadienamides from *Echinacea purpurea*, which interact with the ECS, led to the synthesis of various derivatives and analogs. *N*-alkylcarbamates were tested in U937 cells, showing ultrapotent FAAH inhibition and hyperpotent AEA **1.2** uptake inhibition (Nicolussi et al. 2014a). Some of them (e.g., WOBE492 and WOBE498) demonstrated FAAH-independent AEA **1.2** uptake inhibition in HMC-1 cells with IC_{50} values below 300 nM (Nicolussi et al. 2014a). This led to the identification of (2*E*,4*E*)-*N*-[2-(3,4-dimethoxyphenyl)ethyl] dodeca-2,4-dienamide (WOBE437) **14.2** as a highly potent and selective eCB uptake inhibitor (Chicca et al. 2017).

WOBE437 **14.2** inhibits AEA **1.2** and 2-AG **1.1** uptake in U937 cells with IC_{50} values of 10 ± 8 nM and 283 ± 121 nM, respectively (Chicca et al. 2017). It was tested in Neuro2a mouse neuroblastoma cells, primary rat cortical neurons, and FAAH-deficient HMC-1 cells, showing significant inhibition of AEA **1.2** uptake across all cell lines. WOBE437 **14.2** did not inhibit FAAH, MAGL, α/β-hydrolase domain proteins 6 and 12, or COX-2, nor did it significantly interact with CB_1R/CB_2R, FABP5, or any of 45 relevant CNS-related receptors, transporters, ion channels, and enzymes tested (Chicca et al. 2017). WOBE437 **14.2** was found to be orally bioavailable in C57BL6/J male mice (Reynoso Moreno et al. 2018). In a mouse model of multiple sclerosis-like experimental autoimmune encephalomyelitis WOBE437 **14.2** significantly reduced disease severity and accelerated recovery through CB_1R/CB_2R-dependent mechanisms (Reynoso-Moreno et al. 2021). A structure–activity relationship study on the WOBE437 **14.2** scaffold for cellular

AEA **1.2** uptake inhibition has been published (Mäder et al. 2021). The diazirine-based photoaffinity probe RX-055 **14.3** permanently inhibited the membrane transport of both endocannabinoids and produced targeted anxiolytic effects upon in situ photoactivation in the brain (Chicca et al. 2017). However, using the clickable analog pac-WOBE **14.4** of RX-055 **14.3**, saccharopine dehydrogenase-like oxidoreductase, vesicle amine transport 1, and ferrochelatase were identified as low-affinity (10 μM) off-targets of WOBE437 **14.3** in Neuro-2a cells (Gagestein et al. 2022), highlighting the need for caution in therapeutic use at higher doses.

Currently, a new class of thiazolidinone-based SERIs is being developed for treating psychiatric or neurological disorders and inflammation at Synendos Therapeutics (https://www.synendos.com). Lead molecule SYT510 is supposed to enter phase 1 clinical trials in 2024, although the target protein has not yet been disclosed (Adis Insight Web Page, *Drug report on SYT 510*, https://adisinsight.springer.com/drugs/800061198, accessed November 2024b).

6.2 FABP Inhibitors

So far, the expression of 10 distinct isoforms of the FABP family has been identified, namely FABP1 through FABP9 and FABP12 (Smathers and Petersen 2011). Each of these isoforms is involved in various physiological processes, either individually or in combination with others. FABPs are considered promising therapeutic targets because inhibiting abnormal or excessive lipid signaling could have significant implications for treating a range of conditions and diseases.

In the liver, FABP1 serves as a binding protein for eCBs and phytocannabinoids (pCBs) and plays a role in regulating hepatic eCB levels (Huan Huang et al. 2016). Research involving FABP1 knockout mice indicated a reduced effect of a high-fat diet on brain eCB levels, particularly in male mice, suggesting FABP1's involvement in lipid biosynthesis (Martin et al. 2017). Fatty acid-binding proteins, such as FABP5 and FABP7, are implicated in the trafficking of AEA **1.2**, and their overexpression in cells leads to increased AEA **1.2** uptake and hydrolysis (Kaczocha et al. 2009). Inhibitors like BMS309403 **15.1** can decrease AEA **1.2** uptake, although high concentrations are required for significant inhibition (Fig. 15) (Sulsky et al. 2007; Furuhashi and Hotamisligil 2008). FABP5 is identified as a target for AEA **1.2** uptake inhibitors like OMDM-2, VDM11, and AM404 (Kaczocha et al. 2012; Nicolussi et al. 2014a).

FABP5 also exhibits a strong affinity for AA **1.3**, which challenges its role in AEA **1.2** transport (Beltramo et al. 1997; Hillard et al. 1997; D. Piomelli et al. 1999). The development of specific inhibitors for FABP5, such as SBFI-26 **15.2**, is ongoing, with SBFI-26 **15.2** demonstrating antinociceptive and anti-inflammatory effects in mice (Berger et al. 2012; Kaczocha et al. 2014). SBFI-26 **15.2** binds to FABP5 at the substrate entry portal and the canonical ligand-binding pocket (Hao-Chi Hsu et al. 2017). FABP5 facilitates the hydrolysis of AEA **1.2** to AA **1.3** and transports AA to the nucleus for activation of PPARβ/δ (Yu et al. 2014).

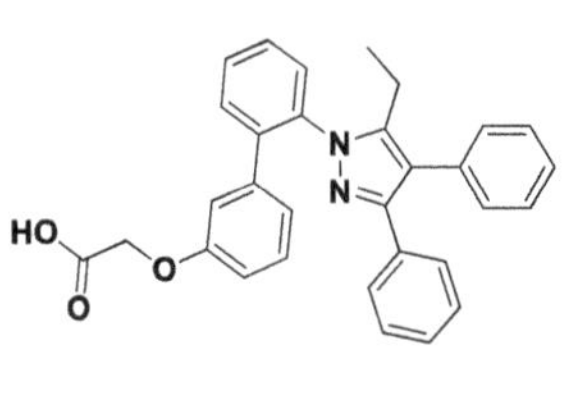

BMS309403 15.1
FABP4/5 inhibitor ✓

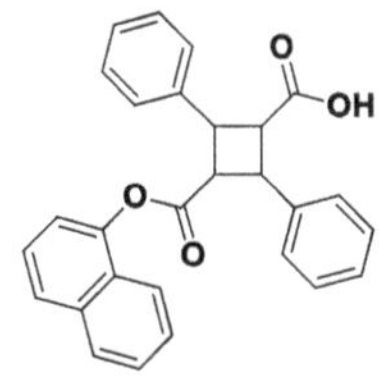

SBFI-26 15.2
Selective FABP5 inhibitor ✓
In vivo efficacy ✓
Phase 1 ✓

Fig. 15 Chemical structures and key characteristics of FABP5 inhibitors

Additionally, FABP5 binds 2-AG **1.1** and acts as a synaptic transporter, influencing retrograde signaling (Haj-Dahmane et al. 2018). Inhibition or absence of FABP5 disrupts depolarization-induced suppression of excitation (DSE) and affects excitatory postsynaptic currents. FABP5 is secreted by astrocytes, indicating its role as an extracellular transporter of 2-AG **1.1**. Deleting FABP5 impairs both tonic and phasic 2-AG **1.1** signaling without affecting CB_1R expression (Fauzan et al. 2022). Conditional knockout of FABP5 in nociceptors increases AEA **1.2** levels, leading to antinociceptive effects mediated by CB_1R (Bogdan et al. 2022). The competition between 2-AG **1.1** and AA **1.3** for binding sites on FABP5 suggests that the effects observed in FABP knockout mice may also be related to AA **1.3** metabolism.

7 Additional Targets Within the "Endocannabinoidome"

The eCBome is an expanded signaling system that includes the ECS and a variety of related molecules and receptors. This broader system was recognized over the past two decades due to two key discoveries. Firstly, the presence of several endogenous congeners of AEA **1.2** and 2-AG **1.1**, such as *N*-acyl-ethanolamines and 2-monoacylglycerols, in tissues and biological fluids. These molecules often share biosynthetic and degradative pathways with AEA **1.2** and 2-AG **1.1** but do not necessarily modulate cannabinoid receptors CB_1R and CB_2R (Vincenzo Di Marzo 2018). Secondly, the identification of plant cannabinoids other than THC **3.1**, such as CBD, cannabigerol, cannabidivarin, and cannabichromene, which interact with various receptors beyond CB_1R and CB_2R (Vincenzo Di Marzo 2018).

The eCBome includes several sub-families of eCB-like molecules such as primary amides of long-chain fatty acids (e.g., oleamide) (Langstein et al. 1996), amides between long-chain fatty acids and amino acids (e.g., *N*-acyl-taurines, -glycines, and -serines) (Susan M. Huang et al. 2001; Milman et al. 2006; Saghatelian et al. 2006). Additional subfamilies consist of amides of long-chain fatty acids with neurotransmitters (e.g., *N*-acyl-dopamines and -serotonins) (Susan

M. Huang et al. 2002; Verhoeckx et al. 2011) and oxidation products of polyunsaturated eCBome mediators, termed the "oxyendocannabinoidome" (oxyeCBome) (Simard et al. 2022).

These eCBome mediators interact with a variety of receptors, including G-protein-coupled receptors (GPCRs) like CB_1R, CB_2R, and orphan GPCRs (e.g., GPR18, GPR55, GPR110, GPR119, and GPR130), or GPCRs that are known to be activated or inhibited by other mediators, such as some serotonin receptors (Morales and Reggio 2017; Muller et al. 2019; Lago-Fernandez et al. 2021). Furthermore, they engage with ligand-activated ion channels, such as TRP channels of the V1-4, A1, and M8 types, and T-type Ca^{2+} (Ca_v3) channels, which have been suggested to be modulated also by other lipid mediators, as well as with amino acid neurotransmitter-activated targets, such as γ-aminobutyric acid or glycine receptors, and nuclear receptors like PPARs. Some eCB-like molecules also interact with eCB metabolic enzymes, influencing levels of AEA **1.2** and 2-AG **1.1** and thereby modulating receptor activity (Petrosino et al. 2019; Jonsson et al. 2001; Petrosino and Di Marzo 2017; Bashashati et al. 2017; Ayoub et al. 2020).

The eCBome offers new therapeutic opportunities and challenges. For example, non-THC cannabinoids and eCBome mediators can modulate receptors beyond CB_1R and CB_2R, potentially widening their medical applications (Bowen et al. 2017; Petrosino and Di Marzo 2017; Wu et al. 2021). However, the promiscuity of eCBome mediators in targeting multiple receptors complicates the development of selective therapeutics. For example, FAAH and MAGL inhibitors may elevate levels of various eCBome mediators, leading to mixed therapeutic outcomes (Tiziana Bisogno et al. 2013; Janssen and Van Der Stelt 2016; Vincenzo Di Marzo and Silvestri 2019). The eCBome interacts significantly with the gut microbiome, which is influenced by diet, medications, and other factors. Gut bacteria can produce eCB-like molecules, expanding the scope of microbe–host communication (De Petrocellis et al. 2009; Cohen et al. 2017; Fang-Yuan Chang et al. 2021). Conversely, the ECS and eCBome can modulate gut microbiota composition and function, affecting metabolic and inflammatory markers (Cluny et al. 2015; Shen et al. 2017; Mehrpouya-Bahrami et al. 2017; Geurts et al. 2015; Dione et al. 2020).

The eCBome encompasses a complex network of mediators and receptors, offering both opportunities and challenges for therapeutic development. Its interaction with the gut microbiome further underscores its potential impact on health and disease. Understanding the specific roles of eCBome components in various conditions is crucial for developing effective treatments for high unmet medical needs. Understanding the specific roles of eCBome components in various conditions is crucial for developing effective treatments for high unmet medical needs. Therefore, deorphanizing associated receptors and generating highly specific chemical tools, with and without reporter units, will be of utmost importance for therapeutically exploiting the largely unexplored elements of the eCBome.

8 Conclusion

Cannabis sativa has been used therapeutically since early civilizations. Key cannabinoids THC **3.1** and CBD were characterized in the 1960s, leading to the discovery of cannabinoid receptors CB_1R and CB_2R and the ECS in the 1990s. The ECS is a complex lipid signaling network involving endogenous ligands like 2-AG **1.1** and AEA **1.2**, which regulates vital functions such as sleep, appetite, or memory formation and has significant therapeutic potential, particularly for neurological disorders. Small molecule-derived pharmacological tools, or chemical probes, targeting key ECS components are crucial for target validation, mechanistic studies, signaling pathway elucidation, phenotypic screening, and drug discovery. These probes selectively interact with specific proteins or pathways, enabling modulation of target activity and observation of biological effects. Created by applying high-throughput screening, medicinal chemistry, and structural biology, effective chemical probes must be selective, potent, and meet criteria such as cell permeability and metabolic stability. Labeled chemical probes carry an additional reporter group, such as a radiolabel or fluorophore. Here reverse design approaches enhance the development of high-quality labeled probes by leveraging accumulated medicinal chemistry knowledge during the development process of the numerous ligands targeting the ECS.

CB_1R and CB_2R, high homologous class A GPCRs are activated by AEA **1.2**, 2-AG **1.1**, and THC **3.1**, but none of these natural ligands are exclusively selective for either receptor. CB_1R is mainly expressed in the CNS and peripheral tissues, while CB_2R is abundant in immune tissues and cells. CB_1R regulates mood, appetite, pain perception, learning, memory, and motor control. Developing CB_1R-targeting drugs is challenging due to their widespread brain distribution and associated side effects. While THC **3.1** and 2-AG **1.1** stimulate appetite, which led to the development of CB_1R inverse agonists like rimonabant for the treatment of obesity, rimonabant was withdrawn from the market due to psychiatric side effects. This prompted a shift to peripherally active agents and selective CB_1R antagonists with reduced central nervous system effects. Advances in imaging techniques like SPECT and PET, along with fluorescently labeled CB_1R probes, have significantly enhanced the understanding of CB_1R localization and function. Recent structural insights have facilitated the design of next-generation ligands to improve receptor activation and antagonist binding. Future research will likely focus on developing safer and more effective CB_1R-targeting therapies, including biased ligands and allosteric modulators.

CB_2R activation generally results in immunosuppressive effects, helping to limit tissue injury in various pathological conditions including neuroinflammation, but it can exacerbate damage in some disease states. Consequently, multiple high-quality CB_2R agonists and inverse agonists have been developed as chemical probes for both in vitro and in vivo applications. Advances in structural biology have facilitated the creation of superior reverse design-based labeled chemical probes for various imaging techniques, providing valuable insights into CB_2R's conformations and functions, and aiding in the design of effective therapeutics. Future research will likely focus on better understanding translational effects and optimizing CB_2R-targeting therapies to balance immunosuppressive benefits with potential risks and will include ortho- and allosteric ligands as well as an emphasis on biased signalling.

AEA **1.2** is synthesized on demand, involving the metallo-β-lactamase NAPE-PLD, which plays a role in energy metabolism. Brain-active inhibitor LEI-401 **7.1** reduces AEA **1.2** levels and affects emotional behavior, representing the best-characterized ligand to date. Advancements in the development of selective and potent inhibitors for NAPE-PLD will be crucial for better understanding its role in energy metabolism and emotional regulation. Hydrolytic degradation of AEA **1.2** occurs through cleavage by FAAH or NAAA. FAAH inhibitors are crucial for increasing endogenous fatty acid amides, offering analgesic effects without the adverse effects associated with direct cannabinoid receptor agonists. Several covalent inhibitors have advanced into clinical trials. Labeled probes, including PET tracers and click probes, have been developed to study FAAH activity and visualize the enzyme under various conditions. Novel labeled chemical probes, along with optionally reversible ligands, could support further mechanistic and translational studies needed to assess the therapeutic potential of FAAH. NAAA is a lysosomal enzyme that hydrolyzes fatty acid ethanolamides and regulates inflammation by activating the nuclear receptor PPARα. Early tool compounds have shown efficacy in reducing inflammation and pain in animal models. Activity-based probes have been developed to study NAAA. Advancing these probes will help to elucidate the enzyme's role in CNS disorders and stimulate new drug discovery programs for therapeutic development.

DAGLα and DAGLβ are crucial for the biosynthesis of the eCB 2-AG **1.1**, with DAGLα primarily responsible for its production in the central nervous system. Numerous covalent DAGL inhibitors and labeled probes have been developed to investigate 2-AG's **1.1** role in processes like food intake, neuroinflammation, and learning and memory regulation. Despite recent advancements, challenges remain in creating highly selective inhibitors that have minimal off-target effects. Future research will therefore focus on enhancing selectivity and specificity to improve therapeutic potential. MAGL degrades the eCB 2-AG **1.1** into AA **1.3**, regulating cannabinoid signaling and playing roles in nociception, learning, addiction, appetite regulation, pain perception, mood, and cancer proliferation. Many covalent and reversible inhibitors have been developed, with some entering clinical trials. Significant progress has been made in developing highly selective inhibitors with minimal off-target effects. These translational efforts toward the clinic application have been strongly supported by PET tracers and other labeled probes. Here reversible inhibitors may offer improved dosing control and safety profiles, enhancing their potential for future therapeutic applications.

Progress in developing inhibitors targeting eCB uptake or trafficking has been slower due to the promiscuity of eCB-binding proteins and the lack of a well-defined target for plasma membrane transport. Despite these challenges, the first potent and selective eCB uptake inhibitors have been identified. These inhibitors do not affect FAAH, MAGL, or other ECS-related enzymes and receptors and have shown efficacy in reducing disease severity in a mouse model of multiple sclerosis, leading to clinical development. Additionally, FABPs are promising therapeutic targets due to their role in lipid signaling, but still remain largely underexplored. Specific FABP5 inhibitors showing antinociceptive and anti-inflammatory effects are currently the most advanced, emphasizing the potential of FABP inhibition in modulating eCB signaling and related physiological processes. Structural information could facilitate the discovery of improved probes and support the exploration of the therapeutic potential of eCB transport modulators for various indications.

Numerous high-quality chemical probes, both labeled and unlabeled, have been developed for CB_1R, CB_2R, FAAH, and MAGL, among others enabling detailed investigation of their pharmacology. Ongoing research aims to create even more refined chemical tools for studying the newer components of the ECS and eCBome. In particular, the eCBome, as an expanded signaling network that encompasses the ECS and additional related molecules beyond, poses a new, still mainly uncharted field of investigation. These include small molecules such as congeners of AEA **1.2** and 2-AG **1.1**, plant cannabinoids like CBD, as well as a variety of receptors such as GPCRs, TRP channels, and nuclear receptors that involve further metabolic enzymes that regulate receptor activity. Therapeutic opportunities may arise here from non-THC cannabinoids and eCBome mediators that can modulate multiple targets and for example interact with the gut microbiome. Future research will aim to deorphanize receptors and develop specific chemical tools to explore and exploit the eCBome in its full breadth to address unmet medical needs.

These efforts will be bolstered by a deeper understanding of the 3D structures of these targets through techniques like cryo-electron microscopy, advanced in silico tools such as computer-aided drug design, virtual screening, and machine learning, as well as insights into biased signaling. These efforts hold the significant promise for improving our translational understanding and thereby accelerating drug discovery processes targeting the ECS and eCBome. Ultimately this will help addressing the high unmet medical needs in the treatment of CNS disorders and many other disease conditions, unlocking the therapeutic potential of the ECS and eCBome.

References

Aaltonen N, Savinainen JR, Ribas CR, Rönkkö J, Kuusisto A, Korhonen J et al (2013) Piperazine and piperidine triazole ureas as ultrapotent and highly selective inhibitors of monoacylglycerol lipase. Chem Biol 20(3):379–390. https://doi.org/10.1016/j.chembiol.2013.01.012

Adis Insight (2024a) Drug report on nabilone. https://adisinsight.springer.com/drugs/800025856. Accessed Nov 2024

Adis Insight (2024b) Drug report on SYT 510. https://adisinsight.springer.com/drugs/800061198. Accessed Nov 2024

Aghazadeh Tabrizi M, Baraldi PG, Borea PA, Varani K (2016) Medicinal chemistry, pharmacology, and potential therapeutic benefits of cannabinoid CB2 receptor agonists. Chem Rev 116(2): 519–560. https://doi.org/10.1021/acs.chemrev.5b00411

Ahn KH, Mahmoud MM, Kendall DA (2012) Allosteric modulator ORG27569 induces CB1 cannabinoid receptor high affinity agonist binding state, receptor internalization, and Gi

protein-independent ERK1/2 kinase activation. J Biol Chem 287(15):12070–12082. https://doi.org/10.1074/jbc.m111.316463
Alexander SPH (2016) Therapeutic potential of cannabis-related drugs. Prog Neuropsychopharmacol Biol Psychiatry 64:157–166. https://doi.org/10.1016/j.pnpbp.2015.07.001
Alexander JP, Cravatt BF (2005) Mechanism of carbamate inactivation of FAAH: implications for the design of covalent inhibitors and in vivo functional probes for enzymes. Chem Biol 12(11): 1179–1187. https://doi.org/10.1016/j.chembiol.2005.08.011
Alhouayek M, Bottemanne P, Subramanian KV, Lambert DM, Makriyannis A, Cani PD et al (2015) N-Acylethanolamine-hydrolyzing acid amidase inhibition increases colon N-palmitoylethanolamine levels and counteracts murine colitis. FASEB J 29(2):650–661. https://doi.org/10.1096/fj.14-255208
Alhouayek M, Masquelier J, Muccioli GG (2018) Lysophosphatidylinositols, from cell membrane constituents to GPR55 ligands. Trends Pharmacol Sci 39(6):586–604. https://doi.org/10.1016/j.tips.2018.02.011
Altamura C, Ventriglia M, Martini MG, Montesano D, Errante Y, Piscitelli F et al (2015) Elevation of plasma 2-arachidonoylglycerol levels in Alzheimer's disease patients as a potential protective mechanism against neurodegenerative decline. J Alzheimers Dis 46(2):497–506
Al-Zoubi R, Morales P, Reggio PH (2019) Structural insights into CB1 receptor biased signaling. Int J Mol Sci 20(8):1837. https://doi.org/10.3390/ijms20081837
Amato G, Khan NS, Maitra R (2019) A patent update on cannabinoid receptor 1 antagonists (2015-2018). Expert Opin Ther Pat 29(4). https://doi.org/10.1080/13543776.2019.1597851
Amin MR, Ali DW (2019) Pharmacology of medical cannabis. Recent advances in cannabinoid physiology and pathology, pp 151–165
Anand P, Whiteside G, Fowler CJ, Hohmann AG (2009) Targeting CB2 receptors and the endocannabinoid system for the treatment of pain. Brain Res Rev 60(1):255–266. https://doi.org/10.1016/j.brainresrev.2008.12.003
Appiah KK, Blat Y, Robertson BJ, Pearce BC, Pedicord DL, Gentles RG et al (2013) Identification of small molecules that selectively inhibit diacylglycerol lipase–α activity. J Biomol Screen 19(4):595–605. https://doi.org/10.1177/1087057113511111
Arrowsmith CH, Audia JE, Austin C, Baell J, Bennett J, Blagg J et al (2015) The promise and peril of chemical probes. Nat Chem Biol 11(8):536–541. https://doi.org/10.1038/nchembio.1867
Artelo Biosciences (2022) Artelo biosciences reports positive pre-clinical results with its novel inhibitor to fatty acid binding protein 5. GlobeNewswire. https://ir.artelobio.com/news-events/press-releases/detail/90/artelo-biosciences-reports-positive-pre-clinical-results. Accessed Nov 2024
Atlas D, Levitzki A (1977) Probing of β-adrenergic receptors by novel fluorescent β-adrenergic blockers. Proc Natl Acad Sci 74(12):5290–5294. https://doi.org/10.1073/pnas.74.12.5290
Atwood BK, Mackie K (2010) CB_2: a cannabinoid receptor with an identity crisis. Br J Pharmacol 160(3):467–479. https://doi.org/10.1111/j.1476-5381.2010.00729.x
Atwood BK, Wager-Miller J, Haskins C, Straiker A, Mackie K (2012) Functional selectivity in CB_2 cannabinoid receptor signaling and regulation: implications for the therapeutic potential of CB_2 ligands. Mol Pharmacol 81(2):250–263. https://doi.org/10.1124/mol.111.074013
Auberson YP, Arimondo PB, Duca M, Essig S, Grether U, Rufer AC et al (2023) EFMC: trends in medicinal chemistry and chemical biology. Chembiochem 24(7):e202200690. https://doi.org/10.1002/cbic.202200690
Aymerich MS, Aso E, Abellanas MA, Tolon RM, Ramos JA, Ferrer I et al (2018) Cannabinoid pharmacology/therapeutics in chronic degenerative disorders affecting the central nervous system. Biochem Pharmacol 157:67–84
Ayoub SM, Smoum R, Farag M, Atwal H, Collins SA, Rock EM et al (2020) Oleoyl alanine (HU595): a stable monomethylated oleoyl glycine interferes with acute naloxone precipitated morphine withdrawal in male rats. Psychopharmacology (Berl) 237(9):2753–2765. https://doi.org/10.1007/s00213-020-05570-4

Baggelaar MP, Chameau PJP, Kantae V, Hummel J, Hsu K-L, Janssen F et al (2015) Highly selective, reversible inhibitor identified by comparative chemoproteomics modulates diacylglycerol lipase activity in neurons. J Am Chem Soc 137(27):8851–8857. https://doi.org/10.1021/jacs.5b04883

Baggelaar MP, Maccarrone M, van der Stelt M (2018) 2-Arachidonoylglycerol: a signaling lipid with manifold actions in the brain. Prog Lipid Res 71:1–17

Balla A, Dong B, Shilpa BM, Vemuri K, Makriyannis A, Pandey SC et al (2018) Cannabinoid-1 receptor neutral antagonist reduces binge-like alcohol consumption and alcohol-induced accumbal dopaminergic signaling. Neuropharmacology 131. https://doi.org/10.1016/j.neuropharm.2017.10.040

Bartholomäus R, Nicolussi S, Baumann A, Rau M, Simão AC, Gertsch J et al (2019) Total synthesis of the endocannabinoid uptake inhibitor Guineensine and SAR studies. ChemMedChem 14(17): 1590–1596. https://doi.org/10.1002/cmdc.201900390

Barutta F, Corbelli A, Mastrocola R, Gambino R, Di Marzo V, Pinach S et al (2010) Cannabinoid receptor 1 blockade ameliorates albuminuria in experimental diabetic nephropathy. Diabetes 59(4):1046–1054. https://doi.org/10.2337/db09-1336

Barutta F, Piscitelli F, Pinach S, Bruno G, Gambino R, Rastaldi MP et al (2011) Protective role of cannabinoid receptor type 2 in a mouse model of diabetic nephropathy. Diabetes 60(9): 2386–2396. https://doi.org/10.2337/db10-1809

Bashashati M, Fichna J, Piscitelli F, Capasso R, Izzo AA, Sibaev A et al (2017) Targeting fatty acid amide hydrolase and transient receptor potential vanilloid-1 simultaneously to modulate colonic motility and visceral sensation in the mouse: a pharmacological intervention with N-arachidonoyl-serotonin (AA−5-HT). Neurogastroenterol Motil 29(12):e13148. https://doi.org/10.1111/nmo.13148

Bell MR, D'Ambra TE, Kumar V, Eissenstat MA, Herrmann JL Jr, Wetzel JR et al (1991) Antinociceptive (aminoalkyl)indoles. J Med Chem 34(3):1099. https://doi.org/10.1021/jm00107a034

Beltramo M, Stella N, Calignano A, Lin S, Makriyannis A, Piomelli D (1997) Functional role of high-affinity anandamide transport, as revealed by selective inhibition. Science 277(5329): 1094–1097

Benito C, Tolón RM, Castillo AI, Ruiz-Valdepeñas L, Martínez-Orgado JA, Fernández-Sánchez FJ et al (2012) β− amyloid exacerbates inflammation in astrocytes lacking fatty acid amide hydrolase through a mechanism involving PPAR-α, PPAR-γ and TRPV1, but not CB1 or CB2 receptors. Br J Pharmacol 166(4):1474–1489

Berger WT, Ralph BP, Kaczocha M, Sun J, Balius TE, Rizzo RC et al (2012) Targeting fatty acid binding protein (FABP) anandamide transporters – a novel strategy for development of anti-inflammatory and anti-nociceptive drugs. PLoS One 7(12):e50968. https://doi.org/10.1371/journal.pone.0050968

Bhuniya D, Kharul RK, Hajare A, Shaikh N, Bhosale S, Balwe S et al (2019) Discovery and evaluation of novel FAAH inhibitors in neuropathic pain model. Bioorg Med Chem Lett 29(2): 238–243. https://doi.org/10.1016/j.bmcl.2018.11.048

Bilsland LG, Dick JR, Pryce G, Petrosino S, Di Marzo V, Baker D et al (2006) Increasing cannabinoid levels by pharmacological and genetic manipulation delays disease progression in SOD1 mice. FASEB J 20(7)

Bisogno T, Melck D, De Petrocellis L, Bobrov MY, Gretskaya NM, Bezuglov VV et al (1998) Arachidonoylserotonin and other novel inhibitors of fatty acid amide hydrolase. Biochem Biophys Res Commun 248(3):515–522. https://doi.org/10.1006/bbrc.1998.8874

Bisogno T, Howell F, Williams G, Minassi A, Cascio MG, Ligresti A et al (2003) Cloning of the first sn1-DAG lipases points to the spatial and temporal regulation of endocannabinoid signaling in the brain. J Cell Biol 163(3):463–468. https://doi.org/10.1083/jcb.200305129

Bisogno T, Mahadevan A, Coccurello R, Chang JW, Allarà M, Chen Y et al (2013) A novel fluorophosphonate inhibitor of the biosynthesis of the endocannabinoid 2-arachidonoylglycerol with potential anti-obesity effects. Br J Pharmacol 169(4):784–793. https://doi.org/10.1111/bph.12013

Blagg J, Workman P (2017) Choose and use your chemical probe wisely to explore cancer biology. Cancer Cell 32(1):9–25

Blankman JL, Cravatt BF (2013) Chemical probes of endocannabinoid metabolism. Pharmacol Rev 65(2):849–871. https://doi.org/10.1124/pr.112.006387

Blankman JL, Simon GM, Cravatt BF (2007) A comprehensive profile of brain enzymes that hydrolyze the endocannabinoid 2-arachidonoylglycerol. Chem Biol 14(12):1347–1356. https://doi.org/10.1016/j.chembiol.2007.11.006

Bluett RJ, Báldi R, Haymer A, Gaulden AD, Hartley ND, Parrish WP et al (2017) Endocannabinoid signalling modulates susceptibility to traumatic stress exposure. Nat Commun 8(1):14782. https://doi.org/10.1038/ncomms14782

Bogdan DM, Studholme K, Dibua A, Gordon C, Kanjiya MP, Yu M et al (2022) FABP5 deletion in nociceptors augments endocannabinoid signaling and suppresses TRPV1 sensitization and inflammatory pain. Sci Rep 12(1). https://doi.org/10.1038/s41598-022-13284-0

Boileau I, Mansouri E, Williams B, Le Foll B, Rusjan P, Mizrahi R et al (2016) Fatty acid amide hydrolase binding in brain of cannabis users: imaging with the novel radiotracer [11C]CURB. Biol Psychiatry 80(9):691–701. https://doi.org/10.1016/j.biopsych.2016.04.012

Bond RA, IJzerman AP (2006) Recent developments in constitutive receptor activity and inverse agonism, and their potential for GPCR drug discovery. Trends Pharmacol Sci 27(2). https://doi.org/10.1016/j.tips.2005.12.007

Bononi G, Poli G, Rizzolio F, Tuccinardi T, Macchia M, Minutolo F et al (2021) An updated patent review of monoacylglycerol lipase (MAGL) inhibitors (2018-present). Expert Opin Ther Pat 31(2):153–168. https://doi.org/10.1080/13543776.2021.1841166

Börner C, Smida M, Höllt V, Schraven B, Kraus J (2009) Cannabinoid receptor type 1- and 2-mediated increase in cyclic AMP inhibits T cell receptor-triggered signaling. J Biol Chem 284(51):35450–35460. https://doi.org/10.1074/jbc.M109.006338

Bouaboula M, Rinaldi M, Carayon P, Carillon C, Delpech B, Shire D et al (1993) Cannabinoid-receptor expression in human leukocytes. Eur J Biochem 214(1):173–180. https://doi.org/10.1111/j.1432-1033.1993.tb17910.x

Bouma J, Broekhuis JD, Van Der Horst C, Kumar P, Ligresti A, Van Der Stelt M et al (2023) Dual allosteric and orthosteric pharmacology of synthetic analog cannabidiol-dimethylheptyl, but not cannabidiol, on the cannabinoid CB2 receptor. Biochem Pharmacol 218:115924. https://doi.org/10.1016/j.bcp.2023.115924

Bowen KJ, Kris-Etherton PM, Shearer GC, West SG, Reddivari L, Jones PJH (2017) Oleic acid-derived oleoylethanolamide: a nutritional science perspective. Prog Lipid Res 67:1–15. https://doi.org/10.1016/j.plipres.2017.04.001

Bradner JE, West N, Grachan ML, Greenberg EF, Haggarty SJ, Warnow T et al (2010) Chemical phylogenetics of histone deacetylases. Nat Chem Biol 6(3):238–243. https://doi.org/10.1038/nchembio.313

Brailoiu GC, Deliu E, Marcu J, Hoffman NE, Console-Bram L, Zhao P et al (2014) Differential activation of intracellular versus Plasmalemmal CB2 cannabinoid receptors. Biochemistry 53(30):4990–4999. https://doi.org/10.1021/bi500632a

Brennecke B, Gazzi T, Atz K, Fingerle J, Kuner P, Schindler T et al (2021) Cannabinoid receptor type 2 ligands: an analysis of granted patents since 2010. Pharm Pat Anal 10(3):111–163. https://doi.org/10.4155/ppa-2021-0002

Bunnage ME, Chekler ELP, Jones LH (2013) Target validation using chemical probes. Nat Chem Biol 9(4):195–199. https://doi.org/10.1038/nchembio.1197

Burch P, Chicca A, Gertsch J, Gademann K (2014) Functionally optimized neuritogenic farinosone C analogs: SAR-study and investigations on their mode of action. ACS Med Chem Lett 5(2):172–177

Burns HD, Van Laere K, Sanabria-Bohórquez S, Hamill TG, Bormans G, Eng W-S et al (2007) [^{18}F]MK-9470, a positron emission tomography (PET) tracer for *in vivo* human PET brain

imaging of the cannabinoid-1 receptor. Proc Natl Acad Sci 104(23):9800–9805. https://doi.org/10.1073/pnas.0703472104
Butini S, Brindisi M, Gemma S, Minetti P, Cabri W, Gallo G et al (2012) Discovery of potent inhibitors of human and mouse fatty acid amide hydrolases. J Med Chem 55(15):6898–6915. https://doi.org/10.1021/jm300689c
Cabral GA, Raborn ES, Griffin L, Dennis J, Marciano-Cabral F (2008) CB_2 receptors in the brain: role in central immune function. Br J Pharmacol 153(2):240–251. https://doi.org/10.1038/sj.bjp.0707584
Cadas H, Di Tomaso E, Piomelli D (1997) Occurrence and biosynthesis of endogenous cannabinoid precursor, *N*-Arachidonoyl phosphatidylethanolamine, in rat brain. J Neurosci 17(4):1226–1242. https://doi.org/10.1523/jneurosci.17-04-01226.1997
Castaneda JT, Harui A, Roth MD (2017) Regulation of cell surface CB2 receptor during human B cell activation and differentiation. J Neuroimmune Pharmacol 12(3):544–554. https://doi.org/10.1007/s11481-017-9744-7
Celorrio M, Fernández-Suárez D, Rojo-Bustamante E, Echeverry-Alzate V, Ramírez MJ, Hillard CJ et al (2016) Fatty acid amide hydrolase inhibition for the symptomatic relief of Parkinson's disease. Brain Behav Immun 57:94–105
Cencioni MT, Chiurchiu V, Catanzaro G, Borsellino G, Bernardi G, Battistini L et al (2010) Anandamide suppresses proliferation and cytokine release from primary human T-lymphocytes mainly via CB2 receptors. PLoS One 5(1). https://doi.org/10.1371/journal.pone.0008688
Centonze D, Finazzi-Agrò A, Bernardi G, Maccarrone M (2007) The endocannabinoid system in targeting inflammatory neurodegenerative diseases. Trends Pharmacol Sci 28(4):180–187. https://doi.org/10.1016/j.tips.2007.02.004
Chambers AP, Vemuri VK, Peng Y, Wood JT, Olszewska T, Pittman QJ et al (2007) A neutral CB1 receptor antagonist reduces weight gain in rat. Am J Physiol Regul Integr Comp Physiol 293(6). https://doi.org/10.1152/ajpregu.00663.2007
Chang JW, Niphakis MJ, Lum KM, Cognetta AB, Wang C, Matthews ML et al (2012) Highly selective inhibitors of monoacylglycerol lipase bearing a reactive group that is bioisosteric with endocannabinoid substrates. Chem Biol 19(5):579–588
Chang JW, Cognetta AB III, Niphakis MJ, Cravatt BF (2013) Proteome-wide reactivity profiling identifies diverse carbamate chemotypes tuned for serine hydrolase inhibition. ACS Chem Biol 8(7):1590–1599. https://doi.org/10.1021/cb400261h
Chang C-P, Huang H-L, Huang J-K, Hung M-S, Wu C-H, Song J-S et al (2019) Fluorine-18 isotope labeling for positron emission tomography imaging. Direct evidence for DBPR211 as a peripherally restricted CB1 inverse agonist. Bioorg Med Chem 27(1). https://doi.org/10.1016/j.bmc.2018.11.043
Chang F-Y, Siuti P, Laurent S, Williams T, Glassey E, Sailer AW et al (2021) Gut-inhabiting Clostridia build human GPCR ligands by conjugating neurotransmitters with diet- and human-derived fatty acids. Nat Microbiol 6(6):792–805. https://doi.org/10.1038/s41564-021-00887-y
Chicca A, Nicolussi S, Bartholomäus R, Blunder M, Aparisi Rey A, Petrucci V et al (2017) Chemical probes to potently and selectively inhibit endocannabinoid cellular reuptake. Proc Natl Acad Sci 114(25):E5006–E5015. https://doi.org/10.1073/pnas.1704065114
Chicca A, Bátora D, Ullmer C, Caruso A, Grüner S, Fingerle J et al (2024) A highly potent, orally bioavailable pyrazole-derived cannabinoid CB2 receptor-selective full agonist for *in vivo* studies. ACS Pharmacol Transl Sci 7(8):2424–2438. https://doi.org/10.1021/acsptsci.4c00269
Choi S-H, Arai AL, Mou Y, Kang B, Yen CC-C, Hallenbeck J et al (2018) Neuroprotective effects of MAGL (monoacylglycerol lipase) inhibitors in experimental ischemic stroke. Stroke 49(3):718–726. https://doi.org/10.1161/STROKEAHA.117.019664
Cisar JS, Weber OD, Clapper JR, Blankman JL, Henry CL, Simon GM et al (2018) Identification of ABX-1431, a selective inhibitor of monoacylglycerol lipase and clinical candidate for treatment of neurological disorders. J Med Chem 61(20):9062–9084. https://doi.org/10.1021/acs.jmedchem.8b00951

Clayton N, Marshall FH, Bountra C, O'Shaughnessy CT (2002) CB1 and CB2 cannabinoid receptors are implicated in inflammatory pain. Pain 96(3):253–260. https://doi.org/10.1016/s0304-3959(01)00454-7

Cluny NL, Keenan CM, Reimer RA, Le Foll B, Sharkey KA (2015) Prevention of diet-induced obesity effects on body weight and gut microbiota in mice treated chronically with Δ9-tetrahydrocannabinol. PLoS One 10(12):e0144270. https://doi.org/10.1371/journal.pone.0144270

Cohen LJ, Esterhazy D, Kim S-H, Lemetre C, Aguilar RR, Gordon EA et al (2017) Commensal bacteria make GPCR ligands that mimic human signalling molecules. Nature 549(7670):48–53. https://doi.org/10.1038/nature23874

Cooper A, Singh S, Hook S, Tyndall JDA, Vernall AJ, Alexander SPH (2017) Chemical tools for studying lipid-binding class A G protein-coupled receptors. Pharmacol Rev 69(3):316–353. https://doi.org/10.1124/pr.116.013243

Cordova T, Ayalon D, Lander N, Mechoulam R, Nir I, Puder M et al (1980) The ovulation blocking effect of cannabinoids: structure-activity relationships. Psychoneuroendocrinology 5(1). https://doi.org/10.1016/0306-4530(80)90009-8

Correa F, Hernangómez-Herrero M, Mestre L, Loría F, Docagne F, Guaza C (2011) The endocannabinoid anandamide downregulates IL-23 and IL-12 subunits in a viral model of multiple sclerosis: evidence for a cross-talk between IL-12p70/IL-23 axis and IL-10 in microglial cells. Brain Behav Immun 25(4):736–749. https://doi.org/10.1016/j.bbi.2011.01.020

Cortez IL, Silva NR, Rodrigues NS, Pedrazzi JFC, Del Bel EA, Mechoulam R et al (2022) HU-910, a CB2 receptor agonist, reverses behavioral changes in pharmacological rodent models for schizophrenia. Prog Neuropsychopharmacol Biol Psychiatry 117:110553. https://doi.org/10.1016/j.pnpbp.2022.110553

Cristino L, Bisogno T, Di Marzo V (2020) Cannabinoids and the expanded endocannabinoid system in neurological disorders. Nat Rev Neurol 16(1):9–29. https://doi.org/10.1038/s41582-019-0284-z

D'Addario C, Di Francesco A, Arosio B, Gussago C, Dell'Osso B, Bari M et al (2012) Epigenetic regulation of fatty acid amide hydrolase in Alzheimer disease. PLoS One 7(6):e39186

Dalzell HC, Uliss DB, Handrick GR, Razdan RK (1981) Hashish. 26. Factors influencing double-bond stability in cannabinoids. J Org Chem 46(5):949–953

D'Ambra TE, Estep KG, Bell MR, Eissenstat MA, Josef KA, Ward SJ et al (1992) Conformationally restrained analogs of pravadoline: nanomolar potent, enantioselective, (aminoalkyl) indole agonists of the cannabinoid receptor. J Med Chem 35(1):124–135

De Laurentiis A, Araujo HA, Rettori V (2014) Role of the endocannabinoid system in the neuroendocrine responses to inflammation. Curr Pharm Des 20(29):4697–4706

de Martin Esteban SR, Benito-Cuesta I, Terradillos I, Martinez-Relimpio AM, Arnanz MA, Ruiz-Perez G et al (2022) Cannabinoid CB2 receptors modulate microglia function and amyloid dynamics in a mouse model of Alzheimer's disease. Front Pharmacol 13:841766. https://doi.org/10.3389/fphar.2022.841766

De Petrocellis L, Deva R, Mainieri F, Schaefer M, Bisogno T, Ciccoli R et al (2009) Chemical synthesis, pharmacological characterization, and possible formation in unicellular fungi of 3-hydroxy-anandamide. J Lipid Res 50(4):658–666. https://doi.org/10.1194/jlr.m800337-jlr200

Deng H, Li W (2020) Monoacylglycerol lipase inhibitors: modulators for lipid metabolism in cancer malignancy, neurological and metabolic disorders. Acta Pharm Sin B 10(4):582–602. https://doi.org/10.1016/j.apsb.2019.10.006

Deng H, Kooijman S, van den Nieuwendijk AMCH, Ogasawara D, van der Wel T, van Dalen F et al (2017) Triazole Ureas act as diacylglycerol lipase inhibitors and prevent fasting-induced refeeding. J Med Chem 60(1):428–440. https://doi.org/10.1021/acs.jmedchem.6b01482

Devane WA, Dysarz FR, Johnson MR, Melvin LS, Howlett AC (1988) Determination and characterization of a cannabinoid receptor in rat brain. Mol Pharmacol 34(5):605–613

Devane WA, Hanus L, Breuer A, Pertwee RG, Stevenson LA, Griffin G et al (1992) Isolation and structure of a brain constituent that binds to the cannabinoid receptor. Science 258(5090):1946. https://doi.org/10.1126/science.1470919

Di Iorio G, Lupi M, Sarchione F, Matarazzo I, Santacroce R, Petruccelli F et al (2013) The endocannabinoid system: a putative role in neurodegenerative diseases. Int J High Risk Behav Addict 2(3):100

Di Marzo V (2018) New approaches and challenges to targeting the endocannabinoid system. Nat Rev Drug Discov 17(9):623–639

Di Marzo V, Fontana A (1995) Anandamide, an endogenous cannabinomimetic eicosanoid:'killing two birds with one stone'. Prostaglandins Leukot Essent Fatty Acids 53(1):1–11

Di Marzo V, Silvestri C (2019) Lifestyle and metabolic syndrome: contribution of the endocannabinoidome. Nutrients 11(8):1956. https://doi.org/10.3390/nu11081956

Di Marzo V, Fontana A, Cadas H, Schinelli S, Cimino G, Schwartz J-C et al (1994) Formation and inactivation of endogenous cannabinoid anandamide in central neurons. Nature 372(6507): 686–691

Dinh TP, Carpenter D, Leslie FM, Freund TF, Katona I, Sensi SL et al (2002) Brain monoglyceride lipase participating in endocannabinoid inactivation. Proc Natl Acad Sci U S A 99(16): 10819–10824. https://doi.org/10.1073/pnas.152334899

Dione N, Lacroix S, Taschler U, Deschênes T, Abolghasemi A, Leblanc N et al (2020) Mgll knockout mouse resistance to diet-induced dysmetabolism is associated with altered gut microbiota. Cells 9(12):2705. https://doi.org/10.3390/cells9122705

Eissenstat MA, Bell MR, D'Ambra TE, Alexander EJ, Daum SJ, Ackerman JH et al (1995) Aminoalkylindoles: structure-activity relationships of novel cannabinoid mimetics. J Med Chem 38(16):3094–3105

Elmes MW, Prentis LE, McGoldrick LL, Giuliano CJ, Sweeney JM, Joseph OM et al (2019) FABP1 controls hepatic transport and biotransformation of Δ9-THC. Sci Rep 9(1). https://doi.org/10.1038/s41598-019-44108-3

El-Remessy AB, Rajesh M, Mukhopadhyay P, Horváth B, Patel V, Al-Gayyar MMH et al (2011) Cannabinoid 1 receptor activation contributes to vascular inflammation and cell death in a mouse model of diabetic retinopathy and a human retinal cell line. Diabetologia 54(6): 1567–1578. https://doi.org/10.1007/s00125-011-2061-4

Everard A, Plovier H, Rastelli M, Van Hul M, De Wouters D'Oplinter A, Geurts L et al (2019) Intestinal epithelial N-acylphosphatidylethanolamine phospholipase D links dietary fat to metabolic adaptations in obesity and steatosis. Nat Commun 10(1). https://doi.org/10.1038/s41467-018-08051-7

Fasia L, Karava V, Siafaka-Kapadai A (2003) Uptake and metabolism of [3H] anandamide by rabbit platelets: lack of transporter? Eur J Biochem 270(17):3498–3506

Fauzan M, Oubraim S, Yu M, Glaser ST, Kaczocha M, Haj-Dahmane S (2022) Fatty acid-binding protein 5 modulates brain endocannabinoid tone and retrograde signaling in the striatum. Front Cell Neurosci 16. https://doi.org/10.3389/fncel.2022.936939

Fazio D, Criscuolo E, Piccoli A, Barboni B, Fezza F, Maccarrone M (2020) Advances in the discovery of fatty acid amide hydrolase inhibitors: what does the future hold? Expert Opin Drug Discovery 15(7):765–778. https://doi.org/10.1080/17460441.2020.1751118

Fernández-Ruiz J (2009) The endocannabinoid system as a target for the treatment of motor dysfunction. Br J Pharmacol 156(7):1029–1040. https://doi.org/10.1111/j.1476-5381.2008.00088.x

Fernández-Ruiz J, Romero J, Velasco G, Tolón RM, Ramos JA, Guzmán M (2007) Cannabinoid CB2 receptor: a new target for controlling neural cell survival? Trends Pharmacol Sci 28(1): 39–45. https://doi.org/10.1016/j.tips.2006.11.001

Fezza F, Bari M, Florio R, Talamonti E, Feole M, Maccarrone M (2014) Endocannabinoids, related compounds and their metabolic routes. Molecules 19(11):17078–17106

Figuerola-Asencio L, Morales P, Zhao P, Hurst DP, Sayed SS, Colón KL et al (2023) Supplementary material for: thienopyrimidine derivatives as GPR55 receptor antagonists: insight into structure-activity relationship. American Chemical Society

Fong TM (2014) Constitutive activity in cannabinoid receptors. Adv Pharmacol 70. https://doi.org/10.1016/B978-0-12-417197-8.00004-3

Fotio Y, Mabou Tagne A, Squire E, Lee H-L, Phillips CM, Chang K et al (2024) NAAA-regulated lipid signaling in monocytes controls the induction of hyperalgesic priming in mice. Nat Commun 15(1). https://doi.org/10.1038/s41467-024-46139-5

Fowler CJ (2013) Transport of endocannabinoids across the plasma membrane and within the cell. FEBS J 280(9):1895–1904

Fowler CJ (2021) The endocannabinoid system – current implications for drug development. J Intern Med 290(1):2–26. https://doi.org/10.1111/joim.13229

Fulp A, Bortoff K, Zhang Y, Snyder R, Fennell T, Marusich JA et al (2013) Peripherally selective diphenyl purine antagonist of the CB1 receptor. J Med Chem 56(20):8066–8072. https://doi.org/10.1021/jm401129n

Furuhashi M, Hotamisligil GS (2008) Fatty acid-binding proteins: role in metabolic diseases and potential as drug targets. Nat Rev Drug Discov 7(6):489–503. https://doi.org/10.1038/nrd2589

Gado F, Di Cesare Mannelli L, Lucarini E, Bertini S, Cappelli E, Digiacomo M et al (2019) Identification of the first synthetic allosteric modulator of the CB_2 receptors and evidence of its efficacy for neuropathic pain relief. J Med Chem 62(1):276–287. https://doi.org/10.1021/acs.jmedchem.8b00368

Gagestein B, Stevens AF, Fazio D, Florea BI, Van Der Wel T, Bakker AT et al (2022) Chemical proteomics reveals off-targets of the anandamide reuptake inhibitor WOBE437. ACS Chem Biol 17(5):1174–1183. https://doi.org/10.1021/acschembio.2c00122

Galiegue S, Mary S, Marchand J, Dussossoy D, Carriere D, Carayon P et al (1995) Expression of central and peripheral cannabinoid receptors in human immune tissues and leukocyte subpopulations. Eur J Biochem 232(1):54–61. https://doi.org/10.1111/j.1432-1033.1995.tb20780.x

Gao Y, Vasilyev DV, Goncalves MB, Howell FV, Hobbs C, Reisenberg M et al (2010) Loss of retrograde endocannabinoid signaling and reduced adult neurogenesis in diacylglycerol lipase knock-out mice. J Neurosci 30(6):2017–2024. https://doi.org/10.1523/jneurosci.5693-09.2010

Garai S, Schaffer PC, Laprairie RB, Janero DR, Pertwee RG, Straiker A et al (2021) Design, synthesis, and pharmacological profiling of cannabinoid 1 receptor allosteric modulators: preclinical efficacy of C2-group GAT211 congeners for reducing intraocular pressure. Bioorg Med Chem 50. https://doi.org/10.1016/j.bmc.2021.116421

García-Martín A, Garrido-Rodríguez M, Navarrete C, Caprioglio D, Palomares B, Demesa J et al (2019) Cannabinoid derivatives acting as dual PPARγ/CB2 agonists as therapeutic agents for systemic sclerosis. Biochem Pharmacol 163:321–334. https://doi.org/10.1016/j.bcp.2019.02.029

Gazzi T, Brennecke B, Atz K, Korn C, Sykes D, Forn-Cuni G et al (2022) Detection of cannabinoid receptor type 2 in native cells and zebrafish with a highly potent, cell-permeable fluorescent probe. Chem Sci 13(19):5539–5545. https://doi.org/10.1039/d1sc06659e

Geurts L, Everard A, Van Hul M, Essaghir A, Duparc T, Matamoros S et al (2015) Adipose tissue NAPE-PLD controls fat mass development by altering the browning process and gut microbiota. Nat Commun 6(1):6495. https://doi.org/10.1038/ncomms7495

Ghidini A, Scalvini L, Palese F, Lodola A, Mor M, Piomelli D (2021) Different roles for the acyl chain and the amine leaving group in the substrate selectivity of *N*-acylethanolamine acid amidase. J Enzyme Inhib Med Chem 36(1):1410–1422. https://doi.org/10.1080/14756366.2021.1912035

Ghosh S, Wise LE, Chen Y, Gujjar R, Mahadevan A, Cravatt BF et al (2013) The monoacylglycerol lipase inhibitor JZL184 suppresses inflammatory pain in the mouse carrageenan model. Life Sci 92(8):498–505. https://doi.org/10.1016/j.lfs.2012.06.020

Gil-Ordóñez A, Martín-Fontecha M, Ortega-Gutiérrez S, López-Rodríguez ML (2018) Monoacylglycerol lipase (MAGL) as a promising therapeutic target. Biochem Pharmacol 157: 18–32. https://doi.org/10.1016/j.bcp.2018.07.036

Godlewski G, Offertáler L, Wagner JA, Kunos G (2009) Receptors for acylethanolamides – GPR55 and GPR119. Prostaglandins Other Lipid Mediat 89(3):105–111. https://doi.org/10.1016/j.prostaglandins.2009.07.001

Gonzalo-Consuegra C, Santos-Garcia I, Garcia-Toscano L, Martin-Baquero R, Rodriguez-Cueto C, Wittwer MB et al (2024) Involvement of CB1 and CB2 receptors in neuroprotective effects of cannabinoids in experimental TDP-43 related frontotemporal dementia using male mice. Biomed Pharmacother 174:116473. https://doi.org/10.1016/j.biopha.2024.116473

Gorelik A, Gebai A, Illes K, Piomelli D, Nagar B (2018) Molecular mechanism of activation of the immunoregulatory amidase NAAA. Proc Natl Acad Sci 115(43):E10032–E10040. https://doi.org/10.1073/pnas.1811759115

Granchi C, Caligiuri I, Minutolo F, Rizzolio F, Tuccinardi T (2017) A patent review of Monoacylglycerol lipase (MAGL) inhibitors (2013-2017). Expert Opin Ther Pat 27(12): 1341–1351. https://doi.org/10.1080/13543776.2018.1389899

Grether U, Foxton RH, Gruener S, Korn C, Kimbara A, Osterwald A et al (2024) RG7774 (Vicasinabin), an orally bioavailable cannabinoid receptor 2 (CB2R) agonist, decreases retinal vascular permeability, leukocyte adhesion, and ocular inflammation in animal models. Front Pharmacol 15. https://doi.org/10.3389/fphar.2024.1426446

Griebel G, Stemmelin J, Lopez-Grancha M, Fauchey V, Slowinski F, Pichat P et al (2018) The selective reversible FAAH inhibitor, SSR411298, restores the development of maladaptive behaviors to acute and chronic stress in rodents. Sci Rep 8(1). https://doi.org/10.1038/s41598-018-20895-z

Gruden G, Barutta F, Kunos G, Pacher P (2016) Role of the endocannabinoid system in diabetes and diabetic complications. Br J Pharmacol 173(7):1116–1127. https://doi.org/10.1111/bph.13226

Guba W, Nazare M, Grether U (2021) Natural compounds and synthetic drugs to target type-2 cannabinoid (CB2) receptor. RSC Drug Discov Ser 76:89–167

Guberman M, Kosar M, Omran A, Carreira EM, Nazaré M, Grether U (2022) Reverse design toward optimized labeled chemical probes – examples from the endocannabinoid system. Chimia 76(5):425–434. https://doi.org/10.2533/chimia.2022.425

Gueye AB, Pryslawsky Y, Trigo JM, Poulia N, Delis F, Antoniou K et al (2016) The CB1 neutral antagonist AM4113 retains the therapeutic efficacy of the inverse agonist Rimonabant for nicotine dependence and weight loss with better psychiatric tolerability. Int J Neuropsychopharmacol 19(12). https://doi.org/10.1093/ijnp/pyw068

Guindon J, Hohmann AG (2008) Cannabinoid CB_2 receptors: a therapeutic target for the treatment of inflammatory and neuropathic pain. Br J Pharmacol 153(2):319–334. https://doi.org/10.1038/sj.bjp.0707531

Guzman M, Lo Verme J, Fu J, Oveisi F, Blazquez C, Piomelli D (2004) Oleoylethanolamide stimulates lipolysis by activating the nuclear receptor peroxisome proliferator-activated receptor α (PPAR-α). J Biol Chem 279(27):27849–27854. https://doi.org/10.1074/jbc.m404087200

Haider A, Kretz J, Gobbi L, Ahmed H, Atz K, Burkler M et al (2019) Structure-activity relationship studies of pyridine-based ligands and identification of a fluorinated derivative for positron emission tomography imaging of cannabinoid type 2 receptors. J Med Chem 62(24): 11165–11181. https://doi.org/10.1021/acs.jmedchem.9b01280

Haider A, Gobbi L, Kretz J, Ullmer C, Brink A, Honer M et al (2020) Identification and preclinical development of a 2,5,6-trisubstituted fluorinated pyridine derivative as a radioligand for the positron emission tomography imaging of cannabinoid type 2 receptors. J Med Chem 63(18): 10287–10306. https://doi.org/10.1021/acs.jmedchem.0c00778

Haj-Dahmane S, Shen R-Y, Elmes MW, Studholme K, Kanjiya MP, Bogdan D et al (2018) Fatty-acid–binding protein 5 controls retrograde endocannabinoid signaling at central glutamate synapses. Proc Natl Acad Sci 115(13):3482–3487. https://doi.org/10.1073/pnas.1721339115

Hajdu Z, Nicolussi S, Rau M, Lorántfy LS, Forgo P, Hohmann J et al (2014) Identification of endocannabinoid system-modulating N-alkylamides from Heliopsis helianthoides var. scabra and Lepidium meyenii. J Nat Prod 77(7):1663–1669

Han S, Thatte J, Buzard DJ, Jones RM (2013) Therapeutic utility of cannabinoid receptor type 2 (CB2) selective agonists. J Med Chem 56(21):8224–8256

Hanuš L, Breuer A, Tchilibon S, Shiloah S, Goldenberg D, Horowitz M et al (1999) HU-308: a specific agonist for CB_2, a peripheral cannabinoid receptor. Proc Natl Acad Sci 96(25): 14228–14233. https://doi.org/10.1073/pnas.96.25.14228

Hattori Y, Aoyama K, Maeda J, Arimura N, Takahashi Y, Sasaki M et al (2019) Design, synthesis, and evaluation of (4R)-1-{3-[2-(18F)Fluoro-4-methylpyridin-3-yl]phenyl}-4-[4-(1,3-thiazol-2-ylcarbonyl)piperazin-1-yl]pyrrolidin-2-one ([18F]T-401) as a novel positron-emission tomography imaging agent for monoacylglycerol lipase. J Med Chem 62(5):2362–2375. https://doi.org/10.1021/acs.jmedchem.8b01576

Hattori Y, Seki C, Maeda J, Nagai Y, Aoyama K, Zhang M-R et al (2021) Quantification of monoacylglycerol lipase and its occupancy by an exogenous ligand in rhesus monkey brains using [18F]T-401 and PET. J Cereb Blood Flow Metab 42(4):656–666. https://doi.org/10.1177/0271678X211058285

Hayes AW, Pressman P, Moser P, Soares-da-Silva P (2020) Regulatory safety pharmacology evaluation of BIA 10-2474. J Pharmacol Toxicol Methods 102:106677. https://doi.org/10.1016/j.vascn.2020.106677

He X-H, Jordan CJ, Vemuri K, Bi G-H, Zhan J, Gardner EL et al (2019) Cannabinoid CB1 receptor neutral antagonist AM4113 inhibits heroin self-administration without depressive side effects in rats. Acta Pharmacol Sin 40(3):365–373. https://doi.org/10.1038/s41401-018-0059-x

He Y, Gobbi LC, Herde AM, Rombach D, Ritter M, Kuhn B et al (2022) Discovery, synthesis and evaluation of novel reversible monoacylglycerol lipase radioligands bearing a morpholine-3-one scaffold. Nucl Med Biol 108-109:24–32. https://doi.org/10.1016/j.nucmedbio.2022.02.002

Hegde VL, Hegde S, Cravatt BF, Hofseth LJ, Nagarkatti M, Nagarkatti PS (2008) Attenuation of experimental autoimmune hepatitis by exogenous and endogenous cannabinoids: involvement of regulatory T cells. Mol Pharmacol 74(1):20–33. https://doi.org/10.1124/mol.108.047035

Hillard CJ, Edgemond WS, Jarrahian A, Campbell WB (1997) Accumulation of *N*-arachidonoylethanolamine (anandamide) into cerebellar granule cells occurs via facilitated diffusion. J Neurochem 69(2):631–638. https://doi.org/10.1046/j.1471-4159.1997.69020631.x

Hillard CJ, Huang H, Vogt CD, Rodrigues BE, Neumann TS, Sem DS et al (2017) Endocannabinoid transport proteins. Elsevier, pp 99–121. https://doi.org/10.1016/bs.mie.2017.06.017

Hollinshead SP, Tidwell MW, Palmer J, Guidetti R, Sanderson A, Johnson MP et al (2013) Selective cannabinoid receptor type 2 (CB2) agonists: optimization of a series of purines leading to the identification of a clinical candidate for the treatment of osteoarthritic pain. J Med Chem 56(14):5722–5733

Honer M, Gobbi L, Martarello L, Comley RA (2014) Radioligand development for molecular imaging of the central nervous system with positron emission tomography. Drug Discov Today 19(12):1936–1944. https://doi.org/10.1016/j.drudis.2014.08.012

Horvath B, Magid L, Mukhopadhyay P, Batkai S, Rajesh M, Park O et al (2012) A new cannabinoid CB2 receptor agonist HU-910 attenuates oxidative stress, inflammation and cell death associated with hepatic ischaemia/reperfusion injury. Br J Pharmacol 165(8):2462–2478. https://doi.org/10.1111/j.1476-5381.2011.01381.x

Hou L, Rong J, Haider A, Ogasawara D, Varlow C, Schafroth MA et al (2021) Positron emission tomography imaging of the endocannabinoid system: opportunities and challenges in radiotracer development. J Med Chem 64(1):123–149. https://doi.org/10.1021/acs.jmedchem.0c01459

Howlett AC (2002) International Union of Pharmacology. XXVII. Classification of cannabinoid receptors. Pharmacol Rev 54(2):161–202. https://doi.org/10.1124/pr.54.2.161

Howlett AC, Johnson MR, Melvin LS, Milne GM (1988) Nonclassical cannabinoid analgetics inhibit adenylate cyclase: development of a cannabinoid receptor model. Mol Pharmacol 33(3): 297–302. https://molpharm.aspetjournals.org/content/molpharm/33/3/297.full.pdf

Howlett AC, Bidaut-Russell M, Devane WA, Melvin LS, Johnson MR, Herkenham M (1990) The cannabinoid receptor: biochemical, anatomical and behavioral characterization. Trends Neurosci 13(10):420. https://doi.org/10.1016/0166-2236(90)90124-s

Howlett AC, Reggio PH, Childers SR, Hampson RE, Ulloa NM, Deutsch DG (2011) BPS publications. Br J Pharmacol 163(7). https://doi.org/10.1111/j.1476-5381.2011.01364.x

Hryhorowicz S, Kaczmarek-Ryś M, Andrzejewska A, Staszak K, Hryhorowicz M, Korcz A et al (2019) Allosteric modulation of cannabinoid receptor 1 – current challenges and future opportunities. Int J Mol Sci 20(23):5874. https://doi.org/10.3390/ijms20235874

Hsu K-L, Tsuboi K, Adibekian A, Pugh H, Masuda K, Cravatt BF (2012) DAGLβ inhibition perturbs a lipid network involved in macrophage inflammatory responses. Nat Chem Biol 8(12): 999–1007. https://doi.org/10.1038/nchembio.1105

Hsu H-C, Tong S, Zhou Y, Elmes MW, Yan S, Kaczocha M et al (2017) The Antinociceptive agent SBFI-26 binds to anandamide transporters FABP5 and FABP7 at two different sites. Biochemistry 56(27):3454–3462. https://doi.org/10.1021/acs.biochem.7b00194

Hua T, Vemuri K, Pu M, Qu L, Han GW, Wu Y et al (2016) Crystal structure of the human cannabinoid receptor CB1. Cell 167(3):750–762.e14. https://doi.org/10.1016/j.cell.2016.10.004

Hua T, Vemuri K, Nikas SP, Laprairie RB, Wu Y, Qu L et al (2017) Crystal structures of agonist-bound human cannabinoid receptor CB1. Nature 547(7664):468–471. https://doi.org/10.1038/nature23272

Hua T, Li X, Wu L, Iliopoulos-Tsoutsouvas C, Wang Y, Wu M et al (2020) Activation and signaling mechanism revealed by cannabinoid receptor-Gi complex structures. Cell 180(4): 655–665.e18. https://doi.org/10.1016/j.cell.2020.01.008

Huang SM, Bisogno T, Petros TJ, Chang SY, Zavitsanos PA, Zipkin RE et al (2001) Identification of a new class of molecules, the arachidonyl amino acids, and characterization of one member that inhibits pain. J Biol Chem 276(46):42639–42644. https://doi.org/10.1074/jbc.m107351200

Huang SM, Bisogno T, Trevisani M, Al-Hayani A, De Petrocellis L, Fezza F et al (2002) An endogenous capsaicin-like substance with high potency at recombinant and native vanilloid VR1 receptors. Proc Natl Acad Sci 99(12):8400–8405. https://doi.org/10.1073/pnas.122196999

Huang H, McIntosh AL, Martin GG, Landrock D, Chung S, Landrock KK et al (2016) FABP1: a novel hepatic endocannabinoid and cannabinoid binding protein. Biochemistry 55(37): 5243–5255. https://doi.org/10.1021/acs.biochem.6b00446

Iannotti FA, Di Marzo V, Petrosino S (2016) Endocannabinoids and endocannabinoid-related mediators: targets, metabolism and role in neurological disorders. Prog Lipid Res 62:107–128. https://doi.org/10.1016/j.plipres.2016.02.002

Ibsen MS, Connor M, Glass M (2017) Cannabinoid CB_1 and CB_2receptor signaling and bias. Cannabis Cannabinoid Res 2(1):48–60. https://doi.org/10.1089/can.2016.0037

Ignatowska-Jankowska BM, Baillie GL, Kinsey S, Crowe M, Ghosh S, Owens RA et al (2015) A cannabinoid CB1 receptor-positive allosteric modulator reduces neuropathic pain in the mouse with no psychoactive effects. Neuropsychopharmacology 40(13):2948–2959. https://doi.org/10.1038/npp.2015.148

Ikeda S, Sugiyama H, Tokuhara H, Murakami M, Nakamura M, Oguro Y et al (2021) Design and synthesis of novel Spiro derivatives as potent and reversible monoacylglycerol lipase (MAGL) inhibitors: bioisosteric transformation from 3-Oxo-3,4-dihydro-2H-benzo[b][1,4]oxazin-6-yl moiety. J Med Chem 64(15):11014–11044. https://doi.org/10.1021/acs.jmedchem.1c00432

Im D-S (2021) GPR119 and GPR55 as receptors for fatty acid ethanolamides, oleoylethanolamide and palmitoylethanolamide. Int J Mol Sci 22(3):1034. https://www.mdpi.com/1422-0067/22/3/1034

Iyer MR, Cinar R, Wood CM, Zawatsky CN, Coffey NJ, Kim KA et al (2022) Synthesis, biological evaluation, and molecular modeling studies of 3,4-diarylpyrazoline series of compounds as potent, nonbrain penetrant antagonists of cannabinoid-1 (CB1R) receptor with reduced lipophilicity. J Med Chem 65(3). https://doi.org/10.1021/acs.jmedchem.1c01836

Izzo AA, Camilleri M (2009) Cannabinoids in intestinal inflammation and cancer. Pharmacol Res 60(2):117–125. https://doi.org/10.1016/j.phrs.2009.03.008

Izzo AA, Sharkey KA (2010) Cannabinoids and the gut: new developments and emerging concepts. Pharmacol Ther 126(1):21–38. https://doi.org/10.1016/j.pharmthera.2009.12.005

Janero DR, Korde A, Makriyannis A (2017) Ligand-assisted protein structure (LAPS): an experimental paradigm for characterizing cannabinoid-receptor ligand-binding domains. Elsevier, pp 217–235. https://doi.org/10.1016/bs.mie.2017.06.022

Janssen FJ, Van Der Stelt M (2016) Inhibitors of diacylglycerol lipases in neurodegenerative and metabolic disorders. Bioorg Med Chem Lett 26(16):3831–3837. https://doi.org/10.1016/j.bmcl.2016.06.076

Janssen FJ, Deng H, Baggelaar MP, Allarà M, van der Wel T, den Dulk H et al (2014) Discovery of glycine sulfonamides as dual inhibitors of sn-1-diacylglycerol lipase α and α/β-hydrolase domain 6. J Med Chem 57(15):6610–6622. https://doi.org/10.1021/jm500681z

Jiang X, Chen S, Zhang Q, Yi C, He J, Ye X et al (2020) Celastrol is a novel selective agonist of cannabinoid receptor 2 with anti-inflammatory and anti-fibrotic activity in a mouse model of systemic sclerosis. Phytomedicine 67:153160. https://doi.org/10.1016/j.phymed.2019.153160

Jiang B-E, Jiang X, Zhang Q, Liang Q, Qiu Z-L, Sun X-B et al (2021) From a designer drug to the discovery of selective cannabinoid type 2 receptor agonists with favorable pharmacokinetic profiles for the treatment of systemic sclerosis. J Med Chem 64(1):385–403. https://doi.org/10.1021/acs.jmedchem.0c01023

Jiang M, Huizenga MCW, Wirt JL, Paloczi J, Amedi A, Van Den Berg RJBHN et al (2023) A monoacylglycerol lipase inhibitor showing therapeutic efficacy in mice without central side effects or dependence. Nat Commun 14(1). https://doi.org/10.1038/s41467-023-43606-3

Jin X-H, Okamoto Y, Morishita J, Tsuboi K, Tonai T, Ueda N (2007) Discovery and characterization of a Ca2+-independent phosphatidylethanolamine N-acyltransferase generating the anandamide precursor and its congeners. J Biol Chem 282(6):3614–3623. https://doi.org/10.1074/jbc.m606369200

Johnson MR, Melvin LS, Althuis TH, Bindra JS, Harbert CA, Milne GM et al (1981) Selective and potent analgetics derived from cannabinoids. J Clin Pharmacol 21(S1):271S–282S. https://doi.org/10.1002/j.1552-4604.1981.tb02605.x

Johnson DS, Stiff C, Lazerwith SE, Kesten SR, Fay LK, Morris M et al (2011) Discovery of PF-04457845: a highly potent, orally bioavailable, and selective urea FAAH inhibitor. ACS Med Chem Lett 2(2):91–96. https://doi.org/10.1021/ml100190t

Johnson MP, Chambers MG, Hollinshead SP, Knopp K, Vasudeva R, Astles PC (2012) A novel selective CB2 agonist, LY2828360 is efficacious in chronic pain models. (Paper presented at the 42nd society for neuroscience annual meeting, New Orleans, LA, USA)

Jonsson KO, Vandevoorde S, Lambert DM, Tiger G, Fowler CJ (2001) Effects of homologues and analogues of palmitoylethanolamide upon the inactivation of the endocannabinoid anandamide. Br J Pharmacol 133(8):1263–1275. https://doi.org/10.1038/sj.bjp.0704199

Jourdan T, Szanda G, Rosenberg AZ, Tam J, Earley BJ, Godlewski G et al (2014) Overactive cannabinoid 1 receptor in podocytes drives type 2 diabetic nephropathy. Proc Natl Acad Sci 111(50):E5420–E5428. https://doi.org/10.1073/pnas.1419901111

Jung K-M, Astarita G, Yasar S, Vasilevko V, Cribbs DH, Head E et al (2012) An amyloid β42-dependent deficit in anandamide mobilization is associated with cognitive dysfunction in Alzheimer's disease. Neurobiol Aging 33(8):1522–1532

Kaczocha M, Haj-Dahmane S (2022) Mechanisms of endocannabinoid transport in the brain. Br J Pharmacol 179(17):4300–4310. https://doi.org/10.1111/bph.15469

Kaczocha M, Glaser ST, Deutsch DG (2009) Identification of intracellular carriers for the endocannabinoid anandamide. Proc Natl Acad Sci 106(15):6375–6380. https://doi.org/10.1073/pnas.0901515106

Kaczocha M, Vivieca S, Sun J, Glaser ST, Deutsch DG (2012) Fatty acid-binding proteins transport N-acylethanolamines to nuclear receptors and are targets of endocannabinoid transport inhibitors. J Biol Chem 287(5):3415–3424. https://doi.org/10.1074/jbc.m111.304907

Kaczocha M, Rebecchi MJ, Ralph BP, Teng Y-HG, Berger WT, Galbavy W et al (2014) Inhibition of fatty acid binding proteins elevates brain anandamide levels and produces analgesia. PLoS One 9(4):e94200. https://doi.org/10.1371/journal.pone.0094200

Kano M, Ohno-Shosaku T, Hashimotodani Y, Uchigashima M, Watanabe M (2009) Endocannabinoid-mediated control of synaptic transmission. Physiol Rev

Karlsson M, Contreras JA, Hellman U, Tornqvist H, Holm C (1997) cDNA cloning, tissue distribution, and identification of the catalytic triad of monoglyceride lipase: evolutionary relationship to esterases, lysophospholipases, and haloperoxidases. J Biol Chem 272(43): 27218–27223. https://doi.org/10.1074/jbc.272.43.27218

Kawahara H, Drew G, Christie M, Vaughan C (2011) Inhibition of fatty acid amide hydrolase unmasks CB1 receptor and TRPV1 channel-mediated modulation of glutamatergic synaptic transmission in midbrain periaqueductal grey. Br J Pharmacol 163(6):1214–1222

Keith JM, Jones WM, Tichenor M, Liu J, Seierstad M, Palmer JA et al (2015) Preclinical characterization of the FAAH inhibitor JNJ-42165279. ACS Med Chem Lett 6(12): 1204–1208. https://doi.org/10.1021/acsmedchemlett.5b00353

Kemble AM, Hornsperger B, Ruf I, Richter H, Benz J, Kuhn B et al (2022) A potent and selective inhibitor for the modulation of MAGL activity in the neurovasculature. PLoS One 17(9): e0268590. https://doi.org/10.1371/journal.pone.0268590

Kenakin T, Kenakin T (2024) Know your molecule: pharmacological characterization of drug candidates to enhance efficacy and reduce late-stage attrition. Nat Rev Drug Discov 23(8) https://doi.org/10.1038/s41573-024-00958-9

Kerbrat A, Ferré J-C, Fillatre P, Ronzière T, Vannier S, Carsin-Nicol B et al (2016) Acute neurologic disorder from an inhibitor of fatty acid amide hydrolase. N Engl J Med 375(18): 1717–1725. https://doi.org/10.1056/NEJMoa1604221

Khajehali E, Malone DT, Glass M, Sexton PM, Christopoulos A, Leach K (2015) Biased Agonism and biased allosteric modulation at the CB_1 cannabinoid receptor. Mol Pharmacol 88(2): 368–379. https://doi.org/10.1124/mol.115.099192

Khurana L, Mackie K, Piomelli D, Kendall DA (2017) Modulation of CB1 cannabinoid receptor by allosteric ligands: pharmacology and therapeutic opportunities. Neuropharmacology 124:3–12. https://doi.org/10.1016/j.neuropharm.2017.05.018

Kindermann M, Roschitzki-Voser H, Caglič D, Repnik U, Miniejew C, Mittl PRE et al (2010) Selective and sensitive monitoring of Caspase-1 activity by a novel bioluminescent activity-based probe. Chem Biol 17(9):999–1007. https://doi.org/10.1016/j.chembiol.2010.07.011

Kirkham TC, Williams CM, Fezza F, Marzo VD (2002) Endocannabinoid levels in rat limbic forebrain and hypothalamus in relation to fasting, feeding and satiation: stimulation of eating by 2-arachidonoyl glycerol. Br J Pharmacol 136(4):550–557

Kishimoto S, Muramatsu M, Gokoh M, Oka S, Waku K, Sugiura T (2005) Endogenous cannabinoid receptor ligand induces the migration of human natural killer cells. J Biochem 137(2): 217–223. https://doi.org/10.1093/jb/mvi021

Kiss LE, Beliaev A, Ferreira HS, Rosa CP, Bonifácio MJ, Loureiro AI et al (2018) Discovery of a potent, Long-acting, and CNS-active inhibitor (BIA 10-2474) of fatty acid amide hydrolase. ChemMedChem 13(20).2177–2188. https://doi.org/10.1002/cmdc.201800393

Kleyer J, Nicolussi S, Taylor P, Simonelli D, Furger E, Anderle P et al (2012) Cannabinoid receptor trafficking in peripheral cells is dynamically regulated by a binary biochemical switch. Biochem Pharmacol 83(10):1393–1412. https://doi.org/10.1016/j.bcp.2012.02.014

Kosar M, Sykes DA, Viray AEG, Vitale RM, Sarott RC, Ganzoni RL et al (2023) Platform reagents enable synthesis of ligand-directed covalent probes: study of cannabinoid receptor 2 in live cells. J Am Chem Soc 145(28):15094–15108. https://doi.org/10.1021/jacs.2c13629

Kosar M, Mach L, Carreira EM, Nazaré M, Pacher P, Grether U (2024a) Patent review of cannabinoid receptor type 2 (CB_2R) modulators (2016-present). Expert Opin Ther Pat 34(8): 665–700. https://doi.org/10.1080/13543776.2024.2368745

Kosar M, Sarott RC, Sykes DA, Viray AEG, Vitale RM, Tomasevic N et al (2024b) Flipping the GPCR switch: structure-based development of selective cannabinoid receptor 2 inverse agonists. ACS Cent Sci 10(5):956–968. https://doi.org/10.1021/acscentsci.3c01461

Krishna Kumar K, Shalev-Benami M, Robertson MJ, Hu H, Banister SD, Hollingsworth SA et al (2019) Structure of a signaling cannabinoid receptor 1-G protein complex. Cell 176(3): 448–458.e12. https://doi.org/10.1016/j.cell.2018.11.040

Lago-Fernandez A, Zarzo-Arias S, Jagerovic N, Morales P (2021) Relevance of peroxisome proliferator activated receptors in multitarget paradigm associated with the endocannabinoid system. Int J Mol Sci 22(3):1001. https://doi.org/10.3390/ijms22031001

Langstein J, Hofstädter F, Schwarz H (1996) Cis-9,10-octadecenoamide, an endogenous sleep-inducing CNS compound, inhibits lymphocyte proliferation. Res Immunol 147(6):389–396. https://doi.org/10.1016/0923-2494(96)82047-5

Laprairie RB, Bagher AM, Kelly MEM, Denovan-Wright EM (2016) Biased type 1 cannabinoid receptor signaling influences neuronal viability in a cell culture model of Huntington disease. Mol Pharmacol 89(3):364–375. https://doi.org/10.1124/mol.115.101980

Lefort C, Roumain M, Van Hul M, Rastelli M, Manco R, Leclercq I et al (2020) Hepatic NAPE-PLD is a key regulator of liver lipid metabolism. Cells 9(5):1247. https://doi.org/10.3390/cells9051247

Leo LM, Abood ME (2021) CB1 cannabinoid receptor signaling and biased signaling. Molecules 26(17):5413. https://doi.org/10.3390/molecules26175413

Li X, Hua T, Vemuri K, Ho J-H, Wu Y, Wu L et al (2019) Crystal structure of the human cannabinoid receptor CB2. Cell 176(3):459–467.e13. https://doi.org/10.1016/j.cell.2018.12.011

Liang L, Takamiya R, Miki Y, Heike K, Taketomi Y, Sugimoto N et al (2022) Group IVE cytosolic phospholipase A_2 limits psoriatic inflammation by mobilizing the anti-inflammatory lipid *N*-acylethanolamine. FASEB J 36(5). https://doi.org/10.1096/fj.202101958r

Lilly (2010, 10–13 January 2010). Company pipeline. Presented at: JP Morgan 28th annual healthcare conference. (Paper presented at the JP Morgan 28th Annual Healthcare Conference, San Francisco, CA, USA)

Lin X, Dhopeshwarkar AS, Huibregtse M, Mackie K, Hohmann AG (2018) Slowly signaling G protein–biased CB_2Cannabinoid receptor agonist LY2828360 suppresses neuropathic pain with sustained efficacy and attenuates morphine tolerance and dependence. Mol Pharmacol 93(2): 49–62. https://doi.org/10.1124/mol.117.109355

Linhorst A, Lübke T (2022) The human Ntn-hydrolase superfamily: structure, functions and perspectives. Cells 11(10):1592. https://doi.org/10.3390/cells11101592

Lo Verme J, Fu J, Astarita G, La Rana G, Russo R, Calignano A et al (2005) The nuclear receptor peroxisome proliferator-activated receptor-α mediates the anti-inflammatory actions of palmitoylethanolamide. Mol Pharmacol 67(1):15–19. https://doi.org/10.1124/mol.104.006353

Long JZ, Li W, Booker L, Burston JJ, Kinsey SG, Schlosburg JE et al (2009) Selective blockade of 2-arachidonoylglycerol hydrolysis produces cannabinoid behavioral effects. Nat Chem Biol 5(1):37–44. https://doi.org/10.1038/nchembio.129

Lu Y, Anderson HD (2017) Cannabinoid signaling in health and disease. Can J Physiol Pharmacol 95(4):311–327

Lucero MY, Gardner SH, Yadav AK, Borri A, Zhao Z, Chan J (2022) Activity-based photoacoustic probes reveal elevated intestinal MGL and FAAH activity in a murine model of obesity. Angew Chem Int Ed 61(44). https://doi.org/10.1002/anie.202211774

Ma M, Bai J, Ling Y, Chang W, Xie G, Li R et al (2016) Monoacylglycerol lipase inhibitor JZL184 regulates apoptosis and migration of colorectal cancer cells. Mol Med Rep 13(3):2850–2856. https://doi.org/10.3892/mmr.2016.4829

Maccarrone M (2020) New tools to interrogate endocannabinoid signalling: from natural compounds to synthetic drugs. The Royal Society of Chemistry

Maccarrone M, Dainese E, Oddi S (2010) Intracellular trafficking of anandamide: new concepts for signaling. Trends Biochem Sci 35(11):601–608. https://doi.org/10.1016/j.tibs.2010.05.008

Maccarrone M, Bab I, Bíró T, Cabral GA, Dey SK, Di Marzo V et al (2015) Endocannabinoid signaling at the periphery: 50 years after THC. Trends Pharmacol Sci 36(5):277–296

Maccarrone M, Di Marzo V, Gertsch J, Grether U, Howlett AC, Hua T et al (2023) Goods and Bads of the endocannabinoid system as a therapeutic target: lessons learned after 30 years. Pharmacol Rev 75(5):885–958. https://doi.org/10.1124/pharmrev.122.000600

Mach L, Omran A, Bouma J, Radetzki S, Sykes DA, Guba W et al (2024) Highly selective drug-derived fluorescent probes for the cannabinoid receptor type 1 (CB1R). J Med Chem 67(14): 11841–11867. https://doi.org/10.1021/acs.jmedchem.4c00465

Mackie K (2008) Signaling via CNS cannabinoid receptors. Mol Cell Endocrinol 286(1–2):S60–S65

Mäder P, Bartholomäus R, Nicolussi S, Baumann A, Weis M, Chicca A et al (2021) Synthesis and biological evaluation of endocannabinoid uptake inhibitors derived from WOBE437. ChemMedChem 16(1):145–154. https://doi.org/10.1002/cmdc.202000153

Magotti P, Bauer I, Igarashi M, Babagoli M, Marotta R, Piomelli D et al (2015) Structure of human N-acylphosphatidylethanolamine-hydrolyzing phospholipase D: regulation of fatty acid Ethanolamide biosynthesis by bile acids. Structure 23(3):598–604. https://doi.org/10.1016/j.str.2014.12.018

Malamas MS, Farah SI, Lamani M, Pelekoudas DN, Perry NT, Rajarshi G et al (2020) Design and synthesis of cyanamides as potent and selective N-acylethanolamine acid amidase inhibitors. Bioorg Med Chem 28(1):115195. https://doi.org/10.1016/j.bmc.2019.115195

Mallat A, Lotersztajn S (2008) Endocannabinoids and liver disease. I. Endocannabinoids and their receptors in the liver. Am J Physiol-Gastrointest Liver Physiol 294(1):G9–G12. https://doi.org/10.1152/ajpgi.00467.2007

Mallat A, Teixeira-Clerc F, Lotersztajn S (2013) Cannabinoid signaling and liver therapeutics. J Hepatol 59(4):891–896. https://doi.org/10.1016/j.jhep.2013.03.032

Manera C, Arena C, Chicca A (2015) Synthetic cannabinoid receptor agonists and antagonists: implication in CNS disorders. Recent Pat CNS Drug Discov 10(2):142–156. https://doi.org/10.2174/1574889810666160519113853

Manning JJ, Green HM, Glass M, Finlay DB (2021) Pharmacological selection of cannabinoid receptor effectors: signalling, allosteric modulation and bias. Neuropharmacology 193:108611. https://doi.org/10.1016/j.neuropharm.2021.108611

Marchalant Y, Brownjohn PW, Bonnet A, Kleffmann T, Ashton JC (2014) Validating antibodies to the cannabinoid CB2 receptor: antibody sensitivity is not evidence of antibody specificity. J Histochem Cytochem 62(6):395. https://doi.org/10.1369/0022155414530995

Marrs WR, Blankman JL, Horne EA, Thomazeau A, Lin YH, Coy J et al (2010) The serine hydrolase ABHD6 controls the accumulation and efficacy of 2-AG at cannabinoid receptors. Nat Neurosci 13(8):951–957. https://doi.org/10.1038/nn.2601

Marsicano G, Lutz B (2006) Neuromodulatory functions of the endocannabinoid system. J Endocrinol Invest 29(3):27

Martella A, Sijben H, Rufer AC, Grether U, Fingerle J, Ullmer C et al (2017) A novel selective inverse agonist of the CB2 receptor as a radiolabeled tool compound for kinetic binding studies. Mol Pharmacol 92(4):389–400. https://doi.org/10.1124/mol.117.108605

Martin GG, Landrock D, Chung S, Dangott LJ, Seeger DR, Murphy EJ et al (2017) *Fabp1* gene ablation inhibits high-fat diet-induced increase in brain endocannabinoids. J Neurochem 140(2): 294–306. https://doi.org/10.1111/jnc.13890

Martín-Couce L, Martín-Fontecha M, Palomares Ó, Mestre L, Cordomí A, Hernangomez M et al (2012) Chemical probes for the recognition of cannabinoid receptors in native systems. Angew Chem Int Ed 51(28):6896–6899. https://doi.org/10.1002/anie.201200467

Matsuda LA, Lolait SJ, Brownstein MJ, Young AC, Bonner TI (1990) Structure of a cannabinoid receptor and functional expression of the cloned cDNA. Nature 346(6284):561–564

Matyas C, Erdelyi K, Trojnar E, Zhao S, Varga ZV, Paloczi J et al (2020) Interplay of liver–heart inflammatory axis and Cannabinoid 2 receptor signaling in an experimental model of hepatic cardiomyopathy. Hepatology 71(4):1391–1407. https://doi.org/10.1002/hep.30916

McAllister LA, Butler CR, Mente S, O'Neil SV, Fonseca KR, Piro JR et al (2018) Discovery of trifluoromethyl glycol carbamates as potent and selective covalent monoacylglycerol lipase (MAGL) inhibitors for treatment of neuroinflammation. J Med Chem 61(7):3008–3026. https://doi.org/10.1021/acs.jmedchem.8b00070

McKinney MK, Cravatt BF (2005) Structure and function of fatty acid amide hydrolase. Annu Rev Biochem 74(1):411–432

Mechoulam R, Gaoni Y (1965) A total synthesis of dl-Δ1-tetrahydrocannabinol, the active constituent of hashish1. J Am Chem Soc 87(14):3273–3275

Mechoulam R, Shvo Y (1963) Hashish? I. Tetrahedron 19(12):2073–2078

Mechoulam R, Ben-Shabat S, Hanus L, Ligumsky M, Kaminski NE, Schatz AR et al (1995) Identification of an endogenous 2-monoglyceride, present in canine gut, that binds to cannabinoid receptors. Biochem Pharmacol 50(1):83–90. https://doi.org/10.1016/0006-2952(95)00109-d

Mehrpouya-Bahrami P, Chitrala KN, Ganewatta MS, Tang C, Murphy EA, Enos RT et al (2017) Blockade of CB1 cannabinoid receptor alters gut microbiota and attenuates inflammation and diet-induced obesity. Sci Rep 7(1). https://doi.org/10.1038/s41598-017-15154-6

Melvin LS, Milne GM, Johnson MR, Subramaniam B, Wilken GH, Howlett AC (1993) Structure-activity relationships for cannabinoid receptor-binding and analgesic activity: studies of bicyclic cannabinoid analogs. Mol Pharmacol 44(5):1008–1015. https://molpharm.aspetjournals.org/content/molpharm/44/5/1008.full.pdf

Melvin LS, Milne GM, Johnson MR, Wilken GH, Howlett AC (1995) Structure-activity relationships defining the ACD-tricyclic cannabinoids: cannabinoid receptor binding and analgesic activity. Drug Des Discov 13(2):155–166. http://europepmc.org/abstract/MED/8872458

Meyers MJ, Long SA, Pelc MJ, Wang JL, Bowen SJ, Schweitzer BA et al (2011) Discovery of novel spirocyclic inhibitors of fatty acid amide hydrolase (FAAH). Part 2. Discovery of 7-azaspiro[3.5]nonane urea PF-04862853, an orally efficacious inhibitor of fatty acid amide hydrolase (FAAH) for pain. Bioorg Med Chem Lett 21(21):6545–6553. https://doi.org/10.1016/j.bmcl.2011.08.048

Migliore M, Pontis S, Fuentes De Arriba AL, Realini N, Torrente E, Armirotti A et al (2016) Second-generation non-covalent NAAA inhibitors are protective in a model of multiple sclerosis. Angew Chem 128(37):11359–11363. https://doi.org/10.1002/ange.201603746

Milano WC, Wild KD, Hui Y, Hubbell CL, Reid LD (1988) PCP, THC, ethanol, and morphine and consumption of palatable solutions. Pharmacol Biochem Behav 31(4):893–897

Milman G, Maor Y, Abu-Lafi S, Horowitz M, Gallily R, Batkai S et al (2006) *N*-arachidonoyl L-serine, an endocannabinoid-like brain constituent with vasodilatory properties. Proc Natl Acad Sci 103(7):2428–2433. https://doi.org/10.1073/pnas.0510676103

Mock ED, Mustafa M, Gunduz-Cinar O, Cinar R, Petrie GN, Kantae V et al (2020) Discovery of a NAPE-PLD inhibitor that modulates emotional behavior in mice. Nat Chem Biol 16(6):667–675. https://doi.org/10.1038/s41589-020-0528-7

Mor M, Rivara S, Lodola A, Plazzi PV, Tarzia G, Duranti A et al (2004) Cyclohexylcarbamic acid 3′- or 4‘-substituted biphenyl-3-yl esters as fatty acid amide hydrolase inhibitors: synthesis, quantitative structure−activity relationships, and molecular modeling studies. J Med Chem 47(21):4998–5008. https://doi.org/10.1021/jm031140x

Morales P, Reggio PH (2017) An update on non-CB1, non-CB2 cannabinoid related G-protein-coupled receptors. Cannabis Cannabinoid Res 2(1):265–273

Morales P, Hernandez-Folgado L, Goya P, Jagerovic N (2016) Cannabinoid receptor 2 (CB2) agonists and antagonists: a patent update. Expert Opin Ther Pat 26(7):843–856. https://doi.org/10.1080/13543776.2016.1193157

Morales P, Goya P, Jagerovic N (2018) Emerging strategies targeting CB2 cannabinoid receptor: biased agonism and allosterism. Biochem Pharmacol 157:8–17. https://doi.org/10.1016/j.bcp.2018.07.031

Morales P, Lago-Fernandez A, Hurst DP, Sotudeh N, Brailoiu E, Reggio PH et al (2020) Therapeutic exploitation of GPR18: beyond the cannabinoids? Miniperspective. J Med Chem 63(23):14216–14227

Muir RK, Guerra M, Bogyo MM (2022) Activity-based diagnostics: recent advances in the development of probes for use with diverse detection modalities. ACS Chem Biol 17(2):281–291. https://doi.org/10.1021/acschembio.1c00753

Mukhopadhyay P, Pan H, Rajesh M, Bátkai S, Patel V, Harvey-White J et al (2010a) CB_1 cannabinoid receptors promote oxidative/nitrosative stress, inflammation and cell death in a murine nephropathy model. Br J Pharmacol 160(3):657–668. https://doi.org/10.1111/j.1476-5381.2010.00769.x

Mukhopadhyay P, Rajesh M, Pan H, Patel V, Mukhopadhyay B, Bátkai S et al (2010b) Cannabinoid-2 receptor limits inflammation, oxidative/nitrosative stress, and cell death in nephropathy. Free Radic Biol Med 48(3):457–467. https://doi.org/10.1016/j.freeradbiomed.2009.11.022

Mulder J, Zilberter M, Pasquare SJ, Alpar A, Schulte G, Ferreira SG et al (2011) Molecular reorganization of endocannabinoid signalling in Alzheimer's disease. Brain 134(4):1041–1060

Muller C, Morales P, Reggio PH (2019) Cannabinoid ligands targeting TRP channels. Front Mol Neurosci 11. https://doi.org/10.3389/fnmol.2018.00487

Müller-Vahl KR, Fremer C, Beals C, Ivkovic J, Loft H, Schindler C (2021) Monoacylglycerol lipase inhibition in Tourette syndrome: a 12-week, randomized, controlled study. Mov Disord 36(10):2413–2418. https://doi.org/10.1002/mds.28681

Müller-Vahl KR, Szejko N, Verdellen C, Roessner V, Hoekstra PJ, Hartmann A et al (2022) European clinical guidelines for Tourette syndrome and other tic disorders: summary statement. Eur Child Adolesc Psychiatry 31(3):377–382. https://doi.org/10.1007/s00787-021-01832-4

Munro S, Thomas KL, Abu-Shaar M (1993) Molecular characterization of a peripheral receptor for cannabinoids. Nature 365(6441):61–65

Nackley AG, Makriyannis A, Hohmann AG (2003) Selective activation of cannabinoid CB2 receptors suppresses spinal fos protein expression and pain behavior in a rat model of inflammation. Neuroscience 119(3):747–757. https://doi.org/10.1016/s0306-4522(03)00126-x

Nawata Y, Yamaguchi T, Fukumori R, Yamamoto T (2019) Inhibition of Monoacylglycerol lipase reduces the reinstatement of methamphetamine-seeking and anxiety-like behaviors in methamphetamine self-administered rats. Int J Neuropsychopharmacol 22(2):165–172. https://doi.org/10.1093/ijnp/pyy086

Nettekoven M, Adam JM, Bendels S, Bissantz C, Fingerle J, Grether U et al (2016) Novel triazolopyrimidine-derived cannabinoid receptor 2 agonists as potential treatment for inflammatory kidney diseases. ChemMedChem 11(2):179–189. https://doi.org/10.1002/cmdc.201500218

Nicolussi S, Gertsch J (2015) Endocannabinoid transport revisited. Vitam Horm 98:441–485

Nicolussi S, Chicca A, Rau M, Rihs S, Soeberdt M, Abels C et al (2014a) Correlating FAAH and anandamide cellular uptake inhibition using N-alkylcarbamate inhibitors: from ultrapotent to hyperpotent. Biochem Pharmacol 92(4):669–689. https://doi.org/10.1016/j.bcp.2014.09.020

Nicolussi S, Viveros-Paredes JM, Gachet MS, Rau M, Flores-Soto ME, Blunder M et al (2014b) Guineensine is a novel inhibitor of endocannabinoid uptake showing cannabimimetic behavioral effects in BALB/c mice. Pharmacol Res 80:52–65. https://doi.org/10.1016/j.phrs.2013.12.010

Niphakis MJ, Cognetta AB III, Chang JW, Buczynski MW, Parsons LH, Byrne F et al (2013) Evaluation of NHS carbamates as a potent and selective class of endocannabinoid hydrolase inhibitors. ACS Chem Nerosci 4(9):1322–1332. https://doi.org/10.1021/cn400116z

Nomura DK, Long JZ, Niessen S, Hoover HS, Ng S-W, Cravatt BF (2010) Monoacylglycerol lipase regulates a fatty acid network that promotes cancer pathogenesis. Cell 140(1):49–61. https://doi.org/10.1016/j.cell.2009.11.027

Nomura DK, Morrison BE, Blankman JL, Long JZ, Kinsey SG, Marcondes MCG et al (2011) Endocannabinoid hydrolysis generates brain prostaglandins that promote neuroinflammation. Science 334(6057):809–813. https://doi.org/10.1126/science.1209200

Oddi S, Fezza F, Pasquariello N, De Simone C, Rapino C, Dainese E et al (2008) Evidence for the intracellular accumulation of anandamide in adiposomes. Cell Mol Life Sci 65(5):840–850. https://doi.org/10.1007/s00018-008-7494-7

Oddi S, Fezza F, Pasquariello N, D'Agostino A, Catanzaro G, De Simone C et al (2009) Molecular identification of albumin and Hsp70 as cytosolic anandamide-binding proteins. Chem Biol 16(6):624–632. https://doi.org/10.1016/j.chembiol.2009.05.004

Oddi S, Fezza F, Catanzaro G, De Simone C, Pucci M, Piomelli D et al (2010) Pitfalls and solutions in assaying anandamide transport in cells. J Lipid Res 51(8):2435–2444. https://doi.org/10.1194/jlr.d004176

Ofek O, Karsak M, Leclerc N, Fogel M, Frenkel B, Wright K et al (2006) Peripheral cannabinoid receptor, CB2, regulates bone mass. Proc Natl Acad Sci 103(3):696–701. https://doi.org/10.1073/pnas.0504187103

Ogasawara D, Deng H, Viader A, Baggelaar MP, Breman A, den Dulk H et al (2016) Rapid and profound rewiring of brain lipid signaling networks by acute diacylglycerol lipase inhibition. Proc Natl Acad Sci 113(1):26–33. https://doi.org/10.1073/pnas.1522364112

Ogura Y, Parsons WH, Kamat SS, Cravatt BF (2016) A calcium-dependent acyltransferase that produces N-acyl phosphatidylethanolamines. Nat Chem Biol 12(9):669–671. https://doi.org/10.1038/nchembio.2127

Okamoto Y, Morishita J, Tsuboi K, Tonai T, Ueda N (2004) Molecular characterization of a phospholipase D generating anandamide and its congeners. J Biol Chem 279(7):5298–5305. https://doi.org/10.1074/jbc.m306642200

Ouali Alami N, Schurr C, Olde Heuvel F, Tang L, Li Q, Tasdogan A et al (2018) NF-κB activation in astrocytes drives a stage-specific beneficial neuroimmunological response in ALS. EMBO J 37(16). https://doi.org/10.15252/embj.201798697

Oyagawa CRM, De La Harpe SM, Saroz Y, Glass M, Vernall AJ, Grimsey NL (2018) Cannabinoid receptor 2 Signalling bias elicited by 2,4,6-trisubstituted 1,3,5-triazines. Front Pharmacol 9. https://doi.org/10.3389/fphar.2018.01202

Pacher P, Gao B (2008) Endocannabinoids and liver disease. III. Endocannabinoid effects on immune cells: implications for inflammatory liver diseases. Am J Physiol Gastrointest Liver Physiol 294(4):G850–G854. https://doi.org/10.1152/ajpgi.00523.2007

Pacher P, Mechoulam R (2011) Is lipid signaling through cannabinoid 2 receptors part of a protective system? Prog Lipid Res 50(2):193–211. https://doi.org/10.1016/j.plipres.2011.01.001

Pacher P, Bátkai S, Kunos G (2006) The endocannabinoid system as an emerging target of pharmacotherapy. Pharmacol Rev 58(3):389–462. https://doi.org/10.1124/pr.58.3.2

Pacher P, Mukhopadhyay P, Mohanraj R, Godlewski G, BáTkai SN, Kunos G (2008) Modulation of the endocannabinoid system in cardiovascular disease. Hypertension 52(4):601–607. https://doi.org/10.1161/hypertensionaha.105.063651

Parker CG, Pratt MR (2020) Click chemistry in proteomic investigations. Cell 180(4):605–632. https://doi.org/10.1016/j.cell.2020.01.025

Paterniti I, Impellizzeri D, Crupi R, Morabito R, Campolo M, Esposito E et al (2013) Molecular evidence for the involvement of PPAR-δ and PPAR-γ in anti-inflammatory and neuroprotective activities of palmitoylethanolamide after spinal cord trauma. J Neuroinflammation 10:20. https://doi.org/10.1186/1742-2094-10-20

Pertwee RG (1999) Pharmacology of cannabinoid receptor ligands. Curr Med Chem 6:635–664

Pertwee RG (2005) Inverse agonism and neutral antagonism at cannabinoid CB1 receptors. Life Sci 76(12). https://doi.org/10.1016/j.lfs.2004.10.025

Pertwee RG (2006) The pharmacology of cannabinoid receptors and their ligands: an overview. Int J Obes (Lond) 30(1):S13–S18

Pertwee RG (2008a) The diverse CB1 and CB2 receptor pharmacology of three plant cannabinoids: Δ9-tetrahydrocannabinol, cannabidiol and Δ9-tetrahydrocannabivarin. Br J Pharmacol 153(2): 199–215. https://doi.org/10.1038/sj.bjp.0707442

Pertwee RG (2008b) Ligands that target cannabinoid receptors in the brain: from THC to anandamide and beyond. Addict Biol 13(2):147–159. https://doi.org/10.1111/j.1369-1600.2008.00108.x

Pertwee RG (2012) Targeting the endocannabinoid system with cannabinoid receptor agonists: pharmacological strategies and therapeutic possibilities. Philos Trans R Soc B Biol Sci 367(1607):3353–3363

Pertwee R, Griffin G, Fernando S, Li X, Hill A, Makriyannis A (1995) AM630, a competitive cannabinoid receptor antagonist. Life Sci 56(23–24):1949–1955. https://doi.org/10.1016/0024-3205(95)00175-6

Pertwee RG, Howlett A, Abood ME, Alexander S, Di Marzo V, Elphick M et al (2010) International Union of Basic and Clinical Pharmacology. LXXIX. Cannabinoid receptors and their ligands: beyond CB1 and CB2. Pharmacol Rev 62(4):588–631

Petracca R, Romeo E, Baggelaar MP, Artola M, Pontis S, Ponzano S et al (2017) Novel activity-based probes for N-acylethanolamine acid amidase. Chem Commun 53(86):11810–11813. https://doi.org/10.1039/c7cc06838g

Petrosino S, Di Marzo V (2017) The pharmacology of palmitoylethanolamide and first data on the therapeutic efficacy of some of its new formulations. Br J Pharmacol 174(11):1349–1365. https://doi.org/10.1111/bph.13580

Petrosino S, Schiano Moriello A, Verde R, Allarà M, Imperatore R, Ligresti A et al (2019) Palmitoylethanolamide counteracts substance P-induced mast cell activation in vitro by stimulating diacylglycerol lipase activity. J Neuroinflammation 16(1). https://doi.org/10.1186/s12974-019-1671-5

Petrucci V, Chicca A, Glasmacher S, Paloczi J, Cao Z, Pacher P et al (2017) Pepcan-12 (RVD-hemopressin) is a CB2 receptor positive allosteric modulator constitutively secreted by adrenals and in liver upon tissue damage. Sci Rep 7(1). https://doi.org/10.1038/s41598-017-09808-8

Piomelli D, Beltramo M, Glasnapp S, Lin SY, Goutopoulos A, Xie X-Q et al (1999) Structural determinants for recognition and translocation by the anandamide transporter. Proc Natl Acad Sci 96(10):5802–5807. https://doi.org/10.1073/pnas.96.10.5802

Piomelli D, Scalvini L, Fotio Y, Lodola A, Spadoni G, Tarzia G et al (2020) N-acylethanolamine acid amidase (NAAA): structure, function, and inhibition. J Med Chem 63(14):7475–7490

Pontis S, Ribeiro A, Sasso O, Piomelli D (2016) Macrophage-derived lipid agonists of PPAR-αas intrinsic controllers of inflammation. Crit Rev Biochem Mol Biol 51(1):7–14. https://doi.org/10.3109/10409238.2015.1092944

Ponzano S, Bertozzi F, Mengatto L, Dionisi M, Armirotti A, Romeo E et al (2013) Synthesis and structure–activity relationship (SAR) of 2-methyl-4-oxo-3-oxetanylcarbamic acid esters, a class of potent N-acylethanolamine acid amidase (NAAA) inhibitors. J Med Chem 56(17): 6917–6934

Porter RF, Szczesniak A-M, Toguri JT, Gebremeskel S, Johnston B, Lehmann C et al (2019) Selective cannabinoid 2 receptor agonists as potential therapeutic drugs for the treatment of endotoxin-induced uveitis. Molecules 24(18):3338. https://doi.org/10.3390/molecules24183338

Portier M, Rinaldi-Carmona M, Pecceu F, Combes T, Poinot-Chazel C, Calandra B et al (1999) SR 144528, an antagonist for the peripheral cannabinoid receptor that behaves as an inverse agonist. J Pharmacol Exp Ther 288(2):582–589

Prevet H, Collins I (2019) Labelled chemical probes for demonstrating direct target engagement in living systems. Future Med Chem 11(10):1195–1224. https://doi.org/10.4155/fmc-2018-0370

Priestley R, Glass M, Kendall D (2017) Functional selectivity at cannabinoid receptors. Elsevier, pp 207–221. https://doi.org/10.1016/bs.apha.2017.03.005

Prokop S, Ábrányi-Balogh P, Barti B, Vámosi M, Zöldi M, Barna L et al (2021) PharmacoSTORM nanoscale pharmacology reveals cariprazine binding on islands of Calleja granule cells. *Nature*. Communications 12(1). https://doi.org/10.1038/s41467-021-26757-z

Punt JM, Van Der Vliet D, Van Der Stelt M (2022) Chemical probes to control and visualize lipid metabolism in the brain. Acc Chem Res 55(22):3205–3217. https://doi.org/10.1021/acs.accounts.2c00521

Qin H, Ruan ZH (2014) The role of monoacylglycerol lipase (MAGL) in the cancer progress. Cell Biochem Biophys 70(1):33–36. https://doi.org/10.1007/s12013-014-9899-2

Rahman SMK, Hussain Z, Morito K, Takahashi N, Sikder MM, Tanaka T et al (2022) Formation of N-acyl-phosphatidylethanolamines by cytosolic phospholipase A2ε in an ex vivo murine model of brain ischemia. Biochim Biophys Acta Mol Cell Biol Lipids 1867(12):159222. https://doi.org/10.1016/j.bbalip.2022.159222

Rajesh M, Bátkai S, Kechrid M, Mukhopadhyay P, Lee W-S, Horváth B et al (2012) Cannabinoid 1 receptor promotes cardiac dysfunction, oxidative stress, inflammation, and fibrosis in diabetic cardiomyopathy. Diabetes 61(3):716–727. https://doi.org/10.2337/db11-0477

Rajesh M, Mukhopadhyay P, Bátkai S, Arif M, Varga ZV, Mátyás C et al (2022) Cannabinoid receptor 2 activation alleviates diabetes-induced cardiac dysfunction, inflammation, oxidative stress, and fibrosis. GeroScience 44(3):1727–1741. https://doi.org/10.1007/s11357-022-00565-9

Rankin L, Fowler CJ (2020) The basal pharmacology of palmitoylethanolamide. Int J Mol Sci 21(21):7942. https://doi.org/10.3390/ijms21217942

Rau M, Nicolussi S, Chicca A, Gertsch J (2016) Assay of endocannabinoid uptake. Springer, New York, pp 191–203. https://doi.org/10.1007/978-1-4939-3539-0_20

Ravinet Trillou C, Delgorge C, Menet C, Arnone M, Soubrie P (2004) CB1 cannabinoid receptor knockout in mice leads to leanness, resistance to diet-induced obesity and enhanced leptin sensitivity. Int J Obes (Lond) 28(4):640–648

Razdan RK (1986) Structure-activity relationships in cannabinoids. Pharmacol Rev 38(2):75–149. https://pharmrev.aspetjournals.org/content/pharmrev/38/2/75.full.pdf

Resat Cinar MRI, Kunos G (2020) The therapeutic potential of second and third generation CB1R antagonists. Pharmacol Ther 208:107477

Reynoso-Moreno I, Najar-Guerrero I, Escareño N, Flores-Soto ME, Gertsch J, Viveros-Paredes JM (2017) An endocannabinoid uptake inhibitor from black pepper exerts pronounced anti-inflammatory effects in mice. J Agric Food Chem 65(43):9435–9442. https://doi.org/10.1021/acs.jafc.7b02979

Reynoso-Moreno I, Chicca A, Flores-Soto ME, Viveros-Paredes JM, Gertsch J (2018) The endocannabinoid reuptake inhibitor WOBE437 is orally bioavailable and exerts indirect polypharmacological effects via different endocannabinoid receptors. Front Mol Neurosci 11. https://doi.org/10.3389/fnmol.2018.00180

Reynoso-Moreno I, Tietz S, Vallini E, Engelhardt B, Gertsch J, Chicca A (2021) Selective endocannabinoid reuptake inhibitor WOBE437 reduces disease progression in a mouse model of multiple sclerosis. ACS Pharmacol Transl Sci 4(2):765–779. https://doi.org/10.1021/acsptsci.0c00214

Reynoso-Moreno I, Rau M, Chicca A, Nicolussi S, Gertsch J (2023) Assay of endocannabinoid uptake. Springer US, pp 329–348. https://doi.org/10.1007/978-1-0716-2728-0_28

Ribeiro A, Pontis S, Mengatto L, Armirotti A, Chiurchiù V, Capurro V et al (2015) A potent systemically active *N*-acylethanolamine acid amidase inhibitor that suppresses inflammation and human macrophage activation. ACS Chem Biol 10(8):1838–1846. https://doi.org/10.1021/acschembio.5b00114

Riether D (2012) Selective cannabinoid receptor 2 modulators: a patent review 2009 – present. Expert Opin Ther Pat 22(5):495–510. https://doi.org/10.1517/13543776.2012.682570

Rinaldi M, Barth F, Casellas P, Congy C, Oustric D, Bell MR et al (1996) Preparation of naphthylcarbonylindoles and analogs as CB2 receptor agonists. Patent FR2735774

Rinaldi-Carmona M, Barth F, Millan J, Derocq J-M, Casellas P, Congy C et al (1998) SR 144528, the first potent and selective antagonist of the CB2 cannabinoid receptor. J Pharmacol Exp Ther 284(2):644–650

Robson P (2001) Therapeutic aspects of cannabis and cannabinoids. Br J Psychiatry 178(2). https://doi.org/10.1192/bjp.178.2.107

Rockwell CE, Snider NT, Thompson JT, Vanden Heuvel JP, Kaminski NE (2006) Interleukin-2 suppression by 2-arachidonyl glycerol is mediated through peroxisome proliferator-activated receptor γ independently of cannabinoid receptors 1 and 2. Mol Pharmacol 70(1):101–111. https://doi.org/10.1124/mol.105.019117

Rouzer CA, Marnett LJ (2011) Endocannabinoid oxygenation by cyclooxygenases, lipoxygenases, and cytochromes P450: cross-talk between the eicosanoid and endocannabinoid signaling pathways. Chem Rev 111(10):5899–5921

Saghatelian A, McKinney MK, Bandell M, Patapoutian A, Cravatt BF (2006) A FAAH-regulated class of *N*-acyl Taurines that activates TRP ion channels. Biochemistry 45(30):9007–9015. https://doi.org/10.1021/bi0608008

Saleh N, Hucke O, Kramer G, Schmidt E, Montel F, Lipinski R et al (2018) Multiple binding sites contribute to the mechanism of mixed agonistic and positive allosteric modulators of the cannabinoid CB1 receptor. Angew Chem 130(10). https://doi.org/10.1002/ange.201708764

Sam AH, Salem V, Ghatei MA (2011) Rimonabant: from RIO to ban. J Obes 2011(1):432607

Sarott RC, Westphal MV, Pfaff P, Korn C, Sykes DA, Gazzi T et al (2020) Development of high-specificity fluorescent probes to enable cannabinoid type 2 receptor studies in living cells. J Am Chem Soc 142(40):16953–16964. https://doi.org/10.1021/jacs.0c05587

Sasso O, Moreno-Sanz G, Martucci C, Realini N, Dionisi M, Mengatto L et al (2013) Antinociceptive effects of the N-acylethanolamine acid amidase inhibitor ARN077 in rodent pain models. Pain 154(3):350–360. https://doi.org/10.1016/j.pain.2012.10.018

Sasso O, Summa M, Armirotti A, Pontis S, De Mei C, Piomelli D (2018) The N-Acylethanolamine acid amidase inhibitor ARN077 suppresses inflammation and pruritus in a mouse model of allergic dermatitis. J Invest Dermatol 138(3):562–569. https://doi.org/10.1016/j.jid.2017.07.853

Schenone M, Dančík V, Wagner BK, Clemons PA (2013) Target identification and mechanism of action in chemical biology and drug discovery. Nat Chem Biol 9(4):232–240. https://doi.org/10.1038/nchembio.1199

Schindler CW, Redhi GH, Vemuri K, Makriyannis A, Le Foll B, Bergman J et al (2016) Blockade of nicotine and cannabinoid reinforcement and relapse by a cannabinoid CB1-receptor neutral antagonist AM4113 and inverse agonist rimonabant in squirrel monkeys. Neuropsychopharmacology 41(9). https://doi.org/10.1038/npp.2016.27

Schlosburg JE, Blankman JL, Long JZ, Nomura DK, Pan B, Kinsey SG et al (2010) Chronic monoacylglycerol lipase blockade causes functional antagonism of the endocannabinoid system. Nat Neurosci 13(9):1113–1119. https://doi.org/10.1038/nn.2616

Schurman LD, Carper MC, Moncayo LV, Ogasawara D, Richardson K, Yu L et al (2019) Diacylglycerol lipase-alpha regulates hippocampal-dependent learning and memory processes in mice. J Neurosci 39(30):5949–5965. https://doi.org/10.1523/jneurosci.1353-18.2019

Schurman LD, Lu D, Kendall DA, Howlett AC, Lichtman AH (2020) Molecular mechanism and cannabinoid pharmacology. Substance use disorders: from etiology to treatment, pp 323–353

Seierstad M, Breitenbucher JG (2008) Discovery and development of fatty acid amide hydrolase (FAAH) inhibitors. J Med Chem 51(23)

Serhan CN, Chiang N, Van Dyke TE (2008) Resolving inflammation: dual anti-inflammatory and pro-resolution lipid mediators. Nat Rev Immunol 8(5):349–361. https://doi.org/10.1038/nri2294

Sgroi S, Romeo E, Fruscia PD, Porceddu PF, Russo D, Realini N et al (2021) Inhibition of N-acylethanolamine-hydrolyzing acid amidase reduces T cell infiltration in a mouse model of multiple sclerosis. Pharmacol Res 172:105816. https://doi.org/10.1016/j.phrs.2021.105816

Shao Z, Yin J, Chapman K, Grzemska M, Clark L, Wang J et al (2016) High-resolution crystal structure of the human CB1 cannabinoid receptor. Nature 540(7634):602–606. https://doi.org/10.1038/nature20613

Shao Z, Yan W, Chapman K, Ramesh K, Ferrell AJ, Yin J et al (2019) Structure of an allosteric modulator bound to the CB1 cannabinoid receptor. Nat Chem Biol 15(12):1199–1205. https://doi.org/10.1038/s41589-019-0387-2

Shen W, Shen M, Zhao X, Zhu H, Yang Y, Lu S et al (2017) Anti-obesity effect of capsaicin in mice fed with high-fat diet is associated with an increase in population of the gut bacterium Akkermansia muciniphila. Front Microbiol 8. https://doi.org/10.3389/fmicb.2017.00272

Shore DM, Reggio PH (2015) The therapeutic potential of orphan GPCRs, GPR35 and GPR55. Front Pharmacol 6:69

Shoup TM, Bonab AA, Wilson AA, Vasdev N (2015) Synthesis and preclinical evaluation of [18F] FCHC for neuroimaging of fatty acid amide hydrolase. Mol Imaging Biol 17(2):257–263. https://doi.org/10.1007/s11307-014-0789-1

Showalter VM, Compton DR, Martin BR, Abood ME (1996) Evaluation of binding in a transfected cell line expressing a peripheral cannabinoid receptor (CB2): identification of cannabinoid receptor subtype selective ligands. J Pharmacol Exp Ther 278(3):989–999

Simard M, Archambault A-S, Lavoie J-PC, Dumais É, Di Marzo V, Flamand N (2022) Biosynthesis and metabolism of endocannabinoids and their congeners from the monoacylglycerol and N-acyl-ethanolamine families. Biochem Pharmacol 205:115261. https://doi.org/10.1016/j.bcp.2022.115261

Sink KS, McLaughlin PJ, Wood JAT, Brown C, Fan P, Vemuri VK et al (2008) The novel cannabinoid CB1 receptor neutral antagonist AM4113 suppresses food intake and food-reinforced behavior but does not induce signs of nausea in rats. Neuropsychopharmacology 33(4):946–955. https://doi.org/10.1038/sj.npp.1301476

Sink KS, Segovia KN, Sink J, Randall PA, Collins LE, Correa M et al (2010) Potential anxiogenic effects of cannabinoid CB1 receptor antagonists/inverse agonists in rats: comparisons between AM4113, AM251, and the benzodiazepine inverse agonist FG-7142. Eur Neuropsychopharmacol 20(2). https://doi.org/10.1016/j.euroneuro.2009.11.002

Slavik R, Grether U, Muller Herde A, Gobbi L, Fingerle J, Ullmer C et al (2015) Discovery of a high affinity and selective pyridine analog as a potential positron emission tomography imaging agent for cannabinoid type 2 receptor. J Med Chem 58(10):4266–4277. https://doi.org/10.1021/acs.jmedchem.5b00283

Slivicki RA, Xu Z, Kulkarni PM, Pertwee RG, Mackie K, Thakur GA et al (2018) Positive allosteric modulation of cannabinoid receptor type 1 suppresses pathological pain without producing tolerance or dependence. Biol Psychiatry 84(10):722–733. https://doi.org/10.1016/j.biopsych.2017.06.032

Slivicki RA, Iyer V, Mali SS, Garai S, Thakur GA, Crystal JD et al (2020) Positive allosteric modulation of CB1 cannabinoid receptor signaling enhances morphine antinociception and attenuates morphine tolerance without enhancing morphine-induced dependence or reward. Front Mol Neurosci 13. https://doi.org/10.3389/fnmol.2020.00054

Smathers RL, Petersen DR (2011) The human fatty acid-binding protein family: Evolutionary divergences and functions. Hum Genomics 5(3):170. https://doi.org/10.1186/1479-7364-5-3-170

Soethoudt M, Grether U, Fingerle J, Grim TW, Fezza F, De Petrocellis L et al (2017) Cannabinoid CB2 receptor ligand profiling reveals biased signalling and off-target activity. Nat Commun 8(1):13958

Soethoudt M, Stolze SC, Westphal MV, van Stralen L, Martella A, van Rooden EJ et al (2018) Selective photoaffinity probe that enables assessment of cannabinoid CB2 receptor expression and ligand engagement in human cells. J Am Chem Soc 140(19):6067–6075. https://doi.org/10.1021/jacs.7b11281

Solorzano C, Zhu C, Battista N, Astarita G, Lodola A, Rivara S et al (2009) Selective *N*-acylethanolamine-hydrolyzing acid amidase inhibition reveals a key role for endogenous

palmitoylethanolamide in inflammation. Proc Natl Acad Sci 106(49):20966–20971. https://doi.org/10.1073/pnas.0907417106
Speers AE, Cravatt BF (2004) Profiling enzyme activities in vivo using click chemistry methods. Chem Biol 11(4):535–546. https://doi.org/10.1016/j.chembiol.2004.03.012
Spiera R, Kuwana M, Khanna D, Hummers L, Frech TM, Stevens W et al (2023) Efficacy and safety of Lenabasum, a cannabinoid type 2 receptor agonist, in a phase 3 randomized trial in diffuse cutaneous systemic sclerosis. Arthritis Rheumatol 75(9):1608–1618. https://doi.org/10.1002/art.42510
Spyridakos D, Mastrodimou N, Vemuri K, Ho TC, Nikas SP, Makriyannis A et al (2022) Blockade of CB1 or activation of CB2 cannabinoid receptors is differentially efficacious in the treatment of the early pathological events in streptozotocin-induced diabetic rats. Int J Mol Sci 24(1):240. https://doi.org/10.3390/ijms24010240
Stephan Röver MA, Bénardeau A, Chaput E, Guba W, Hebeisen P, Mohr S, Nettekoven M, Obst U, Richter WF, Ullmer C, Waldmeier P, Wright MB (2013) 6-Alkoxy-5-aryl-3-pyridinecarboxamides, a new series of bioavailable cannabinoid receptor type 1 (CB1) antagonistsincluding peripherally selective compounds. J Med Chem 56:9874–9895
Sugiura T, Kondo S, Sukagawa A, Nakane S, Shinoda A, Itoh K et al (1995) 2-Arachidonoylglycerol: a possible endogenous cannabinoid receptor ligand in brain. Biochem Biophys Res Commun 215(1):89–97. https://doi.org/10.1006/bbrc.1995.2437
Sulsky R, Magnin DR, Huang Y, Simpkins L, Taunk P, Patel M et al (2007) Potent and selective biphenyl azole inhibitors of adipocyte fatty acid binding protein (aFABP). Bioorg Med Chem Lett 17(12):3511–3515. https://doi.org/10.1016/j.bmcl.2006.12.044
Tam J, Cinar R, Liu J, Godlewski G, Wesley D, Jourdan T et al (2012) Peripheral Cannabinoid-1 receptor inverse agonism reduces obesity by reversing leptin resistance. Cell Metab 16(2). https://doi.org/10.1016/j.cmet.2012.07.002
Tanimura A, Yamazaki M, Hashimotodani Y, Uchigashima M, Kawata S, Abe M et al (2010) The endocannabinoid 2-arachidonoylglycerol produced by diacylglycerol lipase α mediates retrograde suppression of synaptic transmission. Neuron 65(3):320–327. https://www.sciencedirect.com/science/article/pii/S0896627310000474?via%3Dihub
Terry GE, Hirvonen J, Liow J-S, Zoghbi SS, Gladding R, Tauscher JT et al (2010) Imaging and quantitation of cannabinoid CB1 receptors in human and monkey brains using 18F-labeled inverse agonist radioligands. J Nucl Med 51(1):112–120. https://doi.org/10.2967/jnumed.109.067074
Thakur GA, Tichkule R, Bajaj S, Makriyannis A (2009) Latest advances in cannabinoid receptor agonists. Expert Opin Ther Pat 19(12):1647–1673. https://doi.org/10.1517/13543770903436505
Tripathi RKP (2020) A perspective review on fatty acid amide hydrolase (FAAH) inhibitors as potential therapeutic agents. Eur J Med Chem 188:111953. https://doi.org/10.1016/j.ejmech.2019.111953
Trojnar E, Erdelyi K, Matyas C, Zhao S, Paloczi J, Mukhopadhyay P et al (2020) Cannabinoid-2 receptor activation ameliorates hepatorenal syndrome. Free Radic Biol Med 152:540–550. https://doi.org/10.1016/j.freeradbiomed.2019.11.027
Tsang CC, Giudice MG (2016) Nabilone for the management of pain. Pharmacotherapy 36(3). 273–286
Tsuboi K, Takezaki N, Ueda N (2007a) The *N*-acylethanolamine-hydrolyzing acid amidase (NAAA). Chem Biodivers 4(8):1914–1925. https://doi.org/10.1002/cbdv.200790159
Tsuboi K, Zhao L-Y, Okamoto Y, Araki N, Ueno M, Sakamoto H et al (2007b) Predominant expression of lysosomal N-acylethanolamine-hydrolyzing acid amidase in macrophages revealed by immunochemical studies. Biochim Biophys Acta 1771(5):623–632. https://doi.org/10.1016/j.bbalip.2007.03.005
Ueda N, Tsuboi K, Uyama T (2010) N-acylethanolamine metabolism with special reference to N-acylethanolamine-hydrolyzing acid amidase (NAAA). Prog Lipid Res 49(4):299–315. https://doi.org/10.1016/j.plipres.2010.02.003

Uyama T, Ikematsu N, Inoue M, Shinohara N, Jin X-H, Tsuboi K et al (2012) Generation of N-acylphosphatidylethanolamine by members of the phospholipase A/acyltransferase (PLA/AT) family. J Biol Chem 287(38):31905–31919. https://doi.org/10.1074/jbc.m112.368712

Uyama T, Tsuboi K, Ueda N (2017) An involvement of phospholipase A/acyltransferase family proteins in peroxisome regulation and plasmalogen metabolism. FEBS Lett 591(18):2745–2760. https://doi.org/10.1002/1873-3468.12787

Valenzano KJ, Tafesse L, Lee G, Harrison JE, Boulet JM, Gottshall SL et al (2005) Pharmacological and pharmacokinetic characterization of the cannabinoid receptor 2 agonist, GW405833, utilizing rodent models of acute and chronic pain, anxiety, ataxia and catalepsy. Neuropharmacology 48(5):658–672. https://doi.org/10.1016/j.neuropharm.2004.12.008

van der Stelt M, van Kuik JA, Bari M, van Zadelhoff G, Leeflang BR, Veldink GA et al (2002) Oxygenated metabolites of anandamide and 2-arachidonoylglycerol: conformational analysis and interaction with cannabinoid receptors, membrane transporter, and fatty acid amide hydrolase. J Med Chem 45(17):3709–3720

Van Egmond N, Straub VM, van der Stelt M (2021) Targeting endocannabinoid signaling: FAAH and MAG lipase inhibitors. Annu Rev Pharmacol Toxicol 61:441–463. https://doi.org/10.1146/annurev-pharmtox-030220-112741

Van Esbroeck ACM, Janssen APA, Cognetta AB, Ogasawara D, Shpak G, van der Kroeg M et al (2017) Activity-based protein profiling reveals off-target proteins of the FAAH inhibitor BIA 10-2474. Science 356(6342):1084–1087. https://doi.org/10.1126/science.aaf7497

Van Gaal LF, Rissanen AM, Scheen AJ, Ziegler O, Rössner S (2005) Effects of the cannabinoid-1 receptor blocker rimonabant on weight reduction and cardiovascular risk factors in overweight patients: 1-year experience from the RIO-Europe study. Lancet 365(9468):1389–1397

Van Rooden EJ, Kreekel R, Hansen T, Janssen APA, van Esbroeck ACM, den Dulk H et al (2018) Two-step activity-based protein profiling of diacylglycerol lipase. Org Biomol Chem 16(29):5250–5253. https://doi.org/10.1039/C8OB01499J

Varlow C, Boileau I, Wey H-Y, Liang SH, Vasdev N (2020) Classics in neuroimaging: imaging the endocannabinoid pathway with PET. ACS Chem Nerosci 11(13):1855–1862. https://doi.org/10.1021/acschemneuro.0c00305

Veldhuis WB, Van Der Stelt M, Wadman MW, Van Zadelhoff G, Maccarrone M, Fezza F et al (2003) Neuroprotection by the endogenous cannabinoid anandamide and Arvanil against *in vivo* excitotoxicity in the rat: role of vanilloid receptors and lipoxygenases. J Neurosci 23(10):4127–4133. https://doi.org/10.1523/jneurosci.23-10-04127.2003

Verhoeckx KCM, Voortman T, Balvers MGJ, Hendriks HFJ, Wortelboer HM, Witkamp RF (2011) Presence, formation and putative biological activities of N-acyl serotonins, a novel class of fatty-acid derived mediators, in the intestinal tract. Biochim Biophys Acta 1811(10):578–586. https://doi.org/10.1016/j.bbalip.2011.07.008

Wang Y, Casero RA Jr (2006) Mammalian polyamine catabolism: a therapeutic target, a pathological problem, or both? J Biochem 139(1):17–25. https://doi.org/10.1093/jb/mvj021

Wangensteen T, Akselsen H, Holmen J, Undlien D, Retterstøl L (2011) A common haplotype in *NAPEPLD* is associated with severe obesity in a Norwegian population-based cohort (the HUNT study). Obesity 19(3):612–617. https://doi.org/10.1038/oby.2010.219

Wasinka-Kalwa M, Omran A, Mach L, Bouma J, Scipioni L, Li X et al (2024) Visualization of membrane localization and functional state of CB2R pools by matched agonist and inverse agonist probe pairs. American Chemical Society. https://doi.org/10.26434/chemrxiv-2024-q741f

Westphal MV, Sarott RC, Zirwes EA, Osterwald A, Guba W, Ullmer C et al (2020) Highly selective, amine-derived cannabinoid receptor 2 probes. Chem A Eur J 26(6):1380–1387. https://doi.org/10.1002/chem.201904584

Wilkerson JL, Bilbrey JA, Felix JS, Makriyannis A, McMahon LR (2021) Untapped endocannabinoid pharmacological targets: pipe dream or pipeline? Pharmacol Biochem Behav 206:173192

Willmann JK, van Bruggen N, Dinkelborg LM, Gambhir SS (2008) Molecular imaging in drug development. Nat Rev Drug Discov 7(7):591–607. https://doi.org/10.1038/nrd2290

Wong DF, Kuwabara H, Horti A, Kumar A, Brasic J, Ye W et al (2008) PET imaging of cannabinoid CB1 type receptors in healthy humans and patients with schizophrenia using [11C]OMAR. Neuroimage 41:T51. https://doi.org/10.1016/j.neuroimage.2008.04.225

Wong DF, Kuwabara H, Horti AG, Raymont V, Brasic J, Guevara M et al (2010) Quantification of cerebral cannabinoid receptors subtype 1 (CB1) in healthy subjects and schizophrenia by the novel PET radioligand [11C]OMAR. Neuroimage 52(4):1505–1513. https://doi.org/10.1016/j.neuroimage.2010.04.034

Workman P, Collins I (2010) Probing the probes: fitness factors for small molecule tools. Chem Biol 17(6):561–577. https://doi.org/10.1016/j.chembiol.2010.05.013

Wright K, Rooney N, Feeney M, Tate J, Robertson D, Welham M et al (2005) Differential expression of cannabinoid receptors in the human colon: cannabinoids promote epithelial wound healing. Gastroenterology 129(2):437–453. https://doi.org/10.1053/j.gastro.2005.05.026

Wright KL, Duncan M, Sharkey KA (2008) Cannabinoid CB_2 receptors in the gastrointestinal tract: a regulatory system in states of inflammation. Br J Pharmacol 153(2):263–270. https://doi.org/10.1038/sj.bjp.0707486

Wu S, Liao D, Li X, Liu Z, Zhang L, Mo FM et al (2021) Endogenous oleoylethanolamide crystals loaded lipid nanoparticles with enhanced hydrophobic drug loading capacity for efficient stroke therapy. Int J Nanomedicine:8103–8115

Xie X, Li Y, Xu S, Zhou P, Yang L, Xu Y et al (2022) Genetic blockade of NAAA cell-specifically regulates fatty acid ethanolamides (FAEs) metabolism and inflammatory responses. Front Pharmacol 12. https://doi.org/10.3389/fphar.2021.817603

Xing C, Zhuang Y, Xu T-H, Feng Z, Zhou XE, Chen M et al (2020) Cryo-EM structure of the human cannabinoid receptor CB2-Gi signaling complex. Cell 180(4):645–654.e13. https://doi.org/10.1016/j.cell.2020.01.007

Yu S, Levi L, Casadesus G, Kunos G, Noy N (2014) Fatty acid-binding protein 5 (FABP5) regulates cognitive function both by decreasing anandamide levels and by activating the nuclear receptor peroxisome proliferator-activated receptor β/δ (PPARβ/δ) in the brain. J Biol Chem 289(18):12748–12758. https://doi.org/10.1074/jbc.m114.559062

Zanfirescu A, Ungurianu A, Mihai DP, Radulescu D, Nitulescu GM (2021) Targeting monoacylglycerol lipase in pursuit of therapies for neurological and neurodegenerative diseases. https://mdpi-res.com/d_attachment/molecules/molecules-26-05668/article_deploy/molecules-26-05668-v2.pdf?version=1632292879

Zhang Y-M, Greco MN, Macielag MJ, Teleha CA, DesJarlais RL, Tang Y et al (2018) 6-Benzhydryl-4-amino-quinolin-2-ones as potent cannabinoid type 1 (CB1) receptor inverse agonists and chemical modifications for peripheral selectivity. J Med Chem 61(22). https://doi.org/10.1021/acs.jmedchem.8b01467

Zhang L, Butler CR, Maresca KP, Takano A, Nag S, Jia Z et al (2019) Identification and development of an irreversible monoacylglycerol lipase (MAGL) positron emission tomography (PET) radioligand with high specificity. J Med Chem 62(18):8532–8543. https://doi.org/10.1021/acs.jmedchem.9b00847

Zhao P, Abood ME (2013) GPR55 and GPR35 and their relationship to cannabinoid and lysophospholipid receptors. Life Sci 92(8–9):453–457

Zhao L-Y, Tsuboi K, Okamoto Y, Nagahata S, Ueda N (2007) Proteolytic activation and glycosylation of N-acylethanolamine-hydrolyzing acid amidase, a lysosomal enzyme involved in the endocannabinoid metabolism. Biochim Biophys Acta 1771(11):1397–1405. https://doi.org/10.1016/j.bbalip.2007.10.002

Ziring D, Wei B, Velazquez P, Schrage M, Buckley NE, Braun J (2006) Formation of B and T cell subsets require the cannabinoid receptor CB2. Immunogenetics 58(9):714–725. https://doi.org/10.1007/s00251-006-0138-x

Zoja C, Locatelli M, Corna D, Villa S, Rottoli D, Nava V et al (2016) Therapy with a selective cannabinoid receptor type 2 agonist limits albuminuria and renal injury in mice with type 2 diabetic nephropathy. Nephron 132(1):59–69. https://doi.org/10.1159/000442679

Genetic Tools in Rodents to Study Cannabinoid Functions

Krisztina Monory, Inigo Ruiz de Azua, and Beat Lutz

Contents

1 Introduction 146
2 Genetically Engineered Rodent Models 147
2.1 Random Integration of Genes 148
2.2 Targeted Integration of Genes 148
2.3 Site-Specific Recombination Systems 149
2.4 Targeted Genome Editing 153
2.5 Viral Tools 155
3 Understanding the Endocannabinoid System with Genetically Engineered Rodent Models 159
3.1 Diverse Neuronal Types 163
3.2 Glial Cells 173
3.3 Cells Outside the Nervous System 179
3.4 Distinct Expression Sites within the Cell 181
4 Concluding Remarks 183
References 184

The original version of the chapter has been revised. A correction to this chapter can be found at https://doi.org/10.1007/7854_2025_584

K. Monory
Institute of Physiological Chemistry, University Medical Center of the Johannes Gutenberg University, Mainz, Germany
e-mail: monory@uni-mainz.de

I. Ruiz de Azua
Leibniz Institute for Resilience Research (LIR) gGmbH, Mainz, Germany
e-mail: inigo.azua@lir-mainz.de

B. Lutz (✉)
Institute of Physiological Chemistry, University Medical Center of the Johannes Gutenberg University, Mainz, Germany

Leibniz Institute for Resilience Research (LIR) gGmbH, Mainz, Germany
e-mail: blutz@uni-mainz.de

Curr Topics Behav Neurosci (2026) 76: 145–207
https://doi.org/10.1007/7854_2024_550
Published Online: 17 December 2024

Abstract During the past 30 years, the endocannabinoid system (ECS) has emerged as a major signalling system in the mammalian brain regulating neurotransmission in numerous brain regions and in various cell populations. Endocannabinoids are able to regulate specific physiological functions and thus modify their behavioural manifestations and allostatic alterations of the ECS linked to different pathological conditions. As discussed in detail in other chapters of this book, endocannabinoids are involved in learning and memory, stress, and anxiety, feeding, energy balance, development, and ageing. Likewise, many CNS disorders (e.g. schizophrenia, epilepsy, substance use disorders, and multiple sclerosis) are associated with dysregulation of the ECS. Discerning the physiological functions of the synthetic and degrading enzymes of endocannabinoids and their receptors is a challenging task because of their distinct and complex expression patterns. Techniques of genetic engineering have been able to shed light on a number of complex ECS-related tasks during the past years. In this chapter, first, we take a critical look at the toolbox available to researchers who would like to investigate cannabinoid effects using genetic engineering techniques, then we comprehensively discuss genetically modified rodent models in various neuronal and non-neuronal cell populations, both within and outside the nervous system.

Keywords Adeno-associated virus (AAV) · Cannabinoid receptors · Cell-type-specific mutants · Endocannabinoid sensor · Endocannabinoid system · Targeted genome editing · Mouse models

1 Introduction

Research on the functions of various biological systems and on techniques of genetic manipulation and analyses are moving forward in a synergistic fashion. Ever more complex questions are creating a demand for ever more complex models to find answers to them. In the past decades, similarly to other fields, endocannabinoid research has also greatly benefited from the methodological development in fields such as microscopy, in vivo imaging, genetic tools, multi-omics platforms and bioinformatics. Particularly, understanding the intricate mechanisms governing the ECS and its impact on behaviour has been significantly augmented through the application of advanced genetic tools and transgenic animal models. In this review, we embark on a comprehensive journey through the evolving landscape of these tools and rodent models, unravelling their pivotal role in elucidating cellular and behavioural functions related to the ECS.

The remarkable progress in genetic technology has provided researchers with an array of tools to investigate the ECS with unprecedented precision. From site-specific recombination systems to genome editing and viral vector-mediated gene delivery, each methodology offers distinct advantages and considerations. Our initial section offers an insightful exploration of the complex world of genetic manipulation enabling the readers to navigate through it.

In the second part, our focus shifts to cell-type-specific mutants that have, in recent years, revealed much of how the complex tapestry of the ECS influences mammalian physiology. The presented studies show how the ECS extends its reach across diverse neuronal populations, influencing memory, emotional states, homeostasis, and neurogenesis. Beyond neurons, glial cell-specific mutants and those targeting peripheral tissues shed light on the ECS's roles in neuroinflammation, metabolism, and more. Delving deeper, we explore CB1 residing within different cellular compartments, providing perspective on its contributions to endocannabinoid functionality and how all these expression sites eventually shape behaviour.

2 Genetically Engineered Rodent Models

Transgenic technology has widely been recognised as a powerful tool for the analysis of gene function in vivo to study mammalian physiology, development, and a vast array of human diseases. Genetically engineered animals are created by introducing foreign DNA into the animal using recombinant DNA technology. This genetic modification has to be transmitted through the germ line to the offspring so that every cell of the subsequent generations will contain the same modified genetic material. The foreign DNA arrives in the genome by random or targeted integration. Delivery of foreign genetic material can also occur via a recombinant virus, either in a systemic manner or delivered surgically to a limited number of cells in a specific tissue location. Recombinant viruses can be delivered into wild-type or genetically engineered mice enabling more sophisticated genetic modifications, creating neural pathway-, cell-type-, and region-specific mutants.

A huge number of genetic modifications have been generated since the first genetically engineered mouse (Jaenisch and Mintz 1974) was produced. In constitutive knockouts, gene expression is permanently inactivated in all cell types throughout development and thus an animal with a loss-of-function (null allele) is generated. Conditional knockouts make tissue- and/or stage-specific deletion of the gene of interest possible. Knock-in models enable generating animals with a directed mutant allele (e.g. point mutation), leading to altered functions of the targeted gene. Reporter genes, expressed under the control of endogenous regulatory elements permit the precise elucidation of anatomical and subcellular localisation of the gene of interest. Recently, targeted genome editing (e.g. CRISPR/Cas) revolutionised the technology of genomic alterations in animals allowing faster mutant generation at a lower cost. Multiple types of modified viruses are also at the researchers' disposal to deliver genes to specific cells and tissues. The significance of these technologies is emphasised by the two Nobel prizes in the last two decades that were awarded to scientists who developed novel methods to generate genetically modified animals: Mario Capecchi, Martin Evans and Oliver Smithies in 2007 and Jennifer Doudna and Emmanuelle Charpentier in 2020.

In this chapter, we will give an overview of the different genetic engineering techniques that are used for creating animal models, discussing how they are generated, what are they good for, and what their shortcomings are.

2.1 Random Integration of Genes

The first technique to generate genetically modified mice was microinjection of the recombinant DNA into the pronucleus of the zygote, then implantation of the modified zygotes into pseudopregnant mice (Conner 2004). Pronuclear injection results in the integration of the foreign DNA into a random location of the chromosome with relatively low efficiency, compared to other methods. Transcriptionally active sites with a low proportion of heterochromatin, such as the ROSA26 locus on chromosome 6 (Zambrowicz et al. 1997), are permissive for random integration. Importantly, pronuclear microinjection might result in the integration of several copies of the transgene. Although in some cases, this might be a disadvantage, in others, it allows the study of gene dosage effects (Marban and Gearhart 1987). The expression of the transgene is variable and depends on the exact site of the integration (position effect; al-Shawi et al. 1990). Another important consideration is that the random insertion of the DNA construct can disrupt the endogenous gene(s).

Mice that are directly developed from a genetically engineered zygote/embryo are called founders. Founder mice are then crossed with wild-type mice to establish the transgenic line that can be characterised at molecular and phenotypic levels. As each microinjection produces a unique genetic modification (in terms of the integration site, orientation and copy number), each founder is unique and yields a different transgenic line that has to be individually characterised.

Transgenic mice carrying known or suspected disease-causing genes have been very useful in studies of the pathophysiology of various human diseases such as Alzheimer's disease (Games et al. 1995), Huntington's disease (Sathasivam et al. 1999), Parkinson's disease (Allen et al. 2002), and neuroblastoma (Weiss et al. 1997) to name but a few. Reporter genes, such as lacZ (Soriano 1999) or fluorescent proteins (Li et al. 2018), under the control of a specific promoter, facilitated the characterisation of expression patterns of the gene of interest (GOI). Many mouse lines expressing site-specific recombinases (see below), or overexpression models, such as the tissue-specific MAGL overexpression lines (Chon et al. 2012; Jung et al. 2012), were also generated by microinjection/random integration.

2.2 Targeted Integration of Genes

Targeted integration of genes, a technique for introducing specific genetic modifications in mice, has significantly advanced the field of genetically modified animal models. This method, based on homologous recombination, allows for the insertion of expression cassettes into distinct chromosomal locations specified by the experimenter.

To target genes in mice, a linearised targeting construct comprising left and right homology arms, selection cassettes, and the DNA sequence meant to replace or supplement the chromosomal sequence between the homology arms is introduced into mouse embryonic stem cells in culture (Bouabe and Okkenhaug 2013). Following selection steps, cells with the correct recombination are injected into wild-type blastocysts, which are then implanted into pseudopregnant females. The resulting chimeric animals are bred with wild-type mice, and genotyping of the offspring reveals whether they carry the mutation through germ-line transmission. Subsequent breeding of mutant animals allows for molecular and phenotypic characterisation.

Targeted integration facilitates the generation of null or floxed alleles, knock-in models with specific mutations within a GOI, reporter gene expression under endogenous regulatory elements, and humanised mouse models. Notably, the technique allows for the insertion of transcriptional stop cassettes upstream of the translation start codon, enabling tissue-specific excision and gene function restoration in rescue models (see Table 1 for examples of mutants targeting the ECS). While targeted integration offers several advantages, including efficiency, predictability, and reliability, the process involves laborious cloning steps, embryonic stem cell cultures, and an extended timeline for generating genetically engineered animals. Despite its efficacy, targeted integration has limitations, such as potential lethality in early developmental stages and the lack of temporal and spatial selectivity in constitutive knockout models. To address these challenges, conditional gene targeting strategies have been developed, enabling tissue- and/or stage-specific deletion of the gene of interest.

2.3 *Site-Specific Recombination Systems*

Site-specific recombination systems, e.g., primarily from bacteriophages and yeast, enable precise DNA recombination between specific sequences. Classified as tyrosine or serine recombinases based on their active site nucleophile, these systems are essential for in vivo gene expression manipulation. The Cre-loxP system, derived from the P1 bacteriophage, is the most commonly used (Sternberg and Hamilton 1981).

In Cre-loxP recombination, the enzyme Cre recombinase and a DNA sequence containing two loxP sites are required. Recombination between these sites, oriented by an 8 bp spacer, results in excision or inversion of the flanked sequence (Fig. 1a, b). Excision irreversibly removes the sequence flanked by the two loxP sites of the same orientation, while inversion maintains the sequence between two opposing loxP sites, allowing it to flip back and forth. Spatial and temporal gene expression control is achieved by crossing transgenic mice containing in the GOI two loxP sites in the same orientation, flanking, e.g. essential exons, with transgenic mice expressing Cre recombinase under specific regulatory sequences. Various tissue and cell-type-specific Cre mouse lines enable tissue-specific recombination events (Table 2). For temporal control, Cre recombinase fused with a mutated oestrogen receptor ligand-binding domain (called CreERT2) allows inducible recombination

Table 1 Selected mutants targeting the key components of the endocannabinoid system

Gene MGI[a] Ref. Nr.	Null allele	Floxed allele	Additional lines
Cnr1/ 104615	Ledent et al. (1999), Marsicano et al. (2002), Robbe et al. (2002), Zimmer et al. (1999)	Castorena et al. (2021), González-Mariscal et al. (2018), Marcus et al. (2020), Marsicano et al. (2003), Vianna et al. (2012)	Winters et al. (2012): CB1-Tomato reporter line, Ruehle et al. (2013): CB1-Stop, Soria-Gomez et al. (2021): mtCB1-KO, Schneider et al. (2015): rat F238L point mutation, Lutz lab, unpublished: mouse F238L point mutation
Cnr2/ 104650	Buckley et al. (2000), Wotherspoon et al. (2005), López et al. (2018)	Liu et al. (2017), Stempel et al. (2016)	Schmöle et al. (2015): CB2-GFP reporter line, López et al. (2018): CB2-EGFP reporter line
Napepld/ 2140885	Leung et al. (2006), Tsuboi et al. (2011)	Liu et al. (2008), Tsuboi et al. (2011)	
ABHD4/ 1915938	Lee et al. (2015)		
Faah/ 109609	Cravatt et al. (2001)	Lutz lab, unpublished	Mileni et al. (2008): rat humanized FAAH, Dincheva et al. (2015): C385A point mutation
NAAA/ 1914361	Xie et al. (2021)		
Dagla/ 2677061	Gao et al. (2010), Shonesy et al. (2014), Tanimura et al. (2010)	Bluett et al. (2017), Jenniches et al. (2016)	
Daglb/ 2442032	Gao et al. (2010), Tanimura et al. (2010)		
Mgll/ 1346042	Chanda et al. (2010), Taschler et al. (2011)	Viader et al. (2015)	Jung et al. (2012): forebrain overexpression, Chon et al. (2012): intestinal overexpression
ABHD6/ 1913332	Zhao et al. (2014)	Zhao et al. (2014)	

[a] MGI – Mouse Genome Informatics (http://www.informatics.jax.org/)

upon tamoxifen treatment of the mice (Fig. 1c). Codon-improved Cre recombinase (iCre) was developed to mitigate epigenetic silencing (Shimshek et al. 2002).

Advancements extend into using immediate early genes (IEGs), linking gene expression to neuronal activity (Sheng and Greenberg 1990). Cre expression driven by IEG regulatory elements permits activity-dependent recombination. Notably, Arc-CreERT2 and Fos-CreERT2 lines (Denny et al. 2014; Guenthner et al. 2013) have proven invaluable for labelling or manipulating neurons activated during specific conditions, enabling targeted delivery of fluorescent proteins, calcium

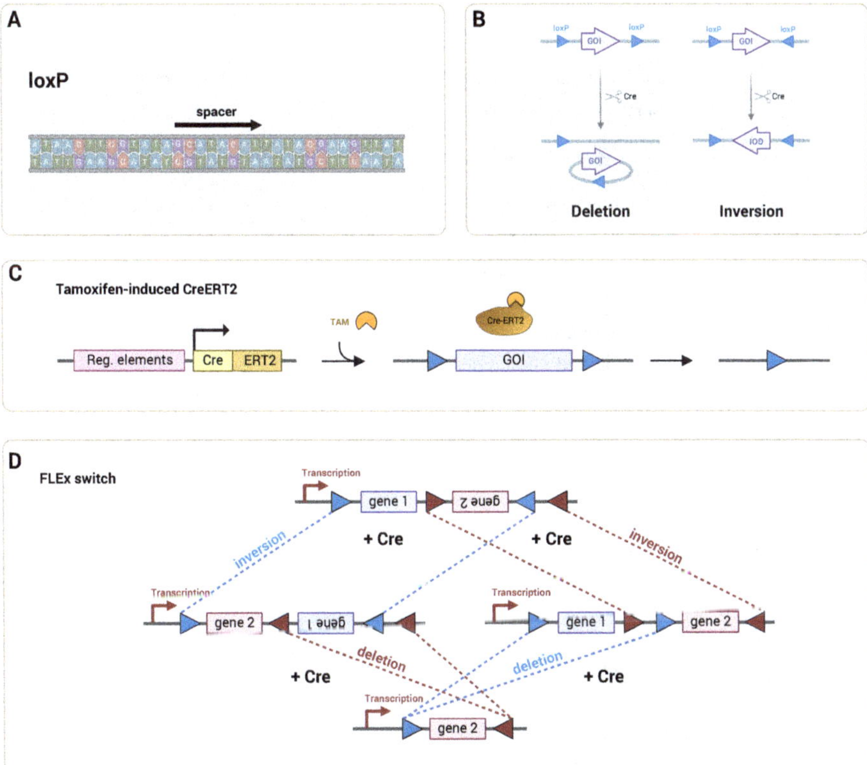

Fig. 1 Principles of the site-specific recombination systems. (**a**) Sequence of the loxP site. The two 13-bp long palindromic sequences are the recognition sites for Cre recombinase and flank the 8-bp long spacer (arrow), which gives the loxP site its orientation. (**b**) Outcomes of the Cre/loxP-mediated recombination depend on the orientation and position of the loxP sites. Deletion of the DNA sequences flanked by the two loxP sites (gene of interest, GOI) occurs if the two loxP sites are oriented in the same direction, while inversion occurs in case the two loxP sites are oriented in opposite directions. (**c**) Ligand-inducible Cre/loxP-mediated recombination provides temporal control over Cre activity. CreERT2 (Cre fused to a mutated oestrogen receptor ligand binding domain) remains inactive until exposed to tamoxifen (TAM). TAM administration to mice activates CreERT2, allowing it to enter the nucleus and initiate recombination at loxP sites, with deletion of the GOI. (**d**) In the FLEx ("flip-excision") switch system (aka DIO, Double-floxed Inverse Orientation), two pairs of loxP sites with different spacer sequences (blue and red triangles, e.g. loxP and lox511, respectively) are incorporated into the genetic construct that may harbour a GOI ("gene 1") and a reporter gene ("gene 2"), or a wild-type and a mutant version of a gene, etc. Cre recombinase can recognise and recombine only loxP variants with the same spacer (here: same colour, e.g. loxP or lox522) and finally mediates a double recombination event, leading in this example to the deletion of gene 1 and the transcriptional activation of gene 2. Created in BioRender. Monory, K. (2024) BioRender.com/m21s977

indicators, chemogenetic or optogenetic tools into these cells. This approach has facilitated the characterisation and manipulation of neurons activated under certain stimuli, such as those activated by chronic social stress, where it has been used to

Table 2 Cre lines used in cannabinoid research. List of Cre lines used to generate mutants of ECS components

Mouse line	Cre recombination pattern	Reference
ACTA-Cre	skeletal muscle cells	Miniou et al. (1999)
AdipoqCreERT2	adipocytes	Sassmann et al. (2010)
albumin-Cre	hepatocytes	Postic et al. (1999)
Arc-CreERT2	activated neurons	Denny et al. (2014)
CaMKII-Cre	principle forebrain neurons, sympathetic neurons	Casanova et al. (2001)
CRF-IRES-Cre	CRH expressing cells including PVN	Taniguchi et al. (2011)
CX3CR1-Cre	monocytes, macrophages, microglial cells	Yona et al. (2013)
DAT-Cre	dopaminergic neurons	Zhuang et al. (2005)
Dbh-Cre	adrenergic and noradrenergic neurons	Parlato et al. (2007)
Dlx5/6 Cre	forebrain GABAergic neurons	Monory et al. (2006)
En1-Cre	developing and mature dopaminergic neurons	Kimmel et al. (2000)
Fabp4-Cre (aka aP2-Cre)	adipocytes	He et al. (2003)
GFAP-Cre	astrocytes	Zhuo et al. (2001)
GFAP-CreERT2	astrocytes	Hirrlinger et al. (2006)
GLAST-CreERT2	astrocytes, radial glial and neuronal stem cells	Mori et al. (2006)
iL1-sglt2-Cre	renal proximal tubule cells	Rubera et al. (2004)
K14-Cre	keratinocytes	Vassar et al. (1989)
LysM-Cre	monocytes/macrophages and neutrophils	Clausen et al. (1999)
MIP-CreERT	pancreatic β-cells	Tamarina et al. (2014)
Nestin-Cre	neuronal and glial precursors	Tronche et al. (1999)
Nestin-CreERT2	neuronal and glial precursors	Lagace et al. (2007)
Nex-Cre	cortical glutamatergic neurons	Goebbels et al. (2006)
NG2-CreERT2	oligodendrocyte precursor cells	Huang et al. (2014)
Pdgfra-CreERT2	oligodendrocyte precursor cells	Rivers et al. (2008)
SF1-Cre	ventromedial hypothalamic nucleus, pituitary, gonad, and adrenal tissue	Dhillon et al. (2006)
Sim-1-Cre	Sim1-expressing tissues including hypothalamus	Balthasar et al. (2005)
Syn-Cre line	differentiated neurons	Zhu et al. (2001)
TPH2-CreERT2	CNS serotonergic neurons	Weber et al. (2009)
Vilin-CreERT2	intestinal epithelial cells	el Marjou et al. (2004)

specifically reduce AEA synthesis by deletion of the AEA synthesising enzyme NAPE-PLD in activated neurons (Tevosian et al. 2023).

To accommodate multiple recombination events, mutant loxP sites with unique asymmetric spacers have been developed (Lee and Saito 1998). By flanking a sequence of interest with two different loxP site sets, a FLEx (flip-excision) switch, also called DIO (Double-floxed Inverse Orientation) is created (Schnütgen et al. 2003). This results in sequence inversion and the removal of one site from each pair of identical recombination sites (Fig. 1d). Combined with activity-dependent recombinases, this technology has become an essential tool for analysing neural pathways and tracing neuronal connections (Gremel et al. 2016). These innovations greatly enhance the versatility of site-specific recombination systems in creating transgenic animal models, providing invaluable insights into the dynamic interplay of biological processes and disease mechanisms.

2.4 Targeted Genome Editing

Genome editing, a groundbreaking technology, has transformed the generation of genetically modified animals. This technique employs specific nuclease enzymes to create double-strand DNA breaks with concomitant genomic modifications (Adli 2018). Genome editing can be applied to any species with known genomic sequences. Among the different nuclease systems, the CRISPR/Cas9 system has gained widespread adoption due to its versatility and efficiency (Fig. 2a). Indeed, CRISPR/Cas9-generated animal models have become favoured over classical transgenic or embryonic stem cell techniques within a short time, and in the past few years, mice with mutations in the ECS were also generated with this technology,

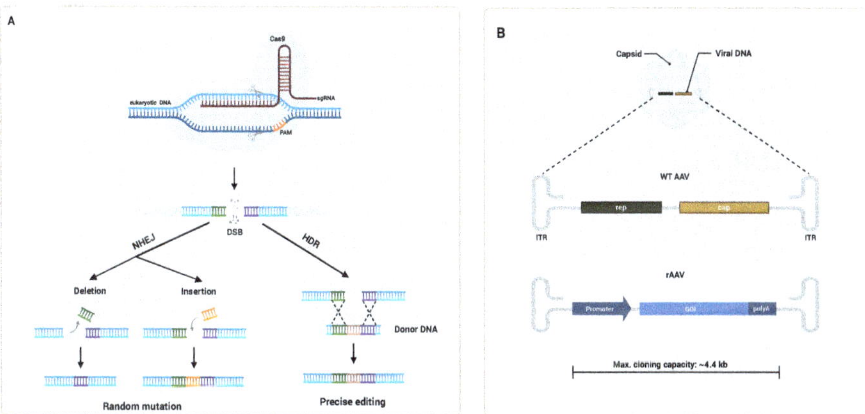

Fig. 2 CRISPR/Cas9 system and adeno-associated virus (AAV). (**a**) Targeted genome editing using CRISPR/Cas9 technology. Synthetic single-stranded guide RNAs (sgRNA) direct Cas9 nuclease to genomic DNA at a specific location with homology with the sgRNA to induce the double-strand break (DSB). The mammalian cell utilises endogenous mechanisms to repair the DSB. One of these mechanisms is the non-homologous end joining (NHEJ), an inaccurate process, which can produce short insertion and deletion mutations of variable lengths at the site of the DSB. In case a donor DNA (ssDNA or dsDNA) is provided, the DSB can be repaired precisely using homology-directed repair (HDR), resulting in the desired genomic alteration (e.g. insertion, replacement, point mutation). (**b**) The AAV virion contains a 4.7 kb ssDNA genome harbouring the rep and cap genes, responsible for capsid formation and replication, respectively. The 145b long inverted terminal repeats (ITRs) contain palindromic sequences allowing them to fold into a hairpin loop conformation. In recombinant AAV (rAAV), all sequences but the two ITRs are replaced with an expression cassette, containing promoter and GOI. Created in BioRender. Monory, K. (2023) BioRender.com/e51f805

such as a new CB1$^{f/f}$ allele (Castorena et al. 2021) and an N-acylethanolamine acid amidase (NAAA) knockout line (Xie et al. 2021).

The potential of CRISPR/Cas9 as a sequence-specific genetic engineering tool was quickly recognised. To create a mutant mouse, the target DNA break site must be determined using genomic search tools (Box 1), followed by the selection of an appropriate Cas9 enzyme. Various Cas9 enzymes from different microbes offer flexibility to the CRISPR system. Mutated Cas9 versions enable activation, repression or labelling of the target sequence (Mali et al. 2013; Cong et al. 2013). The widely used Cas9 from *Streptococcus pyogenes* (spCas9) requires a protospacer adjacent motif (PAM) sequence (5'-NGG-3′) and cleaves the DNA three base pairs upstream from the PAM. For spCas9, a 20 bp guide RNA (gRNA) sequence is selected downstream of the PAM recognition site in the target gene area. An expression vector containing Cas9 and the gRNA, and, in case of the generation of a knock-in, along with a single-stranded donor DNA for the generation of a homologous DNA repair (HDR) event, is assembled by molecular cloning methods and microinjected into pronuclei or electroporated into zygotes. Surviving embryos are then transferred to pseudopregnant mice. Since genetic manipulation occurs at the one-cell stage, resulting embryos are not chimeric, and all cells carry the mutation. Born pups are screened for the desired mutation's presence and fidelity, with correctly targeted founder mice used to establish a colony.

Box 1 Useful Repositories

Mutant mouse repositories

http://www.findmice.org/index
https://mice.jax.org/
https://www.mmrrc.org/
https://www.infrafrontier.eu/emma/
http://www.gensat.org/cre.jsp

Plasmid/vector repositories

https://www.addgene.org/
https://dnasu.org/DNASU/Home.do
https://bccm.belspo.be/genecorner-plasmid-search
https://dna.brc.riken.jp/en/

CRISPR gRNA design tools

http://crispor.tefor.net/
https://www.benchling.com/crispr
https://www.genscript.com/gRNA-design-tool.html
http://www.e-crisp.org/
https://www.atum.bio/eCommerce/cas9/input
http://chopchop.cbu.uib.no/

Although spCas9 is one of the best characterised and most commonly used CRISPR-Cas systems, it comes with certain limitations. Furthermore, longer PAM recognition sequences reduce the likelihood of off-target mutations, while smaller Cas9 molecular weights allow easier packaging into recombinant viral particles for applications discussed further below.

2.5 *Viral Tools*

Recombinant viruses are formidable tools for cell-type-specific gene delivery, a technique routinely used in in vitro or in vivo models to investigate the molecular mechanisms underlying diverse biological functions. Their increasing popularity is due to the fact that rates of gene transfer using viral vectors are far superior to those achieved with non-viral methods. Ideal viral vectors for gene transfer to the CNS are replication-deficient, not pathogenic, easy to produce, stably expressing the transgene, episomally located, able to transduce postmitotic cells and have a large packaging capacity. None of the commonly used viral vectors have all these favourable attributes; for example, adeno-associated viruses (AAV) have limited gene packaging size, and lentiviruses have the potential for recombination leading to replication-competent viruses. Furthermore, recombinant viruses often have to be delivered to the CNS by an invasive surgical process. Nevertheless, viral vectors are comparatively easy to produce, and with their help both precise anatomical and temporal control of the gene expression can be achieved.

2.5.1 Adeno-Associated Virus

Adeno-associated virus (AAV) emerged as a premier vector for gene therapy in both preclinical and clinical research since its discovery in the 1960s alongside adenoviruses (Atchison et al. 1965; Hoggan et al. 1966). Non-pathogenic and boasting minimal immunogenicity, AAV's replication hinges on co-infection with a helper virus, rendering it replication-incompetent on its own. Despite widespread human exposure to AAV, resulting in prevalent immunity (Erles et al. 1999), its utility in gene delivery remains unrivalled due to its benign nature and broad infectivity across species.

AAV's genome is succinct, comprising a single-stranded DNA of about 4.7 kb, encased within capsid proteins (Fig. 2b). The REP and CAP genes orchestrate viral replication and packaging, respectively, but are excised in recombinant AAV (rAAV) to make room for therapeutic or experimental genes.

Recombinant AAVs are produced in HEK cells through co-transfection with a plasmid bearing the gene of interest, and auxiliary plasmids providing structural and replication functions (McClure et al. 2011). The outcome is a batch of AAV particles, specific in their serotype and therefore tissue tropism (Table 3), which are then purified, quantified, and delivered stereotactically to target sites within the brain. A significant recent achievement in AAV delivery is the development of new variants that are able to cross the blood-brain barrier to the CNS following

Table 3 AAV serotypes in the mouse

Serotype	Tropism
AAV1	Muscle, CNS, heart, liver, lungs
AAV2	Heart, CNS, liver, lungs, retina
AAV3	Liver
AAV4	Retina, lungs, kidney
AAV5	Retina, CNS, liver
AAV6	Heart, liver, muscle, retina
AAV7	Liver
AAV8	Muscle, heart, CNS, liver
AAV9	Heart, CNS, liver
AAV10	Muscle, myoblast tissue
AAV11	Muscle, myoblast tissue
AAV12	Salivary glands, muscle
AAV13	Neurons
AAV retrograde (AAV2-retro)	Neuronal pathway specificity
AAV-BBB variants	For transgene delivery across the blood-brain barrier to the CNS following intravenous injection

Data from: Han et al. (2022), Issa et al. (2023), Li and Samulski (2020), Tervo et al. (2016), Drouyer et al. (2024)

intravenous injection (Drouyer et al. 2024). Ideal expression levels following AAV injection are typically reached within 3–4 weeks. Although the modest genome size of the virus limits the length of the insertable genetic material, AAV's stable, episomal persistence in host cells ensures long-term expression of the delivered gene, a trait invaluable for both basic research and therapeutic interventions.

AAV-mediated gene delivery is often used to induce local gene deletion, gene reactivation, gene overexpression, and gene silencing, as well as to deliver biosensors, optogenetic, chemogenetic, or mapping tools (Fig. 3a). With the emergence of genome editing, recombinant viruses are also used for the local delivery of Cas9 and gRNA containing constructs to induce gene knock-out in a cell-type-specific manner (Fig. 3b). A critical innovation in AAV-mediated gene delivery exploits its ability for retrograde transport (Tervo et al. 2016) enabling manipulation of genes within defined neural circuits in a previously unprecedented manner (Fig. 3c).

Accordingly, AAV-mediated gene delivery has been used to investigate region-, cell-type- or circuit-specific roles of the ECS. For instance, this technology was used for the region- and cell-type-specific overexpression of MAGL (Guggenhuber et al. 2015), stress-activated circuit tracing (Marcus et al. 2020), cell-type-specific gene expression silencing (Hu et al. 2022), cell- and organelle type-specific Ca^{2+} sensor delivery (Serrat et al. 2021), circuit-specific CB1 deletion (Kondev et al. 2023) or region- and cell-type-specific mutagenesis (Balsevich et al. 2023) to name but a few.

AAV's adaptation to CRISPR technology has notably advanced the specific reach of genome editing enabling precise circuit- and activity-dependent mutagenesis in the adult mouse brain. Initially hindered by size constraints, the development of

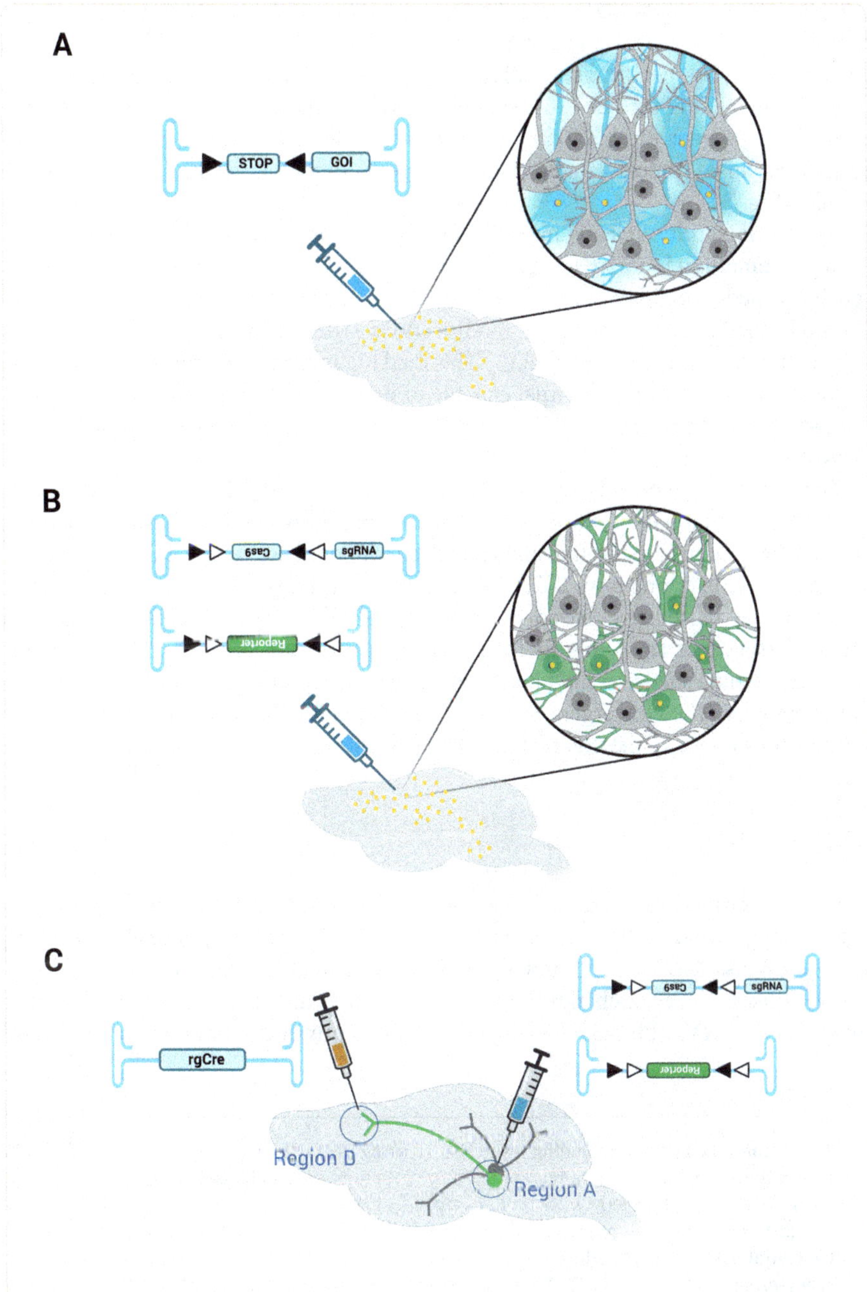

Fig. 3 Examples of AAV-mediated genetic manipulations. (**a**) Region and cell-type-specific expression of GOI. An AAV construct, with a floxed transcriptional STOP cassette preceding the GOI, is injected into a specific brain region of a mouse with specific Cre recombinase expression (yellow dots) leading to the removal of the STOP cassette and to cell-type-specific GOI expression. Only neurons that express Cre (yellow nuclei) are going to express the GOI (bright blue neurons). (**b**) Region- and cell-type-specific modification/inactivation of a GOI. A Cre-dependent Cas9

compact Cas9 variants (such as SaCas9) has facilitated the packaging of the entire CRISPR system within a single AAV vector (Friedland et al. 2015). This innovation enables cell-type-specific gene editing with remarkable efficiency and specificity (Fig. 3b), bypassing the need for separate vectors or transgenic mouse lines (Hunker et al. 2020; Kumar et al. 2018). When paired with retrograde AAV vectors, this system promises precise interrogation of gene functions within specific neural circuits (Fig. 3c).

AAV's versatility extends into the neural circuitry domain, where intersectional strategies employing both Cre- and Flp-dependent systems enable precise pathway-specific genetic manipulations (Fenno et al. 2014). The Flp/Frt mechanism, akin to Cre/loxP, facilitates targeted gene modifications – deletions or expressions – at the intersection of defined neural circuits. This dual recombinase approach culminates in the injection of a retrograde virus, capable of expressing Flp recombinase back to a primary region from its projection target, thereby ensuring pathway-specific gene expression control.

The advent of genetically encoded biosensors, such as GRAB and dLight delivered via AAV, has revolutionised our understanding of neurochemical dynamics in vivo, offering insights into neurotransmitter and neuromodulator activities with unprecedented resolution (Dong et al. 2022; Farrell et al. 2021; Patriarchi et al. 2018; Sun et al. 2018). These biosensors, based on modified GPCRs, illuminate the spatial and temporal aspects of signalling molecule interactions within targeted cell populations or neural circuits, shedding light on the molecular underpinnings of various neurological conditions. The convergence of AAV technology with these sensors opens new avenues for dissecting the complex mechanisms of brain function and disease.

2.5.2 Lentivirus

Lentivirus, derived from *Retroviridae*, is a single-stranded RNA virus known for causing long-duration illnesses, such as HIV/AIDS, in various mammals, including humans (Weiss 1993). Lentiviral vectors, derived mainly from HIV-1, are extensively utilised in research despite their larger size and more complex structure compared to AAVs (Pluta and Kacprzak 2009). Lentiviruses carry a positive-strand

Fig. 3 (continued) construct, including a sgRNA targeting the GOI, is injected into a specific brain region of a mouse expressing Cre recombinase (yellow dots) in a cell-type-specific manner. The Cas9 construct is coinjected with a Cre-dependent reporter construct (e.g. GFP). Upon Cre-mediated recombination processes, in neurons expressing Cre (yellow nuclei), Cas9-mediated genome editing takes place, together with reporter expression (green neurons). (**c**) Pathway-specific modification/inactivation of a GOI. A Cre-dependent Cas9 construct including a sgRNA targeting the GOI, is injected into Region A. The Csa9 construct is coinjected with a Cre-dependent reporter construct (e.g. GFP). Retrograde Cre (rgCre) AAV is injected into Region B. rgCre is taken up by the terminals and travels backward to the cell body. Reaching the cell bodies in Region A, rgCre activates Cas9 and the reporter gene. Only neurons in Region A that project to Region B will carry out Cas9-mediated mutations and will express the reporter gene (green projecting neuron). Created in BioRender. Monory, K. (2024) BioRender.com/f93h849

single-stranded RNA genome, encoding major structural, regulatory, and accessory genes, flanked by two long terminal repeats (LTRs) (Clements and Zink 1996). These vectors maintain cis-acting viral sequences essential for capsid formation, reverse transcription, and genome integration, enabling the efficient delivery of transgenes (Zufferey et al. 1998).

Lentivirus-based vectors are major tools for basic and translational research. A strategic advantage of lentiviruses is that they can efficiently infect nondividing and slowly dividing cells. Furthermore, their broad tropism and low immunogenicity are also important factors why they can effectively deliver genes into a broad selection of cell types. Their relatively large packaging capacity allows for the transfer of more sophisticated constructs. The integration of the viral construct into the host genome enables stable expression or long-term silencing of the GOI. At the same time, integration of the provirus into the host genome can cause mutations by disrupting the gene where it is inserted, which can even lead to malignant cell transformation. Moreover, although its chances are low, the generation of replication-competent viruses cannot be completely excluded either.

Recent studies in the endocannabinoid field have leveraged lentiviral vectors to investigate the role of the ECS in various physiological and pathophysiological paradigms. Thus, lentivirus was utilised to mediate local overexpression of FAAH to investigate anxiety behaviour (Rubino et al. 2008), to screen a lentiviral shRNA library to find novel regulators of anchorage dependant cell death (Simpson et al. 2012), and to induce conditional inactivation of CB1 in neural stem cells (Zimmermann et al. 2018). These application modalities underscore the versatility and utility of lentiviral vectors in elucidating complex biological processes.

3 Understanding the Endocannabinoid System with Genetically Engineered Rodent Models

The ECS is a widely expressed and multifunctional signalling system that is activated on demand following perturbations of the local homeostasis (Lutz 2020). The ECS is composed of the G protein coupled type 1 and type 2 cannabinoid receptors (CB1 and CB2), the endogenous cannabinoids (endocannabinoids, eCBs), such as 2-arachidonoyl glycerol (2-AG) and arachidonoyl ethanolamine (anandamide, AEA), and the enzymatic machinery to synthesise and degrade the eCBs. The main enzymes for synthesis and degradation are N-acyl phosphatidylethanolamine-hydrolysing phospholipase D (NAPE-PLD) and fatty acid amide hydrolase (FAAH) for anandamide, and diacylglycerol lipase (DAGL) α and β and monoacylglycerol lipase (MAGL) for 2-AG (Hillard 2015).

Endocannabinoids regulate multitudes of physiological functions such as memory, emotional states, stress, motor function, energy expenditure, neuronal development, and pain perception (Ligresti et al. 2016). How the ECS can play such a vast array of roles lies in the complexity of its expression, activation, regulation,

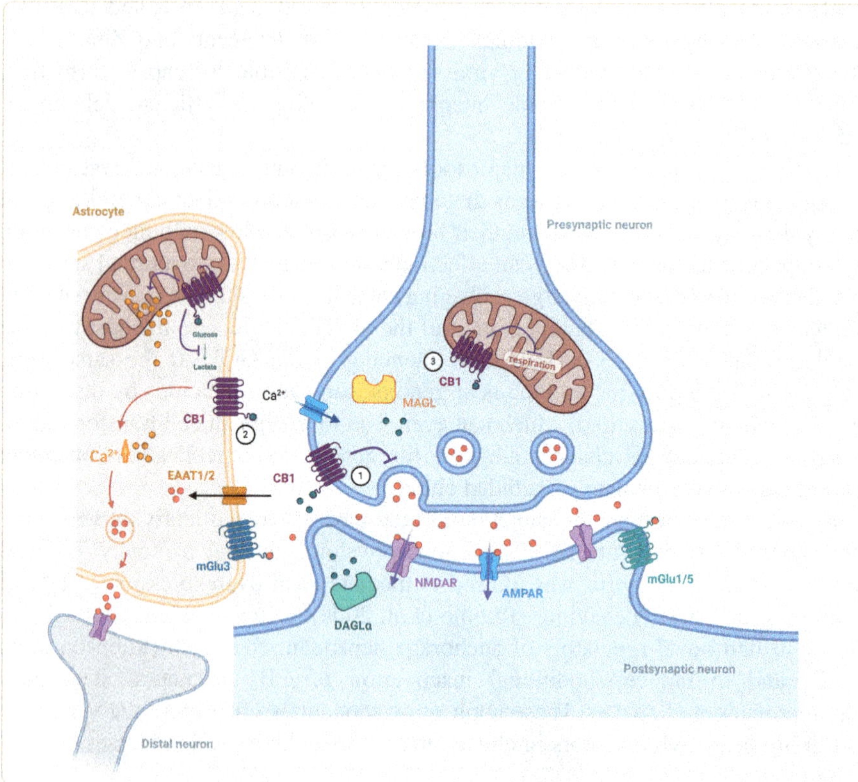

Fig. 4 Schematic view of a synapse with components of the ECS. At the synapse, neurotransmitter release (orange dots) from the presynapse (top) activates receptors (e.g., AMPA and NMDA for glutamatergic transmission) on the postsynapse (bottom). The ensuing Ca^{2+} influx activates endocannabinoid synthesis (2-AG) in the postsynaptic compartment via diacyl glycerol lipase α (DAGLα). Endocannabinoids (green dots) travel retrogradely across the synapse, where they bind to and activate CB1 at the presynapse, thereby decreasing transmitter release (1). CB1 activation on astrocyte plasma membranes initiates an increase in intracellular Ca^{2+} and gliotransmitter release (2). In mitochondria, mtCB1 decreases respiration and regulates cellular Ca^{2+} and lactate levels (3). Bottom left: astrocytic CB1 activation leads to so-called lateral potentiation, i.e. an increased transmission at a more distal neuron. Created in BioRender. Monory, K. (2023) BioRender.com/x29z302

signalling and the actual cellular function that is emerging as a summation of these aspects. Primarily, the ECS is characterised by a specific expression pattern of its constituents and associated proteins (Marsicano and Kuner 2008). CB1, the best-studied protein of the ECS, is present in nearly all regions and cell types of the brain showing especially prominent expression levels in the cerebellum, cortex, hippocampus, striatum, and olfactory bulb. A major function of the eCBs in the brain is the regulation of synaptic neurotransmission (Fig. 4; Noriega-Prieto et al. 2022). Unlike classical neurotransmitters, 2-AG travels from the postsynaptic membrane to the presynaptic one (otherwise known as retrograde transmission; Wilson and Nicoll

2001). The exact mechanisms, of how such lipophilic compounds travel intracellularly or through the synaptic cleft, have not yet been understood, but insights have gradually been gained (Kaczocha and Haj-Dahmane 2021). Retrograde signalling provides an important inhibitory feedback mechanism to the neurons releasing neurotransmitters and plays critical roles in the short- and long-term plasticity of synaptic connections (Winters and Vaughan 2021). CB1 has been shown to modulate transmitter release in glutamatergic, GABAergic, serotonergic, noradrenergic, and cholinergic neurons. Recently, for the first time, retrograde signalling was demonstrated in vivo in CA1 inhibitory synapses with the help of the GRABeCB2.0 endocannabinoid sensor (Dudok et al. 2024). CB1 is predominantly presynaptic, but postsynaptic localisation (Bacci et al. 2004; Maroso et al. 2016; Njoo et al. 2015) as well as expression in intracellular organelles (Bénard et al. 2012; Grimsey et al. 2010) has similarly been shown. The subcellular localisation of the eCB metabolic enzymes was also determined. Consistent with the primary role of 2-AG in retrograde signalling (Gao et al. 2010), DAGLα is found at the somatodendritic domain of neurons, while MAGL is localised at presynaptic terminals (Nyilas et al. 2008). NAPE-PLD has been found both pre- and postsynaptically (Egertová et al. 2008; Nyilas et al. 2008; Reguero et al. 2014). FAAH, on the other hand, is predominantly found in intracellular membranes, such as the endoplasmic reticulum in the somatodendritic compartment of neurons (Gulyás et al. 2004).

CB2, originally termed the peripheral cannabinoid receptor, is strongly expressed in the spleen, thymus, circulating immune cells and various tissue-resident macrophage populations and plays primarily an immunomodulatory function (Turcotte et al. 2016). Although early studies detected no CB2 in the CNS (Munro et al. 1993), it was later shown in microglia (Carlisle et al. 2002) and astrocytes (Molina-Holgado et al. 2002a) although at very low expression levels in a healthy organism. CB2 was reported to have neuroprotective effects in a number of diseases involving neuroinflammation, such as Alzheimer's disease (Ruiz de Martín Esteban et al. 2022), multiple sclerosis (Palazuelos et al. 2008) and amyotrophic lateral sclerosis (Rodríguez-Cueto et al. 2021). In these pathological states, the expression level of CB2 robustly increased (Maresz et al. 2005). The presence of CB2 in neurons, however, has long been and still is the subject of debate. The currently published work posits that functional CB2 is present in certain neuronal populations in various brain regions, such as the brain stem (Van Sickle et al. 2005), ventral tegmental area (Zhang et al. 2014), hippocampus (Stempel et al. 2016) or cortex (Stumpf et al. 2018), with an expression level much lower than CB1.

The eCBs are lipid molecules, and as such, they are thought not to be stored in vesicles; instead, they are synthesised from membrane lipid precursors upon increased Ca^{2+} levels in the cell and degraded by serine hydrolase enzymes to arachidonic acid and other small molecules. However, the metabolism of the eCBs can also occur via parallel pathways (Di Marzo 2018; Liu et al. 2008; Muccioli 2010), Murataeva et al. 2014). The existence of these redundant pathways means, in practical terms, that genetic inactivation of eCB metabolic enzymes might not fully inactivate the targeted process. Moreover, the tissue concentration of 2-AG, an intermediate in lipid metabolism, is hundreds to thousands of times higher than that of anandamide (Tsuboi et al. 2018). The presence of alternative metabolic pathways can influence the availability of eCBs thus modifying their local effect. Furthermore, apart from CB1 and CB2, both AEA and 2-AG as well as palmitoylethanolamide (PEA) can bind to GPR55 and transient receptor potential vanilloid 1 (TRPV1). Nuclear peroxisome proliferator-activated receptors (PPARs) can bind eCBs too; AEA interacts with PPARα and PPARγ, while 2-AG only activates PPARγ (Di Marzo 2018; Iannotti and Vitale 2021). Interestingly, 2-AG and AEA differ in their efficacy; while 2-AG is a full agonist, AEA behaves as a partial agonist at both cannabinoid receptors. The specific outcomes of receptor activation are also influenced by a positive or negative allosteric modulator (PAM or NAM, respectively). In the past years, several endogenous molecules have been found to act as allosteric modulators of the cannabinoid receptors, such as lipoxin A4 (Pamplona et al. 2012) and pregnenolone (Vallée et al. 2014). Apart from these lipids, a group of endogenous cannabinoid peptides (pepcans) has also been discovered, which act as allosteric modulators of the cannabinoid receptors. These peptides are synthesised in an anatomically distinct cell population in the CNS and in the periphery (Hofer et al. 2015). Of the numerous detectable pepcans, two have been shown to have distinct physiological roles: pepcan-23 is the precursor molecule of pepcan-12 (RVD-Hp), which is an endogenous negative allosteric modulator of CB1 and a positive allosteric modulator for CB2 (Bauer et al. 2012; Heimann et al. 2021; Petrucci et al. 2017).

Cannabinoid receptors signal mostly, but not exclusively, through Gi/Go type G proteins (Diez-Alarcia et al. 2016). After activation, receptors that are phosphorylated by G protein receptor kinases can be bound by β-arrestin and internalised (Ibsen et al. 2019). Similarly, downstream intracellular signalling pathways (Leo and Abood 2021; Priestley et al. 2017) also differ depending on the population of receptors investigated. Presynaptic CB1 modulates adenylyl cyclase, voltage-gated calcium channels, K^+ channels, PKA and the ERK pathway. However, it is also known that cannabinoid receptors can form homo- or heterodimers (Oyagawa and Grimsey 2021), for example with dopamine, adenosine, serotonin, or opioid receptors, which can lead to alterations of ligand binding and signalling properties of dimers compared to monomer receptors (Moreno et al. 2018; Rozenfeld et al. 2012).

The description above attempts to provide a basic insight into the complexity of the ECS. It also implies the different levels of technical difficulties faced when trying to decipher eCB functions at specific sites. For instance, the functional role of presynaptic CB1 at GABAergic synapses was revealed relatively early in sufficient detail (Katona et al. 1999; Wilson and Nicoll 2001). This was greatly aided by the very high expression level of CB1 at these sites (Tsou et al. 1999). Remarkably, however, it has been repeatedly demonstrated that low or even very low CB1 expression carries out important physiological functions. In these cases, even demonstrating expression had often been challenging. Deciphering the cell-type-specific physiological role of CB1 would not have been possible without the mutant animal models.

3.1 Diverse Neuronal Types

Glutamatergic and GABAergic neurons are the most abundant excitatory and inhibitory neurons of the mammalian brain. These two neuronal populations act together to modulate neural activity and maintain the homeostasis of neural transmission (Tatti et al. 2017). CB1 expression across cortical and hippocampal regions has been extensively studied and found to be present both in glutamatergic and GABAergic neurons. Specifically, high CB1 expression was observed in a subpopulation of GABAergic interneurons that belong mostly to the cholecystokinin (CCK)-positive subclass (Freund et al. 2003; Marsicano and Lutz 1999; Tsou et al. 1999), while most glutamatergic cells in the cortical areas expressed CB1 at low but detectable levels (Kawamura et al. 2006; Marsicano and Lutz 1999). To specifically decipher the role of CB1 in glutamatergic and GABAergic neurons, conditional knockout mice were generated lacking CB1 in cortical glutamatergic cells and forebrain GABAergic cells, respectively (Monory et al. 2006). To generate these mice, $CB1^{f/f}$ mice (Marsicano et al. 2003) were crossed with Nex-Cre mice (Goebbels et al. 2006) for Glu-CB1-KO and Dlx5/6-Cre mice (Monory et al. 2006) for GABA-CB1-KO. To complement the loss-of-function approach that helped reveal the necessary functions of CB1 at specific sites, cell-type specific rescue models were also generated to restore CB1 expression only in glutamatergic (Glu-CB1-RS; (Ruehle et al. 2013) or GABAergic cells (GABA-CB1-RS; Remmers et al. 2017), respectively, and thereby determine whether re-expression of CB1 in a subset of cells is sufficient for the studied brain functions. Semi-quantification of CB1 receptor immunogold staining showed that in these mutants CB1 re-expression is limited to the targeted cell types and maintains endogenous levels of CB1 (Gutiérrez-Rodríguez et al. 2017).

Activation of CB1 leads to different forms of synaptic plasticity of excitatory and inhibitory neurotransmission. Cell-type-specific mutants were instrumental in specifying the CB1 populations responsible for certain forms of synaptic plasticity. Thus, CB1-dependent depolarisation-induced suppression of inhibition (DSI) was abolished in GABA-CB1-KO mice (Monory et al. 2006), while experiments with GABA-CB1-RS mice showed that this form of plasticity in the hippocampus entirely relies on the expression of CB1 on GABAergic interneurons (Remmers et al. 2017). In contrast, in Glu-CB1-RS mice, DSI was absent, whereas depolarisation-induced suppression of excitation (DSE) was fully rescued (Ruehle et al. 2013). In cortical areas, CB1 on glutamatergic neurons was necessary for the CB1 agonist WIN55,212-2-induced inhibition of glutamatergic transmission (Domenici et al. 2006). WIN55,212-2 effects on long-term depression (LTD), on the other hand, were abolished in GABA-CB1-KO but preserved in CaMK CB1-KO, lacking CB1 in principal forebrain neurons (Azad et al. 2008). The strength of hippocampal long-term potentiation (LTP) is regulated by eCBs acting on both glutamatergic and GABAergic CB1. GABA-CB1-KO mice showed decreased hippocampal LTP, reduced dendritic branching and decreased spine density of CA1 pyramidal neurons, while in Glu-CB1-KO hippocampal LTP was enhanced and

CA1 pyramidal neurons showed an increased branching and an increased spine density (Monory et al. 2015). Interestingly, electrophysiological measurements showed that the eCB tone is significantly stronger in GABAergic cells (Slanina and Schweitzer 2005) than in glutamatergic cells (Roberto et al. 2010). A stronger phasic cannabinoid response might also be related to a more efficiently coupled G protein signalling of CB1 in glutamatergic cells (Steindel et al. 2013).

Regulating neuronal excitability by eCBs conveys a neuroprotective function. Endogenous protection against kainic acid-induced seizures depends on CB1 in glutamatergic hippocampal neurons, while CB1 on GABAergic neurons was dispensable for this function (Monory et al. 2006). ERK phosphorylation and induction of certain immediate early genes by kainic acid in pyramidal cells depends similarly on CB1 in glutamatergic cells (Marsicano et al. 2003). These results were further substantiated by studies with Glu-CB1-RS and GABA-CB1-RS mice, re-expression of CB1 in cortical glutamatergic cells was sufficient to protect against kainic acid-induced epileptiform seizures (Ruehle et al. 2013), while GABAergic rescue of CB1 did not convey protection (Remmers et al. 2017).

An instrumental gap in our understanding of endocannabinoid dynamics in vivo has been recently resolved by the generation of a genetically encoded endocannabinoid sensor (called GRABeCB2.0 sensor, Dong et al. 2022; Farrell et al. 2021; Gunduz-Cinar et al. 2023). The GRABeCB2.0 can detect low micromolar endocannabinoid concentration changes within sub-seconds to seconds (Dong et al. 2022). Thus, it was proven that electrically induced seizures can amplify the endocannabinoid levels by several orders of magnitude in mice expressing the GRABeCB2.0 sensor in CA1 neurons of the hippocampus after AAV2/9-hSyn-GRABeCB2.0 injections (Farrell et al. 2021). The seizure-induced eCB signal was sensitive to changes in 2-AG but not AEA levels after enzymatic manipulation, suggesting that electrical seizures drive excessive 2-AG levels in the hippocampus likely as a neuroprotective signal to suppress seizures. In contrast, kainic acid-induced seizures generated increased AEA but not 2-AG levels in the hippocampus (Marsicano et al. 2003).

Although endocannabinoid-mediated neuroprotection depends on CB1 on glutamatergic neurons, most endocannabinoid-regulated brain functions and behaviours require a coordinated control of both excitatory and inhibitory inputs of the contributing circuits.

The use of different conditional mutants led researchers dissect how the ECS regulates different aspects of stress response. Acute stress caused by forced swim test, for example, leads to increased corticosterone release and immobility in wild-type mice. In CaMK-CB1-KO mice, however, corticosterone secretion was significantly higher, while the behaviour was not different from wild-type littermates, whereas Glu-CB1-KO mice displayed decreased immobility but normal corticosterone release, showing that behavioural and neuroendocrine acute stress coping is mainly influenced by CB1 signalling on different neuronal subpopulations. In this paradigm, GABA-CB1-KO mice did not show altered hormonal or behavioural response (Steiner et al. 2008). Repeated social defeat stress, on the other hand, increased conditioned freezing behaviour in Glu-CB1-KO mice, while locomotor reactivity was particularly reduced in stressed GABA-CB1-KO mice (Dubreucq et al. 2012a).

The ECS exerts opposite functions on the balance between novelty seeking and behavioural inhibition. Both object and social investigation were decreased in Glu-CB1-KO (Häring et al. 2011; Jacob et al. 2009) and increased in GABA-CB1-KO (Häring et al. 2011). Glu-CB1-KO mice also showed reduced interaction with female conspecifics (Terzian et al. 2014). Similarly, interaction with a novel object or new palatable food pellet was decreased in Glu-CB1-KO and increased in GABA-CB1-KO mice (Lafenêtre et al. 2009). Likewise, social and object exploratory behaviour was decreased in GABA-CB1-RS animals and fully rescued in Glu-CB1-RS mice (De Giacomo et al. 2020), showing that CB1 in cortical glutamatergic neurons favours novelty seeking, while in forebrain GABAergic neurons promotes behavioural inhibition.

The ECS is already expressed in the embryonic brain and contributes to the development of the cerebral cortex. Thus, GABAergic interneurons in GABA-CB1-KO show impaired target selection in layer 2/3 of the neocortex with a significantly increased probability of pyramidal cells receiving $VGAT^{+}/VGLUT3^{+}$ inputs (Berghuis et al. 2007). Glu-CB1-KO mice, on the other hand, develop deficits in neuronal progenitor proliferation, axon fasciculation of long-range axons (Mulder et al. 2008), deficits in corticospinal projection neuron development and, as a consequence, abnormal skilled motor function (Díaz-Alonso et al. 2012). These data show that endocannabinoids and CB1 are important axon guidance cues shaping the connectivity of the developing brain.

The response to exogenous cannabinoids strongly depends on the cell-type and cellular compartment of CB1 expression. For instance, both hypothermic and hypolocomotor effects of Δ^9-THC were impaired in Glu-CB1-KO but not in GABA-CB1-KO mice, while the analgesic and cataleptic effects of THC were present in both genotypes (Monory et al. 2007). Correspondingly, THC-induced hypothermia and reduced locomotion were present in Glu-CB1-RS. Interestingly, THC-induced hyperlocomotion in GABA-CB1-RS mice. Neither analgesia nor catalepsy was rescued in either of these mice, indicating that these effects of Δ^9-THC are largely independent of cortical glutamatergic and GABAergic CB1 (De Giacomo et al. 2022). The amnesic-like effects of THC depend on GABAergic CB1 as Δ^9-THC disrupted memory in WT and Glu-CB1-KO but not in GABA-CB1-KO mice in an object recognition test (Puighermanal et al. 2009). A well-known characteristic of cannabinoid treatment is their biphasic effect, i.e. low, and high doses often causing opposite effects. For example, low doses of cannabinoids are anxiolytic, while high doses are anxiogenic. Treatment of Glu- and GABA-CB1-KO mice with 1 or 50μg of the cannabinoid agonist CP 55,940 revealed that activation of CB1 on glutamatergic and GABAergic cells was responsible for the biphasic response of cannabinoids in anxiety. The anxiolytic-like effect of the low dose of cannabinoids was abolished in Glu-CB1-KO mice, while the anxiogenic-like effect under a high-dose treatment was abolished specifically in GABA-CB1-KO mice (Rey et al. 2012). On a similar note, low and high doses of Δ^9-THC have orexigenic and hypophagic effects, respectively, and these effects could also be attributed to the activation of CB1 on cortical glutamatergic or GABAergic neurons (Bellocchio et al. 2010). Thus, orexigenic low dose of Δ^9-THC had no effect on Glu-CB1−KO mice but increased food intake in GABA-CB1−KO mice, whereas the higher dose decreased food intake in Glu-CB1−KO but had no effect in GABA-CB1−KO mice. Furthermore, exogenous THC administration exerted a biphasic effect on fear coping strategies too, with low doses favouring active coping and higher doses promoting passive coping responses in classical fear conditioning, respectively (Metna-Laurent et al. 2012).

Similarly to Δ^9-THC, endocannabinoids can affect food intake through the regulation of GABA and glutamate release as well. Hence, fasting-induced food intake was decreased in Glu-CB1−KO mice and increased in GABA-CB1−KO mice (Bellocchio et al. 2010). Olfaction guides eating behaviour using an array of neuroendocrine molecules (Fadool and Kolling 2020). CB1 is highly expressed in the olfactory bulb, especially on glutamatergic axon terminals innervating olfactory granule cells. Moreover, fasting increased AEA levels in the olfactory bulb (Soria-Gómez et al. 2014). Virally induced CB1 deletion in the olfactory bulb resulted in a hypophagic phenotype, similarly to CB1-KO and Glu-CB1-KO, and virally induced restoration of CB1 expression in the olfactory bulb of mice lacking CB1 (CB1-STOP mice) increased food intake. Olfactory tests showed that cannabinoids increase not only food intake but odour detection as well. Glu-CB1-KO mice exhibited reduced odour perception, food-seeking behaviour and a decreased motivation to obtain palatable food (Ruiz de Azua et al. 2021). Accordingly, Glu-CB1-KO mice showed decreased body weight and food intake when fed with a palatable diet. Rescued CB1 expression in glutamatergic neurons in the olfactory cortex partly restored diet-induced obesity in Glu-CB1-KO mice and improved odour perception. Thus, the endocannabinoid-mediated control of cortical glutamatergic transmission in the olfactory system links olfaction to food intake/feeding behaviour. Furthermore, endocannabinoids also control whole-body energy homeostasis. To complement studies performed with CB1 mutants, transgenic mice were generated overexpressing MAGL in the forebrain (CaMK-MAGL-Tg; Jung et al. 2012). CaMK-MAGL-Tg showed increased MAGL activity and decreased 2-AG levels throughout the adult forebrain. CaMK-MAGL-Tg had reduced body weight, reduced body fat mass, and reduced blood glucose despite eating more and moving less than wild types. Furthermore, CaMK-MGL-Tg showed higher body temperature, higher energy expenditure, and higher mitochondrial complex activity in brown adipose tissue (BAT). These data highlight the fundamental role of 2-AG and CB1 in energy homeostasis.

Adding more complexity to the ECS-related mechanisms that govern whole-body metabolism, a common human polymorphism in FAAH (FAAH C385A) gene has been associated with increased body mass index (Durand et al. 2008; Monteleone et al. 2008; Sipe et al. 2005) although additional studies showed conflicted data (Jensen et al. 2007). FAAH C385A mutation drives a more labile FAAH protein, and a concomitant increase in AEA levels (Sipe et al. 2002). A knock-in mouse model containing FAAH C385A mutation was prone to glucocorticoid-induced hyperphagia and weight gain but not in basal conditions (Balsevich et al. 2023). This evidence suggests that controversial data in humans might be related to individual differences in environmental conditions that dictate the endocrine state. For example, chronic stress and high-fat diet enhance circulating glucocorticoid levels among other appetite signals.

Endocannabinoids play a complex role in fear memory processing, including consolidation, retrieval, reconsolidation, and extinction of fear memory (Mizuno and Matsuda 2021; Ruehle et al. 2012). CB1 is required on both glutamatergic and GABAergic neurons for coordinated control of excitatory and inhibitory inputs within fear-regulating circuits. Thus, CB1 on cortical glutamatergic neurons alleviated whereas CB1 on GABAergic neurons enhanced freezing response to conditioned fear (Llorente-Berzal et al. 2015; Metna-Laurent et al. 2012). At the same time, the deletion of CB1 in GABAergic neurons favoured active coping whereas the deletion restricted to cortical glutamatergic neurons promoted passive coping (Metna-Laurent et al. 2012). Interestingly, the rescue of either cortical glutamatergic (Ruehle et al. 2013) or forebrain GABAergic CB1 (Remmers et al. 2017) was insufficient to restore wild-type behaviour in fear conditioning. Unpredictable threat leads to sustained fear and CB1 on distinct amygdala inputs to BNST determine the fear profile in response to predictable versus unpredictable threat (Lange et al. 2017). Local blockade of CB1 in the BNST led to a phasic, instead of sustained fear response to unpredictable threats. Deleting CB1 from amygdala-BNST projections by injecting AAV-Cre into the basolateral (BA) and central amygdala (CeA) of $CB1^{f/f}$ mice produced the same result. Conversely, AAV-Cre injections to BA/CeA of Stop-CB1 mice locally rescued CB1 expression in amygdala-BNST projections, and this intervention reinstated sustained fear to unpredictable stimuli. Interestingly, both Glu-CB1-RS and GABA-CB1-RS mice showed an intermediate phenotype between phasic and sustained fear. To uncover the importance of CB1 in both GABAergic and glutamatergic neurons, CB1 expression was rescued in GABAergic cells in CeA of Glu-CB1-RS and glutamatergic cells in BA of GABA-CB1-RS mice. AAV-Cre-induced CB1 rescue in amygdala-BNST projections in this manner reinstated sustained fear in both genotypes showing that CB1 in these local subpopulations is necessary and sufficient for sustained fear to unpredictable threat stimuli. The basolateral amygdala (BLA) is a part of the **amygdala** that plays a key role in fear learning and stress-related responses (Li 2019). Recent evidence showed that endocannabinoid signalling in the BLA is important in fear expression using a recently developed selective endocannabinoid biosensor (called GRABeCB2.0) (Dong et al. 2022). Mice expressing the GRABeCB2.0 sensor in BLA neurons showed a transient increase in GRABeCB2.0-related fluorescence when they received a 2 s foot shock (Dong et al. 2022). To elucidate the neuronal circuits underlying, a new study expressed the GRABeCB2.0 sensor (AAV9-hSyn-GRABeCB2.0) in the ventral part of the hippocampus (vHP) whereas changes in fluorescence were recorded in the BLA using a fibre optic (Kondev et al. 2023). In these animals, the foot shock delivery (2 s, 0.7 mA) promoted an increase in GRABeCB2.0 sensor activity in vHP-BLA terminals indicating that stress-induced endocannabinoid release at vHP-BLA synapses. Pharmacological inhibition of DAGL demonstrated that this increase in eCB2.0 sensor activity was related to stress-induced 2-AG release in the BLA. To further investigate whether CB1 in vHP-BLA terminals regulates stress-related responses, a combinatorial viral approach, injecting a retrograde Flp recombinase expressing virus (rgAVV2-Ef1a-mCherry-IRES-Flpo) into BLA and a Flp-dependent Cre virus (AAV5-CMV-fDIO-Cre) in the vHP of the CB1 floxed mice, selectively deleted CB1 in the vHP-BLA pathway. The selective lack of CB1 from vHP-BLA terminals resulted in reduced active coping responses during the restraint stress, tail suspension and forced swim test (Kondev et al. 2023). In summary, these findings revealed an instrumental role of the ECS in the amygdala in fear expression as well as in stress-related passive coping responses.

In line with the role of the ECS in fear and stress-related responses, a growing number of evidence suggest that ECS favours fear extinction and hereby becomes an attractive therapeutic target in general anxiety and post-traumatic stress disorders (PTSD) (Marsicano et al. 2002; Gunduz-Cinar et al. 2013; Lutz et al. 2015; Mayo et al. 2022; Singewald et al. 2023). A recent study shed new light on the temporal dynamics of endocannabinoids during fear extinction in mice using the GRABeCB2.0 sensor (Gunduz-Cinar et al. 2023). The sensor was expressed in the medial PFC (mPFC) by viral delivery (AAV9-hSyn-eCB2.0), and eCB2.0-related signal was recorded in the BLA. Confirming previous evidence, the eCB2.0 signal was increased associated with a foot shock during fear conditioning but also, most strikingly, after tone presentation (CS off period) during extinction training. Notably, this response was attenuated during late extinction training as compared to early extinction training suggesting that the mPCF-BLA endocannabinoid signal is involved in fear extinction learning and, hereby decreased when fear levels are low (late extinction training). To elucidate the putative effect of CB1 at mPFC-BLA synapses in fear extinction, CRISPR/Cas9 gene editing was applied for a selective CB1 loss-of-function in mPFC-BLA neuronal projections using an intersectional viral strategy (Gunduz-Cinar et al. 2023). A Cre-dependent single virus (pAAV-AAV1-CMV-FLEX-SaCas9-U6-sgCNR1) delivered SaCas9 and a sgRNA targeting the CB1 gene into mPFC, whereas a retrograde virus containing Cre recombinase (rgAAV-HSyn-Cre-P2A-dTomato) was injected into BLA. Thus, mice lacking CB1 in the mPFC-BLA circuit showed impairment in fear extinction and a higher anxiety-like behaviour as compared to control mice (Gunduz-Cinar et al. 2023). The application of a new genetic toolbox leads to an unprecedented understanding of endocannabinoid functions in stress-related responses.

Serotonergic neurons form the raphe nuclei, and their axons innervate almost every part of the central nervous system (Conrad et al. 1974). Both serotonin and endocannabinoids were shown to influence various physiological functions and control a wide range of behaviours and emotional states. Therefore, functional interactions between the ECS and the serotonergic system were explored early on. Indeed, cannabinoid agonists were able to decrease the release of serotonin in mouse cortical slices (Nakazi et al. 2000) and serotonergic agonists showed an altered pharmacological profile in CB1 knockout mice (Urigüen et al. 2004). The antidepressant-like properties of CB1 agonists were shown to be mediated by serotonin both in rats (Bambico et al. 2007) and in mice (Häring et al. 2013).

Moreover, blocking AEA degradation in rats resulted in increased firing of dorsal raphe nucleus serotonergic neurons (Gobbi et al. 2005). Consequently, a low but significant portion of serotonergic neurons in the raphe nuclei was shown to express CB1 mRNA (Häring et al. 2007) and expression of CB1 protein was identified on serotonergic terminals in the hippocampus, amygdala (Häring et al. 2007) and frontal cortex (Ferreira et al. 2012).

Mice lacking CB1 specifically in serotonergic neurons were generated crossing the CB1 floxed mice and the TPH2-CreERT2 mice (Weber et al. 2009), where the inducible Cre recombinase expression is under the regulatory control of the mouse gene tryptophan hydroxylase 2 (Tph2), the rate-limiting enzyme for the synthesis of serotonin. TPH2-CB1-KO mice showed increased anxiety and decreased sociability compared to their wild-type littermates (Häring et al. 2015). These mice also showed an altered stress response. Repeated social defeat increased water intake but decreased sucrose intake more in mutants than in wild types (Dubreucq et al. 2012a). TPH2-CB1-KO and their wild-type littermates consumed a comparable amount of food after fasting and their food consumption after CB1 blockade (rimonabant) decreased similarly (Bellocchio et al. 2013), suggesting that CB1 on serotonergic neurons is not necessary for endocannabinoid control of food intake. CB1 on serotonergic neurons was also not responsible for the deficit observed in the object-recognition memory task triggered by acute stress (Busquets-Garcia et al. 2016).

Noradrenergic neurons reside mainly in a small brainstem nucleus, the locus coeruleus (LC; Dahlstroem and Fuxe 1964). Despite its small number of neurons, the LC receives inputs from most brainstem and forebrain regions and sends its projections to almost all cortical and subcortical brain areas (Breton-Provencher et al. 2021). LC neurons release noradrenaline/norepinephrine (NE) to control arousal and a variety of sensory-motor, cognitive, and behavioural functions (Sara 2009).

Although direct evidence has been scarce about the presence of a functional ECS in noradrenergic neurons, ample evidence suggested an interaction between the cannabinoid and noradrenergic systems. Importantly, cannabinoids were shown to modulate noradrenergic transmission in both noradrenergic nuclei and target regions. These interactions were frequently associated with significant functional and behavioural effects, often in a manner relevant to psychiatric disorders such as aversion or anxiety (Carvalho and Van Bockstaele 2012).

Noradrenergic cells were targeted using the Cre recombinase under the regulatory sequences of the dopamine-β-hydroxylase (DBH), the enzyme catalysing the conversion of dopamine to norepinephrine (Parlato et al. 2007). DBH^+ cells are localised in the LC, sympathetic ganglion neurons, and adrenal chromaffin cells (Mercer et al. 1991). DBH-CB1-KO mice were generated by crossing $CB1^{f/f}$ with DBH-Cre mice (Parlato et al. 2007). DBH-CB1-KO mice were insensitive to stress-induced impairment of non-emotional memory consolidation (Busquets-Garcia et al. 2016). Accordingly, in mice where CB1 expression was rescued (Ruehle et al. 2013) specifically in DBH^+ cells (DBH-CB1–RS) stress-induced memory impairment was present, similarly to the wild type. These data show that CB1 in DBH-expressing cells was necessary and sufficient for stress-induced memory impairment. CB1 in DBH$^+$ cells was also found to be an important regulator in obesity and metabolic syndrome as well (Srivastava et al. 2022). Thus, in a diet-induced obesity model, DBH-CB1-KO mice were protected from diet-induced weight gain, while they showed reduced adipogenesis in white adipose tissue and enhanced thermogenesis in brown adipose tissue. DBH-CB1-KO showed also increased plasma NE levels and NE turnover, compared to controls. Furthermore, the mutant mice showed decreased plasma corticosterone levels. Investigating the skeletal phenotype of DBH-CB1-KO mice showed increased bone mass in aged animals but no change in young mice (Deis et al. 2018).

Dopaminergic neurons in the midbrain are the main source of dopamine (DA) in the CNS. Dopaminergic neurons are found mostly in the substantia nigra pars compacta and ventral tegmental area (VTA) and are mainly involved in motivation, reward learning and voluntary movement. Numerous studies showed that cannabinoids and endocannabinoids can regulate DA transmission via multiple mechanisms (Cheer et al. 2003; Covey et al. 2017; Melis et al. 2004; Riegel and Lupica 2004). Based on this evidence, it is generally assumed that CB1 activation in GABAergic interneurons in the VTA or CB1 activation of glutamatergic projections, e.g. from prefrontal cortex to VTA, are responsible for the cannabinoid-modulated reward mechanisms. Recently, a study found that CB1 is expressed in a subset of DA neurons in the middle portion of the VTA using an RNAscope ISH assay (Han et al. 2023), in addition to its expression in midbrain GABAergic and glutamatergic neurons (Han et al. 2017; Harlan et al. 2018). The optogenetic activation of the medial DA neurons of the VTA promotes self-stimulation, whereas selective CB1 deletion from DA neurons reduces optogenetic-stimulating response, which suggests that tonic DA signalling was intrinsically increased in the absence of CB1 in these neurons (Han et al. 2023). Furthermore, the lack of CB1 in DA cells blocked the THC-mediated DA release in NAc, hypolocomotion and anxiogenic effect (Han et al. 2023). An additional study confirmed the CB1 expression in mesencephalic DA (mesDA) neurons in the VTA (Baddenhausen et al. 2024). To elucidate the role of CB1 in mesDA cells, a new conditional CB1 KO mouse line was generated bearing the lack of CB1 in engrailed-1 (En1)-positive cells. En1 is expressed in mesDA precursor and mature dopaminergic neurons. This line, called CB1-EN1-KO, revealed that CB1 lacking in these cells reduced palatable food motivation and increased despair behaviour, without affecting locomotion (Baddenhausen et al. 2024). In summary, these new studies substantiated the role of CB1 in dopaminergic neurons and showed that the ECS regulates DA functions via multiple mechanisms and that the final net effect of exogenous and endogenous cannabinoids may depend on CB1 expression in different neuronal populations.

Hypothalamic neurons The hypothalamus, an assembly of small nuclei situated at the ventral diencephalon, can exert widespread influence over bodily states to maintain homeostasis. The hypothalamus, also a key site for central control of the autonomic nervous system, affects the endocrine-vascular system through its intrinsic connection with the pituitary gland and coordinates them both. Several hypothalamic nuclei display low levels of CB1 mRNA (Marsicano and Lutz 1999) and ample evidence suggests that the ECS in hypothalamic neurons regulate food intake and energy balance (Pagotto et al. 2006) as well as stress hormone secretion under basal and stressful conditions (Steiner and Wotjak 2008).

CB1 was specifically deleted from the paraventricular nucleus (PVN) using a Cre line under the control of the brain patterning transcription factor single-minded 1 (Sim-1-Cre; Balthasar et al. 2005). Sim1-CB1-KO displayed less anxiety and more locomotor activity than their wild-type littermates under basal conditions (Dubreucq et al. 2012b). Repeated social stress, however, increased anxiety and sucrose preference in Sim1-CB1-KO mice. Furthermore, body weight, food intake, lean mass, fat mass and energy expenditure did not differ between Sim1-CB1-KO and Sim1-CB1-WT mice. At the same time, fasting-induced food intake with or without rimonabant treatment was similar in wild-type and Sim1-CB1-KO mice (Bellocchio et al. 2013). Conversely, when animals were fed a high-fat diet, Sim1-CB1-KO showed less body weight, lower fat mass, lower blood insulin and higher energy expenditure, a phenotype that could be rescued by chemical sympathectomy (Cardinal et al. 2015).

Corticotrophin-releasing factor (CRF) is released from PVN into the circulation as part of the hypothalamic-pituitary-adrenal (HPA) axis. CRF and ECS regulate stress responses in an opposite manner and co-localised in several brain regions (Ruat et al. 2021). Therefore, conditional CB1 knockout mice were generated by crossing CB1 floxed (Marsicano et al. 2003) and CRF-IRES-Cre (Taniguchi et al. 2011) mice to selectively delete CB1 in CRF+ neurons. CRF-CB1-KO mice displayed a similar phenotype to control mice, with no effects on fear responses, anxiety-like behaviour or plasma CORT levels. However, mutant mice had a significantly enhanced acoustic startle response, suggesting a hyperarousal trait, but only in male mice and not in female mice (Ruat et al. 2021).

Two independent groups generated mice carrying a deletion of CB1 in steroidogenic factor-1 (SF1) positive neurons of the ventromedial nucleus (VMN): one using ES technology (Cardinal et al. 2015) and one using targeted genome editing (Castorena et al. 2021). Both groups found that SF1-CB1-KO mice on standard diet did not differ from their wild-type littermates in food intake or body weight. However, Cardinal et al. (2015) found that SF1-CB1-KO mice had decreased fat mass, increased glucose tolerance and increased insulin sensitivity. Furthermore, VMN-specific CB1 deletion in these mice increased lipolysis and the metabolic effects of leptin as well. Cardinal et al. (2015) reported increased body weight, fat mass, and food intake in SF1-CB1-KO mice on HFD with leptin resistance and decreased peripheral use of lipid substrates. Castorena et al. (2021), on the other hand, found no differences in body weight, body composition or energy consumption. Fasting-induced hyperphagia was reduced in SF1-CB1-KO mice (Bellocchio et al. 2013). However, similarly to Sim1-CB1-KO mice, rimonabant decreased food intake was not different between wild types and mutants, suggesting that CB1 in the VMN are required for the endogenous control of fasting-induced food intake, but they are dispensable for the hypophagic effect of rimonabant. Likewise, the deletion of CB1 from the VMN did not affect rimonabant-induced weight loss in HFD-fed mice (Castorena et al. 2021).

Adult Neural Stem Cells Adult neurogenesis is a process of generating functional neurons from adult neural precursors (NPs) or neural stem cells (NSCs), which are self-renewing, multipotent cells residing in so-called neurogenic niches in the adult mammalian brain: the subventricular zone (SVZ) in the lateral wall of the lateral ventricles and the subgranular zone (SGZ) in the dentate gyrus of the hippocampus. Newborn neurons generated in the neurogenic niches migrate from the SGZ into the granule cell layer, and from the SVZ to the olfactory bulb and are incorporated into the local circuitry to regulate multiple physiological functions (Ribeiro and Xapelli 2021).

It has already been established that the ECS plays a significant role in CNS development (Harkany et al. 2007), and subsequently, a major role in regulating adult neurogenesis has also emerged. Various studies evidenced that eCBs control basic processes, such as neuronal proliferation (Rubio-Araiz et al. 2008), migration (Oudin et al. 2011; Zhou et al. 2015), and differentiation (Aguado et al. 2006), all of which are also involved in adult neurogenesis. Many of these findings also indicated that increased cannabinoid receptor activity produces neurogenic effects in vitro (Compagnucci et al. 2013) and in vivo (Jiang et al. 2005). Moreover, NSCs were shown to contain a functional endocannabinoid system (Aguado et al. 2005; Oudin et al. 2011).

Studies with transgenic mice further substantiated this notion. CB1-KO mice showed reductions in the number of proliferating cells both in the dentate gyrus and in the subventricular zone (Jin et al. 2004). Furthermore, analysis of in vivo cell fate of proliferating hippocampal NSCs in CB1-KO and FAAH-KO mice suggested that eCBs promote adult hippocampal astrogliogenesis in vivo (Aguado et al. 2006). Interestingly, running-induced neurogenesis in the hippocampus was not affected by CB1 deletion. Even though hippocampal neurogenesis was lower by 40% in CB1−KO mice compared to wild types, a 6-week running protocol increased neurogenesis in both genotypes to a similar extent (Dubreucq et al. 2010). Likewise, deletion of CB2 resulted in a significant decrease in proliferating cells in the dentate gyrus of the hippocampus (Palazuelos et al. 2006). Moreover, CB2 deletion also blocked excitotoxicity-induced adult hippocampal NP proliferation (Palazuelos et al. 2012). In accordance with these data and with previous experiments exploring the effects of DAGL inhibitors (Goncalves et al. 2008), it was found that adult neurogenesis is compromised in the SVZ of DAGLα-KO, but not of DAGLβ-KO mice (Gao et al. 2010). At the same time, both DAGLα-KO and DAGLβ-KO showed reduced cell proliferation in the SGZ.

Even though these studies firmly established the role of the ECS in the physiology of neural progenitors in the adult brain, the question has remained open whether the effects described above are cell-autonomous and/or whether indirectly other cell types play a significant part in cannabinoid regulation of adult neurogenesis. To address the direct function of NSC lineage-specific CB1 in the adult mouse in vivo, NSC-specific conditional CB1 knockout animals were generated using Nestin-CreERT2 transgenic mice expressing tamoxifen-inducible Cre recombinase in neuronal and glial cell precursors in the adult and developing mouse brain (Lagace et al. 2007). To visualise recombined cells, the offspring was additionally crossed with the reporter mouse line ROSA26-floxed-stop-YFP (Srinivas et al. 2001). In the resulting Nes-CB1-KO-YFP mice, proliferation capacity of newborn cells was significantly reduced in the SGZ but not in the SVZ (Zimmermann et al. 2018). NSC-specific CB1 deletion also led to a reduction in the number of newborn neurons and a significantly reduced dendritic length as well as a significantly decreased density of dendritic spines of newborn neurons. Apart from these morphological alterations, Nes-CB1-KO-YFP mice showed functional changes as well: impairments in certain forms of hippocampal synaptic plasticity, deterioration of spatial learning and enhancement of behavioural despair.

To reveal the specific cell types producing 2-AG-induced neurogenesis, DAGLα was specifically deleted from either all neurons or neural progenitors (Schuele et al. 2022). Neuronal-type conditional knockouts were created using the synapsin-Cre line (Zhu et al. 2001), while NSC-specific conditional knockouts were created using GLAST-CreERT2 mice, which carry a tamoxifen-inducible Cre under the control of the glutamate/aspartate transporter, which is expressed in adult astrocytes as well as radial glial cells (Mori et al. 2006). Proliferation of NSCs was significantly decreased in GLAST-DAGLα-KO mice. At the same time, survival of newborn neurons and differentiation was not affected as illustrated by the similar ratio of BrdU positive cells on days 1 and 21 and unchanged relative amounts of neurons and glia in both genotypes, respectively. Interestingly, Syn-DAGLα-KO mice showed neither changes in the proliferation nor in the survival or differentiation of NSPC in the dentate gyrus. Thus, DAGLα, though expressed in NSCs at lower levels than in neurons, is instrumental in producing the endocannabinoids that enhance in an autocrine/paracrine fashion neurogenesis in the adult hippocampus.

3.2 Glial Cells

While the presence of CB1 on distinct neuronal types in the brain has been well established, the presence of the cannabinoid receptors and the eCB synthesising and degrading enzymes on glial cells has been more elusive. The first evidence was reported in in vitro cultured human and rodent glial cells, as the low levels of ECS protein or mRNA expression made it difficult to detect in brain tissue. By now, however, the presence of the ECS in astrocytes, oligodendrocytes, and microglia has been firmly established and their function has been demonstrated using cell-type specific mutant mice.

Astrocytes or astroglia, named after their star-shaped morphology, are the most abundant type of glial cells in the mammalian brain. Astrocytes originate from radial glia (Bayraktar et al. 2014), and they are the primary homeostatic cells of the CNS, which is manifested by a multitude of functions (Verkhratsky et al. 2021). For example, astrocytes provide structural support for neurons, both during development and in the adult brain. They also participate in the formation of the blood-brain barrier, are able to regulate local blood flow by releasing prostaglandins, nitric oxide, or arachidonic acid, and can transport nutrients from the blood to the neurons.

Astrocytes also play a major role in the repair and scarring process after brain injury. These multiple functions are exerted by diverse populations of astrocytes with different molecular identities and distinct morphological features (Miller 2018). Astrocytes cannot generate action potentials, instead, they are activated by rising intracellular Ca^{2+} levels mostly through Gq-coupled GPCRs. Astrocytes communicate through gap junctions that enable a rapid flow of information between these cells. A so-called astroglial perisynaptic sheath enwraps the majority of synapses in the CNS. This special compartment regulates, influences, and assists synaptogenesis, modulates synaptic transmission and plasticity, and plays a role in the metabolism of neurotransmitters (Verkhratsky and Nedergaard 2014). The pre- and postsynaptic compartments together with the astroglial perisynaptic sheath are called the tripartite synapse (Halassa et al. 2007). A crucial astroglial function in the tripartite synapse is the uptake of excess neurotransmitters and K^+ from the synaptic cleft.

The presence of CB1 in astrocytes was first shown in astrocyte cultures (Bouaboula et al. 1995; Sánchez et al. 1998), although early anatomical studies could not unequivocally prove the existence of astrocytic CB1 in the brain (Stella 2004). The first demonstration of astrocytic CB1 in brain tissue came from an electron microscopy study (Rodriguez et al. 2001). Navarrete and Araque, using electrophysiology, Ca^{2+} imaging and immunostaining in mouse hippocampal slices showed that endocannabinoids increase astrocytic Ca^{2+} levels through CB1 activation that is coupled, unlike in neurons, to a PTX-insensitive G protein (Navarrete and Araque 2008). The same group also demonstrated that the endocannabinoids that are released by neuronal depolarisation and locally inhibit transmitter release will potentiate more distant synapses through the activation of astrocytic CB1 (Navarrete and Araque 2010).

Astrocyte-specific CB1 knockout mice (GFAP-CB1-KO) were generated using the tamoxifen-inducible GFAP-Cre mice (Hirrlinger et al. 2006). After tamoxifen treatment, GFAP-CB1R-KO mice displayed a 79% reduction in astrocytic CB1 labelling in the CA1. In these mice, cannabinoid treatment did not disrupt working memory and did not induce long-term depression (LTD) of synaptic strength at hippocampal CA3-CA1 synapses (Han et al. 2012). Deleting CB1 from GABAergic or glutamatergic neurons had no effect on THC-induced LTD, demonstrating that indeed THC-induced impairment of working memory depended on CB1 in astrocytes. Astroglial CB1 also plays a specific role in amygdala synaptic plasticity-related fear behaviour by differentially regulating excitatory and inhibitory synaptic transmission through selective activation of specific signalling pathways (Martin-Fernandez et al. 2017). Neuronal depolarisation in the medial nuclei of the central amygdala (CeM) led to astroglial Ca^{2+} mobilisation and gliotransmitter release, which in turn regulated synaptic transmission in a synapse-specific manner: inhibited excitatory synapses from the basolateral amygdala and enhanced inhibitory synapses from the CeM. Decreased excitatory and increased inhibitory transmission could be blocked by CB1 antagonist and was not present in GFAP-CB1-KO. Remarkably, selective activation of astrocytes in the CeM using an AAV-DREADD construct specifically reduced fear response and enhanced extinction in a fear conditioning paradigm. Astroglial CB1 in the hippocampus were also found to be key determinants of consolidation of object recognition memory by regulating the levels of D-serine, a co-agonist of the NMDA receptor (Robin et al. 2018). Together, these studies show that astroglial CB1 is an active contributor to the modification of synaptic plasticity and behaviour in a region- and circuit-specific manner.

To study the specific role of 2-AG in astrocytes, GFAP-MAGL-KO mice were generated (Grabner et al. 2016) using GFAP-Cre mice (Zhuo et al. 2001). Astrocyte-specific deletion of MAGL was not sufficient to induce changes in body weight, fat/muscle composition, or food consumption. Similarly, respiratory quotient and oxygen consumption were not different from wild-type mice. In the same way, the various behaviours assessed, such as locomotor activity, anxiety-like behaviour, depression-like behaviour, social memory, or spatial memory were not dependent on astroglial 2-AG. Astrocyte-specific MAGL deletion, however, attenuated bacterial toxin lipopolysaccharide (LPS)-induced neuroinflammation, shown by significantly decreased levels of proinflammatory cytokines. Interestingly, reduced neuroinflammation in GFAP-MAGL-KO was not dependent on astrocytic CB1, instead, it was due to the reduced arachidonic acid levels and, consequently, downregulated prostaglandin synthesis. Arachidonic levels are produced after 2-AG degradation by the MAGL enzyme.

DAGLα has also been specifically deleted from astrocytes (Schuele et al. 2021). GLAST-DAGLα-KO mice were generated using the GLAST-CreERT2 mice (Mori et al. 2006) carrying the tamoxifen-inducible Cre-recombinase under the control of the glutamate/aspartate transporter, which is expressed in adult astrocytes as well as radial glial cells at earlier developmental stages. Expression analysis in wild-type mice by in situ hybridisation showed that 50–100% of neurons and 30–70% of astrocytes are DAGLα$^+$ depending on the brain region examined. Female GLAST-DAGLα-KO mice showed significantly increased depressive-like behavioural responses in the nesting behaviour, forced swim and sucrose preference tests. Furthermore, maternal behaviour, measured by the pup retrieval test, was significantly reduced, while levels of oestrogen and progesterone, but not other sex hormones, were significantly increased in GLAST-DAGLα-KO females.

Bidirectional communication between neurons and astrocytes shapes synaptic plasticity and behaviour. It is therefore important to point out that often experiments with cell-type specific mutants not only shed light on the function of the ECS in the targeted cell population but on its modulation of astrocyte-neuron interactions as well. For example, eCBs released from neurons cause a Ca^{2+} increase in astrocytes and this Ca^{2+} increase in response causes a potentiated glutamate response in neurons (Navarrete and Araque 2010; Serrat et al. 2021). This potentiation was produced by cannabinoid regulation of gliotransmitter release, such as D-serine, an NMDAR co-agonist that participates in LTP formation and object recognition memory (Robin et al. 2018). Apart from classical neurotransmitters, retrograde lipid messengers are also coordinately regulated by astrocytes and neurons (Viader et al. 2015). 2-AG controls DSE in the hippocampus, but the level of 2-AG is maintained by neuronal and astrocytic MAGL together. Furthermore, astrocytic MAGL converts 2-AG to prostaglandins as well, connecting eCB metabolism to other lipid signalling molecules. Metabolic cooperation between astrocytes and neurons is another important aspect of how these two cell types influence each other's physiology. The astrocyte-neuron lactate shuttle provides lactate for activated neurons as source energy (Wu et al. 2023), a process endocannabinoids are able to modify and through this have an effect on animal behaviour (Jimenez-Blasco et al. 2020). A recent publication found a remarkable "division of labour" between the two main endocannabinoids (Noriega-Prieto et al. 2023). Measuring synaptic transmission and astrocytic Ca^{2+} mobilisation in CA1 slices of DAGLα-KO, NAPE-PLD-KO, and GFAP-CB1-KO mice, it was shown that 2-AG is signalling to neurons, decreasing neurotransmitter release, while AEA is signalling to astrocytes inducing lateral potentiation through Ca^{2+} mobilisation.

Oligodendrocytes (OLs) are specialised glial cells that generate and maintain the myelin sheath around CNS axons. Oligodendrocytes derive from oligodendrocyte precursor cells (OPCs), which are the major proliferating cells in the adult brain (Dawson et al. 2003). OPCs express platelet-derived growth factor receptor A (PDGFRα) and neural/glial antigen 2 (NG2), both of which are used as markers for this cell type (Nishiyama et al. 1996). Myelination occurs both during development and adult life. Myelin formation and maintenance in the adult brain are highly dynamic, and complex processes that continuously take place to ensure the proper sheathing of axons, resulting either from the generation of new OLs or the active production of new membranes from existing OLs (Kaller et al. 2017). Apart from maintaining existing myelin, activity-dependent myelin remodelling occurs as a response to experience, such as learning, stress or social interaction (Mount and Monje 2017). Moreover, oligodendrocytes and OPCs have increasingly been associated with the pathogenesis of psychiatric disorders, including schizophrenia, depression, and PTSD (Birey et al. 2017). For example, abnormal myelin architecture and reduced OL function have been reported in human post-mortem brains of subjects with several psychiatric disorders (Haroutunian et al. 2014). Likewise, significant changes in terms of myelin content and OPC differentiation have been observed in different rodent models of depression, such as chronic social defeat stress (Lehmann et al. 2017).

Both OPCs and mature OLs were shown to express CB1 and CB2 (Molina-Holgado et al. 2002b) as well as the enzymatic machinery to synthesise and degrade the endocannabinoid 2-AG (Gomez et al. 2010). Studies in the following years have established that the ECS is involved in OPC biology, modulating their survival, proliferation, and differentiation. Thus, stimulation of cannabinoid receptors was shown to have anti-apoptotic effects in OPCs in vivo; and these effects were mediated by the PI3K /Akt signalling, a known pathway for OPC survival (Gomez et al. 2011). Similarly, proper OPC proliferation also depends on endocannabinoid signalling, as both blockades of CB1 or blockade of 2-AG synthesis decreased OPC proliferation (Gomez et al. 2015). CB1 stimulation also affects OPC differentiation: CB1 agonists increase OPC branching and accumulation of myelin basic protein (MBP) (Gomez et al. 2011).

OLs are sensitive to ischemic episodes and cannabinoid agonists were shown to increase OPC proliferation after injury in a rodent model of stroke (Sun et al. 2013). Moreover, administration of 2-AG in a rat model of spinal cord injury prevented the spread of white matter injury by protecting myelin from damage and reducing oligodendrocyte loss (Arevalo-Martin et al. 2010).

The most common demyelinating disease of the central nervous system, multiple sclerosis (MS), seems to be influenced by cannabinoids as well: CB1-KO mice were shown to exhibit more severe symptoms in an experimental model of MS than their wild-type littermates. Using an experimental allergic encephalomyelitis (EAE) model (Pryce et al. 2003), CB1-KO mice showed poor recovery from paralysis and elevated levels of neurodegeneration following inflammatory insults compared to wild-type mice.

In a subsequent study, the role of brain and immune cells as well as the role of CB1 and CB2 were simultaneously investigated in a similar EAE model (Maresz et al. 2007). To this end, two CB1 conditional knockout lines were generated: Nestin-CB1-KO, harbouring a deletion in the central and peripheral nervous system in neuronal and glial cell precursors, giving rise to neurons, astrocytes, and oligodendrocytes (Tronche et al. 1999) and Lck-CB1-KO harbouring a deletion in thymocytes (Hennet et al. 1995). Disease onset and maximal clinical score values were not different in the two mutant lines. However, Δ^9-THC-treatment, which significantly alleviated disease severity and delayed disease onset in wild-type mice, had no effect in CB1-KO and Nestin-CB1-KO mice. At the same time, improved clinical scores were present in Lck-CB1-KO, indicating that CB1 in thymocytes has no protective effect against EAE. To address the role of CB2, CB2-KO mice were also tested in the EAE model. CB2-KO mice showed a higher incidence of the disease, increased clinical score and reduced recovery rate. Given that no conditional CB2 knockouts were available, the role of CB2 on lymphocytes had to be tackled in a different manner. Therefore, T lymphocytes were transplanted from immunised wild-type or CB2-KO mice into naïve wild-type mice to induce EAE. EAE induced by CB2-KO T cells exhibited more severe clinical disease and a higher mortality rate. These data show that CB1 in brain cells as well as CB2 in T lymphocytes protect against EAE and that Δ^9-THC, through CB1 on brain cells, alleviates disease symptoms. Along the same lines, a Δ^9-THC-cannabidiol oromucosal spray (Sativex) is now in clinical use for the management of treatment-resistant MS spasticity (Patti et al. 2016). All in all, one could speculate whether oligodendrocytes play at least a partial role in the Δ^9-THC-mediated alleviation of EAE symptoms. Nevertheless, from these sets of data, it cannot be identified which population of cell or cells in the brain are responsible for the protective effects of THC against EAE.

Recently, CB1 had been specifically deleted from OPCs using NG2-CreERT2 (Huang et al. 2014) and Pdgfra-CreERT2 (Rivers et al. 2008) mouse lines. In the resulting NG2-CB1-KO and Pdgfrα-CB1-KO mice (Sánchez-de la Torre et al. 2022) CB1 deletion was initiated in the early postnatal period (P6-P7). OPC survival was not affected by CB1 deletion, but their differentiation was delayed. Similarly, oligodendrocyte maturation was also affected, and as early as P15 a hypomyelinated phenotype could be observed in the mutants, which persisted to adulthood (P60). OPC-specific CB1 knockouts showed anxiety-like behaviours, motor, and memory impairments. Experiments aimed at uncovering the molecular mechanisms found that CB1 activation inhibited the RhoA/Rho-associated kinase (ROCK) signalling pathway, which is centrally involved in OPC differentiation and OL myelin production (Pedraza et al. 2014) by inducing E3 ubiquitin ligase-dependent RhoA proteasomal degradation. Treating NG2-CB1-WT mice with Δ^9-THC reduced RhoA protein and ROCK activation levels, enhanced OL differentiation and increased myelin-associated protein level. These Δ^9-THC effects were not observed in NG2-CB1-KO mice. Furthermore, by pharmacological blockade of ROCK, it was possible to rescue the dysregulated oligodendrogenesis and myelination phenotype of NG2-CB1-KO as well as the anxiety, motor, and memory deficits. Thus, these data show that Δ^9-THC, through CB1, has a cell-autonomous effect in the modulation of oligodendrogenesis and CNS myelination in vivo.

Remarkably, CB1 localisation in OPCs and OLs in vivo has not been satisfactorily established even after several successful studies showing its functional importance in oligodendrogenesis. CB1 expression in glial cells is low and its detection has been notoriously difficult in this cell population. At the same time, studies employing single-cell RNA sequencing have been able to provide unprecedented in-depth information about gene expression patterns in otherwise difficult-to-characterise cell populations (e.g. Durante et al. 2020; MacDonald et al. 2021). Likewise, oligodendrocytes have been analysed by single-cell transcriptomics as well and data showed that a sizeable proportion of oligodendrocytes expresses CB1, Crip1a, and the endocannabinoid synthesising and degrading enzymes (Marques et al. 2016; http://linnarssonlab.org/oligodendrocytes/). Yet, these data, importantly, lack the spatial component, and if we wish to understand the function of the ECS in OPCs, the "where" matters. A recent publication showed mature oligodendrocytes in the subcortical white matter expressing CB1 on their soma using electron microscopy (Bernal-Chico et al. 2022). Identifying OL somata in the white matter, where very few cells are present, is feasible. However, identifying OLs or OPCs, let alone their processes in grey matter is extremely challenging. This challenging situation could recently be resolved by analysing tissue sections of NG2-EYFP mice (Karram et al. 2008) that express EYFP in $NG2^+$ glia (Manterola et al. 2022). Quantification of double-staining electron microscopy showed that at least 6% of $EYFP^+$ profiles in the adult mouse hippocampus express CB1 in their plasma membrane. This is a low expression level; indeed, CB1 was present at a lower density in OPCs than in glutamatergic cells.

Microglia are the resident innate immune cells as well as the major inflammatory cells of the CNS. In the basal state, they have a small cell body with long, branched processes that they extend and retract in the surrounding parenchyma to safeguard their environment (Nimmerjahn et al. 2005). When a pathogen or a brain injury is detected, microglial cells transform into an activated state characterised by a change in morphology into an amoeboid form. This form is able to release cytokines, prostaglandins and ROS (reactive oxygen species) and clean up invading pathogens, apoptotic cells, or debris by phagocytosis (Kettenmann et al. 2011). Microglial phagocytosis does not exclusively take place in pathological conditions: during development, microglia eliminate cells or cellular substructures, and participate in synaptic pruning (Paolicelli et al. 2011; Wolf et al. 2017).

Isolated resting microglia were shown to express CB1, and their activation was shown to cause substantial upregulation of CB2 receptors (Cabral and Marciano-Cabral 2005), which are normally undetectable in healthy brains. Mice with experimental autoimmune encephalomyelitis express high amounts of CB2 (Maresz et al. 2005) and cannabinoid treatment prevented microglial activation and neurotoxicity in a rodent AD model (Ramírez et al. 2005). Accordingly, post-mortem samples of multiple sclerosis, amyotrophic lateral sclerosis (Yiangou et al. 2006), Huntington's disease (Palazuelos et al. 2009), and AD patients (Ramírez et al. 2005) showed CB1 and CB2 immunostaining together with markers of microglial activation.

Microglia-specific conditional CB1 knockout mice were recently generated (De Meij et al. 2021) using the CX3CR1-Cre mouse line (Yona et al. 2013). In these animals, motor behaviour, anxiety-like behaviour and performance in a memory task were not different compared to wild types. Animals were treated with the bacterial toxin LPS, a potent immune system activator. LPS-induced cytokine production in the brain was significantly stronger in wild types, revealing a proinflammatory role of the microglial CB1. At the same time, sickness behaviour, a set of adaptive behavioural changes that develop in sick individuals during the course of an infection (Dantzer 2001), was more intense in male mutants (i.e., stronger decrease of food intake, more loss of fat mass and body weight). Thus, microglial CB1 was shown to differentially control sickness behaviour in males and females. A model for microglial CB1 loss combined with a chromosomal alteration (16p11dup), representing an increased risk for psychiatric disorders, has recently been created as well (Hasegawa et al. 2023). Adolescent Δ^9-THC exposure in 16p11dup transgenic mice produced microglial abnormalities, reduced excitability of pyramidal tract neurons in mPFC, and social memory deficits. Specifically deleting CB1 from microglia rescued these phenotypes, showing that some important aspects of the deleterious effects of adolescent Δ^9-THC consumption are mediated by microglial CB1.

3.3 Cells Outside the Nervous System

Initially, the periphery was thought to be devoid of CB1. In fact, at first, CB1 had been called the central cannabinoid receptor. At this point, it is evident that CB1 and other elements of the ECS are expressed in various types of peripheral tissues and organs such as muscle, liver, skin, bone, fat, etc. (Maccarrone et al. 2015), and mediate a variety of biological functions. Strikingly, many of the peripheral expression sites participate in the maintenance of the body's energy homeostasis (O'Sullivan et al. 2021; Ruiz de Azua and Lutz 2019). Given that numerous endocannabinoid functions regulate processes such as appetite, energy expenditure, and given that lipid metabolism is a central aspect of the pathophysiology of obesity, considerable research efforts were focussed on elucidating the roles of the ECS at key peripheral sites using cell-type-specific mutants.

CB1 is expressed at a low level in the healthy **liver**; however, the expression level significantly increases in liver disease, a phenomenon that points to the possible involvement of the ECS in the pathophysiology of liver disease (Mallat et al. 2013). Hepatocyte-selective CB1 knockouts (LCB1-KO) were generated using the albumin-Cre line (Postic et al. 1999). LCB1–KO mice fed with a high-fat diet developed obesity similar to wild types, at the same time had less steatosis, hyperglycemia, dyslipidemia, and insulin and leptin resistance than wild-type mice (Osei-Hyiaman et al. 2008). A mouse line with hepatic overexpression of CB1 on a KO background (Liu et al. 2012) corroborated these data, showing that hepatic CB1 is necessary and sufficient for diet-induced insulin resistance. A recent publication, however, questioned the role of CB1 in the development of non-alcoholic fatty liver disease (Wang et al. 2021), as LCB1–KO mice were also reported to be resistant to ethanol-induced steatosis (Jeong et al. 2008) and acute toxin-induced liver injury (Kim et al. 2020). Hepatitis B surface protein (HBs)-transgenic mice spontaneously develop hepatic steatosis. HBs/CB1-KO double transgenic mice, however, had lower body weight, lower liver weight, reduced hepatic triglycerides and lipid droplet accumulation with elevated autophagy than controls (Irungbam et al. 2020). These data show that the ECS has a considerable impact on key processes of chronic liver disease and related metabolic alterations.

CB1 expression has also been shown in **adipose tissue** and cultured adipocytes (Bensaid et al. 2003; Cota et al. 2003). Adipocyte-specific CB1 knockouts were generated using a tamoxifen-inducible adiponectin Cre line (AdipoqCreERT2 line; Sassmann et al. 2010). Ati-CB1–KO had a lower body weight than wild types on a standard diet and were resistant to diet-induced obesity and the ensuing metabolic syndrome. Moreover, HFD-fed Ati-CB1–KO had reduced body fat mass, increased browning of adipose tissue, enhanced energy expenditure, improved insulin sensitivity and an enhanced sympathetic tone in adipose tissue (Ruiz de Azua et al. 2017). Remarkably, inducing CB1 deletion in already obese mice after tamoxifen treatment reversed the obese phenotype, resulted in improved glucose homeostasis, and reduced plasma insulin and leptin levels. Interestingly, recently, an adult pseudo-lean phenotype, caused by low-dose Δ^9-THC exposure during adolescence and characterised by reduced fat mass, increased lean mass and utilisation of fat as fuel, partial resistance to diet-induced obesity and dyslipidaemia, enhanced thermogenesis, and impaired cold- and β-adrenergic receptor-stimulated lipolysis was also dependent on CB1 in adipocytes (Lin et al. 2023). Furthermore, adipocyte-specific NAPE-PLD knockout mice were generated by crossing NAPE-PLD floxed mice with Fabp4-Cre (also called aP2-Cre) mice, in which Cre recombinase is expressed under the control of aP2 (fatty acid binding protein 4) promoter. Fabp4-NAPE-PLD-KO mice developed obesity, adipose tissue inflammation, insulin resistance, glucose intolerance and perturbation of the lipid metabolism (Geurts et al. 2015). These mutant mice also showed reduced adipose tissue browning and decreased thermogenesis. Interestingly, the phenotype is partly mediated by the alteration of gut microbiota as antibiotic treatment improved metabolism in Fabp4-NAPE-PLD-KO mice and gut microbiota transfer to germ-free mice reproduced part of the phenotype.

A number of cell-type-specific conditional mutants were additionally created targeting peripheral organs, many of which were found to have a variety of effects on metabolism. For instance, **pancreatic β-cell** specific CB1 knockouts (β-CB1-KO; González-Mariscal et al. 2018) became more obese on a high-fat/high-sugar diet and had higher insulin secretion than wild types. **Skeletal muscle**-specific CB1-knockout mice (SKM-CB1-KO; González-Mariscal et al. 2019) on a high-fat/high-sugar diet, on the other hand, had increased whole-body lean/fat ratio, increased energy expenditure and increased insulin sensitivity. **Intestinal epithelium**-specific CB1-knockout mice (Int-CB1-KO; Avalos et al. 2020) showed reduced preference for a high-fat/high-sugar diet. When investigating the role of the CB1 in the **kidney**, obese renal proximal tubule cell-specific CB1-knockout mice (RPTC-CB1-KO; Udi et al. 2017) accumulated fewer lipid droplets in the kidney and had higher fatty acid β -oxidation than wild types. A more recent study further established the important metabolic role of proximal tube CB1, showing its critical role in the development of diabetic kidney disease (Hinden et al. 2022). The role of ECS in immune cells was also investigated in obesity and energy homeostasis using **bone marrow** chimeric mice (Miranda et al. 2022). CB1-WT and CB1-KO mice were lethally irradiated and adoptively transferred with donor bone marrow from CB1-WT or CB1-KO mice. CB1-KO mice, whether with wild type or knockout immune cells, were protected from DIO, whereas in mice lacking CB1 in immune cells, obesity, fat mass, caloric intake and glucose intolerance were potentiated. Thus, either as a consequence or cause, CB1 deficiency in immune cells led to increased inflammation in the adipose tissue as well as in the brain. These data highlight the pivotal role of the ECS in regulating a broad spectrum of metabolic effects in peripheral tissues. **Monocyte/macrophage**-specific CB1-KO was generated using the LyaM-Cre line (Wang et al. 2023). CB1 signalling in monocytes and macrophages has a pro-atherosclerotic role through inflammatory cytokine release and mitochondrial oxidative respiration. Interestingly this phenotype is more pronounced in males. **Keratinocyte**-specific CB1-KO, generated using K14-Cre, showed increased and prolonged contact hypersensitivity through increased secretion of proinflammatory cytokines (Gaffal et al. 2013).

3.4 Distinct Expression Sites within the Cell

CB1 is predominantly expressed at axon terminals where it modulates neurotransmitter release by responding to the retrogradely acting endocannabinoids (Noriega-Prieto et al. 2022). A lesser amount of cell surface CB1 modulates neuronal activity by endocannabinoids acting in an autocrine manner (Bacci et al. 2004). However, as already demonstrated in early studies, immunostaining often shows a high amount of CB1 expression intracellularly both in cell cultures (McIntosh et al. 1998) and in brain tissue (Katona et al. 1999). Most of the intracellularly observed receptors are likely being synthesised, going through post-translational modifications, trafficked from the location of synthesis to the axon terminal or getting internalised/recycled.

However, the question arose whether functional CB1 might be present in intracellular organelles as well.

In the past decade, a significant discovery had been the establishment of CB1's presence at **mitochondrial membrane**. Electron microscopy analysis of Glu-CB1-KO and GABA-CB1-KO hippocampi demonstrated that CB1 is present at the outer membranes of approximately 20% of mitochondria both in glutamatergic and in GABAergic cells (Bénard et al. 2012). Most mitochondrial CB1 (mtCB1) was detected in terminals, and just as GABAergic plasma membranes have a higher level of CB1 expression than glutamatergic ones, so do mitochondria in GABAergic cells. CB1 agonists decreased respiration in purified brain mitochondria (Hebert-Chatelain et al. 2014) through Gαi/o proteins, soluble adenylyl cyclase (sAC) and PKA (Hebert-Chatelain et al. 2016). CB1 protein lacking its first 22 N-terminal amino acids (DN22-CB1) does not target mitochondria, a feature that was exploited to study mitochondrion-specific functions of CB1 (Hebert-Chatelain et al. 2016). Treatment of CB1-KO mice intrahippocampally injected with AAV-CB1 with the cannabinoid agonist HU-210 showed an impairment of novel object recognition memory while AAV-DN22-CB1 injected littermates did not, demonstrating that acute cannabinoid-induced memory impairment in mice requires activation of hippocampal mtCB1 (Hebert-Chatelain et al. 2016). To enable further analysis of specific functions of mtCB1, a novel knock-in mouse model lacking mtCB1 was generated replacing the CB1 gene by the coding sequence of the DN22-CB1 protein (DN22-CB1-KI) by homologous recombination (Soria-Gomez et al. 2021). DN22-CB1, apart from not being present in mitochondria, is expressed and distributed the same way as wild-type CB1. DN22-CB1-KI mice responded to cannabinoid agonist treatment with increased latency on the hot plate (antinociceptive effect), but the cataleptic effect of cannabinoids was lost demonstrating that activation of mtCB1 is responsible for the induction of catalepsy by exogenous cannabinoids.

The presence of mtCB1 has been shown in several other cell types as well. For instance, astrocytic mtCB1 has also been detected in the hippocampus (Gutiérrez-Rodríguez et al. 2018) with 11%–13% of the astrocytic mitochondria being CB1 immunopositive. To tackle the function of astroglial mtCB1, mitochondrial function and glucose metabolism were studied in GFAP-CB1-WT and -KO mice (Jimenez-Blasco et al. 2020). Δ^9-THC inhibited the conversion of glucose to lactate and induced impairments in social interaction in GFAP-CB1-WT but not in GFAP-CB1-KO mice. Experiments in cultured astrocytes transfected with wild-type CB1 or DN22-CB1 demonstrated that the effect of Δ^9-THC on lactate depends on mtCB1. Intracerebroventricular (ICV) lactate injection reversed social interaction deficits, demonstrating that astroglial mtCB1 can directly regulate cellular glucose metabolism to eventually modulate behavioural responses. Astroglial mtCB1 was also shown to play an important role in lateral synaptic potentiation, a form of plasticity where neurons that are not in the immediate vicinity of the depolarised neuron get potentiated (Serrat et al. 2021). Cannabinoid agonist WIN55,212-2 increased Ca^{2+} in mitochondria in astrocytes, which in turn potentiated distal neurons. This cannabinoid effect was independent of sAC activation but could be blocked by inhibiting the AKT pathway or the mitochondrial calcium uniporter, the main channel for the entry of calcium into the mitochondrial matrix.

Similarly to brain tissue, about 20% of mitochondria of the skeletal and heart muscle also express CB1. However, in muscle, CB1 density in the plasma membrane is considerably lower than in neurons, thus, most CB1 (60–70%) in muscle tissue is in mitochondria (Mendizabal-Zubiaga et al. 2016) In striated muscle, CB1-KO mice had elevated levels of key enzymes of pyruvate production, and Δ^9-THC treatment inhibited oxygen consumption in isolated wild-type mitochondria. mtCB1 was also detected in about 20% of mitochondria in white adipose tissue (Pagano Zottola et al. 2022). Demonstrating its functional significance, WIN55,212-2 treatment significantly decreased mitochondrial respiration in white adipose tissue isolated from CB1-WT, but not from CB1-KO, Ati-CB1–KO or DN22-CB1 mice. These data are highly relevant because they show that cannabinoid effects on behaviour are partly modulated through the regulation of mtCB1 effects via influencing energy metabolism.

4 Concluding Remarks

Cutting-edge genetic engineering techniques have significantly advanced our understanding of the ECS by generating new transgenic mice. These mutant mice, sometimes in combination with pharmacological, optogenetic, or chemogenetic interventions, allowed us to selectively modify specific components of the ECS and to gain a more comprehensive understanding of the ECS, its diverse roles in different cell types and contexts, and its implications for health and disease.

To further explore the mechanisms of the ECS in the context of the complexity of the brain, circuit-specific genetic manipulations are necessary. At this moment, circuit-specific genetic manipulation is an emerging field in neuroscience, and as numerous valuable tools (such as retrograde and anterograde viral vectors, optogenetics or chemogenetics, inducible genetic tools, trans-synaptic tracing tools) are available, we can anticipate substantial progress in our understanding of neuronal circuits regulated by ECS in the coming years.

One significant feature of the ECS is its presence in a remarkable number of different cell populations with substantially varying levels of expression. During the past few years, highly advanced tools, such as RNAscope, electron microscopy, single-cell transcriptomics or mass spectrometry, made it possible to identify cells with low to very low expression. Remarkably, some sites of low expression (e.g. in astrocytes, oligodendrocytes, mitochondria) proved to be physiologically relevant. With the exceptional sensitivity of these novel methods, however, it will require a multifaceted experimental approach and rigorous data interpretation to identify expression sites with genuine physiological functions.

In recent years, our understanding of the ECS has witnessed remarkable advancements, enabled by increasingly sophisticated tools. Researchers are now capable of introducing known polymorphisms into ECS-related genes using CRISPR-Cas9 technology, allowing for the investigation of genetic variations associated with a gain- or loss-of-function. Additionally, cutting-edge techniques such as cell-type specific manipulation with READR-RNA (Qian et al. 2022), which combines RNA editing with cell-type specificity, provide a level of precision that was previously unattainable. These tools empower scientists to selectively target and modify ECS components within specific cell populations, unveiling intricate details of the ECS's role in various physiological processes. Furthermore, spatial transcriptomics and intersectional genetics (Madisen et al. 2015) will open new dimensions in understanding the spatial distribution of ECS elements within tissues and enable the identification of neural circuits modulated by ECS, offering new insights into its multiple pathophysiological functions.

References

Adli M (2018) The CRISPR tool kit for genome editing and beyond. Nat Commun 9:1911. https://doi.org/10.1038/s41467-018-04252-2

Aguado T, Monory K, Palazuelos J, Stella N, Cravatt B, Lutz B, Marsicano G, Kokaia Z, Guzmán M, Galve-Roperh I (2005) The endocannabinoid system drives neural progenitor proliferation. FASEB J 19:1704–1706. https://doi.org/10.1096/fj.05-3995fje

Aguado T, Palazuelos J, Monory K, Stella N, Cravatt B, Lutz B, Marsicano G, Kokaia Z, Guzmán M, Galve-Roperh I (2006) The endocannabinoid system promotes astroglial differentiation by acting on neural progenitor cells. J Neurosci 26:1551–1561. https://doi.org/10.1523/JNEUROSCI.3101-05.2006

Allen B, Ingram E, Takao M, Smith MJ, Jakes R, Virdee K, Yoshida H, Holzer M, Craxton M, Emson PC, Atzori C, Migheli A, Crowther RA, Ghetti B, Spillantini MG, Goedert M (2002) Abundant Tau filaments and nonapoptotic neurodegeneration in transgenic mice expressing human P301S Tau protein. J Neurosci 22:9340–9351. https://doi.org/10.1523/JNEUROSCI.22-21-09340.2002

Al-Shawi R, Kinnaird J, Burke J, Bishop JO (1990) Expression of a foreign gene in a line of transgenic mice is modulated by a chromosomal position effect. Mol Cell Biol 10:1192–1198. https://doi.org/10.1128/mcb.10.3.1192-1198.1990

Arevalo-Martin A, Garcia-Ovejero D, Molina-Holgado E (2010) The endocannabinoid 2-arachidonoylglycerol reduces lesion expansion and white matter damage after spinal cord injury. Neurobiology of Disease, Neurobiology of developmental disorders involving human higher cognition, vol 38, pp 304–312. https://doi.org/10.1016/j.nbd.2010.02.002

Atchison RW, Casto BC, Hammon WM (1965) Adenovirus-associated defective virus particles. Science 149:754–756. https://doi.org/10.1126/science.149.3685.754

Avalos B, Argueta DA, Perez PA, Wiley M, Wood C, DiPatrizio NV (2020) Cannabinoid CB1 receptors in the intestinal epithelium are required for acute Western-diet preferences in mice. Nutrients 12:2874. https://doi.org/10.3390/nu12092874

Azad SC, Kurz J, Marsicano G, Lutz B, Zieglgänsberger W, Rammes G (2008) Activation of CB1 specifically located on GABAergic interneurons inhibits LTD in the lateral amygdala. Learn Mem 15:143–152. https://doi.org/10.1101/lm.741908

Bacci A, Huguenard JR, Prince DA (2004) Long-lasting self-inhibition of neocortical interneurons mediated by endocannabinoids. Nature 431:312–316. https://doi.org/10.1038/nature02913

Baddenhausen S, Lutz B, Hofmann C (2024) Cannabinoid type-1 receptor signaling in dopaminergic Engrailed-1 expressing neurons modulates motivation and depressive-like behavior. Front Mol Neurosci 17. https://doi.org/10.3389/fnmol.2024.1379889

Balsevich G, Petrie GN, Heinz DE, Singh A, Aukema RJ, Hunker AC, Vecchiarelli HA, Yau H, Sticht M, Thompson RJ, Lee FS, Zweifel LS, Chelikani PK, Gassen NC, Hill MN (2023) A genetic variant of fatty acid amide hydrolase (FAAH) exacerbates hormone-mediated orexigenic feeding in mice. Elife 12:e81919. https://doi.org/10.7554/eLife.81919

Balthasar N, Dalgaard LT, Lee CE, Yu J, Funahashi H, Williams T, Ferreira M, Tang V, McGovern RA, Kenny CD, Christiansen LM, Edelstein E, Choi B, Boss O, Aschkenasi C, Zhang C, Mountjoy K, Kishi T, Elmquist JK, Lowell BB (2005) Divergence of Melanocortin pathways in the control of food intake and energy expenditure. Cell 123:493–505. https://doi.org/10.1016/j.cell.2005.08.035

Bambico FR, Katz N, Debonnel G, Gobbi G (2007) Cannabinoids elicit antidepressant-like behavior and activate serotonergic neurons through the medial prefrontal cortex. J Neurosci 27:11700–11711. https://doi.org/10.1523/JNEUROSCI.1636-07.2007

Bauer M, Chicca A, Tamborrini M, Eisen D, Lerner R, Lutz B, Poetz O, Pluschke G, Gertsch J (2012) Identification and quantification of a new family of peptide endocannabinoids (Pepcans) showing negative allosteric modulation at CB1 receptors. J Biol Chem 287:36944–36967. https://doi.org/10.1074/jbc.M112.382481

Bayraktar OA, Fuentealba LC, Alvarez-Buylla A, Rowitch DH (2014) Astrocyte development and heterogeneity. Cold Spring Harb Perspect Biol 7:a020362. https://doi.org/10.1101/cshperspect.a020362

Bellocchio L, Lafenêtre P, Cannich A, Cota D, Puente N, Grandes P, Chaouloff F, Piazza PV, Marsicano G (2010) Bimodal control of stimulated food intake by the endocannabinoid system. Nat Neurosci 13:281 283. https://doi.org/10.1038/nn.2494

Bellocchio L, Soria-Gómez E, Quarta C, Metna-Laurent M, Cardinal P, Binder E, Cannich A, Delamarre A, Häring M, Martín-Fontecha M, Vega D, Leste-Lasserre T, Bartsch D, Monory K, Lutz B, Chaouloff F, Pagotto U, Guzman M, Cota D, Marsicano G (2013) Activation of the sympathetic nervous system mediates hypophagic and anxiety-like effects of CB_1 receptor blockade. Proc Natl Acad Sci U S A 110:4786–4791. https://doi.org/10.1073/pnas.1218573110

Bénard G, Massa F, Puente N, Lourenço J, Bellocchio L, Soria-Gómez E, Matias I, Delamarre A, Metna-Laurent M, Cannich A, Hebert-Chatelain E, Mulle C, Ortega-Gutiérrez S, Martín-Fontecha M, Klugmann M, Guggenhuber S, Lutz B, Gertsch J, Chaouloff F, López-Rodríguez ML, Grandes P, Rossignol R, Marsicano G (2012) Mitochondrial CB_1 receptors regulate neuronal energy metabolism. Nat Neurosci 15:558–564. https://doi.org/10.1038/nn.3053

Bensaid M, Gary-Bobo M, Esclangon A, Maffrand JP, Le Fur G, Oury-Donat F, Soubrié P (2003) The cannabinoid CB1 receptor antagonist SR141716 increases Acrp30 mRNA expression in adipose tissue of obese fa/fa rats and in cultured adipocyte cells. Mol Pharmacol 63:908–914. https://doi.org/10.1124/mol.63.4.908

Berghuis P, Rajnicek AM, Morozov YM, Ross RA, Mulder J, Urbán GM, Monory K, Marsicano G, Matteoli M, Canty A, Irving AJ, Katona I, Yanagawa Y, Rakic P, Lutz B, Mackie K, Harkany T (2007) Hardwiring the brain: endocannabinoids shape neuronal connectivity. Science 316:1212–1216. https://doi.org/10.1126/science.1137406

Bernal-Chico A, Tepavcevic V, Manterola A, Utrilla C, Matute C, Mato S (2022) Endocannabinoid signaling in brain diseases: emerging relevance of glial cells. Glia. https://doi.org/10.1002/glia.24172

Birey F, Kokkosis AG, Aguirre A (2017) Oligodendroglia-lineage cells in brain plasticity, homeostasis and psychiatric disorders. Curr Opin Neurobiol 47:93–103. https://doi.org/10.1016/j.conb.2017.09.016

Bluett RJ, Báldi R, Haymer A, Gaulden AD, Hartley ND, Parrish WP, Baechle J, Marcus DJ, Mardam-Bey R, Shonesy BC, Uddin MJ, Marnett LJ, Mackie K, Colbran RJ, Winder DG, Patel S (2017) Endocannabinoid signalling modulates susceptibility to traumatic stress exposure. Nat Commun 8:14782. https://doi.org/10.1038/ncomms14782

Bouabe H, Okkenhaug K (2013) A protocol for construction of gene targeting vectors and generation of homologous recombinant embryonic stem cells. Methods Mol Biol 1064:337–354. https://doi.org/10.1007/978-1-62703-601-6_24

Bouaboula M, Bourrié B, Rinaldi-Carmona M, Shire D, Le Fur G, Casellas P (1995) Stimulation of cannabinoid receptor CB1 induces krox-24 expression in human astrocytoma cells. J Biol Chem 270:13973–13980. https://doi.org/10.1074/jbc.270.23.13973

Breton-Provencher V, Drummond GT, Sur M (2021) Locus Coeruleus norepinephrine in learned behavior: anatomical modularity and spatiotemporal integration in targets. Front Neural Circuits 15:638007. https://doi.org/10.3389/fncir.2021.638007

Buckley NE, McCoy KL, Mezey E, Bonner T, Zimmer A, Felder CC, Glass M, Zimmer A (2000) Immunomodulation by cannabinoids is absent in mice deficient for the cannabinoid CB (2) receptor. Eur J Pharmacol 396:141–149. https://doi.org/10.1016/s0014-2999(00)00211-9

Busquets-Garcia A, Gomis-González M, Srivastava RK, Cutando L, Ortega-Alvaro A, Ruehle S, Remmers F, Bindila L, Bellocchio L, Marsicano G, Lutz B, Maldonado R, Ozaita A (2016) Peripheral and central CB1 cannabinoid receptors control stress-induced impairment of memory consolidation. Proc Natl Acad Sci U S A 113:9904–9909. https://doi.org/10.1073/pnas.1525066113

Cabral GA, Marciano-Cabral F (2005) Cannabinoid receptors in microglia of the central nervous system: immune functional relevance. J Leukoc Biol 78:1192–1197. https://doi.org/10.1189/jlb.0405216

Cardinal P, Bellocchio L, Guzmán-Quevedo O, André C, Clark S, Elie M, Leste-Lasserre T, Gonzales D, Cannich A, Marsicano G, Cota D (2015) Cannabinoid type 1 (CB1) receptors on Sim1-expressing neurons regulate energy expenditure in male mice. Endocrinology 156:411–418. https://doi.org/10.1210/en.2014-1437

Carlisle SJ, Marciano-Cabral F, Staab A, Ludwick C, Cabral GA (2002) Differential expression of the CB2 cannabinoid receptor by rodent macrophages and macrophage-like cells in relation to cell activation. Int Immunopharmacol 2:69–82. https://doi.org/10.1016/s1567-5769(01)00147-3

Carvalho AF, Van Bockstaele EJ (2012) Cannabinoid modulation of noradrenergic circuits: implications for psychiatric disorders. Prog Neuropsychopharmacol Biol Psychiatry 38:59–67. https://doi.org/10.1016/j.pnpbp.2012.01.008

Casanova E, Fehsenfeld S, Mantamadiotis T, Lemberger T, Greiner E, Stewart AF, Schütz G (2001) A CamKIIalpha iCre BAC allows brain-specific gene inactivation. Genesis 31:37–42. https://doi.org/10.1002/gene.1078

Castorena CM, Caron A, Michael NJ, Ahmed NI, Arnold AG, Lee J, Lee C, Limboy C, Tinajero AS, Granier M, Wang S, Horton JD, Holland WL, Lee S, Liu C, Fujikawa T, Elmquist JK (2021) CB1Rs in VMH neurons regulate glucose homeostasis but not body weight. Am J Physiol Endocrinol Metab 321:E146–E155. https://doi.org/10.1152/ajpendo.00044.2021

Chanda PK, Gao Y, Mark L, Btesh J, Strassle BW, Lu P, Piesla MJ, Zhang M-Y, Bingham B, Uveges A, Kowal D, Garbe D, Kouranova EV, Ring RH, Bates B, Pangalos MN, Kennedy JD, Whiteside GT, Samad TA (2010) Monoacylglycerol lipase activity is a critical modulator of the tone and integrity of the endocannabinoid system. Mol Pharmacol 78:996–1003. https://doi.org/10.1124/mol.110.068304

Cheer JF, Kendall DA, Mason R, Marsden CA (2003) Differential cannabinoid-induced electrophysiological effects in rat ventral tegmentum. Neuropharmacology 44:633–641. https://doi.org/10.1016/s0028-3908(03)00029-7

Chon S-H, Douglass JD, Zhou YX, Malik N, Dixon JL, Brinker A, Quadro L, Storch J (2012) Overexpression of monoacylglycerol lipase (MGL) in small intestine alters endocannabinoid levels and whole body energy balance, resulting in obesity. PLoS One 7:e43962. https://doi.org/10.1371/journal.pone.0043962

Clausen BE, Burkhardt C, Reith W, Renkawitz R, Förster I (1999) Conditional gene targeting in macrophages and granulocytes using LysMcre mice. Transgenic Res 8:265–277. https://doi.org/10.1023/a:1008942828960

Clements JE, Zink MC (1996) Molecular biology and pathogenesis of animal lentivirus infections. Clin Microbiol Rev 9(1):100–117. https://doi.org/10.1128/CMR.9.1.100

Compagnucci C, Di Siena S, Bustamante MB, Di Giacomo D, Di Tommaso M, Maccarrone M, Grimaldi P, Sette C (2013) Type-1 (CB1) cannabinoid receptor promotes neuronal differentiation and maturation of neural stem cells. PLoS One 8:e54271. https://doi.org/10.1371/journal.pone.0054271

Cong L, Ran FA, Cox D, Lin S, Barretto R, Habib N, Hsu PD, Wu X, Jiang W, Marraffini LA, Zhang F (2013) Multiplex genome engineering using CRISPR/Cas systems. Science 339:819–823. https://doi.org/10.1126/science.1231143

Conner DA (2004) Transgenic mouse production by zygote injection. Curr Protoc Mol Biol. Chapter 23, Unit 23.9. https://doi.org/10.1002/0471142727.mb2309s68

Conrad LC, Leonard CM, Pfaff DW (1974) Connections of the median and dorsal raphe nuclei in the rat: an autoradiographic and degeneration study. J Comp Neurol 156:179–205. https://doi.org/10.1002/cne.901560205

Cota D, Marsicano G, Tschöp M, Grübler Y, Flachskamm C, Schubert M, Auer D, Yassouridis A, Thöne-Reineke C, Ortmann S, Tomassoni F, Cervino C, Nisoli E, Linthorst ACE, Pasquali R, Lutz B, Stalla GK, Pagotto U (2003) The endogenous cannabinoid system affects energy balance via central orexigenic drive and peripheral lipogenesis. J Clin Invest 112:423–431. https://doi.org/10.1172/JCI17725

Covey DP, Mateo Y, Sulzer D, Cheer JF, Lovinger DM (2017) Endocannabinoid modulation of dopamine neurotransmission. Neuropharmacology 124:52–61. https://doi.org/10.1016/j.neuropharm.2017.04.033

Cravatt BF, Demarest K, Patricelli MP, Bracey MH, Giang DK, Martin BR, Lichtman AH (2001) Supersensitivity to anandamide and enhanced endogenous cannabinoid signaling in mice lacking fatty acid amide hydrolase. Proc Natl Acad Sci U S A 98:9371–9376. https://doi.org/10.1073/pnas.161191698

Dahlstroem A, Fuxe K (1964) Evidence for the existence of monoamine-containing neurons in the central nervous system. I. Demonstration of monoamines in the cell bodies of brain stem neurons. Acta Physiol Scand Suppl 232:1–55

Dantzer R (2001) Cytokine-induced sickness behavior: mechanisms and implications. Ann N Y Acad Sci 933:222–234. https://doi.org/10.1111/j.1749-6632.2001.tb05827.x

Dawson MRL, Polito A, Levine JM, Reynolds R (2003) NG2-expressing glial progenitor cells: an abundant and widespread population of cycling cells in the adult rat CNS. Mol Cell Neurosci 24:476–488. https://doi.org/10.1016/s1044-7431(03)00210-0

De Giacomo V, Ruehle S, Lutz B, Häring M, Remmers F (2020) Differential glutamatergic and GABAergic contributions to the tetrad effects of Δ9-tetrahydrocannabinol revealed by cell-type-specific reconstitution of the CB1 receptor. Neuropharmacology 179:108287. https://doi.org/10.1016/j.neuropharm.2020.108287

De Giacomo V, Ruehle S, Lutz B, Häring M, Remmers F (2022) Cell type-specific genetic reconstitution of CB1 receptor subsets to assess their role in exploratory behaviour, sociability, and memory. Eur J Neurosci 55:939–951. https://doi.org/10.1111/ejn.15069

De Meij J, Alfanek Z, Morel L, Decoeur F, Leyrolle Q, Picard K, Carrier M, Aubert A, Séré A, Lucas C, Laforest G, Helbling J-C, Tremblay M-E, Cota D, Moisan M-P, Marsicano G, Layé S, Nadjar A (2021) Microglial cannabinoid type 1 receptor regulates brain inflammation in a sex-specific manner. Cannabis Cannabinoid Res 6:488–507. https://doi.org/10.1089/can.2020.0170

Deis S, Srivastava RK, Ruiz de Azua I, Bindila L, Baraghithy S, Lutz B, Bab I, Tam J (2018) Age-related regulation of bone formation by the sympathetic cannabinoid CB1 receptor. Bone 108:34–42. https://doi.org/10.1016/j.bone.2017.12.018

Denny CA, Kheirbek MA, Alba EL, Tanaka KF, Brachman RA, Laughman KB, Tomm NK, Turi GF, Losonczy A, Hen R (2014) Hippocampal memory traces are differentially modulated by experience, time, and adult neurogenesis. Neuron 83:189–201. https://doi.org/10.1016/j.neuron.2014.05.018

Dhillon H, Zigman JM, Ye C, Lee CE, McGovern RA, Tang V, Kenny CD, Christiansen LM, White RD, Edelstein EA, Coppari R, Balthasar N, Cowley MA, Chua S, Elmquist JK, Lowell

BB (2006) Leptin directly activates SF1 neurons in the VMH, and this action by leptin is required for normal body-weight homeostasis. Neuron 49:191–203. https://doi.org/10.1016/j.neuron.2005.12.021

Di Marzo V (2018) New approaches and challenges to targeting the endocannabinoid system. Nat Rev Drug Discov 17:623–639. https://doi.org/10.1038/nrd.2018.115

Díaz-Alonso J, Aguado T, Wu C-S, Palazuelos J, Hofmann C, Garcez P, Guillemot F, Lu H-C, Lutz B, Guzmán M, Galve-Roperh I (2012) The CB(1) cannabinoid receptor drives corticospinal motor neuron differentiation through the Ctip2/Satb2 transcriptional regulation axis. J Neurosci 32:16651–16665. https://doi.org/10.1523/JNEUROSCI.0681-12.2012

Diez-Alarcia R, Ibarra-Lecue I, Lopez-Cardona ÁP, Meana J, Gutierrez-Adán A, Callado LF, Agirregoitia E, Urigüen L (2016) Biased agonism of three different cannabinoid receptor agonists in mouse brain cortex. Front Pharmacol 7:415. https://doi.org/10.3389/fphar.2016.00415

Dincheva I, Drysdale AT, Hartley CA, Johnson DC, Jing D, King EC, Ra S, Gray JM, Yang R, DeGruccio AM, Huang C, Cravatt BF, Glatt CE, Hill MN, Casey BJ, Lee FS (2015) FAAH genetic variation enhances fronto-amygdala function in mouse and human. Nat Commun 6:6395. https://doi.org/10.1038/ncomms7395

Domenici MR, Azad SC, Marsicano G, Schierloh A, Wotjak CT, Dodt H-U, Zieglgänsberger W, Lutz B, Rammes G (2006) Cannabinoid receptor type 1 located on presynaptic terminals of principal neurons in the forebrain controls glutamatergic synaptic transmission. J Neurosci 26:5794–5799. https://doi.org/10.1523/JNEUROSCI.0372-06.2006

Dong A, He K, Dudok B, Farrell JS, Guan W, Liput DJ, Puhl HL, Cai R, Wang H, Duan J, Albarran E, Ding J, Lovinger DM, Li B, Soltesz I, Li Y (2022) A fluorescent sensor for spatiotemporally resolved imaging of endocannabinoid dynamics in vivo. Nat Biotechnol 40:787–798. https://doi.org/10.1038/s41587-021-01074-4

Drouyer M, Merjane J, Nazareth D, Knight M, Scott S, Liao SHY, Ginn SL, Zhu E, Alexander IE, Lisowski L (2024) Development of CNS tropic AAV1-like variants with reduced liver-targeting following systemic administration in mice. Mol Ther 32:818–836. https://doi.org/10.1016/j.ymthe.2024.01.024

Dubreucq S, Koehl M, Abrous DN, Marsicano G, Chaouloff F (2010) CB1 receptor deficiency decreases wheel-running activity: consequences on emotional behaviours and hippocampal neurogenesis. Exp Neurol 224:106–113. https://doi.org/10.1016/j.expneurol.2010.01.017

Dubreucq S, Matias I, Cardinal P, Häring M, Lutz B, Marsicano G, Chaouloff F (2012a) Genetic dissection of the role of cannabinoid type-1 receptors in the emotional consequences of repeated social stress in mice. Neuropsychopharmacology 37:1885–1900. https://doi.org/10.1038/npp.2012.36

Dubreucq S, Kambire S, Conforzi M, Metna-Laurent M, Cannich A, Soria-Gomez E, Richard E, Marsicano G, Chaouloff F (2012b) Cannabinoid type 1 receptors located on single-minded 1-expressing neurons control emotional behaviors. Neuroscience 204:230–244. https://doi.org/10.1016/j.neuroscience.2011.08.049

Dudok B, Fan LZ, Farrell JS, Malhotra S, Homidan J, Kim DK, Wenardy C, Ramakrishnan C, Li Y, Deisseroth K, Soltesz I (2024) Retrograde endocannabinoid signaling at inhibitory synapses in vivo. Science 383:967–970. https://doi.org/10.1126/science.adk3863

Durand E, Lecoeur C, Delplanque J, Benzinou M, Degraeve F, Boutin P, Marre M, Balkau B, Charpentier G, Froguel P, Meyre D (2008) Evaluating the association of FAAH common gene variation with childhood, adult severe obesity and type 2 diabetes in the French population. Obes Facts 1:305–309. https://doi.org/10.1159/000178157

Durante MA, Kurtenbach S, Sargi ZB, Harbour JW, Choi R, Kurtenbach S, Goss GM, Matsunami H, Goldstein BJ (2020) Single-cell analysis of olfactory neurogenesis and differentiation in adult humans. Nat Neurosci 23:323–326. https://doi.org/10.1038/s41593-020-0587-9

Egertová M, Simon GM, Cravatt BF, Elphick MR (2008) Localization of N-acyl phosphatidylethanolamine phospholipase D (NAPE-PLD) expression in mouse brain: a new perspective on

N-acylethanolamines as neural signaling molecules. J Comp Neurol 506:604–615. https://doi.org/10.1002/cne.21568

el Marjou F, Janssen K-P, Chang BH-J, Li M, Hindie V, Chan L, Louvard D, Chambon P, Metzger D, Robine S (2004) Tissue-specific and inducible Cre-mediated recombination in the gut epithelium. Genesis 39:186–193. https://doi.org/10.1002/gene.20042

Erles K, Sebökovà P, Schlehofer JR (1999) Update on the prevalence of serum antibodies (IgG and IgM) to adeno-associated virus (AAV). J Med Virol 59:406–411. https://doi.org/10.1002/(sici)1096-9071(199911)59:3<406::aid-jmv22>3.0.co;2-n

Fadool DA, Kolling LJ (2020) 3.35 – role of olfaction for eating behavior. In: Fritzsch B (ed) The senses: a comprehensive reference, 2nd edn. Elsevier, Oxford, pp 675–716. https://doi.org/10.1016/B978-0-12-809324-5.23770-X

Farrell JS, Colangeli R, Dong A, George AG, Addo-Osafo K, Kingsley PJ, Morena M, Wolff MD, Dudok B, He K, Patrick TA, Sharkey KA, Patel S, Marnett LJ, Hill MN, Li Y, Teskey GC, Soltesz I (2021) In vivo endocannabinoid dynamics at the timescale of physiological and pathological neural activity. Neuron 109:2398–2403.e4. https://doi.org/10.1016/j.neuron.2021.05.026

Fenno LE, Mattis J, Ramakrishnan C, Hyun M, Lee SY, He M, Tucciarone J, Selimbeyoglu A, Berndt A, Grosenick L, Zalocusky KA, Bernstein H, Swanson H, Perry C, Diester I, Boyce FM, Bass CE, Neve R, Huang ZJ, Deisseroth K (2014) Targeting cells with single vectors using multiple-feature Boolean logic. Nat Methods 11:763–772. https://doi.org/10.1038/nmeth.2996

Ferreira SG, Teixeira FM, Garção P, Agostinho P, Ledent C, Cortes L, Mackie K, Köfalvi A (2012) Presynaptic CB1 cannabinoid receptors control frontocortical serotonin and glutamate release – species differences. Neurochem Int 61:219–226. https://doi.org/10.1016/j.neuint.2012.05.009

Freund TF, Katona I, Piomelli D (2003) Role of endogenous cannabinoids in synaptic signaling. Physiol Rev 83:1017–1066. https://doi.org/10.1152/physrev.00004.2003

Friedland AE, Baral R, Singhal P, Loveluck K, Shen S, Sanchez M, Marco E, Gotta GM, Maeder ML, Kennedy EM, Kornepati AVR, Sousa A, Collins MA, Jayaram H, Cullen BR, Bumcrot D (2015) Characterization of Staphylococcus aureus Cas9: a smaller Cas9 for all-in-one adeno-associated virus delivery and paired nickase applications. Genome Biol 16:257. https://doi.org/10.1186/s13059-015-0817-8

Gaffal E, Cron M, Glodde N, Bald T, Kuner R, Zimmer A, Lutz B, Tüting T (2013) Cannabinoid 1 receptors in keratinocytes modulate proinflammatory chemokine secretion and attenuate contact allergic inflammation. J Immunol 190:4929–4936. https://doi.org/10.4049/jimmunol.1201777

Games D, Adams D, Alessandrini R, Barbour R, Berthelette P, Blackwell C, Carr T, Clemens J, Donaldson T, Gillespie F (1995) Alzheimer-type neuropathology in transgenic mice overexpressing V717F beta-amyloid precursor protein. Nature 373:523–527. https://doi.org/10.1038/373523a0

Gao Y, Vasilyev DV, Goncalves MB, Howell FV, Hobbs C, Reisenberg M, Shen R, Zhang M-Y, Strassle BW, Lu P, Mark L, Piesla MJ, Deng K, Kouranova EV, Ring RH, Whiteside GT, Bates B, Walsh FS, Williams G, Pangalos MN, Samad TA, Doherty P (2010) Loss of retrograde endocannabinoid signaling and reduced adult neurogenesis in diacylglycerol lipase knock-out mice. J Neurosci 30:2017–2024. https://doi.org/10.1523/JNEUROSCI.5693-09.2010

Geurts L, Everard A, Van Hul M, Essaghir A, Duparc T, Matamoros S, Plovier H, Castel J, Denis RGP, Bergiers M, Druart C, Alhouayek M, Delzenne NM, Muccioli GG, Demoulin J-B, Luquet S, Cani PD (2015) Adipose tissue NAPE-PLD controls fat mass development by altering the browning process and gut microbiota. Nat Commun 6:6495. https://doi.org/10.1038/ncomms7495

Gobbi G, Bambico FR, Mangieri R, Bortolato M, Campolongo P, Solinas M, Cassano T, Morgese MG, Debonnel G, Duranti A, Tontini A, Tarzia G, Mor M, Trezza V, Goldberg SR, Cuomo V, Piomelli D (2005) Antidepressant-like activity and modulation of brain monoaminergic transmission by blockade of anandamide hydrolysis. Proc Natl Acad Sci U S A 102:18620–18625. https://doi.org/10.1073/pnas.0509591102

Goebbels S, Bormuth I, Bode U, Hermanson O, Schwab MH, Nave K-A (2006) Genetic targeting of principal neurons in neocortex and hippocampus of NEX-Cre mice. Genesis 44:611–621. https://doi.org/10.1002/dvg.20256

Gomez O, Arevalo-Martin A, Garcia-Ovejero D, Ortega-Gutierrez S, Cisneros JA, Almazan G, Sánchez-Rodriguez MA, Molina-Holgado F, Molina-Holgado E (2010) The constitutive production of the endocannabinoid 2-arachidonoylglycerol participates in oligodendrocyte differentiation. Glia 58:1913–1927. https://doi.org/10.1002/glia.21061

Gomez O, Sanchez-Rodriguez A, Le MQU, Sanchez-Caro C, Molina-Holgado F, Molina-Holgado E (2011) Cannabinoid receptor agonists modulate oligodendrocyte differentiation by activating PI3K/Akt and the mammalian target of rapamycin (mTOR) pathways. Br J Pharmacol 163: 1520–1532. https://doi.org/10.1111/j.1476-5381.2011.01414.x

Gomez O, Sanchez-Rodriguez MA, Ortega-Gutierrez S, Vazquez-Villa H, Guaza C, Molina-Holgado F, Molina-Holgado E (2015) A basal tone of 2-arachidonoylglycerol contributes to early oligodendrocyte progenitor proliferation by activating phosphatidylinositol 3-kinase (PI3K)/AKT and the mammalian target of rapamycin (MTOR) pathways. J Neuroimmune Pharmacol 10:309–317. https://doi.org/10.1007/s11481-015-9609-x

Goncalves MB, Suetterlin P, Yip P, Molina-Holgado F, Walker DJ, Oudin MJ, Zentar MP, Pollard S, Yáñez-Muñoz RJ, Williams G, Walsh FS, Pangalos MN, Doherty P (2008) A diacylglycerol lipase-CB2 cannabinoid pathway regulates adult subventricular zone neurogenesis in an age-dependent manner. Mol Cell Neurosci 38:526–536. https://doi.org/10.1016/j.mcn.2008.05.001

González-Mariscal I, Montoro RA, Doyle ME, Liu Q-R, Rouse M, O'Connell JF, Santa-Cruz Calvo S, Krzysik-Walker SM, Ghosh S, Carlson OD, Lehrmann E, Zhang Y, Becker KG, Chia CW, Ghosh P, Egan JM (2018) Absence of cannabinoid 1 receptor in beta cells protects against high-fat/high-sugar diet-induced beta cell dysfunction and inflammation in murine islets. Diabetologia 61:1470–1483. https://doi.org/10.1007/s00125-018-4576-4

González-Mariscal I, Montoro RA, O'Connell JF, Kim Y, Gonzalez-Freire M, Liu Q-R, Alfaras I, Carlson OD, Lehrmann E, Zhang Y, Becker KG, Hardivillé S, Ghosh P, Egan JM (2019) Muscle cannabinoid 1 receptor regulates Il-6 and myostatin expression, governing physical performance and whole-body metabolism. FASEB J 33:5850–5863. https://doi.org/10.1096/fj.201801145R

Grabner GF, Eichmann TO, Wagner B, Gao Y, Farzi A, Taschler U, Radner FPW, Schweiger M, Lass A, Holzer P, Zinser E, Tschöp MH, Yi C-X, Zimmermann R (2016) Deletion of monoglyceride lipase in astrocytes attenuates lipopolysaccharide-induced neuroinflammation. J Biol Chem 291:913–923. https://doi.org/10.1074/jbc.M115.683615

Gremel CM, Chancey J, Atwood B, Luo G, Neve R, Ramakrishnan C, Deisseroth K, Lovinger DM, Costa RM (2016) Endocannabinoid modulation of orbitostriatal circuits gates habit formation. Neuron 90:1312–1324. https://doi.org/10.1016/j.neuron.2016.04.043

Grimsey NL, Graham ES, Dragunow M, Glass M (2010) Cannabinoid receptor 1 trafficking and the role of the intracellular pool: implications for therapeutics. Biochem Pharmacol 80:1050–1062. https://doi.org/10.1016/j.bcp.2010.06.007

Guenthner CJ, Miyamichi K, Yang HH, Heller HC, Luo L (2013) Permanent genetic access to transiently active neurons via TRAP: targeted recombination in active populations. Neuron 78: 773–784. https://doi.org/10.1016/j.neuron.2013.03.025

Guggenhuber S, Romo-Parra H, Bindila L, Leschik J, Lomazzo E, Remmers F, Zimmermann T, Lerner R, Klugmann M, Pape H-C, Lutz B (2015) Impaired 2-AG signaling in hippocampal glutamatergic neurons: aggravation of anxiety-like behavior and unaltered seizure susceptibility. Int J Neuropsychopharmacol 19:pyv091. https://doi.org/10.1093/ijnp/pyv091

Gulyás AI, Cravatt BF, Bracey MH, Dinh TP, Piomelli D, Boscia F, Freund TF (2004) Segregation of two endocannabinoid-hydrolyzing enzymes into pre- and postsynaptic compartments in the rat hippocampus, cerebellum and amygdala. Eur J Neurosci 20:441–458. https://doi.org/10.1111/j.1460-9568.2004.03428.x

Gunduz-Cinar O, MacPherson KP, Cinar R, Gamble-George J, Sugden K, Williams B, Godlewski G, Ramikie TS, Gorka AX, Alapafuja SO, Nikas SP, Makriyannis A, Poulton R, Patel S, Hariri AR, Caspi A, Moffitt TE, Kunos G, Holmes A (2013) Convergent translational evidence of a role for anandamide in amygdala-mediated fear extinction, threat processing and stress-reactivity. Mol Psychiatry 18:813–823. https://doi.org/10.1038/mp.2012.72

Gunduz-Cinar O, Castillo LI, Xia M, Van Leer E, Brockway ET, Pollack GA, Yasmin F, Bukalo O, Limoges A, Oreizi-Esfahani S, Kondev V, Báldi R, Dong A, Harvey-White J, Cinar R, Kunos G, Li Y, Zweifel LS, Patel S, Holmes A (2023) A cortico-amygdala neural substrate for endocannabinoid modulation of fear extinction. Neuron S0896-6273(23):00482–00488. https://doi.org/10.1016/j.neuron.2023.06.023

Gutiérrez-Rodríguez A, Puente N, Elezgarai I, Ruehle S, Lutz B, Reguero L, Gerrikagoitia I, Marsicano G, Grandes P (2017) Anatomical characterization of the cannabinoid CB1 receptor in cell-type–specific mutant mouse rescue models. J Comp Neurol 525:302–318. https://doi.org/10.1002/cne.24066

Gutiérrez-Rodríguez A, Bonilla-Del Río I, Puente N, Gómez-Urquijo SM, Fontaine CJ, Egaña-Huguet J, Elezgarai I, Ruehle S, Lutz B, Robin LM, Soria-Gómez E, Bellocchio L, Padwal JD, van der Stelt M, Mendizabal-Zubiaga J, Reguero L, Ramos A, Gerrikagoitia I, Marsicano G, Grandes P (2018) Localization of the cannabinoid type-1 receptor in subcellular astrocyte compartments of mutant mouse hippocampus. Glia 66:1417–1431. https://doi.org/10.1002/glia.23314

Halassa MM, Fellin T, Haydon PG (2007) The tripartite synapse: roles for gliotransmission in health and disease. Trends Mol Med 13:54–63. https://doi.org/10.1016/j.molmed.2006.12.005

Han J, Kesner P, Metna-Laurent M, Duan T, Xu L, Georges F, Koehl M, Abrous DN, Mendizabal-Zubiaga J, Grandes P, Liu Q, Bai G, Wang W, Xiong L, Ren W, Marsicano G, Zhang X (2012) Acute cannabinoids impair working memory through astroglial CB1 receptor modulation of hippocampal LTD. Cell 148:1039–1050. https://doi.org/10.1016/j.cell.2012.01.037

Han X, He Y, Bi G-H, Zhang H-Y, Song R, Liu Q-R, Egan JM, Gardner EL, Li J, Xi Z-X (2017) CB1 receptor activation on VgluT2-expressing glutamatergic neurons underlies Δ9-tetrahydrocannabinol (Δ9-THC)-induced aversive effects in mice. Sci Rep 7:12315. https://doi.org/10.1038/s41598-017-12399-z

Han Z, Luo N, Wu Y, Kou J, Ma W, Yang X, Cai Y, Ma L, Han L, Wang X, Qin H, Shi Q, Wang J, Ye C, Lin K, Xu F (2022) AAV13 enables precise targeting of local neural populations. Int J Mol Sci 23:12806. https://doi.org/10.3390/ijms232112806

Han X, Liang Y, Hempel B, Jordan CJ, Shen H, Bi G-H, Li J, Xi Z-X (2023) Cannabinoid CB1 receptors are expressed in a subset of dopamine neurons and underlie cannabinoid-induced aversion, hypoactivity, and anxiolytic effects in mice. J Neurosci 43:373–385. https://doi.org/10.1523/JNEUROSCI.1493-22.2022

Häring M, Marsicano G, Lutz B, Monory K (2007) Identification of the cannabinoid receptor type 1 in serotonergic cells of raphe nuclei in mice. Neuroscience 146:1212–1219. https://doi.org/10.1016/j.neuroscience.2007.02.021

Häring M, Kaiser N, Monory K, Lutz B (2011) Circuit specific functions of cannabinoid CB1 receptor in the balance of investigatory drive and exploration. PLoS One 6:e26617. https://doi.org/10.1371/journal.pone.0026617

Häring M, Grieb M, Monory K, Lutz B, Moreira FA (2013) Cannabinoid CB_1 receptor in the modulation of stress coping behavior in mice: the role of serotonin and different forebrain neuronal subpopulations. Neuropharmacology 65:83–89. https://doi.org/10.1016/j.neuropharm.2012.09.002

Häring M, Enk V, Aparisi Rey A, Loch S, Ruiz De Azua I, Weber T, Bartsch D, Monory K, Lutz B (2015) Cannabinoid type-1 receptor signaling in central serotonergic neurons regulates anxiety-like behavior and sociability. Front Behav Neurosci 9. https://doi.org/10.3389/fnbeh.2015.00235

Harkany T, Guzmán M, Galve-Roperh I, Berghuis P, Devi LA, Mackie K (2007) The emerging functions of endocannabinoid signaling during CNS development. Trends Pharmacol Sci 28: 83–92. https://doi.org/10.1016/j.tips.2006.12.004

Harlan BA, Becker HC, Woodward JJ, Riegel AC (2018) Opposing actions of CRF-R1 and CB1 receptors on VTA-GABAergic plasticity following chronic exposure to ethanol. Neuropsychopharmacology 43:2064–2074. https://doi.org/10.1038/s41386-018-0106-9

Haroutunian V, Katsel P, Roussos P, Davis KL, Altshuler LL, Bartzokis G (2014) Myelination, oligodendrocytes, and serious mental illness. Glia 62:1856–1877. https://doi.org/10.1002/glia.22716

Hasegawa Y, Kim J, Ursini G, Jouroukhin Y, Zhu X, Miyahara Y, Xiong F, Madireddy S, Obayashi M, Lutz B, Sawa A, Brown SP, Pletnikov MV, Kamiya A (2023) Microglial cannabinoid receptor type 1 mediates social memory deficits in mice produced by adolescent THC exposure and 16p11.2 duplication. Nat Commun 14:6559. https://doi.org/10.1038/s41467-023-42276-5

He W, Barak Y, Hevener A, Olson P, Liao D, Le J, Nelson M, Ong E, Olefsky JM, Evans RM (2003) Adipose-specific peroxisome proliferator-activated receptor gamma knockout causes insulin resistance in fat and liver but not in muscle. Proc Natl Acad Sci U S A 100:15712–15717. https://doi.org/10.1073/pnas.2536828100

Hebert-Chatelain E, Reguero L, Puente N, Lutz B, Chaouloff F, Rossignol R, Piazza P-V, Benard G, Grandes P, Marsicano G (2014) Cannabinoid control of brain bioenergetics: exploring the subcellular localization of the CB1 receptor. Mol Metab 3:495–504. https://doi.org/10.1016/j.molmet.2014.03.007

Hebert-Chatelain E, Desprez T, Serrat R, Bellocchio L, Soria-Gomez E, Busquets-Garcia A, Pagano Zottola AC, Delamarre A, Cannich A, Vincent P, Varilh M, Robin LM, Terral G, García-Fernández MD, Colavita M, Mazier W, Drago F, Puente N, Reguero L, Elezgarai I, Dupuy J-W, Cota D, Lopez-Rodriguez M-L, Barreda-Gómez G, Massa F, Grandes P, Bénard G, Marsicano G (2016) A cannabinoid link between mitochondria and memory. Nature 539:555–559. https://doi.org/10.1038/nature20127

Heimann AS, Dale CS, Guimarães FS, Reis RAM, Navon A, Shmuelov MA, Rioli V, Gomes I, Devi LL, Ferro ES (2021) Hemopressin as a breakthrough for the cannabinoid field. Neuropharmacology 183:108406. https://doi.org/10.1016/j.neuropharm.2020.108406

Hennet T, Hagen FK, Tabak LA, Marth JD (1995) T-cell-specific deletion of a polypeptide N-acetylgalactosaminyl-transferase gene by site-directed recombination. Proc Natl Acad Sci U S A 92:12070–12074. https://doi.org/10.1073/pnas.92.26.12070

Hillard CJ (2015) The endocannabinoid signaling system in the CNS: a primer. Int Rev Neurobiol 125:1–47. https://doi.org/10.1016/bs.irn.2015.10.001

Hinden L, Ahmad M, Hamad S, Nemirovski A, Szanda G, Glasmacher S, Kogot-Levin A, Abramovitch R, Thorens B, Gertsch J, Leibowitz G, Tam J (2022) Opposite physiological and pathological mTORC1-mediated roles of the CB1 receptor in regulating renal tubular function. Nat Commun 13:1783. https://doi.org/10.1038/s41467-022-29124-8

Hirrlinger PG, Scheller A, Braun C, Hirrlinger J, Kirchhoff F (2006) Temporal control of gene recombination in astrocytes by transgenic expression of the tamoxifen-inducible DNA recombinase variant CreERT2. Glia 54:11–20. https://doi.org/10.1002/glia.20342

Hofer SC, Ralvenius WT, Gachet MS, Fritschy J-M, Zeilhofer HU, Gertsch J (2015) Localization and production of peptide endocannabinoids in the rodent CNS and adrenal medulla. Neuropharmacology 98:78–89. https://doi.org/10.1016/j.neuropharm.2015.03.021

Hoggan MD, Blacklow NR, Rowe WP (1966) Studies of small DNA viruses found in various adenovirus preparations: physical, biological, and immunological characteristics. Proc Natl Acad Sci U S A 55:1467–1474. https://doi.org/10.1073/pnas.55.6.1467

Hu M, Zhu D, Zhang J, Gao F, Hashem J, Kingsley P, Marnett LJ, Mackie K, Chen C (2022) Enhancing endocannabinoid signalling in astrocytes promotes recovery from traumatic brain injury. Brain 145:179–193. https://doi.org/10.1093/brain/awab310

Huang W, Zhao N, Bai X, Karram K, Trotter J, Goebbels S, Scheller A, Kirchhoff F (2014) Novel NG2-CreERT2 knock-in mice demonstrate heterogeneous differentiation potential of NG2 glia during development. Glia 62:896–913. https://doi.org/10.1002/glia.22648

Hunker AC, Soden ME, Krayushkina D, Heymann G, Awatramani R, Zweifel LS (2020) Conditional single vector CRISPR/SaCas9 viruses for efficient mutagenesis in the adult mouse nervous system. Cell Rep 30:4303–4316.e6. https://doi.org/10.1016/j.celrep.2020.02.092

Iannotti FA, Vitale RM (2021) The endocannabinoid system and PPARs: focus on their signalling crosstalk, action and transcriptional regulation. Cells 10:586. https://doi.org/10.3390/cells10030586

Ibsen MS, Finlay DB, Patel M, Javitch JA, Glass M, Grimsey NL (2019) Cannabinoid CB1 and CB2 receptor-mediated Arrestin translocation: species, subtype, and agonist-dependence. Front Pharmacol 10:350. https://doi.org/10.3389/fphar.2019.00350

Irungbam K, Churin Y, Matono T, Weglage J, Ocker M, Glebe D, Hardt M, Koeppel A, Roderfeld M, Roeb E (2020) Cannabinoid receptor 1 knockout alleviates hepatic steatosis by downregulating perilipin 2. Lab Invest 100:454–465. https://doi.org/10.1038/s41374-019-0327-5

Issa SS, Shaimardanova AA, Solovyeva VV, Rizvanov AA (2023) Various AAV serotypes and their applications in gene therapy: an overview. Cells 12:785. https://doi.org/10.3390/cells12050785

Jacob W, Yassouridis A, Marsicano G, Monory K, Lutz B, Wotjak CT (2009) Endocannabinoids render exploratory behaviour largely independent of the test aversiveness: role of glutamatergic transmission. Genes Brain Behav 8:685–698. https://doi.org/10.1111/j.1601-183X.2009.00512.x

Jaenisch R, Mintz B (1974) Simian virus 40 DNA sequences in DNA of healthy adult mice derived from preimplantation blastocysts injected with viral DNA. Proc Natl Acad Sci U S A 71:1250–1254

Jenniches I, Ternes S, Albayram O, Otte DM, Bach K, Bindila L, Michel K, Lutz B, Bilkei-Gorzo A, Zimmer A (2016) Anxiety, stress, and fear response in mice with reduced endocannabinoid levels. Biol Psychiatry 79:858–868. https://doi.org/10.1016/j.biopsych.2015.03.033

Jensen DP, Andreasen CH, Andersen MK, Hansen L, Eiberg H, Borch-Johnsen K, Jørgensen T, Hansen T, Pedersen O (2007) The functional Pro129Thr variant of the FAAH gene is not associated with various fat accumulation phenotypes in a population-based cohort of 5,801 whites. J Mol Med (Berl) 85:445–449. https://doi.org/10.1007/s00109-006-0139-0

Jeong W, Osei-Hyiaman D, Park O, Liu J, Bátkai S, Mukhopadhyay P, Horiguchi N, Harvey-White J, Marsicano G, Lutz B, Gao B, Kunos G (2008) Paracrine activation of hepatic CB1 receptors by stellate cell-derived endocannabinoids mediates alcoholic fatty liver. Cell Metab 7:227–235. https://doi.org/10.1016/j.cmet.2007.12.007

Jiang W, Zhang Y, Xiao L, Van Cleemput J, Ji S-P, Bai G, Zhang X (2005) Cannabinoids promote embryonic and adult hippocampus neurogenesis and produce anxiolytic- and antidepressant-like effects. J Clin Invest 115:3104–3116. https://doi.org/10.1172/JCI25509

Jimenez-Blasco D, Busquets-Garcia A, Hebert-Chatelain E, Serrat R, Vicente-Gutierrez C, Ioannidou C, Gómez-Sotres P, Lopez-Fabuel I, Resch-Beusher M, Resel E, Arnouil D, Saraswat D, Varilh M, Cannich A, Julio-Kalajzic F, Bonilla-Del Río I, Almeida A, Puente N, Achicallende S, Lopez-Rodriguez M-L, Jollé C, Déglon N, Pellerin L, Josephine C, Bonvento G, Panatier A, Lutz B, Piazza P V, Guzmán M, Bellocchio L, Bouzier-Sore A-K, Grandes P, Bolaños JP, Marsicano G (2020) Glucose metabolism links astroglial mitochondria to cannabinoid effects. Nature 583:603–608. https://doi.org/10.1038/s41586-020-2470-y

Jin K, Xie L, Kim SH, Parmentier-Batteur S, Sun Y, Mao XO, Childs J, Greenberg DA (2004) Defective adult neurogenesis in CB1 cannabinoid receptor knockout mice. Mol Pharmacol 66:204–208. https://doi.org/10.1124/mol.66.2.204

Jung K-M, Clapper JR, Fu J, D'Agostino G, Guijarro A, Thongkham D, Avanesian A, Astarita G, DiPatrizio NV, Frontini A, Cinti S, Diano S, Piomelli D (2012) 2-Arachidonoylglycerol signaling in forebrain regulates systemic energy metabolism. Cell Metab 15:299–310. https://doi.org/10.1016/j.cmet.2012.01.021

Kaczocha M, Haj-Dahmane S (2021) Mechanisms of endocannabinoid transport in the brain. Br J Pharmacol. https://doi.org/10.1111/bph.15469

Kaller MS, Lazari A, Blanco-Duque C, Sampaio-Baptista C, Johansen-Berg H (2017) Myelin plasticity and behaviour-connecting the dots. Curr Opin Neurobiol 47:86–92. https://doi.org/10.1016/j.conb.2017.09.014

Karram K, Goebbels S, Schwab M, Jennissen K, Seifert G, Steinhäuser C, Nave K-A, Trotter J (2008) NG2-expressing cells in the nervous system revealed by the NG2-EYFP-knockin mouse. Genesis 46:743–757. https://doi.org/10.1002/dvg.20440

Katona I, Sperlágh B, Sík A, Kőfalvi A, Vizi ES, Mackie K, Freund TF (1999) Presynaptically located CB1 cannabinoid receptors regulate GABA release from axon terminals of specific hippocampal interneurons. J Neurosci 19:4544–4558

Kawamura Y, Fukaya M, Maejima T, Yoshida T, Miura E, Watanabe M, Ohno-Shosaku T, Kano M (2006) The CB1 cannabinoid receptor is the major cannabinoid receptor at excitatory presynaptic sites in the hippocampus and cerebellum. J Neurosci 26:2991–3001. https://doi.org/10.1523/JNEUROSCI.4872-05.2006

Kettenmann H, Hanisch U-K, Noda M, Verkhratsky A (2011) Physiology of microglia. Physiol Rev 91:461–553. https://doi.org/10.1152/physrev.00011.2010

Kim Y, Gautam S, Aseer KR, Kim J, Chandrasekaran P, Mazucanti CH, Ghosh P, O'Connell JF, Doyle ME, Appleton A, Lehrmann E, Liu Q-R, Egan JM (2020) Hepatocyte cannabinoid 1 receptor nullification alleviates toxin-induced liver damage via NF-κB signaling. Cell Death Dis 11:1–13. https://doi.org/10.1038/s41419-020-03261-8

Kimmel RA, Turnbull DH, Blanquet V, Wurst W, Loomis CA, Joyner AL (2000) Two lineage boundaries coordinate vertebrate apical ectodermal ridge formation. Genes Dev 14:1377–1389

Kondev V, Najeed M, Yasmin F, Morgan A, Loomba N, Johnson K, Adank DN, Dong A, Delpire E, Li Y, Winder D, Grueter BA, Patel S (2023) Endocannabinoid release at ventral hippocampal-amygdala synapses regulates stress-induced behavioral adaptation. Cell Rep 42: 113027. https://doi.org/10.1016/j.celrep.2023.113027

Kumar N, Stanford W, de Solis C, Aradhana N, Abraham ND, Dao T-MJ, Thaseen S, Sairavi A, Gonzalez CU, Ploski JE (2018) The development of an AAV-based CRISPR SaCas9 genome editing system that can be delivered to neurons in vivo and regulated via doxycycline and Cre-recombinase. Front Mol Neurosci 11:413. https://doi.org/10.3389/fnmol.2018.00413

Lafenêtre P, Chaouloff F, Marsicano G (2009) Bidirectional regulation of novelty-induced behavioral inhibition by the endocannabinoid system. Neuropharmacology 57:715–721. https://doi.org/10.1016/j.neuropharm.2009.07.014

Lagace DC, Whitman MC, Noonan MA, Ables JL, DeCarolis NA, Arguello AA, Donovan MH, Fischer SJ, Farnbauch LA, Beech RD, DiLeone RJ, Greer CA, Mandyam CD, Eisch AJ (2007) Dynamic contribution of nestin-expressing stem cells to adult neurogenesis. J Neurosci 27: 12623–12629. https://doi.org/10.1523/JNEUROSCI.3812-07.2007

Lange MD, Daldrup T, Remmers F, Szkudlarek HJ, Lesting J, Guggenhuber S, Ruehle S, Jüngling K, Seidenbecher T, Lutz B, Pape HC (2017) Cannabinoid CB1 receptors in distinct circuits of the extended amygdala determine fear responsiveness to unpredictable threat. Mol Psychiatry 22:1422–1430. https://doi.org/10.1038/mp.2016.156

Ledent C, Valverde O, Cossu G, Petitet F, Aubert JF, Beslot F, Böhme GA, Imperato A, Pedrazzini T, Roques BP, Vassart G, Fratta W, Parmentier M (1999) Unresponsiveness to cannabinoids and reduced addictive effects of opiates in CB1 receptor knockout mice. Science 283:401–404. https://doi.org/10.1126/science.283.5400.401

Lee G, Saito I (1998) Role of nucleotide sequences of loxP spacer region in Cre-mediated recombination. Gene 216:55–65. https://doi.org/10.1016/s0378-1119(98)00325-4

Lee H-C, Simon GM, Cravatt BF (2015) ABHD4 regulates multiple classes of N-acyl phospholipids in the mammalian central nervous system. Biochemistry 54:2539–2549. https://doi.org/10.1021/acs.biochem.5b00207

Lehmann ML, Weigel TK, Elkahloun AG, Herkenham M (2017) Chronic social defeat reduces myelination in the mouse medial prefrontal cortex. Sci Rep 7:46548. https://doi.org/10.1038/srep46548
Leo LM, Abood ME (2021) CB1 cannabinoid receptor signaling and biased signaling. Molecules 26:5413. https://doi.org/10.3390/molecules26175413
Leung D, Saghatelian A, Simon GM, Cravatt BF (2006) Inactivation of N-acyl phosphatidylethanolamine phospholipase D reveals multiple mechanisms for the biosynthesis of endocannabinoids. Biochemistry 45:4720–4726. https://doi.org/10.1021/bi0601631
Li B (2019) Central amygdala cells for learning and expressing aversive emotional memories. Curr Opin Behav Sci 26:40–45. https://doi.org/10.1016/j.cobeha.2018.09.012
Li C, Samulski RJ (2020) Engineering adeno-associated virus vectors for gene therapy. Nat Rev Genet 21:255–272. https://doi.org/10.1038/s41576-019-0205-4
Li S, Chen L, Peng X, Wang C, Qin B, Tan D, Han C, Yang H, Ren X, Liu F, Xu C, Zhou X (2018) Overview of the reporter genes and reporter mouse models. Animal Model Exp Med 1:29–35. https://doi.org/10.1002/ame2.12008
Ligresti A, De Petrocellis L, Di Marzo V (2016) From phytocannabinoids to cannabinoid receptors and endocannabinoids: pleiotropic physiological and pathological roles through complex pharmacology. Physiol Rev 96:1593–1659. https://doi.org/10.1152/physrev.00002.2016
Lin L, Jung K-M, Lee H-L, Le J, Colleluori G, Wood C, Palese F, Squire E, Ramirez J, Su S, Torrens A, Fotio Y, Tang L, Yu C, Yang Q, Huang L, DiPatrizio N, Jang C, Cinti S, Piomelli D (2023) Adolescent exposure to low-dose THC disrupts energy balance and adipose organ homeostasis in adulthood. Cell Metab 35:1227–1241.e7. https://doi.org/10.1016/j.cmet.2023.05.002
Liu J, Wang L, Harvey-White J, Huang BX, Kim H-Y, Luquet S, Palmiter RD, Krystal G, Rai R, Mahadevan A, Razdan RK, Kunos G (2008) Multiple pathways involved in the biosynthesis of anandamide. Neuropharmacology 54:1–7. https://doi.org/10.1016/j.neuropharm.2007.05.020
Liu J, Zhou L, Xiong K, Godlewski G, Mukhopadhyay B, Tam J, Yin S, Gao P, Shan X, Pickel J, Bataller R, O'Hare J, Scherer T, Buettner C, Kunos G (2012) Hepatic cannabinoid receptor-1 mediates diet-induced insulin resistance via inhibition of insulin signaling and clearance in mice. Gastroenterology 142:1218–1228.e1. https://doi.org/10.1053/j.gastro.2012.01.032
Liu Q-R, Canseco-Alba A, Zhang H-Y, Tagliaferro P, Chung M, Dennis E, Sanabria B, Schanz N, Escosteguy-Neto JC, Ishiguro H, Lin Z, Sgro S, Leonard CM, Santos-Junior JG, Gardner EL, Egan JM, Lee JW, Xi Z-X, Onaivi ES (2017) Cannabinoid type 2 receptors in dopamine neurons inhibits psychomotor behaviors, alters anxiety, depression and alcohol preference. Sci Rep 7:17410. https://doi.org/10.1038/s41598-017-17796-y
Llorente-Berzal A, Terzian ALB, di Marzo V, Micale V, Viveros MP, Wotjak CT (2015) 2-AG promotes the expression of conditioned fear via cannabinoid receptor type 1 on GABAergic neurons. Psychopharmacology (Berl) 232:2811–2825. https://doi.org/10.1007/s00213-015-3917-y
López A, Aparicio N, Pazos MR, Grande MT, Barreda-Manso MA, Benito-Cuesta I, Vázquez C, Amores M, Ruiz-Pérez G, García-García E, Beatka M, Tolón RM, Dittel BN, Hillard CJ, Romero J (2018) Cannabinoid CB2 receptors in the mouse brain: relevance for Alzheimer's disease. J Neuroinflammation 15:158. https://doi.org/10.1186/s12974-018-1174-9
Lutz B (2020) Neurobiology of cannabinoid receptor signaling. Dialogues Clin Neurosci 22:207–222. https://doi.org/10.31887/DCNS.2020.22.3/blutz
Lutz B, Marsicano G, Maldonado R, Hillard CJ (2015) The endocannabinoid system in guarding against fear, anxiety and stress. Nat Rev Neurosci 16:705–718. https://doi.org/10.1038/nrn4036
Maccarrone M, Bab I, Bíró T, Cabral GA, Dey SK, Di Marzo V, Konje JC, Kunos G, Mechoulam R, Pacher P, Sharkey KA, Zimmer A (2015) Endocannabinoid signaling at the periphery: 50 years after THC. Trends Pharmacol Sci 36:277–296. https://doi.org/10.1016/j.tips.2015.02.008
MacDonald A, Lu B, Caron M, Caporicci-Dinucci N, Hatrock D, Petrecca K, Bourque G, Stratton JA (2021) Single cell transcriptomics of ependymal cells across age, region and species reveals

cilia-related and metal ion regulatory roles as major conserved ependymal cell functions. Front Cell Neurosci 15:703951. https://doi.org/10.3389/fncel.2021.703951

Madisen L, Garner AR, Shimaoka D, Chuong AS, Klapoetke NC, Li L, van der Bourg A, Niino Y, Egolf L, Monetti C, Gu H, Mills M, Cheng A, Tasic B, Nguyen TN, Sunkin SM, Benucci A, Nagy A, Miyawaki A, Helmchen F, Empson RM, Knöpfel T, Boyden ES, Reid RC, Carandini M, Zeng H (2015) Transgenic mice for intersectional targeting of neural sensors and effectors with high specificity and performance. Neuron 85:942–958. https://doi.org/10.1016/j.neuron.2015.02.022

Mali P, Esvelt KM, Church GM (2013) Cas9 as a versatile tool for engineering biology. Nat Methods 10:957–963. https://doi.org/10.1038/nmeth.2649

Mallat A, Teixeira-Clerc F, Lotersztajn S (2013) Cannabinoid signaling and liver therapeutics. J Hepatol 59:891–896. https://doi.org/10.1016/j.jhep.2013.03.032

Manterola A, Chara JC, Aguado T, Palazuelos J, Matute C, Mato S (2022) Cannabinoid CB1 receptor expression in oligodendrocyte progenitors of the hippocampus revealed by the NG2-EYFP-knockin mouse. Front Neuroanat 16:1030060. https://doi.org/10.3389/fnana.2022.1030060

Marban SL, Gearhart JD (1987) Hyperinsulinemia in transgenic mice carrying multiple copies of the human insulin gene: evidence for a gene dosage effect. Pediatr Res 21:344–344. https://doi.org/10.1203/00006450-198704010-01063

Marcus DJ, Bedse G, Gaulden AD, Ryan JD, Kondev V, Winters ND, Rosas-Vidal LE, Altemus M, Mackie K, Lee FS, Delpire E, Patel S (2020) Endocannabinoid signaling collapse mediates stress-induced Amygdalo-cortical strengthening. Neuron 105:1062–1076.e6. https://doi.org/10.1016/j.neuron.2019.12.024

Maresz K, Carrier EJ, Ponomarev ED, Hillard CJ, Dittel BN (2005) Modulation of the cannabinoid CB2 receptor in microglial cells in response to inflammatory stimuli. J Neurochem 95:437–445. https://doi.org/10.1111/j.1471-4159.2005.03380.x

Maresz K, Pryce G, Ponomarev ED, Marsicano G, Croxford JL, Shriver LP, Ledent C, Cheng X, Carrier EJ, Mann MK, Giovannoni G, Pertwee RG, Yamamura T, Buckley NE, Hillard CJ, Lutz B, Baker D, Dittel BN (2007) Direct suppression of CNS autoimmune inflammation via the cannabinoid receptor CB1 on neurons and CB2 on autoreactive T cells. Nat Med 13:492–497. https://doi.org/10.1038/nm1561

Maroso M, Szabo GG, Kim HK, Alexander A, Bui AD, Lee S-H, Lutz B, Soltesz I (2016) Cannabinoid control of learning and memory through HCN channels. Neuron 89:1059–1073. https://doi.org/10.1016/j.neuron.2016.01.023

Marques S, Zeisel A, Codeluppi S, van Bruggen D, Falcão AM, Xiao L, Li H, Häring M, Hochgerner H, Romanov RA, Gyllborg D, Muñoz-Manchado AB, Manno GL, Lönnerberg P, Floriddia EM, Rezayee F, Ernfors P, Arenas E, Hjerling-Leffler J, Harkany T, Richardson WD, Linnarsson S, Castelo-Branco G (2016) Oligodendrocyte heterogeneity in the mouse juvenile and adult central nervous system. Science. https://doi.org/10.1126/science.aaf6463

Marsicano G, Kuner R (2008) Anatomical distribution of receptors, ligands and enzymes in the brain and in the spinal cord: circuitries and neurochemistry. In: Cannabinoids and the brain. Springer, Boston

Marsicano G, Lutz B (1999) Expression of the cannabinoid receptor CB1 in distinct neuronal subpopulations in the adult mouse forebrain. Eur J Neurosci 11:4213–4225

Marsicano G, Wotjak CT, Azad SC, Bisogno T, Rammes G, Cascio MG, Hermann H, Tang J, Hofmann C, Zieglgänsberger W, Di Marzo V, Lutz B (2002) The endogenous cannabinoid system controls extinction of aversive memories. Nature 418:530–534. https://doi.org/10.1038/nature00839

Marsicano G, Goodenough S, Monory K, Hermann H, Eder M, Cannich A, Azad SC, Cascio MG, Gutiérrez SO, van der Stelt M, López-Rodriguez ML, Casanova E, Schütz G, Zieglgänsberger W, Di Marzo V, Behl C, Lutz B (2003) CB1 cannabinoid receptors and on-demand defense against excitotoxicity. Science 302:84–88. https://doi.org/10.1126/science.1088208

Martin-Fernandez M, Jamison S, Robin LM, Zhao Z, Martín ED, Aguilar J, Benneyworth MA, Marsicano G, Araque A (2017) Synapse-specific astrocyte gating of amygdala-related behavior. Nat Neurosci 20:1540–1548. https://doi.org/10.1038/nn.4649

Mayo LM, Rabinak CA, Hill MN, Heilig M (2022) Targeting the endocannabinoid system in the treatment of posttraumatic stress disorder: a promising case of preclinical-clinical translation? Biol Psychiatry 91:262–272. https://doi.org/10.1016/j.biopsych.2021.07.019

McClure C, Cole KLH, Wulff P, Klugmann M, Murray AJ (2011) Production and titering of recombinant adeno-associated viral vectors. J Vis Exp:e3348. https://doi.org/10.3791/3348

McIntosh HH, Song C, Howlett AC (1998) CB1 cannabinoid receptor: cellular regulation and distribution in N18TG2 neuroblastoma cells. Brain Res Mol Brain Res 53:163–173. https://doi.org/10.1016/s0169-328x(97)00294-5

Melis M, Pistis M, Perra S, Muntoni AL, Pillolla G, Gessa GL (2004) Endocannabinoids mediate presynaptic inhibition of glutamatergic transmission in rat ventral tegmental area dopamine neurons through activation of CB1 receptors. J Neurosci 24:53–62. https://doi.org/10.1523/JNEUROSCI.4503-03.2004

Mendizabal-Zubiaga J, Melser S, Bénard G, Ramos A, Reguero L, Arrabal S, Elezgarai I, Gerrikagoitia I, Suarez J, Rodríguez De Fonseca F, Puente N, Marsicano G, Grandes P (2016) Cannabinoid CB1 receptors are localized in striated muscle mitochondria and regulate mitochondrial respiration. Front Physiol 7:476. https://doi.org/10.3389/fphys.2016.00476

Mercer EH, Hoyle GW, Kapur RP, Brinster RL, Palmiter RD (1991) The dopamine beta-hydroxylase gene promoter directs expression of E. coli lacZ to sympathetic and other neurons in adult transgenic mice. Neuron 7:703–716. https://doi.org/10.1016/0896-6273(91)90274-4

Metna-Laurent M, Soria-Gómez E, Verrier D, Conforzi M, Jégo P, Lafenêtre P, Marsicano G (2012) Bimodal control of fear-coping strategies by CB1 cannabinoid receptors. J Neurosci 32: 7109–7118. https://doi.org/10.1523/JNEUROSCI.1054-12.2012

Mileni M, Johnson DS, Wang Z, Everdeen DS, Liimatta M, Pabst B, Bhattacharya K, Nugent RA, Kamtekar S, Cravatt BF, Ahn K, Stevens RC (2008) Structure-guided inhibitor design for human FAAH by interspecies active site conversion. Proc Natl Acad Sci U S A 105:12820–12824. https://doi.org/10.1073/pnas.0806121105

Miller SJ (2018) Astrocyte heterogeneity in the adult central nervous system. Front Cell Neurosci 12

Miniou P, Tiziano D, Frugier T, Roblot N, Le Meur M, Melki J (1999) Gene targeting restricted to mouse striated muscle lineage. Nucleic Acids Res 27:e27. https://doi.org/10.1093/nar/27.19.e27

Miranda K, Becker W, Busbee PB, Dopkins N, Abdulla OA, Zhong Y, Zhang J, Nagarkatti M, Nagarkatti PS (2022) Yin and yang of cannabinoid CB1 receptor: CB1 deletion in immune cells causes exacerbation while deletion in non-immune cells attenuates obesity. iScience 25:104994. https://doi.org/10.1016/j.isci.2022.104994

Mizuno I, Matsuda S (2021) The role of endocannabinoids in consolidation, retrieval, reconsolidation, and extinction of fear memory. Pharmacol Rep 73:984–1003. https://doi.org/10.1007/s43440-021-00246-y

Molina-Holgado E, Vela JM, Arévalo-Martín A, Almazán C, Molina-Holgado F, Borrell J, Guaza C (2002a) Cannabinoids promote oligodendrocyte progenitor survival: involvement of cannabinoid receptors and phosphatidylinositol-3 kinase/Akt signaling. J Neurosci 22:9742–9753. https://doi.org/10.1523/JNEUROSCI.22-22-09742.2002

Molina-Holgado F, Molina-Holgado E, Guaza C, Rothwell NJ (2002b) Role of CB1 and CB2 receptors in the inhibitory effects of cannabinoids on lipopolysaccharide-induced nitric oxide release in astrocyte cultures. J Neurosci Res 67:829–836. https://doi.org/10.1002/jnr.10165

Monory K, Massa F, Egertová M, Eder M, Blaudzun H, Westenbroek R, Kelsch W, Jacob W, Marsch R, Ekker M, Long J, Rubenstein JL, Goebbels S, Nave K-A, During M, Klugmann M, Wölfel B, Dodt H-U, Zieglgänsberger W, Wotjak CT, Mackie K, Elphick MR, Marsicano G, Lutz B (2006) The endocannabinoid system controls key epileptogenic circuits in the hippocampus. Neuron 51:455–466. https://doi.org/10.1016/j.neuron.2006.07.006

Monory K, Blaudzun H, Massa F, Kaiser N, Lemberger T, Schütz G, Wotjak CT, Lutz B, Marsicano G (2007) Genetic dissection of behavioural and autonomic effects of Delta(9)-tetrahydrocannabinol in mice. PLoS Biol 5:e269. https://doi.org/10.1371/journal.pbio.0050269

Monory K, Polack M, Remus A, Lutz B, Korte M (2015) Cannabinoid CB1 receptor calibrates excitatory synaptic balance in the mouse hippocampus. J Neurosci 35:3842–3850. https://doi.org/10.1523/JNEUROSCI.3167-14.2015

Monteleone P, Tortorella A, Martiadis V, Di Filippo C, Canestrelli B, Maj M (2008) The cDNA 385C to a missense polymorphism of the endocannabinoid degrading enzyme fatty acid amide hydrolase (FAAH) is associated with overweight/obesity but not with binge eating disorder in overweight/obese women. Psychoneuroendocrinology 33:546–550. https://doi.org/10.1016/j.psyneuen.2008.01.004

Moreno E, Chiarlone A, Medrano M, Puigdellívol M, Bibic L, Howell LA, Resel E, Puente N, Casarejos MJ, Perucho J, Botta J, Suelves N, Ciruela F, Ginés S, Galve-Roperh I, Casadó V, Grandes P, Lutz B, Monory K, Canela EI, Lluís C, McCormick PJ, Guzmán M (2018) Singular location and signaling profile of adenosine A2A-cannabinoid CB1 receptor heteromers in the dorsal striatum. Neuropsychopharmacology 43:964–977. https://doi.org/10.1038/npp.2017.12

Mori T, Tanaka K, Buffo A, Wurst W, Kühn R, Götz M (2006) Inducible gene deletion in astroglia and radial glia – a valuable tool for functional and lineage analysis. Glia 54:21–34. https://doi.org/10.1002/glia.20350

Mount CW, Monje M (2017) Wrapped to adapt: experience-dependent myelination. Neuron 95:743–756. https://doi.org/10.1016/j.neuron.2017.07.009

Muccioli GG (2010) Endocannabinoid biosynthesis and inactivation, from simple to complex. Drug Discov Today 15:474–483. https://doi.org/10.1016/j.drudis.2010.03.007

Mulder J, Aguado T, Keimpema E, Barabás K, Ballester Rosado CJ, Nguyen L, Monory K, Marsicano G, Di Marzo V, Hurd YL, Guillemot F, Mackie K, Lutz B, Guzmán M, Lu H-C, Galve-Roperh I, Harkany T (2008) Endocannabinoid signaling controls pyramidal cell specification and long-range axon patterning. Proc Natl Acad Sci U S A 105:8760–8765. https://doi.org/10.1073/pnas.0803545105

Munro S, Thomas KL, Abu-Shaar M (1993) Molecular characterization of a peripheral receptor for cannabinoids. Nature 365:61–65. https://doi.org/10.1038/365061a0

Murataeva N, Straiker A, Mackie K (2014) Parsing the players: 2-arachidonoylglycerol synthesis and degradation in the CNS. Br J Pharmacol 171:1379–1391. https://doi.org/10.1111/bph.12411

Nakazi M, Bauer U, Nickel T, Kathmann M, Schlicker E (2000) Inhibition of serotonin release in the mouse brain via presynaptic cannabinoid CB1 receptors. Naunyn Schmiedebergs Arch Pharmacol 361:19–24. https://doi.org/10.1007/s002109900147

Navarrete M, Araque A (2008) Endocannabinoids mediate neuron-astrocyte communication. Neuron 57:883–893. https://doi.org/10.1016/j.neuron.2008.01.029

Navarrete M, Araque A (2010) Endocannabinoids potentiate synaptic transmission through stimulation of astrocytes. Neuron 68:113–126. https://doi.org/10.1016/j.neuron.2010.08.043

Nimmerjahn A, Kirchhoff F, Helmchen F (2005) Resting microglial cells are highly dynamic surveillants of brain parenchyma in vivo. Science 308:1314–1318. https://doi.org/10.1126/science.1110647

Nishiyama A, Lin XH, Giese N, Heldin CH, Stallcup WB (1996) Co-localization of NG2 proteoglycan and PDGF alpha-receptor on O2A progenitor cells in the developing rat brain. J Neurosci Res 43:299–314. https://doi.org/10.1002/(SICI)1097-4547(19960201)43:3<299::AID-JNR5>3.0.CO;2-E

Njoo C, Agarwal N, Lutz B, Kuner R (2015) The cannabinoid receptor CB1 interacts with the WAVE1 complex and plays a role in actin dynamics and structural plasticity in neurons. PLoS Biol 13:e1002286. https://doi.org/10.1371/journal.pbio.1002286

Noriega-Prieto JA, Kofuji P, Araque A (2022) Endocannabinoid signaling in synaptic function. Glia. https://doi.org/10.1002/glia.24256

Noriega-Prieto JA, Falcón-Moya R, Eraso-Pichot A, Fundazuri UB, Guttipatti P, Belisle L, Rodríguez-Moreno A, van der Stelt M, Cheer J, Marsicano G, Kofuji P, Araque A (2023) Distinct endocannabinoids specifically signal to astrocytes and neurons. bioRxiv. https://doi.org/10.1101/2023.06.13.544877

Nyilas R, Dudok B, Urbán GM, Mackie K, Watanabe M, Cravatt BF, Freund TF, Katona I (2008) Enzymatic machinery for endocannabinoid biosynthesis associated with calcium stores in glutamatergic axon terminals. J Neurosci 28:1058–1063. https://doi.org/10.1523/JNEUROSCI.5102-07.2008

O'Sullivan SE, Yates AS, Porter RK (2021) The peripheral cannabinoid receptor type 1 (CB1) as a molecular target for modulating body weight in man. Molecules 26:6178. https://doi.org/10.3390/molecules26206178

Osei-Hyiaman D, Liu J, Zhou L, Godlewski G, Harvey-White J, Jeong W, Bátkai S, Marsicano G, Lutz B, Buettner C, Kunos G (2008) Hepatic CB1 receptor is required for development of diet-induced steatosis, dyslipidemia, and insulin and leptin resistance in mice. J Clin Invest 118: 3160–3169. https://doi.org/10.1172/JCI34827

Oudin MJ, Gajendra S, Williams G, Hobbs C, Lalli G, Doherty P (2011) Endocannabinoids regulate the migration of subventricular zone-derived neuroblasts in the postnatal brain. J Neurosci 31: 4000–4011. https://doi.org/10.1523/JNEUROSCI.5483-10.2011

Oyagawa CRM, Grimsey NL (2021) Cannabinoid receptor CB1 and CB2 interacting proteins: techniques, progress and perspectives. Methods Cell Biol 166:83–132. https://doi.org/10.1016/bs.mcb.2021.06.011

Pagano Zottola AC, Severi I, Cannich A, Ciofi P, Cota D, Marsicano G, Giordano A, Bellocchio L (2022) Expression of functional cannabinoid Type-1 (CB1) receptor in mitochondria of White adipocytes. Cells 11:2582. https://doi.org/10.3390/cells11162582

Pagotto U, Marsicano G, Cota D, Lutz B, Pasquali R (2006) The emerging role of the endocannabinoid system in endocrine regulation and energy balance. Endocr Rev 27:73–100. https://doi.org/10.1210/er.2005-0009

Palazuelos J, Aguado T, Egia A, Mechoulam R, Guzmán M, Galve-Roperh I (2006) Non-psychoactive CB2 cannabinoid agonists stimulate neural progenitor proliferation. FASEB J 20:2405–2407. https://doi.org/10.1096/fj.06-6164fje

Palazuelos J, Davoust N, Julien B, Hatterer E, Aguado T, Mechoulam R, Benito C, Romero J, Silva A, Guzmán M, Nataf S, Galve-Roperh I (2008) The CB(2) cannabinoid receptor controls myeloid progenitor trafficking: involvement in the pathogenesis of an animal model of multiple sclerosis. J Biol Chem 283:13320–13329. https://doi.org/10.1074/jbc.M707960200

Palazuelos J, Aguado T, Pazos MR, Julien B, Carrasco C, Resel E, Sagredo O, Benito C, Romero J, Azcoitia I, Fernández-Ruiz J, Guzmán M, Galve-Roperh I (2009) Microglial CB2 cannabinoid receptors are neuroprotective in Huntington's disease excitotoxicity. Brain 132:3152–3164. https://doi.org/10.1093/brain/awp239

Palazuelos J, Ortega Z, Díaz-Alonso J, Guzmán M, Galve-Roperh I (2012) CB2 cannabinoid receptors promote neural progenitor cell proliferation via mTORC1 signaling. J Biol Chem 287:1198–1209. https://doi.org/10.1074/jbc.M111.291294

Pamplona FA, Ferreira J, Menezes de Lima O Jr, Duarte FS, Bento AF, Forner S, Villarinho JG, Bellocchio L, Bellochio L, Wotjak CT, Lerner R, Monory K, Lutz B, Canetti C, Matias I, Calixto JB, Marsicano G, Guimarães MZP, Takahashi RN (2012) Anti-inflammatory lipoxin A4 is an endogenous allosteric enhancer of CB1 cannabinoid receptor. Proc Natl Acad Sci U S A 109:21134–21139. https://doi.org/10.1073/pnas.1202906109

Paolicelli RC, Bolasco G, Pagani F, Maggi L, Scianni M, Panzanelli P, Giustetto M, Ferreira TA, Guiducci E, Dumas L, Ragozzino D, Gross CT (2011) Synaptic pruning by microglia is necessary for normal brain development. Science 333:1456–1458. https://doi.org/10.1126/science.1202529

Parlato R, Otto C, Begus Y, Stotz S, Schütz G (2007) Specific ablation of the transcription factor CREB in sympathetic neurons surprisingly protects against developmentally regulated apoptosis. Development 134:1663–1670. https://doi.org/10.1242/dev.02838

Patriarchi T, Cho JR, Merten K, Howe MW, Marley A, Xiong W-H, Folk RW, Broussard GJ, Liang R, Jang MJ, Zhong H, Dombeck D, von Zastrow M, Nimmerjahn A, Gradinaru V, Williams JT, Tian L (2018) Ultrafast neuronal imaging of dopamine dynamics with designed genetically encoded sensors. Science 360:eaat4422. https://doi.org/10.1126/science.aat4422

Patti F, Messina S, Solaro C, Amato MP, Bergamaschi R, Bonavita S, Bossio RB, Morra VB, Costantino GF, Cavalla P, Centonze D, Comi G, Cottone S, Danni M, Francia A, Gajofatto A, Gasperini C, Ghezzi A, Iudice A, Lus G, Maniscalco GT, Marrosu MG, Matta M, Mirabella M, Montanari E, Pozzilli C, Rovaris M, Sessa E, Spitaleri D, Trojano M, Valentino P, Zappia M (2016) Efficacy and safety of cannabinoid oromucosal spray for multiple sclerosis spasticity. J Neurol Neurosurg Psychiatry 87:944–951. https://doi.org/10.1136/jnnp-2015-312591

Pedraza CE, Taylor C, Pereira A, Seng M, Tham C-S, Izrael M, Webb M (2014) Induction of oligodendrocyte differentiation and in vitro myelination by inhibition of rho-associated kinase. ASN Neuro 6:1759091414538134. https://doi.org/10.1177/1759091414538134

Petrucci V, Chicca A, Glasmacher S, Paloczi J, Cao Z, Pacher P, Gertsch J (2017) Pepcan-12 (RVD-hemopressin) is a CB2 receptor positive allosteric modulator constitutively secreted by adrenals and in liver upon tissue damage. Sci Rep 7:9560. https://doi.org/10.1038/s41598-017-09808-8

Pluta K, Kacprzak MM (2009) Use of HIV as a gene transfer vector. Acta Biochim Pol 56:531–595

Postic C, Shiota M, Niswender KD, Jetton TL, Chen Y, Moates JM, Shelton KD, Lindner J, Cherrington AD, Magnuson MA (1999) Dual roles for glucokinase in glucose homeostasis as determined by liver and pancreatic beta cell-specific gene knock-outs using Cre recombinase. J Biol Chem 274:305–315. https://doi.org/10.1074/jbc.274.1.305

Priestley R, Glass M, Kendall D (2017) Functional selectivity at cannabinoid receptors. Adv Pharmacol 80:207–221. https://doi.org/10.1016/bs.apha.2017.03.005

Pryce G, Ahmed Z, Hankey DJR, Jackson SJ, Croxford JL, Pocock JM, Ledent C, Petzold A, Thompson AJ, Giovannoni G, Cuzner ML, Baker D (2003) Cannabinoids inhibit neurodegeneration in models of multiple sclerosis. Brain 126:2191–2202. https://doi.org/10.1093/brain/awg224

Puighermanal E, Marsicano G, Busquets-Garcia A, Lutz B, Maldonado R, Ozaita A (2009) Cannabinoid modulation of hippocampal long-term memory is mediated by mTOR signaling. Nat Neurosci 12:1152–1158. https://doi.org/10.1038/nn.2369

Qian Y, Li J, Zhao S, Matthews EA, Adoff M, Zhong W, An X, Yeo M, Park C, Yang X, Wang B-S, Southwell DG, Huang ZJ (2022) Programmable RNA sensing for cell monitoring and manipulation. Nature 610:713–721. https://doi.org/10.1038/s41586-022-05280-1

Ramírez BG, Blázquez C, Gómez del Pulgar T, Guzmán M, de Ceballos ML (2005) Prevention of Alzheimer's disease pathology by cannabinoids: neuroprotection mediated by blockade of microglial activation. J Neurosci 25:1904–1913. https://doi.org/10.1523/JNEUROSCI.4540-04.2005

Reguero L, Puente N, Elezgarai I, Ramos-Uriarte A, Gerrikagoitia I, Bueno-López J-L, Doñate F, Grandes P (2014) Subcellular localization of NAPE-PLD and DAGL-α in the ventromedial nucleus of the hypothalamus by a preembedding immunogold method. Histochem Cell Biol 141:543–550. https://doi.org/10.1007/s00418-013-1174-x

Remmers F, Lange MD, Hamann M, Ruehle S, Pape H-C, Lutz B (2017) Addressing sufficiency of the CB1 receptor for endocannabinoid-mediated functions through conditional genetic rescue in forebrain GABAergic neurons. Brain Struct Funct 222:3431–3452. https://doi.org/10.1007/s00429-017-1411-5

Rey AA, Purrio M, Viveros M-P, Lutz B (2012) Biphasic effects of cannabinoids in anxiety responses: CB1 and GABA(B) receptors in the balance of GABAergic and glutamatergic neurotransmission. Neuropsychopharmacology. https://doi.org/10.1038/npp.2012.123

Ribeiro FF, Xapelli S (2021) An overview of adult neurogenesis. Adv Exp Med Biol 1331:77–94. https://doi.org/10.1007/978-3-030-74046-7_7

Riegel AC, Lupica CR (2004) Independent presynaptic and postsynaptic mechanisms regulate endocannabinoid signaling at multiple synapses in the ventral tegmental area. J Neurosci 24: 11070–11078. https://doi.org/10.1523/JNEUROSCI.3695-04.2004
Rivers LE, Young KM, Rizzi M, Jamen F, Psachoulia K, Wade A, Kessaris N, Richardson WD (2008) PDGFRA/NG2 glia generate myelinating oligodendrocytes and piriform projection neurons in adult mice. Nat Neurosci 11:1392–1401. https://doi.org/10.1038/nn.2220
Robbe D, Kopf M, Remaury A, Bockaert J, Manzoni OJ (2002) Endogenous cannabinoids mediate long-term synaptic depression in the nucleus accumbens. Proc Natl Acad Sci U S A 99:8384–8388. https://doi.org/10.1073/pnas.122149199
Roberto M, Cruz M, Bajo M, Siggins GR, Parsons LH, Schweitzer P (2010) The endocannabinoid system tonically regulates inhibitory transmission and depresses the effect of ethanol in central amygdala. Neuropsychopharmacology 35:1962–1972. https://doi.org/10.1038/npp.2010.70
Robin LM, da Cruz JFO, Langlais VC, Martin-Fernandez M, Metna-Laurent M, Busquets-Garcia A, Bellocchio L, Soria-Gomez E, Papouin T, Varilh M, Sherwood MW, Belluomo I, Balcells G, Matias I, Bosier B, Drago F, Van Eeckhaut A, Smolders IJ, Georges F, Araque A, Panatier A, Oliet SHR, Oliet SHR, Marsicano G (2018) Astroglial CB1 receptors determine synaptic D-serine availability to enable recognition memory. Neuron 98:935. https://doi.org/10.1016/j.neuron.2018.04.034
Rodriguez JJ, Mackie K, Pickel VM (2001) Ultrastructural localization of the CB1 cannabinoid receptor in mu-opioid receptor patches of the rat Caudate putamen nucleus. J Neurosci 21:823–833. https://doi.org/10.1523/JNEUROSCI.21-03-00823.2001
Rodríguez-Cueto C, Gómez-Almería M, García Toscano L, Romero J, Hillard CJ, de Lago E, Fernández-Ruiz J (2021) Inactivation of the CB2 receptor accelerated the neuropathological deterioration in TDP-43 transgenic mice, a model of amyotrophic lateral sclerosis. Brain Pathol 31:e12972. https://doi.org/10.1111/bpa.12972
Rozenfeld R, Bushlin I, Gomes I, Tzavaras N, Gupta A, Neves S, Battini L, Gusella GL, Lachmann A, Ma'ayan A, Blitzer RD, Devi LA (2012) Receptor heteromerization expands the repertoire of cannabinoid signaling in rodent neurons. PLoS One 7:e29239. https://doi.org/10.1371/journal.pone.0029239
Ruat J, Hartmann A, Heinz DE, Nemcova P, Stoffel R, Deussing JM, Chen A, Wotjak CT (2021) CB1 receptors in corticotropin-releasing factor neurons selectively control the acoustic startle response in male mice. Genes Brain Behav 20:e12775. https://doi.org/10.1111/gbb.12775
Rubera I, Poujeol C, Bertin G, Hasseine L, Counillon L, Poujeol P, Tauc M (2004) Specific Cre/Lox recombination in the mouse proximal tubule. J Am Soc Nephrol 15:2050–2056. https://doi.org/10.1097/01.ASN.0000133023.89251.01
Rubino T, Guidali C, Vigano D, Realini N, Valenti M, Massi P, Parolaro D (2008) CB1 receptor stimulation in specific brain areas differently modulate anxiety-related behaviour. Neuropharmacology 54:151–160. https://doi.org/10.1016/j.neuropharm.2007.06.024
Rubio-Araiz A, Arévalo-Martín A, Gómez-Torres O, Navarro-Galve B, García-Ovejero D, Suetterlin P, Sánchez-Heras E, Molina-Holgado E, Molina-Holgado F (2008) The endocannabinoid system modulates a transient TNF pathway that induces neural stem cell proliferation. Mol Cell Neurosci 38:374–380. https://doi.org/10.1016/j.mcn.2008.03.010
Ruehle S, Rey AA, Remmers F, Lutz B (2012) The endocannabinoid system in anxiety, fear memory and habituation. J Psychopharmacol (Oxf) 26:23–39. https://doi.org/10.1177/0269881111408958
Ruehle S, Remmers F, Romo-Parra H, Massa F, Wickert M, Wörtge S, Häring M, Kaiser N, Marsicano G, Pape H-C, Lutz B (2013) Cannabinoid CB1 receptor in dorsal telencephalic glutamatergic neurons: distinctive sufficiency for hippocampus-dependent and amygdala-dependent synaptic and behavioral functions. J Neurosci 33:10264–10277. https://doi.org/10.1523/JNEUROSCI.4171-12.2013
Ruiz de Azua I, Lutz B (2019) Multiple endocannabinoid-mediated mechanisms in the regulation of energy homeostasis in brain and peripheral tissues. Cell Mol Life Sci 76:1341–1363. https://doi.org/10.1007/s00018-018-2994-6

Ruiz de Azua I, Mancini G, Srivastava RK, Rey AA, Cardinal P, Tedesco L, Zingaretti CM, Sassmann A, Quarta C, Schwitter C, Conrad A, Wettschureck N, Vemuri VK, Makriyannis A, Hartwig J, Mendez-Lago M, Bindila L, Monory K, Giordano A, Cinti S, Marsicano G, Offermanns S, Nisoli E, Pagotto U, Cota D, Lutz B (2017) Adipocyte cannabinoid receptor CB1 regulates energy homeostasis and alternatively activated macrophages. J Clin Invest 127: 4148–4162. https://doi.org/10.1172/JCI83626

Ruiz de Azua I, Martin-Garcia E, Domingo-Rodriguez L, Aparisi Rey A, Pascual Cuadrado D, Islami L, Turunen P, Remmers F, Lutz B, Maldonado R (2021) Cannabinoid CB1 receptor in dorsal telencephalic glutamatergic neurons drives overconsumption of palatable food and obesity. Neuropsychopharmacology 46:982–991. https://doi.org/10.1038/s41386-021-00957-z

Ruiz de Martín Esteban S, Benito-Cuesta I, Terradillos I, Martínez-Relimpio AM, Arnanz MA, Ruiz-Pérez G, Korn C, Raposo C, Sarott RC, Westphal MV, Elezgarai I, Carreira EM, Hillard CJ, Grether U, Grandes P, Grande MT, Romero J (2022) Cannabinoid CB2 receptors modulate microglia function and amyloid dynamics in a mouse model of Alzheimer's disease. Front Pharmacol 13:841766. https://doi.org/10.3389/fphar.2022.841766

Sánchez C, Galve-Roperh I, Rueda D, Guzmán M (1998) Involvement of sphingomyelin hydrolysis and the mitogen-activated protein kinase cascade in the Delta9-tetrahydrocannabinol-induced stimulation of glucose metabolism in primary astrocytes. Mol Pharmacol 54:834–843

Sánchez-de la Torre A, Aguado T, Huerga-Gómez A, Santamaría S, Gentile A, Chara JC, Matute C, Monory K, Mato S, Guzmán M, Lutz B, Galve-Roperh I, Palazuelos J (2022) Cannabinoid CB1 receptor gene inactivation in oligodendrocyte precursors disrupts oligodendrogenesis and myelination in mice. Cell Death Dis 13:585. https://doi.org/10.1038/s41419-022-05032-z

Sara SJ (2009) The locus coeruleus and noradrenergic modulation of cognition. Nat Rev Neurosci 10:211–223. https://doi.org/10.1038/nrn2573

Sassmann A, Offermanns S, Wettschureck N (2010) Tamoxifen-inducible Cre-mediated recombination in adipocytes. Genesis 48:618–625. https://doi.org/10.1002/dvg.20665

Sathasivam K, Hobbs C, Mangiarini L, Mahal A, Turmaine M, Doherty P, Davies SW, Bates GP (1999) Transgenic models of Huntington's disease. Philos Trans R Soc Lond B Biol Sci 354: 963–969. https://doi.org/10.1098/rstb.1999.0447

Schmöle A-C, Lundt R, Gennequin B, Schrage H, Beins E, Krämer A, Zimmer T, Limmer A, Zimmer A, Otte D-M (2015) Expression analysis of CB2-GFP BAC transgenic mice. PLoS One 10:e0138986. https://doi.org/10.1371/journal.pone.0138986

Schneider M, Kasanetz F, Lynch DL, Friemel CM, Lassalle O, Hurst DP, Steindel F, Monory K, Schäfer C, Miederer I, Leweke FM, Schreckenberger M, Lutz B, Reggio PH, Manzoni OJ, Spanagel R (2015) Enhanced functional activity of the cannabinoid type-1 receptor mediates adolescent behavior. J Neurosci 35:13975–13988. https://doi.org/10.1523/JNEUROSCI.1937-15.2015

Schnütgen F, Doerflinger N, Calléja C, Wendling O, Chambon P, Ghyselinck NB (2003) A directional strategy for monitoring Cre-mediated recombination at the cellular level in the mouse. Nat Biotechnol 21:562–565. https://doi.org/10.1038/nbt811

Schuele L-L, Glasmacher S, Gertsch J, Roggan MD, Transfeld J-L, Bindila L, Lutz B, Kolbe C-C, Bilkei-Gorzo A, Zimmer A, Leidmaa E (2021) Diacylglycerol lipase alpha in astrocytes is involved in maternal care and affective behaviors. Glia 69:377–391. https://doi.org/10.1002/glia.23903

Schuele L-L, Schuermann B, Bilkei-Gorzo A, Gorgzadeh S, Zimmer A, Leidmaa E (2022) Regulation of adult neurogenesis by the endocannabinoid-producing enzyme diacylglycerol lipase alpha (DAGLa). Sci Rep 12:633. https://doi.org/10.1038/s41598-021-04600-1

Serrat R, Covelo A, Kouskoff V, Delcasso S, Ruiz-Calvo A, Chenouard N, Stella C, Blancard C, Salin B, Julio-Kalajzić F, Cannich A, Massa F, Varilh M, Deforges S, Robin LM, De Stefani D, Busquets-Garcia A, Gambino F, Beyeler A, Pouvreau S, Marsicano G (2021) Astroglial ER-mitochondria calcium transfer mediates endocannabinoid-dependent synaptic integration. Cell Rep 37:110133. https://doi.org/10.1016/j.celrep.2021.110133

Sheng M, Greenberg ME (1990) The regulation and function of c-fos and other immediate early genes in the nervous system. Neuron 4(4):477–485. https://doi.org/10.1016/0896-6273(90)90106-P

Shimshek DR, Kim J, Hübner MR, Spergel DJ, Buchholz F, Casanova E, Stewart AF, Seeburg PH, Sprengel R (2002) Codon-improved Cre recombinase (iCre) expression in the mouse. Genesis 32:19–26. https://doi.org/10.1002/gene.10023

Shonesy BC, Bluett RJ, Ramikie TS, Báldi R, Hermanson DJ, Kingsley PJ, Marnett LJ, Winder DG, Colbran RJ, Patel S (2014) Genetic disruption of 2-arachidonoylglycerol synthesis reveals a key role for endocannabinoid signaling in anxiety modulation. Cell Rep 9:1644–1653. https://doi.org/10.1016/j.celrep.2014.11.001

Simpson CD, Hurren R, Kasimer D, MacLean N, Eberhard Y, Ketela T, Moffat J, Schimmer AD (2012) A genome wide shRNA screen identifies α/β hydrolase domain containing 4 (ABHD4) as a novel regulator of anoikis resistance. Apoptosis 17:666–678. https://doi.org/10.1007/s10495-012-0723-4

Singewald N, Sartori SB, Reif A, Holmes A (2023) Alleviating anxiety and taming trauma: novel pharmacotherapeutics for anxiety disorders and posttraumatic stress disorder. Neuropharmacology 226:109418. https://doi.org/10.1016/j.neuropharm.2023.109418

Sipe JC, Chiang K, Gerber AL, Beutler E, Cravatt BF (2002) A missense mutation in human fatty acid amide hydrolase associated with problem drug use. Proc Natl Acad Sci U S A 99:8394–8399. https://doi.org/10.1073/pnas.082235799

Sipe JC, Waalen J, Gerber A, Beutler E (2005) Overweight and obesity associated with a missense polymorphism in fatty acid amide hydrolase (FAAH). Int J Obes (Lond) 29:755–759. https://doi.org/10.1038/sj.ijo.0802954

Slanina KA, Schweitzer P (2005) Inhibition of cyclooxygenase-2 elicits a CB1-mediated decrease of excitatory transmission in rat CA1 hippocampus. Neuropharmacology 49:653–659. https://doi.org/10.1016/j.neuropharm.2005.04.019

Soria-Gómez E, Bellocchio L, Reguero L, Lepousez G, Martin C, Bendahmane M, Ruehle S, Remmers F, Desprez T, Matias I, Wiesner T, Cannich A, Nissant A, Wadleigh A, Pape H-C, Chiarlone AP, Quarta C, Verrier D, Vincent P, Massa F, Lutz B, Guzmán M, Gurden H, Ferreira G, Lledo P-M, Grandes P, Marsicano G (2014) The endocannabinoid system controls food intake via olfactory processes. Nat Neurosci 17:407–415. https://doi.org/10.1038/nn.3647

Soria-Gomez E, Pagano Zottola AC, Mariani Y, Desprez T, Barresi M, Bonilla-Del Río I, Muguruza C, Le Bon-Jego M, Julio-Kalajzić F, Flynn R, Terral G, Fernández-Moncada I, Robin LM, Oliveira da Cruz JF, Corinti S, Amer YO, Goncalves J, Varilh M, Cannich A, Redon B, Zhao Z, Lesté-Lasserre T, Vincent P, Tolentino-Cortes T, Busquets-García A, Puente N, Bains JS, Hebert-Chatelain E, Barreda-Gómez G, Chaouloff F, Lohman AW, Callado LF, Grandes P, Baufreton J, Marsicano G, Bellocchio L (2021) Subcellular specificity of cannabinoid effects in striatonigral circuits. Neuron 109:1513–1526.e11. https://doi.org/10.1016/j.neuron.2021.03.007

Soriano P (1999) Generalized lacZ expression with the ROSA26 Cre reporter strain. Nat Genet 21:70–71. https://doi.org/10.1038/5007

Srinivas S, Watanabe T, Lin CS, William CM, Tanabe Y, Jessell TM, Costantini F (2001) Cre reporter strains produced by targeted insertion of EYFP and ECFP into the ROSA26 locus. BMC Dev Biol 1:4. https://doi.org/10.1186/1471-213x-1-4

Srivastava RK, Ruiz de Azua I, Conrad A, Purrio M, Lutz B (2022) Cannabinoid CB1 receptor deletion from Catecholaminergic neurons protects from diet-induced obesity. Int J Mol Sci 23:12635. https://doi.org/10.3390/ijms232012635

Steindel F, Lerner R, Häring M, Ruehle S, Marsicano G, Lutz B, Monory K (2013) Neuron-type specific cannabinoid-mediated G protein signalling in mouse hippocampus. J Neurochem 124:795–807. https://doi.org/10.1111/jnc.12137

Steiner MA, Wotjak CT (2008) Role of the endocannabinoid system in regulation of the hypothalamic-pituitary adrenocortical axis. Prog Brain Res 170:397–432. https://doi.org/10.1016/S0079-6123(08)00433-0

Steiner MA, Marsicano G, Wotjak CT, Lutz B (2008) Conditional cannabinoid receptor type 1 mutants reveal neuron subpopulation-specific effects on behavioral and neuroendocrine stress responses. Psychoneuroendocrinology 33:1165–1170. https://doi.org/10.1016/j.psyneuen.2008.06.004

Stella N (2004) Cannabinoid signaling in glial cells. Glia 48:267–277. https://doi.org/10.1002/glia.20084

Stempel AV, Stumpf A, Zhang H-Y, Özdoğan T, Pannasch U, Theis A-K, Otte D-M, Wojtalla A, Rácz I, Ponomarenko A, Xi Z-X, Zimmer A, Schmitz D (2016) Cannabinoid type 2 receptors mediate a cell type-specific plasticity in the hippocampus. Neuron 90:795–809. https://doi.org/10.1016/j.neuron.2016.03.034

Sternberg N, Hamilton D (1981) Bacteriophage P1 site-specific recombination. I. Recombination between loxP sites. J Mol Biol 150:467–486

Stumpf A, Parthier D, Sammons RP, Stempel AV, Breustedt J, Rost BR, Schmitz D (2018) Cannabinoid type 2 receptors mediate a cell type-specific self-inhibition in cortical neurons. Neuropharmacology 139:217–225. https://doi.org/10.1016/j.neuropharm.2018.07.020

Sun J, Fang Y, Ren H, Chen T, Guo J, Yan J, Song S, Zhang L, Liao H (2013) WIN55,212-2 protects oligodendrocyte precursor cells in stroke penumbra following permanent focal cerebral ischemia in rats. Acta Pharmacol Sin 34:119–128. https://doi.org/10.1038/aps.2012.141

Sun F, Zeng J, Jing M, Zhou J, Feng J, Owen SF, Luo Y, Li F, Wang H, Yamaguchi T, Yong Z, Gao Y, Peng W, Wang L, Zhang S, Du J, Lin D, Xu M, Kreitzer AC, Cui G, Li Y (2018) A genetically encoded fluorescent sensor enables rapid and specific detection of dopamine in flies, fish, and mice. Cell 174:481–496.e19. https://doi.org/10.1016/j.cell.2018.06.042

Tamarina NA, Roe MW, Philipson L (2014) Characterization of mice expressing Ins1 gene promoter driven CreERT recombinase for conditional gene deletion in pancreatic β-cells. Islets 6:e27685. https://doi.org/10.4161/isl.27685

Taniguchi H, He M, Wu P, Kim S, Paik R, Sugino K, Kvitsiani D, Fu Y, Lu J, Lin Y, Miyoshi G, Shima Y, Fishell G, Nelson SB, Huang ZJ (2011) A resource of Cre driver lines for genetic targeting of GABAergic neurons in cerebral cortex. Neuron 71:995–1013. https://doi.org/10.1016/j.neuron.2011.07.026

Tanimura A, Yamazaki M, Hashimotodani Y, Uchigashima M, Kawata S, Abe M, Kita Y, Hashimoto K, Shimizu T, Watanabe M, Sakimura K, Kano M (2010) The endocannabinoid 2-arachidonoylglycerol produced by diacylglycerol lipase alpha mediates retrograde suppression of synaptic transmission. Neuron 65:320–327. https://doi.org/10.1016/j.neuron.2010.01.021

Taschler U, Radner FPW, Heier C, Schreiber R, Schweiger M, Schoiswohl G, Preiss-Landl K, Jaeger D, Reiter B, Koefeler HC, Wojciechowski J, Theussl C, Penninger JM, Lass A, Haemmerle G, Zechner R, Zimmermann R (2011) Monoglyceride lipase deficiency in mice impairs lipolysis and attenuates diet-induced insulin resistance. J Biol Chem 286:17467–17477. https://doi.org/10.1074/jbc.M110.215434

Tatti R, Haley MS, Swanson OK, Tselha T, Maffei A (2017) Neurophysiology and regulation of the balance between excitation and inhibition in neocortical circuits. Biol Psychiatry 81:821–831. https://doi.org/10.1016/j.biopsych.2016.09.017

Tervo DGR, Hwang B-Y, Viswanathan S, Gaj T, Lavzin M, Ritola KD, Lindo S, Michael S, Kuleshova E, Ojala D, Huang C-C, Gerfen CR, Schiller J, Dudman JT, Hantman AW, Looger LL, Schaffer DV, Karpova AY (2016) A designer AAV variant permits efficient retrograde access to projection neurons. Neuron 92:372–382. https://doi.org/10.1016/j.neuron.2016.09.021

Terzian ALB, Micale V, Wotjak CT (2014) Cannabinoid receptor type 1 receptors on GABAergic vs. glutamatergic neurons differentially gate sex-dependent social interest in mice. Eur J Neurosci 40:2293–2298. https://doi.org/10.1111/ejn.12561

Tevosian M, Todorov H, Lomazzo E, Bindila L, Ueda N, Bassetti D, Warm D, Kirischuk S, Luhmann HJ, Gerber S, Lutz B (2023) NAPE-PLD deletion in stress-TRAPed neurons results in

an anxiogenic phenotype. Transl Psychiatry 13:152. https://doi.org/10.1038/s41398-023-02448-9
Tronche F, Kellendonk C, Kretz O, Gass P, Anlag K, Orban PC, Bock R, Klein R, Schütz G (1999) Disruption of the glucocorticoid receptor gene in the nervous system results in reduced anxiety. Nat Genet 23:99–103. https://doi.org/10.1038/12703
Tsou K, Mackie K, Sañudo-Peña MC, Walker JM (1999) Cannabinoid CB1 receptors are localized primarily on cholecystokinin-containing GABAergic interneurons in the rat hippocampal formation. Neuroscience 93:969–975
Tsuboi K, Okamoto Y, Ikematsu N, Inoue M, Shimizu Y, Uyama T, Wang J, Deutsch DG, Burns MP, Ulloa NM, Tokumura A, Ueda N (2011) Enzymatic formation of N-acylethanolamines from N-acylethanolamine plasmalogen through N-acylphosphatidylethanolamine-hydrolyzing phospholipase D-dependent and -independent pathways. Biochim Biophys Acta 1811:565–577. https://doi.org/10.1016/j.bbalip.2011.07.009
Tsuboi K, Uyama T, Okamoto Y, Ueda N (2018) Endocannabinoids and related N-acylethanolamines: biological activities and metabolism. Inflamm Regen 38:28. https://doi.org/10.1186/s41232-018-0086-5
Turcotte C, Blanchet M-R, Laviolette M, Flamand N (2016) The CB2 receptor and its role as a regulator of inflammation. Cell Mol Life Sci 73:4449–4470. https://doi.org/10.1007/s00018-016-2300-4
Udi S, Hinden L, Earley B, Drori A, Reuveni N, Hadar R, Cinar R, Nemirovski A, Tam J (2017) Proximal Tubular Cannabinoid-1 receptor regulates obesity-induced CKD. J Am Soc Nephrol 28:3518–3532. https://doi.org/10.1681/ASN.2016101085
Urigüen L, Pérez-Rial S, Ledent C, Palomo T, Manzanares J (2004) Impaired action of anxiolytic drugs in mice deficient in cannabinoid CB1 receptors. Neuropharmacology 46:966–973. https://doi.org/10.1016/j.neuropharm.2004.01.003
Vallée M, Vitiello S, Bellocchio L, Hébert-Chatelain E, Monlezun S, Martin-Garcia E, Kasanetz F, Baillie GL, Panin F, Cathala A, Roullot-Lacarrière V, Fabre S, Hurst DP, Lynch DL, Shore DM, Deroche-Gamonet V, Spampinato U, Revest J-M, Maldonado R, Reggio PH, Ross RA, Marsicano G, Piazza PV (2014) Pregnenolone can protect the brain from cannabis intoxication. Science 343:94–98. https://doi.org/10.1126/science.1243985
Van Sickle MD, Duncan M, Kingsley PJ, Mouihate A, Urbani P, Mackie K, Stella N, Makriyannis A, Piomelli D, Davison JS, Marnett LJ, Di Marzo V, Pittman QJ, Patel KD, Sharkey KA (2005) Identification and functional characterization of brainstem cannabinoid CB2 receptors. Science 310:329–332. https://doi.org/10.1126/science.1115740
Vassar R, Rosenberg M, Ross S, Tyner A, Fuchs E (1989) Tissue-specific and differentiation-specific expression of a human K14 keratin gene in transgenic mice. Proc Natl Acad Sci U S A 86:1563–1567. https://doi.org/10.1073/pnas.86.5.1563
Verkhratsky A, Nedergaard M (2014) Astroglial cradle in the life of the synapse. Philos Trans R Soc Lond B Biol Sci 369:20130595. https://doi.org/10.1098/rstb.2013.0595
Verkhratsky A, Parpura V, Li B, Scuderi C (2021) Astrocytes: the housekeepers and guardians of the CNS. Adv Neurobiol 26:21–53. https://doi.org/10.1007/978-3-030-77375-5_2
Viader A, Blankman JL, Zhong P, Liu X, Schlosburg JE, Joslyn CM, Liu Q-S, Tomarchio AJ, Lichtman AH, Selley DE, Sim-Selley LJ, Cravatt BF (2015) Metabolic interplay between astrocytes and neurons regulates endocannabinoid action. Cell Rep 12:798–808. https://doi.org/10.1016/j.celrep.2015.06.075
Vianna CR, Donato J, Rossi J, Scott M, Economides K, Gautron L, Pierpont S, Elias CF, Elmquist JK (2012) Cannabinoid receptor 1 in the Vagus nerve is dispensable for body weight homeostasis but required for normal gastrointestinal motility. J Neurosci 32:10331–10337. https://doi.org/10.1523/JNEUROSCI.4507-11.2012
Wang S, Zhu Q, Liang G, Franks T, Boucher M, Bence KK, Lu M, Castorena CM, Zhao S, Elmquist JK, Scherer PE, Horton JD (2021) Cannabinoid receptor 1 signaling in hepatocytes and stellate cells does not contribute to NAFLD. J Clin Invest 131. https://doi.org/10.1172/JCI152242

Wang Y, Li G, Chen B, Shakir G, Volz M, Vorst EPC van der, Maas SL, Muley C, Bartelt A, Li Z, Sachs N, Maegdefessel L, Jahantigh MN, Hristov M, Lacy M, Lutz B, Weber C, Herzig S, Prats RG, Steffens S (2023) Loss of myeloid cannabinoid CB1 receptor confers atheroprotection by reducing macrophage proliferation and immunometabolic reprogramming. https://doi.org/10.1101/2023.04.06.535832

Weber T, Böhm G, Hermann E, Schütz G, Schönig K, Bartsch D (2009) Inducible gene manipulations in serotonergic neurons. Front Mol Neurosci 2:24. https://doi.org/10.3389/neuro.02.024.2009

Weiss RA (1993) How does HIV cause AIDS? Science 260:1273–1279. https://doi.org/10.1126/science.8493571

Weiss WA, Aldape K, Mohapatra G, Feuerstein BG, Bishop JM (1997) Targeted expression of MYCN causes neuroblastoma in transgenic mice. EMBO J 16:2985–2995. https://doi.org/10.1093/emboj/16.11.2985

Wilson RI, Nicoll RA (2001) Endogenous cannabinoids mediate retrograde signalling at hippocampal synapses. Nature 410:588–592. https://doi.org/10.1038/35069076

Winters BL, Vaughan CW (2021) Mechanisms of endocannabinoid control of synaptic plasticity. Neuropharmacology 197:108736. https://doi.org/10.1016/j.neuropharm.2021.108736

Winters BD, Krüger JM, Huang X, Gallaher ZR, Ishikawa M, Czaja K, Krueger JM, Huang YH, Schlüter OM, Dong Y (2012) Cannabinoid receptor 1-expressing neurons in the nucleus accumbens. Proc Natl Acad Sci U S A 109:E2717–E2725. https://doi.org/10.1073/pnas.1206303109

Wolf SA, Boddeke HWGM, Kettenmann H (2017) Microglia in physiology and disease. Annu Rev Physiol 79:619–643. https://doi.org/10.1146/annurev-physiol-022516-034406

Wotherspoon G, Fox A, McIntyre P, Colley S, Bevan S, Winter J (2005) Peripheral nerve injury induces cannabinoid receptor 2 protein expression in rat sensory neurons. Neuroscience 135:235–245. https://doi.org/10.1016/j.neuroscience.2005.06.009

Wu A, Lee D, Xiong W-C (2023) Lactate metabolism, signaling, and function in brain development, synaptic plasticity, angiogenesis, and neurodegenerative diseases. Int J Mol Sci 24:13398. https://doi.org/10.3390/ijms241713398

Xie X, Li Y, Xu S, Zhou P, Yang L, Xu Y, Qiu Y, Yang Y, Li Y (2021) Genetic blockade of NAAA cell-specifically regulates fatty acid ethanolamides (FAEs) metabolism and inflammatory responses. Front Pharmacol 12:817603. https://doi.org/10.3389/fphar.2021.817603

Yiangou Y, Facer P, Durrenberger P, Chessell IP, Naylor A, Bountra C, Banati RR, Anand P (2006) COX-2, CB2 and P2X7-immunoreactivities are increased in activated microglial cells/macrophages of multiple sclerosis and amyotrophic lateral sclerosis spinal cord. BMC Neurol 6:12. https://doi.org/10.1186/1471-2377-6-12

Yona S, Kim K-W, Wolf Y, Mildner A, Varol D, Breker M, Strauss-Ayali D, Viukov S, Guilliams M, Misharin A, Hume DA, Perlman H, Malissen B, Zelzer E, Jung S (2013) Fate mapping reveals origins and dynamics of monocytes and tissue macrophages under homeostasis. Immunity 38:79–91. https://doi.org/10.1016/j.immuni.2012.12.001

Zambrowicz BP, Imamoto A, Fiering S, Herzenberg LA, Kerr WG, Soriano P (1997) Disruption of overlapping transcripts in the ROSA beta geo 26 gene trap strain leads to widespread expression of beta-galactosidase in mouse embryos and hematopoietic cells. Proc Natl Acad Sci U S A 94:3789–3794. https://doi.org/10.1073/pnas.94.8.3789

Zhang H-Y, Gao M, Liu Q-R, Bi G-H, Li X, Yang H-J, Gardner EL, Wu J, Xi Z-X (2014) Cannabinoid CB2 receptors modulate midbrain dopamine neuronal activity and dopamine-related behavior in mice. Proc Natl Acad Sci U S A 111:E5007–E5015. https://doi.org/10.1073/pnas.1413210111

Zhao S, Mugabo Y, Iglesias J, Xie L, Delghingaro-Augusto V, Lussier R, Peyot M-L, Joly E, Taïb B, Davis MA, Brown JM, Abousalham A, Gaisano H, Madiraju SRM, Prentki M (2014) α/β-Hydrolase domain-6-accessible monoacylglycerol controls glucose-stimulated insulin secretion. Cell Metabol 19(6):993–1007. https://doi.org/10.1016/j.cmet.2014.04.003

Zhou Y, Oudin MJ, Gajendra S, Sonego M, Falenta K, Williams G, Lalli G, Doherty P (2015) Regional effects of endocannabinoid, BDNF and FGF receptor signalling on neuroblast motility and guidance along the rostral migratory stream. Mol Cell Neurosci 64:32–43. https://doi.org/10.1016/j.mcn.2014.12.001

Zhu Y, Romero MI, Ghosh P, Ye Z, Charnay P, Rushing EJ, Marth JD, Parada LF (2001) Ablation of NF1 function in neurons induces abnormal development of cerebral cortex and reactive gliosis in the brain. Genes Dev 15:859–876. https://doi.org/10.1101/gad.862101

Zhuang X, Masson J, Gingrich JA, Rayport S, Hen R (2005) Targeted gene expression in dopamine and serotonin neurons of the mouse brain. J Neurosci Methods 143:27–32. https://doi.org/10.1016/j.jneumeth.2004.09.020

Zhuo L, Theis M, Alvarez-Maya I, Brenner M, Willecke K, Messing A (2001) hGFAP-cre transgenic mice for manipulation of glial and neuronal function in vivo. Genesis 31:85–94. https://doi.org/10.1002/gene.10008

Zimmer A, Zimmer AM, Hohmann AG, Herkenham M, Bonner TI (1999) Increased mortality, hypoactivity, and hypoalgesia in cannabinoid CB1 receptor knockout mice. Proc Natl Acad Sci U S A 96:5780–5785. https://doi.org/10.1073/pnas.96.10.5780

Zimmermann T, Maroso M, Beer A, Baddenhausen S, Ludewig S, Fan W, Vennin C, Loch S, Berninger B, Hofmann C, Korte M, Soltesz I, Lutz B, Leschik J (2018) Neural stem cell lineage-specific cannabinoid type-1 receptor regulates neurogenesis and plasticity in the adult mouse hippocampus. Cereb Cortex 28:4454–4471. https://doi.org/10.1093/cercor/bhy258

Zufferey R, Dull T, Mandel RJ, Bukovsky A, Quiroz D, Naldini L, Trono D (1998) Self-inactivating lentivirus vector for safe and efficient in vivo gene delivery. J Virol 72:9873–9880. https://doi.org/10.1128/JVI.72.12.9873-9880.1998

Ageing, Neurodegeneration and the Endocannabinoid System

Javier Fernández-Ruiz, Onintza Sagredo, María Gómez-Ruiz, and Eva de Lago

Contents

1 Endocannabinoids in the CNS: Development, Adulthood and Senescence 210
2 Endocannabinoids and Physiological Ageing in the CNS 214
3 Endocannabinoids and Pathological Ageing in the CNS 218
4 The Ongoing Problem of Limited Efficacy 222
5 The Ongoing Problem of Late Diagnoses 225
6 Why Cannabinoids as Neuroprotectants? 227
7 Concluding Remarks and Future Perspectives 235
References 236

Abstract Numerous studies carried out in the last 30–40 years have strongly demonstrated that the endocannabinoid system exerts important modulatory functions in the central nervous system (CNS). These neuromodulatory functions encompass the whole life of animals, with specific activities during neurodevelopment (prenatal, postnatal and adolescent periods), adulthood and possibly senescence too. However, this is the life stage less investigated in relation with the endocannabinoid

J. Fernández-Ruiz (✉), O. Sagredo, and E. de Lago
Instituto Universitario de Investigación en Neuroquímica, Departamento de Bioquímica y Biología Molecular, Facultad de Medicina, Universidad Complutense, Madrid, Spain

Centro de Investigación Biomédica en Red de Enfermedades Neurodegenerativas (CIBERNED), Madrid, Spain

Instituto Ramón y Cajal de Investigación Sanitaria (IRYCIS), Madrid, Spain
e-mail: jjfr@med.ucm.es; J.Fernandez-Ruiz@ciberned.es

M. Gómez Ruiz
Instituto Universitario de Investigación en Neuroquímica, Departamento de Bioquímica y Biología Molecular, Facultad de Medicina, Universidad Complutense, Madrid, Spain

Centro de Investigación Biomédica en Red de Enfermedades Neurodegenerativas (CIBERNED), Madrid, Spain

Instituto Ramón y Cajal de Investigación Sanitaria (IRYCIS), Madrid, Spain

Departamento de Psicobiología y Metodología en Ciencias del Comportamiento, Facultad de Psicología, Universidad Complutense, Madrid, Spain

Curr Topics Behav Neurosci (2026) 76: 209–248
https://doi.org/10.1007/7854_2025_597
Published Online: 25 July 2025

system to date. In the aged brain, the activity of this system appears to be altered, which contributes to subtle impairments that typically occur during ageing in learning and memory, motor behaviour, social behaviour and other neurobiological functions. Some of the changes in endocannabinoid activity may represent a process to attenuate ageing-related impairment in the brain function, which is consistent with its role as a pro-homeostatic system. An important observation is that these alterations become extreme when normal brain ageing acquires pathological characteristics, as happens in chronic neurodegenerative disorders. This includes the cannabinoid type-1 (CB_1) receptor downregulation or impairment in its signalling and the increase in endocannabinoid-inactivating enzymes, both hypothesised to contribute to pathogenic events. By contrast, elevated levels of endocannabinoids due to a reduced Fatty acid amide hydrolase (FAAH) and monoacyl glycerol lipase (MAGL) expression and the upregulation of cannabinoid type-2 (CB_2) receptors may in turn serve as endogenous pro-homeostatic adaptations against brain impairment. This review synthesises information on: (i) subtle alterations in the endocannabinoid system in the senescent brain in the absence of pathology, with the purpose of demonstrating that these alterations are representative of the extreme changes experienced by this system in the brain pathological ageing; and (ii) the development of neuroprotective therapies based on the pharmacological management of specific endocannabinoid targets to combat neurodegenerative pathologies. Together, research in this area comes at a critical time as global lifespan is increasing, incidence of age-related neurodegenerative disorders is expanding, and the unmet need for efficacious neuroprotective treatments is a public health necessity.

Keywords Cannabinoids · Chronic neurodegenerative disorders · Endocannabinoids · Neurodegeneration · Neuroprotection · Neurorepair · Pathological brain ageing · Physiological brain ageing

1 Endocannabinoids in the CNS: Development, Adulthood and Senescence

Research carried out since the late 1980s and early 1990s aimed to identify the mechanisms of action underlying the pharmacological effects of phytocannabinoids, the active ingredients in the *Cannabis* plant. This facilitated the discovery of a new modulatory system, the so-called endocannabinoid system. Biochemical and pharmacological studies have identified the four types of elements that constitute this modulatory system: (i) the endogenous ligands, so-called endocannabinoids (including arachidonoylethanolamide (AEA, anandamide) and 2-arachidonoylglycerol (2-AG) as the most representative and abundant, but also some minor endocannabinoids), which are lipids acting as messenger molecules to communicate cells located in close vicinity (reviewed in Fernández-Ruiz et al. 2021; Maccarrone et al. 2023); (ii) the molecular machinery responsible for the metabolism of these endocannabinoids, which includes biosynthetic (*N*-arachidonoyl-phosphatidylethanolamine-phospholipase D, NAPE-PLD, and diacylglycerol lipase, DAGL, as the most relevant) and inactivating (fatty acid amide hydrolase, FAAH, and monoacylglycerol lipase, MAGL, as the most active) enzymes (reviewed in Fernández-Ruiz et al. 2021; Maccarrone et al. 2023); (iii) the cell-surface receptors (cannabinoid type-1, CB_1, and type-2, CB_2, receptors as the most classic binding sites) activated by endocannabinoids (and also by some phytocannabinoids) that are responsible for triggering intracellular responses (reviewed in Maccarrone et al. 2023; Gómez-Cañas et al. 2023); and (iv) a possible membrane carrier (although internalisation by endocytosis has been also proposed; Yates and Barker 2009) and intracellular binding proteins for endocannabinoids linked in part to those enzymes responsible for the termination of their activity, then regulating endocannabinoid availability (reviewed in Kaczocha and Haj-Dahmane 2022; Maccarrone et al. 2023; Ciaramellano et al. 2023).

Parallel physiological studies have situated the endocannabinoid system as an important modulator of the homeostatic functions of numerous cells, tissues and organs in the body (reviewed in Maccarrone et al. 2023). Endocannabinoids and their receptors and enzymes are broadly distributed in the body, where they function as an important signalling network. This modulatory function includes the immune (reviewed in Kaminski and Kaplan 2021) and the cardiovascular (reviewed in Karimian Azari et al. 2020) systems, as well as those tissues and organs (e.g. liver, adipose tissue, pancreas, skeletal muscle) involved in the control of energy metabolism (reviewed in Watkins 2019; Mir et al. 2023). The system is also active in several endocrine glands and tissues (reviewed in Meah et al. 2022), reproductive organs (reviewed in Pařízek et al. 2023), bones (reviewed in Xin et al. 2022), gastrointestinal tract (reviewed in Cuddihey et al. 2022) and many other sites in the body (reviewed in Maccarrone et al. 2023). However, the first and possibly the most relevant site for the action of the endocannabinoid system is the nervous system, in particular the CNS. In this system, endocannabinoids exert an important modulatory function at the synapse that is different compared to classic neurotransmitters. Firstly, they act in the inhibitory retrograde control of both excitatory and inhibitory synapses, playing a role in synaptic plasticity (reviewed in Augustin and Lovinger 2018; Winters and Vaughan 2021). Secondly, they play an important function by facilitating glia-neuron communication at the tripartite synapse (reviewed in Covelo et al. 2021; Noriega-Prieto et al. 2023). Such functions enable endocannabinoids to participate in the regulation of numerous neurobiological processes. This includes the control of movement, cognition, learning and memory, control of food intake and appetite, emesis, nociception, neuroendocrine function, and neuronal hyperexcitability, integrity and survival (see Fig. 1 and specific reviews in Fernández-Ruiz 2009 and 2019; Rock and Parker 2016; Hillard et al. 2016; Koch 2017; Fernández-Ruiz et al. 2021; Vega-García et al. 2021; Zieglgänsberger et al. 2022; Figueiredo and Cheer 2023; Chen 2023). Alterations in specific elements of this system in the CNS structures responsible for the control of these neurobiological processes, resulting in dysregulation of the endocannabinoid tone, may elicit, contribute to, or aggravate CNS disorders related to them, such as motor, memory,

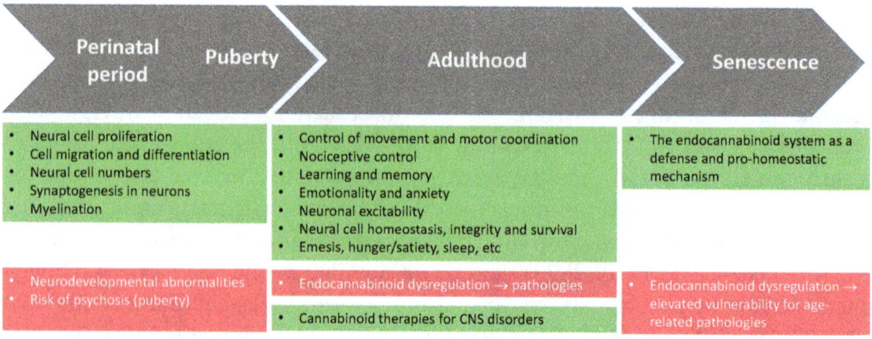

Fig. 1 Diagram summarizing the key modulatory functions exerted by the endocannabinoid system in the CNS during the different phases of life: prenatal, postnatal and peripubertal development, adulthood and senescence

feeding, pain and epileptic disorders (reviewed in Cristino et al. 2020; Fernández-Ruiz et al. 2021). The good news is that numerous studies in the last years are exploring the potential of cannabinoid-based therapies to correct these alterations, serving to alleviate symptoms and/or delay disease progression in these CNS-related disorders (see Fig. 1 and reviews in Fijał and Filip 2016; Fraguas-Sánchez and Torres-Suárez 2018; Cristino et al. 2020; Fernández-Ruiz et al. 2021).

These two activities (acting as inhibitory retrograde messengers and modulating glia-neuron communication) are possibly the two most characteristic functions exerted by the endocannabinoid system in the CNS of mammals (including humans) in the adult state. However, this system is also active (exerting even much more specific activities) at earlier (prenatal, postnatal and peripubertal development; reviewed in Maccarrone et al. 2014; Sagredo et al. 2018; Gomes et al. 2020; Simone et al. 2022) periods in life, as well as, according to more recent experimental evidence (reviewed in Di Marzo et al. 2015), at the senescent brain as will be addressed below (see Fig. 1). In regard to neurodevelopment, elements of this system and evidence of their functioning in the developing brain have been detected in the prenatal period as early as gestational day 14 in rodents (Romero et al. 1997; Berrendero et al. 1998a, 1999; reviewed in Fernández-Ruiz et al. 2000). This correlates with the time at which the phenotypic expression of most neurotransmitters takes place. Of particular interest is the fact that certain endocannabinoid elements (e.g. CB_1 receptors) are transiently expressed in atypical locations (e.g. white matter areas, subventricular zone) compared with their distribution in adult animals (Romero et al. 1997; Berrendero et al. 1998a, 1999; reviewed in Fernández-Ruiz et al. 2000). Such atypical distribution is compatible with a neurotrophic function exerted by the endocannabinoid system during neurodevelopment (reviewed in Fernández-Ruiz et al. 2000). Later, they display substantial changes in their cellular and subcellular location and expression levels up to birth and during the postnatal development (reviewed in Fernández-Ruiz et al. 2000; Harkany et al. 2007). This correlates with their modulatory actions in several development-related events such as the proliferation of neural progenitor cells, their differentiation to the different neural cell lineages and their migration to the final locations (reviewed in Fernández-Ruiz et al. 2000; Harkany et al. 2007). Moreover, in the specific case of developing neurons, endocannabinoids not only contribute to the acquisition of a neuronal identity, but also play a modulatory role on specific neuronal processes such as neuritogenesis, axonal growth, synaptogenesis, network integration and formation of myelin (reviewed in Fernández-Ruiz et al. 2000; Harkany et al. 2007; Díaz-Alonso et al. 2012; Galve-Roperh et al. 2013). They also contribute to reach a balance between cell proliferation and programmed cell death, allowing the accumulation of adequate numbers of neuronal cells during brain development (Aguado et al. 2006). Recent work has indicated that endocannabinoids may even regulate some epigenetic mechanisms (e.g. DNA methylation, histone acetylation) affecting the expression of key genes in the development and function of several neurotransmitter systems, although such effects may be detrimental rather than physiologically beneficial (reviewed in D'Addario et al. 2013; Gomes et al. 2020). Such neuromodulatory function exerted by endocannabinoids and their receptors continues into the peripubertal period (e.g. CB_1 receptor gene expression peaks at the onset of puberty in rodents; Rodríguez de Fonseca et al. 1993), when they participate in the final maturation of specific neuronal circuits (Meyer et al. 2018). Given these important functions of the endocannabinoid system are exerted at key times during the prenatal, early and late postnatal and adolescent neurodevelopment stages, these periods represent the most critical times for the adverse pharmacological effects derived from the consumption of cannabis (Trezza et al. 2008; Moore et al. 2010). By contrast, the onset of some neurodevelopmental disorders (e.g. Dravet, Lennox-Gastaut and Rett syndromes, other epileptic or autism spectrum disorders, hypoxic-ischemic neonatal encephalopathy) also takes place within these periods, but they are, however, responsive to treatment with specific plant-derived cannabinoids (e.g. cannabidiol) (reviewed in Sagredo et al. 2018). Such application is in line with the current direction of the cannabinoid research in all these life periods.

As indicated before, the endocannabinoid system is also active in the aged brain, the major focus of this review, which will be addressed with more detail in subsequent sections. We can anticipate that the activity of this system appears altered (frequently declined, although sometimes elevated depending on the age and/or the CNS structure) in naturally aged mice (Bilkei-Gorzo 2012). Such alterations may contribute to subtle behavioural impairments that typically occur during ageing in certain neurobiological functions (Bilkei-Gorzo 2012), then representing an important risk for the development of aged-related CNS disorders (see Fig. 1). This explains, for example, why naturally aged mice experience progressive learning and memory impairments that can be directly linked with a reduced endocannabinoid activity occurring predominantly in the hippocampus (Piyanova et al. 2015; Bilkei-Gorzo et al. 2017). This also explains why the activation of the CB_1 receptor in aged mice improved these learning and memory deficits (Bilkei-Gorzo et al. 2017; Nidadavolu et al. 2021), whereas this activation produced adverse effects in young animals (reviewed in Hempel and Xi 2022).

It is important to remark that, in all these life periods, endocannabinoid elements have been found to be located in the different types of neural cells (e.g. neurons, astrocytes, oligodendrocytes, microglia, neural progenitor cells, oligodendrocyte precursor cells), even in the blood-brain barrier (BBB). This explains why they are being investigated for their capacity to be activated, inhibited or simply modulated and, therefore, to serve as pharmacological targets to modify endocannabinoid tone, which may result in beneficial effects. This also explains why the pharmacological management of this system is being extensively investigated to develop novel cannabinoid-based therapies, aimed at alleviating symptoms and, in particular, at modifying disease progression, for CNS-related pathologies (and also for peripheral disorders) (reviewed in Fijał and Filip 2016; Maccarrone et al. 2017; Fraguas-Sánchez and Torres-Suárez 2018; Cristino et al. 2020; Fernández-Ruiz et al. 2021). This development is based on: (i) the identification first of those elements of the endocannabinoid system (e.g. classic CB_1 and CB_2 receptors, novel receptors, FAAH and MAGL enzymes, others) that may serve as potential pharmacological targets; and (ii) the identification of those neuropathological conditions that may benefit from these cannabinoid-based therapies. The result of these intense research efforts has been the approval in different countries of some plant-derived or synthetic molecules, formulated as Marinol®, Cesamet®, Sativex® and, more recently, Epidiolex®, for the treatment of vomiting and nausea, anorexia-cachexia syndrome, spasticity in multiple sclerosis, and seizures in infantile refractory epilepsies, respectively (reviewed in Maccarrone et al. 2017; Fraguas-Sánchez and Torres-Suárez 2018). In addition, important regulations for the use of medicinal *Cannabis* have been implemented also in different countries (reviewed in Legare et al. 2022; Hidding et al. 2024).

2 Endocannabinoids and Physiological Ageing in the CNS

Brain ageing is a natural process associated with numerous molecular, subcellular, cellular and tissue changes and adaptations. These include: (i) local reductions in glucose metabolism; (ii) progressively higher generation of reactive oxygen species (ROS) resulting in oxidative stress with higher levels of oxidised biomolecules (DNA, proteins, lipids); (iii) impaired cellular calcium signalling; (iv) defects in protein homeostasis with slower protein synthesis and degradation, which may in turn facilitate the damage of proteins and their accumulation; (v) impairment in other subcellular processes (e.g. unfolded protein response, autophagy, mitochondrial activity, endoplasmic reticulum stress), resulting in non-functional organelles, misfolded proteins and other defects; and (vi) reduced generation of neurotrophic factors, in particular brain-derived neurotrophic factor (BDNF) (see recent reviews in Garaschuk et al. 2018; Zia et al. 2021; Doser and Hoerndli 2021; Ionescu-Tucker and Cotman 2021; Aguilar-Hernández et al. 2023; Bondy 2024). Such molecular and subcellular alterations may result in defects in synaptic plasticity and, in general, in neuronal malfunctioning driven by defective activity at synapses affecting receptors, transporters, ion channels, synthetic enzymes and other synaptic proteins (reviewed in Kumar 2020; Sikora et al. 2021). They can also be accompanied by impaired neurogenesis resulting in defects in neural cell replacement or even abnormal cell proliferation and differentiation events (reviewed in Isaev et al. 2019). In addition, these alterations are associated with elevated glial reactivity involving astrocytes, microglial cells and infiltrated macrophages, resulting in the occurrence of neuroinflammatory events (released proinflammatory cytokines and chemokines that deteriorate neuronal homeostasis and integrity; reviewed in Salat 2011; Stranahan and Mattson 2012; Grimm and Eckert 2017; Rim et al. 2024). For example, astrocytes may become dysfunctional in the aged brain and provide reduced metabolic and trophic support to neurons (García-Matas et al. 2008), whereas microglial cells become activated thereby acquiring a neurotoxic phenotype (Nakanishi and Wu 2009; Luo et al. 2010; von Bernhardi et al. 2010). It is important to identify that these cellular responses are concordant with the changes occurring in the gene expression profile in the brain during ageing, which predominantly affect genes involved in intracellular signalling (e.g. calcium, CREB, etc.), synaptic plasticity (e.g. BDNF), oxidative protection, DNA repair, autophagy, stress responses, inflammation and also mitochondrial genes (Jiang et al. 2001; Blalock et al. 2003; Lu et al. 2004; Tapia-Arancibia et al. 2008). These molecular and cellular changes, when located in specific CNS structures and circuits, may result in several behavioural deficiencies affecting the control of movement, the sensory abilities, the processes of attention and cognition, the memory and learning and some other neurobiological processes (reviewed in Caserta et al. 2009). However, the extent of these ageing-related behavioural alterations is highly variable in the brain areas in mammals, with the frontal and parietal cortex, hippocampus, thalamic structures, nucleus accumbens and putamen being the most ageing-affected areas, and the brainstem nuclei being the less affected (Double et al. 2010; Small et al. 2011). This appears to depend on the particular sensitivities of groups of neurons and circuits in relation with synaptic homeostasis and neuronal plasticity (Mattson 2007; Bano et al. 2011). For example, one of the most characteristic behavioural defects in the aged brain is the decline in memory, associated with a loss of neurons in the hippocampal subiculum and hilus (West 1993) and also in the entorhinal cortex (Simic et al. 2005), in parallel to a reduced neurogenesis in the hippocampal dentate gyrus (Isaev et al. 2019).

This progressive impairment of neurobiological functions that characterises the "natural" process of brain ageing may remain simply as a mere physiological (non-pathological) response. To achieve this, it is necessary that these progressively enhanced "detrimental" molecular and cellular events activated during life may be totally or partly prevented by the activity of some homeostatic defence systems operating in the brain (Bilkei-Gorzo 2012), then limiting the risk of occurrence of debilitating diseases. However, when these pro-homeostatic systems do not operate correctly, this physiological response could progress towards a pathological condition, elevating the incidence of neurodegenerative disorders. This explains the continuum between the changes in "natural" ageing and those found in pathological brain ageing, whose similarities and differences have been extensively reviewed by some authors (reviewed in Bilkei-Gorzo 2012). The endocannabinoid system has been proposed as part of these defence systems (Pacher and Mechoulam 2011; Bilkei-Gorzo 2012; Fernández-Ruiz 2019). Such an idea is based on several experimental observations, for example, that the expression levels and, to a lesser extent, the anatomical distribution of different molecular components of this signalling system change substantially in the brain (and also in peripheral tissues) as a function of age (Di Marzo et al. 2015). This means that the action of potential cannabinoid-based therapies in patients affected by several pathologies may be different depending on age (Di Marzo et al. 2015) resulting, for example, in therapeutic effects visible in aged subjects, which may become inactive (or cause adverse effects) in young individuals. In this sense, studies by Bilkei-Gorzo, Zimmer and coworkers (Bilkei-Gorzo et al. 2005, 2012) have strongly demonstrated that the progressive cognitive decline (e.g. learning and memory impairment) experienced by naturally aged mice can be directly linked to a reduced endocannabinoid activity (lower 2-AG levels, DAGL expression, and CB_1 receptor signalling) occurring in the hippocampus (Piyanova et al. 2013; Piyanova et al. 2015; Bilkei-Gorzo et al. 2017). Such an idea was supported by the fact: (i) that CB_1 receptor-deficient mice, which were able to carry out better several learning and memory tests compared with wildtype animals at the age of 2 months, showed, however, a poorer performance in these tests when they reached 12 months of age and so on, reflecting an early cognitive decline with respect to wild-type mice (reviewed in Di Marzo et al. 2015; Bilkei-Gorzo et al. 2017); and (ii) that, as indicated above, the activation of CB_1 receptors with low doses of Δ^9-tetrahydrocannabinol (Δ^9-THC) chronically given to 12 and 18 month-old mice improved these learning and memory deficits (Paradisi et al. 2006; Bilkei-Gorzo et al. 2017; Nidadavolu et al. 2021). In relation to this last observation, it is important to remark that the same dose of Δ^9-THC given to young mice produced adverse effects (reviewed in Hempel and Xi 2022). This dual response (adverse effects on cognitive abilities in young individuals, but benefits in old subjects) represents an interesting paradox, whose molecular and cellular bases require further investigation. From a general point of view, this may reflect the occurrence of age-dependent differences in the local, on-demand activation of the endocannabinoid system that seeks to correct possible perturbations in brain homeostasis and to restore physiological steady state (Di Marzo et al. 2015). As an example, we can anticipate that such a difference in the response of young and aged mice to Δ^9-THC appears to be possibly related to a different activity (or intensity) of the endocannabinoid system at these two ages in the hippocampus, the key area for these cognitive processes (Bilkei-Gorzo 2012). In the same sense, responses to higher Δ^9-THC doses differ between young (tolerance due to receptor desensitisation) and aged (no effects) individuals (Bilkei-Gorzo et al. 2017). Also, in support of the important role played by the CB_1 receptor in cognitive decline observed in brain ageing, mice with genetic deletion of this receptor experience early, as indicated before, these age-related learning deficits, but this did not happen with motor and sensory abilities (Bilkei-Gorzo et al. 2012). Moreover, this cognitive decline caused by this impairment in the CB_1 receptor was strongly associated with a reduction in BDNF levels (Aso et al. 2008), as well as with an important histopathological deterioration in the hippocampus (e.g. neuronal losses, inflammatory events, deficient neurogenesis; Albayram et al. 2011). Conversely, activation of the CB_1 receptor may be beneficial by preserving memory, as indicated above, which is accompanied by restoration of BDNF levels and protection of hippocampal neurons against ageing-associated impairment (Grigorenko et al. 2002; Butovsky et al. 2005). Further studies demonstrated that CB_1 receptors responsible of both responses (age-related cognitive deficits resulting from its genetic ablation and protection against these defects following their activation) were those located on GABAergic neurons (reviewed in Bilkei-Gorzo 2012). This finding derived from studies in which the elimination of the CB_1 receptor was carried out only in forebrain GABAergic neurons (responses similar to complete knockout mice) or only in glutamatergic neurons (no response) (Albayram et al. 2011). By contrast, the genetic ablation of the CB_2 receptor reduced social memory, but not spatial memory, in aged mice (Komorowska-Müller et al. 2021). In addition, similar levels of lipofuscin accumulation were seen in neurons, but they were elevated in microglial cells which showed a certain profile of activation in these aged CB_2 deficient mice compared to aged wild-type animals (Komorowska-Müller et al. 2021). This would a priori indicate that the CB_2 receptor would play a minor role in brain ageing, at least with the experimental approaches used to date. Interestingly, these approaches identified some important changes in aged CB_2 deficient mice that would occur outside the CNS, for example severe osteoporosis (Ofek et al. 2006; Bab and Zimmer 2008). Similar experiments with mice deficient in endocannabinoid-inactivating enzymes (resulting in elevated levels of anandamide and/or 2-AG) have also been carried out (Ativie et al. 2015), showing delayed/slowed ageing concordant with results obtained in mice deficient in cannabinoid receptors (Bilkei-Gorzo et al. 2012; Komorowska-Müller et al. 2021). Therefore, all these data obtained in laboratory animals collectively support the idea that targeting/activating CB_1 receptors (and perhaps CB_2 receptors too or using FAAH or MAGL inhibitors) in healthy aged individuals may provide benefit against the progressive decline in cognitive functions (e.g. improvement of spatial memory impairment, reduction of glial reactivity, enhanced neurogenesis; Bilkei-Gorzo 2012).

An additional question of interest is the role that the hippocampus appears to play in brain ageing. This CNS structure has been broadly investigated in relation to changes in endocannabinoid ligands, enzymes and receptors in aged rodents (Berrendero et al. 1998b; Romero et al. 1998; Maccarrone et al. 2002; Piyanova et al. 2015; Nidadavolu et al. 2022). In these studies, other CNS structures have also been analysed, demonstrating that ageing-related changes are brain region-specific with the hippocampus likely to be the most-frequently affected, but other regions (e.g. basal ganglia, cerebellum, cerebral cortex, hypothalamus) being also affected (Berrendero et al. 1998b; Romero et al. 1998; Bilkei-Gorzo 2012). However, some studies showed no relevant changes in the hippocampus in ageing (Liu et al. 2003; Wang et al. 2003). Such alterations (e.g. reduced CB_1 receptor binding) were also detected in aged humans (Van Laere et al. 2008). Moreover, these changes appear to vary during ageing progression with the greater alterations occurring in middle age, which are frequently reductions except some few increases at specific middle age periods that may represent transient compensatory responses (Nidadavolu et al. 2022).

Another important aspect in "normal" brain ageing is the occurrence of a reduction in the capability of neurogenic niches to generate new neural cells, in particular new neurons able to replace those affected during senescence (reviewed in Isaev et al. 2019). This declined neurogenic capability is likely due to a reduction in the population of neural stem cells (approximately 50% of reduction in older rodents; Hamilton et al. 2013). Alterations in specific elements of the endocannabinoid system appear to be also involved in this ageing-related response, in particular losses of CB_1/CB_2 receptors presumably occurring at old ages in neurogenic areas (e.g. subventricular zone, dentate gyrus) in rodents. In this sense, previous studies using young animals had demonstrated that: (i) direct activation of these receptors, or indirectly by elevating endocannabinoid levels with FAAH inhibitors, promote neural progenitor cell proliferation and differentiation in the hippocampus (Jiang et al. 2005; Palazuelos et al. 2006; Goncalves et al. 2008), whereas the opposite (e.g. inhibition of endocannabinoid biosynthesis) led to reduced neurogenesis (Gao et al. 2010; Oudin et al. 2011); and (ii) genetic ablation of the CB_1 receptor or DAGL also resulted in defective neurogenesis (Jin et al. 2004; Gao et al. 2010), whereas FAAH-deficient mice again showed an elevated neurogenic capacity (Aguado et al. 2005). Therefore, a presumably declined CB_1 and/or CB_2 receptor function in neurogenic structures of aged animals (something still needing to be confirmed) might explain the reduced capacity of these areas to generate new neurons and also new glial cells in brain ageing.

Hence, all these data appear to indicate that the endocannabinoid system regulates processes involved in ageing, as, in the aged brain, the activity of this system appears to be altered. These changes contribute to subtle impairments that typically occur during ageing in certain neurobiological functions such as learning and memory and, to a lesser extent, in motor behaviour, feeding, emotionality, sleep regulation, social behaviour and others. These changes may be part of the so-called normal ageing, but that may also predispose towards a pathological condition. Some of the alterations identified in the endocannabinoid activity may also represent an attempt to attenuate age-related impairment in the brain function, which agrees with its proposed role as a pro-homeostatic or protective system.

3 Endocannabinoids and Pathological Ageing in the CNS

As was mentioned in the previous section, brain ageing is a natural process that leads to numerous molecular, cellular and tissue changes and adaptations that, to a lesser extent, recapitulate events that are much more intense in magnitude in the so-called "pathological brain ageing" characteristic of chronic neurodegenerative disorders (Salat 2011; Stranahan and Mattson 2012; Grimm and Eckert 2017). As also indicated before, the borderline between the "physiological (healthy)" and the "pathological" brain is extremely diffuse and largely depends on the activity of several endogenous pro-homeostatic systems (e.g. neurotrophic factors, antioxidant defences, regulators of protein homeostasis, etc.). It also depends on a variety of those external factors (e.g. exercise, nutrition, education, etc.) associated with the quality of life. This balance is critical in determining whether the natural process of brain ageing remains as a simply physiological process or progresses to a debilitating pathological condition. An important question affecting this balance is that the progressive elevation in the life expectancy data and global population pyramids observed over the past 50–100 years in the world, in particular in developed countries (see reports from United Nations 2015; World Health Organization 2016; OECD 2017). This has significantly elevated the risk that modest events typically found in natural brain ageing may aggravate and drive physiological processes towards a pathological condition. The result has been a progressive elevation in the incidence of more common chronic neurodegenerative disorders (e.g. Alzheimer's disease (AD) and related dementias, Parkinson's disease (PD), amyotrophic lateral sclerosis (ALS), Huntington's disease (HD) and other polyglutaminopathies (polyQs); see Mazucanti et al. 2015; Fernández-Ruiz 2019). The problem is that the expectation now is that this elevation will continue and even aggravate in coming years, as the predictions for the forthcoming decades indicate (United Nations 2015; World Health Organization 2016; OECD 2017), even extending to less developed countries too. This fact will be determined by two circumstances: (i) developed countries (mainly USA, Canada, most European countries, Japan, Australia, New Zealand) will reach a half-life expectancy close or higher than 90 years (in particular in women) in 2050 (Saba 2015; United Nations 2015; World Health Organization 2016; Fernández-Ruiz 2019); and (ii) people over 65 years will also represent 25–50% of the total population in 2050 (Saba 2015; United Nations 2015; World Health Organization 2016; Fernández-Ruiz 2019). The latter due to a greater life expectancy but also to a lower birth rate typical of developed countries. Therefore, the expectation is that there will be more people in the world, in particular many more old people and much older people, and not only in developed countries; this will be also visible in developing countries (Fernández-Ruiz 2019). These demographic changes are expected to double or triple the incidence of neurodegenerative disorders up to 2050 (Dorsey et al. 2007; Kowal et al. 2013; Rodríguez-Gómez et al. 2014; Kharrazian 2022). This, added to the current absence of effective therapies for neurodegenerative disorders, will become epidemic proportions of elevated incidence of neurodegenerative disorders as the twenty-first century progresses, with their therapeutic control representing a major biomedical challenge that clinical and preclinical researchers will have to address. The question is that the development of neuroprotective therapies has been an important objective from the 90 s decade in the past century, but the success has been terribly limited (see Fig. 2). Thus, despite the large number of therapeutic agents investigated to combat AD (agents that apparently reduce β amyloid production and aggregation, or tau protein aggregation and hyperphosphorylation, or that elevate the clearance of β-amyloid; see Mangialasche et al. 2010; Tobeh and Bruce 2023), only four agents (memantine, a glutamate receptor antagonist, and donepezil, rivastigmine and galantamine that inhibit acetylcholinesterase), all exhibiting a

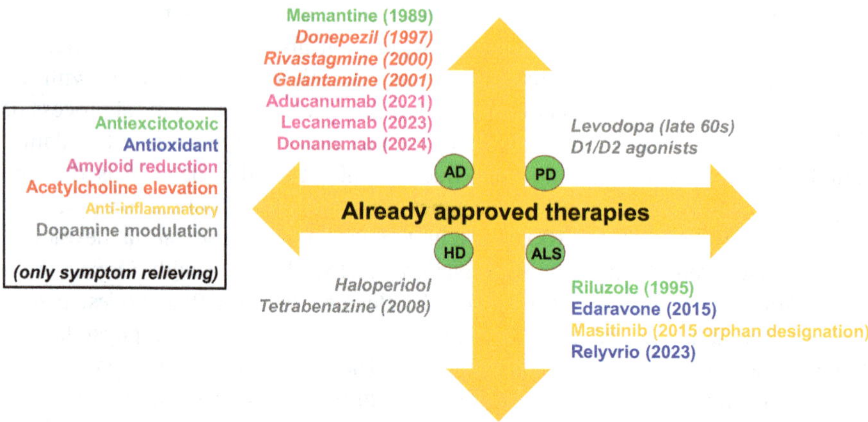

Fig. 2 Current approved therapies for the treatment of the four more important and frequent chronic progressive neurodegenerative disorders. The mechanism(s) by which these therapies are acting was indicated with a specific colour

modest efficacy, were approved, the last one, galantamine, in 2001. They represented the only medicinal arsenal available for AD patients for 20 years up to the recent approval (only by the US-FDA), in 2021 and so on, of the first three immunotherapies against β-amyloid aggregation (reviewed in Wu et al. 2023; Cummings et al. 2024). However, there are numerous concerns on their efficacy (only one has been recently approved by the EMA) and, in particular, on their adverse effects (elevated risk of vascular problems) that will need to be investigated in the follow-up analyses (phase 4) (Shi et al. 2022; Cummings et al. 2024). A similar situation exists for other common neurodegenerative disorders like PD (Olanow et al. 2017; Lorente-Picón and Laguna 2021), ALS (Petrov et al. 2017; Bradford and Rodgers 2024) and HD (Rodrigues and Wild 2018; Huang et al. 2024), in all cases with limited approved therapeutic agents that are active only against symptoms or that have a limited efficacy (see Fig. 2 for a summary). Assuming this present issue, the question is: What are the causes for these high failure rates when attempting to develop efficacious neuroprotective therapies for chronic neurodegenerative disorders? There is no clear response to this question, yet over the years we have been able to identify some important problems that possibly influence the development of such therapies (Fernández-Ruiz 2019):

- Do we know the pharmacological targets? and is polypharmacology preferred to selective compounds?: The answer to the first question is uncertain as yet the feeling is that not all the possible pharmacological targets with neuroprotective potential have been identified and new ones are expected to appear in coming years. In any case, targeting individual elements is not a guarantee of efficacy against neurodegeneration in patients. The issue will almost certainly require the use of combination therapies active at multiple targets, not only those that would respond to selective compounds (see more details later).

- Are these disorders being appropriately reproduced at the experimental level? Are current animal models predictive? and what new models can we use? The relationships between the experimental models used and each of the specific diseases often remain unclear. Indeed, the predictive nature of current animal models is not always ideal, and there is still a clear need for new, more appropriate models (see more details below).
- Are we initiating clinical treatments too late? and do we need earlier biomarkers? Neurodegenerative diseases are particularly susceptible to delays in the initiation of treatments as this is dependent on a previous accurate clinical diagnosis. This has an important influence on the success of many therapies. The problem is that disease diagnosis, which is based on the clinical detection of typical disease symptoms (e.g. dementia in AD, bradykinesia and tremor in PD, etc.), is frequently delayed. This demands urgent identification of early disease biomarkers that may anticipate clinical diagnosis and facilitate earlier onset of treatments (see more details later).
- Only neuroprotection, or neurorepair too? Using only neuroprotective therapy when diagnosis is late and prodromic disease phases are longer may be insufficient. In this context, strategies that also encompass neurorepair should be investigated to ensure progress towards successful treatment of these conditions.
- Are we being sufficiently successful in the translation from basic to clinical science? The answer is clearly not; translation from basic science to clinical practice is more than often inefficient for reasons that have already been mentioned for the previous questions.
- Are pharmacological companies highly enough interested in these disorders? As indicated above, neurodegenerative disorders may become an epidemic in the next years due to the expected elevation in lifespan. However, the idea that these disorders affect a relatively limited percentage of people, whose prognosis may be relatively shorter in most cases, still persist. This possibly influences the interest of pharma companies to research the development of new agents against neurodegenerative events.

Next, we will address with more detail the two major problems derived from these questions, which require urgent solutions with new research activities: (i) the current limited efficacy of therapies for neurodegenerative disorders demands the identification of new targets and the development of new treatments, as well as more predictive preclinical models; and (ii) the delay in disease diagnosis also requires identification of earlier biomarkers for accelerating this diagnosis and the onset of treatments (see Fig. 3). Following these two questions, we will address the advantages that cannabinoid based therapies for these disorders may possibly provide.

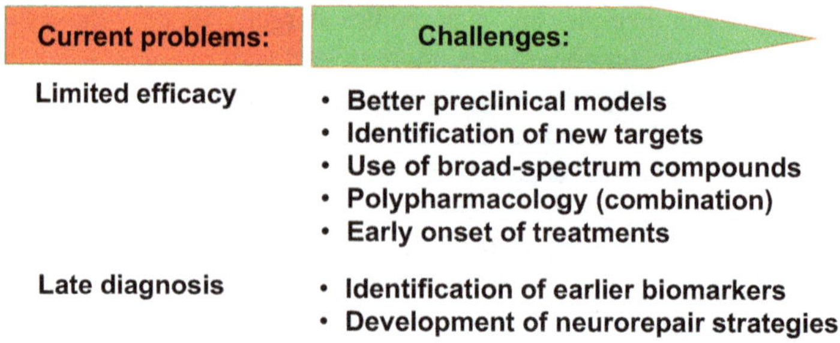

Fig. 3 Diagram showing the most important problems for the treatment of neurodegenerative disorders and the challenges that are currently being addressed to solve these problems

4 The Ongoing Problem of Limited Efficacy

The objective of developing chemical compounds that favour neuronal preservation, rescue, repair and replacement, as disease modifying therapies in preclinical studies, far from having been ignored, has been intensively investigated over the past 20–30 years (Fernández-Ruiz 2019). This includes compounds with antioxidant (e.g. acting as ROS scavengers or as modulators of NRF-2 signalling), anti-inflammatory (e.g. cyclooxygenase-2 (COX-2), inducible nitric oxide synthase (iNOS) or cytokine inhibitors), and anti-excitotoxic (e.g. glutamate receptor antagonists) properties. It also includes neurotrophic factors to promote neuronal survival, caspase inhibitors to manage apoptosis and autophagy enhancers or other strategies (e.g. immunotherapy) to eliminate protein aggregates (reviewed in Sarkar 2013; Baune 2015; Kudelova et al. 2015; Khan et al. 2016; Hoque et al. 2016; Meldolesi 2017; Mallah et al. 2020; Dailah 2022; Esquer et al. 2023; Carles et al. 2024; Bhandari et al. 2024). However, the results obtained have not been as positive as anticipated, with some of these compounds failing in experimental models (e.g. cell-based assays, laboratory animals), and almost every preclinically efficacious agents failing to reproduce these beneficial effects in patients (Berk et al. 2014; Sampaio et al. 2014; Athauda and Foltynie 2015; Müller 2017; Plascencia-Villa and Perry 2020; Wong et al. 2023; Wyse et al. 2024). As indicated above, several reasons may be proposed to explain this repeated failure. One of the most likely is that all these compounds frequently target only one of the different neurotoxic events that kill neurons (e.g. oxidative stress, glial reactivity, excitotoxicity, protein homeostasis dysregulation, etc.), and it is also common that their efficacy is exclusively tested in experimental models that specifically exacerbate each of these mechanisms. This would be, for example, the case of testing antioxidant agents in laboratory animals lesioned with mitochondrial toxins (e.g. 6-hydroxydopamine), anti-inflammatory compounds in animals lesioned with pro-inflammatory toxins (e.g. lipopolysaccharide) or glutamate receptor antagonists in animals lesioned with excitotoxins (e.g. quinolinate, ibotenate). It is likely that they may work against the unique type of insult that is predominant in each of these models, then showing an apparent neuroprotective profile. However, the pathological context in human neurodegenerative disorders is quite different, with neurotoxic insults being diverse and able to interact with each other to kill neurons and glial cells too. In patients, the efficacy can only derive from "polypharmacological approaches", using broad-spectrum agents capable of limiting the different neurotoxic events at the same time or combining more selective agents for each process (Geldenhuys and Van der Schyf 2013; Muhammad et al. 2018; Wang et al. 2019; Gontijo et al. 2020). In this "polypharmacological strategy", it is necessary to consider most of the potential pharmacological targets with neuroprotective activity already identified, but also new examples identified in coming years. For example, in the case of cannabinoid compounds, GPR55, an orphan receptor, which has been recently proposed as a new member of the endocannabinoid-acting receptors (reviewed in Gómez-Cañas et al. 2023), appears to play a promising role in neuroprotection (Celorrio et al. 2017; Martínez-Pinilla et al. 2019; Burgaz et al. 2021; Shen et al. 2022). This role appears to derive in part from its presence and function in microglial cells (Kallendrusch et al. 2013; Minamihata et al. 2020). As will be detailed later, some recent studies are exploring this question.

Another important concern in terms of validating the efficacy of novel neuroprotective agents is related to the currently available laboratory animal models. It is very important to have models that recapitulate all (or at least most) of the pathogenic events operating in a specific neurodegenerative disorder and that have predictive value for use in drug discovery. At present, the available models frequently present problems, which generally limit their value for clinical translation, and frequently too they are used for purposes for which they are not helpful. In other words, some models are adequate for the study of specific pathogenic events, but they present difficulties for being used to determine the efficacy of potential neuroprotective agents, and vice versa. This is the case, for example, for the excitotoxic models (mentioned above) generated by lesions with excitotoxins, pro-oxidant models generated with mitochondrial neurotoxins, etc., which are based on a predominant neurotoxic mechanism, so that these models only reproduce the pathology in part. Such models based on the exacerbation of one neurotoxic mechanism should be used only to investigate the contribution of this event to the disease. However, for the reasons indicated above, they are not useful for drug screening with agents other than those targeting this specific mechanism.

Other problem derives from the use of simple animal models in invertebrates (e.g. flies, nematodes) or vertebrates (e.g. zebrafish), which are not close to humans, but are easier to use and are cheaper. Although they may provide useful data in an initial screen of new compounds, this does not avoid further screens in more complex preclinical models in small mammals (e.g. rodents) and/or in species closer to humans in which new gene editing tools may already be used (e.g. sheep, pigs: Holm et al. 2016). On the other hand, an important part of animal models used for preclinical evaluation of potential neuroprotective agents correspond to transgenic mouse models or, alternatively, inoculation of viral vector containing mutant genes in specific CNS regions, resulting in overexpression of the mutant protein in the whole CNS or in those areas more affected in a specific disease. They are frequently used tools in the case of genetic cases of neurodegenerative pathologies. They are not devoid of limitations; for example, they do not fully reproduce pathological signs, or sometimes, when the disease mutation causes a loss-of-function, the wild-type protein, which is naturally produced by the host animal, may compensate. The more recent development of knock-in models, in which the mutant gene inserted substitutes for the wild-type one, and, in particular, the more recent arrival of the novel gene editing tools may help in improving these experimental tools (Whitelaw et al. 2016; Yang et al. 2016; Zhong et al. 2024). This is the case for CRISPR/Cas9 technology that can facilitate site-specific mutations within the genome, which may serve to generate animals carrying known human disease mutations (Whitelaw et al. 2016; Yang et al. 2016; Zhong et al. 2024). It is also the case of Cre/LoxP-mediated recombination technology (McLellan et al. 2017), which may contribute to making cell- and region-specific and/or time-dependent expression of the mutant protein possible. This originates alterations that occur exclusively in specific cell substrates or CNS areas and/or that are restricted to specific ages or disease stages.

Another important problem derived precisely from this last question is that many of the available models are frequently generated using artificial tools (e.g. transgene technology, viral vector strategies, neurotoxins, CRISPR-Cas9 gene editing, etc.) carried out in laboratory animals, in general, small species (e.g. rodents, fishes, flies, nematodes). Among other consequences, this results in the loss of the natural progression followed by each disease in humans. This is an important characteristic of human neurodegenerative disorders as, irrespective of the genetic or environmental factors that determine their aetiology, they appear spontaneously in patients. Such a feature is extremely difficult to reproduce in experimental models, which, as indicated before, reduces their relevance in terms of translational value towards the clinical scenario. For example, the temporal occurrence of pathogenic events is always missing in neurotoxin-based models (they should be used mainly to screen new drugs in an earlier phase of drug discovery, as indicated before). In other cases (e.g. some transgenic models) the disease is anticipated (or accelerated in progression). In this case, the pathological events appear at very early periods (adolescent or young adulthood periods) compared to the times at which they should appear in humans (senescence occurring at 60–90 years of age), then failing in a critical point for a correct temporal progression of human pathologies. This is the case, for example, of R6/2 mice, a classic model of HD frequently used to screen new neuroprotective agents in this genetic disease (Morton and Howland 2013). However, these mice develop very early (first signs are visible at 4 weeks of age) and progress too rapidly (at 12 weeks of age mice have almost completed the whole deterioration process) through the different pathogenic events occurring in this disease (e.g. huntingtin aggregation, transcriptional dysregulation, motor impairment and neuronal death). It represents a very aggressive model that does not correspond to what is happening in patients. If this were the case, mice should show the first HD-like signs at 16 months of age and progress over at least 6 months and not what is happening in R6/2 (reviewed in Kaye et al. 2021). Therefore, it should be mandatory that experimental models may adjust better to the different time points at which the major neuropathological signs of these diseases occur in humans. A question for future research in this direction would be the use of species in which human-like neurodegenerative disorders develop naturally. This option may represent a more promising approach to progress in the research in such diseases. An interesting option is companion animals, as unlike laboratory rodents, they share the human environment and lifestyle and are exposed to the same pollutants and pathogens (see Mazzatenta et al. 2017). There are already some examples of this, as some dementia-like disorders occurring in domestic animals (e.g. cats, dogs) or even in farm animals (e.g. sheep, cow, pig) (reviewed in Youssef et al. 2016; De Sousa et al. 2023). There is an ALS-like disorder typical of dogs, so-called degenerative myelopathy (Coates and Wininger 2010), and its equivalent described in horses (Divers et al. 1997). These ALS-like disorders share many neuropathological features with human ALS including some common genes as superoxide dismutase-1 (SOD-1) (Coates and Wininger 2010). The use of these species for research in neurodegenerative disorders may afford important benefits: (i) these animal species are phylogenetically closer to humans than rodents; (ii) the disease develops spontaneously as in humans; and (iii) they can be also used to develop new treatments in the veterinary field (De Sousa et al. 2023).

Therefore, the improvement in the experimental models to be used to validate new therapies at the preclinical level appears an important challenge for future research in this field. This improvement should assume issues as the strong complexity of the human CNS, the regional vulnerability apparent in each neurodegenerative disease and the variability in their disease progression along years. Such issues make it extremely difficult to reproduce these diseases well in classic laboratory species (Eaton and Wishart 2017; Kluever and Fornasiero 2021). The objectives should be, first, to better classify the available experimental models in relation to their usefulness for the study of different aspects of these diseases, and, second, to develop more useful and predictive models that reproduce better neurodegenerative pathologies in cells and laboratory animals (reviewed in Menalled and Brunner 2014; Mancuso and Navarro 2015; Koprich et al. 2017; Sasaguri et al. 2017; Fernández-Ruiz 2019).

5 The Ongoing Problem of Late Diagnoses

The second important problem for the clinical development of disease modifying therapies against neurodegeneration is the fact that they are always initiated when patients are in an advanced state of the disease progression, caused by a late diagnosis of their pathological conditions. It is mandatory that patients recruited for clinical trials are clinically diagnosed and, in neurodegenerative disorders, the diagnosis is exclusively based on clinical criteria, namely when classic symptoms for each disease are already visible. However, years of research have demonstrated that, in neurodegenerative disorders, the process of deterioration in the brain function initiates several years before than this deterioration is evident with the occurrence of clinical symptoms (Ugrumov 2020; Jagust et al. 2023). In addition, it is also accepted that an earlier diagnosis based on the identification of early disease biomarkers, and an early onset of treatment based on this fact, may significantly improve the therapeutic efficacy of treatments that were completely inactive when diagnosis was based on classic symptoms (Ugrumov 2020; Haller et al. 2023). A key example in neurodegenerative disorders is the case of PD, a disease having predominantly a sporadic origin. The onset of first motor symptoms (e.g. tremor, akinesia, rigidity, postural instability) in this disease, which are mandatory for the clinical diagnosis, occurs when more than 50% of dopaminergic neurons in the *substantia nigra pars compacta* have already degenerated. This early loss of dopaminergic neurons occurs during the so-called prodromal phase of the disease, which may prolong for years without any parkinsonian sign (Tuominen and Renko 2024; Rees et al. 2024). Similar observations have been obtained in AD and ALS patients and efforts are currently being carried out to detect presymptomatic individuals at risk of progressing towards the symptomatic phase of the disease (Frisoni et al. 2017; Gordon and Meininger 2011, respectively). As it is easy to imagine, in these conditions of late diagnosis, any neuroprotective therapy is doomed to fail. However, the question is whether these failed therapies might have been more active in the case that an earlier diagnosis might have facilitated these therapies being applied to pre-symptomatic patients. This is a key question that highlights the need to identify early biomarkers in patients already affected by a neurodegenerative condition but without symptoms that confirm the diagnosis. Such biomarkers may include: (i) imaging techniques detecting functional (e.g. positron emission tomography (PET)) or structural (e.g. volumetric magnetic resonance imaging (MRI)) anomalies; (ii) neurological recording detecting early neurobehavioral impairment (e.g. Mini-Mental State Examination (MMSE) test for pre-dementia signs; cell phone or computer-aided applications to detect subtle motor defects); and/or (iii) biochemical analysis of biological fluids, tissues or cells to detect variations in different types of biomolecules (e.g. lipids, proteins; reviewed in Delenclos et al. 2016). As an example of the utility of these biomarkers, different parameters in the so-called preclinical stage of AD have been investigated (Frisoni et al. 2017), these including: (i) evaluation of mild cognitive impairment (MCI) using neuropsychological scales (e.g. MMSE), as this is considered to be a pre-Alzheimer state; (ii) analysis of β-amyloid aggregation with PET probes (already available; Franceschi et al. 2022) or in the CSF (Oosthoek et al. 2024); (iii) analysis of synaptic dysfunction in specific CNS areas using PET with fluoro-deoxyglucose or functional MRI (van Oostveen and de Lange 2021); (iv) analysis of tau protein in the CSF (Oosthoek et al. 2024); and (v) analysis of structural brain changes (atrophic areas) using volumetric MRI (Lombardi et al. 2020). It is also important to remark that an important advantage of this type of biomarkers is that they can be recorded using procedures that are poorly invasive. For example, recent studies have identified a number of biomolecules (e.g. GFAP protein) that are affected in neurodegenerative conditions (dementia cases). Changes in GFAP may have a reflect in their levels in CSF, whose collection needs a highly invasive procedure, but these changes are also evident in blood samples which makes the collection much more easy for patients (Sánchez-Juan et al. 2024). The development of these and novel biomarkers that may be identified in coming years will help to achieve an earlier diagnosis, to identify subjects at risk of developing the disease, and even to personalise the disease in subgroups of patients. This will enable neuroprotective treatments to be initiated earlier and improve the chances that they may be efficacious.

Assuming that clinical diagnosis in neurodegenerative disorders currently occurs when the degree of neuronal injury is already established, an important objective of present treatments is that disease modifying therapies are accompanied by additional neurorepair strategies aimed at restoring the neural cells that were lost during the prodromal stages (Ziemka-Nałęcz and Zalewska 2012; Demuth et al. 2017; Gao et al. 2022; Zhao et al. 2024). This represents an additional research challenge for all neurodegenerative disorders, but particular for PD and others with longer prodromal periods such as AD and HD (Rosser and Svendsen 2014; Mu et al. 2024; Pérot et al. 2024; Kluge and Iranzo 2024). The case of PD is possibly the one that has received the most attention, with the search for neuronal cell replacement strategies with the use of different cell sources that can potentially acquire dopaminergic characteristics (e.g. chromaffin cells, foetal dopaminergic neurons, neural stem cells reprogrammed from human fibroblasts; Stoker et al. 2017; Fernández-Ruiz 2019; Park et al. 2024). However, some problems and side effects are associated with these procedures, so that these strategies have progressed towards the potentiation of those processes involved in natural cell replacement from neurogenic niches (Farzanehfar 2016; Salmina et al. 2021). However, our understanding of these processes is still limited.

6 Why Cannabinoids as Neuroprotectants?

As indicated above, a second objective of this review is to update the information available on the potential of cannabinoids and other modulators of the endocannabinoid system as neuroprotectants against chronic neurodegeneration. As also indicated in the first part of the review, this potential derives from the homeostatic role played by the endocannabinoid system in brain ageing. This is based on four important advantages that cannabinoids may afford in comparison with other neuroprotective agents (see Fig. 4): (i) that cannabinoids have a pleiotropic profile exerting broad-spectrum activities against different neurotoxic stimuli (e.g. excitotoxicity, oxidative stress, glial reactivity, protein aggregation, etc.) that kill neurons; (ii) that they may also promote neuronal cell replacement as the endocannabinoid system is also active in adult brain neurogenesis, an important aspect of neurodegenerative disorders having a long prodromal period; (iii) that these neuroprotective and neurorepair effects are possible because the location of those elements of the endocannabinoid system that are targeted by cannabinoids in neural cell substrates and CNS structures that play a key role in the regulation of neuronal homeostasis, integrity and survival; and (iv) that, collectively, these properties reflect the role played by the endocannabinoid system as a possible endogenous protective system against neurotoxic insults (e.g. protein dysregulation, oxidative stress, excitotoxicity, inflammatory events, trauma, etc.) other than those (e.g. pathogens) controlled by the immune system.

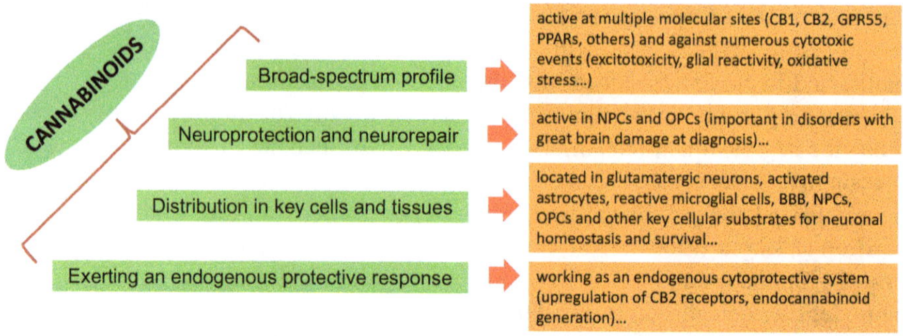

Fig. 4 Diagram showing the advantages of cannabinoids and, in general, modulators of the endocannabinoid system as disease modifying agents compared to other available and in development therapies

As regards the first of these advantages, the first experimental evidence that cannabinoids may offer neuroprotection was obtained in the 1990s when a series of in vitro studies demonstrated the capability of the phytocannabinoids Δ^9-THC and CBD to increase neuronal cell survival against oxidative stress (Hampson et al. 1998). These beneficial effects appeared to be exerted by cannabinoid receptor-independent mechanisms related to an action as ROS scavengers (Hampson et al. 1998). Such antioxidant activity was simply related to the particular chemical structure of these two plant-derived cannabinoids and would be available for similar phytocannabinoids (e.g. cannabinol (CBN), cannabidiolic acid (CBDA), cannabidivarin (CBDV), Δ^9-tetrahydrocannabidivarin (Δ^9-THCV) and cannabigerol (CBG); Marsicano et al. 2002). Additional studies demonstrated that these properties were also evident in some synthetic cannabinoids (e.g. nabilone, levonantradol, dexanabinol; Marsicano et al. 2002). These chemical properties of certain plant-derived or synthetic cannabinoids also included their metal-ion chelating activity, which may also contribute to their antioxidant properties (Kessiova et al. 2006). In parallel, a second series of studies was obtained with the above-mentioned dexanabinol (also known as HU-211), which is the inactive enantiomer (it has no activity at classic cannabinoid receptors) of the potent cannabinoid receptor agonist HU-210. These studies provided evidence that dexanabinol was neuroprotective, by acting as a non-selective antagonist of NMDA glutamate receptors (Feigenbaum et al. 1989), in experimental models of ischemia and brain trauma (Eshhar et al. 1993; Shohami et al. 1993). Interestingly, these preclinical studies progressed to further clinical trials, although, unfortunately, these benefits could not be replicated in patients affected by traumatic brain injury (Pop 2000; Maas et al. 2006). This pioneering work carried out in the 1990s was followed in subsequent decades by numerous studies that have strongly demonstrated the potential of cannabinoids to preserve not only neurons, but also astrocytes, oligodendrocytes and even their precursor cells. Such cytoprotective profile was demonstrated against different pro-oxidative insults, but also to conditions of excitotoxicity, glial-driven inflammation and protein aggregation that jeopardize neural cell homeostasis and integrity (reviewed in Fernández-Ruiz et al. 2015a, b; Fernández-Ruiz 2019). Collectively, all this work has contributed to identify the canonical pharmacological targets within the endocannabinoid system (e.g. cannabinoid receptors, endocannabinoid enzymes), as well as the non-canonical elements or other targets outside the endocannabinoid system, which are available for certain cannabinoids to afford cytoprotection. These targets are located in distinct neural cell types critical for cell survival or neuronal replacement (e.g. neurons, different glial cells, neural precursor cells), so that their activation or inhibition may be beneficial for promoting cell integrity and survival. This experimental evidence has contributed to highlight the promising neuroprotective profile of cannabinoids (reviewed in Fernández-Ruiz et al. 2015a, b; Fernández-Ruiz 2019). It has situated cannabinoids and the endocannabinoid system itself in a promising position as potential neuroprotective therapies in patients affected by chronic progressive neurodegenerative disorders, as well as accidental brain damage (e.g. stroke, brain trauma, spinal injury), with notable advantages when compared to other neuroprotective agents studied, as indicated before.

These advantages are related first to their wide range of action (reviewed in Fernández-Ruiz et al. 2015a, b; Fernández-Ruiz 2019). In the case of cannabinoids, a single molecule, or a combination of two or more cannabinoids can be active against excitotoxicity, glia-driven inflammation, oxidative stress, protein aggregation and other neurotoxic mechanisms that damage neuronal homeostasis and integrity. It is well known that these pathological events work in a cooperative manner to kill neurons and other neural cells, so that, as indicated before, to protect these cells it will be necessary to have a "multi-target strategy" with broad-spectrum agents or combinations of compounds. This is something that cannabinoids have been found to afford. The research carried out in our laboratory on HD is a good example of this concept (reviewed in Fernández-Ruiz et al. 2015a, b, 2017; Fernández-Ruiz 2019, where the original information can be found), with CBD, CBG and other antioxidant phytocannabinoids being active in a model of HD priming oxidative damage (e.g. 3-nitropropionate-lesioned rodents). In other models (e.g. R6/2, quinolinate-lesioned mice), cannabinoids that selectively activate the CB_1 receptor also afford neuroprotection (reviewed in Fernández-Ruiz et al. 2015a, b, 2017; Fernández-Ruiz 2019, where the original information can be found). In addition, compounds that selectively activate the CB_2 receptor also displayed activity in both models and they provided neuroprotection in rats lesioned with malonate, a model with strong glial reactivity and activation of the apoptotic machinery (reviewed in Fernández-Ruiz et al. 2015a, b, 2017; Fernández-Ruiz 2019, where the original information can be found). Lastly, cannabinoids that were not active at CB_1/CB_2 receptors, but that target the peroxisome proliferator-activated receptor-γ (PPAR-γ) were also beneficial in R6/2 and 3-nitropropionate-lesioned mice (reviewed in Fernández-Ruiz et al. 2015a, b, 2017; Fernández-Ruiz 2019, where the original information can be found). Collectively, these results support the idea that therapies for HD patients should be based on antioxidant cannabinoids acting through canonical cannabinoid receptors (CB_1 and CB_2 receptors) and non-classical targets (e.g. PPAR-γ), or on a combination of cannabinoids ensuring all these activities (reviewed in Fernández-Ruiz et al. 2015a, b, 2017; Fernández-Ruiz 2019). The preclinical studies in ALS drive to similar conclusions with contribution of cannabinoid receptor-independent mechanisms, activation of CB_1/CB_2 receptors and also PPAR-γ receptors. This has been observed against the numerous neurotoxic insults that contribute to kill motor neurons in different experimental models of this disease (e.g. mutant SOD1 mice, TDP-43 transgenic mice, NSC34 cells, etc.; reviewed in Rodríguez-Cueto et al. 2021a). Studies in experimental models of AD, FTD, PD and other disorders indicate that similar approaches might also be beneficial, using cannabinoids with a broad activity profile (reviewed in Fernández-Ruiz et al. 2015a, b, 2017; Fernández-Ruiz 2019).

What explanations can be proposed for this broad-spectrum activity of cannabinoids? Definitely, their pleiotropic profile, which, as indicated before, on one side, is related to their direct actions on elements of the endocannabinoid system (e.g. CB_1/CB_2 receptors, inactivating enzymes, GPR55), and on the other side, it involves targets not directly related to endocannabinoid signalling (reviewed in Fernández-Ruiz et al. 2015a, b, 2017; Fernández-Ruiz 2019; Rodríguez-Cueto et al. 2021a). In this sense, as indicated in the diagram presented in Fig. 5:

(i) Certain antioxidant phytocannabinoids (e.g. Δ^9-THC, CBN, CBD, CBDA, CBDV, Δ^9-THCV, CBG) and some synthetic cannabinoids (e.g. nabilone, levonantradol, dexanabinol), as indicated before, may exert cannabinoid receptor-independent antioxidant actions, which may contribute to their neuroprotective effects (Blebea et al. 2024). However, further studies aimed at exploring these antioxidant properties have also suggested a possible mediation of nuclear receptors of the PPAR family, or the transcription factor Nrf-2

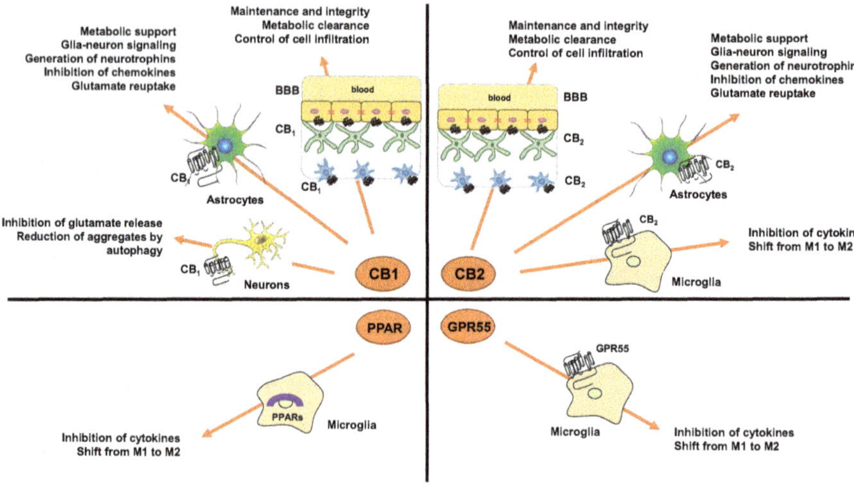

Fig. 5 Summary of the molecular and cellular mechanisms that underlie in neuroprotective effects of different cannabinoids acting through the four most important cannabinoid receptors

that regulates, among other processes, the inducible antioxidant response (Fernández-Ruiz 2019).

(ii) Cannabinoids may reduce excitotoxic damage by activating CB_1 receptors, which have a presynaptic location on glutamatergic terminals, resulting in inhibition of excessive glutamate release, thereby avoiding excitotoxic damage (Marsicano et al. 2003). The same result can be obtained by promoting glutamate reuptake in glial cells, something that cannabinoids could also facilitate given the presence of CB_1 receptors on astrocytes present at glutamatergic synapses, although this possibility would need to be confirmed. The CB_1 receptor-mediated anti-excitotoxic effect of certain cannabinoids may also derive from the activation of a pool of these receptors located post-synaptically in neurons containing glutamate receptors (e.g. NMDA receptors). However, this has been a controversial issue (reviewed in Fernández-Ruiz et al. 2015a, b, Fernández-Ruiz 2019). This activation may facilitate the closing of voltage-dependent calcium channels, thereby controlling the intracellular levels of calcium and avoiding the overactivation of calcium-dependent destructive pathways that occurs in post-synaptic neurons in glutamatergic synapses (reviewed in Fernández-Ruiz et al. 2015a, b; Fernández-Ruiz 2019). Therefore, all these effects, irrespective of the pre- and post-synaptic neuronal or astroglial location of CB_1 receptors at glutamatergic synapses, may serve to preserve the correct synapse homeostasis and avoid the consequences of excessive excitatory transmission.

(iii) Cannabinoids can also target CB_1 receptors to activate autophagy, which may serve to eliminate damaged proteins in cell aggregates thereby improving neuronal homeostasis and integrity. That the CB_1 receptor may be active against damaged proteins is supported by the location of this receptor at intracellular compartments (e.g. lysosomes, endosomes, mitochondria; Rozenfeld and Devi 2008; Brailoiu et al. 2011; Fernández-Moncada et al. 2024). It is also supported by data obtained in cancer studies, in which a CB_1 receptor-dependent activation of autophagy has been found as part of the antitumoral properties of certain cannabinoids (Ellert-Miklaszewska et al. 2021). In fact, CB_1 receptor-deficient mice showed altered autophagy and lysosome function, which led to the idea of autophagy dysregulation derived from a reduced CB_1 receptor signalling in the ageing brain (reviewed in Bilkei-Gorzo 2012).

(iv) Cannabinoids, which are able to activate the CB_2 receptor, the orphan receptor GPR55, the nuclear receptors of the PPAR family and, in certain conditions, the CB_1 receptor too, have been found to function through some of these receptors in attenuating glial toxicity for neurons. This also depends on both astrocytes and microglial cells, in particular, when these cells become activated, serving as key cell substrates for neuroprotection. For example, reactive astrocytes express not only CB_2 but also CB_1 receptors, and both receptors are able to work in conjunction to regulate astrocyte functions and, particularly, their influence on neuronal homeostasis, integrity and survival, either positively or negatively (reviewed in Fernández-Ruiz et al. 2015a, b; Stella 2010). These effects include the supply of metabolic substrates (e.g. lactate, ketone bodies) to neurons, but also a contribution in glutamate clearance, as indicated above. They also include the generation of neurotrophins, anti-inflammatory mediators (e.g. interleukin-10, interleukin-1 receptor antagonist) or pro-survival factors (e.g. transforming growth factor-β) for neurons, or a reduction in the production of harmful factors (e.g. chemokines). All of these effects are useful to improve neuronal homeostasis and to potentially rescue damaged neurons (Molina-Holgado et al. 2003; Arévalo-Martín et al. 2008; Duarte et al. 2012; Fernández-Ruiz et al. 2017; Fernández-Moncada et al. 2024). The endocannabinoid signalling is also active in microglial cells. When they are activated, a frequent CB_2 receptor up-regulation is seen both in vitro and in vivo and in all neurodegenerative conditions (Benito et al. 2003; Cabral and Griffin-Thomas 2009; Cassano et al. 2017; Duffy et al. 2021). These microglial CB_2 receptors play an important role in the proliferation of microglia and their migration to lesion sites, as well as in the transformation of proinflammatory M1 cells towards the anti-inflammatory M2 phenotype. Such benefits appear to derive from the CB_2 receptor-mediated inhibition of the transcription factor NFκB and its downstream activators of proinflammatory gene expression (e.g. iNOS, COX-2, metalloproteases, tumour necrosis factor-α (TNF-α), other proinflammatory cytokines; see Stella 2010; Zoppi et al. 2014). This can also be elicited by cannabinoids that activate GPR55 and/or PPAR-γ, consistent with the broad profile of these compounds (reviewed in Fernández-Ruiz et al. 2017).

(v) Cannabinoids have also been found to contribute to the preservation of blood-brain barrier integrity and function, which facilitate, among others, the blood supply to the CNS (Hagan et al. 2022). This is essential in neuroprotection having important benefits in neurodegenerative conditions. This function is under the control of both CB_1 and CB_2 receptor-mediated signals (Fujii et al. 2014), although cannabinoids could also be active at non-endocannabinoid receptors (e.g. some serotonin receptor types) (Haj-Dahmane and Shen 2011). Both CB_1 and CB_2 receptors are located in specific cells that are part of the brain-blood barrier (e.g. astrocytes, perivascular microglial cells and endothelial cells; Hagan et al. 2022), where they help maintain the integrity of tight junctions, inhibit leukocyte infiltration and facilitate β-amyloid clearance (Vendel and de Lange 2014). They are also located in the brain microvasculature, where they help to reduce endothelin-1-induced vasoconstriction and restore blood supply to the injured brain (Mechoulam and Shohami 2007; Choi et al. 2013).

The second advantage of cannabinoids deals with their potential for cell neurorepair and neuro-replacement. This is still a possibility yet to be conclusively preclinically demonstrated and, of course, clinically developed and that will require much more research. Anyway, it is something that also situates cannabinoids in an advantageous position. This question derives, as indicated above, from the fact that clinical diagnosis in most neurodegenerative disorders, in particular those of sporadic origin, is established late as it is mainly based on the detection of clinical symptoms following standardised scales (see previous comments). This means that diagnosis comes at a time when clinical symptoms become evident. However, this frequently occurs when the extent of neuronal injury is already quite advanced due to the long prodromal phases in which these symptoms are absent. In such conditions, the potential treatments would need not only to preserve the surviving neurons, but also to promote the replacement from neurogenic niches of those neurons (or other neural cells) lost during the prodromal stage of the disease. Interestingly, this is something that the pharmacological manipulation of endocannabinoid signalling may facilitate, as this modulatory system appears to play a role in the regulation of proliferation, maturation and/or differentiation processes of neural precursor cells to neurons or glial cells (Huerga-Gómez et al. 2021). This possibility is supported by the presence of the CB_1 and CB_2 receptors, FAAH activity and other elements of the endocannabinoid system in neural precursor cells, which may differentiate into neurons, astrocyte or oligodendrocytes (reviewed in Galve-Roperh et al. 2013; Valeri and Mazzon 2021). Importantly, the activation of these two classic cannabinoid receptors or the inhibition of FAAH has been associated with alterations in such processes affecting the generation of new neurons or glial cells (Ilyasov et al. 2018). Assuming this possibility, it is possible to imagine a cannabinoid-based therapy addressed to potentially repair lesioned areas (reviewed in Galve-Roperh et al. 2013; Valeri and Mazzon 2021), which would add to the neuroprotective properties of cannabinoids. In fact, there is already evidence that the activation of CB_2 receptors identified in oligodendrocytes and their precursor cells (OPCs) may play a key role in these glial cells that are in turn crucial for neurons (e.g. remyelination, reviewed in Arévalo-Martín et al. 2008).

The two above-mentioned advantages of cannabinoids, which are based on years of research, reflect that there is sufficient experimental evidence on the benefits that can be achieved with these compounds in relation to neuroprotection and, to a lesser extent, with neuro-replacement. This has positioned these compounds close to clinical development as novel and efficacious therapies for the treatment of patients affected by neurodegenerative disorders. It is important to remark that these neuroprotective and neurorepair effects are made possible by the specific location of those elements of the endocannabinoid system that are targeted by cannabinoids in neural cell substrates and CNS structures that are crucial for the preservation, rescue, repair and/or replacement of neurons and other neural cells. This represents the third advantage of cannabinoids compared to other neuroprotective agents (see Fig. 4). As will be addressed later, such location is consistent with the idea that the endocannabinoid system plays a role as an endogenous protective system against inflammatory, excitotoxic, infectious, traumatic or oxidant stimuli that potentially damage the CNS (Pacher and Mechoulam 2011; Fernández-Ruiz et al. 2015a, b; Fernández-Ruiz 2019). This is likely the reason for the active elements of this system being located in these key neural cells and CNS structures. Other types of locations would not be compatible with such a physiological role. This includes their location in specific neuronal subpopulations (e.g. glutamatergic neurons or glutamatergic synapses) but also in astrocytes, resting and reactive microglia, perivascular microglial cells, oligodendrocytes, OPCs and NPCs, and even in key CNS structures like the blood-brain barrier. In these key locations, endocannabinoids and their receptors can selectively control the activity of these cells (or structures), particularly in processes related to survival, protection and/or repair (Galve-Roperh et al. 2013; Fernández-Ruiz et al. 2015a; Fernández-Ruiz 2019). As mentioned in the beginning of this paragraph, such particular cell and tissue location is completely compatible with the idea, proposed nearly a decade ago (Pacher and Mechoulam 2011), that the endocannabinoid signalling may function as an endogenous protective system in response to conditions of inflammation, trauma, oxidative stress, excitotoxicity, protein and cell organelle aggregation, and others that could damage the CNS. Such protective function would be distinct from the defence against pathogens that is exerted by the immune system, and would cover distinct types of cytotoxic insults of extreme relevance in the nervous system due to their particular functions and their significant vulnerabilities (Pacher and Mechoulam 2011; Fernández-Ruiz 2019). The hypothesis of the endocannabinoid system serving as a defensive pro-homeostatic system derives even from the data presented in the first part of this review in relation to the role of this system in brain ageing (Bilkei-Gorzo 2012). Thus, its subtle impairment during senescence has been found to be critical for the possibility that the natural brain ageing process evolves into a pathological condition. This is also supported by the responses experienced by specific elements of this signalling system in physiological brain ageing and, in particular, with the pathological deterioration of CNS structures associated with the different neurodegenerative disorders. This may be the case of the up-regulation of CB_2 receptors occurring in reactive glial cells, as well as of the elevated generation of endocannabinoids by these and other neural cells (Pacher and Mechoulam 2011). If this is the case, it would be extremely important to investigate how the pharmacological enhancement of these responses (e.g. blocking endocannabinoid inactivation, administering endocannabinoids, selectively activating the CB_2 receptor; reviewed in Mechoulam and Shohami 2007; Pacher and Mechoulam 2011; Fernández-Ruiz et al. 2015a, b; Fernández-Ruiz 2019) may potentially benefit the disease progression in these disorders. As indicated above, this was particularly evident in the case of CB_2 receptor up-regulation occurring in glial elements (microglia and astrocytes) in parallel to their reactivity in response to damaging stimuli, whose potentiation with selective CB_2 receptor agonists have been found to be positive in experimental AD (Benito et al. 2003; Aso and Ferrer 2016; Magham et al. 2021), PD (Gómez-Gálvez et al. 2016; Basile and Mazzon 2022), HD (Sagredo et al. 2009; Palazuelos et al. 2009), ALS (Yiangou et al. 2006; Bilsland and Greensmith 2008; Espejo-Porras et al. 2015, 2019; Rodríguez-Cueto et al. 2021a, b), and others neurodegenerative disorders (reviewed in Aymerich et al. 2018; Fernández-Ruiz 2019). By contrast, some controversial data have been published in FTD based on Tau protein (Galán-Ganga et al. 2021) compared to TDP-43-dependent FTD (Gonzalo-Consuegra et al. 2024). In parallel, induction of experimental PD, HD or ALS in CB_2 receptor-deficient mice was followed by an earlier onset and/or aggravation in the pathological phenotype (Sagredo et al. 2009; Palazuelos et al. 2009; Gómez-Gálvez et al. 2016; Rodríguez-Cueto et al. 2021b). However, it is also possible that some responses experienced by the endocannabinoid signalling (e.g. reduced CB_1 receptor signalling, elevated FAAH activity) reflect a dysregulation of this system rather than a pro-homeostatic response, the resulting instrumental in the pathogenesis of these disorders (Fagan and Campbell 2014; reviewed in Fernández-Ruiz 2019). In these conditions, rather than being promoted, these responses would need better to be pharmacologically corrected, for example, with FAAH inhibition (Hwang et al. 2010). The choice will depend on the type of disease, its stage and some other conditions, in part related to the different elements of the endocannabinoid system.

7 Concluding Remarks and Future Perspectives

As indicated at the beginning of this review, our purpose was first to collect all information generated in the last years on the changes experienced by the endocannabinoid system during the natural brain ageing and to compare these changes with those occurring during the pathological ageing that the brain experiences in neurodegenerative disorders. This comparison demonstrates that alterations in natural brain ageing are a copy in miniature of the extreme changes that take place in neurodegenerative conditions. It also demonstrates the important role that the endocannabinoid system plays as a pro-homeostatic system, as well as how their dysregulation may contribute to move this natural process to the pathological condition. A second objective was to update the current knowledge about cannabinoids as neuroprotectant agents and to remark on their advantages (broad-spectrum activity covering most neurotoxic events that kill neurons and the capacity to promote neural cell replacement in pathologies with long prodromal periods) compared to other investigated agents. These advantages likely derive from two characteristics that provide the endocannabinoid system a strong capacity to serve for the development of novel disease-modifying therapies against neurodegeneration: (i) that those endocannabinoid elements useful as pharmacological targets are located in cell substrates and CNS structures that are crucial for the preservation, rescue, repair and replacement of neurons; and (ii) that this particular location reflects that the endocannabinoid system may work as an endogenous protective system against those neurotoxic insults (excitotoxicity, oxidative stress, glial reactivity and protein dysregulation) that interact to deteriorate neural cell homeostasis and integrity in neurodegenerative disorders. Such findings derive from an intense experimental work carried out in recent years using predominantly preclinical models, which have positioned cannabinoid-based neuroprotective therapies in a solid and promising position to progress towards a true clinical application. This is likely the challenge for the forthcoming years, to move the issue to the clinical scenario with the purpose to validate in patients the expectations generated by cannabinoids in experimental models of neurodegeneration disorders. However, it is also important to remark that this progression will also need improvements in the diagnosis of neurodegenerative pathologies (identification of earlier biomarkers that anticipate this diagnosis) and in the experimental models used in preclinical drug development. Without a progression in these two issues, cannabinoids may suffer of similar problems to reach the clinical scenario than other neuroprotectant agents investigated so far.

Acknowledgements This work was supported by grants from CIBERNED (CB06/05/0089), MICINN (PID2021-128906OB-I00) with FEDER funds and Michael J. Fox Foundation (MJFF-022552). Authors are indebted to all those who participated in the studies carried out by his group that were mentioned in this review and to Yolanda García-Movellán for administrative assistance.

Competing Interests' Statement Authors declare they have no competing interests in relation with this submission.

References

Aguado T, Monory K, Palazuelos J, Stella N, Cravatt B, Lutz B et al (2005) The endocannabinoid system drives neural progenitor proliferation. FASEB J 19:1704–1706

Aguado T, Palazuelos J, Monory K, Stella N, Cravatt B, Lutz B et al (2006) The endocannabinoid system promotes astroglial differentiation by acting on neural progenitor cells. J Neurosci 26: 1551–1561

Aguilar-Hernández L, Alejandre R, César Morales-Medina J, Iannitti T, Flores G (2023) Cellular mechanisms in brain aging: focus on physiological and pathological aging. J Chem Neuroanat 128:102210

Albayram O, Alferink J, Pitsch J, Piyanova A, Neitzert K, Poppensieker K et al (2011) Role of CB1 cannabinoid receptors on GABAergic neurons in brain aging. Proc Natl Acad Sci USA 108: 11256–11261

Arévalo-Martín A, García-Ovejero D, Gómez O, Rubio-Araiz A, Navarro-Galve B, Guaza C et al (2008) CB2 cannabinoid receptors as an emerging target for demyelinating diseases: from neuroimmune interactions to cell replacement strategies. Br J Pharmacol 153:216–225

Aso E, Ozaita A, Valdizán EM, Ledent C, Pazos A, Maldonado R, Valverde O (2008) BDNF impairment in the hippocampus is related to enhanced despair behavior in CB1 knockout mice. J Neurochem 105:565–572

Aso E, Ferrer I (2016) CB2 cannabinoid receptor as potential target against Alzheimer's disease. Front Neurosci 10:243

Athauda D, Foltynie T (2015) The ongoing pursuit of neuroprotective therapies in Parkinson's disease. Nat Rev Neurol 11:25–40

Ativie F, Albayram O, Bach K, Pradier B, Zimmer A, Bilkei-Gorzo A (2015) Enhanced microglial activity in FAAH(−/−) animals. Life Sci 138:52–56

Augustin SM, Lovinger DM (2018) Functional relevance of endocannabinoid-dependent synaptic plasticity in the central nervous system. ACS Chem Neurosci 9:2146–2161

Aymerich MS, Aso E, Abellanas MA, Tolon RM, Ramos JA, Ferrer I et al (2018) Cannabinoid pharmacology/therapeutics in chronic degenerative disorders affecting the central nervous system. Biochem Pharmacol 157:67–84

Bab I, Zimmer A (2008) Cannabinoid receptors and the regulation of bone mass. Br J Pharmacol 153:182–188

Bano D, Agostini M, Melino G, Nicotera P (2011) Ageing, neuronal connectivity and brain disorders: an unsolved ripple effect. Mol Neurobiol 43:124–130

Basile MS, Mazzon E (2022) The role of cannabinoid type 2 receptors in Parkinson's disease. Biomedicine 10:2986

Baune BT (2015) Inflammation and neurodegenerative disorders: is there still hope for therapeutic intervention? Curr Opin Psychiatry 28:148–154

Benito C, Nuñez E, Tolon RM, Carrier EJ, Rábano A, Hillard CJ et al (2003) Cannabinoid CB2 receptors and fatty acid amide hydrolase are selectively overexpressed in neuritic plaque-associated glia in Alzheimer's disease brains. J Neurosci 23:11136–11141

Berk C, Paul G, Sabbagh M (2014) Investigational drugs in Alzheimer's disease: current progress. Expert Opin Investig Drugs 23:837–846

Berrendero F, García-Gil L, Hernández ML, Romero J, Cebeira M, de Miguel R et al (1998a) Localization of mRNA expression and activation of signal transduction mechanisms for cannabinoid receptor in rat brain during fetal development. Development 125:3179–3188

Berrendero F, Romero J, García-Gil L, Suárez I, De la Cruz P, Ramos JA et al (1998b) Changes in cannabinoid receptor binding and mRNA levels in several brain regions of aged rats. Biochim Biophys Acta 1407:205–214

Berrendero F, Sepe N, Ramos JA, Di Marzo V, Fernández-Ruiz J (1999) Analysis of cannabinoid receptor binding and mRNA expression and endogenous cannabinoid contents in the developing rat brain during late gestation and early postnatal period. Synapse 33:181–191

Bhandari UR, Danish SM, Ahmad S, Ikram M, Nadaf A, Hasan N et al (2024) New opportunities for antioxidants in amelioration of neurodegenerative diseases. Mech Ageing Dev 221:111961

Bilkei-Gorzo A, Racz I, Valverde O, Otto M, Michel K, Sastre M et al (2005) Early age-related cognitive impairment in mice lacking cannabinoid CB1 receptors. Proc Natl Acad Sci USA 102: 15670–15675

Bilkei-Gorzo A (2012) The endocannabinoid system in normal and pathological brain ageing. Philos Trans R Soc Lond Ser B Biol Sci 367:3326–3341

Bilkei-Gorzo A, Drews E, Albayram Ö, Piyanova A, Gaffal E, Tueting T et al (2012) Early onset of aging-like changes is restricted to cognitive abilities and skin structure in $Cnr1^{-}/^{-}$ mice. Neurobiol Aging 33:200.e11–200.e22

Bilkei-Gorzo A, Albayram O, Draffehn A, Michel K, Piyanova A, Oppenheimer H et al (2017) A. A chronic low dose of Δ^9-tetrahydrocannabinol (THC) restores cognitive function in old mice. Nat Med 23:782–787

Bilsland LG, Greensmith L (2008) The endocannabinoid system in amyotrophic lateral sclerosis. Curr Pharm Des 14:2306–2316

Blalock EM, Chen KC, Sharrow K, Herman JP, Porter NM, Foster TC et al (2003) Gene microarrays in hippocampal aging: statistical profiling identifies novel processes correlated with cognitive impairment. J Neurosci 23:3807–3819

Blebea NM, Pricopie AI, Vlad RA, Hancu G (2024) Phytocannabinoids: exploring pharmacological profiles and their impact on therapeutical use. Int J Mol Sci 25:4204

Bondy SC (2024) Mitochondrial dysfunction as the major basis of brain aging. Biomolecules 14: 402

Bradford D, Rodgers KE (2024) Advancements and challenges in amyotrophic lateral sclerosis. Front Neurosci 18:1401706

Brailoiu GC, Oprea TI, Zhao P, Abood ME, Brailoiu E (2011) Intracellular cannabinoid type 1 (CB1) receptors are activated by anandamide. J Biol Chem 286:29166–29174

Burgaz S, García C, Gonzalo-Consuegra C, Gómez-Almería M, Ruiz-Pino F, Unciti JD et al (2021) Preclinical investigation in neuroprotective effects of the GPR55 ligand VCE-006.1 in experimental models of Parkinson's disease and amyotrophic lateral sclerosis. Molecules 26:7643

Butovsky E, Juknat A, Goncharov I, Elbaz J, Eilam R, Zangen A et al (2005) In vivo up-regulation of brain-derived neurotrophic factor in specific brain areas by chronic exposure to Delta-tetrahydrocannabinol. J Neurochem 93:802–811

Cabral GA, Griffin-Thomas L (2009) Emerging role of the cannabinoid receptor CB2 in immune regulation: therapeutic prospects for neuroinflammation. Expert Rev Mol Med 11:e3

Carles A, Freyssin A, Perin-Dureau F, Rubinstenn G, Maurice T (2024) Targeting *N*-methyl-d-aspartate receptors in neurodegenerative diseases. Int J Mol Sci 25:3733

Caserta MT, Bannon Y, Fernandez F, Giunta B, Schoenberg MR, Tan J (2009) Normal brain aging clinical, immunological, neuropsychological, and neuroimaging features. Int Rev Neurobiol 84: 1–19

Cassano T, Calcagnini S, Pace L, De Marco F, Romano A, Gaetani S (2017) Cannabinoid receptor 2 signaling in neurodegenerative disorders: from pathogenesis to a promising therapeutic target. Front Neurosci 11:30

Celorrio M, Rojo-Bustamante E, Fernández-Suárez D, Sáez E, Estella-Hermoso de Mendoza A, Müller CE et al (2017) GPR55: a therapeutic target for Parkinson's disease? Neuropharmacology 125:319–332

Chen C (2023) Inhibiting degradation of 2-arachidonoylglycerol as a therapeutic strategy for neurodegenerative diseases. Pharmacol Ther 244:108394

Choi IY, Ju C, Anthony Jalin AM, Lee DI, Prather PL, Kim WK (2013) Activation of cannabinoid CB2 receptor-mediated AMPK/CREB pathway reduces cerebral ischemic injury. Am J Pathol 182:928–939

Ciaramellano F, Fanti F, Scipioni L, Maccarrone M, Oddi S (2023) Endocannabinoid metabolism and transport as drug targets. Methods Mol Biol 2576:201–211

Coates JR, Wininger FA (2010) Canine degenerative myelopathy. Vet Clin North Am Small Anim Pract 40:929–950

Covelo A, Eraso-Pichot A, Fernández-Moncada I, Serrat R, Marsicano G (2021) CB1R-dependent regulation of astrocyte physiology and astrocyte-neuron interactions. Neuropharmacology 195: 108678

Cristino L, Bisogno T, Di Marzo V (2020) Cannabinoids and the expanded endocannabinoid system in neurological disorders. Nat Rev Neurol 16:9–29

Cuddihey H, MacNaughton WK, Sharkey KA (2022) Role of the endocannabinoid system in the regulation of intestinal homeostasis. Cell Mol Gastroenterol Hepatol 14:947–963

Cummings JL, Osse AML, Kinney JW, Cammann D, Chen J (2024) Alzheimer's disease: combination therapies and clinical trials for combination therapy development. CNS Drugs 38:613–624

D'Addario C, Di Francesco A, Pucci M, Finazzi Agrò A, Maccarrone M (2013) Epigenetic mechanisms and endocannabinoid signalling. FEBS J 280:1905–1917

Dailah HG (2022) Potential of therapeutic small molecules in apoptosis regulation in the treatment of neurodegenerative diseases: an updated review. Molecules 27:7207

Delenclos M, Jones DR, McLean PJ, Uitti RJ (2016) Biomarkers in Parkinson's disease: advances and strategies. Parkinsonism Relat Disord 22:S106–S110

Demuth HU, Dijkhuizen RM, Farr TD, Gelderblom M, Horsburgh K, Iadecola C et al (2017) Recent progress in translational research on neurovascular and neurodegenerative disorders. Restor Neurol Neurosci 35:87–103

De Sousa AA, Rigby Dames BA, Graff EC, Mohamedelhassan R, Vassilopoulos T, Charvet CJ (2023) Going beyond established model systems of Alzheimer's disease: companion animals provide novel insights into the neurobiology of aging. Commun Biol 6:655

Di Marzo V, Stella N, Zimmer A (2015) Endocannabinoid signalling and the deteriorating brain. Nat Rev Neurosci 16:30–42

Díaz-Alonso J, Guzmán M, Galve-Roperh I (2012) Endocannabinoids via CB_1 receptors act as neurogenic niche cues during cortical development. Philos Trans R Soc Lond Ser B Biol Sci 367:3229–3241

Divers TJ, Mohammed HO, Cummings JF (1997) Equine motor neuron disease. Vet Clin North Am Equine Pract 13:97–105

Dorsey ER, Constantinescu R, Thompson JP, Biglan KM, Holloway RG, Kieburtz K et al (2007) Projected number of people with Parkinson disease in the most populous nations, 2005 through 2030. Neurology 68:384–386

Doser RL, Hoerndli FJ (2021) Regulation of neuronal excitability by reactive oxygen species and calcium signaling: insights into brain aging. Curr Res Neurobiol 2:100012

Double KL, Reyes S, Werry EL, Halliday GM (2010) Selective cell death in neurodegeneration: why are some neurons spared in vulnerable regions? Prog Neurobiol 92:316–329
Duarte JM, Ferreira SG, Carvalho RA, Cunha RA, Köfalvi A (2012) CB_1 receptor activation inhibits neuronal and astrocytic intermediary metabolism in the rat hippocampus. Neurochem Int 60:1–8
Duffy SS, Hayes JP, Fiore NT, Moalem-Taylor G (2021) The cannabinoid system and microglia in health and disease. Neuropharmacology 190:108555
Eaton SL, Wishart TM (2017) Bridging the gap: large animal models in neurodegenerative research. Mamm Genome 28:324–337
Ellert-Miklaszewska A, Ciechomska IA, Kaminska B (2021) Synthetic cannabinoids induce autophagy and mitochondrial apoptotic pathways in human glioblastoma cells independently of deficiency in *TP53* or *PTEN* tumor suppressors. Cancers (Basel) 13:419
Eshhar N, Striem S, Biegon A (1993) HU-211, a non-psychotropic cannabinoid, rescues cortical neurones from excitatory amino acid toxicity in culture. Neuroreport 5:237–240
Espejo-Porras F, Piscitelli F, Verde R, Ramos JA, Di Marzo V, de Lago E et al (2015) Changes in the endocannabinoid signaling system in CNS structures of TDP-43 transgenic mice: relevance for a neuroprotective therapy in TDP-43-related disorders. J Neuroimmune Pharmacol 10:233–244
Espejo-Porras F, García-Toscano L, Rodríguez-Cueto C, Santos-García I, de Lago E, Fernandez-Ruiz J (2019) Targeting glial cannabinoid CB_2 receptors to delay the progression of the pathological phenotype in TDP-43 (A315T) transgenic mice, a model of amyotrophic lateral sclerosis. Br J Pharmacol 176:1585–1600
Esquer A, Blanc F, Collongues N (2023) Immunotherapies targeting amyloid and tau protein in Alzheimer's disease: should we move away from diseases and focus on biological targets? A systematic review and expert opinion. Neurol Ther 12:1883–1907
Fagan SG, Campbell VA (2014) The influence of cannabinoids on generic traits of neurodegeneration. Br J Pharmacol 171:1347–1360
Farzanehfar P (2016) Towards a better treatment option for Parkinson's disease: a review of adult neurogenesis. Neurochem Res 41:3161–3170
Feigenbaum JJ, Bergmann F, Richmond SA, Mechoulam R, Nadler V, Kloog Y et al (1989) Nonpsychotropic cannabinoid acts as a functional N-methyl-D-aspartate receptor blocker. Proc Natl Acad Sci USA 86:9584–9587
Fernández-Moncada I, Rodrigues RS, Fundazuri UB, Bellocchio L, Marsicano G (2024) Type-1 cannabinoid receptors and their ever-expanding roles in brain energy processes. J Neurochem 168:693–703
Fernández-Ruiz J, Berrendero F, Hernández ML, Ramos JA (2000) The endogenous cannabinoid system and brain development. Trends Neurosci 23:14–20
Fernández-Ruiz J (2009) The endocannabinoid system as a target for the treatment of motor dysfunction. Br J Pharmacol 156:1029–1040
Fernández-Ruiz J, Romero J, Ramos JA (2015a) Endocannabinoids and neurodegenerative disorders: Parkinson's disease, Huntington's chorea, Alzheimer's disease, and others. Handb Exp Pharmacol 231:233–259
Fernández-Ruiz J, Moro MA, Martínez-Orgado J (2015b) Cannabinoids in neurodegenerative disorders and stroke/brain trauma: from preclinical models to clinical applications. Neurotherapeutics 12:793–806
Fernández-Ruiz J, Gómez-Ruiz M, García C, Hernández M, Ramos JA (2017) Modeling neurodegenerative disorders for developing cannabinoid-based neuroprotective therapies. Methods Enzymol 593:175–198
Fernández-Ruiz J (2019) The biomedical challenge of neurodegenerative disorders: an opportunity for cannabinoid-based therapies to improve on the poor current therapeutic outcomes. Br J Pharmacol 176:1370–1383
Fernández-Ruiz J, Sagredo O, de Lago E (2021) Chapter 2 – Phytocannabinoids *versus* endocannabinoids. A modern view of the endocannabinoid system. In: Maccarrone M

(ed) New tools to interrogate endocannabinoid signalling: from natural compounds to synthetic drugs. Royal Society of Chemistry, pp 10–47

Figueiredo A, Cheer JF (2023) Endocannabinoid regulation of hippocampus-dependent memory. Exp Neurol 364:114384

Fijał K, Filip M (2016) Clinical/therapeutic approaches for cannabinoid ligands in central and peripheral nervous system diseases: mini review. Clin Neuropharmacol 39:94–101

Fraguas-Sánchez AI, Torres-Suárez AI (2018) Medical use of cannabinoids. Drugs 78:1665–1703

Franceschi AM, Petrover DR, Giliberto L, Clouston SAP, Gordon ML (2022) Semiquantitative approach to amyloid positron emission tomography interpretation in clinical practice. World J Nucl Med 22:15–21

Frisoni GB, Boccardi M, Barkhof F, Blennow K, Cappa S, Chiotis K et al (2017) Strategic roadmap for an early diagnosis of Alzheimer's disease based on biomarkers. Lancet Neurol 16:661–676

Fujii M, Sherchan P, Krafft PR, Rolland WB, Soejima Y, Zhang JH (2014) Cannabinoid type 2 receptor stimulation attenuates brain edema by reducing cerebral leukocyte infiltration following subarachnoid hemorrhage in rats. J Neurol Sci 342:101–106

Galve-Roperh I, Chiurchiù V, Díaz-Alonso J, Bari M, Guzmán M, Maccarrone M (2013) Cannabinoid receptor signaling in progenitor/stem cell proliferation and differentiation. Prog Lipid Res 52:633–650

Gao Y, Vasilyev DV, Goncalves MB, Howell FV, Hobbs C, Reisenberg M et al (2010) Loss of retrograde endocannabinoid signaling and reduced adult neurogenesis in diacylglycerol lipase knock-out mice. J Neurosci 30:2017–2024

Gao Z, Pang Z, Chen Y, Lei G, Zhu S, Li G et al (2022) Restoring after central nervous system injuries: neural mechanisms and translational applications of motor recovery. Neurosci Bull 38: 1569–1587

Galán-Ganga M, Rodríguez-Cueto C, Merchán-Rubira J, Hernández F, Ávila J, Posada-Ayala M et al (2021) Cannabinoid receptor CB2 ablation protects against TAU induced neurodegeneration. Acta Neuropathol Commun 9:90

Garaschuk O, Semchyshyn HM, Lushchak VI (2018) Healthy brain aging: interplay between reactive species, inflammation and energy supply. Ageing Res Rev 43:26–45

García-Matas S, Gutierrez-Cuesta J, Coto-Montes A, Rubio-Acero R, Díez-Vives C, Camins A et al (2008) Dysfunction of astrocytes in senescence-accelerated mice SAMP8 reduces their neuroprotective capacity. Aging Cell 7:630–640

Geldenhuys WJ, Van der Schyf CJ (2013) Rationally designed multi-targeted agents against neurodegenerative diseases. Curr Med Chem 20:1662–1672

Gomes TM, Dias da Silva D, Carmo H, Carvalho F, Silva JP (2020) Epigenetics and the endocannabinoid system signaling: an intricate interplay modulating neurodevelopment. Pharmacol Res 162:105237

Gómez-Cañas M, Rodríguez-Cueto C, Satta V, Hernández-Fisac I, Navarro E, Fernández-Ruiz J (2023) Endocannabinoid-binding receptors as drug targets. Methods Mol Biol 2576:67–94

Gómez-Gálvez Y, Palomo-Garo C, Fernández-Ruiz J, García C (2016) Potential of the cannabinoid CB(2) receptor as a pharmacological target against inflammation in Parkinson's disease. Prog Neuro-Psychopharmacol Biol Psychiatry 64:200–208

Goncalves MB, Suetterlin P, Yip P, Molina-Holgado F, Walker DJ, Oudin MJ et al (2008) A diacylglycerol lipase-CB2 cannabinoid pathway regulates adult subventricular zone neurogenesis in an age-dependent manner. Mol Cell Neurosci 38:526–536

Gontijo VS, Viegas FPD, Ortiz CJC, de Freitas SM, Damasio CM, Rosa MC et al (2020) Molecular hybridization as a tool in the design of multi-target directed drug candidates for neurodegenerative diseases. Curr Neuropharmacol 18:348–407

Gonzalo-Consuegra C, Santos-García I, García-Toscano L, Martín-Baquero R, Rodríguez-Cueto C, Wittwer MB et al (2024) Involvement of CB_1 and CB_2 receptors in neuroprotective effects of cannabinoids in experimental TDP-43 related frontotemporal dementia using male mice. Biomed Pharmacother 174:116473

Gordon PH, Meininger V (2011) How can we improve clinical trials in amyotrophic lateral sclerosis? Nat Rev Neurol 7:650–654

Grigorenko E, Kittler J, Clayton C, Wallace D, Zhuang S, Bridges D et al (2002) Assessment of cannabinoid induced gene changes: tolerance and neuroprotection. Chem Phys Lipids 121:257–266

Grimm A, Eckert A (2017) Brain aging and neurodegeneration: from a mitochondrial point of view. J Neurochem 143:418–431

Hagan K, Varelas P, Zheng H (2022) Endocannabinoid system of the blood-brain barrier: current understandings and therapeutic potentials. Cannabis Cannabinoid Res 7:561–568

Haj-Dahmane S, Shen RY (2011) Modulation of the serotonin system by endocannabinoid signaling. Neuropharmacology 61:414–420

Haller S, Jäger HR, Vernooij MW, Barkhof F (2023) Neuroimaging in dementia: more than typical Alzheimer's disease. Radiology 308:e230173

Hamilton LK, Joppé SE, Cochard LM, Fernandes KJ (2013) Aging and neurogenesis in the adult forebrain: what we have learned and where we should go from here. Eur J Neurosci 37:1978–1986

Hampson AJ, Grimaldi M, Axelrod J, Wink D (1998) Cannabidiol and (−)Δ^9-tetrahydrocannabinol are neuroprotective antioxidants. Proc Natl Acad Sci USA 95:8268–8273

Harkany T, Guzmán M, Galve-Roperh I, Berghuis P, Devi LA, Mackie K (2007) The emerging functions of endocannabinoid signaling during CNS development. Trends Pharmacol Sci 28: 83–92

Hempel B, Xi ZX (2022) Receptor mechanisms underlying the CNS effects of cannabinoids: CB_1 receptor and beyond. Adv Pharmacol 93:275–333

Hidding U, Mainka T, Buhmann C (2024) Therapeutic use of medical Cannabis in neurological diseases: a clinical update. J Neural Transm (Vienna) 131:117–126

Hillard CJ, Beatka M, Sarvaideo J (2016) Endocannabinoid signaling and the hypothalamic-pituitary-adrenal axis. Compr Physiol 7:1–15

Holm IE, Alstrup AK, Luo Y (2016) Genetically modified pig models for neurodegenerative disorders. J Pathol 238:267–287

Hoque A, Hossain MI, Ameen SS, Ang CS, Williamson N, Ng DC et al (2016) A beacon of hope in stroke therapy-Blockade of pathologically activated cellular events in excitotoxic neuronal death as potential neuroprotective strategies. Pharmacol Ther 160:159–179

Huang C, Zheng X, Yan S, Zhang Z (2024) Advances in clinical therapies for Huntington's disease and the promise of multi-targeted/functional drugs based on Clinicaltrials.gov. Clin Pharmacol Ther 116:1452

Huerga-Gómez A, Aguado T, Sánchez-de la Torre A, Bernal-Chico A, Matute C, Mato S et al (2021) Δ^9-Tetrahydrocannabinol promotes oligodendrocyte development and CNS myelination in vivo. Glia 69:532–545

Hwang J, Adamson C, Butler D, Janero DR, Makriyannis A, Bahr BA (2010) Enhancement of endocannabinoid signaling by fatty acid amide hydrolase inhibition: a neuroprotective therapeutic modality. Life Sci 86:615–623

Ilyasov AA, Milligan CE, Pharr EP, Howlett AC (2018) The endocannabinoid system and oligodendrocytes in health and disease. Front Neurosci 12:733

Ionescu-Tucker A, Cotman CW (2021) Emerging roles of oxidative stress in brain aging and Alzheimer's disease. Neurobiol Aging 107:86–95

Isaev NK, Stelmashook EV, Genrikhs EE (2019) Neurogenesis and brain aging. Rev Neurosci 30: 573–580

Jagust WJ, Teunissen CE, DeCarli C (2023) The complex pathway between amyloid β and cognition: implications for therapy. Lancet Neurol 22:847–857

Jiang CH, Tsien JZ, Schultz PG, Hu Y (2001) The effects of aging on gene expression in the hypothalamus and cortex of mice. Proc Natl Acad Sci USA 98:1930–1934

Jiang W, Zhang Y, Xiao L, Van Cleemput J, Ji SP, Bai G et al (2005) Cannabinoids promote embryonic and adult hippocampus neurogenesis and produce anxiolytic- and antidepressant-like effects. J Clin Invest 115:3104–3116

Jin K, Xie L, Kim SH, Parmentier-Batteur S, Sun Y, Mao XO et al (2004) Defective adult neurogenesis in CB1 cannabinoid receptor knockout mice. Mol Pharmacol 66:204–208

Kaczocha M, Haj-Dahmane S (2022) Mechanisms of endocannabinoid transport in the brain. Br J Pharmacol 179:4300–4310

Kallendrusch S, Kremzow S, Nowicki M, Grabiec U, Winkelmann R, Benz A et al (2013) The G protein-coupled receptor 55 ligand l-α-lysophosphatidylinositol exerts microglia-dependent neuroprotection after excitotoxic lesion. Glia 61:1822–1831

Kaminski NE, Kaplan BLF (2021) Immunomodulation by cannabinoids: current uses, mechanisms, and identification of data gaps to be addressed for additional therapeutic application. Adv Pharmacol 91:1–59

Karimian Azari E, Kerrigan A, O'Connor A (2020) Naturally occurring cannabinoids and their role in modulation of cardiovascular health. J Diet Suppl 17:625–650

Kaye J, Reisine T, Finkbeiner S (2021) Huntington's disease mouse models: unraveling the pathology caused by CAG repeat expansion. Fac Rev 10:77

Kessiova M, Alexandrova A, Georgieva A, Kirkova M, Todorov S (2006) In vitro effects of CB1 receptor ligands on lipid peroxidation and antioxidant defense systems in the rat brain. Pharmacol Rep 58:870–875

Khan TA, Hassan I, Ahmad A, Perveen A, Aman S, Quddusi S et al (2016) Recent updates on the dynamic association between oxidative stress and neurodegenerative disorders. CNS Neurol Disord Drug Targets 15:310–320

Kharrazian D (2022) Functional medicine approaches to neurodegeneration. Phys Med Rehabil Clin N Am 33:733–743

Kluever V, Fornasiero EF (2021) Principles of brain aging: status and challenges of modeling human molecular changes in mice. Ageing Res Rev 72:101465

Kluge A, Iranzo A (2024) Biofluid detection of pathological α-synuclein in the prodromal phase of synucleinopathies. J Parkinsons Dis 14:S323

Koch M (2017) Cannabinoid receptor signaling in central regulation of feeding behavior: a mini-review. Front Neurosci 11:293

Komorowska-Müller JA, Rana T, Olabiyi BF, Zimmer A, Schmöle AC (2021) Cannabinoid receptor 2 alters social memory and microglial activity in an age-dependent manner. Molecules 26:5984

Koprich JB, Kalia LV, Brotchie JM (2017) Animal models of α-synucleinopathy for Parkinson disease drug development. Nat Rev Neurosci 18:515–529

Kowal SL, Dall TM, Chakrabarti R, Storm MV, Jain A (2013) The current and projected economic burden of Parkinson's disease in the United States. Mov Disord 28:311–318

Kudelova J, Fleischmannova J, Adamova E, Matalova E (2015) Pharmacological caspase inhibitors: research towards therapeutic perspectives. J Physiol Pharmacol 66:473–482

Kumar A (2020) Calcium signaling during brain aging and its influence on the hippocampal synaptic plasticity. Adv Exp Med Biol 1131:985–1012

Legare CA, Raup-Konsavage WM, Vrana KE (2022) Therapeutic potential of cannabis, cannabidiol, and cannabinoid-based pharmaceuticals. Pharmacology 107:131–149

Liu P, Bilkey DK, Darlington CL, Smith PF (2003) Cannabinoid CB1 receptor protein expression in the rat hippocampus and entorhinal, perirhinal, postrhinal and temporal cortices: regional variations and age-related changes. Brain Res 979:235–239

Lombardi G, Crescioli G, Cavedo E, Lucenteforte E, Casazza G, Bellatorre AG et al (2020) Structural magnetic resonance imaging for the early diagnosis of dementia due to Alzheimer's disease in people with mild cognitive impairment. Cochrane Database Syst Rev 3:CD009628

Lorente-Picón M, Laguna A (2021) New avenues for Parkinson's disease therapeutics: disease-modifying strategies based on the gut microbiota. Biomolecules 11:433

Lu T, Pan Y, Kao SY, Li C, Kohane I, Chan J, Yankner BA (2004) Gene regulation and DNA damage in the ageing human brain. Nature 429:883–891
Luo XG, Ding JQ, Chen SD (2010) Microglia in the aging brain: relevance to neurodegeneration. Mol Neurodegener 5:12
Maas AI, Murray G, Henney H 3rd, Kassem N, Legrand V, Mangelus M et al (2006) Efficacy and safety of dexanabinol in severe traumatic brain injury: results of a phase III randomised, placebo-controlled, clinical trial. Lancet Neurol 5:38–45
Maccarrone M, Valverde O, Barbaccia ML, Castañé A, Maldonado R, Ledent C et al (2002) Age-related changes of anandamide metabolism in CB1 cannabinoid receptor knockout mice: correlation with behaviour. Eur J Neurosci 15:1178–1186
Maccarrone M, Guzmán M, Mackie K, Doherty P, Harkany T (2014) Programming of neural cells by (endo)cannabinoids: from physiological rules to emerging therapies. Nat Rev Neurosci 15: 786–801
Maccarrone M, Maldonado R, Casas M, Henze T, Centonze D (2017) Cannabinoids therapeutic use: what is our current understanding following the introduction of THC, THC: CBD oromucosal spray and others? Expert Rev Clin Pharmacol 10:443–455
Maccarrone M, Di Marzo V, Gertsch J, Grether U, Howlett AC, Hua T et al (2023) Goods and bads of the endocannabinoid system as a therapeutic target: lessons learned after 30 years. Pharmacol Rev 75:885–958
Magham SV, Thaggikuppe Krishnamurthy P, Shaji N, Mani L, Balasubramanian S (2021) Cannabinoid receptor 2 selective agonists and Alzheimer's disease: an insight into the therapeutic potentials. J Neurosci Res 99:2888–2905
Mallah K, Couch C, Borucki DM, Toutonji A, Alshareef M, Tomlinson S (2020) Anti-inflammatory and neuroprotective agents in clinical trials for CNS disease and injury: where do we go from here? Front Immunol 11:2021
Mancuso R, Navarro X (2015) Amyotrophic lateral sclerosis: current perspectives from basic research to the clinic. Prog Neurobiol 133:1–26
Mangialasche F, Solomon A, Winblad B, Mecocci P, Kivipelto M (2010) Alzheimer's disease: clinical trials and drug development. Lancet Neurol 9:702–716
Marsicano G, Moosmann B, Hermann H, Lutz B, Behl C (2002) Neuroprotective properties of cannabinoids against oxidative stress: role of the cannabinoid receptor CB_1. J Neurochem 80: 448–456
Marsicano G, Goodenough S, Monory K, Hermann H, Eder M, Cannich A et al (2003) CB1 cannabinoid receptors and on-demand defense against excitotoxicity. Science 302:84–88
Martínez-Pinilla E, Aguinaga D, Navarro G, Rico AJ, Oyarzábal J, Sánchez-Arias JA et al (2019) Targeting CB_1 and GPR55 endocannabinoid receptors as a potential neuroprotective approach for Parkinson's disease. Mol Neurobiol 56:5900–5910
Mattson MP (2007) Mitochondrial regulation of neuronal plasticity. Neurochem Res 32:707–715
Mazucanti CH, Cabral-Costa JV, Vasconcelos AR, Andreotti DZ, Scavone C, Kawamoto EM (2015) Longevity pathways (mTOR, SIRT, insulin/IGF-1) as key modulatory targets on aging and neurodegeneration. Curr Top Med Chem 15:2116–2138
Mazzatenta A, Carluccio A, Robbe D, Giulio CD, Cellerino A (2017) The companion dog as a unique translational model for aging. Semin Cell Dev Biol 70:141–153
McLellan MA, Rosenthal NA, Pinto AR (2017) Cre-loxP-mediated recombination: general principles and experimental considerations. Curr Protoc Mouse Biol 7:1–12
Meah F, Lundholm M, Emanuele N, Amjed H, Poku C, Agrawal L, Emanuele MA (2022) The effects of cannabis and cannabinoids on the endocrine system. Rev Endocr Metab Disord 23: 401–420
Mechoulam R, Shohami E (2007) Endocannabinoids and traumatic brain injury. Mol Neurobiol 36: 68–74
Meldolesi J (2017) Neurotrophin receptors in the pathogenesis, diagnosis and therapy of neurodegenerative diseases. Pharmacol Res 121:129–137

Menalled L, Brunner D (2014) Animal models of Huntington's disease for translation to the clinic: best practices. Mov Disord 29:1375–1390

Meyer HC, Lee FS, Gee DG (2018) The role of the endocannabinoid system and genetic variation in adolescent brain development. Neuropsychopharmacology 43:21–33

Minamihata T, Takano K, Moriyama M, Nakamura Y (2020) Lysophosphatidylinositol, an endogenous ligand for G protein-coupled receptor 55, has anti-inflammatory effects in cultured microglia. Inflammation 43:1971–1987

Mir HD, Giorgini G, Di Marzo V (2023) The emerging role of the endocannabinoidome-gut microbiome axis in eating disorders. Psychoneuroendocrinology 154:106295

Molina-Holgado F, Pinteaux E, Moore JD, Molina-Holgado E, Guaza C, Gibson RM et al (2003) Endogenous interleukin-1 receptor antagonist mediates anti-inflammatory and neuroprotective actions of cannabinoids in neurons and glia. J Neurosci 23:6470–6474

Moore NL, Greenleaf AL, Acheson SK, Wilson WA, Swartzwelder HS, Kuhn CM (2010) Role of cannabinoid receptor type 1 desensitization in greater tetrahydrocannabinol impairment of memory in adolescent rats. J Pharmacol Exp Ther 335:294–301

Morton AJ, Howland DS (2013) Large genetic animal models of Huntington's disease. J Huntingtons Dis 2:3–19

Mu Y, Chang KX, Chen YF, Yan K, Wang CX, Hua Q (2024) Diagnosis of Alzheimer's disease: towards accuracy and accessibility. J Biol Methods 11:e99010010

Muhammad J, Khan A, Ali A, Fang L, Yanjing W, Xu Q et al (2018) Network pharmacology: exploring the resources and methodologies. Curr Top Med Chem 18:949–964

Müller T (2017) Investigational agents for the management of Huntington's disease. Expert Opin Investig Drugs 26:175–185

Nakanishi H, Wu Z (2009) Microglia-aging: roles of microglial lysosome- and mitochondria-derived reactive oxygen species in brain aging. Behav Brain Res 201:1–7

Nidadavolu P, Bilkei-Gorzo A, Krämer M, Schürmann B, Palmisano M, Beins EC et al (2021) Efficacy of Δ^9-tetrahydrocannabinol (THC) alone or in combination with a 1:1 ratio of cannabidiol (CBD) in reversing the spatial learning deficits in old mice. Front Aging Neurosci 13:718850

Nidadavolu P, Bilkei-Gorzo A, Effah F, Leidmaa E, Schürmann B, Berger M et al (2022) Dynamic changes in the endocannabinoid system during the aging process: focus on the middle-age crisis. Int J Mol Sci 23:10254

Noriega-Prieto JA, Kofuji P, Araque A (2023) Endocannabinoid signaling in synaptic function. Glia 71:36–43

OECD (2017) Health at a glance. OECD

Ofek O, Karsak M, Leclerc N, Fogel M, Frenkel B, Wright K et al (2006) Peripheral cannabinoid receptor, CB2, regulates bone mass. Proc Natl Acad Sci USA 103:696–701

Olanow CW, Kieburtz K, Katz R (2017) Clinical approaches to the development of a neuroprotective therapy for PD. Exp Neurol 298:246–251

Oosthoek M, Vermunt L, de Wilde A, Bongers B, Antwi-Berko D, Scheltens P et al (2024) Utilization of fluid-based biomarkers as endpoints in disease-modifying clinical trials for Alzheimer's disease: a systematic review. Alzheimers Res Ther 16:93

Oudin MJ, Gajendra S, Williams G, Hobbs C, Lalli G, Doherty P (2011) Endocannabinoids regulate the migration of subventricular zone-derived neuroblasts in the postnatal brain. J Neurosci 31: 4000–4011

Pacher P, Mechoulam R (2011) Is lipid signaling through cannabinoid 2 receptors part of a protective system? Prog Lipid Res 50:193–211

Palazuelos J, Aguado T, Egia A, Mechoulam R, Guzmán M, Galve-Roperh I (2006) Non-psychoactive CB2 cannabinoid agonists stimulate neural progenitor proliferation. FASEB J 20:2405–2407

Palazuelos J, Aguado T, Pazos MR, Julien B, Carrasco C, Resel E, Sagredo O, Benito C, Romero J, Azcoitia I, Fernández-Ruiz J, Guzmán M, Galve-Roperh I (2009) Microglial CB2 cannabinoid receptors are neuroprotective in Huntington's disease excitotoxicity. Brain 132:3152–3164

Paradisi A, Oddi S, Maccarrone M (2006) The endocannabinoid system in ageing: a new target for drug development. Curr Drug Targets 7:1539–1552

Pařízek A, Suchopár J, Laštůvka Z, Alblová M, Hill M, Dušková M (2023) The endocannabinoid system and its relationship to human reproduction. Physiol Res 72:S365–S380

Park TY, Jeon J, Cha Y, Kim KS (2024) Past, present, and future of cell replacement therapy for parkinson's disease: a novel emphasis on host immune responses. Cell Res 34:479–492

Pérot JB, Brouillet E, Flament J (2024) The contribution of preclinical magnetic resonance imaging and spectroscopy to Huntington's disease. Front Aging Neurosci 16:1306312

Petrov D, Mansfield C, Moussy A, Hermine O (2017) ALS clinical trials review: 20 years of failure. Are we any closer to registering a new treatment? Front Aging Neurosci 9:68

Piyanova A, Albayram O, Rossi CA, Farwanah H, Michel K, Nicotera P et al (2013) Loss of CB1 receptors leads to decreased cathepsin D levels and accelerated lipofuscin accumulation in the hippocampus. Mech Ageing Dev 134:391–399

Piyanova A, Lomazzo E, Bindila L, Lerner R, Albayram O, Ruhl T et al (2015) Age-related changes in the endocannabinoid system in the mouse hippocampus. Mech Ageing Dev 150:55–64

Plascencia-Villa G, Perry G (2020) Status and future directions of clinical trials in Alzheimer's disease. Int Rev Neurobiol 154:3–50

Pop E (2000) Dexanabinol Pharmos. Curr Opin Investig Drugs 1:494–503

Rees RN, Noyce AJ, Schrag AE (2024) Identification of prodromal Parkinson disease: we may be able to but should we? Neurology 102:e209394

Rim C, You MJ, Nahm M, Kwon MS (2024) Emerging role of senescent microglia in brain aging-related neurodegenerative diseases. Transl Neurodegener 13:10

Rock EM, Parker LA (2016) Cannabinoids as potential treatment for chemotherapy-induced nausea and vomiting. Front Pharmacol 7:221

Rodrigues FB, Wild EJ (2018) Huntington's disease clinical trials corner: February 2018. J Huntingtons Dis 7:89–98

Rodríguez-Cueto C, García-Toscano L, Santos-García I, Gómez-Almería M, Gonzalo-Consuegra C, Espejo-Porras F et al (2021a) Targeting the CB_2 receptor and other endocannabinoid elements to delay disease progression in amyotrophic lateral sclerosis. Br J Pharmacol 178:1373–1387

Rodríguez-Cueto C, Gómez-Almería M, García Toscano L, Romero J, Hillard CJ, de Lago E et al (2021b) Inactivation of the CB_2 receptor accelerated the neuropathological deterioration in TDP-43 transgenic mice, a model of amyotrophic lateral sclerosis. Brain Pathol 31:e12972

Rodríguez de Fonseca F, Ramos JA, Bonnin A, Fernández-Ruiz J (1993) Presence of cannabinoid binding sites in the brain from early postnatal ages. Neuroreport 4:135–138

Rodríguez-Gómez O, Palacio-Lacambra ME, Palasí A, Ruiz-Laza A, Boada-Rovira M (2014) Prevention of Alzheimer's disease: a global challenge for next generation neuroscientists. J Alzheimers Dis 42:S515–S523

Romero J, García-Palomero E, Berrendero F, García-Gil L, Hernández ML, Ramos JA et al (1997) Atypical location of cannabinoid receptors in white matter areas during rat brain development. Synapse 26:317–323

Romero J, Berrendero F, Garcia-Gil L, de la Cruz P, Ramos JA, Fernández-Ruiz J (1998) Loss of cannabinoid receptor binding and messenger RNA levels and cannabinoid agonist-stimulated [35S]guanylyl-5'O-(thio)-triphosphate binding in the basal ganglia of aged rats. Neuroscience 84:1075–1083

Rosser A, Svendsen CN (2014) Stem cells for cell replacement therapy: a therapeutic strategy for HD? Mov Disord 29:1446–1454

Rozenfeld R, Devi LA (2008) Regulation of CB1 cannabinoid receptor trafficking by the adaptor protein AP-3. FASEB J 22:2311–2322

Saba L (2015) Imaging neurodegenerative disorders. Oxford University Press

Sagredo O, González S, Aroyo I, Pazos MR, Benito C, Lastres-Becker I, Romero JP, Tolón RM, Mechoulam R, Brouillet E, Romero J, Fernández-Ruiz J (2009) Cannabinoid CB2 receptor agonists protect the striatum against malonate toxicity: relevance for Huntington's disease. Glia 57:1154–1167

Sagredo O, Palazuelos J, Gutierrez-Rodriguez A, Satta V, Galve-Roperh I, Martínez-Orgado J (2018) Cannabinoid signalling in the immature brain: encephalopathies and neurodevelopmental disorders. Biochem Pharmacol 157:85–96

Salat DH (2011) The declining infrastructure of the aging brain. Brain Connect 1:279–293

Salmina AB, Kapkaeva MR, Vetchinova AS, Illarioshkin SN (2021) Novel approaches used to examine and control neurogenesis in Parkinson's disease. Int J Mol Sci 22:9608

Sampaio C, Borowsky B, Reilmann R (2014) Clinical trials in Huntington's disease: interventions in early clinical development and newer methodological approaches. Mov Disord 29:1419–1428

Sánchez-Juan P, Valeriano-Lorenzo E, Ruiz-González A, Pastor AB, Rodrigo Lara H, López-González F et al (2024) Serum GFAP levels correlate with astrocyte reactivity, post-mortem brain atrophy and neurofibrillary tangles. Brain 147:1667–1679

Sarkar S (2013) Regulation of autophagy by mTOR-dependent and mTOR-independent pathways: autophagy dysfunction in neurodegenerative diseases and therapeutic application of autophagy enhancers. Biochem Soc Trans 41:1103–1130

Sasaguri H, Nilsson P, Hashimoto S, Nagata K, Saito T, De Strooper B et al (2017) APP mouse models for Alzheimer's disease preclinical studies. EMBO J 36:2473–2487

Shen SY, Yu R, Li W, Liang LF, Han QQ, Huang HJ et al (2022) The neuroprotective effects of GPR55 against hippocampal neuroinflammation and impaired adult neurogenesis in CSDS mice. Neurobiol Dis 169:105743

Shi M, Chu F, Zhu F, Zhu J (2022) Impact of anti-amyloid-β monoclonal antibodies on the pathology and clinical profile of Alzheimer's disease: a focus on aducanumab and lecanemab. Front Aging Neurosci 14:870517

Shohami E, Novikov M, Mechoulam R (1993) A nonpsychotropic cannabinoid, HU-211, has cerebroprotective effects after closed head injury in the rat. J Neurotrauma 10:109–119

Sikora E, Bielak-Zmijewska A, Dudkowska M, Krzystyniak A, Mosieniak G, Wesierska M et al (2021) Cellular senescence in brain aging. Front Aging Neurosci 13:646924

Simic G, Bexheti S, Kelovic Z, Kos M, Grbic K, Hof PR et al (2005) Hemispheric asymmetry, modular variability and age-related changes in the human entorhinal cortex. Neuroscience 130:911–925

Simone JJ, Green MR, McCormick CM (2022) Endocannabinoid system contributions to sex-specific adolescent neurodevelopment. Prog Neuro-Psychopharmacol Biol Psychiatry 113:110438

Small SA, Schobel SA, Buxton RB, Witter MP, Barnes CA (2011) A pathophysiological framework of hippocampal dysfunction in ageing and disease. Nat Rev Neurosci 12:585–601

Stella N (2010) Cannabinoid and cannabinoid-like receptors in microglia, astrocytes, and astrocytomas. Glia 58:1017–1030

Stoker TB, Blair NF, Barker RA (2017) Neural grafting for Parkinson's disease: challenges and prospects. Neural Regen Res 12:389–392

Stranahan AM, Mattson MP (2012) Recruiting adaptive cellular stress responses for successful brain ageing. Nat Rev Neurosci 13:209–216

Tapia-Arancibia L, Aliaga E, Silhol M, Arancibia S (2008) New insights into brain BDNF function in normal aging and Alzheimer disease. Brain Res Rev 59:201–220

Tobeh NS, Bruce KD (2023) Emerging Alzheimer's disease therapeutics: promising insights from lipid metabolism and microglia-focused interventions. Front Aging Neurosci 15:1259012

Tuominen RK, Renko JM (2024) Biomarkers of Parkinson's disease in perspective of early diagnosis and translation of neurotrophic therapies. Basic Clin Pharmacol Toxicol 135:271

Trezza V, Cuomo V, Vanderschuren LJ (2008) Cannabis and the developing brain: insights from behavior. Eur J Pharmacol 585:441–452

Ugrumov M (2020) Development of early diagnosis of Parkinson's disease: illusion or reality? CNS Neurosci Ther 26:997–1009

United Nations (2015) World population prospects: the 2015 revision. UN

Valeri A, Mazzon E (2021) Cannabinoids and neurogenesis: the promised solution for neurodegeneration? Molecules 26:6313
Van Laere K, Goffin K, Casteels C, Dupont P, Mortelmans L, de Hoon J et al (2008) Gender-dependent increases with healthy aging of the human cerebral cannabinoid-type 1 receptor binding using [(18)F]MK-9470 PET. NeuroImage 39:1533–1541
van Oostveen WM, de Lange ECM (2021) Imaging techniques in Alzheimer's disease: a review of applications in early diagnosis and longitudinal monitoring. Int J Mol Sci 22:2110
von Bernhardi R, Tichauer JE, Eugenín J (2010) Aging-dependent changes of microglial cells and their relevance for neurodegenerative disorders. J Neurochem 112:1099–1114
Vega-García A, Feria-Romero I, García-Juárez A, Munguia-Madera AC, Montes-Aparicio AV, Zequeida-Muñoz E et al (2021) Cannabinoids: a new perspective on epileptogenesis and seizure treatment in early life in basic and clinical studies. Front Behav Neurosci 14:610484
Vendel E, de Lange EC (2014) Functions of the CB1 and CB2 receptors in neuroprotection at the level of the blood-brain barrier. NeuroMolecular Med 16:620–642
Wang L, Liu J, Harvey-White J, Zimmer A, Kunos G (2003) Endocannabinoid signaling via cannabinoid receptor 1 is involved in ethanol preference and its age-dependent decline in mice. Proc Natl Acad Sci USA 100:1393–1398
Wang T, Liu XH, Guan J, Ge S, Wu MB, Lin JP et al (2019) Advancement of multi-target drug discoveries and promising applications in the field of Alzheimer's disease. Eur J Med Chem 169: 200–223
Watkins BA (2019) Diet, endocannabinoids, and health. Nutr Res 70:32–39
West MJ (1993) Regionally specific loss of neurons in the aging human hippocampus. Neurobiol Aging 14:287–293
Whitelaw CB, Sheets TP, Lillico SG, Telugu BP (2016) Engineering large animal models of human disease. J Pathol 238:247–256
Winters BL, Vaughan CW (2021) Mechanisms of endocannabinoid control of synaptic plasticity. Neuropharmacology 197:108736
Wong C, Gregory JM, Liao J, Egan K, Vesterinen HM, Ahmad Khan A et al (2023) Systematic, comprehensive, evidence-based approach to identify neuroprotective interventions for motor neuron disease: using systematic reviews to inform expert consensus. BMJ Open 13:e064169
World Health Organization (2016) Life expectancy increased by 5 years since 2000, but health inequalities persist. WHO
Wu W, Ji Y, Wang Z, Wu X, Li J, Gu F et al (2023) The FDA-approved anti-amyloid-β monoclonal antibodies for the treatment of Alzheimer's disease: a systematic review and meta-analysis of randomized controlled trials. Eur J Med Res 28:544
Wyse RK, Isaacs T, Barker RA, Cookson MR, Dawson TM, Devos D et al (2024) Twelve years of drug prioritization to help accelerate disease modification trials in Parkinson's disease: the International Linked Clinical Trials Initiative. J Parkinsons Dis 14:657–666
Xin Y, Tang A, Pan S, Zhang J (2022) Components of the endocannabinoid system and effects of cannabinoids against bone diseases: a mini-review. Front Pharmacol 12:793750
Yang W, Tu Z, Sun Q, Li XJ (2016) CRISPR/Cas9: implications for modeling and therapy of neurodegenerative diseases. Front Mol Neurosci 9:30
Yates ML, Barker EL (2009) Organized trafficking of anandamide and related lipids. Vitam Horm 81:25–53
Yiangou Y, Facer P, Durrenberger P, Chessell IP, Naylor A, Bountra C et al (2006) COX-2, CB2 and P2X7-immunoreactivities are increased in activated microglial cells/macrophages of multiple sclerosis and amyotrophic lateral sclerosis spinal cord. BMC Neurol 6:12
Youssef SA, Capucchio MT, Rofina JE, Chambers JK, Uchida K, Nakayama H et al (2016) Pathology of the aging brain in domestic and laboratory animals, and animal models of human neurodegenerative diseases. Vet Pathol 53:327–348
Zhao J, Liu S, Xiang X, Zhu X (2024) Versatile strategies for adult neurogenesis: avenues to repair the injured brain. Neural Regen Res 19:774–780

Zhong MZ, Peng T, Duarte ML, Wang M, Cai D (2024) Updates on mouse models of Alzheimer's disease. Mol Neurodegener 19:23
Zia A, Pourbagher-Shahri AM, Farkhondeh T, Samarghandian S (2021) Molecular and cellular pathways contributing to brain aging. Behav Brain Funct 17:6
Zieglgänsberger W, Brenneisen R, Berthele A, Wotjak CT, Bandelow B, Tölle TR et al (2022) Chronic pain and the endocannabinoid system: smart lipids – a novel therapeutic option? Med Cannabis Cannabinoids 5:61–75
Ziemka-Nałęcz M, Zalewska T (2012) Endogenous neurogenesis induced by ischemic brain injury or neurodegenerative diseases in adults. Acta Neurobiol Exp 72:309–324
Zoppi S, Madrigal JL, Caso JR, García-Gutiérrez MS, Manzanares J, Leza JC et al (2014) Regulatory role of the cannabinoid CB2 receptor in stress-induced neuroinflammation in mice. Br J Pharmacol 171:2814–2826

Pharmacotherapy for Cannabis Use Disorder: Preclinical and Clinical Models

Margaret F. Bedillion, Catherine F. Moore, Elise M. Weerts, Caroline A. Arout, Hannah M. Harris, and Margaret Haney

Contents

1 Introduction: Epidemiology of Cannabis Use Disorder 250
1.1 Current Patterns of Cannabis Use 250
2 Neurobiology of Cannabis 251
3 Preclinical Models of Cannabis Use 252
3.1 Interoceptive/"Subjective" Effects: Drug Discrimination 254
3.2 Positive Reinforcing Effects: Cannabinoid Self-Administration 254
3.3 Rewarding Effects: Conditioned Place Preference 257
3.4 Extinction and Reinstatement (Relapse) Models 258
3.5 Cannabinoid Withdrawal 259
3.6 Intracranial Self-Stimulation Models: Reward Function 260
4 Medication Effects: Preclinical Models 261
4.1 Cannabidiol 261
4.2 Cannabinoid Receptor Competitive Antagonists and Biased Antagonists 271
4.3 Other Candidate Medication Agents 272
5 Human Laboratory Models of Cannabis Use Disorder 273
5.1 Positive Subjective Effects 273
5.2 Positive Reinforcing Effects: Cannabinoid Self-Administration 273
5.3 Cannabis Withdrawal 274
6 Medication Effects: Human Laboratory Models and Clinical Trials 274
6.1 Abstinence Initiation: Reducing the Positive Subjective and Reinforcing Effects of Cannabis 275
6.2 Relapse Prevention: Reducing the Negative Reinforcing Effects of Cannabis 277
7 Summary 281
References 284

M. F. Bedillion (✉), C. A. Arout, H. M. Harris, and M. Haney
Department of Psychiatry, Columbia University Irving Medical Center and New York State Psychiatric Institute, New York, NY, USA
e-mail: margaret.bedillion@nyspi.columbia.edu

C. F. Moore and E. M. Weerts
Division of Behavioral Biology, Department of Psychiatry and Behavioral Sciences, Johns Hopkins University School of Medicine, Baltimore, MD, USA

Curr Topics Behav Neurosci (2026) 76: 249–296
https://doi.org/10.1007/7854_2025_598
Published Online: 05 August 2025

Abstract Cannabis use disorder (CUD) is increasingly prevalent, and there are no FDA-approved medications to facilitate its treatment. Research on the development of CUD pharmacotherapies lags compared to other substance use disorders (SUDs), partly due to the difficulty in establishing robust preclinical models of cannabinoid self-administration. This chapter evaluates the translational pipeline, from preclinical and human laboratory models to randomized controlled clinical trials for CUD, reviewing medication effects observed within these models and discussing the challenges in translating preclinical and human laboratory findings, particularly regarding efficacy endpoints (e.g., prolonged abstinence versus a reduction in use). The chapter concludes by addressing future directions for improving translational validity and optimizing therapeutic development for CUD.

Keywords Cannabinoids · Cannabis Use Disorder · Clinical Models · Pharmacotherapy · Preclinical Models · THC

1 Introduction: Epidemiology of Cannabis Use Disorder

1.1 Current Patterns of Cannabis Use

Cannabis is one of the most commonly used addictive drugs, with 2.5% of the global population reporting past-year use (World Health Organization n.d.). In the United States (U.S.), 47% of individuals aged 12 years or older reported lifetime cannabis use (Substance Abuse, Administration MHS 2024). Although at present (2025), cannabis remains a federally illicit substance in the U.S. under the Controlled Substances Act of 1970, 39 states have legalized medicinal use, and 24 of these states have also legalized recreational ("adult") use. These policy changes parallel long-term trends in cannabis use within the U.S. Between 1992 and 2022, the number of individuals reporting daily or near-daily use (defined as use on 21 or more days in the past month) increased 15-fold from 0.9 million to 17.7 million (Caulkins 2024). While most individuals do not develop clinically significant problems related to their cannabis use, a subset engages in patterns of chronic, heavy use, which increases the risk of more psychosocial and behavioral consequences and developing a cannabis use disorder (CUD) (Callaghan et al. 2020; Wittemann et al. 2021; Buckner et al. 2015; Chan et al. 2021; Millar et al. 2021). The Diagnostic and Statistical Manual of Mental Disorders (DSM-5) defines CUD as the presence of clinically significant impairment or distress evidenced by at least 2 of the 11 criteria, such as cravings, withdrawal symptoms, and using larger amounts than intended, within a 12-month period (Hasin et al. 2013).

In the U.S., 19.4 million people reported past-year CUD (Substance Abuse, Administration MHS 2024), and 16.5% of individuals with CUD received past-year substance use treatment (Choi et al. 2024). This low engagement in treatment may be attributed to several barriers, including overconfidence in the ability to stop on one's own, minimization of problems associated with cannabis use, stigma surrounding substance use disorders (SUDs), and limited awareness of available treatment options (Kerridge et al. 2017). Among those seeking treatment, evidence-based behavioral and psychotherapies, including cognitive behavioral therapy (Budney et al. 2006; Copeland et al. 2001), motivational enhancement therapy (Budney et al. 2000), motivational interviewing (Zaharakis et al. 2024), and contingency management (Budney et al. 2015; Schuster et al. 2016; Litt et al. 2008; Carroll et al. 2006), have shown effectiveness in reducing cannabis use, yet most individuals relapse to regular cannabis use and are unable to achieve sustained abstinence (Flórez-Salamanca et al. 2013).

To date, there are no Food and Drug Administration (FDA)-approved medications for CUD, and developing efficacious pharmacotherapy remains a goal of researchers and clinicians in the field of addiction treatment. Several promising pharmacological agents for CUD have been examined including cannabinoid type 1 (CB1) receptor agonists (e.g., dronabinol, nabilone, and nabiximols), a signaling specific inhibitor of CB1 receptors, fatty acid amide hydrolase (FAAH) inhibitors, and non-cannabinoid medications (e.g., naltrexone, quetiapine, and varenicline). A recent study examining treatment preferences among individuals who use cannabis found that 80% favored a combination of medication and behavioral therapy (Lile et al. 2023), supporting the importance of this avenue of research. As the rates of cannabis use and CUD continue to rise, there is a clear need for diverse treatment options that include safe and effective pharmacotherapies.

2 Neurobiology of Cannabis

Cannabis is often categorized as Δ-9-tetrahydrocannabinol (THC)-dominant or cannabidiol (CBD)-dominant, but it also contains other cannabinoids, as well as volatile oils or terpenes, which impart distinct odors and flavors (Reimann-Philipp et al. 2020; Spindle et al. 2019a). THC is the primary cannabinoid of interest in this chapter, as it is the main psychoactive component of cannabis and is associated with the development of CUD. Cannabis/cannabinoids act primarily on the endocannabinoid system (ECS), which comprises two major receptor types, CB1 and CB2 receptors, their naturally occurring endogenous ligands, 2-arachidonoylglycerol (2-AG) and arachidonoylethanolamide (AEA), and the enzymes FAAH and monoacylglycerol lipase (MAGL).

CB1 and CB2 receptors belong to the superfamily of G-protein-coupled receptors and can indirectly modulate glutamate, gamma-aminobutyric acid (GABA), opioid, serotonin, and dopamine signaling (Cohen et al. 2019). CB1 receptors are widely distributed throughout the brain, with a high density in regions involved in reward, mood, stress regulation, motor function, habit learning, and cognition (Mecha et al. 2016; Wong et al. 2010). Correspondingly, the ECS regulates a wide array of processes including motivation for natural rewards (e.g., food, sex, social behavior, etc.), memory, anxiety, pain, and stress response (Parsons and Hurd 2015). THC, a partial agonist at CB1 and CB2 receptors, exerts wide-ranging effects on these processes. Notably, the psychoactive and abuse-related effects of cannabis and THC are primarily mediated via CB1 receptors (Cooper and Haney 2008; Huestis et al. 2001; Tanda and Goldberg 2003).

Repeated exposure to cannabis/THC significantly impacts the ECS. Based on positron emission topography (PET) studies, individuals who regularly use cannabis have lower levels of FAAH, the AEA metabolizing enzyme (Boileau et al. 2016) and unexpectedtly also have lower levels of AEA in cerebrospinal fluid (Morgan et al. 2013). Circulating levels of 2-AG appears to be inversely correlated with the amount of cannabis used (Kearney-Ramos et al. 2023). PET studies investigating CB1 receptor availability in individuals who use cannabis daily have shown a downregulation of CB1 receptors in both female (Spindle et al. 2021) and male participants when compared to controls (Ceccarini et al. 2015; D'Souza et al. 2016a, b). These receptor levels recover with continued abstinence.

Preclinical data are largely consistent with clinical data. Repeated THC administration to animals downregulates CB1 receptors (Breivogel et al. 1999; Sim-Selley 2003), but is associated with higher levels of AEA in the limbic forebrain, and decreases in AEA and 2-AG levels in the striatum (Di Marzo et al. 2000; González et al. 2004). Circulating endocannabinoid concentrations are lower in animals administered THC (Hillard 2018), as in humans. Overall, there is still much to learn about the functional consequences of disrupting the ECS with regular cannabis use.

3 Preclinical Models of Cannabis Use

Laboratory animal models allow for the evaluation of drug effects under tightly controlled experimental conditions (e.g., drug history, parameters of drug and dose exposure, conditions of access, and environmental context) that are not possible in human volunteers. While preclinical studies cannot model all aspects of SUDs, there are well-established procedures for investigating drug abuse liability and behavioral processes associated with various substances of misuse (Corley et al. 2024). Models include those in which drug is administered by the experimenter (i.e., noncontingent) and those in which drug consumption is controlled by the animal (i.e., contingent). For example, noncontingent drug exposure models include drug discrimination to assess interoceptive drug effects, place conditioning to assess the rewarding and aversive drug effects, and chronic drug exposure models to assess drug tolerance, physical dependence, and the associated withdrawal syndrome when drug use is abruptly stopped. Contingent models include drug self-administration to assess drug taking and reinforcement and related models of drug seeking and reinstatement that attempt to model aspects of relapse to drug taking (Fig. 1).

A key consideration in cannabinoid research is the route of administration used in the model, as it significantly influences drug pharmacokinetics and pharmacodynamics. In humans, inhalation of smoked or vaporized cannabis, the most common

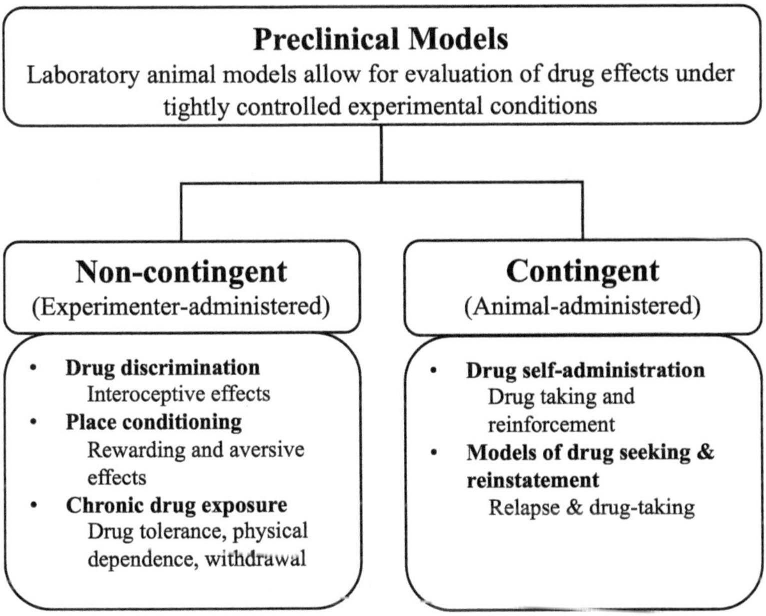

Fig. 1 Preclinical models of drug administration to assess cannabis use disorder

method of use, is characterized by fast absorption through the lungs resulting in almost immediate detectable drug levels in plasma (Huestis et al. 1992; Ohlsson et al. 1986), which peak after 5–10 min and decrease to near nondetectable levels (<5 ng/ml) within 3 h (Huestis et al. 1992; Schwope et al. 2011). Recently, inhalation models in rodents have been developed using Volcano® type vaporizers, meter-dose inhalers, and e-cigarette technology to more closely mimic human use (Bruijnzeel et al. 2016; Javadi-Paydar et al. 2018; Lichtman et al. 2000; Manwell et al. 2014a, b; Nguyen et al. 2016; Wilson et al. 2002). In rats, vapor exposure to THC also results in peak plasma concentrations occurring rapidly following administration (e.g., ~10–15 min) in a concentration-dependent manner (Javadi-Paydar et al. 2018, 2019). In contrast, oral administration of THC/cannabis results in slower and more variable absorption (e.g., peak ~2 h post administration) compared to inhalation, due in part to first-pass metabolism in the liver. This results in markedly lower levels of THC in plasma, but high concentrations of the psychoactive metabolite 11-OH-THC compared to inhalation (Huestis 2007; Lemberger et al. 1973; Zagzoog et al. 2024). In rodents, peak behavioral effects and blood cannabinoid concentrations occur several hours (1.5–2) after administration, similar to humans (Zagzoog et al. 2024; Spindle et al. 2020; Pryor et al. 1977). Preclinical models often administer THC by intraperitoneal (IP) injection, as it is a technically simple route of administration and produces significant detectable plasma THC concentrations that are comparable to concentrations seen in humans (e.g., 40–200 ng/ml) (Huestis et al. 1992). Following an IP injection, THC levels in the blood peak approximately 30–60 min post-dosing, with detectable levels in the blood lasting for hours (Baglot et al. 2021; Britch et al. 2017; Hložek et al. 2017). Notably, the peak plasma THC concentrations (C_{max}) achieved via IP administration is comparable in magnitude to those observed with inhalation, both in rodent and human models, yet the duration of elevated plasma THC levels is typically longer. For example, a recent comparison of the pharmacokinetics between inhalation models and IP injection noted that, at doses that produced similar levels of THC in the blood, IP administration resulted in dramatically higher levels of the active metabolite 11-OH-THC in the blood and brain (Baglot et al. 2021). IP administration could produce a much more robust and sustained activation of brain CB1 receptors, via THC and 11-OH-THC, that produce markedly different physiological, psychological, and behavioral effects compared to inhalation. Thus, the choice of route of administration is particularly important for preclinical cannabinoid research and its translational relevance to human use.

3.1 Interoceptive/"Subjective" Effects: Drug Discrimination

Drug discrimination, a behavioral pharmacology procedure used to assess the interoceptive effects of drugs, models subjective drug effects in humans, such as feel "good effects" and "high," which can provide insights into abuse-related effects of a drug. This has been recommended as a primary method for preclinical evaluation of cannabinoids for abuse liability by U.S. federal agencies (FDA 2017). Given the difficulty establishing reliable models of THC self-administration, drug discrimination studies provide an alternative method for assessing pharmacological actions of THC that model both the abuse liability and subjective effects of cannabinoids (Balster and Prescott 1992).

3.2 Positive Reinforcing Effects: Cannabinoid Self-Administration

Drug self-administration is the "gold standard" for demonstration of drug reinforcement and abuse liability. Drug reinforcement, as with any reinforcement, refers to the process by which drug presentation, contingent upon a behavioral response, increases the likelihood of that behavioral response (Young and Herling 1986). Animals are trained to emit a behavior (e.g., nose-poke or lever press) that results in drug delivery. Reinforcement is demonstrated if the behavioral response associated with drug delivery is maintained and exceeds responding for the drug vehicle. The ability of a drug to serve as a reinforcer involves a dynamic behavioral process that is influenced by the state of the individual as well as the context in which a drug is available for consumption. Multiple variables can influence the rate and pattern of drug-maintained behavior including the speed of drug delivery, the onset and duration of drug effects (i.e., pharmacokinetics), and the schedule of reinforcement under which drug is available, including the duration of drug access and time between each drug delivery.

Historically, self-administration studies of THC and other cannabinoids in laboratory animals have used intravenous (IV) drug delivery, which allows for rapid drug onset. Initial studies looking to establish IV THC self-administration in laboratory animals primarily used simple fixed ratio schedules of reinforcement and short duration of daily drug access (e.g., 1-h sessions) (Panlilio et al. 2010; Tanda et al. 2000). Early studies in rats had difficulty in demonstrating self-administration of IV THC using these standard procedures (doses ranging from 0.00625 to 0.05 mg/kg/infusion) (Reinaldo and Singer 1980; Takahashi and Singer 1979; van Ree et al. 1978). Similarly, several early studies using rhesus monkeys found that responding for THC did not supersede vehicle (Harris et al. 1974; Kaymakcalan 1973) and were deemed "negative." More recent studies of IV THC self-administration in rodents have also shown levels of drug intake that are not greater than vehicle or saline controls (Wakeford et al. 2017; Stringfield and Torregrossa 2021). However, better lever discrimination was observed in THC-administering animals compared to vehicle-administering controls and classic biphasic dose response curves to THC were observed (Spencer et al. 2018). Animals self-administering THC achieve physiologically relevant doses of THC, show relevant behavioral effects of THC, and exhibit long-term neurobiological changes, such as downregulation of CB1 receptors (Stringfield and Torregrossa 2021; Spencer et al. 2018). Interestingly, THC is self-administered intracranially into the cerebral ventricles (Braida et al. 2004), the ventral tegmental area, and the shell of the nucleus accumbens in rats (Zangen et al. 2006).

Studies in squirrel monkeys have so far provided the strongest evidence of the reinforcing effects of IV THC using an animal model (Tanda et al. 2000; Justinova et al. 2003). As detailed in a comprehensive review (Panlilio et al. 2010), these studies modified several key aspects of the procedure that likely contributed to its success over prior attempts. This included the use of very low unit THC doses (2–4 μg/kg/injection), use of a vehicle that enhanced THC solubility, rapid injection speed, exposure to a full range of THC doses, and frequent vehicle substitution early in training to increase sensitivity and discrimination of doses. The authors determined that these changes in procedural parameters were critical to establish THC self-injection. Moreover, reinforcement was more robust when using more complex schedules of reinforcement (e.g., fixed ratio (FR) 10, second-order schedule) (Tanda et al. 2000; Justinova et al. 2003; John et al. 2017).

While THC self-administration has proven difficult to achieve, selective high affinity CB1 receptor agonists, which have greater efficacy than THC, are self-administered IV in a dose-dependent manner (Deiana et al. 2007; Fattore et al. 2007; Martellotta et al. 1998; Lefever et al. 2014). Self-administration is attenuated by pretreatment with CB1 receptor antagonists, confirming that the reinforcing effects are CB1 receptor-mediated (Lefever et al. 2014; Fattore et al. 2001). These models of CB1 receptor agonist self-administration have been useful for screening potential medications for CUD and determining the neurobiological mechanisms of cannabinoid reinforcement (Mendizabal et al. 2006). However, THC has distinct properties from full CB1 receptor agonists. For example, rats will self-administer the CB1 agonist WIN 55,212-2 (0.01 mg/kg/infusion), but substitution with THC (0.003–0.1 mg/kg/infusion) did not maintain responding, demonstrating lower reinforcing efficacy of THC compared with WIN 55,212-2 (Lefever et al. 2014). Thus, models using THC rather than CB1 receptor agonists likely provide better translational validity when it comes to understanding the neurobiological mechanisms of cannabis reinforcement.

Over the last decade, efforts have been made to establish models that use vaporized cannabis or cannabinoid exposure and, more recently, THC self-administration (Freels et al. 2020, 2024; Pickens et al. 1978; Glodosky et al. 2020; Cooper et al. 2021). The speed of vapor delivery of THC may facilitate reinforcement and the necessary contingent associations that develop with operant tasks and bypass the limitations of IV methods, such as catheterization and cannabinoid solubility issues in vehicle solutions suitable for IV injection. Importantly, vapor 'doses' do not necessarily reflect the amount of drug received by the animal, as location, respiratory rate, and air flow can influence the pulmonary deposition and systemic bioavailability. To account for this, blood levels of THC and its metabolites are measured to verify exposure.

Recently, the first instance of cannabis vapor self-administration in male rats was demonstrated (Freels et al. 2020). Under FR4 and progressive ratio (PR) schedules of reinforcement, greater responding for vapor deliveries was observed in rats self-administering a THC-rich cannabis vapor compared to rats responding for CBD-rich cannabis vapor or the vehicle vapor. Rats showed discrimination (>2:1 active vs. inactive nose-poke responding) for both THC- and CBD-rich cannabis extract vapor but not for vehicle vapor. Rats administered a greater amount of vapor deliveries early in the session, consistent with the loading dose phenomenon seen in other drug self-administration models (Pickens et al. 1978). Motivation to obtain THC-rich cannabis extract vapor was higher than vehicle or CBD-rich cannabis extract vapor, as evidenced by higher breakpoints under the PR schedule. The reinforcing efficacy of THC-rich cannabis extract was CB1 receptor-mediated, as pretreatment with the CB1 receptor antagonist AM251 reduced responding. Further, response-contingent delivery of vaporized cannabis extracts produced behavioral and physiological changes that were consistent with passive vapor models; namely, reduced locomotor activity, increased daily food intake, and plasma THC and CBD levels that were consistent with amount of vapor deliveries obtained (Freels et al. 2020). Recent studies from this group have replicated this initial finding that rats will voluntarily administer cannabis vapor, with delivery-dependent increases in drug plasma levels as well as behavioral and biological changes in cannabis vapor-administering groups (Freels et al. 2024; Glodosky et al. 2020), although THC and vehicle vapor maintained similar levels of responding. In these more recent studies, animals were maintained on an FR1 schedule versus an FR4 or PR schedule as in the initial study, where responding for THC vapor was higher than for vehicle (Freels et al. 2020).

Recently, Cooper and colleagues (Cooper et al. 2021) evaluated the reinforcing effects of aerosolized THC and other CB1 receptor agonists in rhesus macaques using a food-maintained responding procedure. Monkeys were trained to puff on a stem for aerosol drug or vehicle delivery, and puffing behavior was reinforced with candy under a PR schedule. In this study, full CB1 receptor agonists functioned as weak reinforcers, but inhaled aerosolized THC did not maintain responding above vehicle levels.

Overall, cannabinoid self-administration in animal models is less robust than for other addictive drugs. Instead of assuming a failure of translation, researchers should think critically about redefining "success" in these models and adapt existing models to account for the complex and unique pharmacology of cannabis and cannabinoids. Successful models should demonstrate sustained responding for drug and administration of biologically relevant doses. Given the long duration of action of many cannabinoids, future studies should optimize self-administration paradigms to reflect this – for example, by incorporating intermittent access, extended session durations, longer inter-infusion intervals, or more complex schedules (e.g., second-order schedules) that result in lower total drug intake. Further, researchers can continue to refine these models through genetic (selective breeding, modification) and environmental manipulations that could improve their ability to capture key features of problematic cannabis use and CUD. Future studies could also focus on optimizing reinforcement conditions, drug access schedules, and cue reactivity to enhance the translational value of these models for understanding vulnerability to CUD and evaluating potential treatments.

3.3 Rewarding Effects: Conditioned Place Preference

Place conditioning is a commonly used Pavlovian conditioning procedure to assess the positive (rewarding) or negative (aversive) motivational effects of addictive drugs in rodents, non-human primates, and humans (Bardo et al. 1999; Bardo and Neisewander 1986; Borges et al. 2015; Childs and Wit de 2009; Collier and Echevarria 2013; Cunningham et al. 2006). The place conditioning apparatus used for rodents is typically a 2-sided or 3-sided chambered compartment with dividers between sides; each side of the box has distinct environmental cues (e.g., floor texture, color). Animals are treated with a drug/dose and placed in one side of the compartment (CS+); on alternate trials, drug vehicle administrations are paired with the other side of the compartment (CS-). After repeated drug-CS+ or vehicle-CS-pairings, the divider is removed, and drug-free animals are allowed to move freely between compartments. Conditioned place preference (CPP) is established if animals spend more time in the drug-paired chamber on the drug-free test. Conditioned place aversion (CPA) is established if the animal avoids the drug-paired compartment during the drug-free test. Using place conditioning procedures, most addictive drugs, depending on the dose, support significant CPP or CPA.

In tests of the conditioned effects of systemically injected THC in rodents, low doses (0.075–0.75 mg/kg, IP) produce preference, while higher doses (5–6 mg/kg, IP) produce aversion (Braida et al. 2004). The aversion seen at higher doses reflects a dysphoric effect of THC and CB1 receptor agonists, which may be overcoming any rewarding effects of the drug (McGregor et al. 1996; Parker and Gillies 1995). If a primer dose of THC is given prior to CPP conditioning (i.e., pre-exposure), it can shift preference for THC-paired compartments (Valjent and Maldonado 2000), supporting a theory of an initial dysphoric effect of THC upon first administration, particularly with injection routes of administration. Pretreatment with CBD prior to THC conditioning (1:1, 10 mg/kg) also can reverse the CPA seen with THC-alone (Klein et al. 2011). Administration of CB1 receptor antagonists block CPP and CPA produced by THC and other cannabinoid agonists (Braida et al. 2001; Chaperon et al. 1998).

Several studies have demonstrated CPP to vaporized THC in rats. In one study, rats exposed to THC vapor (30 min exposure, 200 mg/ml) displayed CPP, with female rats requiring more THC vapor exposure to produce CPP compared with males (Moore et al. 2024). Vapor concentrations that produced CPP also resulted in significant levels of THC in blood (~25–50 ng/ml) (Moore et al. 2024), concentrations that are comparable to those detected in humans reporting positive subjective drug effects after vaporizing cannabis (Spindle et al. 2019b). Similarly, a separate study also observed CPP in male rats exposed to 10 mg of vaporized THC across 10 min (Manwell et al. 2014a). To date, no studies using vaporized THC exposure have produced CPA, which may reflect a route of administration difference compared with IP administration. However, additional studies using vaporized THC would be needed to determine whether CPA is observed at higher doses.

3.4 Extinction and Reinstatement (Relapse) Models

Extinction and reinstatement are key processes used to study the persistence of drug-seeking behavior and to model features of drug relapse (Shaham et al. 2003). In extinction, a learned behavior (e.g., responding on the drug-associated lever) decreases or is eliminated when the reinforcing stimulus (e.g., the drug or associated cues) is no longer available. In the context of operant self-administration, drug is not delivered following an operant response, and in the context of CPP, exposure to the previously drug paired environment is not preceded by drug exposure. Reinstatement refers to the return of previously extinguished behavior (i.e., active responses or time spent in drug-paired chamber), which can be triggered by exposure to specific cues, stress, or the drug itself.

In squirrel monkeys that self-administer THC, classic reinstatement occurs with the presentation of a small, noncontingent dose of THC (Haney et al. 2023). Yet in rodents, as validation of self-administration models of THC and cannabis are considerably behind when compared to other addictive drugs, so are models of reinstatement. There is some evidence that stimuli paired with THC serve as conditioned reinforcers and may maintain seeking behavior. As described above, rats showed CPP to vaporized THC. When tested under extinction conditions, this preference was more persistent in females compared with males (Moore et al. 2024).

Rats showed increases in lever responding during an extinction test after a 7-day forced abstinence period from daily IV self-administration of THC + CBD (Spencer et al. 2018). Further, rats showed reinstatement following reintroduction of drug-paired cues contingent on lever pressing (Spencer et al. 2018). Likewise, in a study by Freels and colleagues (Freels et al. 2020), rats showed classic increases in responding during extinction tests.

3.5 *Cannabinoid Withdrawal*

The assessment of physical dependence in laboratory animals typically involves chronic drug administration and assessment of withdrawal symptoms when drug administration is abruptly discontinued (i.e., abstinence withdrawal) or after acute administration of a receptor-specific antagonist, which displaces the agonist from the receptor, immediately resulting in withdrawal (i.e., precipitated withdrawal). For abstinence withdrawal, significant effects on behavior should differ from both the effects of the drug and no drug controls and be *time limited* (e.g., return to baseline). For precipitated withdrawal, significant effects on behavior must also differ from effects of the antagonist and its vehicle (control).

The evaluation of THC/cannabis physical dependence and abstinence withdrawal in laboratory animals has several challenges. First, many symptoms of cannabis withdrawal in humans are subjective and mood related. Second, THC is highly lipophilic, stored in fat, and then slowly released. The redistribution and slow release of THC from tissues results in a slower onset and time course of withdrawal symptoms to subside when chronic THC/cannabis use is abruptly discontinued. Therefore, demonstration of a THC abstinence withdrawal syndrome in laboratory animals has been variable, and when found, the magnitude of withdrawal effects appears relatively minor (Lichtman and Martin 2002; Trexler et al. 2018). In contrast, antagonist-precipitated withdrawal in rats and nonhuman primates treated with THC is more robust and reliable and is the predominantly used model (Aceto et al. 1996; Cook et al. 1998; Wilson et al. 2006; Kesner et al. 2022; Marusich et al. 2014).

In rats, precipitated and spontaneous withdrawal symptoms include wet-dog shakes, facial rubs, horizontal and vertical activity, forepaw fluttering, chewing, tongue rolling, paw shakes, head shakes, retropulsion, myoclonic spasms, front paw treading, and ptosis (Aceto et al. 1996). In addition to somatic symptoms, deficits in early-stage cognition including disrupted prepulse inhibition and affective responses (increased startle reactivity) are observed immediately after administration of the antagonist (Marusich et al. 2014). Mice exposed repeatedly to cannabis smoke exhibit similar precipitated withdrawal effects as THC injected mice (Wilson et al. 2006). Recent studies have examined more complex behaviors and neurobiological markers during spontaneous withdrawal and have observed that mice show sleep disturbances and increased irritability, as well as altered striatal dopamine release during abstinence (Kesner et al. 2022).

Another strategy to model the subjective aspects of withdrawal is to examine changes in ongoing behavior suppressed by withdrawal (e.g., operant responding) (Beardsley et al. 1986). This strategy can be particularly useful for medications development to examine whether novel compounds can restore behavior suppressed after abrupt abstinence or as precipitated with an antagonist. For example, under a PR schedule of operant reinforcement for sucrose, precipitated withdrawal from THC reduces the motivation for sucrose (breakpoint) in mice (Eckard et al. 2020).

The interoceptive (subjective) effects of a dysphoric withdrawal state can also be assessed in animals using drug discrimination procedures, as described above. Rhesus monkeys that were maintained on the long-acting CB1 receptor agonist AM2389 were trained to discriminate rimonabant from saline (Kangas et al. 2013). When AM2389 was discontinued, subjects exhibited the discriminative stimulus effects of rimonabant, demonstrating that spontaneous withdrawal substituted for this rimonabant-precipitated withdrawal (Kangas et al. 2013). This procedure allows for testing medications that can alter the subjective experience associated with cannabis withdrawal.

Notably, there is growing evidence that females may be more vulnerable than males to developing dependence after chronic cannabinoid exposure (Cooper and Craft 2018a), and this is also observed in preclinical models. Females are more sensitive to the antinociceptive, hypothermic, and locomotor effects of acute IP injections of THC and synthetic cannabinoids (Cooper and Craft 2018a; Tseng and Craft 2001; Wakley et al. 2014). Female rats develop tolerance faster and at lower cannabinoid doses than males (Wakley et al. 2014; Wiley 2003) and discontinuation of chronic THC administration produces more anxiety-related behaviors in female rats compared with males (Harte-Hargrove and Dow-Edwards 2012).

3.6 Intracranial Self-Stimulation Models: Reward Function

Intracranial self-stimulation (ICSS) is an operant conditioning technique where animals can self-administer direct electrical stimulation to brain regions that underlie reward, such as the medial forebrain bundle of the mesolimbic dopamine system. ICSS can be leveraged to assess how these drugs modify the threshold intensity and frequency required for this electrical stimulation to maintain its reinforcing properties (Kornetsky et al. 1979). Drugs with rewarding properties generally facilitate ICSS, meaning they enhance dopamine system activity and lower the stimulation threshold required for reinforcement. Consistent with this, CB1 receptor agonists have been shown to increase sensitivity to electrical brain stimulation (enhanced ICSS), while the CB1 receptor antagonist/inverse agonist rimonabant produces the opposite effect (Gardner et al. 1988; Lepore et al. 1996; Xi et al. 2008). Interestingly, higher doses of THC have also been shown to increase brain stimulation thresholds, indicative of a biphasic effect (Katsidoni et al. 2013; Tanda 2016). Additionally, ICSS can be used to study withdrawal-related deficits in reward system activity, characterized by increased stimulation thresholds following cessation of chronic cannabinoid exposure (Ravula et al. 2019). Therefore, ICSS can be a valuable tool for evaluating potential pharmacotherapies, including assessing how candidate treatments modulate cannabinoid-induced or cannabinoid withdrawal-induced alterations in ICSS thresholds.

4 Medication Effects: Preclinical Models

A goal of developing animal models of cannabis or THC administration is to establish a medications development pipeline for treatment of CUD in humans. Due in part to the difficulty in establishing parameters that result in self-administration of cannabinoids in animals, there has been limited preclinical research in developing new medications to reduce cannabis intake. However, each of the models described above can be useful in testing potential CUD medications and/or uncovering therapeutic targets through understanding the neurobiology of cannabinoid subjective/interoceptive effects, reward, reinforcement, and withdrawal. Table 1 summarizes findings from preclinical experiments, human laboratory studies, and/or clinical trials for medications that have been evaluated as potential pharmacotherapies for CUD. For brevity, only compounds with supporting preclinical data and evidence of testing in humans are included. Several additional therapeutic strategies (e.g., neutral CB1 receptor antagonists, adenosine A2A receptor antagonists, and alpha-7 nicotinic acetylcholine receptor antagonists) have shown promise in preclinical models (Justinova et al. 2011; Schindler et al. 2016; Solinas et al. 2007); however, these studies are not included in the table or discussed in detail in this chapter, as they have not yet progressed to human testing.

4.1 Cannabidiol

The effects of CBD on THC/cannabis reward, reinforcement, and withdrawal have been tested across a range of species. CBD engages with a diverse set of molecular targets, including G-protein coupled receptors, such as serotonin-1A (5-HT_{1A}) and opioid receptors (Kathmann et al. 2006), and ion channels, such as transient receptor potential V1 (TRPV1) (McPartland et al. 2015; Russo and Marcu 2017). More recent evidence also indicates that CBD acts as a negative allosteric modulator at the CB1 receptors (Laprairie et al. 2015; Pandey et al. 2025). CBD's effects in combination with THC or cannabis, however, have been variable, depending on the dose, dosing regimen, and outcome of interest. For example, when THC and CBD are co-administered orally, "drug-drug" pharmacokinetic/pharmacodynamic interactions have been observed, such that high dose CBD increases THC's subjective drug effects (Zamarripa et al. 2023); these effects were associated with inhibition of THC clearance via cytochrome P450 (CYP) enzymes (Bansal et al. 2023).

Table 1 Summary of studies examining medications for cannabis use disorder

Preclinical				Human Laboratory				Clinical Trials			
Sample	Dose	Findings		Sample	Dose	Findings		Sample	Dose	Findings	
Cannabidiol											
Hiltunen and Jarbe (1986)											
Sprague Dawley rats $N = 14M$	0.3, 1.0 mg/kg, IP	↑	Time course of THC discriminative stimulus effects								
Vann et al. (2008)											
Long-Evans rats $N = 7M$	1–30 mg/kg IP	↔	THC discriminative stimulus effects								
McMahon (2016)				Haney et al. (2016)				Freeman et al. (2020)			
Rhesus monkeys $N = 6M$	17.8 mg/kg IV	↑	THC potency	$N = 31$: 17M, 14F	200, 400, 800 mg PO	↔	Self-admin Abuse liability	$N = 70$: 50M, 20F *Phase 2a RCT*	400, 800 mg/day PO	↓	Use days/wk
										↓	Withdrawal
Wakeford et al. (2017)											
Long-Evans rats $N = 5$: 3M, 2F	1:1 or 1:10 ratio (mean 0.41 (±0.07) mg/kg/IV for 1:1; 0.041 mg/kg (±0.007) for 1:10	↔	THC self-admin								
Navarrete et al. (2018)											
C57BL/6J mice $N = 180M$	5–20 mg/kg IP	↓	CP-55,940 Withdrawal-induced motor & anxiety-like behavior								

FAAH inhibitors: URB597, PF-04457845										
Justinova et al. (2008)				D'Souza et al. (2019)				Clinical Trial Data Not Yet Published (PF-04457845)		
Squirrel monkeys $N = 23$	0.3 mg/kg IV (URB597)	↔	Reinforcing effects of THC	$N = 70$M	4 mg/day (PF-04457845) PO	↓	Withdrawal Use/day			
Stewart and McMahon 2011										
Rhesus monkeys $N = 9$: 5M, 4F	0.32–3.2 mg/kg IV (URB597)	↔	Discriminative stimulus effects of THC	–	–	–				
Schlosburg et al. (2009)										
C57BL/6 J mice $N = 8$M/group	10 mg/kg SC (URB597)	↓	Precipitated THC withdrawal	–	–	–				
Rimonabant										
Braida et al. (2001)				Huestis et al. (2001)				–	–	–
Wistar rats $N = 8$M	0.5 mg/kg IP	↓	Conditioned rewarding effects of THC	$N = 63$M	1, 3, 10, 30, 90 mg PO (escalating dose)	↓	Abuse liability (90 mg)			
Justinova et al. (2008)				Huestis et al. (2007)						
Squirrel monkeys $N = 5$M	0.3 mg/kg IM	↓	IV THC self-admin	$N = 42$M	40 mg/day PO, 90 mg PO single dose	↓	Peak abuse liability (40 mg, day 8)			
						↔	Peak abuse liability (40 mg, 90 mg, day 15)			

(continued)

Table 1 (continued)

Preclinical				Human Laboratory				Clinical Trials			
Sample	Dose	Findings		Sample	Dose	Findings		Sample	Dose	Findings	
Wiley et al. (2017)											
Sprague-Dawley rats $N = 12$: 4M, 8F	1 ml/kg IP	↓	THC discriminative stimulus effects								
AEF0117											
Haney et al. (2023)				Haney et al. (2023)				Phase 2b Clinical Trial: Multisite Data Not Yet Published			
Mice $N = 13$/group	5, 15 ug/kg PO	↓	WIN 55,212–2 IV self-admin	$N = 29$: 28M, 1F	0.6 mg, 1 mg/day PO	↓	Self-admin (1 mg) Abuse liability (0.6, 1 mg)				
Squirrel monkeys $N = 4$	1.5–50 ug/kg PO	↓	IV THC self-admin								
Naltrexone											
Justinova et al. (2004)				Haney et al. (2015)				Notzon et al. (2018)			
Squirrel monkeys $N = 7M$	0.03–0.1 mg/kg IV	↓	THC self-admin	$N = 23$: 18M, 5F	50 mg/day PO	↓	Self-admin Abuse liability	$N = 12$: 7M, 5F *Open-label pilot RCT*	380 mg IM	↓	Use days/wk
										⟷	Grams, Dollars THC urine
Solinas and Goldberg (2005)											
Sprague-Dawley rats $N = 10M$	1–10 mg/kg IP	⟷	Discriminative stimulus effects of THC								

Wakley and Craft (2011)											
Sprague-Dawley rats $N = 18M$	0.32–3.2 mg/kg SC	↔	Discriminative stimulus effects of THC								
Tiagabine											
				Wesley et al. (2018)							
–	–	–		$N = 12$: 9M, 3F	12 mg/day PO	↑	Craving Worsened sleep	–	–	–	
						↔	Self-admin Abuse liability				
Pregabalin											
				Lile et al. (2021)							
Mice $N = 10M$/group	40 mg/kg PO	↓	Spontaneous CP-55,940 withdrawal	$N = 27$: 23M, 4F	300 mg, 450 mg/day PO	↔	Self-admin				
Dronabinol											
–	–	–		Haney et al. (2008)				Levin et al. (2011)			
				$N = 8M$	60 mg/day PO	↑	Abuse liability	$N = 156$: 128M, 28F *Phase 2 RCT*	40 mg/day PO	↓	Continuous abstinence (wks 7,8)
						↔	Withdrawal			↔	Withdrawal
						↓	"Relapse"				
				Schlienz et al. (2018)							
				$N = 13$: 10M, 3F	120–240 mg/day PO	↑	Abuse liability				
						↓	Withdrawal				

(continued)

Table 1 (continued)

Preclinical			Human Laboratory			Clinical Trials		
Sample	Dose	Findings	Sample	Dose	Findings	Sample	Dose	Findings
Lofexidine (+ Dronabinol)								
–	–	–	Haney et al. (2008)			Levin et al. (2016)		
			$N = 8$M	2.4 mg/day (Lofex) PO 60 mg/day (Dro) PO	↑ Sedation Abstinence-related anorexia Improved sleep	$N = 122$: 84M, 38F *Phase 2/3* *RCT*	1.8 mg/day (Lofex) 60 mg/day (Dro)	↓ Abstinence (Lofex-Dro)
					↓ "Relapse" (Lofex)			
					↔ "Relapse" Withdrawal Craving (Lofex-Dro)			
Zolpidem-XR (+ Nabilone)								
–	–	–	Vandrey et al. (2011)			Lee et al. (2024)		
			$N = 20$: 16M, 4F	12.5 mg/day (Zol) PO	↑ Sleep quality	$N = 127$: 79M, 58F *Phase 2/3* *RCT*	12.5 mg/day (Zol) PO	↓ Withdrawal-related sleep disturbance
					↔ Withdrawal Craving			↔ Urine toxicology
			Herrmann et al. (2016)					
			$N = 11$M	12.5 mg/day (Zol) PO 3 mg/day (Nab) PO	↓ Withdrawal-related sleep disruptions (Zol, Zol-Nab)			
					↓ "Relapse" (Zol-Nab)			
			Haney et al. (2013a, b)					
			$N = 11$: 8M, 3F	6 mg/day (Nab) PO	↓ "Relapse" Withdrawal			

Nabiximols										
–	–	–	Trigo et al. (2016)				Allsop et al. (2014)			
			$N = 9$: 8M, 1F	108 mg THC :100 mg CBD	↓	Withdrawal	$N = 51$: 39M, 12F *Phase* *3 RCT*	86.4 mg THC: 80 mg CBD	↔	Withdrawal Craving
					↓	Craving			↔	Self- reported use
							Lintzeris et al. (2019)			
							$N = 128$: 98M, 30F *Phase* *3 RCT*	86.4 mg THC: 80 mg CBD	↓	Use days
Buspirone										
–	–	–	–	–	–		McRae-Clark et al. (2009, 2015)			
							$N = 50$: 45M, 5F *Phase* *4 RCT* (2009) $N = 175$: 134M, 41F *Phase* *4 RCT* (2015)	60 mg/day (flexible dose) PO	↔	Urine drug screen Self- reported use
Lorcaserin										
–	–	–	Arout et al. (2021)				Brezing et al. (2024)			
			$N = 15$: 11M, 4F	20 mg PO	↓	"Relapse" Craving	$N = 8$ *Open-* *label* *Pilot RCT*	20 mg/day (XR) PO	↓	Use days/ wk

(continued)

Table 1 (continued)

Preclinical			Human Laboratory				Clinical Trials			
Sample	Dose	Findings	Sample	Dose	Findings		Sample	Dose	Findings	
Baclofen										
–	–	–	Haney et al. (2010)				–	–	–	
			$N = 10$M	60 mg, 90 mg/day PO	↔	"Relapse" Withdrawal				
Divalproex										
–	–	–	Haney et al. (2004)				Levin et al. (2004)			
			$N = 7$: 6M, 1F	1,500 mg/day PO	↑	"Irritable" "On edge" "Anxious" "Sleepy"	$N = 25$: 23M, 2F *Pilot RCT*	1,500 mg/day PO	↔	Abstinence Cannabis use
Celecoxib										
–	–	–	Haney et al. (2022)				–	–	–	
			$N = 15$: 12M, 3F	400 mg PO	↔	"Relapse" Withdrawal				
					↑	Craving				
Nefazodone										
–	–	–	Haney et al. (2003)				Carpenter et al. (2009)			
			$N = 7$: 6M, 1F	450 mg/day PO	↓	"Anxious" "Muscle pain"	$N = 66$: 56M, 10F *Phase 2 RCT*	600 mg PO	↔	Withdrawal Use/wk
					↔	"'Irritable" "Miserable" Sleep quality				
Buproprion-SR										
–	–	–	Haney et al. (2001)				Carpenter et al. (2009)			
			$N = 10$: 8M, 2F	300 mg/day PO	↑	Withdrawal	$N = 70$: 53M, 17F *Phase 2 RCT*	300 mg PO	↔	Withdrawal Use/wk

Mirtazapine										
–	–	–	Haney et al. (2010)				–	–	–	
			$N = 10$M	30 mg/day PO	↔	"Relapse" Abuse liability				
					↑	Withdrawal				
Atomoxetine										
–	–	–	–	–	–		Tirado et al. (2008)			
							$N = 13$: 10M, 3F *Open-label RCT*	25, 40, 80 mg/day PO (flexible dose)	↔	Days abstinent
							McRae-Clark et al. (2010)			
							$N = 38$: 29M, 9F *Phase 2 RCT*	Up to 100 mg/day PO (flexible dose)	↔	Avg use
Guanfacine										
–	–	–	Haney et al. (2019)				Dakwar et al. (2020)			
			$N = 15$: 13M, 2F	2 mg/day (IR) PO	↓	"Irritable"	$N = 22$: 14M, 8F *Open-label Pilot RCT*	4 mg (titrate-ER) PO	↑	Avg grams Use days/wk
					↔	Improved sleep				
					↓	"Relapse"				
			Holst et al. (2019)							
			$N = 7$: 3M, 4F	3 mg/day (IR) PO	↔	Withdrawal Craving Sleep				

(continued)

Table 1 (continued)

Preclinical			Human Laboratory			Clinical Trials		
Sample	Dose	Findings	Sample	Dose	Findings	Sample	Dose	Findings
Quetiapine								
–	–	–	Cooper et al. (2013)			Mariani et al. (2021)		
			$N = 14$: 12M, 2F	200 mg/day PO	↑ "Relapse" Craving Sleep Food intake	$N = 130$: 102M, 28F *Phase 2 RCT*	300 mg/day O	↓ Withdrawal Amount spent # use days
Varenicline (+ Nabilone)								
–	–	–	Herrmann et al. (2019)			McRae-Clark et al. (2021)		
			$N = 46$: 38M, 8F	2 mg (Varen) PO 8 mg (Nab) PO	↔ "Relapse"Withdrawal (Varen)	$N = 72$: 49M, 23F *Phase 2 Pilot RCT*	2 mg/day (Varen) PO	↓ Withdrawal % days used Use sess/ day
					↔ Withdrawal			
					↓ "Relapse" (Varen-Nab)			

Note: Animal doses are not directly equivalent to human doses and are typically calculated based on body surface area; for example, human of ~6.2 for rats or ~12.3 for mice to obtain equivalent mg/kg doses. = no change in outcome.

Human laboratory studies enroll individuals not seeking treatment for their cannabis use and clinical trials typically enroll those interested in post abstinence; *CPP* conditioned place preference, *ER* extended-release, *F* female, *IM* intramuscular, *IP* intraperitoneal, intravenous, *M* male, *PO* oral administration, *RCT* randomized clinical trial, *SC* subcutaneous, *THC* Δ-9-tetrahydrocannabinol

In a drug discrimination study using rats, CBD (0.3–30 mg/kg, IP) did not alter the discriminative stimulus effects of THC (Vann et al. 2008). In contrast, CBD (17.8 mg/kg, IV) increased the potency of THC (0.1 mg/kg, IV) in a drug-discrimination study using rhesus monkeys (McMahon 2016). CBD (30 mg/kg, IP) lengthened the time course of the discriminative stimulus effects of low doses of THC (0.3 and 1.0 mg/kg, IP) in rats (Hiltunen and Jarbe 1986). The one study, to our knowledge, that has examined the effect of CBD on THC self-administration found that CBD pretreatment did not impact THC IV self-administration across a range of CBD:THC dose ratios (Wakeford et al. 2017). In a mouse model of cannabinoid (CP,55,940) withdrawal, CBD blocked some somatic symptoms associated with withdrawal and normalized the cannabinoid withdrawal-induced increase in anxiety-like behavior (Navarrete et al. 2018).

Together, these preclinical findings suggest that while CBD's effects on THC reward and reinforcement are mixed, they generally point toward an enhancement; however, CBD may attenuate withdrawal- and anxiety-related behaviors. Further research is needed to clarify CBD's potential utility for treating CUD.

4.2 *Cannabinoid Receptor Competitive Antagonists and Biased Antagonists*

CB1 receptor antagonists and inverse agonists (e.g., rimonabant) attenuate a range of cannabinoid effects and early evidence has shown that repeated THC administration produces dependence (Aceto et al. 1995). Specifically, rimonabant decreases self-administration of IV THC (Fattore et al. 2001; Justinova et al. 2008), reduces motivation for cannabis vapor puffs (Freels et al. 2020), blocks THC's discriminative stimulus effects (Wiley et al. 2017), and reduces the conditioned rewarding effects of THC (Braida et al. 2001; Chaperon et al. 1998). However, with repeated THC administration, CB1 receptor antagonists/inverse agonists also precipitate cannabinoid withdrawal symptoms, including physical symptoms (i.e., wet dog shakes) and anxiety-like behavior (Tanda and Goldberg 2003; Aceto et al. 1996; Cook et al. 1998; Rubino et al. 1998) suggesting these compounds would have limited utility for patients with CUD. In addition, rimonabant produced serious adverse effects in a clinical trial (described below).

An alternative approach is to alter, rather than block, the effects of cannabis at the CB1 receptor. The steroid hormone pregnenolone alters the effect of CB1 receptor agonists in a signaling-specific manner by binding to a specific intracellular site on the CB1 receptor, distinct from the binding site of THC and other CB1 receptor agonists (Vallée et al. 2014). As pregnenolone is rapidly metabolized, Aelis Pharma developed a new pharmacological agent, AEF0117, which is a non-metabolizable pregnenolone derivative. AEF0117 was shown to antagonize a range of THC-related behaviors in animals, including hypolocomotion and hyperphagia (i.e., increased appetite), as well as to reduce THC-induced dopamine release in the nucleus accumbens. Further, pretreatment with AEF0117 reduced self-administration of the CB1 receptor agonist WIN55,212-2 in mice and reduced self-administration of THC in squirrel monkeys (Haney et al. 2023). AEF0117 also reduced reinstatement of THC seeking in monkeys after THC-reinforced responding was extinguished (Haney et al. 2023). Unlike rimonabant, AEF0117 alone produced none of the behavioral effects associated with CB1 antagonists that likely contribute to their poor tolerability, such as reduced food intake and increased anxiety-like behavior.

4.3 Other Candidate Medication Agents

A potential therapeutic strategy to reduce symptoms of cannabis withdrawal is to increase endocannabinoid levels by targeting FAAH or MAGL, the enzymes responsible for the degradation of the endogenous cannabinoid ligands AEA and 2-AG, respectively. In animals, acute administration of either a FAAH inhibitor (URB597) or MAGL inhibitor (JZL184) significantly attenuated rimonabant-precipitated withdrawal signs (Schlosburg et al. 2009). In contrast to its promising effects on withdrawal, URB597 did not significantly modify the discriminative stimulus effects of THC (Stewart and McMahon 2011), nor did it alter the reinforcing effects of THC in nonhuman primates (Justinova et al. 2008).

Preclinical studies have shown that the mixed opioid antagonist naltrexone can reduce the discriminative stimulus effects of THC. Neither norbinaltorphimine (a selective kappa-opioid receptor antagonist) nor naltrindole (a selective delta-opioid antagonist) had any effect, suggesting that THC's interoceptive effects are specifically mediated by mu-opioid receptor activity (Solinas and Goldberg 2005). However, other preclinical studies using drug discrimination have not shown an effect of naltrexone (Wakley and Craft 2011), indicating opioid-receptor antagonist modulation of the discriminative-stimulus effects of cannabinoids may be complex and depend on multiple factors (e.g., antagonist dose and cannabinoid history).

Opioid antagonists have also shown promise in reducing the reinforcing effects of cannabinoids. Naloxone reduced self-administration of CB1 receptor agonists in mice (Navarro et al. 2001) and naltrexone reduced IV THC self-administration in squirrel monkeys (Justinova et al. 2004). Naloxone also reduced CPP for a CB1 receptor agonist (Braida et al. 2001). The mechanism of opioid-receptor antagonist effects on reducing cannabinoid reward and reinforcement appears to occur via blocking CB1 receptor activation-induced dopamine release in mesolimbic areas, as shown by in vivo microdialysis studies (Tanda et al. 1997). However, naloxone was also able to precipitate withdrawal symptoms in rats chronically exposed to cannabinoids (Navarro et al. 2001), although this was not replicated in human studies (see below).

5 Human Laboratory Models of Cannabis Use Disorder

Like preclinical studies, key features of CUD can be modeled in placebo-controlled human laboratory studies enrolling nontreatment-seeking volunteers: cannabis's abuse-related subjective effects, cannabis self-administration, and cannabis withdrawal. These models allow for researchers to control cannabis administration (e.g., varying doses, blinding procedures, and cued, or ad-libitum smoking protocols), assess discrete behavioral aspects of cannabis use (e.g., self-administration, withdrawal symptoms, and positive subjective effects), and test novel pharmacotherapies for CUD while controlling for naturalistic cannabis and other drug use. The following section briefly reviews human laboratory models that examine discrete behaviors relevant to CUD.

5.1 *Positive Subjective Effects*

Abuse liability is defined as the potential of a drug, in this case cannabis, to elicit positive subjective effects that may reinforce use (FDA 2017). Early research showed that among a sample of individuals who regularly use cannabis, administration of smoked cannabis (1.8–3.6% THC) or oral THC (2.5–30 mg) increased ratings such as "good drug effect," "high" "liking," and "willingness to take cannabis again," compared to placebo (Chait and Zacny 1992; Haney et al. 1999a, b). Repeated daily exposure to active cannabis or THC administration resulted in diminished subjective effects, suggesting the development of tolerance to the abuse-related effects of both smoked cannabis and oral THC (Haney et al. 1999a, b; Haney et al. 1997a). Tolerance to the effects of cannabis may lead to increased consumption as individuals seek to achieve the desired effects that were previously experienced with lower levels of use. Given that positive subjective responses have been shown to correspond with the development of CUD (Scherrer et al. 2009; Zeiger et al. 2010), one approach to developing a pharmacotherapy for CUD is to reduce cannabis's positive effects.

5.2 *Positive Reinforcing Effects: Cannabinoid Self-Administration*

The choice to self-administer drugs is influenced by both the nature of available alternative reinforcers, such as social interaction, food, or monetary rewards, and the relative cost of those reinforcers. In order to assess the contribution of these factors on cannabis self-administration, researchers have manipulated the THC concentration and "cost" of an alternative reinforcer. For example, in a 16-day residential study among individuals who use cannabis, Haney and colleagues (1997b) showed that THC (0.0, 2.2, or 3.9% w/w) was the essential reinforcing component of cannabis self-administration. Moreover, findings from this study showed that cannabis self-administration can be influenced by monetary contingencies, with participants opting not to smoke cannabis when an opportunity to earn additional money was presented (Haney et al. 1997b). Further research from this laboratory supports these findings, demonstrating that, in a separate sample, the opportunity to earn money served as an effective alternative to cannabis self-administration (Ward et al. 1997). Recent work examining the effects of oral THC maintenance on cannabis self-administration showed that under forced-choice (drug vs. money) conditions, active oral THC significantly reduced self-administration (5.7% THC) (Schlienz et al. 2018). Findings from these studies demonstrates that the positive-reinforcing effects of cannabis are malleable and sensitive to changes in THC concentrations and contingencies to earn money.

5.3 *Cannabis Withdrawal*

There are also negative reinforcing effects of cannabis that may maintain cannabis use, particularly among those who use daily (Cooper and Haney 2008). When cannabis use is abruptly reduced or ceased, withdrawal symptoms may begin within 24 h, typically peak between 2 and 6 days and last up to 1–2 weeks. Cannabis withdrawal comprises negative mood (e.g., irritability, anger, aggression, and anxiety), sleep difficulty, physical symptoms (e.g., sweating, fever, chills, headache, shakiness/tremors, and abdominal pain) and decreased food intake (Budney et al. 2001, 2003, 2007; Haney et al. 1999a, 2004; Hart et al. 2002). Administering smoked cannabis or subthreshold doses of oral THC during abstinence alleviates withdrawal symptoms (Haney et al. 1999a, b; Hart et al. 2002). These findings highlight the clinical significance of cannabis withdrawal symptoms in the development of potential pharmacotherapies, as they often serve as a precursor to cannabis relapse and drive persistent patterns of use.

6 Medication Effects: Human Laboratory Models and Clinical Trials

Although there are no FDA-approved medications to treat CUD, several medication approaches have shown some promise in human laboratory studies. In the following sections, we briefly review the effects of medications on factors likely to contribute to the maintenance of daily cannabis use, organized by outcome and pharmacological targets (Table 1). For a more detailed review of pharmacotherapy candidates for CUD, see (Alayoubi et al. 2024; Brezing and Levin 2018).

6.1 Abstinence Initiation: Reducing the Positive Subjective and Reinforcing Effects of Cannabis

6.1.1 Cannabidiol

One approach to facilitating drug treatment is by reducing the positive reinforcing effects of the substance. CBD, a non-intoxicating constituent of the cannabis plant, has been tested using this approach. Although early studies showed that oral CBD reduced oral THC's subjective effects (Zuardi et al. 1982; Dalton et al. 1976), subsequent studies have reported no such effect (Juckel et al. 2007; Roser et al. 2009; Morgan et al. 2010; Hindocha et al. 2015). Haney and colleagues (2016) conducted a study to determine whether oral CBD pretreatment reduced the positive subjective and reinforcing effects of smoked cannabis. Yet, among nontreatment-seeking individuals who frequently used cannabis (>4 times per week), cannabis's positive subjective and reinforcing effects did not vary as a function of CBD pretreatment (200 mg, 400 mg, 800 mg) relative to placebo (Haney et al. 2016). These data are consistent with the negative findings in preclinical studies (Table 1) and suggest that acute, oral CBD administration may have limited therapeutic utility as a pharmacotherapy for CUD.

By contrast, a phase 2a, double-blind, placebo-controlled, randomized, adaptive Bayesian trial showed that maintenance on CBD (800 mg), rather than pretreatment, significantly reduced cannabis use across a 4-week period compared to placebo (Freeman et al. 2020). Notably, the treatment effects observed in this study were small: participants used 0.27–0.48 fewer days/week in the CBD condition relative to placebo, so the clinical relevance of this finding is unclear; the authors note the study was not designed to estimate efficacy. Further studies are needed.

6.1.2 Cannabinoid Receptor Competitive Antagonists and Biased Antagonists

Reducing cannabis use by directly blocking or altering its rewarding effects at the cannabinoid receptor is another treatment option for CUD. Rimonabant, a selective orthosteric CB1 receptor antagonist (Rinaldi-Carmona et al. 1994, 1995), had shown promising clinical results for weight loss and metabolic disorders (Pi-Sunyer et al. 2006) and tobacco smoking cessation (Cahill and Ussher 2011). Human laboratory studies showed that rimonabant, when administered in a single high dose or repeated low doses, attenuated the subjective effects of smoked cannabis (Huestis et al. 2001, 2007). However, rimonabant has been associated with serious adverse psychiatric effects (e.g., anxiety, depression, and suicidality) and was removed from the market, precluding further investigation of its use as a treatment for CUD.

As described above, AEF0117 is a pregnenolone derivative that alters the effects of CB1 receptor agonists by binding to an intracellular site on the CB1 receptor (distinct from THC's binding site) that does not produce the myriad adverse effects of competitive antagonists like rimonabant. In a phase 2a trial among nontreatment-seeking volunteers with CUD, AEF0117 significantly reduced cannabis's positive subjective effects by 19% (0.6 mg) and 38% (1 mg) and reduced cannabis self-administration (1 mg) compared to placebo (Haney et al. 2023). These findings, suggesting that AEF0117 may be a tolerable and an efficacious treatment option for CUD, led to a large multisite phase 2b clinical trial for the treatment of CUD (results not yet published).

6.1.3 Other Candidate Medication Agents

Naltrexone is an opioid receptor antagonist that is FDA-approved for the treatment of opioid use disorder and alcohol use disorder. As discussed above, opioid antagonists reduce the reinforcing effects of CB1 receptor agonists in preclinical studies (Braida et al. 2001; Justinova et al. 2004), suggesting that naltrexone may hold some promise as an effective pharmacotherapy for CUD. Although acute naltrexone pretreatment has been shown to increase the positive subjective effects of cannabis in individuals who use cannabis (Haney 2007; Cooper and Haney 2010), naltrexone maintenance (50 mg) over 4–6 weeks significantly reduced active cannabis self-administration and its positive subjective cannabis effects (Haney et al. 2015). An open-label clinical trial testing injectable naltrexone (380 mg, IM) demonstrated the safety and feasibility of this approach in CUD patients (Notzon et al. 2018). Future placebo-controlled randomized clinical trials are needed to determine the efficacy of naltrexone as a treatment for CUD.

Tiagabine, a GABA reuptake inhibitor, has also been examined for treating CUD. Wesley and colleagues (Wesley et al. 2018) showed that tiagabine (12 mg) did not significantly affect most of cannabis's subjective effects or cannabis self-administration and, in some, worsened measures of sleep and cannabis craving. The authors concluded these findings do not support the continued evaluation of tiagabine as a pharmacotherapy for CUD. Similarly, pregabalin, another anticonvulsant medication, did not reduce cannabis self-administration relative to placebo in the human laboratory (Lile et al. 2021).

6.1.4 Cannabinoid Receptor Agonists

Dronabinol (oral THC), a cannabinoid receptor agonist, has been examined as a pharmacotherapy for CUD in the context of ongoing cannabis use. Schlienz and colleagues (Schlienz et al. 2018) tested whether high doses of dronabinol (120–240 mg) impacted subjective cannabis effects and rates of smoked cannabis self-administration under progressive ratio conditions. They showed that maintenance on dronabinol significantly increased self-reported ratings of "feel a good drug effect" compared to placebo. Moreover, both doses of dronabinol reduced cannabis withdrawal, craving, and self-administration, compared to placebo (Schlienz et al. 2018). To note, the dronabinol doses used within this study were substantially higher than those tested in other studies that failed to shift cannabis use with dronabinol under abstinent conditions (see below).

6.2 *Relapse Prevention: Reducing the Negative Reinforcing Effects of Cannabis*

6.2.1 Cannabinoid Receptor Agonists

Both inpatient and outpatient studies among nontreatment-seeking samples have shown that dronabinol administered during cannabis abstinence dose-dependently decreased withdrawal symptoms, relative to placebo (Schlienz et al. 2018; Haney et al. 2004; Budney et al. 2007; Vandrey et al. 2013). Yet, in a laboratory model of "relapse" (operationally defined as a return to cannabis self-administration at financial cost after a period of abstinence), dronabinol (60 mg) reduced symptoms of cannabis withdrawal without altering cannabis relapse (Haney et al. 2008). Similarly, dronabinol (40 mg) also reduced cannabis withdrawal symptoms and improved retention in patients seeking treatment for their cannabis use, yet did not significantly alter rates of continuous (2 weeks) abstinence relative to placebo (dronabinol: 17.7%; placebo: 15.6%; (Levin et al. 2011)).

Dronabinol has also been tested in combination with lofexidine, an α2-adrenergic receptor agonist, based on preclinical data showing that α2-adrenergic receptor agonists decrease symptoms of precipitated cannabis withdrawal (Lichtman and Martin 2002). In the human laboratory, dronabinol (60 mg) combined with lofexidine (2.4 mg) decreased both symptoms of withdrawal and cannabis "relapse" (Haney et al. 2008). However, in the clinic, dronabinol (60 mg) combined with lofexidine (1.8 mg) failed to increase rates of continuous (3 weeks) abstinence relative to placebo (Levin et al. 2016). To note, unlike the laboratory study, many patients in the clinical study were unable to tolerate the target medication doses, which may have contributed to the discordant findings.

Other cannabinoid receptor agonists such as nabilone and nabiximols have been investigated as treatments for CUD. Nabilone, a synthetic cannabinoid analogue of THC with greater bioavailability, a longer duration of action, and clearer dose-linearity than dronabinol, may be a superior medication (Balter and Haney 2017; Bedi et al. 2013). Nabilone, alone and in combination with zolpidem (a GABAergic receptor agonist used to treat insomnia), has been shown to reverse characteristic symptoms of cannabis withdrawal, including worsened mood, disrupted sleep, and decreased food intake, and to decrease cannabis "relapse" (Haney et al. 2013a; Herrmann et al. 2016). Nabilone has not yet been tested in a clinical population.

Nabiximols, an oromucosal spray composed of a 1:1 ratio of THC (2.7 mg/actuation) and CBD (2.5 mg/actuation), reduces cannabis withdrawal symptoms in both treatment-seeking (Allsop et al. 2014) and nontreatment-seeking samples (Trigo et al. 2016). However, there have been inconsistent findings regarding the effectiveness of nabiximols to reduce cannabis use. Initial work by Allsop and colleagues (Allsop et al. 2014) found that nabiximols (86.4 mg THC: 80 mg CBD per day) did not promote long-term reductions in weekly cannabis use following medication cessation. More recent work has shown that nabiximols in combination with a psychosocial intervention significantly reduced the number of days cannabis was used over a 12-week period (days of cannabis use: 53 days with placebo vs 35 days with nabiximols; (Lintzeris et al. 2019)). Overall, these studies suggest that CB1 receptor agonist medications show promise as potential treatments for CUD by consistently reducing symptoms of cannabis withdrawal and reducing measures of cannabis use in the laboratory and in the clinic under certain conditions.

6.2.2 Serotonergic Agents

Buspirone is a $5HT_{1A}$ partial agonist that is commonly used to treat anxiety. A clinical study evaluating the efficacy of buspirone (60 mg) as a treatment for CUD has shown that this medication was not more efficacious than placebo in reducing cannabis use (McRae-Clark et al. 2015). More recently, a human laboratory study testing lorcaserin (10 mg), a serotonin-2c ($5HT_{2c}$) agonist, in a nontreatment-seeking sample, showed that it significantly reduced cannabis craving during abstinence and reduced a laboratory measure of cannabis relapse compared to placebo (Arout et al. 2021). In a 10-week, outpatient, open-label clinical trial in treatment seekers, lorcaserin (target dose of 20 mg extended-release daily) reduced the average frequency of cannabis use days per week by 1.74 days relative to baseline, suggesting the medication warrants further controlled study (Brezing et al. 2024). Notably, in 2020, the FDA issued a Drug Safety Communication that requested lorcaserin be withdrawn due to its potential to increase cancer risk (Sharretts et al. 2020), leaving the future unclear for $5HT_{2c}$-receptor agonists for the treatment of CUD.

6.2.3 GABAergic Agents

Zolpidem (12.5 mg; extended-release), a non-benzodiazepine $GABA_A$ receptor agonist, reduced cannabis withdrawal-related disruptions in sleep (Herrmann et al. 2016; Vandrey et al. 2011). Zolpidem alone did not decrease cannabis relapse in the laboratory setting, although zolpidem combined with nabilone (6 mg/day) did decrease cannabis relapse (Herrmann et al. 2016). Similarly, a 12-week clinical trial in treatment seekers showed that zolpidem-extended release (12.5 mg) attenuated abstinence-induced sleep disturbance early in treatment but did not affect cannabis abstinence rates compared to placebo (Lee et al. 2024). These findings suggest zolpidem alone may not be an effective pharmacotherapy for treating CUD.

Baclofen, a $GABA_B$ receptor agonist used to treat spasticity and muscle pain, and divalproex, an anticonvulsant commonly used to prevent and control seizures associated with epilepsy, have both been examined as potential treatments for CUD. Neither baclofen (60 mg, 90 mg/day) nor divalproex (1,500 mg/day) reduced symptoms of cannabis withdrawal in the human laboratory (Haney et al. 2004, 2010); divalproex also did not alter cannabis "relapse" (Haney et al. 2004). Moreover, divalproex (1,500 mg/day) did not improve rates of abstinence in a randomized clinical trial (RCT) (Levin et al. 2004).

6.2.4 Other Candidate Medication Agents

Other candidate medications agents with varied mechanisms of action have been examined as potential pharmacotherapy for CUD, including a FAAH inhibitor (PF-04457845), celecoxib, guanfacine, bupropion, mirtazapine, nefazodone, quetiapine, and varenicline.

The rationale for testing a FAAH-inhibitor and the COX-2 inhibitor, celecoxib, was to see if potentiating endocannabinoid levels through inhibition of their enzymatic metabolism would reduce cannabis withdrawal and use. D'Souza and colleagues (D'Souza et al. 2019) showed that the long-acting, time-dependent, and selective FAAH inhibitor, PF-04457845 (4 mg), reduced cannabis withdrawal symptoms and lowered self-reported cannabis use at 4 weeks in men. These data supported a recent multi-site randomized clinical trial for CUD; results have not yet been published. Celecoxib, used for osteoarthritis and rheumatoid arthritis, was also hypothesized to increase endocannabinoid levels and thereby reduce cannabis withdrawal and "relapse" in the human laboratory. However, celecoxib (200 mg BID) did not show significant effects on cannabis withdrawal or "relapse," and in fact, increased cannabis craving compared to placebo, suggesting it does not hold promise for the treatment of CUD (Haney et al. 2022).

Antidepressants have also been explored as potential treatments for CUD. These include bupropion, an atypical antidepressant that inhibits norepinephrine and dopamine reuptake and is FDA-approved for tobacco smoking cessation; mirtazapine, a tetracyclic antidepressant that antagonizes the adrenergic α2-autoreceptors and -heteroreceptors and blocks serotonin 5-HT_2 and 5-HT_3 receptors; nefazodone, which antagonizes 5-HT_2 receptors and inhibits 5-HT uptake, serotonin; and atomoxetine, a selective norepinephrine reuptake inhibitor commonly prescribed for attention deficit hyperactivity disorder (ADHD). Human laboratory studies have found mixed results regarding the effects of these antidepressant medications on cannabis withdrawal symptoms and self-administration. Among nontreatment-seeking samples, bupropion-sustained release (SR) (300 mg) worsened withdrawal symptoms (Haney et al. 2001), while nefazodone (450 mg) decreased ratings of some cannabis withdrawal symptoms (e.g., "anxious" and "muscle pain") (Haney et al. 2003), and mirtazapine (30 mg) improved sleep and food intake during cannabis abstinence but otherwise did not reduce other symptoms of withdrawal, or reduce "relapse" (Haney et al. 2010). Similarly, neither bupropion-SR (150 mg BID) nor nefazodone (300 mg BID) improved symptoms of cannabis withdrawal or reduced cannabis use in a clinical trial (Carpenter et al. 2009). Moreover, studies have shown that atomoxetine results in clinically significant gastrointestinal side effects and does not have a promising role in the treatment of CUD (Tirado et al. 2008; McRae-Clark et al. 2010).

More recently, studies have examined guanfacine, an α2A-adrenergic receptor agonist that is FDA-approved for the treatment of ADHD and hypertension (Posey and Mcdougle 2007). In nontreatment-seeking cannabis smokers, Haney and colleagues (2019) found that immediate-release guanfacine (2 mg) at bedtime significantly reduced ratings of irritability and improved objective sleep measures during cannabis abstinence relative to placebo but did not affect cannabis "relapse". However, in a separate sample of nontreatment-seeking cannabis smokers, immediate-release guanfacine (3 mg) did not show significant effects on withdrawal, craving, or sleep (Holst et al. 2019). An outpatient open-label pilot trial showed that extended-release guanfacine, gradually titrated to a dose of 4 mg (or highest dose tolerated), significantly reduced the average daily amount of cannabis use and cannabis use days over the course of 8 weeks (Dakwar et al. 2020). Taken together, there are signals that guanfacine, particularly the extended-release formulation, may have promise for CUD, but further study is needed.

Quetiapine is an atypical antipsychotic that is commonly used to treat schizophrenia and bipolar disorder and as an adjunct treatment for major depressive disorder (Daly and Trivedi 2007; Riedel et al. 2007). Quetiapine is an antagonist at serotonin type 2A, dopamine type 2, histamine type 1, and adrenergic receptors. A human laboratory study showed that quetiapine (200 mg) improved sleep quality, increased caloric intake, and decreased weight loss during cannabis abstinence. Quetiapine also increased cannabis craving and self-administration, suggesting it does not show promise as a pharmacotherapy for CUD (Cooper et al. 2013). However, there are intriguing clinical findings at a higher dose that suggest quetiapine may warrant further study. A 12-week randomized double-blind placebo-controlled trial showed that quetiapine (300 mg nightly) increased the likelihood of a transition from heavy cannabis use (5–7 days per week) to moderate use (2–4 days per week) (Mariani et al. 2014). The authors note that the clinical significance of these results was likely to be modest, and future trials should consider examining the combined effects of quetiapine and behavioral interventions (e.g., contingency management).

Researchers have also examined varenicline, an nAChR partial agonist of the α4β2 subtype and a full agonist of the α7 subtype, which is an FDA-approved medication for tobacco smoking cessation (Gonzales et al. 2006; Eisenberg et al. 2008), as an approach to treat CUD. In a nontreatment-seeking sample of individuals who co-used tobacco and cannabis, varenicline (1 mg BID) alone facilitated tobacco cessation and lessened negative mood but only reduced symptoms of cannabis withdrawal when combined with nabilone (4 mg BID) (Herrmann et al. 2019). In a placebo-controlled 6-week pilot trial, varenicline (1 mg BID), added to a brief MET intervention, significantly decreased self-reported cannabis withdrawal, percentage of days used, and use sessions per day during treatment compared to baseline (McRae-Clark et al. 2021). This preliminary data provide support for further investigation of varenicline in larger clinical trials.

7 Summary

Research on the development of pharmacotherapies for CUD has lagged relative to that of most other SUDs. One challenge in this field has been establishing robust preclinical models of cannabinoid self-administration, which impacts the translational pipeline for novel targets. Existing preclinical models with cannabis or cannabinoids, adapted from those used for drugs such as cocaine or opioids, do not recapitulate key behavioral characteristics of a model of addictive behavior (e.g., escalation, high levels of intake). This is an evolving field, however, with continuing advances in preclinical models of vaporized cannabis administration showing promise relative to IV self-administration models better suited to fast-acting, non-lipophilic drugs.

Human laboratory models, using smoked or vaporized cannabis, can capture key features of CUD, including cannabis's abuse-related subjective effects, cannabis withdrawal, and cannabis self-administration under conditions of abstinence (a model of "relapse"), and under non-abstinent conditions (a model of abstinence initiation). A wide range of medications have been tested using these models, with a wide range of outcomes, i.e., increased and decreased cannabis withdrawal and cannabis self-administration.

The most essential value of preclinical and human laboratory models focusing on developing pharmacotherapies for CUD is their predictive validity (Haney and Spealman 2008). Although defining predictive validity is difficult in lieu of a medication with demonstrated efficacy, there is evidence of consistent effects from preclinical through human laboratory to clinic. One such example is AEF0117, which reduced cannabis self-administration in non-human primates and nontreatment-seeking cannabis smokers (Haney et al. 2023) and reduced cannabis use (without promoting abstinence) in treatment-seeking patients with moderate CUD (in preparation).

Discrepancies between preclinical and human laboratory models and RCTs also provide valuable opportunities for both backward and forward translation to improve the predictive validity of these models. For example, in the human laboratory, the combination of dronabinol and lofexidine significantly reduced cannabis withdrawal and "relapse" (Haney et al. 2008), but failed to produce sustained cannabis abstinence in an RCT, the primary endpoint (Levin et al. 2011). As noted, patients in the RCT could not tolerate the doses deemed efficacious in the inpatient human laboratory, indicating that human laboratory models need to prioritize testing medication conditions (e.g., dose, frequency of dose administration) that are feasible and well-tolerated in an outpatient setting. Interestingly, a secondary analysis of the RCT showed that dronabinol and lofexidine did not promote abstinence but did significantly shift CUD patients in the clinic from heavy to moderate cannabis smokers, demonstrating some consistency with the human laboratory findings when similar endpoints are evaluated (Levin et al. 2016).

This finding highlights the important possibility that some positive human laboratory findings do not parallel RCTs because the latter's definition of sustained abstinence from cannabis may be a problematic definition of medication efficacy. A significant proportion of those seeking treatment for their cannabis use are not seeking abstinence, so it is no surprise that every well-powered RCT requiring sustained abstinence from cannabis has failed. The primary definition of efficacy does not reflect the goal of many patients seeking treatment for their cannabis use.

In fact, there are important signals of efficacy in clinical trials looking at reductions in cannabis use rather than abstinence as the primary endpoints. Quetiapine shifted heavy cannabis use (5–7 days/week) to moderate cannabis use (2–4 days/week; (Mariani et al. 2021)). Similarly, as mentioned above, large RCTs for CUD that failed to demonstrate sustained abstinence showed that starting at week 5 of medication maintenance, patients shifted from heavy use (5–7 days/week) to moderate use (2–4 days/week) relative to placebo (Levin et al. 2016). Lintzeris and colleagues (2019) similarly showed that nabiximols, an oral mucosal spray containing THC and CBD, reduced the number of days cannabis was used out of the 84-day study from 53 days in the placebo group to 35 days in the nabiximols group. Finally, maintenance on CBD alone also reduced cannabis use more than placebo (although the effect was small: < half a day/week (Freeman et al. 2020).

Patients are still using a significant amount of cannabis in these studies, albeit less than in the placebo condition, so understanding the clinical significance of these reductions in use is essential (Loflin et al. 2020). A recent secondary analysis of several RCTs, seeking to understand how much of a reduction in cannabis use is needed to see improvements in functional outcomes, showed that reductions of 50% in cannabis use days and 75% in cannabis use amounts (episodes or grams of use) were associated with reductions in cannabis-related problems and improvements in clinician-rated severity (McClure et al. 2024). Showing that a substantial reduction in cannabis use has clinically meaningful consequences, in lieu of abstinence, and provides clear and testable hypotheses for future RCTs.

Overall, there have been a vast range of outcome measures in CUD studies (Lee et al. 2019), highlighting the need for standardized measurements outcomes. A recent international consortium that sought to identify a global consensus on standard measurement for outcomes in addiction (Black et al. 2024) proposed a series of patient-centered outcome measures for substance-related disorders that may have particular relevance to a more bottom-up approach to cannabis treatment, i.e., being guided by what patients consider a meaningful change in their cannabis use rather than having clinicians dictate what an effective treatment looks like. This consortium, which included individuals with lived experience, recommended specific quality of life ratings, i.e., Substance Use Recovery Evaluator (Neale et al. 2016), that capture issues of particular relevance to substance use (e.g., coping and social support).

Regulatory (e.g., FDA) and funding bodies (e.g., NIDA) appear open to a discussion of non-abstinent endpoints (Kiluk et al. 2019; Compton and Volkow 2024). The FDA has approved a treatment for alcohol use disorder that targets a reduction in heavy drinking days (not abstinence) because there are clinically meaningful health outcomes to reduced heavy drinking. This is more difficult to demonstrate for cannabis, which has less physiological toxicity than alcohol. In lieu of objective, physiological endpoints, it is likely that outcomes will have to be subjective, i.e., patient satisfaction; improvements in quality of life, consistent with a movement toward patient-reported outcomes. Kiluk and colleagues (2019) make a compelling argument that measures that capture the negative psychosocial and health consequences of drug use, as opposed to overall functioning, are critical to capturing meaningful effects of reductions in use.

Additional Issues to Consider in Translating Preclinical and Human Laboratory Findings

Additional issues to consider when defining endpoints that define efficacy include: (1) *CUD severity*: Nontreatment-seeking individuals who use cannabis daily in human laboratory studies, almost by definition, do not view cannabis as causing significant problems in their lives, so they typically meet criteria for moderate, not severe, CUD. Yet patients seeking treatment are more likely to meet criteria for severe CUD, particularly if they are agreeing to take an experimental medication for weeks to months. Severe CUD would suggest that their cannabis use is more refractory and less responsive to interventions. On the other hand, there would be a low placebo response increasing the likelihood of a significant medication effect (see Brezing and Levin 2018); (2) *Gender*: Although not the focus of this review, sex differences exist in the response to cannabis and cannabinoids across species (Cooper and Craft 2018b) and to medications tested for the treatment of CUD (see Brezing and Levin 2018; Brezing et al. 2024); (3) *Co-morbidities*: Those with a psychiatric diagnosis (Budney et al. 2019), alcohol and other SUDs (Hayaki et al. 2016), including tobacco use disorder (Haney et al. 2013b; McClure et al. 2018), fare worse in cannabis treatment. Human laboratory studies typically exclude participants who meet criteria for alcohol use disorder, anxiety, and depression, for example, whereas not all clinical trials do so. This impacts how medications alter cannabis use; (4) *Patient goals*: Evaluating the treatment goal of patients (abstinence versus reduction in use) and changes in specific aspects of quality of life that are important to patients needs to be integrated into measures of efficacy. Defining efficacy as abstinence is problematic when it is not what many patients are seeking in treatment (McClure et al. 2024; Lee et al. 2019), increasing the likelihood that most RCTs for CUD result in a false negative. The secondary analysis of large RCTs, demonstrating that several medications reduce the number of cannabis use days and amount of cannabis used, support integration of these meaningful, self-reported metrics of reduction as endpoints in future trials (Levin et al. 2016, 2021).

To conclude, while CUD medication development remains in its early stages, the tools we have – combined with continued refinement of preclinical models and clinical research strategies – are paving the way for more effective treatments. By accelerating translational research and optimizing our experimental approaches, we can increase throughput and improve the likelihood of identifying viable pharmacotherapies for people with CUD.

References

Aceto MD, Scates SM, Lowe JA, Martin BR (1995) Cannabinoid precipitated withdrawal by the selective cannabinoid receptor antagonist, SR 141716A. Eur J Pharmacol 282:R1–R2. Available from: https://www.sciencedirect.com/science/article/pii/001429999500447S

Aceto DM, Scates MS, Lowe AJ et al (1996) Dependence on delta 9-tetrahydrocannabinol: studies on precipitated and abrupt withdrawal. J Pharmacol Exp Ther 278(3):1290–1295

Alayoubi M, Henry BA, Cahill CM, Cooper ZD (2024) Exploring novel pharmacotherapy candidates for cannabis use disorder: uncovering promising agents on the horizon by mechanism of action. Drugs. https://doi.org/10.1007/s40265-024-02098-1

Allsop DJ, Copeland J, Lintzeris N, Dunlop AJ, Montebello M, Sadler C et al (2014) Nabiximols as an agonist replacement therapy during cannabis withdrawal: a randomized clinical trial. JAMA Psychiatry 71:281–291

Arout CA, Cooper ZD, Reed SC, Foltin RW, Comer SD, Levin FR et al (2021) 5HT-2C agonist lorcaserin decreases cannabis self-administration in daily cannabis smokers. Addict Biol 26: e12993

Baglot SL, Hume C, Petrie GN, Aukema RJ, Lightfoot SHM, Grace LM et al (2021) Pharmacokinetics and central accumulation of delta-9-tetrahydrocannabinol (THC) and its bioactive metabolites are influenced by route of administration and sex in rats. Sci Rep:11

Balster RL, Prescott WR (1992) Δ9-tetrahydrocannabinol discrimination in rats as a model for cannabis intoxication. Neurosci Biobehav Rev 16:55–62. Available from: https://www.sciencedirect.com/science/article/pii/S014976340580051X

Balter RE, Haney M (2017) Chapter 85: The synthetic analog of Δ9-Tetrahydrocannabinol (THC): Nabilone. Pharmacology and clinical application. In: Preedy VR (ed) Handbook of cannabis and related pathologies. Academic, San Diego, pp 821–827. Available from: https://www.sciencedirect.com/science/article/pii/B9780128007563000995

Bansal S, Zamarripa AC, Spindle RT et al (2023) Evaluation of cytochrome P450-mediated cannabinoid-drug interactions in healthy adult participants. Clin Pharmacol Ther 114(3): 693–703

Bardo TM, Neisewander LJ (1986) Single-trial conditioned place preference using intravenous morphine. Pharmacol Biochem Behav 25(5):1101–1105

Bardo TM, Valone MJ, Bevins AR (1999) Locomotion and conditioned place preference produced by acute intravenous amphetamine: role of dopamine receptors and individual differences in amphetamine self-administration. Psychopharmacology 143(1):39–46

Beardsley MP, Balster LR, Harris SL (1986) Dependence on tetrahydrocannabinol in rhesus monkeys. J Pharmacol Exp Ther 239(2):311–319

Bedi G, Cooper ZD, Haney M (2013) Subjective, cognitive and cardiovascular dose-effect profile of nabilone and dronabinol in marijuana smokers. Addict Biol 18:872–881

Black N, Chung S, Tisdale C, Fialho LS, Aramrattana A, Assanangkornchai S et al (2024) An international, multidisciplinary consensus set of Patient-centered outcome measures for substance-related and addictive disorders. J Clin Med 13:2154

Boileau I, Mansouri E, Williams B, Le Foll B, Rusjan P, Mizrahi R et al (2016) Fatty acid amide hydrolase binding in brain of cannabis users: imaging with the novel radiotracer [11C]CURB. Biol Psychiatry 80:691–701

Borges CA, Duarte BR, Nogueira L et al (2015) Temporal and dose-dependent differences in simultaneously-induced cocaine hypervigilance and conditioned-place-preference in marmoset monkeys. Drug Alcohol Depend 148:188–194

Braida D, Pozzi M, Cavallini R et al (2001) Conditioned place preference induced by the cannabinoid agonist CP 55,940: interaction with the opioid system. Neuroscience 104(4):923–926

Braida D, Iosuè S, Pegorini S, Sala M (2004) Δ 9-tetrahydrocannabinol-induced conditioned place preference and intracerebroventricular self-administration in rats. Eur J Pharmacol 506:63–69

Breivogel CS, Childers SR, Deadwyler SA, Hampson RE, Vogt LJ, Sim-Selley LJ (1999) Chronic △9-tetrahydrocannabinol treatment produces a time-dependent loss of cannabinoid receptors

and cannabinoid receptor-activated G proteins in rat brain. J Neurochem 73:2447–2459. https://doi.org/10.1046/j.1471-4159.1999.0732447.x
Brezing CA, Levin FR (2018) The current state of pharmacological treatments for cannabis use disorder and withdrawal. Neuropsychopharmacology (Nature Publishing Group) 43:173–194
Brezing CA, Sibai M, Choi CJ, Mitra S, Mariani JJ, Naqvi N et al (2024) Open-label pilot of lorcaserin (a serotonin 2C-receptor agonist) for cannabis use disorder. J Subst Use 29:487–494
Britch SC, Wiley JL, Yu Z, Clowers BH, Craft RM (2017) Cannabidiol-Δ9-tetrahydrocannabinol interactions on acute pain and locomotor activity. Drug Alcohol Depend 175:187–197
Bruijnzeel AW, Qi X, Guzhva LV, Wall S, Deng JV, Gold MS et al (2016) Behavioral characterization of the effects of cannabis smoke and anandamide in rats. PLoS One 11
Buckner JD, Zvolensky MJ, Crosby RD, Wonderlich SA, Ecker AH, Richter A (2015) Antecedents and consequences of cannabis use among racially diverse cannabis users: an analysis from ecological momentary assessment. Drug Alcohol Depend 147:20–25
Budney AJ, Higgens ST, Radonovich KJ, Novy PL (2000) Adding voucher-based incentives to coping skills and motivational enhancement improves outcomes during treatment for marijuana dependence. J Consult Clin Psychol 68:1051–1061
Budney AJ, Hughes JR, Moore BA, Novy PL (2001) Marijuana abstinence effects in marijuana smokers maintained in their home environment. Arch Gen Psychiatry 58:917–924
Budney AJ, Moore BA, Vandrey RG, Hughes JR (2003) The time course and significance of cannabis withdrawal. J Abnorm Psychol 112:393–402
Budney AJ, Moore BA, Rocha HL, Higgins ST (2006) Clinical trial of abstinence-based vouchers and cognitive-behavioral therapy for cannabis dependence. J Consult Clin Psychol 74:307–316
Budney AJ, Vandrey RG, Hughes JR, Moore BA, Bahrenburg B (2007) Oral delta-9-tetrahydrocannabinol suppresses cannabis withdrawal symptoms. Drug Alcohol Depend 86:22–29
Budney AJ, Stanger C, Tilford JM, Scherer EB, Brown PC, Li Z et al (2015) Computer-assisted behavioral therapy and contingency management for cannabis use disorder. Psychol Addict Behav 29:501–511
Budney AJ, Sofis MJ, Borodovsky JT (2019) An update on cannabis use disorder with comment on the impact of policy related to therapeutic and recreational cannabis use. Eur Arch Psychiatry Clin Neurosci (Dr. Dietrich Steinkopff Verlag GmbH and Co. KG):73–86
Cahill K, Ussher MH (2011) Cannabinoid type 1 receptor antagonists for smoking cessation. Cochrane Database Syst Rev (John Wiley and Sons Ltd) 2011(3):CD005353
Callaghan RC, Sanches M, Kish SJ (2020) Quantity and frequency of cannabis use in relation to cannabis-use disorder and cannabis-related problems. Drug Alcohol Depend 217:108271. Available from: https://www.sciencedirect.com/science/article/pii/S0376871620304361
Carpenter KM, McDowell D, Brooks DJ, Cheng WY, Levin FR (2009) A preliminary trial: double-blind comparison of nefazodone, bupropion-SR, and placebo in the treatment of cannabis dependence. Am J Addict 18:53–64
Carroll KM, Easton CJ, Nich C, Hunkele KA, Neavins TM, Sinha R et al (2006) The use of contingency management and motivational/skills-building therapy to treat young adults with marijuana dependence. J Consult Clin Psychol 74(5):955–966
Caulkins JP (2024) Changes in self-reported cannabis use in the United States from 1979 to 2022. Addiction 119:1648–1652
Ceccarini J, Kuepper R, Kemels D, van Os J, Henquet C, Van Laere K (2015) [18F]MK-9470 PET measurement of cannabinoid CB1 receptor availability in chronic cannabis users. Addict Biol 20:357–367. https://doi.org/10.1111/adb.12116
Chait LD, Zacny JP (1992) Reinforcing and subjective effects of oral Ag-THC and smoked marijuana in humans. Psychopharmacology. https://doi.org/10.1007/BF02245145
Chan GCK, Becker D, Butterworth P, Hines L, Coffey C, Hall W et al (2021) Young-adult compared to adolescent onset of regular cannabis use: a 20-year prospective cohort study of later consequences. Drug Alcohol Rev 40:627–636
Chaperon F, Soubrie P, Puech JA et al (1998) Involvement of central cannabinoid (CB1) receptors in the establishment of place conditioning in rats. Psychopharmacology 135(4):324–332

Childs E, Wit de H (2009) Amphetamine-induced place preference in humans. Biol Psychiatry 65(10):900–904

Choi NG, Moore J, Choi BY (2024) Cannabis use disorder and substance use treatment among U.S. adults. J Subst Use Addict Treat 167:209486

Cohen K, Weizman A, Weinstein A (2019) Modulatory effects of cannabinoids on brain neurotransmission. Eur J Neurosci 50:2322–2345. https://doi.org/10.1111/ejn.14407

Collier DA, Echevarria JD (2013) The utility of the zebrafish model in conditioned place preference to assess the rewarding effects of drugs. Behav Pharmacol 24(5–6):375–383

Compton WM, Volkow ND (2024) Reductions in substance use as outcome targets for treatment development. Am J Psychiatry 181:947–948. Available from: http://www.ncbi.nlm.nih.gov/pubmed/39482950

Cook AS, Lowe AJ, Martin RB (1998) CB1 receptor antagonist precipitates withdrawal in mice exposed to Delta9-tetrahydrocannabinol. J Pharmacol Exp Ther 285(3):1150–1156

Cooper DZ, Craft MR (2018a) Sex-dependent effects of cannabis and Cannabinoids: a translational perspective. Neuropsychopharmacology 43(1):34–51

Cooper ZD, Craft RM (2018b) Sex-dependent effects of cannabis and Cannabinoids: a translational perspective. Neuropsychopharmacology (Nature Publishing Group) 43:34–51

Cooper ZD, Haney M (2008) Cannabis reinforcement and dependence: role of the cannabinoid CB1 receptor. Addict Biol:188–195

Cooper ZD, Haney M (2010) Opioid antagonism enhances marijuana's effects in heavy marijuana smokers. Psychopharmacology 211:141–148

Cooper ZD, Foltin RW, Hart CL, Vosburg SK, Comer SD, Haney M (2013) A human laboratory study investigating the effects of quetiapine on marijuana withdrawal and relapse in daily marijuana smokers. Addict Biol 18:993–1002

Cooper DZ, Evans MS, Foltin WR (2021) Self-administration of inhaled delta-9-tetrahydrocannabinol and synthetic cannabinoids in non-human primates. Exp Clin Psychopharmacol 29(2): 137–146

Copeland J, Swift W, Roffman R, Stephens R (2001) A randomized controlled trial of brief cognitive-behavioral interventions for cannabis use disorder. J Subst Abuse Treat 21(2):55–64

Corley C, Craig A, Sadek S, Marusich JA, Chehimi SN, White AM et al (2024) Enhancing translation: a need to leverage complex preclinical models of addictive drugs to accelerate substance use treatment options. Pharmacol Biochem Behav 243:1–28

Cunningham LC, Gremel MC, Groblewski AP (2006) Drug-induced conditioned place preference and aversion in mice. Nat Protoc 1(4):1662–1670

D'Souza D, Cortes-Briones JA, Ranganathan M, Thurnauer H, Creatura G, Surti T et al (2016a) rapid changes in CB1 receptor availability in cannabis dependent males after abstinence from cannabis. Biol Psychiatry Cogn Neurosci Neuroimaging. 1(1):60–67

D'Souza DC, Cortes-Briones JA, Ranganathan M, Thurnauer H, Creatura G, Surti T et al (2016b) Rapid changes in cannabinoid 1 receptor availability in cannabis-dependent male subjects after abstinence from cannabis. Biol Psychiatry Cogn Neurosci Neuroimaging 1:60–67. Available from: https://www.sciencedirect.com/science/article/pii/S2451902215000099

D'Souza DC, Cortes-Briones J, Creatura G, Bluez G, Thurnauer H, Deaso E et al (2019) Efficacy and safety of a fatty acid amide hydrolase inhibitor (PF-04457845) in the treatment of cannabis withdrawal and dependence in men: a double-blind, placebo-controlled, parallel group, phase 2a single-site randomised controlled trial. Lancet Psychiatry 6:35–45

Dakwar E, Mahony A, Choi CJ, Pavlicova M, Brooks D, Mariani JP et al (2020) Guanfacine extended-release for cannabis use disorder: a pilot feasibility trial. Am J Drug Alcohol Abuse 46:44–48

Dalton W, Martz R, Lemberger L, Rodda B, Forney R (1976) Influence of cannabidiol on delta-9-tetrahydrocannabinol effects. Clin Pharmacol Ther 19:300–309

Daly EJ, Trivedi MH (2007) A review of quetiapine in combination with antidepressant therapy in patients with depression. Neuropsychiatr Dis Treat. https://doi.org/10.2147/ndt.s1862

Deiana S, Fattore L, Spano SM et al (2007) Strain and schedule-dependent differences in the acquisition, maintenance and extinction of intravenous cannabinoid self-administration in rats. Neuropharmacology 52(2):646–654

Di Marzo V, Berrendero F, Bisogno T, González S, Cavaliere P, Romero J et al (2000) Enhancement of anandamide formation in the limbic forebrain and reduction of endocannabinoid contents in the striatum of Δ9-tetrahydrocannabinol-tolerant rats. J Neurochem 74:1627–1635. https://doi.org/10.1046/j.1471-4159.2000.0741627.x

Eckard LM, Trexler RK, Kotson TB et al (2020) Precipitated Delta9-THC withdrawal reduces motivation for sucrose reinforcement in mice. Pharmacol Biochem Behav 195:172966

Eisenberg MJ, Filion KB, Yavin D, Bélisle P, Mottillo S, Joseph L et al (2008) Pharmacotherapies for smoking cessation: a meta-analysis of randomized controlled trials. CMAJ Can Med Assoc J 179:135–144

Fattore L, Cossu G, Martellotta MC et al (2001) Intravenous self-administration of the cannabinoid CB1 receptor agonist WIN 55,212-2 in rats. Psychopharmacology 156(4):410–416

Fattore L, Spano MS, Altea S, Angius F, Fadda P, Fratta W (2007) Cannabinoid self-administration in rats: sex differences and the influence of ovarian function. Br J Pharmacol 152:795–804

FDA (2017) Assessment of abuse potential of drugs: guidance for industry. U.S. Food and Drug Administration. Available from: https://www.fda.gov/regulatory-information/search-fda-guidance-documents/assessment-abuse-potential-drugs

Flórez-Salamanca L, Secades-Villa R, Budney AJ, García-Rodríguez O, Wang S, Blanco C (2013) Probability and predictors of cannabis use disorders relapse: results of the National Epidemiologic Survey on Alcohol and Related Conditions (NESARC). Drug Alcohol Depend 132:127–133

Freels GT, Baxter-Potter NL, Lugo MJ et al (2020) Vaporized cannabis extracts have reinforcing properties and support conditioned drug-seeking behavior in rats. J Neurosci 40(9):1897–1908

Freels GT, Westbrook RS, Zamberletti E et al (2024) Sex differences in response-contingent cannabis vapor administration during adolescence mediate enduring effects on behavioral flexibility and prefrontal microglia activation in rats. Cannabis Cannabinoid Res 9(4):e1184–e1196

Freeman TP, Hindocha C, Baio G, Shaban NDC, Thomas EM, Astbury D et al (2020) Cannabidiol for the treatment of cannabis use disorder: a phase 2a, double-blind, placebo-controlled, randomised, adaptive Bayesian trial. Lancet Psychiatry 7:865–874

Gardner LE, Paredes W, Smith D et al (1988) Facilitation of brain stimulation reward by delta 9-tetrahydrocannabinol. Psychopharmacology 96(1):142–144

Glodosky CN, Cuttler C, Freels GT et al (2020) Cannabis vapor self-administration elicits sex- and dose-specific alterations in stress reactivity in rats. Neurobiol Stress 13:100260

Gonzales D, Rennard SI, Nides M, Oncken C, Azoulay S, Billing CB et al (2006) Varenicline, an α4β2 nicotinic acetylcholine receptor partial agonist, vs sustained-release bupropion and placebo for smoking Cessation A randomized controlled trial. JAMA 296:47–55. https://doi.org/10.1001/jama.296.1.47

González S, Fernández-Ruiz J, Di MV, Hernández M, Arévalo C, Nicanor C et al (2004) Behavioral and molecular changes elicited by acute administration of SR141716 to Δ9-tetrahydrocannabinol-tolerant rats: an experimental model of cannabinoid abstinence. Drug Alcohol Depend 74:159–170. Available from: https://www.sciencedirect.com/science/article/pii/S0376871604000055

Haney M (2007) Opioid antagonism of cannabinoid effects: differences between marijuana smokers and nonmarijuana smokers. Neuropsychopharmacology 32:1391–1403

Haney M, Spealman R (2008) Controversies in translational research: drug self-administration. Psychopharmacology 199:403–419

Haney M, Comer SD, Foltin RW, Fischman MW (1997a) Behavioral contingencies modulate alprazolam self-administration by humans. Behav Pharmacol 8. Available from: https://journals.lww.com/behaviouralpharm/fulltext/1997/02000/behavioral_contingencies_modulate_alprazolam.9.aspx

Haney M, Comer SD, Ward AS, Foltin RW, Fischman MW (1997b) Factors influencing marijuana self-administration by humans. Behav Pharmacol 8:101–112. Available from: https://www.safetylit.org/citations/index.php?citationIds%5B%5D=citjournalarticle_253118_38

Haney M, Ward AS, Comer SD, Foltin RW, Fischman MW (1999a) Abstinence symptoms following smoked marijuana in humans. Psychopharmacology. https://doi.org/10.1007/s002130050849

Haney M, Ward AS, Comer SD, Foltin RW, Fischman MW (1999b) Abstinence symptoms following oral THC administration to humans. Psychopharmacology. https://doi.org/10.1007/s002130050848

Haney M, Ward AS, Comer SD, Hart CL, Foltin RW, Fischman MW (2001) Bupropion SR worsens mood during marijuana withdrawal in humans. Psychopharmacology 155:171–179

Haney M, Hart CL, Ward AS, Foltin RW (2003) Nefazodone decreases anxiety during marijuana withdrawal in humans. Psychopharmacology 165:157–165

Haney M, Hart CL, Vosburg SK, Nasser J, Bennett A, Zubaran C et al (2004) Marijuana withdrawal in humans: effects of Oral THC or divalproex. Neuropsychopharmacology 29:158–170

Haney M, Hart CL, Vosburg SK, Comer SD, Reed SC, Foltin RW (2008) Effects of THC and lofexidine in a human laboratory model of marijuana withdrawal and relapse. Psychopharmacology 197:157–168

Haney M, Hart CL, Vosburg SK, Comer SD, Reed SC, Cooper ZD et al (2010) Effects of baclofen and mirtazapine on a laboratory model of marijuana withdrawal and relapse. Psychopharmacology 211:233–244

Haney M, Cooper ZD, Bedi G, Vosburg SK, Comer SD, Foltin RW (2013a) Nabilone decreases marijuana withdrawal and a laboratory measure of marijuana relapse. Neuropsychopharmacology 38:1557–1565

Haney M, Bedi G, Cooper ZD, Glass A, Vosburg SK, Comer SD et al (2013b) Predictors of marijuana relapse in the human laboratory: robust impact of tobacco cigarette smoking status. Biol Psychiatry 73:242–248

Haney M, Ramesh D, Glass A, Pavlicova M, Bedi G, Cooper ZD (2015) Naltrexone Maintenance Decreases Cannabis Self-Administration and Subjective Effects in Daily Cannabis Smokers. Neuropsychopharmacology: official publication of the American College of Neuropsychopharmacology 40(11):2489–2498. https://doi.org/10.1038/npp.2015.108

Haney M, Malcolm RJ, Babalonis S, Nuzzo PA, Cooper ZD, Bedi G et al (2016) Oral Cannabidiol does not Alter the subjective, reinforcing or cardiovascular effects of smoked cannabis. Neuropsychopharmacology 41:1974–1982

Haney M, Cooper ZD, Bedi G, Herrmann E, Comer SD, Reed SC et al (2019) Guanfacine decreases symptoms of cannabis withdrawal in daily cannabis smokers. Addict Biol 24:707–716

Haney M, Bedi G, Cooper ZD, Herrmann ES, Reed SC, Foltin RW et al (2022) Impact of cyclooxygenase-2 inhibition on cannabis withdrawal and circulating endocannabinoids in daily cannabis smokers. Addict Biol 27:e13183

Haney M, Vallée M, Fabre S, Collins Reed S, Zanese M, Campistron G et al (2023) Signaling-specific inhibition of the CB1 receptor for cannabis use disorder: phase 1 and phase 2a randomized trials. Nat Med 29:1487–1499

Harris RT, Waters W, McLendon D (1974) Evaluation of reinforcing capability of delta-9-tetrahydrocannabinol in rhesus monkeys. Psychopharmacologia 37:23–29. https://doi.org/10.1007/BF00426679

Hart CL, Ward AS, Haney M, Comer SD, Foltin RW, Fischman MW (2002) Comparison of smoked marijuana and oral Δ9-tetrahydrocannabinol in humans. Psychopharmacology 164:407–415

Harte-Hargrove CL, Dow-Edwards LD (2012) Withdrawal from THC during adolescence: sex differences in locomotor activity and anxiety. Behav Brain Res 231(1):48–59

Hasin DS, O'Brien CP, Auriacombe M, Borges G, Bucholz K, Budney A et al (2013) DSM-5 criteria for substance use disorders: recommendations and rationale. Am J Psychiatry 170:834–851

Hayaki J, Anderson BJ, Stein MD (2016) Dual cannabis and alcohol use disorders in young adults: problems magnified. Subst Abus 37:579–583

Herrmann ES, Cooper ZD, Bedi G, Ramesh D, Reed SC, Comer SD et al (2016) Effects of zolpidem alone and in combination with nabilone on cannabis withdrawal and a laboratory model of relapse in cannabis users. Psychopharmacology 233:2469–2478

Herrmann ES, Cooper ZD, Bedi G, Ramesh D, Reed SC, Comer SD et al (2019) Varenicline and nabilone in tobacco and cannabis co-users: effects on tobacco abstinence, withdrawal and a laboratory model of cannabis relapse. Addict Biol 24:765–776

Hillard CJ (2018) Circulating endocannabinoids: from whence do they come and where are they going? Neuropsychopharmacology (Nature Publishing Group):155–172

Hiltunen JA, Jarbe UT (1986) Interactions between delta 9-tetrahydrocannabinol and cannabidiol as evaluated by drug discrimination procedures in rats and pigeons. Neuropharmacology 25(2): 133–142

Hindocha C, Freeman TP, Schafer G, Gardener C, Das RK, Morgan CJA et al (2015) Acute effects of delta-9-tetrahydrocannabinol, cannabidiol and their combination on facial emotion recognition: a randomised, double-blind, placebo-controlled study in cannabis users. Eur Neuropsychopharmacol 25:325–334

Hložek T, Uttl L, Kadeřábek L, Balíková M, Lhotková E, Horsley RR et al (2017) Pharmacokinetic and behavioural profile of THC, CBD, and THC+CBD combination after pulmonary, oral, and subcutaneous administration in rats and confirmation of conversion in vivo of CBD to THC. Eur Neuropsychopharmacol 27:1223 1237. Available from: https://www.sciencedirect.com/science/article/pii/S0924977X17309835

Holst M, Mathai DS, Patel MM, Rodgman C, Keller J, Hussain MZ, De La Garza R, 2nd, Kosten TR, Verrico CD (2019) Use of Guanfacine for Cannabis Use Disorder and Related Symptomology. The American journal on addictions 28(6):455–464. https://doi.org/10.1111/ajad.12959

Huestis MA (2007) Human cannabinoid pharmacokinetics. Chem Biodivers:1770–1804

Huestis MA, Henningfield JE, Cone EJ (1992) Blood cannabinoids. I. Absorption of THC and formation of 11-OH-THC and THCCOOH during and after smoking marijuana*. J Anal Toxicol 16:276–282. https://doi.org/10.1093/jat/16.5.276

Huestis MA, Gorelick DA, Heishman SJ, Preston KL, Nelson RA, Moolchan ET et al (2001) Blockade of effects of smoked marijuana by the CB1-selective cannabinoid receptor antagonist SR141716. Arch Gen Psychiatry 58:322–328. https://doi.org/10.1001/archpsyc.58.4.322

Huestis MA, Boyd SJ, Heishman SJ, Preston KL, Bonnet D, Le Fur G et al (2007) Single and multiple doses of rimonabant antagonize acute effects of smoked cannabis in male cannabis users. Psychopharmacology 194:505–515

Javadi-Paydar M, Nguyen JD, Kerr TM, Grant Y, Vandewater SA, Cole M et al (2018) Effects of Δ9-THC and cannabidiol vapor inhalation in male and female rats. Psychopharmacology 235: 2541–2557

Javadi-Paydar M, Creehan KM, Kerr TM, Taffe MA (2019) Vapor inhalation of cannabidiol (CBD) in rats. Pharmacol Biochem Behav 184

John WS, Martin TJ, Nader MA (2017) Behavioral determinants of cannabinoid self-Administration in old World Monkeys. Neuropsychopharmacology 42:1522 1530

Juckel G, Roser P, Nadulski T, Stadelmann AM, Gallinat J (2007) Acute effects of Δ9-tetrahydrocannabinol and standardized cannabis extract on the auditory evoked mismatch negativity. Schizophr Res 97:109–117

Justinova Z, Tanda G, Redhi GH, Goldberg SR (2003) Self administration of Δ9-tetrahydrocannabinol (THC) by drug naive squirrel monkeys. Psychopharmacology 169:135–140. https://doi.org/10.1007/s00213-003-1484-0

Justinova Z, Tanda G, Munzar P, Goldberg SR (2004) The opioid antagonist naltrexone reduces the reinforcing effects of Δ9-tetrahydrocannabinol (THC) in squirrel monkeys. Psychopharmacology 173:186–194

Justinova Z, Mangieri RA, Bortolato M, Chefer SI, Mukhin AG, Clapper JR et al (2008) Fatty acid amide hydrolase inhibition heightens anandamide signaling without producing reinforcing effects in primates. Biol Psychiatry 64:930–937

Justinova Z, Ferre S, Redhi HG et al (2011) Reinforcing and neurochemical effects of cannabinoid CB1 receptor agonists, but not cocaine, are altered by an adenosine A2A receptor antagonist. Addict Biol 16(3):405–415

Kangas DB, Delatte SM, Vemuri KV et al (2013) Cannabinoid discrimination and antagonism by CB1 neutral and inverse agonist antagonists. J Pharmacol Exp Ther 344(3):561–567

Kathmann M, Flau K, Redmer A et al (2006) Cannabidiol is an allosteric modulator at mu- and delta-opioid receptors. Naunyn Schmiedeberg's Arch Pharmacol 372(5):354–361

Katsidoni V, Kastellakis A, Panagis G (2013) Biphasic effects of Delta9-tetrahydrocannabinol on brain stimulation reward and motor activity. Int J Neuropsychopharmacol 16(10):2273–2284

Kaymakcalan S (1973) Tolerance to and dependence on cannabis. Bull Narc 25:39–47

Kearney-Ramos T, Herrmann ES, Belluomo I, Matias I, Vallée M, Monlezun S et al (2023) The relationship between circulating endogenous Cannabinoids and the effects of smoked cannabis. Cannabis Cannabinoid Res 8:1069–1078

Kerridge BT, Mauro PM, Chou SP, Saha TD, Pickering RP, Fan AZ et al (2017) Predictors of treatment utilization and barriers to treatment utilization among individuals with lifetime cannabis use disorder in the United States. Drug Alcohol Depend 181:223–228

Kesner JA, Mateo Y, Abrahao PK et al (2022) Changes in striatal dopamine release, sleep, and behavior during spontaneous Delta-9-tetrahydrocannabinol abstinence in male and female mice. Neuropsychopharmacology 47(8):1537–1549

Kiluk BD, Fitzmaurice GM, Strain EC, Weiss RD (2019) What defines a clinically meaningful outcome in the treatment of substance use disorders: reductions in direct consequences of drug use or improvement in overall functioning? Addiction 114:9–15. https://doi.org/10.1111/add.14289

Klein C, Karanges E, Spiro A et al (2011) Cannabidiol potentiates Delta(9)-tetrahydrocannabinol (THC) behavioural effects and alters THC pharmacokinetics during acute and chronic treatment in adolescent rats. Psychopharmacology 218(2):443–457

Kornetsky C, Esposito UR, McLean S et al (1979) Intracranial self-stimulation thresholds: a model for the hedonic effects of drugs of abuse. Arch Gen Psychiatry 36(3):289–292

Laprairie BR, Bagher MA, Kelly EM et al (2015) Cannabidiol is a negative allosteric modulator of the cannabinoid CB1 receptor. Br J Pharmacol 172(20):4790–4805

Lee DC, Schlienz NJ, Peters EN, Dworkin R, Turk D, Strain E et al (2019) Systematic review of outcome domains and measures used in psychosocial and pharmacological treatment trials for cannabis use disorder. Drug Alcohol Depend (Elsevier Ireland Ltd) 194:500–517

Lee DC, Schlienz NJ, Herrmann ES, Martin EL, Leoutsakos J, Budney AJ et al (2024) Randomized controlled trial of zolpidem as a pharmacotherapy for cannabis use disorder. J Subst Use Addict Treat 156:209180

Lefever WT, Marusich AJ, Antonazzo RK et al (2014) Evaluation of WIN 55,212-2 self-administration in rats as a potential cannabinoid abuse liability model. Pharmacol Biochem Behav 118:30–35

Lemberger L, Martz R, Rodda B, Forney R, Rowe H (1973) Comparative pharmacology of A-tetrahydrocannabinol and its metabolite, 11-OH-A9-tetrahydrocannabinol. J Clin Invest 52:2411–2417

Lepore M, Liu X, Savage V et al (1996) Genetic differences in delta 9-tetrahydrocannabinol-induced facilitation of brain stimulation reward as measured by a rate-frequency curve-shift electrical brain stimulation paradigm in three different rat strains. Life Sci 58(25):PL365-372

Levin FR, McDowell D, Evans SM, Nunes E, Akerele E, Donovan S et al (2004) Pharmacotherapy for marijuana dependence: a double-blind, placebo-controlled pilot study of divalproex sodium. Am J Addict 13:21–32

Levin FR, Mariani JJ, Brooks DJ, Pavlicova M, Cheng W, Nunes EV (2011) Dronabinol for the treatment of cannabis dependence: a randomized, double-blind, placebo-controlled trial. Drug Alcohol Depend 116:142–150

Levin FR, Mariani JJ, Pavlicova M, Brooks D, Glass A, Mahony A et al (2016) Dronabinol and lofexidine for cannabis use disorder: a randomized, double-blind, placebo-controlled trial. Drug Alcohol Depend 159:53–60

Levin FR, Mariani JJ, Choi CJ, Basaraba C, Brooks DJ, Brezing CA et al (2021) Non-abstinent treatment outcomes for cannabis use disorder. Drug Alcohol Depend 225

Lichtman HA, Martin RB (2002) Marijuana withdrawal syndrome in the animal model. J Clin Pharmacol 42(S1):20S–27S

Lichtman AH, Peart J, Poklis JL, Bridgen DT, Razdan RK, Wilson DM et al (2000) Pharmacological evaluation of aerosolized cannabinoids in mice. Eur J Pharmacol 399:141–149. Available from: https://www.sciencedirect.com/science/article/pii/S0014299900003216

Lile JA, Alcorn JL, Hays LR, Kelly TH, Stoops WW, Wesley MJ et al (2021) Influence of pregabalin maintenance on cannabis effects and related behaviors in daily cannabis users. Exp Clin Psychopharmacol. https://doi.org/10.1037/pha0000464

Lile JA, Turner BW, Cox DH, Bonn-Miller MO, Katz NR, Shellenberg TP et al (2023) Cannabis use disorder treatment preferences: a pilot survey in current users of cannabis. J Addict Med 17: E87–E93

Lintzeris N, Bhardwaj A, Mills L, Dunlop A, Copeland J, McGregor I et al (2019) Nabiximols for the treatment of cannabis dependence: a randomized clinical trial. JAMA Intern Med 179:1242–1253

Litt MD, Kadden RM, Kabela Cormier E, Petry NM (2008) Coping skills training and contingency management treatments for marijuana dependence: exploring mechanisms of behavior change. Addiction 103:638–648

Loflin MJE, Kiluk BD, Huestis MA, Aklin WM, Budney AJ, Carroll KM et al (2020) The state of clinical outcome assessments for cannabis use disorder clinical trials: a review and research agenda. Drug Alcohol Depend (Elsevier Ireland Ltd). https://doi.org/10.1016/j.drugalcdep.2020.107993

Manwell LA, Charchoglyan A, Brewer D, Matthews BA, Heipel H, Mallet PE (2014a) A vapourized Δ(9)-tetrahydrocannabinol (Δ(9)-THC) delivery system part I: development and validation of a pulmonary cannabinoid route of exposure for experimental pharmacology studies in rodents. J Pharmacol Toxicol Methods 70:120–127. Available from: http://europepmc.org/abstract/MED/24973534

Manwell LA, Ford B, Matthews BA, Heipel H, Mallet PE (2014b) A vapourized Δ9-tetrahydrocannabinol (Δ9-THC) delivery system part II: comparison of behavioural effects of pulmonary versus parenteral cannabinoid exposure in rodents. J Pharmacol Toxicol Methods 70:112–119. Available from: https://www.sciencedirect.com/science/article/pii/S1056871914002214

Mariani JJ, Pavlicova M, Mamczur AK, Bisaga A, Nunes EV, Levin FR (2014) Open-label pilot study of quetiapine treatment for cannabis dependence. Am J Drug Alcohol Abuse 40:280–284

Mariani JJ, Pavlicova M, Jean Choi C, Basaraba C, Carpenter KM, Mahony AL et al (2021) Quetiapine treatment for cannabis use disorder. Drug Alcohol Depend 218:108366

Martellotta CM, Cossu G, Fattore L et al (1998) Self administration of the cannabinoid receptor agonist WIN 55,212-2 in drug-naive mice. Neuroscience 85(2):327–330

Marusich AJ, Lefever WT, Antonazzo RK et al (2014) Evaluation of sex differences in cannabinoid dependence. Drug Alcohol Depend. https://doi.org/10.1016/j.drugalcdep.2014.01.019

McClure EA, Baker NL, Sonne SC, Ghitza UE, Tomko RL, Montgomery LT et al (2018) Tobacco use during cannabis cessation: use patterns and impact on abstinence in a National Drug Abuse Treatment Clinical Trials Network study. Drug Alcohol Depend 192:59–66

McClure EA, Neelon B, Tomko RL, Gray KM, McRae-Clark AL, Baker NL (2024) Association of Cannabis use Reduction with Improved Functional Outcomes, an exploratory aggregated analysis from seven cannabis use disorder treatment trials to extract data-driven cannabis reduction metrics. Am J Psychiatry 181:988–996. Available from: https://psychiatryonline.org/doi/10.1176/appi.ajp.20230508

McGregor SI, Issakidis NC, Prior G (1996) Aversive effects of the synthetic cannabinoid CP 55,940 in rats. Pharmacol Biochem Behav 53(3):657–664

McMahon LR (2016) Enhanced discriminative stimulus effects of Δ9-THC in the presence of cannabidiol and 8-OH-DPAT in rhesus monkeys. Drug Alcohol Depend 165:87–93

McPartland MJ, Duncan M, Marzo DV et al (2015) Are cannabidiol and Delta(9)-tetrahydrocannabivarin negative modulators of the endocannabinoid system? A systematic review. Br J Pharmacol 172(3):737–753

McRae-Clark AL, Carter RE, Killeen TK, Carpenter MJ, White KG, Brady KT (2010) A placebo-controlled trial of atomoxetine in marijuana-dependent individuals with attention deficit hyperactivity disorder. Am J Addict 19:481–489

McRae-Clark AL, Baker NL, Gray KM, Killeen TK, Wagner AM, Brady KT et al (2015) Buspirone treatment of cannabis dependence: a randomized, placebo-controlled trial. Drug Alcohol Depend 156:29–37

McRae-Clark AL, Gray KM, Baker NL, Sherman BJ, Squeglia L, Sahlem GL et al (2021) Varenicline as a treatment for cannabis use disorder: a placebo-controlled pilot trial. Drug Alcohol Depend 229:109111

Mecha M, Carrillo-Salinas FJ, Feliú A, Mestre L, Guaza C (2016) Microglia activation states and cannabinoid system: therapeutic implications. Pharmacol Ther 166:40–55. Available from: https://www.sciencedirect.com/science/article/pii/S0163725816301012

Mendizabal V, Zimmer A, Maldonado R (2006) Involvement of kappa/dynorphin system in WIN 55,212-2 self-administration in mice. Neuropsychopharmacology 31(9):1957–1966

Millar SR, Mongan D, Smyth BP, Perry IJ, Galvin B (2021) Relationships between age at first substance use and persistence of cannabis use and cannabis use disorder. BMC Public Health 21:997

Moore FC, Davis MC, Sempio C et al (2024) Delta(9)-tetrahydrocannabinol vapor exposure produces conditioned place preference in male and female rats. Cannabis Cannabinoid Res 9(1):111–120

Morgan CJA, Schafer G, Freeman TP, Curran HV (2010) Impact of cannabidiol on the acute memory and psychotomimetic effects of smoked cannabis: naturalistic study. Br J Psychiatry 197:285–290

Morgan CJA, Page E, Schaefer C, Chatten K, Manocha A, Gulati S et al (2013) Cerebrospinal fluid anandamide levels, cannabis use and psychotic-like symptoms. Br J Psychiatry 202:381–382. Available from: https://www.cambridge.org/core/product/6C4E16919C53C3444C90AA53C76DC85D

Navarrete F, Aracil-Fernández A, Manzanares J (2018) Cannabidiol regulates behavioural alterations and gene expression changes induced by spontaneous cannabinoid withdrawal. Br J Pharmacol 175:2676–2688

Navarro M, Carrera MRA, Fratta W, Valverde O, Cossu G, Fattore L et al (2001) Functional interaction between opioid and cannabinoid receptors in drug self-administration. J Neurosci 21(14):5344–5350

Neale J, Vitoratou S, Finch E, Lennon P, Mitcheson L, Panebianco D et al (2016) Development and validation of 'sure': a patient reported outcome measure (prom) for recovery from drug and alcohol dependence. Drug Alcohol Depend 165:159–167

Nguyen JD, Aarde SM, Vandewater SA, Grant Y, Stouffer DG, Parsons LH et al (2016) Inhaled delivery of Δ9-tetrahydrocannabinol (THC) to rats by e-cigarette vapor technology. Neuropharmacology 109:112–120

Notzon DP, Kelly MA, Choi CJ, Pavlicova M, Mahony AL, Brooks DJ et al (2018) Open-label pilot study of injectable naltrexone for cannabis dependence. Am J Drug Alcohol Abuse 44:619–627

Ohlsson A, Lindgren J-E, Andersson S, Agurell S, Gillespie H, Hollister LE (1986) Single-dose kinetics of deuterium-labelled cannabidiol in man after smoking and intravenous administration. Biomed Environ Mass Spectrom 13:77–83. https://doi.org/10.1002/bms.1200130206

Pandey P, Zagzoog A, Laprairie BR et al (2025) Determination of the negative allosteric binding site of Cannabidiol at the CB1 receptor: a combined computational and site-directed mutagenesis study. ACS Chem Neurosci 16(3):311–328

Panlilio LV, Justinova Z, Goldberg SR (2010) Animal models of cannabinoid reward. Br J Pharmacol (John Wiley and Sons Inc.):499–510

Parker AL, Gillies T (1995) THC-induced place and taste aversions in Lewis and Sprague-Dawley rats. Behav Neurosci 109(1):71–78

Parsons LH, Hurd YL (2015) Endocannabinoid signalling in reward and addiction. Nat Rev Neurosci (Nature Publishing Group):579–594

Pickens R, Meisch AR, Thompson T (1978) Drug self-administration: an analysis of the reinforcing effects of drugs. In: Drugs of abuse. Springer, Boston, pp 1–37

Pi-Sunyer FX, Aronne LJ, Heshmati HM, Devin J, Rosenstock J (2006) RIO-North America study group for the. Effect of Rimonabant, a Cannabinoid-1 receptor blocker, on weight and Cardiometabolic risk factors in overweight or obese PatientsRIO-North America: a randomizedff controlled trial. JAMA 295:761–775. https://doi.org/10.1001/jama.295.7.761

Posey DJ, Mcdougle CJ (2007) Guanfacine and Guanfacine extended release: treatment for ADHD and related disorders. CNS Drug Rev (Blackwell Publishing Inc). https://doi.org/10.1111/j.1527-3458.2007.00026.x

Pryor GT, Husain S, Mitoma C (1977) Influence of fasting on the absorption and effects of δ9-tetrahydrocannabinol after oral administration in sesame oil. Pharmacol Biochem Behav 6:331–341. Available from: https://www.sciencedirect.com/science/article/pii/0091305777900338

Ravula A, Chandasana H, Jagnarine D et al (2019) Pharmacokinetic and Pharmacodynamic characterization of tetrahydrocannabinol-induced cannabinoid dependence after chronic passive cannabis smoke exposure in rats. Cannabis Cannabinoid Res 4(4):240–254

Reimann-Philipp U, Speck M, Orser C, Johnson S, Hilyard A, Turner H et al (2020) Cannabis Chemovar nomenclature misrepresents chemical and genetic diversity; survey of variations in chemical profiles and genetic markers in Nevada medical cannabis samples. Cannabis Cannabinoid Res 5:215–230

Reinaldo NT, Singer G (1980) Effects of body weight levels on cannabis self-injection. Pharmacol Biochem Behav 13:877–881. Available from: https://www.sciencedirect.com/science/article/pii/0091305780902221

Riedel M, Müller N, Strassnig M, Spellmann I, Severus E, Möller H-J (2007) Quetiapine in the treatment of schizophrenia and related disorders. Neuropsychiatr Dis Treat. https://doi.org/10.2147/nedt.2007.3.2.219

Rinaldi-Carmona M, Barth F, Héaulme M, Shire D, Calandra B, Congy C et al (1994) SR141716A, a potent and selective antagonist of the brain cannabinoid receptor. FEBS Lett 350:240–244

Rinaldi-Carmona M, Barth F, Heaulme M, Alonso R, Shire D, Congy C et al (1995) Biochemical and pharmacological characterisation of SR141716A, the first potent and selective brain cannabinoid receptor antagonist. Life Sci 56(23-24):1941–1947

Roser P, Gallinat J, Weinberg G, Juckel G, Gorynia I, Stadelmann AM (2009) Psychomotor performance in relation to acute oral administration of Δ9-tetrahydrocannabinol and standardized cannabis extract in healthy human subjects. Eur Arch Psychiatry Clin Neurosci 259:284–292

Rubino T, Patrini G, Massi P et al (1998) Cannabinoid-precipitated withdrawal: a time-course study of the behavioral aspect and its correlation with cannabinoid receptors and G protein expression. J Pharmacol Exp Ther 285(2):813–819

Russo BE, Marcu J (2017) Cannabis pharmacology: the usual suspects and a few promising leads. Adv Pharmacol 80:67–134

Scherrer JF, Grant JD, Duncan AE, Sartor CE, Haber JR, Jacob T et al (2009) Subjective effects to cannabis are associated with use, abuse and dependence after adjusting for genetic and environmental influences. Drug Alcohol Depend 105:76–82

Schindler WC, Redhi HG, Vemuri K et al (2016) Blockade of nicotine and cannabinoid reinforcement and relapse by a cannabinoid CB1-receptor neutral antagonist AM4113 and inverse agonist Rimonabant in squirrel monkeys. Neuropsychopharmacology 41(9):2283–2293

Schlienz NJ, Lee DC, Stitzer ML, Vandrey R (2018) The effect of high-dose dronabinol (oral THC) maintenance on cannabis self-administration. Drug Alcohol Depend 187:254–260

Schlosburg JE, Carlson BLA, Ramesh D, Abdullah RA, Long JZ, Cravatt BF et al (2009) Inhibitors of endocannabinoid-metabolizing enzymes reduce precipitated withdrawal responses in THC-dependent mice. AAPS J 11:342–352

Schuster RM, Hanly A, Gilman J, Budney A, Vandrey R, Evins AE (2016) A contingency management method for 30-days abstinence in non-treatment seeking young adult cannabis users. Drug Alcohol Depend 167:199–206

Schwope DM, Scheidweiler KB, Huestis MA (2011) Direct quantification of cannabinoids and cannabinoid glucuronides in whole blood by liquid chromatography–tandem mass spectrometry. Anal Bioanal Chem 401:1273–1283. https://doi.org/10.1007/s00216-011-5197-7

Shaham Y, Shalev U, Lu L et al (2003) The reinstatement model of drug relapse: history, methodology and major findings. Psychopharmacology 168(1–2):3–20

Sharretts J, Galescu O, Gomatam S, Andraca-Carrera E, Hampp C, Yanoff L (2020) Cancer risk associated with Lorcaserin – the FDA'S review of the CAMELLIA-TIMI 61 trial. N Engl J Med 383:1000–1002

Sim-Selley LJ (2003) Regulation of cannabinoid CB1 receptors in the central nervous system by chronic Cannabinoids. Crit Rev Trade Neurobiol 15:91

Solinas M, Goldberg SR (2005) Involvement of mu-, delta- and kappa-opioid receptor subtypes in the discriminative-stimulus effects of delta-9-tetrahydrocannabinol (THC) in rats. Psychopharmacology 179:804–812

Solinas M, Scherma M, Fattore L et al (2007) Nicotinic alpha 7 receptors as a new target for treatment of cannabis abuse. J Neurosci 27(21):5615–5620

Spencer S, Neuhofer D, Chioma VC, Garcia-Keller C, Schwartz DJ, Allen N et al (2018) A model of Δ9-tetrahydrocannabinol self-administration and reinstatement that alters synaptic plasticity in nucleus Accumbens. Biol Psychiatry 84:601–610

Spindle TR, Bonn-Miller MO, Vandrey R (2019a) Changing landscape of cannabis: novel products, formulations, and methods of administration. Curr Opin Psychol 30:98–102, Elsevier BV

Spindle TR, Cone EJ, Schlienz NJ, Mitchell JM, Bigelow GE, Flegel R et al (2019b) Acute pharmacokinetic profile of smoked and vaporized cannabis in human blood and Oral fluid. J Anal Toxicol 43:233–258. https://doi.org/10.1093/jat/bky104

Spindle TR, Cone EJ, Herrmann ES, Mitchell JM, Flegel R, LoDico C et al (2020) Pharmacokinetics of cannabis brownies: a controlled examination of Δ9-tetrahydrocannabinol and metabolites in blood and Oral fluid of healthy adult males and females. J Anal Toxicol 44:661–671

Spindle TR, Kuwabara H, Eversole A, Nandi A, Vandrey R, Antoine DG et al (2021) Brain imaging of cannabinoid type I (CB1) receptors in women with cannabis use disorder and male and female healthy controls. Addict Biol 26

Stewart JL, McMahon LR (2011) The fatty acid amide hydrolase inhibitor URB 597: interactions with anandamide in rhesus monkeys. Br J Pharmacol 164:655–666

Stringfield SJ, Torregrossa MM (2021) Intravenous self-administration of delta-9-THC in adolescent rats produces long-lasting alterations in behavior and receptor protein expression. Psychopharmacology 238:305–319

Substance Abuse, Administration MHS (2024) Key substance use and mental health indicators in the United States: results from the 2023 National Survey on Drug Use and Health. Available from: https://www.samhsa.gov/data/report/2023-nsduh-annual-national-report

Takahashi RN, Singer G (1979) Self-administration of Δ9-tetrahydrocannabinol by rats. Pharmacol Biochem Behav 11:737–740. Available from: https://www.sciencedirect.com/science/article/pii/0091305779902740

Tanda G (2016) Preclinical studies on the reinforcing effects of cannabinoids. A tribute to the scientific research of Dr Steve Goldberg. Psychopharmacology (Berl) 233(10):1845–1866

Tanda G, Goldberg SR (2003) Cannabinoids: reward, dependence, and underlying neurochemical mechanisms – a review of recent preclinical data. Psychopharmacology:115–134

Tanda G, Pontieri EF, Chiara D, G. (1997) Cannabinoid and heroin activation of mesolimbic dopamine transmission by a common mu1 opioid receptor mechanism. Science 276(5321): 2048–2050

Tanda G, Munzar P, Goldberg SR (2000) Self-administration behavior is maintained by the psychoactive ingredient of marijuana in squirrel monkeys. Nat Neurosci 3:1073–1074. https://doi.org/10.1038/80577

Tirado CF, Goldman M, Lynch K, Kampman KM, Obrien CP (2008) Atomoxetine for treatment of marijuana dependence: a report on the efficacy and high incidence of gastrointestinal adverse events in a pilot study. Drug Alcohol Depend 94:254–257

Trexler RK, Nass RS, Crowe SM et al (2018) Novel behavioral assays of spontaneous and precipitated THC withdrawal in mice. Drug Alcohol Depend 191:14–24

Trigo JM, Lagzdins D, Rehm J, Selby P, Gamaleddin I, Fischer B et al (2016) Effects of fixed or self-titrated dosages of Sativex on cannabis withdrawal and cravings. Drug Alcohol Depend 161:298–306

Tseng HA, Craft MR (2001) Sex differences in antinociceptive and motoric effects of cannabinoids. Eur J Pharmacol 430(1):41–47

Valjent E, Maldonado R (2000) A behavioural model to reveal place preference to delta 9-tetrahydrocannabinol in mice. Psychopharmacology 147(4):436–438

Vallée M, Vitiello S, Bellocchio L, Hebert-Chatelain E, Monlezun S, Martin-Garcia E et al (2014) Pregnenolone can protect the brain from cannabis intoxication. Science (1979) 343:94–98

van Ree JM, Slangen JL, de Wied D (1978) Intravenous self-administration of drugs in rats. J Pharmacol Exp Ther 204:547–557

Vandrey R, Smith MT, McCann UD, Budney AJ, Curran EM (2011) Sleep disturbance and the effects of extended-release zolpidem during cannabis withdrawal. Drug Alcohol Depend 117: 38–44

Vandrey R, Stitzer ML, Mintzer MZ, Huestis MA, Murray JA, Lee D (2013) The dose effects of short-term dronabinol (oral THC) maintenance in daily cannabis users. Drug Alcohol Depend 128:64–70

Vann ER, Gamage FT, Warner AJ et al (2008) Divergent effects of cannabidiol on the discriminative stimulus and place conditioning effects of Delta(9)-tetrahydrocannabinol. Drug Alcohol Depend 94(1–3):191–198

Wakeford AGP, Wetzell BB, Pomfrey RL, Clasen MM, Taylor WW, Hempel BJ et al (2017) The effects of cannabidiol (CBD) on Δ9-tetrahydrocannabinol (thc) self-administration in male and female Long-Evans rats. Exp Clin Psychopharmacol 25:242–248

Wakley AA, Craft RM (2011) THC-methadone and THC-naltrexone interactions on discrimination, antinociception, and locomotion in rats. Behav Pharmacol 22:489–497

Wakley AA, Wiley LJ, Craft MR (2014) Sex differences in antinociceptive tolerance to delta-9-tetrahydrocannabinol in the rat. Drug Alcohol Depend 143:22–28

Ward AS, Comer SD, Haney M, Foltin RW, Fischman MW (1997) The effects of a monetary alternative on marijuana self-administration. Behav Pharmacol:8. Available from: https://journals.lww.com/behaviouralpharm/fulltext/1997/08000/the_effects_of_a_monetary_alternative_on_marijuana.1.aspx

Wesley MJ, Westgate PM, Stoops WW, Kelly TH, Hays LR, Lile JA (2018) Influence of tiagabine maintenance on cannabis effects and related behaviors in daily cannabis users. Exp Clin Psychopharmacol 26:310–319

Wiley LJ (2003) Sex-dependent effects of delta 9-tetrahydrocannabinol on locomotor activity in mice. Neurosci Lett 352(2):77–80

Wiley JL, Lefever TW, Marusich JA, Craft RM (2017) Comparison of the discriminative stimulus and response rate effects of Δ9-tetrahydrocannabinol and synthetic cannabinoids in female and male rats. Drug Alcohol Depend 172:51–59

Wilson DM, Peart J, Martin BR, Bridgen DT, Byron PR, Lichtman AH (2002) Physiochemical and pharmacological characterization of a Δ9-THC aerosol generated by a metered dose inhaler. Drug Alcohol Depend 67:259–267. Available from: https://www.sciencedirect.com/science/article/pii/S0376871602000789

Wilson MD, Varvel AS, Harloe PJ et al (2006) SR 141716 (Rimonabant) precipitates withdrawal in marijuana-dependent mice. Pharmacol Biochem Behav 85(1):105–113

Wittemann M, Brielmaier J, Rubly M, Kennel J, Werler F, Schmitgen MM et al (2021) Cognition and cortical thickness in heavy cannabis users. Eur Addict Res 27:115–122

Wong DF, Kuwabara H, Horti AG, Raymont V, Brasic J, Guevara M et al (2010) Quantification of cerebral cannabinoid receptors subtype 1 (CB1) in healthy subjects and schizophrenia by the novel PET radioligand [11C]OMAR. NeuroImage 52:1505–1513

World Health Organization (n.d.) Cannabis. World Health Organization. Available from: https://www.who.int/teams/mental-health-and-substance-use/alcohol-drugs-and-addictive-behaviours/drugs-psychoactive/cannabis

Xi XZ, Spiller K, Pak CA et al (2008) Cannabinoid CB1 receptor antagonists attenuate cocaine's rewarding effects: experiments with self-administration and brain-stimulation reward in rats. Neuropsychopharmacology 33(7):1735–1745

Young AM, Herling S (1986) Drugs as Reinforcers: studies in laboratory animals. In: Goldberg SR, Stolerman IP (eds) Behavioral analysis of drug dependence. Academic, New York, pp 9–68

Zagzoog A, Halter K, Jones AM, Bannatyne N, Cline J, Wilcox A et al (2024) The intoxication equivalency of 11-Hydroxy-tetrahydrocannabinol relative to tetrahydrocannabinol. J Pharmacol Exp Ther 391:194–205. https://doi.org/10.1124/jpet.123.001998

Zaharakis N, Coatsworth JD, Riggs NR, Radford A, Rayburn S, Mennis J et al (2024) Treating young adult cannabis use disorder with text message-delivered peer network counseling. Contemp Clin Trials 144

Zamarripa AC, Spindle RT, Surujunarain R et al (2023) Assessment of orally administered Delta9-tetrahydrocannabinol when Coadministered with Cannabidiol on Delta9-tetrahydrocannabinol pharmacokinetics and pharmacodynamics in healthy adults: a randomized clinical trial. JAMA Netw Open 6(2):e2254752

Zangen A, Solinas M, Ikemoto S, Goldberg SR, Wise RA (2006) Two brain sites for cannabinoid reward. J Neurosci 26:4901–4907

Zeiger JS, Haberstick BC, Corley RP, Ehringer MA, Crowley TJ, Hewitt JK et al (2010) Subjective effects to marijuana associated with marijuana use in community and clinical subjects. Drug Alcohol Depend 109:161–166

Zuardi AW, Shirakawa I, Finkelfarb E, Karniol IG (1982) Psycho phar macolog y action of Cannabidiol on the anxiety and other effects produced by a 9-THC in Normal subjects. Psychopharmacology. https://doi.org/10.1007/BF00432554

Cannabinoid-based Pharmacology for the Management of Substance Use Disorders

M. Á. Luján, Y. Kim, L. Y. Zhang, and J. F. Cheer

Contents

1 Introduction ... 298
1.1 Substance Use Disorders ... 298
1.2 The Brain Reward Circuitry and the Endocannabinoid System in Brief ... 299
2 The Cannabinoid-Based *pharmacopoeia* ... 302
3 Opioids ... 304
4 Cocaine and Other Psychostimulants ... 307
5 Nicotine ... 313
6 Alcohol ... 317
7 Cannabis ... 320
8 Discussion ... 323
References ... 325

Abstract In the last two decades, the endocannabinoid system has emerged as a crucial modulator of motivation and emotional processing. Due to its widespread neuroanatomical distribution and characteristic retrograde signaling nature, cannabinoid type I receptors and their endogenous ligands finely orchestrate somatic and axon terminal activity of dopamine neurons. Owing to these unique features, this signaling system is a promising pharmacological target to ameliorate dopamine-mediated drug-seeking behaviors while circumventing the adverse side effects of, for instance, dopaminergic antagonists. Despite considerable preclinical efforts, an agreement on the efficacy of endocannabinoid-targeting compounds for treating drug substance use disorders in humans has not been reached. In the following chapter, we will summarize preclinical and clinical evidence addressing the therapeutic potential of cannabinoids and endocannabinoid-targeting compounds in substance use disorders. To bridge the gap between animal and clinical research, we capitalize on studies evaluating the impact of endocannabinoid-targeting compounds in relevant settings, such as the management of drug relapse. Finally, we discuss the

M. Á. Luján, Y. Kim, L. Y. Zhang, and J. F. Cheer (✉)
Department of Neurobiology, University of Maryland, School of Medicine, Baltimore, MD, USA
e-mail: JCheer@som.umaryland.edu

Curr Topics Behav Neurosci (2026) 76: 297–338
https://doi.org/10.1007/7854_2024_551
Published Online: 16 January 2025

therapeutic potential of novel cannabinoid compounds that hold promise for treating substance use disorders.

Keywords Cannabinoids · Relapse · Substance use disorder

1 Introduction

1.1 Substance Use Disorders

In the United States, substance use disorders have a lifetime prevalence of 10% in the general population, amounting to more than 23 million adults with problematic drug use (Ignaszewski 2021). Opioids, primarily of a synthetic nature, account for more than half of all drug-involved overdose deaths (National Center for Health Statistics 2021). Substance use and overdose mortality rates were amplified during the COVID-19 pandemic, which strained the healthcare system and exposed vulnerabilities for individuals with these mental health disorders (Ornell et al. 2020; Ignaszewski 2021). A case in point, in 2021, there were over 100,000 drug overdose deaths in the United States (Ahmad et al. 2022). Despite the manifold and devastating consequences of substance use disorders, only a minority of this at-risk population seeks or receives substance use treatment. This is due, in part, to the scarcity of clinically validated pharmaceuticals that offer robust treatment retention and suppression of renewed drug use.

The challenge for developing effective pharmacotherapies for substance use disorders lies in the etiology of the disease itself. Substance use disorder is a chronic brain disease characterized by the compulsive seeking and taking of drugs, with a diminished capacity to exert self-control over drug use despite its harmful consequences. At the core of this malfunction resides a chronically unbalanced brain system that disproportionally potentiates dopaminergic encoding of the drug and its environmental cues, as well as dysregulated neurocircuitry required to sustain executive control (Volkow and Morales 2015). Furthermore, these disturbances change as drug use progresses and this has crucial implications for formulating all-encompassing therapies (Longo et al. 2016). The current consensus is that initial drug use is characterized by a period of drug consumption and intoxication driven by positive reinforcement processes, known as the binge/intoxication stage (Wise and Bozarth 1987). While animal studies frequently utilize pharmacological approaches to suppress the rewarding effects of a given drug, this type of intervention may be of limited clinical utility as individuals typically are not evaluated nor diagnosed for substance use disorder during the binge/intoxication phase. With prolonged drug use, negative reinforcement mechanisms intensify and sustain drug-seeking behavior as dysphoric, negative emotional consequences of withdrawal deter maintenance of abstinence (Koob and Le Moal 1979, 2001). Animal studies aimed at pharmacologically dampening negative reinforcement of drug-seeking are highly relevant as most people who use drugs for a long time (i.e., tobacco smokers) are identified in this stage of addiction. However, individuals are often diagnosed with substance use disorder characterized, in part, by a ubiquitous state of preoccupation to obtain the drug and at the cost of extensive personal interference. This excessive desire for the drug, commonly called "craving," permeates nearly every aspect of the subject's life, who is found in an unrelenting cycle of attempts to maintain abstinence. During this stage, relapse to drug-seeking behaviors is triggered by exposure to drug-associated cues, stressful events, and/or dysphoric emotional states (Everitt and Robbins 2016; Luján et al. 2023b). Individuals consuming alcohol, cocaine, nicotine, or cannabis delay seeking treatment only after years of use and several unsuccessful abstinence attempts (Marín-Navarrete et al. 2014). Despite preclinical efforts to explore pharmacological avenues to diminish the impact of relapse-precipitating events, few FDA-approved medications are validated for the treatment of abstinence maintenance and relapse prevention (Volkow et al. 2019). Along with a reduction in drug use (McCann et al. 2015), achieving lasting abstinence remains the primary treatment goal of substance use disorders research. This chapter highlights recent advances in the research and development of promising cannabinoid-based pharmacotherapies that demonstrate potential reducing drug relapse rates.

1.2 The Brain Reward Circuitry and the Endocannabinoid System in Brief

Many substances with addictive potential abnormally increase mesolimbic dopamine (DA) neurotransmission, leading to greater encoding of both drug reinforcers and their associated environmental cues (Sulzer 2011). DA neurons in the ventral tegmental area (VTA) project to cortical and limbic structures and are prominently involved in motivated behavior (Salamone and Correa 2012). VTA DA cells integrate and transmit information about environmental associations and action-outcome events by encoding incentive salience, value of resource allocation, and reward prediction errors (Schultz 2015; Berke 2018; Kim et al. 2020). DA neurotransmission is further integrated in the striatum, particularly in the nucleus accumbens (NAc), where it is integrated with efferent signaling from the amygdala, prefrontal cortex (PFC), and hippocampus, among other brain structures. In terminal regions, DA release is not only determined by DA cell body firing, but also by independent synaptic systems and neuromodulators including endocannabinoids (eCBs) (Peters et al. 2021a).

In the midbrain, eCBs influence DA function and motivated behavior via their actions at cannabinoid 1 receptors (CB1R), which indirectly disinhibit DA neurons (Cheer et al. 2000, 2004; Pillolla et al. 2007). In the VTA, DA neurons are equipped with the molecular machinery required for eCBs synthesis. Here, CB1R signaling is more relevant at gamma-aminobutyric acid (GABA), rather than glutamate, releasing pre-synaptic terminals interfacing DA neurons (Riegel and Lupica 2004).

Remarkably, CB1R are not found in DA terminals (Julian et al. 2003). DA neuron activation promotes "on demand" de novo synthesis and retrograde release of 2-arachidonoylglycerol (2AG), the primary eCB and CB1R ligand found in the central nervous system, to nearby CB1R-expressing synaptic terminals (Riegel and Lupica 2004; Lupica and Riegel 2005). This unique receptor positioning explains why blocking CB1R reduces NAc DA release in response to a variety of natural and drug reinforcers in rats (Cheer et al. 2007; Wang et al. 2015; Covey et al. 2016). Conversely, direct activation of CB1R rescues extinguished reward-seeking behavior (De Vries et al. 2001). Notably, pharmacological manipulations of endogenous 2AG concentration and concomitant CB1R activation in operant tasks primarily impact cue-evoked striatal DA release while reward-evoked DA signaling remains unaltered (Oleson et al. 2012; Covey et al. 2021). The modulation of DA transients by 2AG, however, is only patent during learning acquisition, but not after consolidation (Wenzel et al. 2018). The current working model posits that VTA 2AG is necessary to facilitate learning about reward-predictive cues and thus instigate the shift of accumbal DA phasic release towards earlier predictors (Luján et al. 2023a) (Fig. 1). In this sense, an important question remains: if CB1R signaling is primarily involved in acquisition of action-outcome learning, how does CB1R activation reinstate well-consolidated reward-seeking behaviors (Serrano and Parsons 2011)? The answer may lie in the ability of eCB-CB1R neurotransmission to gate reward-seeking behaviors not only during acquisition, but also in response to representations of reward-predictive cues that characteristically follow extinction learning.

A comprehensive understanding of the mechanistic role of the eCB, anandamide (AEA), in the neurobiology of motivated behavior, remains lacking (Scherma et al. 2019). Different observations support that the elevation of AEA concentration in the central nervous system dampens effortful motivation (Oleson et al. 2012). AEA is a partial agonist ligand of CB1R, and it is currently hypothesized that AEA and 2AG engage in competitive antagonism of CB1R (Howlett and Mukhopadhyay 2000). AEA, compared to 2AG, has a broader pharmacodynamic repertoire, acting as a partial agonist of transient receptor potential vanilloid (TRPV) and GPR55 receptors, while also activating peroxisome proliferator-activated receptors (PPARs) (Pertwee 2008). As substrates for AEA and other eCBs, these receptors are part of the "expanded eCB system" (Cristino et al. 2020). These differences between receptor affinity and selectivity underlie the divergent effects that the two eCBs have on drug-motivated behavior. Recent advancements enabling in vivo monitoring of 2AG and AEA dynamics will be key to better ascertaining the specific roles of these two eCB moieties on motivation (Dong et al. 2021; Kondev et al. 2023; Singh et al. 2023).

In striatal synapses, eCBs participate in the terminal control of DA release (Cachope et al. 2012). As Mateo et al. (2017) found, this two-step mechanism requires the participation of cholinergic and glutamatergic terminals in the striatum. Activating striatal cholinergic interneurons results in acetylcholine (ACh) release which in turn facilitates glutamatergic release from ACh receptor-expressing terminals (Collins et al. 2016). Then, glutamate binds to AMPA receptors found on DA varicosities, potentiating DA release (Kosillo et al. 2016). Importantly, glutamatergic terminals in the striatum express CB1R and thus 2AG binding results

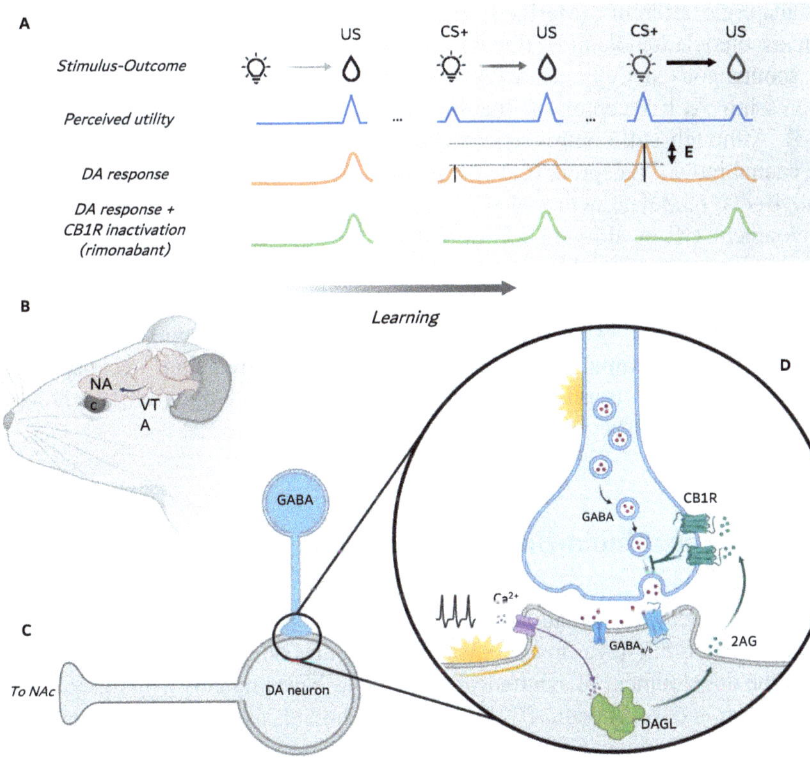

Fig. 1 Proposed eCB orchestration of DA encoding of reward-predictive cues. (**A**) Transfer of DA response from the unconditioned stimulus (US, reward) to the earliest predictor (CS+, cue). Downregulation of CB1R signaling (e.g., rimonabant administration) suppresses DA activity evoked by cue presentation, but not by the reward (Oleson et al. 2012). (**B**) Principal anatomical components of the brain DA response during CS-US associative learning. Both DA cell firing in the VTA and DA release in the NAc follow similar dynamics during the encoding of rewards and reward-predictive stimuli, but local, postsynaptic influences exist. (**C**) Increased DA cell firing in response to conditioned stimuli is posited to arise from downscaled GABA neurotransmission-impinging VTA DA neurons. (**D**) DA neuron high-frequency phasic burst results in an increase in intracellular calcium levels. Calcium-activated, eCB-producing enzymes (such as DAGL) synthesize 2AG, which in turn is released from the post-synaptic compartment to the presynaptic terminal. 2AG retrogradely activates CB1R preferentially located on GABA afferents. Gi/o-coupled CB1R activation suppresses GABA release onto VTA DA cells. (**E**) This series of synaptic events is thought to mediate the net increase in DA encoding of reward-predictive cues during learning. Ultimately, this mechanism has the potential to account for the efficacy of eCB-CB1R pharmacological manipulations to modulate the reinstatement of drug reward-seeking behaviors in response to predictive cues

in the inhibition of cortical glutamate terminals (Mateo et al. 2017). The net action of this eCB mechanism results in the inhibition of ACh-facilitated DA release in the striatum. Other lipid substrates constituting the "expanded eCB system" are also proposed to regulate DA neuron activity (Melis and Pistis 2012). For instance, AEA drives DA cell activation in the midbrain by activating TRPV1 receptors on glutamatergic afferents (Marinelli et al. 2003). In contrast, N-acylethanolamines such as oleoylethanolamide (OEA) and palmitoylethanolamide (PEA), decrease the spontaneous activity of VTA DA cells through a PPARα mechanism by suppressing ACh receptor facilitation of DA cell activation (Melis et al. 2010, 2013). Although still controversial (Morales and Bonci 2012), evidence indicates that cannabinoid 2 receptors (CB2R), mainly expressed by microglial cells (Munro et al. 1993), modulate neuronal signaling in the NAc and thus impact drug abuse development (Xi et al. 2011). How CB2R influence neuronal signaling and DA transmission remains unclear, but few preclinical studies show that CB2R agonists may reduce some motivational effects of different drugs of abuse (Navarrete et al. 2021). In conclusion, extensive research supports a variety of eCB-mediated mechanisms that can potentially be targeted to finetune DA signaling, opening a therapeutic window to treat motivational disorders stemming from pathological alterations in the mesocorticolimbic system (Luján et al. 2021).

2 The Cannabinoid-Based *pharmacopoeia*

The crystal structure of ligand-bound CB1R (Hua et al. 2017, 2020; Shao et al. 2019) has provided atomic-level insight into the molecular docking dynamics, which has fueled the development of synthetic, cannabinoid-based ligands with clinical potential (Ramesh and Rosenbaum 2021). These compounds can be broadly grouped into four categories: (**i**) direct CB1/2R ligands, (**ii**) inhibitors of eCB-degrading enzymes, (**iii**) ligands of the expanded-eCB system and (**iv**) multi-target compounds (mainly phytocannabinoids) that modulate non-eCB signaling.

Among the direct CB1/2 receptor ligands studied for their anti-relapse properties, classic **CB1R receptor antagonists** are at the forefront of cannabinoid-based pharmacotherapies. As previously stated, 2AG-CB1R signaling is potentiated by exposure to drugs and subsequently drives incentive salience. For this reason, CB1R receptor antagonists such as rimonabant were hypothesized to reduce drug relapse rates and thereby suffice as a treatment for substance use disorders. However, its adverse psychiatric side effects (e.g., anxiety, depression, suicidality), prompted its withdrawal from the pharmaceutical market (Sam et al. 2011). Extensive research demonstrated that the side effects of rimonabant were due to its **CB1R receptor inverse agonist** profile (Le Foll et al. 2009), urging the need for synthesizing and validating alternative CB1R receptor blockers that would exert fewer unwanted side effects. Recently, the National Institute of Drug Abuse listed CB1R receptor neutral antagonists and negative allosteric modulators as top priority targets for developing new medications to address the current opioid crisis (Rasmussen et al. 2019). **Neutral antagonists**, unlike inverse agonists, do not modify ligand-independent control of CB1R's second messenger signaling (Greasley and Clapham 2006; Bergman et al. 2008) and thus offer a safer therapeutic profile (Järbe et al. 2008; Nguyen et al. 2019). **Negative allosteric modulators (NAMs)** emerged as an alternative to modulate eCB dynamics exclusively in the presence of endogenous ligands (Morales and Jagerovic 2020). By binding to their allosteric site, NAMs decrease the probability of CB1R conformational changes upon the binding of the orthosteric ligand.

Alternatively, eCB neurotransmission can be pharmacologically manipulated by targeting the lipid metabolic machinery, thereby preserving ligand-unbound CB1R signaling. **Fatty acid amide hydrolase (FAAH) inhibitors**, like URB597, have been consistently used to decrease AEA degradation and thereby increase AEA concentration in the brain and plasma, which has the advantage of not only modulating CB1R, but also CB2R, PPARα/γ, and TRPV1 receptors. Inhibition of FAAH also elevates PEA and OEA concentrations (Sagheddu et al. 2019). Unfortunately, pharmacological manipulation of eCB metabolism has proven difficult as off-target effects may occur (Kerbrat et al. 2016). Enhancing eCB tone via FAAH inhibition can be of great therapeutic efficacy, but this intervention requires cautious consideration to avoid the lipidome-related neurological complications displayed by the FAAH inhibitor BIA10-2474 (van Esbroeck et al. 2017). Otherwise, **AEA transport inhibitors**, like VDM11 or AM404, may serve as safer therapeutic alternatives to FAAH blockers. Pharmacological strategies targeting 2AG metabolism have not been as thoroughly developed and are still poorly understood. While **MAGL inhibitors**, which increase 2AG levels, expectedly increase cue-induced reinstatement of drug-seeking (Oleson et al. 2012; Trigo and Le Foll 2016), **DAGL inhibitors** (responsible for the generation of 2AG (Wenzel et al. 2018)) may induce more off-target effects than FAAH inhibitors due to their upstream placement in the lipid synthesis pathway.

The cannabinoid-based portfolio extends to **phytocannabinoids**, a diverse class of naturally occurring chemical constituents derived from the *Cannabis sativa* plant. A hallmark feature of phytocannabinoids is their ability to influence multitarget actions on an array of neurobiological substrates beyond those of the eCB system (Turner et al. 2017). **Cannabidiol (CBD)** has attracted substantial interest as a pharmacological agent to treat motivational disorders (Zlebnik and Cheer 2016a). This non-intoxicating phytocannabinoid has a promiscuous pharmacological profile that combines most of the desirable eCB-mediated effects mentioned previously (Galaj and Xi 2021). CBD inhibits FAAH activity and blocks the transmembrane transport of AEA (Bisogno et al. 2001), and is posited to act as a NAM at CB1R and CB2R (Tham et al. 2019; Jakowiecki et al. 2021) while also activating PPARγ, GPR55, and TRPV1 receptors (De Petrocellis et al. 2011). Beyond the eCB system, CBD activates neuronal Kv7 channels (Zhang et al. 2022), interacts with $5HT_{1A}$ receptors (Zanelati et al. 2010), and blocks the adenosine transporter ENT1 (Carrier et al. 2006) as well as oxytocin 1 receptors (Vitale et al. 2021). Another cannabis constituent relevant for the management of drug relapse is the terpene β-caryophyllene (**BCP**), a food additive abundantly found in essential oils from black pepper, cloves, or thyme (Sharma et al. 2016). Unlike CBD, BCP elicits a full agonist action on CB2R and is a potent antagonist of α7-nACh receptors (Hashiesh et al. 2021). Lastly, another important *Cannabis s.* plant constituent is the phytocannabinoid **Δ^9-tetrahydrocannabivarin (THCV)**. THCV is an interesting pharmacological candidate to mitigate relapse given its combined CB1R neutral antagonist and CB2R agonist properties (McPartland et al. 2015).

The unique properties of the eCB system in modulating mesolimbic DA dynamics open a therapeutic window to tackle the reinstatement of drug-seeking behaviors. In the following sections, we review evidence from clinical and animal studies showing the anti-relapse effects of cannabinoid-based agents upon each of the major classes of substances of abuse.

3 Opioids

Opioids are one of the most lethal classes of drugs of misuse, and the development of new pharmacotherapies preventing relapse to opioid use and overdose remains more critical than ever (Ignaszewski 2021). Mounting research on opioid pharmacology and the eCB system has elucidated a series of maladaptive changes that underlie their excessive seeking (Parolaro et al. 2010; Laviolette 2017). Interestingly, endogenous opioids and cannabinoids similarly modulate DA neuron excitability by inhibiting GABAergic interneurons within the VTA (Spanagel 2020). A recent study by Zhang et al. (2021) suggests that adaptations of the eCB system following repeated morphine exposure synergistically amplify opioid-induced disinhibition of VTA DA neurons. α,β-Hydrolase-domain-containing protein 6 (ABHD6) also hydrolyses 2AG and its activity is even higher than that of MAGL in some brain structures (Baggelaar et al. 2017). In rats, sustained morphine exposure downregulates ABHD6 expression in the VTA, a mechanism that may drive opioid-mediated long-term synaptic depression in VTA inhibitory interneurons (Zhang et al. 2021). Previous research has also shown that repeated morphine exposure increases *Cnr1* mRNA levels (Jin et al. 2014). Accordingly, studies assessing the reinforcing efficacy of opioid drugs (mainly heroin and morphine) demonstrate that pharmacological facilitation of eCB-CB1R signaling enhances opioid-seeking behavior (Panlilio and Justinova 2018).

Growing evidence, especially from preclinical models, demonstrates the promising utility of CB1R antagonists to tackle opioid relapse (Table 1). Seminal work by De Vries et al. (2003) showed that the CB1R antagonist **rimonabant (SR141716A)** attenuates the reinstatement of heroin-seeking precipitated by drug-paired cues or heroin priming. These findings were corroborated by additional studies showing that administration of rimonabant reduced drug-induced reinstatement of heroin-seeking in rats (Fattore et al. 2003, 2005). Later, it was demonstrated that local CB1R antagonism in the NAc, PFC, but not the basolateral amygdala, accounted for the effects of systemic exposure to rimonabant in the reinstatement of heroin-seeking (Alvarez-Jaimes et al. 2008). Disruption of eCB-CB1R signaling may also prevent opioid relapse triggered by CB1R activation (Fattore et al. 2011), a promising outcome considering that cannabis-opioid co-use is present in at least 25% of hospitalized patients (Rogers et al. 2019). The use of classic rimonabant-like CB1R antagonists, as described previously, is currently not clinically viable.

Table 1 Table summary of the studies assessing the effects of cannabinoid agents in the reduction of reinstatement of **opioid** drug seeking; –, no change found; ↑, increase; ↓, decrease

Compound class	Compound	Experimental measurement	Effect	Species	Reference
CB1/2R ligands	Rimonabant (SR141716)	Cue- and drug-induced **heroin** seeking	↓	Rats	De Vries et al. (2003)
		Drug-induced reinstatement of heroin-seeking behavior	↓	Rats	Fattore et al. (2003, 2005)
		Cannabinoid-induced reinstatement of heroin seeking	↓	Rats	Fattore et al. (2011)
		Cue-induced reinstatement of heroin seeking	↓	Rats	Alvarez-Jaimes et al. (2008)
	AM6527 (CB1R neutral antagonist)	Drug-induced reinstatement of heroin-seeking behavior	↓	Rats	Soler-Cedeño et al. (2024)
Inhibitors of eCB metabolism/ transport	SA57 (FAAH/ MAGL inhibitor)	Cue-induced reinstatement of heroin-seeking behavior	–	Mice	Wilkerson et al. (2017)
Ligands of the “expanded-eCB” system	SB366791 (TRPV1 antagonist)	Drug-induced reinstatement of morphine seeking	↓	Rats	Ma et al. (2017)
	Pioglitazone (PPARγ agonist)	Stress and drug-induced reinstatement of heroin seeking	↓	Rats	de Guglielmo et al. (2017)
		Cue-induced reinstatement of heroin seeking	–		
		Opioid-positive urine specimens during buprenorphine taper	–	Humans	Schroeder et al. (2018)
Phytocannabinoids	CBD	Cue-induced reinstatement of heroin seeking	↓	Rats	Ren et al. (2009)
		Drug-induced reinstatement of heroin seeking	–		
		Craving and anxiety induced by the presentation of heroin cues	↓	Humans	Hurd et al. (2019)

Alternative pharmacological approaches that circumvent the challenges of CB1R inverse agonism have been pursued, but the number of studies specifically addressing opioid relapse are few with limited results. The novel CB1R **neutral antagonist AM6527** was recently described to dose-dependently inhibit heroin self-administration and reinstatement in rats (Soler-Cedeño et al. 2024). Alternatively, the **dual FAAH/MAGL inhibitor SA57** did not affect the reinstatement of heroin-seeking behavior elicited by cues, despite reducing motivation for heroin and increasing its antinociceptive effects (Wilkerson et al. 2017).

Other studies have focused on non-CB1R targets within the eCB system. TRPV1 receptors directly interact with mu-opioid 1 receptors, the main substrate for opioid drugs, and prevent agonist-induced receptor internalization from the cell membrane (Scherer et al. 2017). Consequently, the **TRPV1 antagonist SB366791** was shown to reduce the priming-induced reinstatement of morphine-seeking behavior in rats (Ma et al. 2017). Inactivation of TRPV1 receptors has the additional benefit of reducing anxiety-like behaviors and chronic pain (Morales-Lázaro et al. 2016) and could help manage the relapse-triggering effects of stress and pain-related allodynia during abstinence. Indeed, Ma et al. (2017) showed that administration of SB366791 for relapse-preventing treatment was accompanied by a reduction of anxiety-like behaviors. These results warrant further investigation into the therapeutic effects of TRPV1 receptor antagonists, specifically in hampering stress-induced relapse to opioid use. PPAR have also shown promising potential in reducing certain types of opioid relapse. $PPAR_\gamma$ specifically regulate morphine-induced activation of DA neurons and DA release in the NAc shell (De Guglielmo et al. 2015). Although the precise mechanism is unknown, growing evidence supports the hypothesis that $PPAR_\gamma$ may facilitate VTA GABA interneuron signaling (Domi et al. 2016). As such, the **$PPAR_\gamma$ agonist, pioglitazone**, which is used to treat type 2 diabetes, blunts stress- and priming-induced reinstatement of heroin-seeking behaviors in rats (De Guglielmo et al. 2015). However, heroin-seeking reinstatement by associated cues was not affected. In humans, the effectiveness of pioglitazone remains an open question. A randomized, placebo-controlled clinical trial among participants receiving oral pioglitazone or placebo before, during, and after a buprenorphine taper reported no overall differences between treatment groups with trends of exacerbated opioid withdrawal scores and increase of rescue medications in the pioglitazone arm during the post-taper phase (Schroeder et al. 2018). Lastly, the ability of **CBD** to prevent relapse to opioid use has also been tested. CBD, which activates $PPAR_\gamma$ and is a partial agonist of TRPV1 channels (Turner et al. 2017), reduced cue-induced reinstatement of heroin-seeking in rats (Ren et al. 2009). These effects were accompanied by the normalization of imbalances in glutamatergic signaling [α-amino-3-hydroxy-5-methyl-4-isoxazole propionic acid (AMPA) receptor subunit composition], which is theorized to occur during periods of abstinence preceding relapse. While opioid abstinence or relapse rates after CBD treatment were not assessed, administration of CBD to drug-abstinent individuals with heroin use disorder showed that CBD significantly reduced cue-induced heroin cravings (Hurd et al. 2019).

Overall, initial studies yielded promising results regarding the utility of cannabinoid pharmacological manipulations to mitigate opioid relapse (De Vries et al. 2001, 2003; Fattore et al. 2003, 2005; Alvarez-Jaimes et al. 2008). However, research in this direction seems to have stalled in recent years, probably owing to the adverse side effects of rimonabant in humans. On this front, further studies are required to specifically develop and validate safer pharmacological treatments to curtail CB1R signaling (e.g., neutral antagonists, NAMs). Alternative targets in the eCB system may also open interesting avenues for scientific inquiry, especially on TRPV1 receptor antagonists, which have yet to be tested in clinical trials among opioid-abstinent individuals.

4 Cocaine and Other Psychostimulants

Cocaine use is the leading non-opioid cause of drug overdose death in the US (Center for Disease Control 2020). In 2017, 6.2% of 18–25 aged people in the US reported using cocaine in the last year, compared to 0.6% using heroin (Center for Behavioral Health Statistics and Quality 2017). Unfortunately, cocaine popularity has rebounded after many years of decline (Hswen et al. 2021) and, despite extensive research on the neurobiology of cocaine use disorder (Luján et al. 2023b), no FDA-approved drugs currently exist for the prevention of relapse. Extensive evidence in the literature supports the notion that cocaine-induced disruptions in eCB plasticity result in the concurrent loss of inhibitory control over DA neurons in the VTA and dysregulation of excitatory signaling in the NAc (Zlebnik and Cheer 2016b; Peters et al. 2021b). In the VTA, cocaine exposure increases the burst firing of DA cells in a CB1R-dependent manner (Cheer et al. 2004, 2007; Wang et al. 2015). Consistent with these phenomena, cocaine mobilizes 2AG via local inhibition of norepinephrine uptake and enhances short-term (Lupica and Riegel 2005; Wang et al. 2015) and long-term (Liu et al. 2005; Pan et al. 2008a, b) eCB-mediated inhibition of GABAergic synapses in the VTA. In the NAc, acute and forced (Fourgeaud et al. 2004), as well as chronic and voluntary (McCutcheon et al. 2011), cocaine exposure abolishes eCB long-term depression of excitatory synapses (eCB-LTDe). Strengthening of excitatory synapses in the NAc is required for craving incubation and facilitates reinstatement of cocaine-seeking behaviors (Wolf 2016). Hence, counteracting cocaine-induced loss of eCB-LTDe in the NAc may serve as a viable therapeutic strategy to maintain abstinence from cocaine use.

There is a growing body of preclinical work assessing the impact of cannabinoid-based drugs on the animal models of relapse to cocaine use. As shown in Table 2, the trends are generally encouraging but not without some notable inconsistencies. Twenty years ago, another seminal paper by de Vries et al. (2001) showed that systemic administration of **CB1R antagonists** blunts reinstatement of cocaine-seeking triggered by cues and cocaine priming. The relapse-triggering effects of stress, however, remained unaltered by CB1R antagonist administration, a finding that has been consistently replicated (Kupferschmidt et al. 2012; McReynolds et al. 2016; Luján et al. 2022). The extent to which CB1R inactivation attenuates priming-induced reinstatement of cocaine-seeking is variable based on the animal model and molecular composition, dosage, and route of administration of the compound. Reduction of drug-induced reinstatement is consistently achieved with relatively high doses of rimonabant (1, 3.2 mg/kg, *i.p.*) in rats (De Vries et al. 2001, Jing et al. 2014), but not in squirrel monkeys (0.3 mg/kg, *i.p.*) (Schindler et al. 2016b). The neutral **antagonist AM6527** effectively reduces cocaine reinstatement induced by a

Table 2 Table summary of the studies assessing the effects of cannabinoid agents in the reduction of reinstatement of **cocaine/psychostimulant** seeking; –, no change found; ↑, increase; ↓, decrease

Compound class	Compound	Experimental measurement	Effect	Species	Reference
CB1/2R ligands	Rimonabant	Cue- and drug-induced reinstatement of cocaine seeking	↓	Rats	De Vries et al. (2001)
		Stress-induced reinstatement of cocaine seeking	–		
		Cue- and drug-induced reinstatement of cocaine or METH seeking	↓	Rats	Jing et al. (2014)
		Cue-induced reinstatement of cocaine seeking	↓	Mice	Ward et al. (2009)
			↓	Squirrel monkeys	Schindler et al. (2016a)
		Priming-induced reinstatement of cocaine seeking	–		
		Cue- and drug-induced reinstatement of METH seeking	↓	Rats	Anggadiredja et al. (2004)
	AM251	Stress-potentiation of drug-induced cocaine seeking reinstatement	↓	Rats	McReynolds et al. (2016)
		Footshock-induced reinstatement of cocaine seeking	–		
		Drug-induced reinstatement of cocaine seeking	–		
		Cue-induced reinstatement of cocaine seeking	↓	Rats	Adamczyk et al. (2012)
		Cocaine-induced reinstatement of cocaine-seeking behavior	↓	Rats	Adamczyk et al. (2012)
			↓	Rats	Xi et al. (2006)
		Stress- and drug-induced reinstatement of cocaine seeking	–	Rats	Kupferschmidt et al. (2012)
	AM4113 (CB1R neutral antagonist)	Cue- and drug-induced reinstatement of cocaine seeking	↓	Squirrel monkeys	Schindler et al. (2016a)

(continued)

Table 2 (continued)

Compound class	Compound	Experimental measurement	Effect	Species	Reference
		Cue-, stress- and drug-induced reinstatement of cocaine seeking	–	Mice	Luján et al. (2022)
	Org27569 (CB1R NAM)	Drug- and cue-induced reinstatement of cocaine or METH	↓	Rats	Jing et al. (2014)
	PIMSR (CB1R NAM)	Cue-induced reinstatement of cocaine	↓	Rats	Galaj et al. (2022)
	SR144528 (CB2 antagonist)	Cocaine-induced reinstatement of cocaine-seeking behavior	↓	Rats	Adamczyk et al. (2012)
		Cue-induced reinstatement of cocaine-seeking	–		
Inhibitors of eCB metabolism/ transport	URB597 (FAAH inhibitor)	Cue- and stress-triggered reinstatement of cocaine seeking	↓	Rats	Chauvet et al. (2015)
	AM404 (AEA transport inhibitor)	Priming-induced reinstatement of cocaine-seeking behavior	↑	Squirrel monkeys	Schindler et al. (2016b)
	Diclofenac (COX1/2 inhibitor)	Cue- and drug-triggered reinstatement of METH seeking	↓	Rats	Anggadiredja et al. (2004)
Ligands of the "expanded-eCB" system	SB366791 (TRPV1 antagonist)	Cocaine-induced reinstatement of cocaine-seeking behavior	↓	Rats	Adamczyk et al. (2012)
		Cue-induced reinstatement of cocaine-seeking	–		
	Pioglitazone (PPARγ agonist)	Cocaine cue reactivity	↓	Rats	Miller et al. (2018a, b)
		Cocaine cravings	↓	Humans	Schmitz et al. (2017)
Phytocannabinoids	CBD	Cue-induced reinstatement of cocaine seeking	↓	Mice	Luján et al. (2018, 2022)
		Drug-induced reinstatement of cocaine seeking	–		

(continued)

Table 2 (continued)

Compound class	Compound	Experimental measurement	Effect	Species	Reference
		Stress-induced reinstatement of cocaine seeking	↑		
		Cocaine craving incubation	–	Mice	Alegre-Zurano et al. (2022)
		Cue- and stress-induced reinstatement	↓	Rats	Gonzalez-Cuevas et al. (2018)
		Drug-induced reinstatement of cocaine seeking	–	Rats	Mahmud et al. (2017)
			↓		Hay et al. (2018)
		Crack-cocaine cravings	–	Humans	de Meneses-Gaya et al. (2020)
		Cue-induced cocaine cravings	–	Humans	Mongeau-Pérusse et al. (2021)
		Time-to-cocaine relapse			
	THC	Drug-induced reinstatement of METH seeking	↓	Rats	Anggadiredja et al. (2004)
		Cue-induced reinstatement of METH seeking	↑		
	BCP	Drug- and cue-induced reinstatement of METH seeking	↓	Rats	He et al. (2021)

cocaine *i.p.* priming (Soler-Cedeño et al. 2024). In contrast, a low dose of the **neutral antagonist AM4113** (5 mg/kg, *i.p.*) fails to alter the reinstatement of cocaine-seeking elicited by cocaine priming in mice (Luján et al. 2022). CB1R inverse agonists, such as AM251, do not influence drug-induced reinstatement administered by systemic or intracerebroventricular injection (Kupferschmidt et al. 2012; McReynolds et al. 2016). However, local NAc infusion of AM251 suppresses drug-induced reinstatement (Xi et al. 2006).

Owing to the prominent role of eCB-CB1R neurotransmission in the encoding of reward-paired incentive salience (Luján et al. 2023a), studies addressing the effects of rimonabant on **cue**-induced reinstatement of cocaine-seeking are considerably less discrepant (De Vries et al. 2001; Anggadiredja et al. 2004; Ward et al. 2009; Jing et al. 2014; Schindler et al. 2016b). This reduction can be recapitulated using the inverse agonist AM251 (Adamczyk et al. 2012), but not the neutral antagonist AM4113 (Luján et al. 2022). Importantly, the **CB1R NAMs Org27569** and

PIMSR dose-dependently prevent the reinstatement of cocaine-seeking behavior triggered by cues in rats (Jing et al. 2014; Galaj et al. 2022) without exerting adverse side effects of CB1R inverse agonists. These data justify the need for further investigation but confirm the potential of CB1R NAM as a potentially safe pharmacotherapy for the prevention of cue-induced psychostimulant relapse. Unfortunately, no clinical data investigating the effects of CB1R inactivation in individuals with cocaine use disorder are available at present. Clinical studies assessing and validating CB1R inactivation during periods of abstinence may serve as one of the most promising cannabinoid-based strategies for the sustained prevention of relapse to psychostimulant use triggered by associated stimuli.

The **FAAH inhibitor URB597**, despite increasing AEA/CB1R signaling, is not intrinsically rewarding, and is regarded as a viable option for reducing relapse to psychostimulant use. Indeed, the effects of URB597 resemble those of CB1R antagonists: it reduces cue- and stress-induced reinstatement of cocaine-seeking (Chauvet et al. 2015). The precise mechanism by which FAAH inhibition recapitulates the effects of CB1R inactivation in animal models of drug relapse is unclear. One possibility is that AEA, a partial agonist of CB1R, outcompetes 2AG binding, thus reducing net 2AG-CB1R signaling. URB597 may also exert off-target effects. Justinova et al. (2008a) showed that URB597 results in an increase of midbrain OEA levels, which activates PPARα receptors and reduces DA cell excitability (Melis et al. 2010, 2013). Similar results were observed in methamphetamine-seeking animals by inhibiting the alternative AEA/2AG metabolizing enzymes cyclooxygenase 1/2 with **diclofenac** (Anggadiredja et al. 2004). The effects of AEA are also terminated by its reuptake into neurons (Cristino et al. 2020). The specific mechanisms of this process are still a matter of debate, but fatty acid-binding proteins (FABPs) have been proposed to serve as active transporters of AEA in neurons (Kaczocha et al. 2009). FABP inhibitors may circumvent key issues related to pharmacological manipulations of eCB metabolism (e.g., unspecific alteration of the neuronal lipidome). While significant progress has been made to assess the therapeutic potential of FABP inhibitors against other drugs (e.g., nicotine), Schindler et al. (2016b) showed that a single injection of the **AEA uptake inhibitor AM404** triggered the reinstatement of cocaine-seeking behavior.

As seen with opioid drugs, substances targeting the "expanded-eCB system" could be of significant utility in tackling psychostimulant relapse. The **TRPV1 antagonist, SB366791,** reduced priming-induced reinstatement of cocaine-seeking behavior, while relapse rates elicited by cues remained unchanged (Adamczyk et al. 2012). More studies are needed to further elucidate the role of TRPV1 antagonists as therapeutic candidates for preventing drug-induced reinstatement of drug-seeking behaviors and as modulators of vanilloid channels in the neurocircuitry of motivated behavior. The **activation of PPAR$_\gamma$** has been similarly explored. Pioglitazone, a PPAR$_\gamma$ agonist, decreases reactivity towards cocaine paired cues in rats (Miller et al. 2018b) and reduces cocaine cravings in humans (Schmitz et al. 2017). In the latter report, activation of PPAR$_\gamma$ was accompanied by an increase in white matter integrity, highlighting one of the potentially protective, neuroplastic effects of pioglitazone (Schmitz et al. 2017). **CBD** also targets FAAH, PPAR and TRPV1 receptors and, accordingly, its anti-relapse properties have been of great interest in the psychostimulant research field (Calpe-López et al. 2019; Alegre-Zurano et al. 2022). Evidence, however, is not yet fully consistent. As seen with other cannabinoid-based treatments, CBD seems to be particularly effective in reducing cocaine-seeking behaviors reinstated by cues in rodents (Gonzalez-Cuevas et al. 2018; Luján et al. 2022). Although the precise mechanism of action remains to be understood, data indicate that the protective effect of CBD employs intact CB1R signaling during memory retrieval (Luján et al. 2022), suggesting that CBD treatment during abstinence introduces neuroplastic changes in CB1R neurotransmission that are required to prevent cue-elicited relapse. Intriguingly, the same study suggested that the opposite effect is observed for the stress-induced reinstatement of cocaine-seeking. That is, CBD facilitates reinstatement induced by stressful foot shocks in mice, an effect that was abolished when a CB1R antagonist was administered during reinstatement (Luján et al. 2022). The nature of CB1R-related neuroplastic changes during CBD treatment was investigated in the same study, revealing an overall increase in CB1R protein levels in different brain structures of the mesocorticolimbic system in CBD-treated mice. Previous studies have reported long-lasting changes induced by CBD in critical CB1R-dependent neuroplastic processes including increased adult hippocampal neurogenesis (Luján et al. 2018, 2020; Luján and Valverde 2020), reduced neuroinflammation (Campos et al. 2017; Luján et al. 2018) and enhanced autophagy (Hosseinzadeh et al. 2016). Altogether, converging evidence suggests that CBD-induced neuroplasticity enables CB1R signaling to control the retrieval of drug-seeking behaviors more prominently. Based on these promising preclinical data, Mongeau-Pérusse et al. (2021) conducted a phase II double-blind, randomized, placebo-controlled trial in which recently abstinent individuals with cocaine use disorder were treated with CBD. CBD neither prevented cocaine use relapse nor cue-induced cravings. These findings are in accordance with another pilot trial reporting that CBD had no effect in reducing cravings for crack cocaine (de Meneses-Gaya et al. 2020). Several factors could contribute to the stark contrast observed between the clinical and pre-clinical data. Mongeau-Pérusse et al. acknowledged several limitations that may have hindered their ability to detect the positive effects of CBD in decreasing cocaine relapse or craving, including small sample size, no direct supervision of medication intake, and high attrition rates (Mongeau-Pérusse et al. 2021). In addition, animal models of drug relapse, particularly rodent models, may lack the predictive validity they are usually credited for. The complex pharmacology of CBD may also explain species-specific divergencies. Considering the myriad of targets for CBD in the brain, it is safe to assume that unexpected pharmacodynamic interactions may uniquely occur in rodents but not in humans, and vice versa.

In addition to CBD, the terpene **BCP**, used as a food additive, features a similarly appealing pharmacodynamic profile. A recent investigation of BCP's therapeutic potential in a rat relapse model has shown that BCP decreases cue-induced reinstatement of methamphetamine-seeking and inhibits methamphetamine-primed reinstatement of drug-seeking, suggesting that CB2R and non-CB2R mechanisms are involved in these protective properties. (He et al. 2021). Inactivation of α7-nAChR by BCP (Hashiesh et al. 2021), which modulates DA terminal release and DA cell excitability (Maex et al. 2014), could explain these effects.

In conclusion, cannabinoid-based pharmacotherapies for the management of cocaine/psychostimulant relapse have the potential to effectively prevent reinstatement of drug-seeking induced by drug-cue exposure. Classic and neutral CB1R antagonists, CB1R NAMs, as well as inhibitors of eCB metabolism are promising contenders, but clinical validation of their safety and efficacy is lacking. Phytocannabinoids, namely CBD and BCP, hold promise as safe and well-tolerated compounds (Elsaid et al. 2019) but preclinical evidence of their efficacy in preventing relapse to psychostimulant use is mixed or insufficient, respectively. Lastly, modulators of the "expanded-eCB system" may be useful to palliate drug-induced reinstatement of psychostimulant-seeking (via TRPV1 antagonism).

5 Nicotine

Tobacco smoking is the world's leading cause of preventable death (World Health Organization 2017). The reinforcing properties of nicotine hinder smoking cessation and make it even harder to curb the associated mortality. Unlike other substance use disorders, various FDA-approved medicines to promote smoking cessation are available: nicotine replacement therapy, varenicline, and bupropion. (Giulietti et al. 2020). Nevertheless, current treatment options show limited efficacy and warrant further investigation for developing innovative pharmacological strategies to treat smoking dependence. The eCB system is tightly implicated in the reinforcing actions of nicotine, and exposure to this alkaloid induces profound changes in the eCB system (Scherma et al. 2016). Several studies conducted in rat models have shown amplified eCB-CB1R signaling upon chronic exposure to nicotine (Werling et al. 2009; Buczynski et al. 2013; Reisiger et al. 2014). Accordingly, compounds that upregulate eCB levels facilitate nicotine reinforcement, while those that disrupt eCB signaling lower nicotine's rewarding effects (Saravia et al. 2021). These strong relationships are preserved in the context of reinstatement of nicotine-seeking behaviors (Table 3).

As seen with other drugs of abuse, **inactivation of CB1R** has shown to attenuate reinstatement of nicotine-seeking followed by exposure to cues, stressful stimuli, and nicotine priming (Forget et al. 2009; de Bruin et al. 2011; Gueye et al. 2016; Schindler et al. 2016a). In some studies, these effects were similarly achieved by the **neutral antagonist, AM4113,** in non-human primates (Gueye et al. 2016; Schindler et al. 2016a), which constitute the most compelling results on its clinical utility to date. Similarly, diverse **inhibitors of AEA degradation, or reuptake**, consistently impair priming- and cue-induced reinstatement of nicotine-seeking behaviors in rats (Scherma et al. 2008; Forget et al. 2009, 2016; Gamaleddin et al. 2011, 2013) and squirrel monkeys (Justinova et al. 2015). Accordingly, inactivation of 2AG hydrolysis by the **MAGL inhibitor, JZL184,** enhances cue induced reinstatement of nicotine-seeking (Trigo and Le Foll 2016). In humans, results from three

Table 3 Table summary of the studies assessing the effects of cannabinoid agents in the reduction of reinstatement of **nicotine** seeking; –, no change found; ↑, increase; ↓, decrease

Compound class	Compound	Experimental measurement	Effect	Species	Reference
CB1/2R ligands	Rimonabant SR141716A	Priming- and cue-induced reinstatement of nicotine-seeking behavior	↓	Squirrel monkeys	Schindler et al. (2016a)
			↓	Rats	Forget et al. (2009)
		Smoking abstinence rates	↓	Humans	STRATUS EU, US, META; pooled review in Robinson et al. (2018)
	AM630 (CB2R antagonist)	Priming- and cue-induced reinstatement of nicotine-seeking behavior	–	Rats	Gamaleddin et al. (2012a, b)
	AM1241 (CB2R agonist)				
	SLV330 (CB1R antagonist)	Cue-induced reinstatement of nicotine-seeking behavior	↓	Rats	de Bruin et al. (2011)
	AM4113 (CB1R neutral antagonist)	Cue-, priming- and stress-induced reinstatement of nicotine-seeking behavior	↓	Squirrel monkeys	Gueye et al. (2016)
		Priming- and cue-induced reinstatement of nicotine-seeking behavior	↓	Squirrel monkeys	Schindler et al. (2016a)
Inhibitors of 2AG/AEA metabolism or transport	URB597 (FAAH inhibitor)	Nicotine-induced reinstatement of nicotine seeking	↓	Rats	Scherma et al. (2008)
		Priming- and cue-induced reinstatement of nicotine-seeking behavior	↓	Rats	Forget et al. (2009)
				Squirrel monkeys	Justinova et al. (2015)
		Cue-induced reinstatement of nicotine-seeking behavior	↓	Rats	Forget et al. (2016)
	URB694 (FAAH inhibitor)	Priming- and cue-induced reinstatement of nicotine-seeking behavior	↓	Squirrel monkeys	Justinova et al. (2015)

(continued)

Table 3 (continued)

Compound class	Compound	Experimental measurement	Effect	Species	Reference
	JZL184 (MAGL inhibitor)	Cue-induced reinstatement of nicotine-seeking behavior	↑	Mice	Trigo and Le Foll (2016)
	AM404 (AEA reuptake inhibitor)	Priming- and cue-induced reinstatement of nicotine-seeking behavior	↓	Rats	Gamaleddin et al. (2013)
	VDM11 (AEA reuptake inhibitor)				Gamaleddin et al. (2011)
Ligands of the "expanded-eCB" system	Clofibrate (PPARa agonist)	Cue- and drug-induced reinstatement of nicotine-seeking	↓	Squirrel monkeys	Panlilio et al. (2012)
	WY14643 (PPARa agonist)		↓	Rats, squirrel monkeys	Mascia et al. (2011)
	Fenofibrate (PPARa agonist)	Smoking abstinence rates	–	Humans	Perkins et al. (2016)
	Gemfibrozil (PPARa agonist)	Smoking abstinence rates and cue reactivity	–	Humans	Gendy et al. (2018)
Phytocannabinoids	THCV	Cue-induced and nicotine-induced relapse to nicotine-seeking behavior	↓	Rats	Xi et al. (2019)
	BCP	Cue-induced reinstatement of nicotine-seeking	↓	Rats	He et al. (2020)
	CBD	Tobacco craving and attentional bias to cigarette cues	↓	Humans	Hindocha et al. (2018)
		Smoking cravings	–	Humans	Morgan et al. (2013)

randomized, placebo-controlled trials (STRATUS EU, US, and META) supported rimonabant as a modestly effective aid for smoking cessation (Robinson et al. 2018). Although work in this direction was halted due to rimonabant's toxic side effects, these results inform and spur the development of alternative antagonists with improved tolerability, such as AM4113, which have yet to enter clinical trials.

Clinical and preclinical data has not been obtained among PPARα agonists. As previously mentioned, activation of PPARα abolishes nicotine-induced α7-nAChR facilitation of DA cell activation and decreases spontaneous activity of VTA DA cells (Melis et al. 2010, 2013). In line with this mechanism, studies in rat and non-human primate models supported the efficacy of the **PPARα agonists, clofibrate, and WY14643,** in lessening cue- and priming-induced reinstatement of nicotine seeking (Mascia et al. 2011; Panlilio et al. 2012). These promising reports were later followed by two pilot clinical trials using other FDA-approved PPARα agonists. Unfortunately, neither fenofibrate nor gemfibrozil affected smoking abstinence or reactivity towards smoking cues in adult smokers intending to quit smoking (Perkins et al. 2016; Gendy et al. 2018). Substance-specific differences could account for the divergent effects observed between PPARα agonists in animal and human studies. Nevertheless, PPARα agonists have a safe pharmacological profile and a suitable mechanism of action, which warrants further clinical investigation into the relapse-preventing effects of WY14643 and various fibrates. The **phytocannabinoid CBD** has also been reported to inhibit the function of α7-nAChRs (Mahgoub et al. 2013). CBD treatment eases nicotine withdrawal symptoms, reducing abstinence-induced hyperalgesia (Smith et al. 2021), cognitive impairments, and neuroinflammation (Saravia et al. 2019). Surprisingly, preclinical work assessing the potential of CBD in reducing the reinstatement of nicotine-seeking behaviors remains sparse. Two clinical pilot trials administered CBD to tobacco smokers and evaluated relapse-predictive correlates such as self-reported feelings of craving and attentional biases to cues suggestive of cigarette smoking. In one trial, ad hoc inhaled CBD use reduced the number of cigarettes smoked by almost 40%, but did not affect craving among participants (Morgan et al. 2013). In the other study, orally ingested CBD reduced attentional bias to cigarette cues and reduced craving (Hindocha et al. 2018). The disparity between these two studies may depend on the dose, as cravings were only reduced with the higher dose (400 vs. 800 mg) and route of CBD administration (inhaled vs. oral). These data are encouraging, but more work is certainly needed. Alternatively, the terpene **BCP** and phytocannabinoid **THCV** may offer clinical promise, as they have both attenuated cue- and nicotine-induced reinstatement of nicotine-seeking behavior in rats (He et al. 2020) (Xi et al. 2019).

Altogether, a growing number of studies support the use of cannabinoid-based pharmacological strategies to prevent relapse to nicotine consumption. While initial enthusiasm for CB1R antagonists was tempered by the detection of the psychiatric side effects in rimonabant-like compounds, progress continues to be made in exploring alternative approaches targeting CB1R inactivation. Among promising candidates, the neutral antagonist AM4113 currently stands out, in our opinion, due to its encouraging preclinical findings and a well-known neurobiological rationale for the blockade of CB1R signaling. Alongside CB1R antagonists, phytocannabinoids and fibrates open new lanes on the highway for cannabinoid-based drug development, though their clinical safety and efficacy remain to be validated.

6 Alcohol

According to the National Survey on Drug Use and Health (NSDUH 2021), 10% of people aged 12 or older (28.3 million individuals) in the US suffered from alcohol use disorder in the past year. Overall, it is estimated that alcohol use is responsible for 5.1% of the global burden of disease (Hammer et al. 2018). Medications for the management of alcohol use disorders are available (naltrexone, acamprosate, and disulfiram), but with limited efficacy in preventing relapse (Morley et al. 2021). As seen with previous types of drugs of abuse, alcohol exerts part of its reinforcing and deleterious effects through the eCB system in the brain. Acute alcohol exposure results in the CB1R-dependent activation of VTA DA neurons (Perra et al. 2005; Cheer et al. 2007). Withdrawal from chronic alcohol vapor exposure further increases eCB-CB1R-mediated inhibition of GABAergic transmission in the VTA (Harlan et al. 2018). Alcohol intake increases eCB levels in the striatum both in rats (Vinod et al. 2012) and in postmortem tissue from patients with alcohol use disorder (Vinod et al. 2010). Additionally, chronic alcohol treatment reduces striatal *Cnr1* mRNA levels (Ortiz et al. 2004; Cippitelli et al. 2005), possibly as a compensatory mechanism. The existing data makes a strong case for the development and use of cannabinoid-based pharmacotherapies to tackle alcohol-induced disturbances, namely by mitigating the reinforcing effects of alcohol and thereby preventing relapse to alcohol use.

Early research on the relapse-preventing effects of **CB1R antagonists** was, again, auspicious (Table 4). Rimonabant was deployed and found to effectively attenuate alcohol-seeking behaviors in response to cues (Cippitelli et al. 2005; Economidou et al. 2006, 2007). These effects were also observed using other classic antagonists/ inverse agonists, such as SR147778 (Gessa et al. 2005) and SLV330 (de Bruin et al. 2011). However, this effect was not accompanied by a reduction of stress-induced reinstatement (Economidou et al. 2006, 2007). It is hypothesized that intact CB1R neurotransmission is necessary for the adaptive regulation of stress-induced anxiety-like behaviors (Haller et al. 2004; Marcus et al. 2020) and thus, disruption of CB1R signaling may negate the relapse-preventing properties of rimonabant in stressful situations (Economidou et al. 2007). However, at the time, a large human clinical trial was conducted to assess the potential effect of rimonabant on preventing relapse in recently detoxified alcohol-dependent individuals. Soyka et al. (2008), however, reported no differences between placebo and rimonabant-treated groups.

As cannabinoid-based drug development efforts steered away from rimonabant due to its adverse drug effects, eCB enzymatic inhibitors emerged into the limelight as promising candidates. It was initially found in animal models that the **FAAH inhibitor URB597** does not attenuate cue- or stress-induced reinstatement of alcohol-seeking (Cippitelli et al. 2008). However, Zhou et al. (2017) later suggested that pharmacological elevation of AEA tone decreases relapse-like behavior in a different preclinical model of alcohol relapse. In their study, relapse was modeled after the alcohol deprivation effect (conventionally referred to as ADE). The ADE paradigm refers to the increase in alcohol intake from baseline drinking levels after a

Table 4 Table summary of the studies assessing the effects of cannabinoid agents in the reduction of reinstatement of **alcohol** seeking; –, no change found; ↑, increase; ↓, decrease

Compound class	Compound	Experimental measurement	Effect	Species	Reference
CB1/2R ligands	Rimonabant	Cue-induced reinstatement of ethanol-seeking behavior	↓	Rats	Cippitelli et al. (2005); Economidou et al. (2006, 2007); Getachew et al. (2011)
		Stress-induced reinstatement of ethanol-seeking behavior	–	Rats	Economidou et al. (2006, 2007)
		Alcohol Deprivation Effect (ADE)	↓	Rats	Serra et al. (2002); Getachew et al. (2011); Vinod et al. (2012)
		Alcohol relapse rates	–	Humans	Soyka et al. (2008)
		Abstinence duration			
		Percentage of drinking days			
	SR147778 (CB1R antagonist)	Alcohol Deprivation Effect (ADE)	↓	Rats	Gessa et al. (2005)
	SLV330 (CB1R antagonist)	Cue-induced reinstatement of ethanol-seeking behavior	↓	Rats	de Bruin et al. (2011)
Inhibitors of eCB metabolism/ transport	URB597 (FAAH inhibitor)	Alcohol Deprivation Effect (ADE)	↓	Mice	Zhou et al. (2017)
		Cue-, footshock stress- and yohimbine-induced alcohol relapse	–	Rats	Cippitelli et al. (2008)
	DO34 (DAGL inhibitor)	Reinstatement of alcohol drinking	↓	Mice	Winters et al. (2021)
Ligands of the "expanded-eCB" system	Pioglitazone (PPARγ agonist)	Stress-induced reinstatement of ethanol-seeking	↓	Rats	Stopponi et al. (2011, 2013)
		Cue-induced reinstatement of ethanol-seeking	–		
	WY14643 (PPARα agonist)	Cue-induced reinstatement of ethanol-seeking	↓	Rats	Bilbao et al. (2016)

(continued)

Table 4 (continued)

Compound class	Compound	Experimental measurement	Effect	Species	Reference
Phytocannabinoids	CBD	Cue- and stress-induced reinstatement of ethanol-seeking	↓	Rats	Gonzalez-Cuevas et al. (2018)

period of temporary deprivation and is considered a more robust measure of relapse-like drinking behavior in rats (Vengeliene et al. 2014). No clinical study has yet been conducted to demonstrate the safety and efficacy of URB597 in alcohol-abstinent individuals. Human studies report that a loss-of-function FAAH mutation correlates with increased binge consumption of alcohol and dependence severity (Sipe et al. 2002; Sloan et al. 2018). Additionally, treatment-seeking patients with alcohol use disorder present increased levels of AEA and reduced FAAH levels (Best et al. 2020). A more robust mechanistic understanding of how FAAH inhibition bolsters relapse prevention must be obtained before deploying FAAH inhibitors for clinical trials. Recently, a report by Winters et al. (2021) alternatively proposed that targeting 2AG levels in the brain may be a useful strategy to reduce relapse to alcohol. The **DAGL inhibitor, DO34,** was given during abstinence to mice with a history of alcohol drinking. By inhibiting the rate-limiting enzyme involved in the synthesis of 2-AG, DO34 induced a marked decrease in 2AG levels (but not AEA) and prevented the reinstatement of alcohol drinking (Winters et al. 2021). In accordance with the hypothesized mechanism, pharmacologically suppressed 2AG-CB1R signaling prevented alcohol-induced suppression of GABA transmission in the VTA. These data are promising and warrant future studies on the utility of DAGL inhibitors for the prevention of alcohol relapse.

As mentioned previously, it is unlikely that pharmacological manipulation of eCB-CB1R signaling alone can effectively modulate the relapse-triggering effects of stress in alcohol-consuming animals (Economidou et al. 2006, 2007). This is a gap that other pharmacological approaches may very well fill. The **PPARγ agonist pioglitazone** specifically prevents the reinstatement of alcohol-seeking induced by stress, but not cues (Stopponi et al. 2011, 2013). This outcome is in concordance with its effects observed in heroin-dependent rats (de Guglielmo et al. 2017) and suggests that pioglitazone may be a promising cannabinoid-based candidate for preventing stress-induced relapse to multiple drugs of abuse. Lastly, preclinical studies have also shown that **CBD** may reduce relapse to alcohol drinking. Gonzalez-Cuevas et al. (2018) utilized an elegant method of transdermal delivery of CBD to treat alcohol-abstinent rats, showing that CBD attenuated the cue- and stress-induced reinstatement of alcohol-seeking. These encouraging findings have spurred a currently ongoing clinical study evaluating the use of CBD for the treatment of alcohol use disorder (NCT04873453), focusing on cue-reactivity and alcohol craving.

In summary, an array of candidates informs the development of eCB and cannabinoid-based pharmacotherapies for the management of alcohol relapse. Though much preclinical and clinical work remains to be done, at present, we highlight DAGL inhibitors, CBD, and PPARy agonists as the three main contenders of this quest.

7 Cannabis

The incidence of cannabis use disorder is on the rise (SAMHSA 2019) and there is currently no FDA-approved pharmacotherapy for the treatment of cannabis use disorder (Connor et al. 2021). Thus, the generation of effective pharmacotherapies for this syndrome is a top priority for the national research agenda. Utilizing cannabinoid-derived compounds to treat cannabis use disorder may seem counter-intuitive. However, the diverse pharmacology of isolated phytocannabinoids, and other cannabinoid-based substances, provides a gamut of agonist replacement and relapse-preventing strategies that may be clinically viable (Hill et al. 2022). Clinical data has significantly informed drug development, which is due, in part, to the Schedule I status of cannabis and the experimental challenges of training rodents to self-administer THC. In humans, cannabis use leads to a downregulation of CB1R in limbic and cortical structures (Hirvonen et al. 2012; Ceccarini et al. 2015) and normalizes after drug use cessation (D'Souza et al. 2016). As with alcohol, FAAH activity in the brain is negatively correlated with acute and chronic cannabis use (Boileau et al. 2016). The dopaminergic disturbances related to cannabis and THC are in line with that of other substances of misuse (Ferland and Hurd 2020). Acute THC induces DA release in the striatum (Bossong et al. 2008), followed by a deficit in drug-evoked DA transmission after chronic use (Thiruchselvam et al. 2017). Animal studies substantiate these findings. The reported downregulation of CB1R is consistent with the THC-evoked receptor desensitization characterized both in vivo and ex vivo (Breivogel et al. 1999; Burston et al. 2010). This agrees with evidence from animal models showing that the increase of DA release after THC use is related to the suppression of GABAergic VTA interneuron function via activation of CB1R (Oleson et al. 2021). Finally, animal studies have elucidated the synaptic and cellular alterations induced by chronic THC exposure in key brain regions linked to diminished cognitive control, such as the PFC. Specifically, a low-dose THC treatment reduces dendritic arborization and spine number in PFC pyramidal neurons in rats (Miller et al. 2018a). Despite significant debate about the neuropsychological consequences of cannabis use, human and animal evidence converge to a common point: chronic cannabis use can manifest the same structural, neurochemical, and cellular disturbances found in individuals suffering from other substance use disorders.

Early research using a rat model of voluntary cannabinoid intake suggested that the **CB1R antagonist rimonabant** attenuates the drug-induced reinstatement of cannabinoid-seeking behavior (Spano et al. 2004) (Table 5). This first suitable rat

Table 5 Table summary of the studies assessing the effects of cannabinoid agents in the reduction of reinstatement of **THC** drug seeking; –, no change found; ↑, increase; ↓, decrease

Compound class	Compound	Experimental measurement	Effect	Species	Reference
CB1/2R ligands	Rimonabant SR141716A	Cannabinoid-seeking behavior triggered by WIN 55,212-2 or heroin priming	↓	Rats	Spano et al. (2004)
		THC- and morphine-induced reinstatement of THC-seeking	↓	Squirrel monkeys	Justinova et al. (2008b)
		Priming- and cue-induced reinstatement of THC-seeking behavior	↓	Squirrel monkeys	Schindler et al. (2016a)
	AM4113 (CB1R neutral antagonist)	Priming- and cue-induced reinstatement of THC-seeking behavior	↓	Squirrel monkeys	Schindler et al. (2016a)
	AEF0117 (CB1R biased NAM)	Priming-induced reinstatement of THC-seeking behavior	↓	Squirrel monkeys	Haney et al. (2023)
		Cannabis self-administration	↓	Humans	Haney et al. (2023)
Inhibitors of eCB metabolism/ transport	AM404 (AEA transport inhibitor)	Priming-induced reinstatement of THC-seeking behavior	↑	Squirrel monkeys	Schindler et al. (2016b)
	URB597 (FAAH inhibitor)	Reinstatement of THC seeking behavior	–	Squirrel monkeys	Justinova et al. (2008a)
	PF-0447845 (FAAH inhibitor)	Cannabis use after forced abstinence	↓	Humans	D'Souza et al. (2019)
Phytocannabinoids	THC (Dronabinol)	Cannabis craving during withdrawal	↓	Humans	Haney et al. (2004); Budney et al. (2007)
		Cannabis abstinence rates	–	Humans	Haney et al. (2008); Levin et al. (2011)
	THC analog (Nabilone)	Marijuana "laboratory" relapse	↓	Humans	Haney et al. (2013)
	Nabiximols (THC/CBD)	Cannabis craving during withdrawal	↓	Humans	Allsop et al. (2014); Trigo et al. (2018)
			–	Humans	Trigo et al. (2016)
		Cannabis abstinence rates	–	Humans	Trigo et al. (2018)

(continued)

Table 5 (continued)

Compound class	Compound	Experimental measurement	Effect	Species	Reference
	CBD	Cannabis abstinence rates	↓	Humans	Freeman et al. (2020)
		Attentional bias to drug cues	↓	Humans	Morgan et al. (2010)

model of cannabinoid abuse utilized the CB1R/CB2R agonist WIN55,212-2 as a substitute for THC, which is not readily self-administered by rats. Later reports expanded upon these results, in support of the relapse-preventing effects of rimonabant in monkeys self-administering THC (Justinova et al. 2008b; Schindler et al. 2016a). In addition to reducing cannabinoid-induced reinstatement, CB1R antagonism also curtailed relapse triggered by opioid priming (morphine, heroin) (Spano et al. 2004; Justinova et al. 2008b). To date, only two placebo-controlled laboratory studies have explored the effects of rimonabant in humans with cannabis use disorder (Brezing and Levin 2018). For example, non-treatment-seeking participants administered with rimonabant experienced dampened sensations of euphoria elicited by THC (Huestis et al. 2001, 2007). Further work demonstrated a similar effect by the **CB1R neutral antagonist AM4113** (Schindler et al. 2016a), though evidence on this drug remains lacking. More recently, a phase 1 and phase 2a randomized trial demonstrated the promising efficacy of the CB1R-**biased NAM AEF0117** to treat cannabis use disorder (Haney et al. 2023). AEF0117 is a novel compound derived from the neurosteroid pregnenolone, both of which allosterically inhibit CB1 receptor signaling and reduce THC's reinforcing effects (Vallée et al. 2014; Vallée 2016; Haney et al. 2023) without affecting ligand-unbound receptor function. Interestingly, pregnenolone and **AEF0117** act as **biased NAMs**, blocking THC-induced CB1 receptor ERK1/2 recruitment without impacting cAMP signaling (Vallée et al. 2014; Haney et al. 2023). Unlike pregnenolone, **AEF0117** displays favorable bioavailability, a longer half-life, and is well-tolerated in humans. These properties led to a Phase 2a randomized clinical trial in individuals with cannabis use disorder, showing reduced cannabis self-administration following **AEF0117** treatment (Haney et al. 2023).

The use of **FAAH inhibitors** for cannabis use disorder (Boileau et al. 2016), has been conducted in several human and non-human primate studies. Given previous evidence with other drugs (Galaj and Xi 2019) and FAAH polymorphisms (Tyndale et al. 2007), D'Souza et al. (2019) administered an FAAH inhibitor, PF-04457843, to participants in an outpatient setting. Although relapse was not directly assessed, subjects receiving PF-04457843 exhibited decreased cannabis use after a forced 8-day abstinence period (D'Souza et al. 2019). Whether FAAH inhibitors may be clinically deployable for preventing relapse to cannabis use remains to be determined. It should be noted that levels of FAAH binding in the brain are lower in cannabis users during the early stages of withdrawal, a phenomenon that is related to cannabis intake and impulsivity (Boileau et al. 2016). In this context, further inhibition of FAAH could boost impulsivity and be counterproductive for the goal of maintaining abstinence. A considerable number of human studies have made use of **plant-derived CB1R agonists**, such as THC (dronabinol), THC analogs (nabilone), and THC/CBD combinations (nabiximols), to palliate relapse to cannabis use (Table 5) (Werneck et al. 2018). Agonist replacement therapy with dronabinol dose-dependently suppresses cannabis cravings during withdrawal (Haney et al. 2004; Budney et al. 2007). In contrast, when abstinence rates from cannabis are evaluated, the positive effects previously seen with dronabinol administration are not observed (Haney et al. 2008; Levin et al. 2011). To overcome these discrepancies, a subsequent study evaluated the THC analog nabilone, which offers improved bioavailability and a more linear, dose-dependent effect (Haney et al. 2013). Nabilone blunted cannabis relapse in a laboratory setting, with no observed adverse effects (Haney et al. 2013). More research in outpatient settings is warranted, but the evidence on the relapse-preventing effects exerted by THC analogs is promising thus far. The efficacy of nabiximols is compromised when THC is mixed with CBD. For example, nabiximols decreased cannabis cravings during abstinence (Allsop et al. 2014; Trigo et al. 2018, but see Trigo et al. 2016) but did not affect abstinent rates in clinical settings (Trigo et al. 2018). Pharmacological interactions between THC and CBD are poorly understood. On one hand, CBD inhibits the activity of cytochrome P450 (Jiang et al. 2013), therefore improving the bioavailability of THC. Yet, CBD is also known to counteract the intoxicating effects of THC, such as paranoia and subjective euphoria (Niesink and van Laar 2013), through the modulation of common signaling cascades (Hudson et al. 2019). In this case, it remains to be determined whether CBD's promiscuous pharmacology also suppresses the relapse-preventing effects of THC. Administration of **CBD** alone, however, has yielded different results than that of nabiximols. As a follow-up to a successful experiment demonstrating that CBD can reduce attentional bias to cannabis-associated cues (Morgan et al. 2010), Freeman et al. (2020) carried out a phase 2 clinical trial aimed to identify CBD's effective doses for the treatment of cannabis use disorder. Their findings revealed increased cannabis abstinence rates during and after CBD treatment (Freeman et al. 2020). A CBD dose of 400 mg was deemed effective, well tolerated, and achieved satisfactory treatment adherence. Along with **AEF0117**, clinical evidence posits CBD as the best cannabinoid-based compound to combat cannabis relapse to date.

8 Discussion

The research reviewed here highlights the remarkable role that eCB neuropharmacology has played in informing the etiopathology and medication development for treating patients with substance use disorder by promoting abstinence and preventing relapse behaviors. As evidence continues to accumulate, the potential therapeutic advantages of modulating eCB-CB1R neurotransmission are compelling. Novel targets and ligands, inspired by recent advances in our understanding of the "expanded-eCB system," as well as isolated phytocannabinoids, such as CBD, have been invigorating clinical research and pharmaceutical development of cannabinoid-based anti-relapse therapeutics.

Future work will need to address several open questions. **First**, a better understanding of the considerable gap existing between animal and human literature is of capital importance. We have noted important discrepancies between the clinical and preclinical data addressing the protective actions of (i) CBD in the relapse to psychostimulants, (ii) fibrates for the relapse to nicotine, and (iii) rimonabant in alcohol use disorder. Technical and conceptual limitations in rodent models of drug relapse impose significant constraints that hinder the predictive validity of the preclinical findings (Sanchis-Segura and Spanagel 2006). Of course, we must not rule out the possibility of species-specific differences in the neurobiology of the eCB system, and as such, sustained investment in applied human research will remain critical for the field of addiction medicine. **Second**, we currently lack sufficient understanding of the implications of cannabinoid-based pharmacology in stress responses during relapse. From our review of the literature, we have identified a notably high number of studies reporting the limited efficacy of drug candidates interacting with the eCB system that ameliorated stress-induced reinstatement. This stands in contrast to the consensus that cannabinoid-based compounds offer the concomitant advantage of normalizing stress response abnormalities that arise during dysphoric states exhibited during periods of abstinence. eCBs are integral regulators of the stress response (Morena et al. 2016), and a clearer delineation of their stress-related effects is needed to inform clinically appropriate innovation of eCB-derived therapies. Novel evidence is starting to show that if these conditions are not systematically evaluated, adverse outcomes may arise and undermine the clinical potential of otherwise promising compounds such as CBD (Luján et al. 2022).

Emerging cannabinoid-based approaches utilizing TRPV1 receptor antagonists may be especially useful for the management of drug- and stress-induced reinstatement of opioid-seeking behaviors, a feature that is lacking among classical cannabinoid compounds. Inactivation of TRPV1 receptors has been found to induce hyperthermia, however, which has precluded many TRPV1 antagonists from advancing to clinical trials in humans. Yet, novel formulations of TRPV1 antagonists such as AG1529 or PAC-14028 (Asivatrep), are devoid of hyperthermic side effects and may offer a safer pharmacological profile (Nikolaeva-Koleva et al. 2021). The CB1R neutral antagonist, AM4113, appears to be an outstanding candidate for preventing relapse to nicotine use, based on mounting preclinical data on its tolerability and efficacy. Further clinical validation of this compound will likely yield invaluable insight on the feasibility of deploying CB1R-blocking medications on patients with substance use disorders at large. In this area, the CB1R-biased NAM AEF0117 shows great promise in reducing cannabis reinforcement in humans and is well-positioned to deliver exciting results in the near future. Lastly, the discovery of novel DAGL inhibitors illuminates new avenues for pharmaceutical development that more effectively attenuates 2AG-CB1R signaling and thereby circumvents the drawbacks inherent to ligands that directly block CB1R. The promising study of

Winters et al. (2021) has already demonstrated the potential of DAGL inhibition in preventing relapse to alcohol use.

In conclusion, this chapter highlights that cannabinoid receptor interacting compounds, modulators of eCB metabolism, ligands of the expanded-eCB system, and broadly-acting phytocannabinoids can offer a diverse *pharmacopoeia* of potential candidates readily available to inform medication development for the long-term prevention of relapse among patients diagnosed with substance use disorders.

References

Adamczyk P, Miszkiel J, McCreary AC, Filip M, Papp M, Przegaliński E (2012) The effects of cannabinoid CB1, CB2 and vanilloid TRPV1 receptor antagonists on cocaine addictive behavior in rats. Brain Res 1444:45–54

Ahmad F, Rossen L, Sutton P (2022) Provisional drug overdose death counts. National Center for Health Statistics

Alegre-Zurano L, Berbegal-Sáez P, Luján MÁ, Cantacorps L, Martín-Sánchez A, García-Baos A et al (2022) Cannabidiol decreases motivation for cocaine in a behavioral economics paradigm but does not prevent incubation of craving in mice. Biomed Pharmacother 148:112708

Allsop DJ, Copeland J, Lintzeris N, Dunlop AJ, Montebello M, Sadler C et al (2014) Nabiximols as an agonist replacement therapy during cannabis withdrawal: a randomized clinical trial. JAMA Psychiatry 71:281–291

Alvarez-Jaimes L, Polis I, Parsons LH (2008) Attenuation of cue-induced heroin-seeking behavior by cannabinoid CB1 antagonist infusions into the nucleus accumbens core and prefrontal cortex, but not basolateral amygdala. Neuropsychopharmacology 33:2483–2493

Anggadiredja K, Nakamichi M, Hiranita T, Tanaka H, Shoyama Y, Watanabe S et al (2004) Endocannabinoid system modulates relapse to methamphetamine seeking: possible mediation by the arachidonic acid Cascade. Neuropsychopharmacology 29:1470–1478

Baggelaar MP, Van Esbroeck ACM, Van Rooden EJ, Florea BI, Overkleeft HS, Marsicano G et al (2017) Chemical proteomics maps brain region specific activity of endocannabinoid hydrolases. ACS Chem Biol 12:852–861

Bergman J, Delatte MS, Paronis CA, Vemuri K, Thakur GA, Makriyannis A (2008) Some effects of CB1 antagonists with inverse agonist and neutral biochemical properties. Physiol Behav 93: 666–670

Berke JD (2018) What does dopamine mean? Nat Neurosci 21:787–793

Best LM, Williams B, Le Foll B, Mansouri E, Bazinet RP, Lin L et al (2020) Lower brain fatty acid amide hydrolase in treatment-seeking patients with alcohol use disorder: a positron emission tomography study with [C-11]CURB. Neuropsychopharmacology 45:1289–1296

Bilbao A, Serrano A, Cippitelli A, Pavón FJ, Giuffrida A, Suárez J, García-Marchena N, Baixeras E, Gómez de Heras R, Orio L, Alén F, Ciccocioppo R, Cravatt BF, Parsons LH, Piomelli D, Rodríguez de Fonseca F (2016) Role of the saticty factor oleoylethanolamide in alcoholism. Addict Biol 21(4):859–872. https://doi.org/10.1111/adb.12276. Epub 2015 Jun 2. PMID: 26037332; PMCID: PMC4668242

Bisogno T, Hanuš L, De Petrocellis L, Tchilibon S, Ponde DE, Brandi I et al (2001) Molecular targets for cannabidiol and its synthetic analogues: effect on vanilloid VR1 receptors and on the cellular uptake and enzymatic hydrolysis of anandamide. Br J Pharmacol 134:845–852

Boileau I, Mansouri E, Williams B, Le Foll B, Rusjan P, Mizrahi R et al (2016) Fatty acid amide hydrolase binding in brain of cannabis users: imaging with the novel radiotracer [11C]CURB. Biol Psychiatry 80:691–701

Bossong MG, Van Berckel BNM, Boellaard R, Zuurman L, Schuit RC, Windhorst AD et al (2008) Δ9-tetrahydrocannabinol induces dopamine release in the human striatum. Neuropsychopharmacology 34:759–766

Breivogel CS, Childers SR, Deadwyler SA, Hampson RE, Vogt LJ, Sim-Selley LJ (1999) Chronic Δ9-tetrahydrocannabinol treatment produces a time-dependent loss of cannabinoid receptors and cannabinoid receptor-activated G proteins in rat brain. J Neurochem 73:2447–2459

Brezing CA, Levin FR (2018) The current state of pharmacological treatments for cannabis use disorder and withdrawal. Neuropsychopharmacology 43:173–194

Buczynski MW, Polis IY, Parsons LH (2013) The volitional nature of nicotine exposure alters anandamide and oleoylethanolamide levels in the ventral tegmental area. Neuropsychopharmacology 38:574–584

Budney AJ, Vandrey RG, Hughes JR, Moore BA, Bahrenburg B (2007) Oral delta-9-tetrahydrocannabinol suppresses cannabis withdrawal symptoms. Drug Alcohol Depend 86:22–29

Burston JJ, Wiley JL, Craig AA, Selley DE, Sim-Selley LJ (2010) Regional enhancement of cannabinoid CB 1 receptor desensitization in female adolescent rats following repeated Δ 9-tetrahydrocannabinol exposure. Br J Pharmacol 161:103–112

Cachope R, Mateo Y, Mathur BN, Irving J, Wang HL, Morales M et al (2012) Selective activation of cholinergic interneurons enhances accumbal phasic dopamine release: setting the tone for reward processing. Cell Rep 2:33–41

Calpe-López C, García-Pardo MP, Aguilar MA (2019) Cannabidiol treatment might promote resilience to cocaine and methamphetamine use disorders: a review of possible mechanisms. Molecules 24:2583

Campos AC, Fogaça MV, Scarante FF, Joca SRL, Sales AJ, Gomes FV et al (2017) Plastic and neuroprotective mechanisms involved in the therapeutic effects of Cannabidiol in psychiatric disorders. Front Pharmacol 8:269

Carrier EJ, Auchampach JA, Hillard CJ (2006) Inhibition of an equilibrative nucleoside transporter by cannabidiol: a mechanism of cannabinoid immunosuppression. Proc Natl Acad Sci USA 103(20):7895–7900

Ceccarini J, Kuepper R, Kemels D, Van Os J, Henquet C, Van Laere K (2015) [18F]MK-9470 PET measurement of cannabinoid CB1 receptor availability in chronic cannabis users. Addict Biol 20:357–367

Center for Behavioral Health Statistics and Quality (2017) National survey on drug use and health, 2017 (NSDUH-2017-DS0001). SAMHDA

Center for Disease Control (2020) Multiple cause of death data on CDC WONDER. Center for Disease Control

Chauvet C, Nicolas C, Thiriet N, Lardeux MV, Duranti A, Solinas M (2015) Chronic stimulation of the tone of endogenous anandamide reduces Cue- and stress-induced relapse in rats. Int J Neuropsychopharmacol 18:pyu025–pyu025

Cheer JF, Kendall DA, Marsden CA (2000) Cannabinoid receptors and reward in the rat: a conditioned place preference study. Psychopharmacology 151:25–30

Cheer JF, Wassum KM, Heien MLAV, Phillips PEM, Wightman RM (2004) Cannabinoids enhance subsecond dopamine release in the nucleus accumbens of awake rats. J Neurosci 24: 4393–4400

Cheer JF, Wassum KM, Sombers LA, Heien MLAV, Ariansen JL, Aragona BJ et al (2007) Phasic dopamine release evoked by abused substances requires cannabinoid receptor activation. J Neurosci 27:791–795

Cippitelli A, Bilbao A, Hansson AC, Del Arco I, Sommer W, Heilig M et al (2005) Cannabinoid CB1 receptor antagonism reduces conditioned reinstatement of ethanol-seeking behavior in rats. Eur J Neurosci 21:2243–2251

Cippitelli A, Cannella N, Braconi S, Duranti A, Tontini A, Bilbao A et al (2008) Increase of brain endocannabinoid anandamide levels by FAAH inhibition and alcohol abuse behaviours in the rat. Psychopharmacology 198:449–460

Collins AL, Aitken TJ, Greenfield VY, Ostlund SB, Wassum KM (2016) Nucleus Accumbens acetylcholine receptors modulate dopamine and motivation. Neuropsychopharmacology 41: 2830–2838
Connor JP, Stjepanović D, Le Foll B, Hoch E, Budney AJ, Hall WD (2021) Cannabis use and cannabis use disorder. Nat Rev Dis Prim 7:1–24
Covey DP, Bunner KD, Schuweiler DR, Cheer JF, Garris PA (2016) Amphetamine elevates nucleus accumbens dopamine via an action potential-dependent mechanism that is modulated by endocannabinoids. Eur J Neurosci 43:1661–1673
Covey DP, Hernandez E, Luján MÁ, Cheer JF (2021) Chronic augmentation of endocannabinoid levels persistently increases dopaminergic encoding of reward cost and motivation. J Neurosci 41:6946–6953
Cristino L, Bisogno T, Di Marzo V (2020) Cannabinoids and the expanded endocannabinoid system in neurological disorders. Nat Rev Neurol 16:9–29
D'Souza DC, Cortes-Briones JA, Ranganathan M, Thurnauer H, Creatura G, Surti T et al (2016) Rapid changes in cannabinoid 1 receptor availability in cannabis-dependent male subjects after abstinence from cannabis. Biol Psychiatry Cogn Neurosci Neuroimaging 1:60–67
D'Souza DC, Cortes-Briones J, Creatura G, Bluez G, Thurnauer H, Deaso E et al (2019) Efficacy and safety of a fatty acid amide hydrolase inhibitor (PF-04457845) in the treatment of cannabis withdrawal and dependence in men: a double-blind, placebo-controlled, parallel group, phase 2a single-site randomised controlled trial. Lancet Psychiatry 6:35–45
de Bruin NMWJ, Lange JHM, Kruse CG, Herremans AH, Schoffelmeer ANM, van Drimmelen M et al (2011) SLV330, a cannabinoid CB(1) receptor antagonist, attenuates ethanol and nicotine seeking and improves inhibitory response control in rats. Behav Brain Res 217:408–415
De Guglielmo G, Melis M, De Luca MA, Kallupi M, Li HW, Niswender K et al (2015) PPARγ activation attenuates opioid consumption and modulates mesolimbic dopamine transmission. Neuropsychopharmacology 40:927–937
de Guglielmo G, Kallupi M, Scuppa G, Demopulos G, Gaitanaris G, Ciccocioppo R (2017) Pioglitazone attenuates the opioid withdrawal and vulnerability to relapse to heroin seeking in rodents. Psychopharmacology 234:223–234
de Meneses-Gaya C, Crippa JA, Hallak JE, Miguel AQ, Laranjeira R, Bressan RA et al (2020) Cannabidiol for the treatment of crack-cocaine craving: an exploratory double-blind study. Braz J Psychiatry 43:467–476
De Petrocellis L, Ligresti A, Moriello AS, Allarà M, Bisogno T, Petrosino S et al (2011) Effects of cannabinoids and cannabinoid-enriched cannabis extracts on TRP channels and endocannabinoid metabolic enzymes. Br J Pharmacol 163:1479–1494
De Vries TJ, Shaham Y, Homberg JR, Crombag H, Schuurman K, Dieben J et al (2001) A cannabinoid mechanism in relapse to cocaine seeking. Nat Med 7:1151–1154
De Vries TJ, Homberg JR, Binnekade R, Raasø H, Schoffelmeer ANM (2003) Cannabinoid modulation of the reinforcing and motivational properties of heroin and heroin-associated cues in rats. Psychopharmacology 168:164–169
Domi E, Uhrig S, Soverchia L, Spanagel R, Hansson AC, Barbier E et al (2016) Genetic deletion of neuronal PPARγ enhances the emotional response to acute stress and exacerbates anxiety: an effect reversed by Rescue of Amygdala PPARγ function. J Neurosci 36:12611
Dong A, He K, Dudok B, Farrell JS, Guan W, Liput DJ et al (2021) A fluorescent sensor for spatiotemporally resolved imaging of endocannabinoid dynamics in vivo. Nat Biotechnol 40: 787–798
Economidou D, Mattioli L, Cifani C, Perfumi M, Massi M, Cuomo V et al (2006) Effect of the cannabinoid CB1 receptor antagonist SR-141716A on ethanol self-administration and ethanol-seeking behaviour in rats. Psychopharmacology 183:394–403
Economidou D, Mattioli L, Ubaldi M, Lourdusamy A, Soverchia L, Hardiman G et al (2007) Role of cannabinoidergic mechanisms in ethanol self-administration and ethanol seeking in rat adult offspring following perinatal exposure to ??9-tetrahydrocannabinol. Toxicol Appl Pharmacol 223:73–85

Elsaid S, Kloiber S, Le Foll B (2019) Effects of cannabidiol (CBD) in neuropsychiatric disorders: a review of pre-clinical and clinical findings. Prog Mol Biol Transl Sci 167:25–75

Everitt BJ, Robbins TW (2016) Drug addiction: updating actions to habits to compulsions ten years on. Annu Rev Psychol 67:23–50

Fattore L, Spano MS, Cossu G, Deiana S, Fratta W (2003) Cannabinoid mechanism in reinstatement of heroin-seeking after a long period of abstinence in rats. Eur J Neurosci 17:1723–1726

Fattore L, Spano S, Cossu G, Deiana S, Fadda P, Fratta W (2005) Cannabinoid CB(1) antagonist SR 141716A attenuates reinstatement of heroin self-administration in heroin-abstinent rats. Neuropharmacology 48:1097–1104

Fattore L, Spano MS, Melis V, Fadda P, Fratta W (2011) Differential effect of opioid and cannabinoid receptor blockade on heroin-seeking reinstatement and cannabinoid substitution in heroin-abstinent rats. Br J Pharmacol 163:1550–1562

Ferland J-MN, Hurd YL (2020) Deconstructing the neurobiology of cannabis use disorder. Nat Neurosci 23:600–610

Forget B, Coen KM, Le Foll B (2009) Inhibition of fatty acid amide hydrolase reduces reinstatement of nicotine seeking but not break point for nicotine self-administration--comparison with CB(1) receptor blockade. Psychopharmacology 205:613–624

Forget B, Guranda M, Gamaleddin I, Goldberg SR, Le Foll B (2016) Attenuation of cue-induced reinstatement of nicotine seeking by URB597 through cannabinoid CB1 receptor in rats. Psychopharmacology 233:1823–1828

Fourgeaud L, Mato S, Bouchet D, Hémar A, Worley PF, Manzoni OJ (2004) A single in vivo exposure to cocaine abolishes endocannabinoid-mediated long-term depression in the nucleus accumbens. J Neurosci 24:6939–6945

Freeman TP, Hindocha C, Baio G, Shaban NDC, Thomas EM, Astbury D et al (2020) Cannabidiol for the treatment of cannabis use disorder: a phase 2a, double-blind, placebo-controlled, randomised, adaptive Bayesian trial. Lancet Psychiatry 7(10):865–874

Galaj E, Xi Z-X (2019) Potential of cannabinoid receptor ligands as treatment for substance use disorders. CNS Drugs 33:1001–1030

Galaj E, Xi ZX (2021) Possible receptor mechanisms underlying cannabidiol effects on addictive-like behaviors in experimental animals. Int J Mol Sci 22:1–14

Galaj E, Hempel B, Moore A, Klein B, Bi GH, Gardner EL et al (2022) Therapeutic potential of PIMSR, a novel CB1 receptor neutral antagonist, for cocaine use disorder: evidence from preclinical research. Transl Psychiatry 12:286

Gamaleddin I, Guranda M, Goldberg SR, Le Foll B (2011) The selective anandamide transport inhibitor VDM11 attenuates reinstatement of nicotine seeking behaviour, but does not affect nicotine intake. Br J Pharmacol 164:1652–1660

Gamaleddin I, Wertheim C, Zhu AZ, Coen KM, Vemuri K, Makryannis A, Goldberg SR, Le Foll B (2012a) Cannabinoid receptor stimulation increases motivation for nicotine and nicotine seeking. Addict Biol 17(1):47–61. https://doi.org/10.1111/j.1369-1600.2011.00314.x. Epub 2011 Apr 26

Gamaleddin I, Zvonok A, Makriyannis A, Goldberg SR, Le Foll B. (2012b) Effects of a selective cannabinoid CB2 agonist and antagonist on intravenous nicotine self administration and reinstatement of nicotine seeking. PLoS One 7(1):e29900. https://doi.org/10.1371/journal.pone.0029900. Epub 2012 Jan 26. PMID: 22291896; PMCID: PMC3266883

Gamaleddin I, Guranda M, Scherma M, Fratta W, Makriyannis A, Vadivel SK et al (2013) AM404 attenuates reinstatement of nicotine seeking induced by nicotine-associated cues and nicotine priming but does not affect nicotine- and food-taking. J Psychopharmacol 27:564–571

Gendy MNS, Di Ciano P, Kowalczyk WJ, Barrett SP, George TP, Heishman S et al (2018) Testing the PPAR hypothesis of tobacco use disorder in humans: a randomized trial of the impact of gemfibrozil (a partial PPARα agonist) in smokers. PLoS One 13:e0201512

Gessa GL, Serra S, Vacca G, Carai MAM, Colombo G (2005) Suppressing effect of the cannabinoid CB1 receptor antagonist, SR147778, on alcohol intake and motivational properties of alcohol in alcohol-preferring sP rats. Alcohol Alcohol 40:46–53

Getachew B, Hauser SR, Dhaher R, Katner SN, Bell RL, Oster SM, McBride WJ, Rodd ZA (2011) CB1 receptors regulate alcohol-seeking behavior and alcohol self-administration of alcohol-preferring (P) rats. Pharmacol Biochem Behav 97(4):669–75. https://doi.org/10.1016/j.pbb.2010.11.006. Epub 2010 Nov 24. PMID: 21110997; PMCID: PMC4393646

Giulietti F, Filipponi A, Rosettani G, Giordano P, Iacoacci C, Spannella F et al (2020) Pharmacological approach to smoking cessation: an updated review for daily clinical practice. High Blood Press Cardiovasc Prev 27:349–362

Gonzalez-Cuevas G, Martin-Fardon R, Kerr TM, Stouffer DG, Parsons LH, Hammell DC et al (2018) Unique treatment potential of cannabidiol for the prevention of relapse to drug use: preclinical proof of principle. Neuropsychopharmacology 43:2036–2045

Greasley PJ, Clapham JC (2006) Inverse agonism or neutral antagonism at G-protein coupled receptors: a medicinal chemistry challenge worth pursuing? Eur J Pharmacol 553:1–9

Gueye AB, Pryslawsky Y, Trigo JM, Poulia N, Delis F, Antoniou K et al (2016) The CB 1 neutral antagonist AM4113 retains the therapeutic efficacy of the inverse agonist Rimonabant for nicotine dependence and weight loss with better psychiatric tolerability. Int J Neuropsychopharmacol 19:pyw068

Haller J, Varga B, Ledent C, Barna I, Freund TF (2004) Context-dependent effects of CB1 cannabinoid gene disruption on anxiety-like and social behaviour in mice. Eur J Neurosci 19:1906–1912

Hammer JH, Parent MC, Spiker DA, World Health Organization (2018) Global status report on alcohol and health 2018

Haney M, Hart CL, Vosburg SK, Nasser J, Bennett A, Zubaran C et al (2004) Marijuana withdrawal in humans: effects of oral THC or divalproex. Neuropsychopharmacology 29:158–170

Haney M, Hart CL, Vosburg SK, Comer SD, Reed SC, Foltin RW (2008) Effects of THC and Lofexidine in a human laboratory model of marijuana withdrawal and relapse. Psychopharmacology 197:157

Haney M, Cooper ZD, Bedi G, Vosburg SK, Comer SD, Foltin RW (2013) Nabilone decreases marijuana withdrawal and a laboratory measure of marijuana relapse. Neuropsychopharmacology 38:1557

Haney M, Vallée M, Fabre S, Collins Reed S, Zanese M, Campistron G et al (2023) Signaling-specific inhibition of the CB1 receptor for cannabis use disorder: phase 1 and phase 2a randomized trials. Nat Med 29:1487–1499

Harlan BA, Becker HC, Woodward JJ, Riegel AC (2018) Opposing actions of CRF-R1 and CB1 receptors on VTA-GABAergic plasticity following chronic exposure to ethanol. Neuropsychopharmacology 43:2064–2074

Hashiesh HM, Sharma C, Goyal SN, Sadek B, Jha NK, Al Kaabi J et al (2021) A focused review on CB2 receptor-selective pharmacological properties and therapeutic potential of β-caryophyllene, a dietary cannabinoid. Biomed Pharmacother 140:111639

Hay GL, Baracz SJ, Everett NA, Roberts J, Costa PA, Arnold JC, McGregor IS, Cornish JL (2018) Cannabidiol treatment reduces the motivation to self-administer methamphetamine and methamphetamine-primed relapse in rats. J Psychopharmacol 32(12):1369–1378. https://doi.org/10.1177/0269881118799954. Epub 2018 Sep 27

He Y, Galaj E, Bi GH, Wang XF, Gardner E, Xi ZX (2020) β-Caryophyllene, a dietary terpenoid, inhibits nicotine taking and nicotine seeking in rodents. Br J Pharmacol 177:2058

He XH, Galaj E, Bi GH, He Y, Hempel B, Wang YL et al (2021) β-caryophyllene, an FDA-approved food additive, inhibits methamphetamine-taking and methamphetamine-seeking behaviors possibly via CB2 and non-CB2 receptor mechanisms. Front Pharmacol 12:2363

Hill KP, Gold MS, Nemeroff CB, McDonald W, Grzenda A, Widge AS et al (2022) Risks and benefits of cannabis and cannabinoids in psychiatry. Am J Psychiatry 179:98–109

Hindocha C, Freeman TP, Grabski M, Stroud JB, Crudgington H, Davies AC et al (2018) Cannabidiol reverses attentional bias to cigarette cues in a human experimental model of tobacco withdrawal. Addiction 113:1696–1705

Hirvonen J, Goodwin RS, Li CT, Terry GE, Zoghbi SS, Morse C et al (2012) Reversible and regionally selective downregulation of brain cannabinoid CB1 receptors in chronic daily cannabis smokers. Mol Psychiatry 17:642–649

Hosseinzadeh M, Nikseresht S, Khodagholi F, Naderi N, Maghsoudi N (2016) Cannabidiol post-treatment alleviates rat epileptic-related behaviors and activates hippocampal cell autophagy pathway along with antioxidant defense in chronic phase of pilocarpine-induced seizure. J Mol Neurosci 58:432–440

Howlett AC, Mukhopadhyay S (2000) Cellular signal transduction by anandamide and 2-arachidonoylglycerol. Chem Phys Lipids 108:53–70

Hswen Y, Zhang A, Brownstein JS (2021) Estimating the incidence of cocaine use and mortality with music lyrics about cocaine. Npj Digital Medicine 4:1–6

Hua T, Vemuri K, Nikas SP, Laprairie RB, Wu Y, Qu L et al (2017) Crystal structures of agonist-bound human cannabinoid receptor CB1. Nature 547:468–471

Hua T, Li X, Wu L, Iliopoulos-Tsoutsouvas C, Wang Y, Wu M et al (2020) Activation and signaling mechanism revealed by cannabinoid receptor-Gi complex structures. Cell 180:655–665.e18

Hudson R, Renard J, Norris C, Rushlow WJ, Laviolette SR (2019) Cannabidiol counteracts the psychotropic side-effects of Δ-9-tetrahydrocannabinol in the ventral hippocampus through bidirectional control of erk1-2 phosphorylation. J Neurosci 39:8762–8777

Huestis MA, Gorelick DA, Heishman SJ, Preston KL, Nelson RA, Moolchan ET et al (2001) Blockade of effects of smoked marijuana by the CB1-selective cannabinoid receptor antagonist SR141716. Arch Gen Psychiatry 58:322–328

Huestis MA, Boyd SJ, Heishman SJ, Preston KL, Bonnet D, Le Fur G et al (2007) Single and multiple doses of rimonabant antagonize acute effects of smoked cannabis in male cannabis users. Psychopharmacology 194:505

Hurd YL, Spriggs S, Alishayev J, Winkel G, Gurgov K, Kudrich C et al (2019) Cannabidiol for the reduction of Cue-induced craving and anxiety in drug-abstinent individuals with heroin use disorder: a double-blind randomized placebo-controlled trial. Am J Psychiatry 176:911–922

Ignaszewski MJ (2021) The epidemiology of drug abuse. J Clin Pharmacol 61:S10–S17

Jakowiecki J, Abel R, Orzeł U, Pasznik P, Preissner R, Filipek S (2021) Allosteric modulation of the CB1 cannabinoid receptor by Cannabidiol – a molecular modeling study of the N-terminal domain and the allosteric-Orthosteric coupling. Molecules 26:2456

Järbe TUC, LeMay BJ, Olszewska T, Vemuri VK, Wood JT, Makriyannis A (2008) Intrinsic effects of AM4113, a putative neutral CB1 receptor selective antagonist, on open-field behaviors in rats. Pharmacol Biochem Behav 91:84–90

Jiang R, Yamaori S, Okamoto Y, Yamamoto I, Watanabe K (2013) Cannabidiol is a potent inhibitor of the catalytic activity of cytochrome P450 2C19. Drug Metab Pharmacokinet 28:332–338

Jin L, Pan L, Guo Y, Zheng Y, Nie Z, Zhu R (2014) Expression and localization of cannabinoid receptor 1 in rats' brain treated with acute and repeated morphine. Acta Neurobiol Exp (Wars) 74:288–297

Jing L, Qiu Y, Zhang Y, Li J-X (2014) Effects of the cannabinoid CB1 receptor allosteric modulator ORG 27569 on reinstatement of cocaine- and methamphetamine-seeking behavior in rats. Drug Alcohol Depend 143:251–256

Julian MD, Martin AB, Cuellar B, Rodriguez De Fonseca F, Navarro M, Moratalla R et al (2003) Neuroanatomical relationship between type 1 cannabinoid receptors and dopaminergic systems in the rat basal ganglia. Neuroscience 119:309–318

Justinova Z, Mangieri RA, Bortolato M, Chefer SI, Mukhin AG, Clapper JR et al (2008a) Fatty acid amide hydrolase inhibition heightens anandamide signaling without producing reinforcing effects in primates. Biol Psychiatry 64:930–937

Justinova Z, Munzar P, Panlilio LV, Yasar S, Redhi GH, Tanda G et al (2008b) Blockade of THC-seeking behavior and relapse in monkeys by the cannabinoid CB(1)-receptor antagonist rimonabant. Neuropsychopharmacology 33:2870–2877

Justinova Z, Panlilio LV, Moreno-Sanz G, Redhi GH, Auber A, Secci ME et al (2015) Effects of fatty acid amide hydrolase (FAAH) inhibitors in non-human primate models of nicotine reward and relapse. Neuropsychopharmacology 40:2185–2197

Kaczocha M, Glaser ST, Deutsch DG (2009) Identification of intracellular carriers for the endocannabinoid anandamide. Proc Natl Acad Sci USA 106:6375–6380

Kerbrat A, Ferré J-C, Fillatre P, Ronzière T, Vannier S, Carsin-Nicol B et al (2016) Acute neurologic disorder from an inhibitor of fatty acid amide hydrolase. N Engl J Med 375:1717–1725

Kim HGR, Malik AN, Mikhael JG, Bech P, Tsutsui-Kimura I, Sun F et al (2020) A unified framework for dopamine signals across timescales. Cell 183:1600–1616.e25

Kondev V, Najeed M, Yasmin F, Morgan A, Loomba N, Johnson K et al (2023) Endocannabinoid release at ventral hippocampal-amygdala synapses regulates stress-induced behavioral adaptation. Cell Rep 42:113027

Koob GF, Le Moal M (1979) Drug abuse: hedonic homeostatic dysregulation. Science 278:52–58

Koob GF, Le Moal M (2001) Drug addiction, dysregulation of reward, and allostasis. Neuropsychopharmacology 24:97–129

Kosillo P, Zhang Y-F, Threlfell S, Cragg SJ (2016) Cortical control of striatal dopamine transmission via striatal cholinergic interneurons. Cereb Cortex 26:4160–4169

Kupferschmidt DA, Klas PG, Erb S (2012) Cannabinoid CB 1 receptors mediate the effects of corticotropin-releasing factor on the reinstatement of cocaine seeking and expression of cocaine-induced behavioural sensitization. Br J Pharmacol 167:196–206

Laviolette SR (2017) Cannabinoid regulation of opiate motivational processing in the mesolimbic system: the integrative roles of amygdala, prefrontal cortical and ventral hippocampal input pathways. Curr Opin Behav Sci 13:46–54

Le Foll B, Gorelick DA, Goldberg SR (2009) The future of endocannabinoid-oriented clinical research after CB1 antagonists. Psychopharmacology 205:171–174

Levin FR, Mariani JJ, Brooks DJ, Pavlicova M, Cheng W, Nunes EV (2011) Dronabinol for the treatment of cannabis dependence: a randomized, double-blind, placebo-controlled trial. Drug Alcohol Depend 116:142

Liu QS, Pu L, Poo MM (2005) Repeated cocaine exposure in vivo facilitates LTP induction in midbrain dopamine neurons. Nature 437(7061):1027–1031

Longo DL, Volkow ND, Koob GF, McLellan AT (2016) Neurobiologic advances from the brain disease model of addiction. N Engl J Med 374:363–371

Luján MÁ, Valverde O (2020) The pro-neurogenic effects of Cannabidiol and its potential therapeutic implications in psychiatric disorders. Front Behav Neurosci 14:109

Luján MÁ, Castro-Zavala A, Alegre-Zurano L, Valverde O (2018) Repeated Cannabidiol treatment reduces cocaine intake and modulates neural proliferation and CB1R expression in the mouse hippocampus. Neuropharmacology 143:163–175

Luján MÁ, Cantacorps L, Valverde O (2020) The pharmacological reduction of hippocampal neurogenesis attenuates the protective effects of cannabidiol on cocaine voluntary intake. Addict Biol 25:e12778

Luján MÁ, Cheer JF, Melis M (2021) Choosing the right drug: status and future of endocannabinoid research for the prevention of drug-seeking reinstatement. Curr Opin Pharmacol 56:29–38

Luján M, Alegre-Zurano L, Martín-Sánchez A, Cantacorps L, Valverde O (2022) CB1 receptor antagonist AM4113 reverts the effects of cannabidiol on cue and stress-induced reinstatement of cocaine-seeking behaviour in mice. Prog Neuro-Psychopharmacol Biol Psychiatry 113:110462

Luján MÁ, Covey DP, Young-Morrison R, Zhang L, Kim A, Morgado F et al (2023a) Mobilization of endocannabinoids by midbrain dopamine neurons is required for the encoding of reward prediction. Nat Commun 14:7545

Luján MÁ, Oliver BL, Young-Morrison R, Engi SA, Zhang L-Y, Wenzel JM et al (2023b) A multivariate regressor of patterned dopamine release predicts relapse to cocaine. Cell Rep 42:112553

Lupica CR, Riegel AC (2005) Endocannabinoid release from midbrain dopamine neurons: a potential substrate for cannabinoid receptor antagonist treatment of addiction. Neuropharmacology 48:1105–1116

Ma SX, Kwon SH, Seo JY, Hwang JY, Hong SI, Kim HC et al (2017) Impairment of opiate-mediated behaviors by the selective TRPV1 antagonist SB366791. Addict Biol 22:1817–1828

Maex R, Grinevich VP, Grinevich V, Budygin E, Bencherif M, Gutkin B (2014) Understanding the role α7 nicotinic receptors play in dopamine efflux in nucleus Accumbens. ACS Chem Neurosci 5:1032

Mahgoub M, Keun-Hang SY, Sydorenko V, Ashoor A, Kabbani N, Al Kury L et al (2013) Effects of cannabidiol on the function of α7-nicotinic acetylcholine receptors. Eur J Pharmacol 720: 310–319

Mahmud A, Gallant S, Sedki F, D'Cunha T, Shalev U (2017) Effects of an acute cannabidiol treatment on cocaine selfadministration and cue-induced cocaine seeking in male rats. J Psychopharmacol 31(1):96–104. https://doi.org/10.1177/0269881116667706. Epub 2016 Sep 26

Marcus DJ, Bedse G, Gaulden AD, Ryan JD, Kondev V, Winters ND et al (2020) Endocannabinoid signaling collapse mediates stress-induced Amygdalo-cortical strengthening. Neuron 105: 1062–1076.e6

Marinelli S, Di Marzo V, Berretta N, Matias I, Maccarrone M, Bernardi G et al (2003) Presynaptic facilitation of glutamatergic synapses to dopaminergic neurons of the rat substantia Nigra by endogenous stimulation of Vanilloid receptors. J Neurosci 23:3136–3144

Marín-Navarrete R, Templos-Nuñez L, Eliosa-Hernández A, Villalobos-Gallegos L, Fernández-Mondragón J, Pérez-López A et al (2014) Characteristics of a treatment-seeking population in outpatient addiction treatment centers in Mexico. Subst Use Misuse 49:1784–1794

Mascia P, Pistis M, Justinova Z, Panlilio LV, Luchicchi A, Lecca S et al (2011) Blockade of nicotine reward and reinstatement by activation of alpha-type peroxisome proliferator-activated receptors. Biol Psychiatry 69:633–641

Mateo Y, Johnson KA, Covey DP, Atwood BK, Wang H, Zhang S et al (2017) Endocannabinoid actions on cortical terminals orchestrate local modulation of dopamine release in the nucleus Accumbens. Neuron 96:1112–1126.e5

McCann DJ, Ramey T, Skolnick P (2015) Outcome measures in medication trials for substance use disorders. Curr Treat Options Psychiatry 2:113–121

McCutcheon JE, Wang X, Tseng KY, Wolf ME, Marinelli M (2011) Calcium-permeable AMPA receptors are present in nucleus Accumbens synapses after prolonged withdrawal from cocaine self-administration but not experimenter-administered cocaine. J Neurosci 31:5737–5743

McPartland JM, Duncan M, Di Marzo V, Pertwee RG (2015) Are cannabidiol and Δ 9 -tetrahydrocannabivarin negative modulators of the endocannabinoid system? A systematic review. Br J Pharmacol 172:737–753

McReynolds JR, Doncheck EM, Vranjkovic O, Ganzman GS, Baker DA, Hillard CJ et al (2016) CB1 receptor antagonism blocks stress-potentiated reinstatement of cocaine seeking in rats. Psychopharmacology 233:99–109

Melis M, Pistis M (2012) Hub and switches: endocannabinoid signalling in midbrain dopamine neurons. Philos Trans R Soc Lond B Biol Sci 367:3276–3285

Melis M, Carta S, Fattore L, Tolu S, Yasar S, Goldberg SR et al (2010) Peroxisome proliferator-activated receptors-alpha modulate dopamine cell activity through nicotinic receptors. Biol Psychiatry 68:256–264

Melis M, Scheggi S, Carta G, Madeddu C, Lecca S, Luchicchi A et al (2013) PPAR regulates cholinergic-driven activity of midbrain dopamine neurons via a novel mechanism involving 7 nicotinic acetylcholine receptors. J Neurosci 33:6203–6211

Miller ML, Chadwick B, Dickstein DL, Purushothaman I, Egervari G, Rahman T et al (2018a) Adolescent exposure to Δ9-tetrahydrocannabinol alters the transcriptional trajectory and dendritic architecture of prefrontal pyramidal neurons. Mol Psychiatry 24:588–600

Miller WR, Fox RG, Stutz SJ, Lane SD, Denner L, Cunningham KA et al (2018b) PPARγ agonism attenuates cocaine cue reactivity. Addict Biol 23:55–68

Mongeau-Pérusse V, Brissette S, Bruneau J, Conrod P, Dubreucq S, Gazil G et al (2021) Cannabidiol as a treatment for craving and relapse in individuals with cocaine use disorder: a randomized placebo-controlled trial. Addiction 116:2431–2442

Morales M, Bonci A (2012) Hooking CB2 receptor into drug abuse? Nature Publishing Group

Morales P, Jagerovic N (2020) Novel approaches and current challenges with targeting the endocannabinoid system. Expert Opin Drug Discov 00:1–14

Morales-Lázaro SL, Llorente I, Sierra-Ramírez F, López-Romero AE, Ortíz-Rentería M, Serrano-Flores B et al (2016) Inhibition of TRPV1 channels by a naturally occurring omega-9 fatty acid reduces pain and itch. Nat Commun 7:1–12

Morena M, Patel S, Bains JS, Hill MN (2016) Neurobiological interactions between stress and the endocannabinoid system. Neuropsychopharmacology 41:80–102

Morgan CJA, Freeman TP, Schafer GL, Curran HV (2010) Cannabidiol attenuates the appetitive effects of Δ9- tetrahydrocannabinol in humans smoking their chosen cannabis. Neuropsychopharmacology 35:1879–1885

Morgan CJA, Das RK, Joye A, Curran HV, Kamboj SK (2013) Cannabidiol reduces cigarette consumption in tobacco smokers: preliminary findings. Addict Behav 38:2433–2436

Morley KC, Perry CJ, Watt J, Hurzeler T, Leggio L, Lawrence AJ et al (2021) New approved and emerging pharmacological approaches to alcohol use disorder: a review of clinical studies. Expert Opin Pharmacother 22:1291–1303

Munro S, Thomas KL, Abu-Shaar M (1993) Molecular characterization of a peripheral receptor for cannabinoids. Nature 365:61–65

National Center for Health Statistics (2021) National drug-involved overdose deaths by specific category. National Center for Health Statistics

Navarrete F, García-Gutiérrez MS, Gasparyan A, Navarro D, Manzanares J (2021) CB2 receptor involvement in the treatment of substance use disorders. Biomolecules 11:1556

Nguyen T, Thomas BF, Zhang Y (2019) Overcoming the psychiatric side effects of the cannabinoid CB1 receptor antagonists: current approaches for therapeutics development. Curr Top Med Chem 19:1418–1435

Niesink RJM, van Laar MW (2013) Does Cannabidiol protect against adverse psychological effects of THC? Front Psych 4:130

Nikolaeva-Koleva M, Butron L, González-Rodríguez S, Devesa I, Valente P, Serafini M et al (2021) A capsaicinoid-based soft drug, AG1529, for attenuating TRPV1-mediated histaminergic and inflammatory sensory neuron excitability. Sci Rep 11:1–18

NSDUH (2021) 2020 National Survey of drug use and health (NSDUH) detailed tables. NSDUH

Oleson EB, Beckert MV, Morra JT, Lansink CS, Cachope R, Abdullah RA et al (2012) Endocannabinoids shape Accumbal encoding of Cue-motivated behavior via CB1 receptor activation in the ventral tegmentum. Neuron 73:360–373

Oleson EB, Hamilton LR, Gomez DM (2021) Cannabinoid modulation of dopamine release during motivation, periodic reinforcement, exploratory behavior, habit formation, and attention. Front Synaptic Neurosci 13:29

Ornell F, Moura HF, Scherer JN, Pechansky F, Kessler FHP, von Diemen L (2020) The COVID-19 pandemic and its impact on substance use: implications for prevention and treatment. Psychiatry Res 289:113096

Ortiz S, Oliva JM, Pérez-Rial S, Palomo T, Manzanares J (2004) Chronic ethanol consumption regulates cannabinoid CB1 receptor gene expression in selected regions of rat brain. Alcohol Alcohol 39:88–92

Pan B, Hillard CJ, Liu QS (2008a) D2 dopamine receptor activation facilitates endocannabinoid-mediated long-term synaptic depression of GABAergic synaptic transmission in midbrain dopamine neurons via cAMP-protein kinase A signaling. J Neurosci 28:14018–14030

Pan B, Hillard CJ, Liu QS (2008b) Endocannabinoid signaling mediates cocaine-induced inhibitory synaptic plasticity in midbrain dopamine neurons. J Neurosci 28:1385–1397

Panlilio LV, Justinova Z (2018) Preclinical studies of cannabinoid reward, treatments for cannabis use disorder, and addiction-related effects of cannabinoid exposure. Neuropsychopharmacology 43:116

Panlilio LV, Justinova Z, Mascia P, Pistis M, Luchicchi A, Lecca S et al (2012) Novel use of a lipid-lowering fibrate medication to prevent nicotine reward and relapse: preclinical findings. Neuropsychopharmacology 37:1838–1847

Parolaro D, Rubino T, Vigano D, Massi P, Guidali C, Realini N (2010) Cellular mechanisms underlying the interaction between cannabinoid and opioid system. Curr Drug Targets 11:393–405

Perkins KA, Karelitz JL, Michael VC, Fromuth M, Conklin CA, Roy Chengappa KN et al (2016) Initial evaluation of Fenofibrate for efficacy in aiding smoking abstinence. Nicotine Tob Res 18: 74–78

Perra S, Pillolla G, Melis M, Muntoni AL, Gessa GL, Pistis M (2005) Involvement of the endogenous cannabinoid system in the effects of alcohol in the mesolimbic reward circuit: electrophysiological evidence in vivo. Psychopharmacology 183:368–377

Pertwee RG (2008) Ligands that target cannabinoid receptors in the brain: from THC to anandamide and beyond. Addict Biol 13:147–159

Peters KZ, Cheer JF, Tonini R (2021a) Modulating the neuromodulators: dopamine, serotonin, and the endocannabinoid system. Trends Neurosci 44:464–477

Peters KZ, Zlebnik NE, Cheer JF (2021b) Cannabis exposure during adolescence: a uniquely sensitive period for neurobiological effects. In: International review of neurobiology. Academic Press, pp 95–120

Pillolla G, Melis M, Perra S, Muntoni AL, Gessa GL, Pistis M (2007) Medial forebrain bundle stimulation evokes endocannabinoid-mediated modulation of ventral tegmental area dopamine neuron firing in vivo. Psychopharmacology 191:843–853

Ramesh K, Rosenbaum DM (2021) Molecular basis for ligand modulation of the cannabinoid CB 1 receptor. Br J Pharmacol 179(14):3487–3495

Rasmussen K, White DA, Acri JB (2019) NIDA's medication development priorities in response to the opioid crisis: ten most wanted. Neuropsychopharmacology 44:657–659

Reisiger AR, Kaufling J, Manzoni O, Cador M, Georges F, Caillé S (2014) Nicotine self-administration induces CB1-dependent LTP in the bed nucleus of the stria terminalis. J Neurosci 34:4285–4292

Ren Y, Whittard J, Higuera-Matas A, Morris CV, Hurd YL (2009) Cannabidiol, a nonpsychotropic component of cannabis, inhibits Cue-induced heroin seeking and normalizes discrete mesolimbic neuronal disturbances. J Neurosci 29:14764–14769

Riegel AC, Lupica CR (2004) Independent presynaptic and postsynaptic mechanisms regulate endocannabinoid signaling at multiple synapses in the ventral tegmental area. J Neurosci 24: 11070–11078

Robinson JD, Cinciripini PM, Karam-Hage M, Aubin HJ, Dale LC, Niaura R et al (2018) Pooled analysis of three randomized, double-blind, placebo controlled trials with rimonabant for smoking cessation. Addict Biol 23:291–303

Rogers AH, Bakhshaie J, Buckner JD, Orr MF, Paulus DJ, Ditre JW et al (2019) Opioid and cannabis co-use among adults with chronic pain: relations to substance misuse, mental health, and pain experience. J Addict Med 13:287–294

Sagheddu C, Scherma M, Congiu M, Fadda P, Carta G, Banni S et al (2019) Inhibition of N-acylethanolamine acid amidase reduces nicotine-induced dopamine activation and reward. Neuropharmacology 144:327–336

Salamone JD, Correa M (2012) The mysterious motivational functions of mesolimbic dopamine. Neuron 76:470–485

Sam AH, Salem V, Ghatei MA (2011) Rimonabant: from RIO to ban. J Obes 2011:1–4

SAMHSA (2019) 2019 NSDUH detailed tables | CBHSQ data. SAMHSA

Sanchis-Segura C, Spanagel R (2006) Behavioural assessment of drug reinforcement and addictive features in rodents: an overview. Addict Biol 11:2–38

Saravia R, Ten-Blanco M, Grande MT, Maldonado R, Berrendero F (2019) Anti-inflammatory agents for smoking cessation? Focus on cognitive deficits associated with nicotine withdrawal in male mice. Brain Behav Immun 75:228–239

Saravia R, Ten-Blanco M, Pereda-Pérez I, Berrendero F (2021) New insights in the involvement of the endocannabinoid system and natural cannabinoids in nicotine dependence. Int J Mol Sci 22:13316

Scherer PC, Zaccor NW, Neumann NM, Vasavda C, Barrow R, Ewald AJ et al (2017) TRPV1 is a physiological regulator of μ-opioid receptors. Proc Natl Acad Sci USA 114:13561–13566

Scherma M, Panlilio LV, Fadda P, Fattore L, Gamaleddin I, Le Foll B et al (2008) Inhibition of anandamide hydrolysis by Cyclohexyl Carbamic acid 3′-Carbamoyl-3-yl Ester (URB597) reverses abuse-related behavioral and neurochemical effects of nicotine in rats. J Pharmacol Exp Ther 327:482–490

Scherma M, Muntoni AL, Melis M, Fattore L, Fadda P, Fratta W et al (2016) Interactions between the endocannabinoid and nicotinic cholinergic systems: preclinical evidence and therapeutic perspectives. Psychopharmacology 233:1765–1777

Scherma M, Masia P, Satta V, Fratta W, Fadda P, Tanda G (2019) Brain activity of anandamide: a rewarding bliss? Acta Pharmacol Sin 40:309–323

Schindler CW, Redhi GH, Vemuri K, Makriyannis A, Le Foll B, Bergman J et al (2016a) Blockade of nicotine and cannabinoid reinforcement and relapse by a cannabinoid CB1-receptor neutral antagonist AM4113 and inverse agonist rimonabant in squirrel monkeys. Neuropsychopharmacology 41:2283–2293

Schindler CW, Scherma M, Redhi GH, Vadivel SK, Makriyannis A, Goldberg SR et al (2016b) Self-administration of the anandamide transport inhibitor AM404 by squirrel monkeys. Psychopharmacology 233:1867–1877

Schmitz JM, Green CE, Hasan KM, Vincent J, Suchting R, Weaver MF et al (2017) PPAR-gamma agonist pioglitazone modifies craving intensity and brain white matter integrity in patients with primary cocaine use disorder: a double-blind randomized controlled pilot trial. Addiction 112:1861–1868

Schroeder JR, Phillips KA, Epstein DH, Jobes ML, Furnari MA, Kennedy AP et al (2018) Assessment of pioglitazone and proinflammatory cytokines during buprenorphine taper in patients with opioid use disorder. Psychopharmacology 235(10):2957–2966

Schultz W (2015) Neuronal reward and decision signals: from theories to data. Physiol Rev 95:853–951

Serra S, Brunetti G, Pani M, Vacca G, Carai MA, Gessa GL, Colombo G (2002) Blockade by the cannabinoid CB(1) receptor antagonist, SR 141716, of alcohol deprivation effect in alcohol-preferring rats. Eur J Pharmacol 443(1-3):95–97. https://doi.org/10.1016/s0014-2999(02)01594-7

Serrano A, Parsons LH (2011) Endocannabinoid influence in drug reinforcement, dependence and addiction-related behaviors. Pharmacol Ther 132:215–241

Shao Z, Yan W, Chapman K, Ramesh K, Ferrell AJ, Yin J et al (2019) Structure of an allosteric modulator bound to the CB1 cannabinoid receptor. Nat Chem Biol 15:1199–1205

Sharma C, Al Kaabi JM, Nurulain SN, Goyal S, Amjad Kamal M, Ojha S (2016) Polypharmacological properties and therapeutic potential of β-Caryophyllene: a dietary phytocannabinoid of pharmaceutical promise. Curr Pharm Des 22:3237–3264

Singh S, Sarroza D, English A, McGrory M, Dong A, Zweifel L, et al. (2023) Pharmacological characterization of the endocannabinoid sensor GRABeCB2.0. https://home.liebertpub.com/can

Sipe JC, Chiang K, Gerber AL, Beutler E, Cravatt BF (2002) A missense mutation in human fatty acid amide hydrolase associated with problem drug use. Proc Natl Acad Sci USA 99:8394–8399

Sloan ME, Gowin JL, Yan J, Schwandt ML, Spagnolo PA, Sun H et al (2018) Severity of alcohol dependence is associated with the fatty acid amide hydrolase Pro129Thr missense variant. Addict Biol 23:474

Smith LC, Tieu L, Suhandynata RT, Boomhower B, Hoffman M, Sepulveda Y et al (2021) Cannabidiol reduces withdrawal symptoms in nicotine-dependent rats. Psychopharmacology 238:2201–2211

Soler-Cedeño O, Alton H, Bi GH, Linz E, Ji L, Makriyannis A et al (2024) AM6527, a neutral CB1 receptor antagonist, suppresses opioid taking and seeking, as well as cocaine seeking in rodents without aversive effects. Neuropsychopharmacology 49:1678–1688

Soyka M, Koller G, Schmidt P, Lesch OM, Leweke M, Fehr C et al (2008) Cannabinoid receptor 1 blocker rimonabant (SR 141716) for treatment of alcohol dependence: results from a placebo-controlled, double-blind trial. J Clin Psychopharmacol 28:317–324

Spanagel R (2020) Cannabinoids and the endocannabinoid system in reward processing and addiction: from mechanisms to interventions. Dialogues Clin Neurosci 22:241–250

Spano MS, Fattore L, Cossu G, Deiana S, Fadda P, Fratta W (2004) CB1 receptor agonist and heroin, but not cocaine, reinstate cannabinoid-seeking behaviour in the rat. Br J Pharmacol 143: 343

Stopponi S, Somaini L, Cippitelli A, Cannella N, Braconi S, Kallupi M et al (2011) Activation of nuclear PPARγ receptors by the antidiabetic agent pioglitazone suppresses alcohol drinking and relapse to alcohol seeking. Biol Psychiatry 69:642–649

Stopponi S, de Guglielmo G, Somaini L, Cippitelli A, Cannella N, Kallupi M et al (2013) Activation of PPARγ by pioglitazone potentiates the effects of naltrexone on alcohol drinking and relapse in msP rats. Alcohol Clin Exp Res 37:1351–1360

Sulzer D (2011) How addictive drugs disrupt presynaptic dopamine neurotransmission. Neuron 69: 628–649

Tham M, Yilmaz O, Alaverdashvili M, Kelly MEM, Denovan-Wright EM, Laprairie RB (2019) Allosteric and orthosteric pharmacology of cannabidiol and cannabidiol-dimethylheptyl at the type 1 and type 2 cannabinoid receptors. Br J Pharmacol 176:1455–1469

Thiruchselvam T, Malik S, Le Foll B (2017) A review of positron emission tomography studies exploring the dopaminergic system in substance use with a focus on tobacco as a co-variate. Am J Drug Alcohol Abuse 43:197–214

Trigo JM, Le Foll B (2016) Inhibition of monoacylglycerol lipase (MAGL) enhances cue-induced reinstatement of nicotine-seeking behavior in mice. Psychopharmacology 233:1815–1822

Trigo JM, Lagzdins D, Rehm J, Selby P, Gamaleddin I, Fischer B et al (2016) Effects of fixed or self-titrated dosages of Sativex on cannabis withdrawal and cravings. Drug Alcohol Depend 161:298–306

Trigo JM, Soliman A, Quilty LC, Fischer B, Rehm J, Selby P et al (2018) Nabiximols combined with motivational enhancement/cognitive behavioral therapy for the treatment of cannabis dependence: a pilot randomized clinical trial. PLoS One 13:e0190768

Turner SE, Williams CM, Iversen L, Whalley BJ (2017) Molecular pharmacology of phytocannabinoids. In: Progress in the chemistry of organic natural products. Springer, Cham, pp 61–101

Tyndale RF, Payne JI, Gerber AL, Sipe JC (2007) The fatty acid amide hydrolase C385A (P129T) missense variant in cannabis users: studies of drug use and dependence in Caucasians. Am J Med Genet B Neuropsychiatr Genet 144B:660–666

Vallée M (2016) Neurosteroids and potential therapeutics: focus on pregnenolone. J Steroid Biochem Mol Biol 160:78–87

Vallée M, Vitiello S, Bellocchio L, Hébert-Chatelain E, Monlezun S, Martin-Garcia E et al (2014) Pregnenolone can protect the brain from cannabis intoxication. Science 1979(343):94–98

van Esbroeck ACM, Janssen APA, Cognetta AB, Ogasawara D, Shpak G, van der Kroeg M et al (2017) Activity-based protein profiling reveals off-target proteins of the FAAH inhibitor BIA 10-2474. Science 1979(356):1084–1087

Vengeliene V, Bilbao A, Spanagel R (2014) The alcohol deprivation effect model for studying relapse behavior: a comparison between rats and mice. Alcohol 48:313–320

Vinod KY, Kassir SA, Hungund BL, Cooper TB, Mann JJ, Arango V (2010) Selective alterations of the CB1 receptors and the fatty acid amide hydrolase in the ventral striatum of alcoholics and suicides. J Psychiatr Res 44:591–597

Vinod KY, MacCioni P, Garcia-Gutierrez MS, Femenia T, Xie S, Carai MAM et al (2012) Innate difference in the endocannabinoid signaling and its modulation by alcohol consumption in alcohol-preferring sP rats. Addict Biol 17:62–75

Vitale RM, Iannotti FA, Moriello AS, Tunisi L, Piscitelli F, Savopoulos R et al (2021) Identification and characterization of Cannabidiol as an OX1R antagonist by computational and in vitro functional validation. Biomolecules 11:1134

Volkow ND, Morales M (2015) The brain on drugs: from reward to addiction. Cell 162:712–725

Volkow ND, Michaelides M, Baler R (2019) The neuroscience of drug reward and addiction. Physiol Rev 99:2115–2140

Wang H, Treadway T, Covey DP, Cheer JF, Lupica CR (2015) Cocaine-induced endocannabinoid mobilization in the ventral tegmental area. Cell Rep 12:1997–2008

Ward SJ, Rosenberg M, Dykstra LA, Walker EA (2009) The CB1 antagonist rimonabant (SR141716) blocks cue-induced reinstatement of cocaine seeking and other context and extinction phenomena predictive of relapse. Drug Alcohol Depend 105:248–255

Wenzel JM, Oleson EB, Gove WN, Cole AB, Gyawali U, Dantrassy HM et al (2018) Phasic dopamine signals in the nucleus Accumbens that cause active avoidance require endocannabinoid mobilization in the midbrain. Curr Biol 28:1392–1404.e5

Werling LL, Reed SC, Wade D, Izenwasser S (2009) Chronic nicotine alters cannabinoid-mediated locomotor activity and receptor density in periadolescent but not adult male rats. Int J Dev Neurosci 27:263

Werneck MA, Kortas GT, de Andrade AG, Castaldelli-Maia JM (2018) A systematic review of the efficacy of cannabinoid agonist replacement therapy for cannabis withdrawal symptoms. CNS Drugs 32:1113–1129

Wilkerson JL, Ghosh S, Mustafa M, Abdullah RA, Niphakis MJ, Cabrera R et al (2017) The endocannabinoid hydrolysis inhibitor SA-57: intrinsic antinociceptive effects, augmented morphine-induced antinociception, and attenuated heroin seeking behavior in mice. Neuropharmacology 114:156–167

Winters ND, Bedse G, Astafyev AA, Patrick TA, Altemus M, Morgan AJ et al (2021) Targeting diacylglycerol lipase reduces alcohol consumption in preclinical models. J Clin Invest 131: e146861

Wise RA, Bozarth MA (1987) A psychomotor stimulant theory of addiction. Psychol Rev 94:469–492

Wolf ME (2016) Synaptic mechanisms underlying persistent cocaine craving. Nat Rev Neurosci 17: 351–365

World Health Organization (2017) WHO report on the global tobacco epidemic, 2017: monitoring tobacco use and prevention policies. WHO

Xi ZX, Gilbert JG, Peng XQ, Pak AC, Li X, Gardner EL (2006) Cannabinoid CB1 receptor antagonist AM251 inhibits cocaine-primed relapse in rats: role of glutamate in the nucleus accumbens. J Neurosci 26:8531–8536

Xi Z-X, Peng X Q, Li X, Song R, Zhang H-Y, Liu Q-R et al (2011) Brain cannabinoid CB (2) receptors modulate cocaine's actions in mice. Nat Neurosci 14:1160–1166

Xi ZX, Muldoon P, Wang XF, Bi GH, Damaj MI, Lichtman AH et al (2019) Δ8-Tetrahydrocannabivarin has potent anti-nicotine effects in several rodent models of nicotine dependence. Br J Pharmacol 176:4773

Zanelati TV, Biojone C, Moreira FA, Guimaraes FS, Joca SRL (2010) Antidepressant like effects of cannabidiol in mice: possible involvement of 5-HT 1A receptors. Br J Pharmacol 159:122–128

Zhang H, Lipinski AA, Liktor-Busa E, Smith AF, Moutal A, Khanna R et al (2021) The effects of repeated morphine treatment on the endogenous cannabinoid system in the ventral tegmental area. Front Pharmacol 12:632757

Zhang H-XB, Heckman L, Niday Z, Jo S, Fujita A, Shim J et al (2022) Cannabidiol activates neuronal Kv7 channels. elife 11:e73246
Zhou Y, Schwartz BI, Giza J, Gross SS, Lee FS, Kreek MJ (2017) Blockade of alcohol escalation and "relapse" drinking by pharmacological FAAH inhibition in male and female C57BL/6J mice. Psychopharmacology 234:2955–2970
Zlebnik NE, Cheer JF (2016a) Beyond the CB1 receptor: is Cannabidiol the answer for disorders of motivation? Annu Rev Neurosci 39:1–17
Zlebnik NE, Cheer JF (2016b) Drug-induced alterations of endocannabinoid-mediated plasticity in brain reward regions. J Neurosci 36:10230–10238

The Role of Cannabinoids and the Endocannabinoid System in the Treatment and Regulation of Nausea and Vomiting

Erin M. Rock and Linda A. Parker

Contents

1 Introduction 340
2 Cannabinoids in Animal Models of Nausea and Vomiting 341
2.1 THC and THC-Like Synthetics in Animal Models of CINV 343
2.2 THC-Induced Conditioned Gaping: A Model of Cannabinoid Hyperemesis Syndrome 349
2.3 CBD in Animal Models of CINV 349
2.4 Minor Cannabinoids in Animal Models of CINV 354
2.5 Manipulations of eCBs in Animal Models of CINV 356
2.6 N-Acyl Glycines in Animal Models of CINV 362
3 Cannabinoids in Human CINV Patients 363
3.1 Exogenous Cannabinoids in Human CINV Patients 363
3.2 Cannabinoid Treatments Producing Nausea and Vomiting in Human Patients 366
4 Endocannabinoid Levels in Humans Experiencing Nausea and Vomiting 367
4.1 Cannabinoids and eCBs in Human Patients: Conclusions 370
5 General Conclusions 370
References 372

Abstract Despite using the recommended anti-emetic treatments, control of nausea and vomiting is still an unmet need for cancer patients undergoing chemotherapy treatment. Few properly controlled clinical trials have evaluated the potential of exogenously administered cannabinoids or manipulations of the endogenous cannabinoid (eCB) system to treat nausea and vomiting. In this chapter, we explore the pre-clinical and human clinical trial evidence for the potential of exogenous cannabinoids and manipulations of the eCB system to reduce nausea and vomiting. Although there are limited high-quality human clinical trials, pre-clinical evidence suggests that cannabinoids and manipulations of the eCB system have anti-nausea/anti-emetic potential. The pre-clinical anti-nausea/anti-emetic evidence highlights

E. M. Rock and L. A. Parker (✉)
Department of Psychology and Collaborative Neuroscience Graduate Program, University of Guelph, Guelph, ON, Canada
e-mail: parkerl@uoguelph.ca

Curr Topics Behav Neurosci (2026) 76: 339–387
https://doi.org/10.1007/7854_2024_554
Published Online: 31 December 2024

the need for further evaluation of cannabinoids and manipulations of eCBs and other fatty acid amides in clinical trials.

Keywords Chemotherapy-induced nausea and vomiting · Conditioned gaping · Endocannabinoid · Delta-9-tetrahydrocannabinol · Cannabidiol

1 Introduction

Cannabis has been used as a medicine for centuries (see Mechoulam and Hanus 2001 for review) but was not evaluated for its ability to relieve chemotherapy-induced nausea and vomiting (CINV) until the 1970s. Since then, few clinical trials have compared medical cannabis or the individual components in cannabis with the currently recommended anti-emetic regimen in patients receiving chemotherapy, nor evaluated it as a possible adjunct treatment. Data from pre-clinical studies suggest that cannabis and some of its components may reduce nausea and vomiting.

The natural products derived from the cannabis plant (cannabinoids) include the psychoactive delta-9-tetrahydrocannabinol (THC) and the non-intoxicating cannabidiol (CBD), as well as other minor cannabinoids, such as cannabigerol (CBG). THC was first identified in the 1960s by Yechiel Gaoni and Raphael Mechoulam (Gaoni and Mechoulam 1964). This led to the search and eventual discovery of its endogenous target, the endogenous cannabinoid (eCB) system. This system consists of the cannabinoid receptors 1 and 2 (CB_1 and CB_2) (Devane et al. 1988; Munro et al. 1993), their endogenous ligands (eCBs) including anandamide (AEA)(Devane et al. 1992) and 2-arachidonoylglycerol (2-AG) (Mechoulam et al. 1995), as well as their degradative enzymes including fatty acid amide hydrolase (FAAH) (Cravatt et al. 1996) and monoacylglycerol lipase (MAGL) (Dinh et al. 2002a, b). The eCB system is now regarded as an important modulatory system involved in many aspects of health and disease states. Research is in the early stages of understanding how it modulates many aspects of daily functioning.

The most studied cannabinoid, THC is a partial agonist at the CB_1 and CB_2 receptors (Pertwee 2008), and its activity at the CB_1 receptor is thought to underlie its psychoactive effects. CBD (unlike THC) is not intoxicating, likely due to its low affinity for the CB_1 receptor. CBD has been termed a promiscuous ligand because of its activity at many receptors which may actually underlie its widespread therapeutic effects (Crippa et al. 2018). CBD is a non-competitive CB_1 receptor antagonist, a CB_2 receptor inverse agonist (Thomas et al. 2007) and a negative allosteric modulator of the CB_1 receptor (Laprairie et al. 2015). CBD is an agonist at both the nuclear peroxisome proliferator-activated receptor-γ (PPAR-γ), and the transient receptor potential cation channel subfamily V member 1 (TRPV1) and TRPV2 receptors (Bisogno et al. 2001). CBD also acts as an indirect agonist of the CB_1 receptor by blocking anandamide uptake and inhibiting its breakdown by FAAH in rodents (Bisogno et al. 2001) but may not inhibit FAAH in humans (Criscuolo et al.

2020). CBD's mechanism of action that is of greatest relevance to nausea and vomiting is its activation of the serotonin 1A (5-HT$_{1A}$) receptor. Here, CBD may be a positive allosteric modulator because it can enhance direct activation of the 5-HT$_{1A}$ receptor (Rock et al. 2012; Russo et al. 2005). CBG is a partial agonist of CB$_1$ and CB$_2$ receptors (Zagzoog et al. 2020) and acts as an antagonist at the 5-HT$_{1A}$ receptor (Cascio et al. 2010).

In this chapter, we review findings on the potential of exogenous cannabinoids and manipulations of the eCB system to reduce nausea and vomiting, from pre-clinical animal research to human clinical trials.

Chemotherapy patients experience *acute* (occurring within 24 h post-treatment) and/or *delayed* (occurring more than 24 h post-treatment and up to 5 days) nausea and vomiting as distressing side effects of treatment (Fiore and Gralla 1984; Navari and Aapro 2016). If improperly managed, *anticipatory* nausea and vomiting – a conditioned response upon re-exposure to the chemotherapy clinic – can develop (Morrow 1982). Guidelines for the management of highly emetogenic chemotherapy agents recommend the combination of a serotonin 3 (5-HT$_3$) receptor antagonist (such as ondansetron [OND], dolasetron, and granisetron), a neurokinin-1 (NK1) receptor antagonist (such as aprepitant, fosaprepitant, and rolapitant), and corticosteroids (such as dexamethasone) (Hesketh et al. 2017). Even with this standard treatment, up to 40% of patients still report nausea and/or vomiting following chemotherapy (Dranitsaris et al. 2017). As an example, in a survey of 168 cancer patients that received chemotherapy, 71% of patients indicated they experienced nausea and 26% reported experiencing vomiting (Hernandez Torres et al. 2015), despite receiving the standard anti-emetic treatment. None of the currently available treatments are effective in managing anticipatory nausea (Roscoe et al. 2011), with sedating benzodiazepines often prescribed (Malik et al. 1995; Razavi et al. 1993). Indeed, nausea control remains to be an unmet need for these patients. Therefore, cannabinoids may have efficacy for this indication based on considerable evidence from preclinical animal models reviewed next.

2 Cannabinoids in Animal Models of Nausea and Vomiting

The *Suncus murinus* (house musk shrew) and *Cryptotis parva* (least shrew) are often used to study emesis as they vomit to toxins such as nicotine (Matsuki et al. 1988, 1990; Nakayama et al. 2005; Parker et al. 2009; Rock et al. 2012; Torii et al. 1991), the chemotherapeutic agent cisplatin (Darmani 1998, 2001b; Lau et al. 2005; Matsuki et al. 1988, 1990; Parker et al. 2009; Ray et al. 2009; Rock et al. 2012, Sam et al. 2003; Torii et al. 1991), or lithium chloride (LiCl) (e.g. Parker et al. 2004). Ferrets, cats, dogs, and pigeons also vomit to toxins such as cisplatin or morphine 6 glucuronide (M6G) (McCarthy and Borison 1984; Sharkey et al. 2007; Van Sickle et al. 2001, 2003). These species have been used to study the neurobiology of vomiting and to evaluate the effects of cannabinoids and manipulations of the eCB system on vomiting.

The literature on CINV patients suggests that females may be at a higher risk than males for developing CINV (Mosa et al. 2020), although this has been attributed to a greater likelihood for females to self-report poor health symptoms (Barsky et al. 2001; Boerma et al. 2016). Few studies have examined sex differences in vomiting using emetic species such as shrews, frogs, and ferrets. In house musk shrews subjected to horizontal motion, more emetic episodes have been reported in females (Javid and Naylor 1999), or in males (Brame and Lucot 2011; Matsuki et al. 1997), with others reporting no sex difference (Ito et al. 2003; Kaji et al. 1990; Ueno et al. 1988). One study has reported no sex differences in treefrogs subjected to parabolic flight manoeuvres (Naitoh et al. 2001). No sex differences have been found in female and male house musk shrews in vomiting produced by the toxins nicotine, cisplatin, LiCl, radiation, isoflurane, or copper sulphate (Horn et al. 2012, 2013; Ito et al. 2003; Matsuki et al. 1997; O'Brien et al. 2013; Parker et al. 2009; Rock et al. 2020a; Rock and Parker 2015). Finally, no sex differences have been reported in cisplatin-induced vomiting in ferrets (Percie du Sert et al. 2011). Very few studies have examined whether the anti-emetic efficacy of cannabinoids or manipulations of the eCB system differ between male and female animals (reviewed below).

The use of pre-clinical animal models has allowed an understanding of the neurobiology of vomiting (Hornby 2001), but that of nausea is still being discovered and may be more complex (see Zhong et al. 2021 for a recent review). While vomiting is mediated by brainstem structures (Hornby 2001), nausea seems to require activation of forebrain structures (Horn 2008; Napadow et al. 2013; Sanger and Andrews 2006). Indeed, CB_1 and CB_2 receptors have been identified in ferret brainstem structures implicated in emesis (Van Sickle et al. 2001, 2005) where cannabinoids exert their anti-emetic effects. In addition, FAAH and MAGL have been detected in ferret and rat brainstem structures (Suárez et al. 2010; Van Sickle et al. 2001).

The selective and reliable rodent model of nausea-induced conditioned gaping has revealed some brain regions that may be key in mediating nausea and has also allowed for the evaluation of candidate anti-nausea compounds. Although rats cannot vomit, they show conditioned gaping reactions when exposed to a flavour that has been previously paired with an illness-inducing agent such as LiCl (Grill and Norgren 1978) a rodent model of acute nausea. Indeed, only emetic drugs produce conditioned gaping in rats, and anti-emetic treatments (including cannabinoids and manipulations of the eCB system) block conditioned gaping (Parker 2014).

Rats also display conditioned gaping if they are returned to a context that was previously paired with illness-inducing LiCl, similar to the development of anticipatory nausea in humans (Limebeer et al. 2008). Like human anticipatory nausea patients, treatment with a 5-HT_3 receptor antagonist does not reduce contextually elicited conditioned gaping in rats (Limebeer et al. 2006; Rock et al. 2014). Furthermore, although treatment with a benzodiazepine (a common treatment for human anticipatory nausea patients) reduced contextually elicited conditioned gaping in rats, it produced sedative effects (Rock et al. 2014). Thus, the rodent model of contextually elicited conditioned gaping can be used as a reliable model of anticipatory nausea.

Using the acute conditioned gaping model, the Parker lab has accumulated much evidence to suggest that the interoceptive insular cortex (IIC) is a critical region mediating nausea. In brief, elevated 5-HT in the IIC triggers nausea, which can be suppressed by treatments that reduce 5-HT (Limebeer et al. 2018). Indeed, intra-IIC infusion of a 5-HT_3 receptor antagonist, ondansetron, reduced LiCl-induced gaping, while intra-IIC infusion of a 5-HT_3 receptor agonist produced it (Tuerke et al. 2012). In addition, LiCl elevates 5-HT in the IIC and treatments that reduce conditioned gaping also reduce 5-HT release in the IIC (Limebeer et al. 2018).

The conditioned gaping model has also been used to detect whether a compound might produce nausea as a side effect of its administration. Indeed, compounds such as selective serotonin reuptake inhibitors (which produce nausea as a side effect in humans) also produce conditioned gaping in rats (Limebeer et al. 2009; Tuerke et al. 2009). In 2006, the CB_1 receptor antagonist, rimonabant (SR141716), was approved for the treatment of obesity in Europe but was withdrawn worldwide in 2008 because of its side effects, including severe depression and suicidal thoughts, with nausea also reported in some patients (de Mattos Viana et al. 2009; Despres 2009). Rimonabant is an antagonist/inverse agonist of the CB_1 receptor (Landsman et al. 1997; Mato et al. 2002; McLaughlin et al. 2006). Its inverse agonist activity is thought to underlie nausea produced by rimonabant (and the CB_1 receptor inverse agonist AM251). Rimonabant and AM251 have been shown to produce conditioned gaping (Sink et al. 2008), that is not produced by equivalent doses of the neutral CB_1 receptor antagonists AM4113 (Rock et al. 2013b; Sink et al. 2008) or AM6545 (Cluny et al. 2010). These findings suggest that neutral CB_1 receptor antagonists may be effective anti-obesity therapeutics, devoid of nauseating side effects. This also highlights the importance of dose selection when using AM251 or SR141716 to block anti-nausea/anti-emetic effects to demonstrate CB_1 receptor mediation. Low doses of these compounds must be used that on their own do not produce nausea or vomiting.

Here we review the pre-clinical evidence for the potential of cannabinoids and eCB manipulations to reduce vomiting as well as acute and anticipatory nausea. We also discuss the mechanism of action for these effects when they have been reported.

2.1 *THC and THC-Like Synthetics in Animal Models of CINV*

2.1.1 Toxin-Induced Vomiting

The effects of THC and THC-like synthetics have been evaluated for their ability to reduce toxin-induced vomiting in several species. Please refer to Table 1 for a summary of these experiments. In the least shrew, lower doses of THC ($\lesssim$10 mg/kg, i.p.) reduced vomiting to cisplatin (Darmani 2001b; Ray et al. 2009; Wang et al. 2009), substance P (Darmani et al. 2019) and dopamine D2/D3 receptor agonists (Darmani and Crim 2005), while a higher dose of THC (20 mg/kg, i.p.) was

Table 1 Effects of THC and THC-like synthetics on toxin-induced vomiting in animals

Compound	Species	Effective dose (s) tested	Mechanism of action	Toxin	Reference
THC	Least shrew	≤10 mg/kg, i.p.	CB_1 receptor	Cisplatin	Darmani (2001b), Darmani et al. (2014), Ray et al. (2009), Wang et al. (2009)
				Substance P	Darmani et al. (2019)
				Dopamine D2/D3 receptor agonists	Darmani and Crim (2005)
		20 mg/kg, i.p.	CB_1 receptor	Serotonin	Darmani and Johnson (2004)
				SR141716	Darmani (2001a)
				Radiation	Darmani et al. (2007)
	House musk shrew	2.5–20 mg/kg, i.p.	CB_1 receptor	LiCl	Parker et al. (2004)
			Not evaluated	Cisplatin	Kwiatkowska et al. (2004)
	Ferret	0.5, 1 mg/kg, i.p.	CB_1 receptor	Cisplatin	Van Sickle et al. (2003)
				M6G	Van Sickle et al. (2001)
	Squirrel monkey	0.1 mg/kg, i.m.	CB_1 receptor	Nicotine	Wooldridge et al. (2020)
			Not evaluated	LiCl	Wooldridge et al. (2020)
THC + TRPV1 agonist	Least Shrew	0.5–10 mg/kg + 1, 5 μg/kg, i.p.[a]	Not evaluated	Cisplatin	Darmani et al. (2014)
THC + $5\text{-}HT_3$ antagonist	House musk shrew	1.25 + 0.02 mg/kg, i.p.[a]	Not evaluated	Cisplatin	Kwiatkowska et al. (2004)
	Least shrew	0.25, 0.5 + 0.025–0.25 mg/kg, i.p. [a]	Not evaluated	Cisplatin	Wang et al. (2009)
THC + CBD	House musk shrew	1 + 2.5 mg/kg, i.p.[a]	Not evaluated	LiCl	Rock and Parker (2015)
THC + CBDA	House musk shrew	1 + 0.05 mg/kg, i.p.[a]	Not evaluated	LiCl	Rock and Parker (2015)
Dronabinol	Ferret	0.2–5 mg/kg, s.c	Not evaluated	Morphine	Mori et al. (2012)

(continued)

Table 1 (continued)

Compound	Species	Effective dose (s) tested	Mechanism of action	Toxin	Reference
Nabilone	Cat	25–0.1 mg/kg, i.v.)	Not evaluated	Cisplatin	London et al. (1979), McCarthy and Borison (1981)
			Not evaluated	Apomorphine	London et al. (1979)
HU-210	Pigeon	12.5–50 μg/kg, i.v.	Not evaluated	Cisplatin	Ferrari et al. (1999)
			Not evaluated	Emetine	Ferrari et al. (1999)
HU-211	Pigeon	2.5 mg/kg, s.c.	Not evaluated	Cisplatin	Feigenbaum et al. (1989)
THCA	House musk shrew	0.05, 0.5 mg/kg, i.p.	CB_1 receptor	LiCl	Rock et al. (2013a)

[a] Enhanced suppression of vomiting when compounds combined

necessary to reduce vomiting induced by serotonin (Darmani and Johnson 2004), the CB_1 receptor antagonist/inverse agonist SR141716 (Darmani 2001a), and radiation (Darmani et al. 2007). Similarly in the house musk shrew, THC (2.5–20 mg/kg, i.p.) reduced LiCl-, cisplatin-, and motion-induced vomiting (Cluny et al. 2008; Kwiatkowska et al. 2004; Parker et al. 2004). These anti-emetic effects of THC were shown to be CB_1 receptor-mediated in the least shew (Darmani 2001b; Darmani and Crim 2005; Darmani et al. 2007; Ray et al. 2009) and house musk shrew (Cluny et al. 2008; Kwiatkowska et al. 2004; Parker et al. 2004). Despite the evidence for the anti-emetic effects of THC, there is a report of THC-induced vomiting in the house musk shrew at the higher doses of 3, 5 and 10 mg/kg (i.p.) (Cluny et al. 2008). This suggests a biphasic effect of THC's role in vomiting in this species.

THC (0.5, 1 mg/kg, i.p.) reduced cisplatin-, and M6G-induced vomiting in ferrets (Van Sickle et al. 2001, 2003) and THC (0.1 mg/kg, intramuscular, [i.m.]) suppressed nicotine- and LiCl-induced vomiting in squirrel monkeys (Wooldridge et al. 2020). These effects were mediated by agonism of the CB_1 receptor (Van Sickle et al. 2001, 2003; Wooldridge et al. 2020).

Similar anti-emetic effects have been reported with THC-like synthetics. Dronabinol (0.2–5 mg/kg, s.c) reduced morphine-induced vomiting in ferrets (Mori et al. 2012). In cats, nabilone (25–100 μg/kg, i.v.) reduced cisplatin- and apomorphine-induced vomiting (McCarthy and Borison 1981), but was ineffective against nicotine-induced vomiting in cats (London et al. 1979) and cisplatin-induced vomiting in dogs (Gylys et al. 1979). At these doses of nabilone, severe locomotor impairment and ataxia were observed in these animals. The potent cannabinoid agonist HU-210 (12.5–50 μg/kg, i.v.) suppressed cisplatin- and emetine-induced vomiting in pigeons, but produced sedation (Ferrari et al. 1999), while its synthetic enantiomer HU-211 (Dexanabinol; 2.5 mg/kg, s.c.) suppressed cisplatin-induced vomiting in pigeons (Feigenbaum et al. 1989) without psychoactive effects (Mechoulam et al. 1988).

THC's acidic precursor tetrahydrocannabinolic acid (THCA), present in fresh cannabis and decarboxylated upon heating or drying of the plant, does not exert psychotomimetic activity when administered to rhesus monkeys, mice, and dogs (Grunfeld and Edery 1969). THCA's lack of psychoactivity may be due to THCA's limited access to the central nervous system (Moreno-Sanz 2016). THCA (0.05, 0.5 mg/kg, i.p.), reduced LiCl-induced vomiting through a CB_1 receptor mechanism (Rock et al. 2013a).

Combinations of THC with other anti-emetic compounds have also been evaluated in shrews. In the least shrew, an enhanced anti-emetic effect occurred when low doses of THC (0.5, 2.5, 5, 10 mg/kg, i.p.) were combined with low doses of the TRPV1 receptor agonist resiniferatoxin (Darmani et al. 2014). In addition, low doses of THC (0.25 and 0.5 mg/kg, i.p.) in combination with low doses of tropisetron (a 5-HT_3 receptor antagonist) showed an enhancement in the reduction of vomiting, with no such enhanced effect when combined with dexamethasone in the least shrew (Wang et al. 2009). In the house musk shrew, there was an enhanced anti-emetic effect for combined ineffective doses of the 5-HT_3 receptor antagonist OND and THC against cisplatin-induced vomiting (Kwiatkowska et al. 2004). These results complement human findings that THC and THC-like compounds have anti-emetic properties through their action at the CB_1 receptor and may have added efficacy when combined with other anti-emetic treatments. Use of THC at lower doses may be preferable because it is devoid of intoxicating effects at such doses.

2.1.2 Acute Nausea

Using the acute nausea model of LiCl-induced conditioned gaping, the effects of THC and THC-like synthetics have been evaluated. Please refer to Table 2 for a summary of these studies.

THC (0.5, 1, 2.5, 3, 10 mg/kg, i.p; 2.5, 10, 20 mg/kg, i.g.) reduced cyclophosphamide- (Limebeer and Parker 1999), and LiCl-induced conditioned gaping in rats (Limebeer et al. 2012; Parker and Mechoulam 2003; Parker et al. 2003; Rock et al. 2013b, 2015a, 2016b). The synthetic CB_1 receptor full agonist HU-210 (0.001–0.01 mg/kg, i.p.) potently reduced LiCl-induced conditioned gaping (Parker and Mechoulam 2003; Parker et al. 2003) and its intracranial administration to the IIC reduced conditioned gaping through a CB_1 receptor-mediated effect (Limebeer et al. 2012). Most recently, our group has demonstrated that THC (0.5, 1 mg/kg, i.p.) reduced LiCl-induced conditioned gaping in a similar manner in both male and female rats (Rock et al. 2022).

THCA (0.01 0.05, 0.5 mg/kg, i.p.) reduced conditioned gaping (Rock et al. 2013a, 2020c). Although the anti-vomiting effect of THCA in the shrew was mediated by its effect on the CB_1 receptor (Rock et al. 2013a), the anti-nausea effect

Table 2 Effects of THC and THC-like synthetics on acute and anticipatory nausea in male rats

Compound	Effective dose(s) tested	Mechanism (s) of action	Reference
Acute nausea			
THC	0.5, 1, 2.5, 3, 10 mg/kg, i.p	CB_1 receptor	Limebeer and Parker (1999), Limebeer et al. (2012), Parker and Mechoulam (2003), Parker et al. (2003), Rock et al. (2015a), Rock et al. (2022), (2013b), (2020c)
	2.5, 10, 20 mg/kg, i.g.	Not evaluated	Rock et al. (2016b)
THC + CBD	0.1 + 0.1 mg/kg, i.p.	CB_1 + $5\text{-}HT_{1A}$ receptors	Rock et al. (2020c)
THC + CBDA	0.01 mg/kg + 0.01 μg/kg, 0.1 mg/kg + 0.1 μg/kg, i.p.[a] 1 mg/kg + 1 μg/kg, 10 mg/ kg + 10 μg/kg, i.p.	Not evaluated	Rock et al. (2015a)
	0.5 mg/kg + 0.5 μg/kg, 1 mg/kg + 1 μg/kg, i.g.[a] 2.5 mg/kg + 2.5 μg/kg, 10 mg/kg + 10 μg/kg, 20 mg/kg + 20 μg/kg, i.g.	Not evaluated	Rock et al. (2016b)
HU-210	0.001–0.01 mg/kg, i.p.	CB_1 receptor	Parker and Mechoulam (2003), Parker et al. (2003)
	0.001, 0.01 μg into the IIC	CB_1 receptor	Limebeer et al. (2012)
THCA	0.01 0.05, 0.5 mg/kg, i.p.	PPARα	Rock et al. (2013a), (2020c)
THCA + CBDA	1 + 0.01 μg/kg, i.p.[a]	PPARα + $5\text{-}HT_{1A}$ receptors	Rock et al. (2020c)
Anticipatory nausea			
THC	0.5, 1, 10 mg/kg, i.p.	Not evaluated	Limebeer et al. (2006), Rock et al. (2014), (2015a)
	1, 2.5, 10 mg/kg, i.g.	Not evaluated	Rock et al. (2016b)
THC + CBDA	1 mg/kg + 1 μg/kg, 10 mg/ kg + 10 μg/kg, i.p.	Not evaluated	Rock et al. (2015a)
	0.1 mg/kg + 0.1 μg/kg, 1 mg/kg + 1 μg/kg, i.g.	Not evaluated	Rock et al. (2016b)
THCA	0.05 mg/kg, i.p.	CB_1 receptor	Rock et al. (2013a)
THCA + CBDA	5 + 0.1 μg/kg, i.p.[a]	CB_1 + $5\text{-}HT_{1A}$ receptors	Rock et al. (2014)

[a] Enhanced suppression of nausea when compounds combined

of THCA was mediated by its action at the proliferator-activated receptor alpha (PPARα). Administration of a peroxisome PPARα antagonist (MK886) blocked THCA's anti-nausea effect (Rock et al. 2020c). In vitro studies suggest that THCA is a weak FAAH inhibitor (De Petrocellis et al. 2013). Thus, THCA's anti-nausea effect may be a result of elevated levels of the fatty acid ethanolamides oleoylethanolamide (OEA) and palmitoylethanolamide (PEA), which are high-affinity PPARα agonists (Sun et al. 2007). This is consistent with the finding in cyclic vomiting patients that showed elevations of OEA and PEA (Venkatesan et al. 2016), presumably as a protective mechanism during the sick phase (discussed later). This suggests that OEA and PEA may be anti-nausea compounds.

Thus, consistent with human trials, THC (likely via activation of presynaptic CB_1 receptors to reduce 5-HT release) and THC-like synthetics have anti-nausea effects in the rat conditioned gaping model of acute nausea, with THC exerting similar anti-nausea effects in both male and female rats. In addition, the non-intoxicating THCA also seems to reduce acute nausea via action at PPARα.

2.1.3 Anticipatory Nausea

The effects of THC and THCA have been evaluated in the contextually elicited gaping model. Please refer to Table 2 for a summary of these findings. THC (0.5, 1, 10 mg/kg, i.p.; 1, 2.5, 10 mg/kg, i.g.) was effective in this rodent model of anticipatory nausea (Limebeer et al. 2006; Rock et al. 2014, 2015a, 2016b), although the high dose of 10 mg/kg (i.p.) THC suppressed activity, suggesting a sedative effect.

In addition, THCA (0.05 mg/kg, i.p.) reduced contextually-elicited gaping through a CB_1 receptor-mediated mechanism but did not induce hypothermia or suppress locomotion, indicating a non-CB_1 agonist-like effect (Rock et al. 2013a). In vitro studies suggest that THCA is a weak FAAH inhibitor in rodents (De Petrocellis et al. 2013). Thus, THCA may exert an indirect effect on CB_1 receptors via elevation of AEA levels to reduce anticipatory nausea, or its action as an FAAH inhibitor may be elevating OEA and PEA levels to act on PPARα to reduce acute nausea. There are conflicting results about the ability of THCA to bind to CB_1 and CB_2 receptors (see Moreno-Sanz 2016 for review), which may be due to differing purity (synthetic vs plant-derived) or its degree of decarboxylation into THC. Further research carefully monitoring the stability of THCA is needed to fully understand its mechanisms.

THC (at lower doses) or THCA may be effective in reducing anticipatory nausea. To date, we have examined the acute anti-nausea effects of THC in female rats, but no such studies have been conducted using the anticipatory nausea model.

2.2 THC-Induced Conditioned Gaping: A Model of Cannabinoid Hyperemesis Syndrome

THC can have dose-dependent biphasic effects on nausea. Repeated, high-dose administration of THC (5 and 10 mg/kg, i.p.) produces conditioned gaping, while a known anti-nausea dose of THC (0.5 mg/kg) does not (DeVuono et al. 2018). This THC-induced conditioned gaping is CB_1 receptor-mediated (DeVuono et al. 2018). When examining the brains of rats exposed to high dose THC (10 mg/kg), an upregulation of genes related to MAGL in the hypothalamus was seen, suggesting that THC may promote nausea via dysregulation of eCB function in the hypothalamus (DeVuono et al. 2018). This effect of high dose THC in rats may serve as a model of Cannabinoid Hyperemesis Syndrome (CHS) in humans (see Sect. 3.2 below; for an excellent review see DeVuono et al. 2024).

Like with CHS patients, the typical anti-emetic agent OND did not interfere with THC-induced conditioned gaping, but treatments that interfere with the stress response (including the MAGL inhibitor MJN110, the FAAH inhibitor URB597 and chlordiazepoxide) did interfere with the THC-induced conditioned gaping response (DeVuono et al. 2020b). Furthermore, THC produced significantly higher levels of the stress hormone corticosterone at 10 mg/kg than at the anti-nausea dose of 0.5 mg/kg THC (DeVuono et al. 2020b) suggesting that this THC-induced nausea may be due to activation of the stress response. Like THC, JWH-018, a synthetic full agonist of the CB_1 receptor, potently produced conditioned gaping with repeated treatment that was accompanied by elevated levels of corticosterone (DeVuono et al. 2020a). The effect of JWH-018 was also mediated by the CB_1 receptor.

Administration of CBD (5 mg/kg, i.p.) reduced THC-induced conditioned gaping through a 5-HT_{1A} receptor-mediated effect (DeVuono et al. 2022). Furthermore, CBD (5 mg/kg, i.p.) reversed the THC-induced increase in corticosterone (DeVuono et al. 2022). Together, these findings support the hypothesis that THC-induced nausea, and CHS, results from a dysregulated stress response (DeVuono et al. 2020b) and that CBD or manipulations of the eCB system may be potential treatments. These findings also suggest that the use of CBD-rich cannabis strains may help to alleviate or prevent the development of CHS. Further research utilising this model might help to better understand CHS and potential treatments for this disorder.

2.3 CBD in Animal Models of CINV

2.3.1 Toxin-Induced Vomiting

To date, the toxin-induced anti-emetic effects of CBD and CBDA have only been evaluated in the house musk shrew. Please refer to Table 3 for a summary of these studies.

Table 3 Effects of CBD and CBDA on vomiting in shrews

Compound	Effective dose (s) tested	Mechanism(s) of action	Reference
Toxin-induced vomiting			
CBD	5, 10 mg/kg i.p., s.c.	5-HT_{1A} receptor	Kwiatkowska et al. (2004), Parker et al. (2004), Rock et al. (2011), (2012)
	5 mg/kg, s.c., daily for 7 days	Not evaluated	Rock et al. (2020b)
CBD + THC	2.5 + 1 mg/kg, i.p.[a]	Not evaluated	Rock and Parker (2015)
CBDA	0.1, 0.5 mg/kg, i.p.	Not evaluated	Bolognini et al. (2013)
CBDA + THC	0.05 + 1 mg/kg, i.p.[a]	Not evaluated	Rock and Parker (2015)
Motion-induced vomiting			
CBDA	0.1, 0.5 mg/kg, i.p.	Not evaluated	Bolognini et al. (2013)

[a] Enhanced suppression of vomiting when compounds combined

In the house musk shrew, CBD produces a biphasic effect on toxin-induced vomiting. At low doses (5, 10 mg/kg, i.p., s.c.) CBD suppresses (Kwiatkowska et al. 2004; Parker et al. 2004; Rock et al. 2011, 2012), but at high doses (25, 40 mg/kg, i.p.) CBD potentiates (but does not on its own produce) toxin-induced vomiting (Kwiatkowska et al. 2004; Parker et al. 2004). The anti-emetic effect of CBD is 5-HT_{1A} receptor-mediated (Rock et al. 2012). Repeated administration of CBD (5 mg/kg, s.c.) daily for 7 days retains its anti-emetic efficacy suggesting that tolerance does not develop to this effect (Rock et al. 2020b). These findings suggest that although CBD does have anti-emetic efficacy, this effect occurs at a limited dose range, perhaps limiting its clinical utility.

The acidic precursor of CBD, cannabidiolic acid (CBDA; 0.1 and 0.5 mg/kg, i.p.) potently reduced toxin-induced vomiting (Bolognini et al. 2013) with recent unpublished data suggesting that, unlike CBD, at a high dose (20 mg/kg, i.p.) CBDA does not potentiate LiCl-induced vomiting. CBDA similarly reduced LiCl-induced vomiting in male and female shrews (unpublished data). In addition, combined ineffective doses of THC (1 mg/kg, i.p.) and CBD (2.5 mg/kg, i.p.) or CBDA (0.05 mg/kg, i.p.) resulted in an enhanced anti-emetic effect (Rock and Parker 2015), suggesting synergism of these compounds in reducing vomiting. Importantly, no sex differences were found for the effects of these doses of THC, CBD, CBDA or their combinations in toxin-induced vomiting (Rock and Parker 2015). These findings suggest that CBD and CBDA reduce toxin-induced vomiting, likely acting through 5-HT_{1A} receptors. CBD has a biphasic effect with low/moderate doses suppressing vomiting, but higher doses enhancing toxin-induced vomiting in shrews. In contrast, CBDA does not seem to enhance toxin-induced vomiting and has anti-emetic effects at lower doses. This suggests that lower doses of CBD (or even lower doses of CBDA) should be investigated as a treatment (or adjunct treatment with other anti-emetics) of toxin-induced vomiting.

Also in the house musk shrew, CBDA (0.1 and 0.5 mg/kg, i.p.) reduced motion-induced vomiting (Bolognini et al. 2013), but CBD (0.5, 1, 2, 5, 10, 20 and 40 mg/kg, i.p.) did not (Cluny et al. 2008). The differential effectiveness of CBD in toxin- vs motion-induced vomiting may be because of the differing neuronal pathways involved in these two types of emesis or differing magnitudes of the sickness they produce. Motion-induced emesis may be more severe and a more potent anti-emetic agent such as CBDA may therefore be necessary to overcome it.

2.3.2 Acute Nausea

CBD, CBDA and the more stable methyl ester version of CBDA (CBDA-ME, also called EPM 301 or HU-580) have anti-nausea effects in a rodent model of acute nausea (see Table 4 for a summary of these studies). CBD (0.5, 1, 5 mg/kg, i.p, 2.5, 5 mg/kg, s.c.) reduced LiCl-induced conditioned gaping (Parker and Mechoulam 2003; Parker et al. 2002; Rock et al. 2011, 2012, 2020b, c), and high doses (20, 40 mg/kg (i.p.) did not promote conditioned gaping (Rock et al. 2020c). CBD's (5, 20 mg/kg, s.c. and 1, 5, 20 mg/kg, i.p.) acute anti-nausea effects were similar in male and female rats (Rock et al. 2020b, 2022). Repeated treatment with CBD (daily for 7 days or weekly for 4 weeks) maintained its anti-nausea properties through a 5-HT_{1A} receptor mechanism of action (Rock et al. 2020b) without the development of tolerance. Enhanced anti-nausea effects occurred when an ineffective dose of CBD (0.5 mg/kg, s.c. or 0.1 mg/kg, i.p.) was combined with an ineffective dose of the 5-HT_{1A} agonist 8-OH-DPAT (Rock et al. 2012) or an ineffective dose of THC (0.1 mg/kg, i.p.) (Rock et al. 2020c). Therefore, very low doses of CBD can be combined with low doses of other anti-nausea compounds to effectively reduce acute nausea.

CBD's acute anti-nausea effects are mediated by its action at the 5-HT_{1A} receptor. When systemically administered (s.c) or microinfused into the dorsal raphe nucleus (a brain region that produces forebrain 5-HT and where somatodendritic 5-HT_{1A} autoreceptors are located), CBD's anti-nausea effects were blocked by 5-HT_{1A} receptor antagonist (Rock et al. 2012). Projections from the dorsal raphe nucleus reach the interoceptive insular cortex (IIC), a critical forebrain region for nausea. Indeed, LiCl selectively elevated 5-HT in the IIC and CBD (5 mg/kg, i.p.) reduced this elevation of 5-HT (Limebeer et al. 2018). Taken together, these findings indicate that the anti-nausea effects of CBD occur by its indirect action at 5-HT_{1A} autoreceptors in the dorsal raphe nucleus. CBD does not exhibit direct agonist action on the 5-HT_{1A} receptor but instead exerts its effect by augmenting the action of endogenous 5-HT at this receptor (Rock et al. 2012) to reduce the firing of 5-HT neurons projecting to the IIC. Elevation of 5-HT in the IIC is a trigger for nausea, which CBD acts to reduce.

CBD's acidic precursor CBDA (0.5, 1, 5, 10, 100 µg/kg, i.p.; 2.5, 10, 20 µg/kg, i.g.) potently reduced conditioned gaping in rats (Bolognini et al. 2013; Rock et al. 2015a, 2016b; Rock and Parker 2013a), through a 5-HT_{1A} receptor-mediated effect (Bolognini et al. 2013). CBDA (1 µg/kg, s.c.) maintained its anti-nausea efficacy

Table 4 Effects of CBD, CBDA and CBDA-ME on acute and anticipatory nausea in male rats

Compound	Effective dose(s) tested	Mechanism (s) of action	Reference
Acute nausea			
CBD	0.5, 1, i.p.; 2.5 mg/kg, s.c.; 5 mg/kg i.p., s.c., 20 mg/kg, i.p., s.c.	5-HT_{1A} receptor	Parker and Mechoulam (2003), Parker et al. (2002), Rock et al. (2011), (2012), (2020b, c), (2022)
	5 mg/kg, s.c., daily for 7 days or weekly for 4 weeks	5-HT_{1A} receptor	Rock et al. (2020b)
	10 μg into the dorsal raphe nucleus	5-HT_{1A} receptor	Rock et al. (2012)
CBD + THC	0.1 + 0.1 mg/kg, i.p.[a]	5-HT_{1A} + CB_1 receptors	Rock et al. (2020c)
CBD + 8-OH-DPAT	0.5 + 0.005 mg/kg, s.c.[a] 2.5 + 0.05 mg/kg, s.c.	Not evaluated	Rock et al. (2012)
CBDA	0.5, 1, 5, 10, 100 μg/kg, i. p.; 1 μg/kg, s.c	5-HT_{1A} receptor	Bolognini et al. (2013), Rock et al. (2015a), (2020b), (2022), Rock and Parker (2013a)
	2.5, 10, 20 μg/kg, i.g.	Not evaluated	Rock et al. (2016b)
	1 μg/kg, s.c., daily for 7 days or weekly for 4 weeks	5-HT_{1A} receptor	Rock et al. (2020b)
CBDA + THC	0.01 μg/kg + 0.01 mg/kg, 0.1 μg/kg + 0.1 mg/kg, i.p.[a] 1 μg/kg + 1 mg/kg, 10 μg/kg + 10 mg/kg, i.p.	Not evaluated	Rock et al. (2015a)
	0.5 μg/kg + 0.5 mg/kg, 1 μg/kg + 1 mg/kg, i.g.[a] 2.5 μg/kg + 2.5 mg/kg, 10 μg/kg + 10 mg/kg, 20 μg/kg + 20 mg/kg, i.g.	Not evaluated	Rock et al. (2016b)
CBDA + THCA	0.01 + 1 μg/kg, i.p.	5-HT_{1A} + PPARα receptors	Rock et al. (2020c)
CBDA + OND	0.1 + 1 μg/kg, i.p.[a] 0.5 + 10 μg/kg, i.p.	Not evaluated	Rock and Parker (2013a)
CBDA + MCP	0.1 μg/kg + 0.3 mg/kg, i.p.[a]	Not evaluated	Rock and Parker (2013b)
CBDA-ME	0.1, 1 μg/kg, i.p.	5-HT_{1A} receptor	Pertwee et al. (2018)
	1 μg/kg, s.c., daily for 7 days or weekly for 4 weeks	5-HT_{1A} receptor	Rock et al. (2020b)

(continued)

Table 4 (continued)

Compound	Effective dose(s) tested	Mechanism (s) of action	Reference
Anticipatory Nausea			
CBD	1, 5 mg/kg, i.p.	Not evaluated	Rock et al. (2008)
CBDA	1, 10, 100 μg/kg, i.p.	$5\text{-}HT_{1A}$ receptor	Bolognini et al. (2013), Rock et al. (2014), (2015a)
	10 μg/kg, i.g.	Not evaluated	Rock et al. (2016b)
CBDA + THCA	0.1 μg/kg CBDA + 5 μg/kg THCA, i.p.	$5\text{-}HT_{1A}$ + CB_1 receptors	Rock et al. (2014)
CBDA + THC	1 μg/kg + 1 mg/kg, 10 μg/kg + 10 mg/kg, i.p.	Not evaluated	Rock et al. (2015a)
	0.1 μg/kg + 0.1 mg/kg, 1 μg/kg + 1 mg/kg, i.g.	Not evaluated	Rock et al. (2016b)
CBDA-ME	0.01, 0.1 μg/kg, i.p.	$5\text{-}HT_{1A}$ receptor	Pertwee et al. (2018)

[a] Enhanced suppression of nausea when compounds combined

when administered daily (for 7 days) or weekly (for 4 treatments) (Rock et al. 2020b). Indeed, CBDA (0.01, 0.1, 1 μg/kg, i.p.) was similarly effective in reducing conditioned gaping in male and female rats (Rock et al. 2022). Enhanced anti-nausea effects have been demonstrated when ineffective doses of CBDA (0.01, 0.1 μg/kg, i.p.) were combined with an ineffective dose of anti-emetic agents such as metoclopramide (MCP; an anti-emetic drug pre-dating $5\text{-}HT_3$ receptor antagonists) (Rock and Parker 2013b), THC (Rock et al. 2015a, 2016b), THCA (Rock et al. 2020c), or a low dose of OND (Rock and Parker 2013a).

The potent and more stable CBDA-ME (0.1, 1 μg/kg, i.p.) also reduced acute nausea through a $5\text{-}HT_{1A}$ receptor mechanism (Pertwee et al. 2018). In addition, with repeated daily dosing and weekly dosing (for 4 weeks) CBDA-ME (1 μg/kg, s.c.) maintained its anti-nausea efficacy, suggesting a lack of tolerance to this effect (Rock et al. 2020b).

Together, these findings suggest that CBD reduced acute nausea via indirect $5\text{-}HT_{1A}$ receptor activation, while its acidic precursor demonstrated greater potency; however, because it is rapidly decarboxylated, this instability limits CBDA's clinical utility. In contrast, the more stable CBDA-ME is even more potent, achieving anti-nausea effects at the ultralow dose of 0.1 μg/kg. Importantly, the acute anti-nausea effect of these compounds is maintained over repeated treatments, without the development of tolerance. Importantly, the anti nausea effects of CBD and CBDA are similar in male and female rats

2.3.3 Anticipatory Nausea

CBD, CBDA and CBDA-ME reduced anticipatory nausea (see Table 4 for a summary of these studies). CBD (1, 5 mg/kg, i.p.) reduced contextually elicited conditioned gaping in rats (Rock et al. 2008). Potently, CBDA (1, 10, 100 μg/kg, i. p.; 10 μg/kg i.g.) and CBDA-ME (0.01, 0.1 μg/kg, i.p.) reduced contextually elicited conditioned gaping (Bolognini et al. 2013; Pertwee et al. 2018; Rock et al. 2014, 2015a, 2016b) through a 5-HT_{1A} receptor-mediated effect (Bolognini et al. 2013; Pertwee et al. 2018). When ineffective doses of CBDA were co-administered with THC or THCA there was an enhanced suppression of anticipatory nausea (Rock et al. 2014, 2015a, 2016b).

Taken together, these results indicate an anti-nausea effect of CBD, with increased potency for CBDA. CBDA-ME demonstrated the greatest potency in rats. These compounds effectively reduce anticipatory nausea, a condition for which there are no selective treatments. Extremely low doses of CBD (or CBDA) can be co-administered with other anti-emetic compounds for synergistic suppression of nausea and vomiting, likely avoiding any biphasic effects that might occur at higher doses. Recent work by Anderson et al. (2021) has highlighted the pharmacokinetic interactions between cannabinoids within a full-spectrum cannabis extract. This group found that CBDA plasma concentrations were much higher (14-fold higher) when administered in a cannabis extract than when administered alone. Furthermore, their work suggested that this increased CBDA concentration resulted from interactions at the ATP-binding cassette (ABC) transporter breast cancer resistance protein (BCRP) in the intestinal lumen, which facilitates the movement of drugs across membranes and limits systemic absorption. These findings suggest that CBDA is a substrate of BCRP, and its efflux was inhibited by THC, leading to increased CBDA plasma concentrations (Anderson et al. 2021). Therefore, the administration of cannabis extracts (which contain THC) may provide a method to enhance CBDA concentrations, which could have therapeutic implications.

In addition, in vitro anti-cancer effects have been demonstrated for CBD and CBDA in cancer cell survival and colony-forming assays. CBD or CBDA treatment reduced migration/invasion and viability of cancer cells and even improved the cytotoxicity of anti-cancer drugs (Go et al. 2020; Henry et al. 2021; Inkol et al. 2021; Patel et al. 2021; Takeda et al. 2012). These pre-clinical findings may help to inform planned clinical trials assessing CBD (as well as CBDA and CBDA-ME) in oncology patients experiencing CINV.

2.4 *Minor Cannabinoids in Animal Models of CINV*

Although over 100 cannabinoids have been identified, most are not present in substantial amounts in the cannabis plant (Hanuš et al. 2016). These minor cannabinoids are much less well studied than THC and CBD, with only

Table 5 Effect of the minor cannabinoids on acute nausea in male rats

Compound	Effective dose(s) tested	Mechanism(s) of action	Reference
Acute nausea			
CBDV	200 mg/kg, i.p.	Not evaluated	Rock et al. (2013b)
CBG	1 mg/kg, i.p.	Not evaluated	Rock et al. (2011)
THCV	10 mg/kg, i.p.	Not evaluated	Rock et al. (2013b)

tetrahydrocannabivarin (THCV), cannabidivarin (CBDV), and cannabigerol (CBG) having been evaluated in pre-clinical models of nausea and/or vomiting (refer to Table 5 for a summary of these studies).

Interestingly, CBG seems to act in opposition to CBD on nausea/vomiting. Treatment with CBG (1, 5, and 10 mg/kg, i.p.) blocked the CBD-induced suppression of conditioned gaping in rats and LiCl-induced vomiting in shrews (Rock et al. 2011), suggesting that CBG opposes CBD's action at the 5-HT_{1A} receptor. Interestingly, at 1 mg/kg CBG reduced LiCl-induced conditioned gaping in rats, suggesting an anti-nausea effect at this low dose (Rock et al. 2011). This dose-dependent effect may be due to CBG's varied mechanisms of action. CBG (pM to low nM concentrations) stimulates GTPγS binding to mouse brain membranes, and this effect disappears at higher concentrations (1 μM) to be replaced by its action as a 5-HT_{1A} receptor antagonist (Cascio et al. 2010). CBG is also a highly potent alpha2-adrenoceptor agonist (Cascio et al. 2010), with other alpha2-adrenoceptor agonists such as clonidine known for their ability to reduce postoperative nausea and vomiting (see Hsu et al. 2019 for review). A recent survey of those using CBG-predominant (>50% CBG) products ($n = 127$) reported that 76% of respondents indicated that CBG-predominant product use improved their nausea and was more effective than conventional medications (Russo et al. 2022a). No subjects reported vomiting as a side-effect of its use or during withdrawal (Russo et al. 2022a). The recent increase in the availability of cannabis strains containing high percentages of CBG and CBG-predominant products suggests that further research on CBG is timely (see Nachnani et al. 2021 for a review of CBG's effects).

The minor cannabinoids THCV (10 mg/kg, i.p.) and CBDV (200 mg/kg, i.p.) suppressed LiCl-induced conditioned gaping, without producing conditioned gaping on their own, suggesting no CB_1 receptor inverse agonist activity (Rock et al. 2013b). No mechanisms of action for these anti-nausea effects have yet been investigated. THCV has been shown to act as a partial agonist at the CB_1 receptor (Zagzoog et al. 2020), as an agonist at the CB_2 receptor (Bolognini et al. 2010), and as an agonist at transient receptor potential (TRP) channels (De Petrocellis et al. 2011). Also of interest, THCV has been shown to act as a neutral CB_1 receptor antagonist (Thomas et al. 2005) that may have anti-obesity properties (Riedel et al. 2009).In contrast, CBDV does not bind to the CB_1 receptor but does act at the CB_2 receptor (Navarro et al. 2020; Zagzoog et al. 2020). The anti-nausea effect of CBDV may be occurring via action at the CB_2 receptor. Indeed, the selective CB_2 agonist HU 308 effectively reduced conditioned gaping (Rock et al. 2016a). Further research is needed to understand the potential anti-nausea properties of these minor cannabinoids and the multitude of effects that these cannabinoids may exert when administered together in cannabis.

2.5 *Manipulations of eCBs in Animal Models of CINV*

The eCBs are rapidly degraded upon their "on-demand" release. For this reason, exogenous administration of AEA or 2-AG may not remain available for a sufficient time to exert therapeutic effects. In contrast, FAAH or MAGL inhibitors can elevate levels of the fatty acids, AEA (as well as OEA and PEA) or 2-AG respectively, by up to 24 h (Ahn et al. 2009; Long et al. 2009a). While AEA and 2-AG act at the cannabinoid receptors, OEA and PEA instead display high affinity for the PPARα receptors (Fu et al. 2003). Enzyme inhibition may be a highly selective manipulation to determine the site-specific increased levels of eCBs and other fatty acids (such as OEA and PEA) that alleviate nausea and vomiting in animal models.

2.5.1 Effects of eCBs and Their Enzyme Inhibitors on Vomiting

There is conflicting evidence about the role of 2-AG in emesis which may be a result of species differences, the use of differing emetic agents, observation timelines, or the definition of an "emetic episode" (an event characterised by retching that involves the expulsion of solids/fluids from the mouth vs an event characterised by retching with no passage of material from the mouth). Please refer to Table 6 for a summary of the studies outlining the anti-emetic role of 2-AG, AEA and MAGL and FAAH inhibition.

Increased levels of 2-AG (but not AEA) were detected in least shrew brains following administration of the chemotherapeutic agent cisplatin, with higher levels detected at 30 min post-cisplatin than at 60 min (Darmani et al. 2005). Indeed, vomiting began 15–30 min post-cisplatin injection, with most vomiting occurring from 30 to 60 min of cisplatin exposure (Darmani 1998). This suggests that brain 2-AG levels (and not AEA levels) may be mobilised rapidly (within 30 min), as a protective mechanism in least shrews against this toxic chemotherapeutic agent.

A report by Zheng et al. (2014) indicated that rats subjected to acceleration (motion sickness) had lower 2-AG levels than controls (but no difference in levels of AEA) and showed decreased mRNA levels of CB_1 receptors in the dorsal vagal complex (a classic emetic brainstem region) and stomach. Like the human studies demonstrating reduced 2-AG levels during motion-sickness (Chouker et al. 2010; Strewe et al. 2012), rats subjected to acceleration also showed reduced 2-AG levels. These findings suggest that illness is associated with reduced 2-AG (but not AEA) levels. As a protective mechanism, 2-AG may be mobilised.

At moderate to high doses, 2-AG (2.5, 5, 10, 20 mg/kg, i.p.) itself has been reported to produce vomiting in the least shrew (Darmani 2002; Darmani et al. 2005), and this 2-AG-induced vomiting was attenuated by treatment with THC

Table 6 Effect of eCB system manipulation on toxin-induced vomiting in animals

Compound	Species	Effective dose (s) tested	Mechanism (s) of action	Toxin	Reference
2-AG	Ferret	1, 2 mg/kg, i.p.	CB_1, CB_2, TRPV1 receptors	M6G	Sharkey et al. (2007), Van Sickle et al. (2001), (2005)
	House musk shrew	2, 5 mg/kg, i.p.	Not evaluated	LiCl	Sticht et al. (2013)
MJN110	House musk shrew	10, 20 mg/kg, i.p.	CB_1 receptor	LiCl	Parker et al. (2015)
AEA	Ferret	1 mg/kg, i. p.	CB_1, TRPV1 receptors	M6G	Sharkey et al. (2007)
mAEA	Ferret	3 mg/kg, i. p.	CB_1 receptor	M6G	Van Sickle et al. (2001)
	Squirrel monkey	10 mg/kg, i.m.	CB_1 receptor	nicotine	Wooldridge et al. (2020)
			Not evaluated	LiCl	Wooldridge et al. (2020)
URB597	Ferret	5 mg/kg, i. p.	CB_1, TRPV1 receptors	M6G	Sharkey et al. (2007)
	House musk shrew	0.9 mg/kg, i.p.	Not evaluated	cisplatin	Parker et al. (2009)
			CB_1 receptor	nicotine	Parker et al. (2009)
URB597 + AEA	House musk shrew	0.9 + 5 mg/kg, i. p.	Not evaluated	cisplatin	Parker et al. (2009)

(1, 2.5, 5 mg/kg, i.p.), AEA (5 mg/kg, i.p.), methanandamide (a stable methylated AEA analog, 10 mg/kg, i.p.) but not CBD (10, 20 mg/kg) (Darmani 2002). Administration of 2-AG (1, 2, 5, 10 mg/kg, i.p.) itself did not produce vomiting in the house musk shrew (Sticht et al. 2013). Indeed, pre-treatment with 2-AG (2, 5 mg/kg, i.p.) reduced LiCl-induced vomiting in the house musk shrew (Sticht et al. 2013), but this effect was not CB_1 receptor-mediated (Sticht et al. 2013). Similarly, 2-AG (1, 2 mg/kg, i.p.) reduced M6G-induced emesis in ferrets (Sharkey et al. 2007; Van Sickle et al. 2005), through a CB_1, CB_2 (Van Sickle et al. 2005) and TRPV1 (Sharkey et al. 2007) receptor-mediated effect.

Due to the rapid degradation of exogenous 2-AG, MAGL inhibition may be a more selective and sustained anti-emetic treatment. MJN110 (10, 20 mg/kg, i.p.) reduced LiCl-induced vomiting in house musk shrews through a CB_1 receptor-mediated effect (Parker et al. 2015) and tissue analysis of the shrew brain confirmed that MJN110 selectively inhibited MAGL (Parker et al. 2015).

In ferrets, AEA (1 mg/kg, i.p.), and the global FAAH inhibitor URB597 (5 mg/kg, i.p.) reduced M6G-induced emesis, with AEA's effect mediated by TRPV1 and URB597's effect mediated by TRPV1 as well as CB_1 receptors (Sharkey et al. 2007). Also in ferrets, methanandamide (3 mg/kg, i.p.) reduced M6G-induced vomiting, via a CB_1 receptor-mediated effect (Van Sickle et al. 2001). Similarly, in the house musk shrew, URB597 (0.9 mg/kg, i.p.) attenuated cisplatin- and nicotine-induced vomiting and the combination of URB597+AEA (0.9 mg/kg + 5 mg/kg, i.p.) suppressed cisplatin-induced vomiting through a CB_1 receptor-mediated effect (Parker et al. 2009). Importantly, no sex differences were found in the anti-emetic effects of URB597 or the combination of URB597+AEA in toxin-induced vomiting (Parker et al. 2009). Finally, in squirrel monkeys methanandamide (10 mg/kg, i.m.) blocked nicotine- and LiCl-induced emesis in a CB_1-mediated manner (Wooldridge et al. 2020). In contrast, URB597 (5, 10 mg/kg, i.p.) was ineffective in reducing apomorphine-induced vomiting in ferrets (Percie du Sert et al. 2010) and cisplatin-, or apomorphine-induced vomiting in the least shrew (Darmani et al. 2005). These differential findings for URB597's effectiveness may be due to differences in the magnitude of emesis induced by the various emetic agents used. For example, in ferrets, baseline vomiting to apomorphine was 5.3 ± 1.0 vomits (vs 1.8 ± 0.2 for M6G). Perhaps apomorphine-induced emesis may have been too severe for FAAH inhibition to overcome it. Also of note is the pretreatment interval between the administration of the FAAH inhibitor and the toxin. Although FAAH inhibition is rapid (within 15 min) after URB597 administration, significant elevation of AEA (and other fatty acids) does not occur until at least 2 h post-injection (Kathuria et al. 2003). In the Percie du Sert (2010) and the Darmani (2005) studies, URB597 was administered 30 or 10 min prior to the toxin, which may not have been sufficient time for AEA levels to adequately increase to the level necessary to reduce vomiting.

Together, these findings suggest that MAGL and FAAH inhibition exert anti-emetic effects in several species and these effects seem to occur via action at CB_1, CB_2, and TRPV1 receptors in brainstem structures involved in emesis. CB_1 receptors, FAAH and TRPV1 receptors have been localised in ferret brainstem structures implicated in emesis (Sharkey et al. 2007; Van Sickle et al. 2001). Conflicting results have been found about eCB levels present in animals while experiencing illness, which may be driven by species differences, differential emetic agents used, and, the timeline at which tissue or plasma is sampled. Further research is needed to clarify the role of eCBs in vomiting.

2.5.2 Effects of eCBs and Their Enzyme Inhibitors in Acute Nausea

The conditioned gaping response has been utilised to determine the anti-nausea effects of manipulations of the eCB system and to further investigate the mechanisms and critical brain regions that mediate nausea in rats. Evidence suggests that FAAH and MAGL inhibition exert their anti-nausea effects via distinct receptor systems with differing critical regions identified. For a summary of these findings please refer to Table 7.

Levels of 2-AG (but not AEA) were selectively elevated in the rat brain IIC during an episode of LiCl-induced nausea (Sticht et al. 2016), suggesting a mobilisation of 2-AG as a protective mechanism (as discussed above). Administration of the MAGL inhibitor MJN110 (10, 20 mg/kg, i.p.) reduced acute nausea

Table 7 Effect of FAAH and MAGL inhibition on acute and anticipatory nausea in male rats

	Acute nausea	Anticipatory nausea
Systemic administration		
FAAH Inhibitors	↓ PPARα mediated (Rock et al. 2015b)	↓ CB_1 mediated (Rock et al. 2008, 2015b)
MAGL Inhibitors	↓ CB_1 mediated (Parker et al. 2015)	↓ CB_1 mediated (Parker et al. 2015)
Central administration		
Intra-IIC administration		
FAAH Inhibitors	– (Sticht et al. 2016)	– (Limebeer et al. 2016)
MAGL Inhibitors	↓ CB_1 mediated (Sticht et al. 2016)	↓ CB_1 mediated (Limebeer et al. 2016)
Intra-VP administration		
FAAH Inhibitors	↓ PPARα mediated (Rock et al. 2019)	Not yet evaluated

↓, reduced; –, no effect

through a CB_1 receptor-mediated effect, with increased levels of 2-AG detected in the IIC (Parker et al. 2015). Furthermore, administration of MJN110 (10 mg/k, i.p.) was able to prevent the elevation of 5-HT in the IIC in response to an injection of LiCl (Limebeer et al. 2018). Finally, the MAGL inhibitor AM4301 (20 mg/kg, i.p.) also reduced acute nausea through a CB_1 receptor-mediated effect (Parker et al. 2016).

To investigate the brain region mediating 2-AG's acute anti-nausea effect in rats, MAGL inhibitors were microinfused into the IIC. MJN110 (0.5, 2 μg) and AM4301 (2 μg) reduced acute nausea through a CB_1 receptor mechanism (Parker et al. 2016; Sticht et al. 2016). Inhibition of MAGL robustly increased levels of 2-AG in the IIC and reduced IIC c-Fos immunoreactivity in rats treated with LiCl (Sticht et al. 2016). Taken together, these findings suggest that 2-AG's acute anti-nausea effect is mediated by CB_1 receptors and that the IIC is a critical brain region.

The anti-nausea effect of AEA has been investigated with the use of selective FAAH inhibitors in rats. The global FAAH inhibitors PF-3845 (10 mg/kg), AM4303 (20 mg/kg) and URB597 (0.3 mg/kg, i.p.) suppressed acute nausea (Cross-Mellor et al. 2007; Parker et al. 2016; Rock et al. 2015b). The anti-nausea effect of PF-3845 was mediated via PPARα, rather than CB_1 receptors (Rock et al. 2015b). In line with this finding, systemic administration or intra-IIC infusion of PF-3845 did not alter AEA levels in the IIC but did elevate levels of OEA and PEA in the IIC (Sticht et al. 2016). Intra-IIC inhibition of FAAH with PF-3845 (2 μg) or URB597 (0.01 μg) was without effect (Sticht et al. 2016). Together these findings suggest that the anti-nausea effects of FAAH inhibition occur via PPARα receptors, and that the IIC may not be the primary critical brain region involved. This also suggests that the anti-nausea effects of FAAH inhibition are driven by the elevation of levels of OEA and PEA (which act at PPARα) and not AEA (which acts at the CB_1 receptor).

Rock et al. (2017) used the peripherally restricted FAAH inhibitor URB937, to further elucidate the role of FAAH inhibition in acute nausea. Administration of URB937 (1, 3 mg/kg, i.p.) reduced acute nausea through activation of PPARα (Rock et al. 2017), and administration of a PPARα agonist similarly reduced acute nausea (Rock et al. 2017). Furthermore, URB937 significantly suppressed FAAH, while elevating levels of OEA and PEA in the area postrema (Rock et al. 2017), a brainstem circumventricular organ that has been shown to be critical in the establishment of conditioned gaping (Eckel and Ossenkopp 1996; Ossenkopp and Eckel 1995). Taken together, these findings suggest that the anti-nausea effect of FAAH inhibition may be mediated by PPARα receptors (via PEA and OEA). The area postrema may act as the initial central relay station which eventually activates a more anterior brain region to reduce nausea.

One such anterior region of interest is the ventral pallidum. Both human neuroimaging and rodent studies have highlighted the ventral pallidum as an important region in processing aversive stimuli (e.g. Ho and Berridge 2014; Royet et al. 2016). Intra-ventral pallidum infusion of PF-3845 reduced acute nausea through PPARα but not CB_1 receptor activation (Rock et al. 2019). Similarly, microinfusion of a PPARα agonist to this region also reduced acute nausea (Rock et al. 2019). These findings suggest the acute anti-nausea effects of FAAH inhibition involve the ventral pallidum. This effect is likely due to the elevation of OEA and/or PEA (which have a high affinity for PPARα receptors), rather than AEA (which acts at CB_1 receptors and is a weak PPARα agonist (Sun et al. 2007).

Compounds that inhibit both FAAH and MAGL have also been synthesised and a few of these have been tested in the acute nausea model. The dual FAAH/MAGL inhibitor AM4302 (20 mg/kg, i.p.) reduced acute nausea (Parker et al. 2016) and intra-IIC infusion of the dual FAAH/MAGL inhibitor JZL195 (10 μg) reduced conditioned gaping (Sticht et al. 2016). As JZL195's anti-nausea effect occurred when directly infused into the IIC, this suggests that its anti-nausea effects may be primarily due to the MAGL inhibition action of this compound.

Taken together, these findings suggest the acute anti-nausea effects of MAGL inhibition are mediated by 2-AG acting at CB_1 receptors and the IIC is a critical brain region. In contrast, FAAH inhibition is likely mediated by OEA and PEA acting at PPARα receptors with the ventral pallidum being a critical region involved. This role for OEA and PEA in reducing acute nausea is consistent with the finding in cyclic vomiting patients that show elevations of OEA and PEA, which are presumably mobilised as a protective mechanism during the sick phase. Randomised, controlled trials have evaluated PEA for treating pain and OEA as an anti-obesity treatment and have been well-tolerated in patients (Artukoglu et al. 2017; Laleh et al. 2019). No trials involving these compounds for CINV have yet been conducted.

2.5.3 Effects of eCBs and Their Enzyme Inhibitors in Anticipatory Nausea

The anti-nausea effects of the administration of exogenous eCBs and their enzyme inhibitors have also been examined in the rat model of anticipatory nausea. Exogenous administration of 2-AG (1.25 mg/kg, i.p.) was ineffective in reducing contextually elicited gaping (Limebeer et al. 2014). This finding is consistent with the rapid breakdown of endogenously administered eCBs, limiting their use to a very brief therapeutic time window. Administration of the MAGL inhibitor MJN110 (10, 20 mg/kg, i.p.) did effectively reduce anticipatory nausea, with corresponding increased IIC levels of 2-AG (Parker et al. 2015). Similarly, the MAGL inhibitor AM4301 (10, 20 mg/kg, i.p.) reduced anticipatory nausea with elevated levels of 2-AG (but not AEA) in the IIC (Parker et al. 2016). Intra-IIC infusion of MJN110 (2 μg) reduced anticipatory nausea through a CB_1 receptor-mediated effect (Limebeer et al. 2016). These findings suggest that MAGL inhibition's elevation of 2-AG effectively reduced anticipatory nausea via activation of CB_1 receptors and the IIC is a critical region involved.

Exogenous administration of AEA (5 mg/kg, i.p.) suppressed contextually elicited gaping (Limebeer et al. 2014), through a CB_1 receptor-mediated effect. The FAAH inhibitors URB597 (0.3 and 10 mg/kg), PF-3845 (10 and 20 mg/kg) or AM4303 (20 mg/kg, i.p.) reduced contextually-elicited gaping via CB_1 receptors, but not via PPARα (Rock et al. 2008, 2015b), with IIC elevation of AEA but not 2-AG) (Parker et al. 2016). However, intra-IIC infusion of PF-3845 (2 μg) did not reduce anticipatory nausea (Limebeer et al. 2016). These results suggested that FAAH inhibition reduces anticipatory nausea by the action of AEA at CB_1 receptors, but that the IIC may not be the primary critical brain region involved. Using the peripherally restricted FAAH inhibitor, URB937, anticipatory nausea was reduced, and this was accompanied by elevated levels of AEA in the area postrema (Rock et al. 2017).

The dual FAAH/MAGL inhibitor JZL195 (10 mg/kg, i.p.) suppressed contextually-elicited gaping via CB_1 (but not CB_2) receptors, and this effect was enhanced by the co-administration of AEA (5 mg/kg, i.p.) or 2-AG (1.25 mg/kg, i.p.) (Limebeer et al. 2014). Furthermore, whole-brain lipid analysis showed that JZL195 increased levels of AEA, PEA and OEA, but not 2-AG) (Limebeer et al. 2014). Similarly, the dual FAAH/MAGL inhibitor AM4302 (5, 10, 20 mg/kg, i.p.) reduced anticipatory nausea through a CB_1 receptor-mediated effect (Parker et al. 2016). Indeed, levels of 2-AG and AEA were elevated in the IIC in those rats treated with AM4302 (Parker et al. 2016). Although the combined blockade of FAAH and MAGL seems promising, it does seem to produce sedation as a side effect. The dual FAAH/MAGL inhibitor JZL195 (20 mg/kg, i.p.) and the MAGL inhibitor JZL184 (40 mg/kg), but not the FAAH inhibitor PF-3845 (10 mg/kg, i.p.) have been shown to reduce locomotor activity in mice (Long et al. 2009b), suggesting a CB_1 agonist-like sedative effect. This sedative side-effect may be therapeutically undesirable.

Together, these findings suggest that while MAGL inhibition is likely elevating 2-AG to reduce anticipatory nausea through its action on CB_1 receptors with the IIC being a critical region, FAAH inhibition is likely reducing anticipatory nausea through elevation of AEA which acts on CB_1 receptors located in the area postrema (Van Sickle et al. 2003). The area postrema might be the initial central site of action producing the suppression of anticipatory nausea. Studies have not yet evaluated more anterior brain regions (such as the ventral pallidum) that might also be activated during anticipatory nausea. Unlike the FAAH inhibitor PF-3845, the dual FAAH/MAGL inhibitor JZL195, and the MAGL inhibitor JZL184 produce sedative effects which might limit their clinical utility.

2.6 N-Acyl Glycines in Animal Models of CINV

A family of lipid signalling molecules, the N-acyl glycines, are newly recognised neuromodulators that are structurally like AEA. These lipids include oleoyl glycine (OlGly) and oleoyl alanine (OlAla). In collaboration with Raphael Mechoulam's group, we have evaluated their anti-nausea and anti-emetic properties.

OlGly has been shown to interfere with LiCl-induced vomiting in house musk shrews (Rock et al. 2020a). Importantly, OlGly was similarly effective in both male and female shrews (Rock et al. 2020a). OlGly also reduced LiCl-induced conditioned gaping in rats (Rock et al. 2020a).

Because OlGly is rapidly deactivated, OlAla, a more stable methylated version of OlGly was developed. In house musk shrews, although both OlGly and OlAla (1, 5 mg/kg) suppressed LiCl-induced vomiting, OlAla was superior in its degree of suppression (Rock et al. 2021). In rats, at the high dose of 20 mg/kg (i.p.) OlGly reduced LiCl-induced conditioned gaping but also produced sedative-like effects, whereas OlAla (1, 5, 20 mg/kg, i.p.) effectively reduced LiCl-induced conditioned gaping at a wider dose range, without any sedation (Rock et al. 2021). OlAla (1, 5, 20 mg/kg, i.p.) reduced LiCl-induced conditioned gaping in male and female rats in a similar manner (Rock et al. 2022). These findings suggest that both OlAla and OlGly reduce LiCl-induced gaping in rats and LiCl-induced vomiting in shrews, with OlAla being more effective.

Although the mechanism(s) of action for these compounds to reduce LiCl-induced nausea have not yet been investigated, their ability to reduce illness behaviours associated with morphine withdrawal has been shown to be mediated by both CB_1 and PPARα receptors (Ayoub et al. 2020; Rock et al. 2020a). This is consistent with the in vitro findings that OlAla inhibits FAAH and activates PPARα (Ayoub et al. 2020). These findings suggest that these fatty acid amides may be worth further investigation for their anti-nausea/anti-emetic effects. Their dual action at PPARα and CB_1 receptors may boost their therapeutic potential. Indeed, Calignano et al. (1998) have previously reported that PEA and AEA act synergistically to alleviate pain in animal models.

3 Cannabinoids in Human CINV Patients

Considerable preclinical evidence and the failure of current treatments to properly manage CINV in human patients suggest that cannabinoids and manipulations of the eCB system may provide relief for nausea and vomiting in these patients. I Indeed, survey data from those using medical cannabis to manage their CINV suggests that these patients are experiencing benefits. One such survey of 45 gynecologic cancer patients prescribed medical cannabis indicates that over 70% of patients reported improvements in nausea and/or vomiting (Fehniger et al. 2021). Despite these promising self-reports, properly controlled, randomised clinical trials are needed to objectively evaluate the effectiveness of cannabinoids and manipulations of the eCB system in CINV.

3.1 Exogenous Cannabinoids in Human CINV Patients

Oral or smoked cannabis have been shown to be effective in reducing CINV (Bar-Lev Schleider et al. 2018; Ofir et al. 2019; Vinciguerra et al. 1988). Two orally available THC-like synthetics, nabilone and dronabinol, are approved for CINV, after first-line anti-emetics have failed. A report by The National Academy of Sciences (2017) concluded that there was sufficient evidence for the effectiveness of oral cannabinoids in the treatment of CINV. This conclusion was based on the available studies involving THC and THC-like synthetics (reviewed below).

3.1.1 THC and THC-like Synthetics in CINV Patients

The first medically approved use of THC in the United States by the Food and Drug Administration (FDA) was for treatment of nausea and vomiting. Synthetic THC was approved for use in treating CINV in both the United States (Dronabinol) and Canada (Marinol) in 1985 with several reports in the literature regarding the effectiveness of THC (Chang et al. 1979; Frytak et al. 1979; Kluin-Neleman et al. 1979; Lucas and Laszlo 1980; McCabe et al. 1988; Neidhart et al. 1981; Orr and McKernan 1981; Orr et al. 1980; Sallan et al. 1975; Sweet et al. 1981) and THC-like synthetics (Ahmedzai et al. 1983; Chan et al. 1987; Dalzell et al. 1986; Einhorn et al. 1981; Elder and Knoderer 2015; Herman et al. 1979; Johansson et al. 1982; Levitt 1982; Maida et al. 2008; Meiri et al. 2007; Niamatali et al. 1984; Niederle et al. 1986; Niiranen and Mattson 1985; Polito et al. 2018; Pomeroy et al. 1986; Steele et al. 1980; Wada et al. 1982) in reducing CINV (see Tramer et al. 2001 for review). At higher doses, THC administration can produce psychoactive effects that may limit its clinical utility (e.g. Noyes et al. 1975).

To our knowledge, only one published clinical trial has explicitly assessed the effect of dronabinol (10 mg every 6 h, beginning 24 h prior to and continued for

24 hr after the last dose of chemotherapy for 6 days total) on anticipatory nausea in chemotherapy patients (Lane et al. 1991). This study also assessed its ability to reduce acute nausea and vomiting (occurring over 6 days of treatment) (Lane et al. 1991). All patients had received prior chemotherapy and 50% had previously experienced 2–10 episodes of nausea and vomiting. Thirty percent of those in the dronabinol group experienced anticipatory nausea. Seventy percent of those in the dronabinol group experienced two or fewer episodes of nausea or vomiting, with 40% experiencing no nausea and vomiting over the 6 days. Over 70% reported side effects including dizziness and drowsiness. The effectiveness of dronabinol is difficult to interpret in this study because there was no placebo control group included. Most patients included in this study had prior experience with CINV and most were receiving highly emetogenic chemotherapy regimens. It is possible that the magnitude of nausea and vomiting may have been too great to overcome with dronabinol in this difficult-to-treat group of patients. Given that these promising improvements were seen in this difficult-to-treat group, further properly controlled trials are needed.

The NIH database of registered clinical trials (https://clinicaltrials.gov) lists one multicenter, randomised, double-blind, placebo-controlled phase III clinical trial (NCT03984214) currently recruiting patients to evaluate the efficacy and safety of dronabinol in chemotherapy-induced and tumour-related symptoms in pancreatic cancer patients. This study aims to determine optimal dosing, employing a *titration period* (4 weeks) from 2.5 mg up to 30 mg per day with the dose increased by 2.5 mg every other day. Next, during the *maintenance period* (12 weeks) the individually tolerated maximum dose (up to 30 mg) from the end of the titration period will be continued for 12 weeks. During the *tapering period* (2 weeks) the dose will be decreased every other day by 5 mg, with a safety follow-up occurring 4 weeks after the end of treatment. The results of this study will help to inform clinicians and patients about dronabinol's optimal dosing for cancer patients in improving quality of life measures, including nausea and vomiting.

In some cases, the undesirable psychoactive effects of THC have caused patients to move towards the use of cannabis strains with lower THC content. THC and CBD are derived from the same precursor cannabigerolic acid, so as the THC content in a cannabis strain drops, the CBD content correspondingly rises. In recent years, CBD products have gained popularity for both recreational and medicinal use, but there is limited data to support the benefits reported by consumers, including its efficacy in CINV.

3.1.2 CBD in CINV Patients

Only one case report has been published with the non-intoxicating cannabinoid CBD (100–450 mg/day, oral capsule, for two years, presumably in addition to typical antiemetics) used in two patients undergoing chemotherapy and radiation (Dall'Stella et al. 2019). When taking CBD during their cancer treatment period, two male patients with gliomas reported little nausea with no evidence of disease progression for at least 2 years. This occurred without any significant abnormalities in blood counts/plasma biochemistry. Future randomised placebo-controlled trials with a larger number of patients are needed to properly examine CBD's effects on CINV.

A recently completed non-randomised, parallel assignment intervention study (NCT04585841) examined the effect of CBD (300 mg/day) on lean body mass in cancer patients receiving chemotherapy, presumably in addition to typical antiemetic treatments. The secondary measures of change in nausea and vomiting, and quality of life parameters were also examined. No results are yet available from this trial. Currently recruiting, is a study (NCT06137027) evaluating the effect of CBD oil (5.5 mg sublingually, given two to three hours preoperatively) on the incidence of postoperative nausea and vomiting following gynecological surgery.

Oral CBD (Epidiolex) is FDA-approved for seizures associated with Lennox-Gastaut syndrome, Dravet syndrome, or tuberous sclerosis complex. In a recently completed trial, Epidiolex was administered for 12 weeks in recurrent prostate cancer patients who had undergone surgery or radiation. As a secondary outcome, this open-label, phase I study (NCT04428203) also evaluated changes in quality of life, including the symptoms of nausea and vomiting. No results are available currently.

3.1.3 THC + CBD in CINV Patients

Sativex (Nabiximols) is an oromucosal cannabis extract spray containing THC and CBD (2.7 mg THC + 2.5 mg CBD/spray) as well as minor cannabinoids. Regulatory approval has been obtained in numerous countries for its use as a treatment for Multiple Sclerosis (MS) spasticity (muscle stiffness or spasms). Because this product contains both THC and CBD (along with other minor cannabinoids) it is difficult to determine whether its effects are THC- and/or CBD-dependent. A pilot, phase II randomised double-blind placebo-controlled trial evaluated Sativex® (mean daily dose 13 mg THC + 12 mg CBD for 4 days, beginning within 2 h post-chemotherapy administration) as an adjunct treatment (consisting of a corticosteroid with a 5-HT_3 receptor antagonist or metoclopramide) in 16 chemotherapy patients, with favourable results (Duran et al. 2010). No delayed vomiting was reported in 71% of the Sativex group (22% in placebo) and no delayed nausea was reported in 57% of the Sativex group (22% in placebo). No differences occurred in the acute period. This study suggests anti-emetic and anti-nausea benefits with Sativex as an adjunct to typical anti-emetic treatment but should be interpreted with caution due to the limited sample size.

A phase II multicenter, randomised, double-blind, placebo-controlled trial evaluated oral THC + CBD (2.5 mg THC + 2.5 mg CBD capsule, three times daily, from days -1 to 5, and 1 cycle of matching placebo in a crossover design) extract for the prevention of CINV (Grimison et al. 2020). The addition of the cannabis extract to the anti-emetic regimen reduced vomiting and nausea (Grimison et al. 2020). Adverse events such as sedation, dizziness, and disorientation were reported, but most patients (85%) reported that they preferred the extract over placebo. Based on these results, this group is continuing recruitment for the phase III parallel component of the trial.

In contrast, a multicenter, randomised, double-blind, placebo-controlled, parallel-group study examined the efficacy of oral extracts containing THC only or THC: CBD in patients with cancer-related pain. Patients reported more nausea and vomiting with THC:CBD compared with placebo (but no difference in the THC-only extract) (Johnson et al. 2010). Because no cannabinoid quantification was completed in these patients, it is difficult to determine whether this increased nausea and vomiting was due to the CBD present in the extract or due to CBD altering the metabolism of THC by partially inhibiting the cytochrome P450 enzyme, the enzyme responsible for the breakdown of THC.

A recent randomised, double-blinded, crossover and placebo-controlled trial examined the effects of THC:CBD (1:1) oil in gynecologic cancer patients, with doses ranging between 0.25 and 0.4 mg (Sukpiriyagul et al. 2023). In the first two days following chemotherapy treatment, those patients receiving the THC:CBD oil (+ standard antiemetic agents) reported less nausea in comparison to those who received only standard antiemetic. There were no differences in nausea scores between groups on days 3 and 4 post-chemotherapy, but on day 5, the THC:CBD oil patients reported less nausea than controls. These findings suggest that THC: CBD oil may help to alleviate acute and delayed nausea.

3.2 *Cannabinoid Treatments Producing Nausea and Vomiting in Human Patients*

Paradoxically, nausea and/or vomiting can occur as adverse events associated with cannabinoid treatment. Indeed there are reports of high doses of THC (Vaziri et al. 1981) and synthetic cannabinoid agonists such as "Spice" (Forrester et al. 2012; Tait et al. 2016) producing nausea and/or vomiting in humans. Co-occurring with the increased rates of high-THC content cannabis is the newly recognised clinical condition of cannabinoid hyperemesis syndrome (discussed below). A recent review of the adverse effects of nabiximols indicated that in large (>100 randomised patients) placebo-controlled trials in cancer pain patients, nausea was a treatment-related adverse event in 10% of patients receiving nabiximols vs 7% of patients receiving placebo (Prieto González and Vila Silván 2021). Certainly, factors such as dosing, disease progression and differences in the reporting of adverse events can make comparisons between studies quite difficult.

CBD treatment can also be associated with the symptom of vomiting. In a study of 607 treatment-resistant epilepsy patients taking CBD, no vomiting was reported in those in a low-dose group of CBD (2–10 mg/kg/day), but the incidence of vomiting increased with higher doses (above 40 mg/kg/day) (Szaflarski et al. 2018). These findings suggest a dose-dependent effect of CBD in the promotion of vomiting (as is also seen in preclinical studies described above). During these trials, patients remain on their prescribed anti-epileptic medications so it is difficult to determine whether CBD (perhaps at higher doses) is producing vomiting, whether it might be interacting with an adjunct treatment, or both. In addition, the content of the CBD product being taken cannot always be verified. This is an important consideration given the finding that only 30% of 84 online CBD and hemp oil products examined were accurately labelled for CBD and THC content (Bonn-Miller et al. 2017). Indeed, in a recent analysis of CBD products labelled as "THC Free," 5 out of 21 (24%) of the products contained THC ranging from 0.015–0.656 mg/ml (Johnson et al. 2022). A recent meta-analysis of adverse events associated with CBD treatment did not find significantly increased odds of experiencing nausea or vomiting (Chesney et al. 2020).

3.2.1 THC and Cannabinoid Hyperemesis Syndrome (CHS)

Associated with prolonged cannabis use, particularly with the use of high-THC strains, is a clinical disorder known as cannabinoid hyperemesis syndrome (CHS) (see DeVuono et al. 2024 for review). CHS is characterised by cyclical episodes of nausea and vomiting accompanied by severe abdominal pain (Allen et al. 2004). These patients do not respond to typical anti-emetic drugs (Richards 2017), but topical application of abdominal capsaicin cream (Dezieck et al. 2017; Kum et al. 2021), and taking hot showers can improve symptoms (Wallace et al. 2007). A recent study of 28 CHS patients revealed mutations in genes affecting neurotransmitters, the eCB system, and the cytochrome P450 complex associated with the metabolism of cannabinoids (Russo et al. 2022b). These genetic mutations may be a diagnostic tool to identify those who may be more susceptible or predisposed to developing CHS (Russo et al. 2022b). Further research is needed to better understand and treat CHS. As discussed in Sect. 2.2 above, preclinical evidence suggests that high dose THC and other CB_1 agonists produces nausea and vomiting in animal models (for review see DeVuono and Parker 2020).

4 Endocannabinoid Levels in Humans Experiencing Nausea and Vomiting

Evidence suggests that although exogenous administration of eCBs is undesirable because they are quickly degraded (Di Marzo et al. 1994), targeted manipulations that inhibit the degradative enzymes, thereby selectively elevating levels of eCBs, may be a useful therapeutic approach. No clinical trials have yet investigated the anti-nausea/anti-emetic potential of manipulations of the eCB system (such as increased concentration of AEA and 2-AG via inhibition of FAAH or MAGL), although BIA 10–2474, a FAAH inhibitor did reach a Phase I clinical trial, with patients reporting nausea as a side effect. This trial was suspended when one participant was hospitalised and declared brain dead after receiving BIA 10-2474. Others displayed neurological signs and were hospitalised. In the single ascending dose part of the trial, one instance of nausea was reported in a patient in the 40 mg group, and one instance of nausea was also reported in a patient in the 50 mg group of the multiple ascending dose part of the trial (Rocha et al. 2021). The side-effect of nausea was also reported in dogs at the high dose (100 mg/kg/day, administered for 4 weeks) (Hardisty et al. 2020) and in monkeys at the high dose (125 mg/kg, administered for 4 days) (Weber et al. 2020). BIA 10-2474 potently inhibits FAAH, but it also interacts with several off-target lipid processing enzymes with high levels of exposure (Bonifácio et al. 2020). Indeed, using activity-based protein profiling, van Esbroeck et al. (2017) demonstrated that it targets lipolytic enzymes as well. This might lead to a disruption of neuronal lipid metabolism. Although this data suggests that these off-target effects may have contributed to its neurotoxicity, it remains unclear what the implication of these off-targets may have had in the clinical trial patients, and if they contributed to the serious side effects that were seen. Evidence suggests that these serious adverse events associated with BIA 10-2474 were not due to pharmacological inhibition of FAAH, rather they were due to properties specific to this drug. Since this time, other FAAH inhibitors (JNJ-42165279, ASP3652 and PF-04457845) have been assessed in clinical trials, without serious adverse effects (D'Souza et al. 2019; Schmidt et al. 2021; Takizawa et al. 2020). In fact, the side effects of nausea and vomiting were either reported as not present in the treatment groups or reported at the same incidence (or less) as that of the placebo control groups.

Changes in eCB levels have been measured in those experiencing nausea and/or vomiting following anaesthesia and during motion sickness. They have also been examined in those with cyclic vomiting syndrome (CVS) and hyperemesis gravidarum.

Evidence from the few studies that have been conducted suggests that postoperative nausea may be associated with reduced eCB levels. Decreased AEA levels were detected in patients exposed to the anaesthesia sevoflurane, which commonly produces postoperative nausea (Schelling et al. 2006). In line with this finding, a prospective genotype-blinded observational study of 259 healthy children undergoing tonsillectomy found significant associations between FAAH polymorphisms and an increased risk for postoperative nausea (Sadhasivam et al. 2015), although causality needs to be further studied. These findings suggest a key role for AEA in the experience of postoperative nausea.

Those experiencing motion sickness may have reduced eCB levels. Low blood levels of AEA and 2-AG have been shown in those experiencing severe motion sickness during parabolic flight manoeuvres on the International Space Station (Chouker et al. 2010; Strewe et al. 2012). These reduced eCB levels were associated with increased salivary cortisol levels (Chouker et al. 2010). Furthermore, those astronauts experiencing chronic (6 months), but well-tolerated exposure to parabolic flights, showed increased blood eCBs, which returned to baseline at the end of the flight (Strewe et al. 2012). In addition, lower CB_1 receptor mRNA expression in leucocytes was seen in those experiencing motion sickness (Chouker et al. 2010). These findings suggest that eCB levels are correlated with the experience of motion sickness and that eCBs may be recruited as a protective mechanism to offset motion-induced decreases in eCB levels. Stress may be an associated factor mediating this response. That is a rapid decline in AEA may be driving the initiation of the stress response, with its termination occurring due to a more sluggish increase in 2-AG signalling (see Morena et al. 2016 for review).

CVS, characterised by episodes of severe vomiting that have no apparent cause, may be associated with alterations in the fatty acids oleoylethanolamide (OEA) and palmitoylethanolamide (PEA). Like AEA, both these fatty acids are substrates of FAAH. eCBs and associated fatty acids were examined in 22 cyclic vomiting patients in their well and sick phases (Venkatesan et al. 2016). In comparison to matched controls, there were no differences in levels of AEA and 2-AG, although there was a trend towards an increase in these levels during the sick phase. Compared to the well phase, there were elevated levels of OEA, PEA, and salivary cortisol during the sick phase. Furthermore, PEA and nausea scores in the sick phase were positively correlated. In a more recent study, CVS patients were examined only during their well phase, with CVS patients ($n = 10$) showing higher 2-AG levels (compared to healthy controls), suggesting that 2-AG may be protective for nausea and vomiting (Bashashati et al. 2023). AEA, OEA, and PEA levels were similar between these CVS patients and healthy controls.

Hyperemesis gravidarum is characterised by severe nausea and vomiting during pregnancy. Measured levels of AEA, OEA and PEA were not altered in hyperemesis gravidarum patients ($n = 15$) vs healthy pregnant controls ($n = 30$) around seven weeks of gestation (Gebeh et al. 2014). These levels were not necessarily measured during the sick phase for these pregnant women. Therefore, OEA and PEA may only be mobilised in the sick phase, as a protective mechanism.

Although the data is limited, eCBs seem to be important neuromodulators involved in the experience of nausea and vomiting, with decreased levels of AEA and/or 2-AG evident with nausea-inducing manipulations or conditions associated with nausea and/or vomiting. Further research is needed to clarify how eCBs and related fatty acids are involved in the experience of nausea and vomiting and how manipulations of the eCB system could manage these symptoms. The timepoint at which these eCBs are measured (i.e. during the sick vs well phase) may be particularly important in revealing their role in nausea and vomiting. Examining the role of eCBs (and related fatty acids) in nausea and vomiting and the interconnection with the stress response may be important (see DeVuono et al. 2024). It is also possible that the stress-induced downregulation of CB_1 receptor expression may result in the mobilisation of OEA and PEA as a protective mechanism. The relationship between the CB_1 receptor and PPARα in reducing nausea and vomiting is not well understood. However, it has been reported that the co-administration of PEA and AEA act synergistically to alleviate pain in animal models (Calignano et al. 1998).

4.1 Cannabinoids and eCBs in Human Patients: Conclusions

Several studies have examined the effect of THC and synthetic THC such as dronabinol, mostly with favourable effects on nausea and vomiting in CINV patients, but at higher doses patients do report unpleasant psychoactive effects. In contrast, there is only one published case study with the non-intoxicating CBD showing improved reduction of nausea. A few studies have examined the effects of cannabis extracts containing THC and CBD, with mixed results. At high doses, THC and CBD may cause the side-effect of nausea and/or vomiting, with prolonged use of high-THC content cannabis developing into CHS for some individuals, who may be predisposed to its development.

The field is only beginning to unravel the role that eCBs and manipulations of this system play in modulating nausea and vomiting. Reduced levels of eCBs seem to be associated with the experience of nausea and/or vomiting, with mobilisation of these neuromodulators to restore homeostasis. Elevations of OEA and PEA have been reported during the sick phase in cyclic vomiting patients, presumably as a protective mechanism, but further research is needed to understand this relationship.

There is a scarcity of high-quality blinded, randomised control trials evaluating the effects of cannabinoids and modulations of the eCB system in mediating nausea and vomiting. These important trials are needed to better understand optimal dosing and whether the treatments will remain effective over extended durations. More attention needs to be devoted to assessing the control of nausea as a separate outcome from that of vomiting, with particular attention to whether the symptom (s) are occurring during the acute, delayed, or anticipatory stage. In addition, understanding possible adverse effects associated with cannabinoid use is critical. Pre-clinical animal studies can therefore help to screen potential compounds, with attention to the timepoints of their administration that may be worthy of further clinical investigation.

5 General Conclusions

Treatment of chemotherapy-induced nausea is an unmet need for many patients. Cannabinoids and manipulations of the eCB system are promising adjunct treatment options. The oral THC-like synthetics nabilone and dronabinol are approved for CINV, but intoxication can limit their clinical utility. Although no clinical trials have evaluated CBD in CINV patients, some planned or in-progress trials will evaluate measures of nausea and vomiting in cancer patients being treated with CBD. Clinical trials are currently evaluating FAAH inhibitors for cannabis use disorder and withdrawal. No such trials are examining the effect of manipulations of the eCB system on CINV, although a few studies have measured levels of eCBs during nausea-provoking manipulations. These studies, although preliminary, do suggest that decreased levels of eCBs and the fatty acids OEA and PEA are correlated with the experience of nausea. The recruitment of this system may be protective. This pattern is consistent with findings from animal studies.

Paradoxically, treatment with cannabinoids can dose-dependently produce nausea and/or vomiting, with the condition of Cannabinoid Hyperemesis Syndrome – cyclical episodes of nausea and vomiting associated with prolonged high-THC cannabis use – gaining popular media attention. In a rodent model of THC-induced nausea, the high dose of THC that produces nausea also dysregulates the stress system. Further research utilising this model might help to better understand the aetiology of CHS and to identify candidate treatments (in addition to cessation of cannabis use).

Due to the scarcity of properly controlled clinical trials with cannabinoids and manipulations of the eCBs system on CINV, we must rely on the pre-clinical research to inform us about potential treatments that may have clinical utility. These models have also helped to elucidate their mechanisms of action and the critical brain regions involved.

In animal models, THC and THC-like synthetics dose-dependently reduce toxin-induced vomiting in a variety of species via action at the CB_1 receptor. In rats, THC also dose-dependently reduces acute and anticipatory nausea. THC is likely exerting its effects by activation of presynaptic CB_1 receptors, resulting in suppression of neurotransmitter release, presumably 5-HT. THCA's anti-emetic and anticipatory nausea effects seem to be CB_1 receptor-mediated (although it did not induce hypothermia or hypolocomotion like typical CB_1 agonists), while its acute nausea effects seem to be PPARα receptor-mediated. As THCA is a weak FAAH inhibitor, the elevation of AEA may be acting at CB_1 receptors to reduce vomiting and anticipatory nausea, while the elevation of OEA and PEA may be acting at PPARα to reduce acute nausea.

CBD dose-dependently reduces toxin-induced vomiting in shrews, as well as acute and anticipatory nausea, with CBDA and CBDA-ME demonstrating even greater potency. CBD, CBDA, and CBDA-ME seem to be reducing nausea and vomiting through activation of the $5\text{-}HT_{1A}$ receptor, enhancing the ability of endogenous 5-HT to bind to these receptors, preventing the further release of 5-HT to forebrain structures such as the IIC.

Enhanced anti-emetic and anti-nausea effects have been demonstrated when ineffective doses of cannabinoids are combined. As these phytocannabinoids can have distinct mechanisms of action, their combined suppressive effects are likely due to their separate mechanisms of action working simultaneously. Utilising very low doses of cannabinoids can eliminate the likelihood of undesirable side effects. Clinical trials utilising single or combined cannabinoids as adjunct treatments with existing anti-emetic regimens are needed for CINV patients.

Although little research has focused on the anti-emetic/anti-nausea of the minor cannabinoids, evidence suggests that THCV and CBDV reduce acute nausea in rats, although their mechanisms of action have not been evaluated. Interestingly, CBG seems to act in opposition to CBD, blocking its anti-emetic and acute nausea effects. This may be problematic for nausea and vomiting control, given the rise in popularity of CBG-predominant cannabis products.

Anti-emetic effects are seen following exogenous administration of 2-AG, AEA, and inhibition of FAAH or MAGL, through CB_1 and/or TRPV1 receptors. FAAH inhibition seems to exert its effects in acute nausea by elevating OEA and PEA to activate PPARα, with the ventral pallidum as a key brain area. Its effect in anticipatory nausea occurs through AEA elevation via CB_1 receptor activation, with the area postrema being a key brain area involved. Through MAGL inhibition 2-AG is elevated to reduce acute and anticipatory via activation of the CB_1 receptor, with the IIC being a critical area involved in both. Although dual FAAH/MAGL inhibitors reduce acute and anticipatory nausea, they do seem to be associated with sedation which might be therapeutically undesirable.

These findings support growing evidence to suggest that differing mechanisms regulate anticipatory nausea and acute nausea. For example, the administration of THCA (which acts to inhibit FAAH) or selective FAAH inhibitors like PF-3845, reduce acute nausea by elevating OEA and PEA to act on PPARα. In contrast, their effects on anticipatory nausea occur by elevating AEA to activate CB_1 receptors. Alternatively, these presumed differential mechanisms may simply be due to differences between these experimental paradigms. During the acute nausea paradigm, the cannabinoid or eCB manipulation occurs during the establishment of the taste-illness association (i.e. the compound is given as a pre-treatment to the lithium chloride-induced illness during conditioning). In contrast, during the anticipatory nausea paradigm, it occurs during the expression of an already established context-illness association (i.e. the compound is given at the test). Further research is needed to tease apart this distinction.

Both newly recognised neuromodulators OlGly and OlAla reduce vomiting and acute nausea. OlAla does have a wider effective dose range, devoid of sedative effects. Because these fatty acid amides exert their effects through dual action at CB_1 receptors and PPARα they may be particularly promising for the development of new treatments.

There are certainly many more questions to answer regarding the role of cannabinoids in the regulation of nausea and vomiting. Although no sex differences have been reported for CBD's acute anti-nausea effects, further studies examining the anti-nausea and anti-emetic effects of cannabinoids are needed, including their mechanism(s) of action. Repeated dosing of cannabinoids should also be better evaluated, with attention to sex differences in the development of tolerance.

References

Ahmedzai S, Carlyle DL, Calder IT, Moran F (1983) Anti-emetic efficacy and toxicity of nabilone, a synthetic cannabinoid, in lung cancer chemotherapy. Br J Cancer 48:657–663. https://doi.org/10.1038/bjc.1983.247

Ahn K, Johnson DS, Mileni M, Beidler D, Long JZ, McKinney MK, Weerapana E, Sadagopan N, Liimatta M, Smith SE, Lazerwith S, Stiff C, Kamtekar S, Bhattacharya K, Zhang Y, Swaney S, Van Becelaere K, Stevens RC, Cravatt BF (2009) Discovery and characterization of a highly selective FAAH inhibitor that reduces inflammatory pain. Chem Biol 16:411–420. https://doi.org/10.1016/j.chembiol.2009.02.013

Allen JH, de Moore GM, Heddle R, Twartz JC (2004) Cannabinoid hyperemesis: cyclical hyperemesis in association with chronic cannabis abuse. Gut 53:1566–1570. https://doi.org/10.1136/gut.2003.036350

Anderson LL, Etchart MG, Bahceci D, Golembiewski TA, Arnold JC (2021) Cannabis constituents interact at the drug efflux pump BCRP to markedly increase plasma cannabidiolic acid concentrations. Sci Rep 11:14948. https://doi.org/10.1038/s41598-021-94212-6

Artukoglu BB, Beyer C, Zuloff-Shani A, Brener E, Bloch MH (2017) Efficacy of palmitoylethanolamide for pain: a meta-analysis. Pain Physician 20:353–362

Ayoub SM, Smoum R, Farag M, Atwal H, Collins SA, Rock EM, Limebeer CL, Piscitelli F, Iannotti FA, Lichtman AH, Leri F, Di Marzo V, Mechoulam R, Parker LA (2020) Oleoyl alanine (HU595): a stable monomethylated oleoyl glycine interferes with acute naloxone precipitated morphine withdrawal in male rats. Psychopharmacology (Berl) 237:2753–2765. https://doi.org/10.1007/s00213-020-05570-4

Bar-Lev Schleider L, Mechoulam R, Lederman V, Hilou M, Lencovsky O, Betzalel O, Shbiro L, Novack V (2018) Prospective analysis of safety and efficacy of medical cannabis in large unselected population of patients with cancer. Eur J Intern Med 49:37–43. https://doi.org/10.1016/j.ejim.2018.01.023

Barsky AJ, Peekna HM, Borus JF (2001) Somatic symptom reporting in women and men. J Gen Intern Med 16:266–275. https://doi.org/10.1046/j.1525-1497.2001.00229.x

Bashashati M, Bradshaw HB, Johnson CT, Zuckerman MJ, Sarosiek J, McCallum RW, Sarosiek I (2023) Plasma endocannabinoids and cannabimimetic fatty acid derivatives are altered in cyclic vomiting syndrome: the effects of sham feeding. J Invest Med 71:821–829. https://doi.org/10.1177/10815589231196591

Bisogno T, Hanus L, De Petrocellis L, Tchilibon S, Ponde DE, Brandi I, Moriello AS, Davis JB, Mechoulam R, Di Marzo V (2001) Molecular targets for cannabidiol and its synthetic analogues: effect on vanilloid VR1 receptors and on the cellular uptake and enzymatic hydrolysis of anandamide. Br J Pharmacol 134:845–852. https://doi.org/10.1038/sj.bjp.0704327

Boerma T, Hosseinpoor AR, Verdes E, Chatterji S (2016) A global assessment of the gender gap in self-reported health with survey data from 59 countries. BMC Public Health 16:675. https://doi.org/10.1186/s12889-016-3352-y

Bolognini D, Costa B, Maione S, Comelli F, Marini P, Di Marzo V, Parolaro D, Ross RA, Gauson LA, Cascio MG, Pertwee RG (2010) The plant cannabinoid Delta9-tetrahydrocannabivarin can decrease signs of inflammation and inflammatory pain in mice. Br J Pharmacol 160:677–687. https://doi.org/10.1111/j.1476-5381.2010.00756.x

Bolognini D, Rock EM, Cluny NL, Cascio MG, Limebeer CL, Duncan M, Stott CG, Javid FA, Parker LA, Pertwee RG (2013) Cannabidiolic acid prevents vomiting in Suncus murinus and nausea-induced behaviour in rats by enhancing 5-HT1A receptor activation. Br J Pharmacol 168:1456–1470. https://doi.org/10.1111/bph.12043

Bonifácio MJ, Sousa F, Aires C, Loureiro AI, Fernandes-Lopes C, Pires NM, Palma PN, Moser P, Soares da-Silva P (2020) Preclinical pharmacological evaluation of the fatty acid amide hydrolase inhibitor BIA 10-2474. Br J Pharmacol 177:2123–2142. https://doi.org/10.1111/bph.14973

Bonn-Miller MO, Loflin MJE, Thomas BF, Marcu JP, Hyke T, Vandrey R (2017) Labeling Accuracy of Cannabidiol Extracts Sold Online. JAMA 318:1708–1709. https://doi.org/10.1001/jama.2017.11909

Brame RE, Lucot JB (2011) Guamanian Suncus murinus responsiveness to emetic stimuli and the antiemetic effects of 8-OH-DPAT. Pharmacol Biochem Behav 99:381–384. https://doi.org/10.1016/j.pbb.2011.05.012

Calignano A, La Rana G, Giuffrida A, Piomelli D (1998) Control of pain initiation by endogenous cannabinoids. Nature 394:277–281. https://doi.org/10.1038/28393

Cascio MG, Gauson LA, Stevenson LA, Ross RA, Pertwee RG (2010) Evidence that the plant cannabinoid cannabigerol is a highly potent alpha2-adrenoceptor agonist and moderately potent

5HT1A receptor antagonist. Br J Pharmacol 159:129–141. https://doi.org/10.1111/j.1476-5381.2009.00515.x
Chan HS, Correia JA, MacLeod SM (1987) Nabilone versus prochlorperazine for control of cancer chemotherapy-induced emesis in children: a double-blind, crossover trial. Pediatrics 79:946–952
Chang AE, Shiling DJ, Stillman RC, Goldberg NH, Seipp CA, Barofsky I, Simon RM, Rosenberg SA (1979) Delata-9-tetrahydrocannabinol as an antiemetic in cancer patients receiving high-dose methotrexate. A prospective, randomized evaluation. Ann Intern Med 91:819–824. https://doi.org/10.7326/0003-4819-91-6-819
Chesney E, Oliver D, Green A, Sovi S, Wilson J, Englund A, Freeman TP, McGuire P (2020) Adverse effects of cannabidiol: a systematic review and meta-analysis of randomized clinical trials. Neuropsychopharmacology. https://doi.org/10.1038/s41386-020-0667-2
Chouker A, Kaufmann I, Kreth S, Hauer D, Feuerecker M, Thieme D, Vogeser M, Thiel M, Schelling G (2010) Motion sickness, stress and the endocannabinoid system. PLoS One 5: e10752. https://doi.org/10.1371/journal.pone.0010752
Cluny NL, Naylor RJ, Whittle BA, Javid FA (2008) The effects of cannabidiol and tetrahydrocannabinol on motion-induced emesis in Suncus murinus. Basic Clin Paharmacol Toxicol 103:150–156. https://doi.org/10.1111/j.1742-7843.2008.00253.x
Cluny NL, Vemuri VK, Chambers AP, Limebeer CL, Bedard H, Wood JT, Lutz B, Zimmer A, Parker LA, Makriyannis A, Sharkey KA (2010) A novel peripherally restricted cannabinoid receptor antagonist, AM6545, reduces food intake and body weight, but does not cause malaise, in rodents. Br J Pharmacol 161:629–642. https://doi.org/10.1111/j.1476-5381.2010.00908.x
Cravatt BF, Giang DK, Mayfield SP, Boger DL, Lerner RA, Gilula NB (1996) Molecular characterization of an enzyme that degrades neuromodulatory fatty-acid amides. Nature 384:83–87. https://doi.org/10.1038/384083a0
Crippa JA, Guimarães FS, Campos AC, Zuardi AW (2018) Translational investigation of the therapeutic potential of cannabidiol (CBD): toward a new age. Front Immunol 9:2009. https://doi.org/10.3389/fimmu.2018.02009
Criscuolo E, De Sciscio ML, Fezza F, Maccarrone M (2020) In silico and in vitro analysis of major cannabis-derived compounds as fatty acid amide hydrolase inhibitors. Molecules 26:48. https://doi.org/10.3390/molecules26010048
Cross-Mellor SK, Ossenkopp KP, Piomelli D, Parker LA (2007) Effects of the FAAH inhibitor, URB597, and anandamide on lithium-induced taste reactivity responses: a measure of nausea in the rat. Psychopharmacology (Berl) 190:135–143. https://doi.org/10.1007/s00213-006-0589-7
D'Souza DC, Cortes-Briones J, Creatura G, Bluez G, Thurnauer H, Deaso E, Bielen K, Surti T, Radhakrishnan R, Gupta A, Gupta S, Cahill J, Sherif MA, Makriyannis A, Morgan PT, Ranganathan M, Skosnik PD (2019) Efficacy and safety of a fatty acid amide hydrolase inhibitor (PF-04457845) in the treatment of cannabis withdrawal and dependence in men: a double-blind, placebo-controlled, parallel group, phase 2a single-site randomised controlled trial. Lancet Psychiatry 6:35–45. https://doi.org/10.1016/s2215-0366(18)30427-9
Dall'Stella PB, Docema MFL, Maldaun MVC, Feher O, Lancellotti CLP (2019) Case report: clinical outcome and image response of two patients with secondary high-grade glioma treated with chemoradiation, PCV, and cannabidiol. Front Oncol 8. https://doi.org/10.3389/fonc.2018.00643
Dalzell AM, Bartlett H, Lilleyman JS (1986) Nabilone: an alternative antiemetic for cancer chemotherapy. Arch Dis Child 61:502–505. https://doi.org/10.1136/adc.61.5.502
Darmani NA (1998) Serotonin 5-HT3 receptor antagonists prevent cisplatin-induced emesis in Cryptotis parva: a new experimental model of emesis. J Neural Transm 105:1143–1154. https://doi.org/10.1007/s007020050118
Darmani NA (2001a) Delta(9)-tetrahydrocannabinol and synthetic cannabinoids prevent emesis produced by the cannabinoid CB(1) receptor antagonist/inverse agonist SR 141716A. Neuropsychopharmacology 24:198–203. https://doi.org/10.1016/S0893-133X(00)00197-4

Darmani NA (2001b) Delta-9-tetrahydrocannabinol differentially suppresses cisplatin-induced emesis and indices of motor function via cannabinoid CB(1) receptors in the least shrew. Pharmacol Biochem Behav 69:239–249

Darmani NA (2002) The potent emetogenic effects of the endocannabinoid, 2-AG (2-arachidonoylglycerol) are blocked by delta(9)-tetrahydrocannabinol and other cannnabinoids. J Pharmacol Exp Ther 300:34–42. https://doi.org/10.1124/jpet.300.1.34

Darmani NA, Belkacemi L, Zhong W (2019) Δ(9)-THC and related cannabinoids suppress substance P-induced neurokinin NK(1)-receptor-mediated vomiting via activation of cannabinoid CB(1) receptor. Eur J Pharmacol 865:172806. https://doi.org/10.1016/j.ejphar.2019.172806

Darmani NA, Chebolu S, Zhong W, Trinh C, McClanahan B, Brar RS (2014) Additive antiemetic efficacy of low-doses of the cannabinoid CB(1/2) receptor agonist Δ(9)-THC with ultralow-doses of the vanilloid TRPV1 receptor agonist resiniferatoxin in the least shrew (Cryptotis parva). Eur J Pharmacol 722:147–155. https://doi.org/10.1016/j.ejphar.2013.08.051

Darmani NA, Crim JL (2005) Delta-9-tetrahydrocannabinol differentially suppresses emesis versus enhanced locomotor activity produced by chemically diverse dopamine D2/D3 receptor agonists in the least shrew (Cryptotis parva). Pharmacol Biochem Behav 80:35–44. https://doi.org/10.1016/j.pbb.2004.10.019

Darmani NA, Janoyan JJ, Crim J, Ramirez J (2007) Receptor mechanism and antiemetic activity of structurally-diverse cannabinoids against radiation-induced emesis in the least shrew. Eur J Pharmacol 563:187–196. https://doi.org/10.1016/j.ejphar.2007.01.093

Darmani NA, Johnson JC (2004) Central and peripheral mechanisms contribute to the antiemetic actions of delta-9-tetrahydrocannabinol against 5-hydroxytryptophan-induced emesis. Eur J Pharmacol 488:201–212. https://doi.org/10.1016/j.ejphar.2004.02.018

Darmani NA, McClanahan BA, Trinh C, Petrosino S, Valenti M, Di Marzo V (2005) Cisplatin increases brain 2-arachidonoylglycerol (2-AG) and concomitantly reduces intestinal 2-AG and anandamide levels in the least shrew. Neuropharmacology 49:502–513. https://doi.org/10.1016/j.neuropharm.2005.04.007

de Mattos VB, Prais HA, Daker MV (2009) Melancholic features related to rimonabant. Gen Hosp Psychiatry 31:583–585. https://doi.org/10.1016/j.genhosppsych.2008.12.009

De Petrocellis L, Ligresti A, Moriello AS, Allara M, Bisogno T, Petrosino S, Stott CG, Di Marzo V (2011) Effects of cannabinoids and cannabinoid-enriched Cannabis extracts on TRP channels and endocannabinoid metabolic enzymes. Br J Pharmacol 163:1479–1494. https://doi.org/10.1111/j.1476-5381.2010.01166.x

De Petrocellis L, Ligresti A, Schiano Moriello A, Iappelli M, Verde R, Stott CG, Cristino L, Orlando P, Di Marzo V (2013) Non-THC cannabinoids inhibit prostate carcinoma growth in vitro and in vivo: pro-apoptotic effects and underlying mechanisms. Br J Pharmacol 168:79–102. https://doi.org/10.1111/j.1476-5381.2012.02027.x

Despres JP (2009) Pleiotropic effects of rimonabant: clinical implications. Curr Pharm Des 15:553–570. https://doi.org/10.2174/138161209787315666

Devane WA, Dysarz FA 3rd, Johnson MR, Melvin LS, Howlett AC (1988) Determination and characterization of a cannabinoid receptor in rat brain. Mol Pharmacol 34:605–613

Devane WA, Hanus L, Breuer A, Pertwee RG, Stevenson LA, Griffin G, Gibson D, Mandelbaum A, Etinger A, Mechoulam R (1992) Isolation and structure of a brain constituent that binds to the cannabinoid receptor. Science 258:1946–1949. https://doi.org/10.1126/science.1470919

DeVuono MV, Hrelja KM, Petrie GN, Limebeer CL, Rock EM, Hill MN, Parker LA (2020a) Nausea-induced conditioned gaping reactions in rats produced by high-dose synthetic cannabinoid, JWH-018. Cannabis Cannabinoid Res 5:298–304. https://doi.org/10.1089/can.2019.0103

DeVuono MV, Hrelja KM, Sabaziotis L, Rajna A, Rock EM, Limebeer CL, Mutch DM, Parker LA (2018) Conditioned gaping produced by high dose Δ(9)-tetrahydracannabinol: dysregulation of the hypothalamic endocannabinoid system. Neuropharmacology 141:272–282. https://doi.org/10.1016/j.neuropharm.2018.08.039

DeVuono MV, La Caprara O, Petrie GN, Limebeer CL, Rock EM, Hill MN, Parker LA (2022) Cannabidiol interferes with establishment of δ 9 - tetrahydrocannabinol-induced Nausea through a 5-HT 1A mechanism. Cannabis Cannabinoid Res 7:58–64. https://doi.org/10.1089/can.2020.0083

DeVuono MV, La Caprara O, Sullivan MT, Bath A, Petrie GN, Limebeer CL, Rock EM, Hill MN, Parker LA (2020b) Role of the stress response and the endocannabinoid system in Δ(9)-tetrahydrocannabinol (THC)-induced nausea. Psychopharmacology (Berl) 237:2187–2199. https://doi.org/10.1007/s00213-020-05529-5

DeVuono MV, Parker LA (2020) Cannabinoid hyperemesis syndrome: a review of potential mechanisms. Cannabis Cannabinoid Res 5:132–144. https://doi.org/10.1089/can.2019.0059

DeVuono MV, Venkatesan T, Hillard CJ (2024) Endocannabinoid signaling in stress, nausea, and vomiting. Neurogastroenterol Motil. https://doi.org/10.1111/nmo.14911

Dezieck L, Hafez Z, Conicella A, Blohm E, O'Connor MJ, Schwarz ES, Mullins ME (2017) Resolution of cannabis hyperemesis syndrome with topical capsaicin in the emergency department: a case series. Clin Toxicol 55:908–913. https://doi.org/10.1080/15563650.2017.1324166

Di Marzo V, Fontana A, Cadas H, Schinelli S, Cimino G, Schwartz JC, Piomelli D (1994) Formation and inactivation of endogenous cannabinoid anandamide in central neurons. Nature 372:686–691. https://doi.org/10.1038/372686a0

Dinh TP, Carpenter D, Leslie FM, Freund TF, Katona I, Sensi SL, Kathuria S, Piomelli D (2002a) Brain monoglyceride lipase participating in endocannabinoid inactivation. Proc Natl Acad Sci U S A 99:10819–10824. https://doi.org/10.1073/pnas.152334899

Dinh TP, Freund TF, Piomelli D (2002b) A role for monoglyceride lipase in 2-arachidonoylglycerol inactivation. Chem Phys Lipids 121:149–158. https://doi.org/10.1016/s0009-3084(02)00150-0

Dranitsaris G, Molassiotis A, Clemons M, Roeland E, Schwartzberg L, Dielenseger P, Jordan K, Young A, Aapro M (2017) The development of a prediction tool to identify cancer patients at high risk for chemotherapy-induced nausea and vomiting. Ann Oncol 28:1260–1267. https://doi.org/10.1093/annonc/mdx100

Duran M, Perez E, Abanades S, Vidal X, Saura C, Majem M, Arriola E, Rabanal M, Pastor A, Farre M, Rams N, Laporte JR, Capella D (2010) Preliminary efficacy and safety of an oromucosal standardized cannabis extract in chemotherapy-induced nausea and vomiting. Br J Clin Pharmacol 70:656–663. https://doi.org/10.1111/j.1365-2125.2010.03743.x

Eckel LA, Ossenkopp KP (1996) Area postrema mediates the formation of rapid, conditioned palatability shifts in lithium-treated rats. Behav Neurosci 110:202–212

Einhorn LH, Nagy C, Furnas B, Williams SD (1981) Nabilone: an effective antiemetic in patients receiving cancer chemotherapy. J Clin Pharmacol 21:64s–69s. https://doi.org/10.1002/j.1552-4604.1981.tb02576.x

Elder JJ, Knoderer HM (2015) Characterization of dronabinol usage in a pediatric oncology population. J Pediatr Pharmacol Ther 20:462–467. https://doi.org/10.5863/1551-6776-20.6.462

Fehniger J, Brodsky AL, Kim A, Pothuri B (2021) Medical marijuana utilization in gynecologic cancer patients. Gynecol Oncol Rep 37:100820

Feigenbaum JJ, Richmond SA, Weissman Y, Mechoulam R (1989) Inhibition of cisplatin-induced emesis in the pigeon by a non-psychotropic synthetic cannabinoid. Eur J Pharmacol 169:159–165. https://doi.org/10.1016/0014-2999(89)90828-5

Ferrari F, Ottani A, Giuliani D (1999) Cannabimimetic activity in rats and pigeons of HU 210, a potent antiemetic drug. Pharmacol Biochem Behav 62:75–80. https://doi.org/10.1016/s0091-3057(98)00114-2

Fiore JJ, Gralla RJ (1984) Pharmacologic treatment of chemotherapy-induced nausea and vomiting. Cancer Invest 2:351–361. https://doi.org/10.3109/07357908409040312

Forrester MB, Kleinschmidt K, Schwarz E, Young A (2012) Synthetic cannabinoid and marijuana exposures reported to poison centers. Hum Exp Toxicol 31:1006–1011. https://doi.org/10.1177/0960327111421945

Frytak S, Moertel CG, O'Fallon JR, Rubin J, Creagan ET, O'Connell MJ, Schutt AJ, Schwartau NW (1979) Delta-9-tetrahydrocannabinol as an antiemetic for patients receiving cancer chemotherapy. A comparison with prochlorperazine and a placebo. Ann Intern Med 91:825–830. https://doi.org/10.7326/0003-4819-91-6-825

Fu J, Gaetani S, Oveisi F, Lo Verme J, Serrano A, Rodriguez De Fonseca F, Rosengarth A, Luecke H, Di Giacomo B, Tarzia G, Piomelli D (2003) Oleylethanolamide regulates feeding and body weight through activation of the nuclear receptor PPAR-alpha. Nature 425:90–93. https://doi.org/10.1038/nature01921

Gaoni Y, Mechoulam R (1964) Isolation, structure, and partial synthesis of an active constituent of Hashish. J Am Chem Soc 86:1646–1647. https://doi.org/10.1021/ja01062a046

Gebeh AK, Willets JM, Marczylo TH, Konje JC (2014) Plasma anandamide and related n-acylethanolamide levels are not elevated in pregnancies complicated by hyperemesis gravidarum. J Matern Fetal Neonatal Med 27:954–959. https://doi.org/10.3109/14767058.2013.847413

Go YY, Kim SR, Kim DY, Chae SW, Song JJ (2020) Cannabidiol enhances cytotoxicity of anticancer drugs in human head and neck squamous cell carcinoma. Sci Rep 10:20622. https://doi.org/10.1038/s41598-020-77674-y

Grill HJ, Norgren R (1978) The taste reactivity test. I. Mimetic responses to gustatory stimuli in neurologically normal rats. Brain Res 143:263–279. https://doi.org/10.1016/0006-8993(78)90568-1

Grimison P, Mersiades A, Kirby A, Lintzeris N, Morton R, Haber P, Olver I, Walsh A, McGregor I, Cheung Y, Tognela A, Hahn C, Briscoe K, Aghmesheh M, Fox P, Abdi E, Clarke S, Della-Fiorentina S, Shannon J, Gedye C, Begbie S, Simes J, Stockler M (2020) Oral THC:CBD cannabis extract for refractory chemotherapy-induced nausea and vomiting: a randomised, placebo-controlled, phase II crossover trial. Ann Oncol 31:1553–1560. https://doi.org/10.1016/j.annonc.2020.07.020

Grunfeld Y, Edery H (1969) Psychopharmacological activity of the active constituents of hashish and some related cannabinoids. Psychopharmacologia 14:200–210. https://doi.org/10.1007/BF00404218

Gylys JA, Doran KM, Buyniski JP (1979) Antagonism of cisplatin induced emesis in the dog. Res Commun Chem Pathol Pharmacol 23:61–68

Hanuš LO, Meyer SM, Muñoz E, Taglialatela-Scafati O, Appendino G (2016) Phytocannabinoids: a unified critical inventory. Nat Prod Rep 33:1357–1392. https://doi.org/10.1039/c6np00074f

Hardisty JF, Harris SB, Hayes WA, Weber K (2020) Oral repeated-dose toxicity studies of BIA 10-2474 in beagle dogs. Regul Toxicol Pharmacol 111:104555. https://doi.org/10.1016/j.yrtph.2019.104555

Henry JG, Shoemaker G, Prieto JM, Hannon MB, Wakshlag JJ (2021) The effect of cannabidiol on canine neoplastic cell proliferation and mitogen-activated protein kinase activation during autophagy and apoptosis. Vet Comp Oncol 19:253–265. https://doi.org/10.1111/vco.12669

Herman TS, Einhorn LH, Jones SE, Nagy C, Chester AB, Dean JC, Furnas B, Williams SD, Leigh SA, Dorr RT, Moon TE (1979) Superiority of nabilone over prochlorperazine as an antiemetic in patients receiving cancer chemotherapy. N Engl J Med 300:1295–1297. https://doi.org/10.1056/NEJM197906073002302

Hernandez Torres C, Mazzarello S, Ng T, Dranitsaris G, Hutton B, Smith S, Munro A, Jacobs C, Clemons M (2015) Defining optimal control of chemotherapy-induced nausea and vomiting-based on patients' experience. Support Care Cancer 23:3341–3359. https://doi.org/10.1007/s00520-015-2801-y

Hesketh PJ, Bohlke K, Kris MG (2017) Antiemetics: American Society of Clinical Oncology Clinical Practice Guideline Update Summary. J Oncol Pract 13:825–830. https://doi.org/10.1200/jop.2017.026351

Ho CY, Berridge KC (2014) Excessive disgust caused by brain lesions or temporary inactivations: mapping hotspots of the nucleus accumbens and ventral pallidum. Eur J Neurosci 40:3556–3572. https://doi.org/10.1111/ejn.12720

Horn CC (2008) Why is the neurobiology of nausea and vomiting so important? Appetite 50:430–434. https://doi.org/10.1016/j.appet.2007.09.015

Horn CC, Meyers K, Pak D, Nagy A, Apfel CC, Williams BA (2012) Post-anesthesia vomiting: impact of isoflurane and morphine on ferrets and musk shrews. Physiol Behav 106:562–568. https://doi.org/10.1016/j.physbeh.2012.03.031

Horn CC, Wang H, Estival L, Meyers K, Magnusson MS (2013) Novel dynamic measures of emetic behavior in musk shrews. Auton Neurosci 179:60–67. https://doi.org/10.1016/j.autneu.2013.07.006

Hornby PJ (2001) Central neurocircuitry associated with emesis. Am J Med 111(Suppl 8A):106S–112S. https://doi.org/10.1016/s0002-9343(01)00849-x

Hsu YP, Chu KC, Bai CH, Huang CJ, Chen C, Hsu CW (2019) Safety and efficacy of clonidine on postoperative vomiting and pain in pediatric ophthalmic surgery: a systematic review and meta-analysis. Paediatr Anaesth 29:1011–1023. https://doi.org/10.1111/pan.13712

Inkol JM, Hocker SE, Mutsaers AJ (2021) Combination therapy with cannabidiol and chemotherapeutics in canine urothelial carcinoma cells. PLoS One 16:e0255591. https://doi.org/10.1371/journal.pone.0255591

Ito H, Nishibayashi M, Kawabata K, Maeda S, Seki M, Ebukuro S (2003) Induction of Fos protein in neurons in the medulla oblongata after motion- and X-irradiation-induced emesis in musk shrews (Suncus murinus). Auton Neurosci 107:1–8. https://doi.org/10.1016/s1566-0702(03)00026-2

Javid FA, Naylor RJ (1999) Variables of movement amplitude and frequency in the development of motion sickness in Suncus murinus. Pharmacol Biochem Behav 64:115–122. https://doi.org/10.1016/s0091-3057(99)00066-0

Johansson R, Kilkku P, Groenroos M (1982) A double-blind, controlled trial of nabilone vs. prochlorperazine for refractory emesis induced by cancer chemotherapy. Cancer Treat Rev 9 Suppl B:25–33. https://doi.org/10.1016/s0305-7372(82)80032-7

Johnson E, Kilgore M, Babalonis S (2022) Cannabidiol (CBD) product contamination: quantitative analysis of Δ(9)-tetrahydrocannabinol (Δ(9)-THC) concentrations found in commercially available CBD products. Drug Alcohol Depend 237:109522. https://doi.org/10.1016/j.drugalcdep.2022.109522

Johnson JR, Burnell-Nugent M, Lossignol D, Ganae-Motan ED, Potts R, Fallon MT (2010) Multicenter, double-blind, randomized, placebo-controlled, parallel-group study of the efficacy, safety, and tolerability of THC:CBD extract and THC extract in patients with intractable cancer-related pain. J Pain Symptom Manage 39:167–179. https://doi.org/10.1016/j.jpainsymman.2009.06.008

Kaji T, Saito H, Ueno S, Matsuki N (1990) Comparison of various motion stimuli on motion sickness and acquisition of adaptation in Suncus Murinus. Jikken Dobutsu 39:75–79. https://doi.org/10.1538/expanim1978.39.1_75

Kathuria S, Gaetani S, Fegley D, Valino F, Duranti A, Tontini A, Mor M, Tarzia G, La Rana G, Calignano A, Giustino A, Tattoli M, Palmery M, Cuomo V, Piomelli D (2003) Modulation of anxiety through blockade of anandamide hydrolysis. Nat Med 9:76–81. https://doi.org/10.1038/nm803

Kluin-Neleman JC, Neleman FA, Meuwissen OJ, Maes RA (1979) Delta 9-tetrahydrocannabinol (THC) as an antiemetic in patients treated with cancerchemotherapy; a double-blind cross-over trial against placebo. Vet Hum Toxicol 21:338–340

Kum V, Bell A, Fang W, VanWert E (2021) Efficacy of topical capsaicin for cannabinoid hyperemesis syndrome in a pediatric and adult emergency department. Am J Emerg Med 49:343–351. https://doi.org/10.1016/j.ajem.2021.06.049

Kwiatkowska M, Parker LA, Burton P, Mechoulam R (2004) A comparative analysis of the potential of cannabinoids and ondansetron to suppress cisplatin-induced emesis in the Suncus murinus (house musk shrew). Psychopharmacology (Berl) 174:254–259. https://doi.org/10.1007/s00213-003-1739-9

Laleh P, Yaser K, Alireza O (2019) Oleoylethanolamide: a novel pharmaceutical agent in the management of obesity-an updated review. J Cell Physiol 234:7893–7902. https://doi.org/10.1002/jcp.27913

Landsman RS, Burkey TH, Consroe P, Roeske WR, Yamamura HI (1997) SR141716A is an inverse agonist at the human cannabinoid CB1 receptor. Eur J Pharmacol 334:R1–R2. https://doi.org/10.1016/s0014-2999(97)01160-6

Lane M, Vogel CL, Ferguson J, Krasnow S, Saiers JL, Hamm J, Salva K, Wiernik PH, Holroyde CP, Hammill S (1991) Dronabinol and prochlorperazine in combination for treatment of cancer chemotherapy-induced nausea and vomiting. J Pain Symptom Manage 6:352–359

Laprairie RB, Bagher AM, Kelly ME, Denovan-Wright EM (2015) Cannabidiol is a negative allosteric modulator of the cannabinoid CB1 receptor. Br J Pharmacol 172:4790–4805. https://doi.org/10.1111/bph.13250

Lau AH, Rudd JA, Yew DT (2005) Action of ondansetron and CP-99,994 on cisplatin-induced emesis and locomotor activity in Suncus murinus (house musk shrew). Behav Pharmacol 16: 605–612. https://doi.org/10.1097/00008877-200512000-00002

Levitt M (1982) Nabilone vs. placebo in the treatment of chemotherapy-induced nausea and vomiting in cancer patients. Cancer Treat Rev 9 Suppl B:49–53. https://doi.org/10.1016/s0305-7372(82)80036-4

Limebeer CL, Abdullah RA, Rock EM, Imhof E, Wang K, Lichtman AH, Parker LA (2014) Attenuation of anticipatory nausea in a rat model of contextually elicited conditioned gaping by enhancement of the endocannabinoid system. Psychopharmacology (Berl) 231:603–612. https://doi.org/10.1007/s00213-013-3282-7

Limebeer CL, Hall G, Parker LA (2006) Exposure to a lithium-paired context elicits gaping in rats: a model of anticipatory nausea. Physiol Behav 88:398–403. https://doi.org/10.1016/j.physbeh.2006.04.014

Limebeer CL, Krohn JP, Cross-Mellor S, Litt DE, Ossenkopp KP, Parker LA (2008) Exposure to a context previously associated with nausea elicits conditioned gaping in rats: a model of anticipatory nausea. Behav Brain Res 187:33–40. https://doi.org/10.1016/j.bbr.2007.08.024

Limebeer CL, Litt DE, Parker LA (2009) Effect of 5-HT3 antagonists and a 5-HT(1A) agonist on fluoxetine-induced conditioned gaping reactions in rats. Psychopharmacology (Berl) 203:763–770. https://doi.org/10.1007/s00213-008-1421-3

Limebeer CL, Parker LA (1999) Delta-9-tetrahydrocannabinol interferes with the establishment and the expression of conditioned rejection reactions produced by cyclophosphamide: a rat model of nausea. Neuroreport 10:3769–3772

Limebeer CL, Rock EM, Mechoulam R, Parker LA (2012) The anti-nausea effects of CB1 agonists are mediated by an action at the visceral insular cortex. Br J Pharmacol 167:1126–1136. https://doi.org/10.1111/j.1476-5381.2012.02066.x

Limebeer CL, Rock EM, Puvanenthirarajah N, Niphakis MJ, Cravatt BF, Parker LA (2016) Elevation of 2-AG by monoacylglycerol lipase inhibition in the visceral insular cortex interferes with anticipatory nausea in a rat model. Behav Neurosci 130:261–266. https://doi.org/10.1037/bne0000132

Limebeer CL, Rock EM, Sharkey KA, Parker LA (2018) Nausea-induced 5-HT release in the interoceptive insular cortex and regulation by monoacylglycerol lipase (MAGL) inhibition and cannabidiol. eNeuro 5. https://doi.org/10.1523/eneuro.0256-18.2018

London SW, McCarthy LE, Borison HL (1979) Suppression of cancer chemotherapy-induced vomiting in the cat by nabilone, a synthetic cannabinoid. Proc Soc Exp Biol Med 160:437–440. https://doi.org/10.3181/00379727-160-40465

Long JZ, Li W, Booker L, Burston JJ, Kinsey SG, Schlosburg JE, Pavon FJ, Serrano AM, Selley DE, Parsons LH, Lichtman AH, Cravatt BF (2009a) Selective blockade of 2-arachidonoylglycerol hydrolysis produces cannabinoid behavioral effects. Nat Chem Biol 5: 37–44. https://doi.org/10.1038/nchembio.129

Long JZ, Nomura DK, Vann RE, Walentiny DM, Booker L, Jin X, Burston JJ, Sim-Selley LJ, Lichtman AH, Wiley JL, Cravatt BF (2009b) Dual blockade of FAAH and MAGL identifies behavioral processes regulated by endocannabinoid crosstalk in vivo. Proc Natl Acad Sci U S A 106:20270–20275. https://doi.org/10.1073/pnas.0909411106

Lucas VS Jr, Laszlo J (1980) Delta 9-tetrahydrocannabinol for refractory vomiting induced by cancer chemotherapy. JAMA 243:1241–1243

Maida V, Ennis M, Irani S, Corbo M, Dolzhykov M (2008) Adjunctive nabilone in cancer pain and symptom management: a prospective observational study using propensity scoring. J Support Oncol 6:119–124

Malik IA, Khan WA, Qazilbash M, Ata E, Butt A, Khan MA (1995) Clinical efficacy of lorazepam in prophylaxis of anticipatory, acute, and delayed nausea and vomiting induced by high doses of cisplatin. A prospective randomized trial. Am J Clin Oncol 18:170–175. https://doi.org/10.1097/00000421-199504000-00017

Mato S, Pazos A, Valdizán EM (2002) Cannabinoid receptor antagonism and inverse agonism in response to SR141716A on cAMP production in human and rat brain. Eur J Pharmacol 443:43–46. https://doi.org/10.1016/s0014-2999(02)01575-3

Matsuki N, Torii Y, Muto M, Ueno S, Saito H (1990) Emesis induced by cisplatin and its block by serotonergic (5HT3) antagonists in Suncus murinus. Eur J Pharmacol 183:1968–1969. https://doi.org/10.1016/0014-2999(90)92319-E

Matsuki N, Ueno S, Kaji T, Ishihara A, Wang CH, Saito H (1988) Emesis induced by cancer chemotherapeutic agents in the Suncus murinus: a new experimental model. Jpn J Pharmacol 48:303–306. https://doi.org/10.1254/jjp.48.303

Matsuki N, Wang CH, Okada F, Tamura M, Ikegaya Y, Lin SC, Hsu YN, Chaung LJ, Chen SJ, Saito H (1997) Male/female differences in drug-induced emesis and motion sickness in Suncus murinus. Pharmacol Biochem Behav 57:721–725. https://doi.org/10.1016/s0091-3057(96)00389-9

McCabe M, Smith FP, Macdonald JS, Woolley PV, Goldberg D, Schein PS (1988) Efficacy of tetrahydrocannabinol in patients refractory to standard antiemetic therapy. Invest New Drugs 6:243–246. https://doi.org/10.1007/bf00175407

McCarthy LE, Borison HL (1981) Antiemetic activity of N-methyllevonantradol and nabilone in cisplatin-treated cats. J Clin Pharmacol 21:30S–37S

McCarthy LE, Borison HL (1984) Cisplatin-induced vomiting eliminated by ablation of the area postrema in cats. Cancer Treat Rep 68:401–404

McLaughlin PJ, Qian L, Wood JT, Wisniecki A, Winston KM, Swezey LA, Ishiwari K, Betz AJ, Pandarinathan L, Xu W, Makriyannis A, Salamone JD (2006) Suppression of food intake and food-reinforced behavior produced by the novel CB1 receptor antagonist/inverse agonist AM 1387. Pharmacol Biochem Behav 83:396–402. https://doi.org/10.1016/j.pbb.2006.02.022

Mechoulam R, Ben-Shabat S, Hanus L, Ligumsky M, Kaminski NE, Schatz AR, Gopher A, Almog S, Martin BR, Compton DR et al (1995) Identification of an endogenous 2-monoglyceride, present in canine gut, that binds to cannabinoid receptors. Biochem Pharmacol 50:83–90. https://doi.org/10.1016/0006-2952(95)00109-d

Mechoulam R, Feigenbaum JJ, Lander N, Segal M, Järbe TU, Hiltunen AJ, Consroe P (1988) Enantiomeric cannabinoids: stereospecificity of psychotropic activity. Experientia 44:762–764. https://doi.org/10.1007/bf01959156

Mechoulam R, Hanus L (2001) The cannabinoids: an overview. Therapeutic implications in vomiting and nausea after cancer chemotherapy, in appetite promotion, in multiple sclerosis and in neuroprotection. Pain Res Manag 6:67–73. https://doi.org/10.1155/2001/183057

Meiri E, Jhangiani H, Vredenburgh JJ, Barbato LM, Carter FJ, Yang HM, Baranowski V (2007) Efficacy of dronabinol alone and in combination with ondansetron versus ondansetron alone for delayed chemotherapy-induced nausea and vomiting. Curr Med Res Opin 23:533–543. https://doi.org/10.1185/030079907x167525

Morena M, Patel S, Bains JS, Hill MN (2016) Neurobiological interactions between stress and the endocannabinoid system. Neuropsychopharmacology 41:80–102. https://doi.org/10.1038/npp.2015.166

Moreno-Sanz G (2016) Can you pass the acid test? Critical review and novel therapeutic perspectives of δ(9)-tetrahydrocannabinolic acid A. Cannabis Cannabinoid Res 1:124–130. https://doi.org/10.1089/can.2016.0008

Mori T, Shibasaki M, Abe M, Udagawa Y, Suzuki T (2012) Effects of dronabinol on morphine-induced dopamine-related behavioral effects in animals. Synapse 66:931–937. https://doi.org/10.1002/syn.21586

Morrow GR (1982) Prevalence and correlates of anticipatory nausea and vomiting in chemotherapy patients. J Natl Cancer Inst 68:585–588

Mosa ASM, Hossain AM, Lavoie BJ, Yoo I (2020) Patient-related risk factors for chemotherapy-induced nausea and vomiting: a systematic review. Front Pharmacol 11:329. https://doi.org/10.3389/fphar.2020.00329

Munro S, Thomas KL, Abu-Shaar M (1993) Molecular characterization of a peripheral receptor for cannabinoids. Nature 365:61–65. https://doi.org/10.1038/365061a0

Nachnani R, Raup-Konsavage WM, Vrana KE (2021) The pharmacological case for cannabigerol. J Pharmacol Exp Ther 376:204–212. https://doi.org/10.1124/jpet.120.000340

Naitoh T, Wassersug RJ, Yamashita N (2001) Factors influencing the susceptibility of anurans to motion sickness. J Comp Physiol A 187:105–113. https://doi.org/10.1007/s003590100181

Nakayama H, Yamakuni H, Higaki M, Ishikawa H, Imazumi K, Matsuo M, Mutoh S (2005) Antiemetic activity of FK1052, a 5-HT3- and 5-HT4-receptor antagonist, in Suncus murinus and ferrets. J Pharmacol Sci 98:396–403. https://doi.org/10.1254/jphs.fpj05001x

Napadow V, Sheehan JD, Kim J, Lacount LT, Park K, Kaptchuk TJ, Rosen BR, Kuo B (2013) The brain circuitry underlying the temporal evolution of nausea in humans. Cereb Cortex 23:806–813. https://doi.org/10.1093/cercor/bhs073

National Academies of Sciences E, and Medicine, Division HaM, Practice BoPHaPH, Agenda CotHEoMAERaR (2017) The health effects of cannabis and cannabinoids: the current state of evidence and recommendations for research. The National Academics Press, Washington

Navari RM, Aapro M (2016) Antiemetic prophylaxis for chemotherapy-induced nausea and vomiting. N Engl J Med 374:1356–1367. https://doi.org/10.1056/NEJMra1515442

Navarro G, Varani K, Lillo A, Vincenzi F, Rivas-Santisteban R, Raïch I, Reyes-Resina I, Ferreiro-Vera C, Borea PA, Sánchez de Medina V, Nadal X, Franco R (2020) Pharmacological data of cannabidiol- and cannabigerol-type phytocannabinoids acting on cannabinoid CB(1), CB(2) and CB(1)/CB(2) heteromer receptors. Pharmacol Res 159:104940. https://doi.org/10.1016/j.phrs.2020.104940

Neidhart JA, Gagen MM, Wilson HE, Young DC (1981) Comparative trial of the antiemetic effects of THC and haloperidol. J Clin Pharmacol 21:38s–42s. https://doi.org/10.1002/j.1552-4604.1981.tb02571.x

Niamatali C, Fallon SD, Egan EL (1984) Nabilone in the management of prochlorperazine resistant cancer chemotherapy induced emesis. Ir Med J 77:276–277

Niederle N, Schütte J, Schmidt CG (1986) Crossover comparison of the antiemetic efficacy of nabilone and alizapride in patients with nonseminomatous testicular cancer receiving cisplatin therapy. Klin Wochenschr 64:362–365. https://doi.org/10.1007/bf01728184

Niiranen A, Mattson K (1985) A cross-over comparison of nabilone and prochlorperazine for emesis induced by cancer chemotherapy. Am J Clin Oncol 8:336–340. https://doi.org/10.1097/00000421-198508000-00013

Noyes R Jr, Brunk SF, Avery DA, Canter AC (1975) The analgesic properties of delta-9-tetrahydrocannabinol and codeine. Clin Pharmacol Ther 18:84–89. https://doi.org/10.1002/cpt197518184

O'Brien LD, Limebeer CL, Rock EM, Bottegoni G, Piomelli D, Parker LA (2013) Anandamide transport inhibition by ARN272 attenuates nausea-induced behaviour in rats, and vomiting in shrews (Suncus murinus). Br J Pharmacol 170:1130–1136. https://doi.org/10.1111/bph.12360

Ofir R, Bar-Sela G, Weyl Ben-Arush M, Postovsky S (2019) Medical marijuana use for pediatric oncology patients: single institution experience. Pediatr Hematol Oncol 36:255–266. https://doi.org/10.1080/08880018.2019.1630537

Orr LE, McKernan JF (1981) Antiemetic effect of delta 9-tetrahydrocannabinol in chemotherapy-associated nausea and emesis as compared to placebo and compazine. J Clin Pharmacol 21:76S–80S. https://doi.org/10.1002/j.1552-4604.1981.tb02578.x

Orr LE, McKernan JF, Bloome B (1980) Antiemetic effect of tetrahydrocannabinol. Compared with placebo and prochlorperazine in chemotherapy-associated nausea and emesis. Arch Intern Med 140:1431–1433

Ossenkopp KP, Eckel LA (1995) Toxin-induced conditioned changes in taste reactivity and the role of the chemosensitive area postrema. Neurosci Biobehav Rev 19:99–108. https://doi.org/10.1016/0149-7634(94)00024-u

Parker LA (2014) Conditioned flavor avoidance and conditioned gaping: rat models of conditioned nausea. Eur J Pharmacol 722:122–133. https://doi.org/10.1016/j.ejphar.2013.09.070

Parker LA, Kwiatkowska M, Burton P, Mechoulam R (2004) Effect of cannabinoids on lithium-induced vomiting in the Suncus murinus (house musk shrew). Psychopharmacology (Berl) 171: 156–161. https://doi.org/10.1007/s00213-003-1571-2

Parker LA, Limebeer CL, Rock EM, Litt DL, Kwiatkowska M, Piomelli D (2009) The FAAH inhibitor URB-597 interferes with cisplatin- and nicotine-induced vomiting in the Suncus murinus (house musk shrew). Physiol Behav 97:121–124. https://doi.org/10.1016/j.physbeh.2009.02.014

Parker LA, Limebeer CL, Rock EM, Sticht MA, Ward J, Turvey G, Benchama O, Rajarshi G, Wood JT, Alapafuja SO, Makriyannis A (2016) A comparison of novel, selective fatty acid amide hydrolase (FAAH), monoacyglycerol lipase (MAGL) or dual FAAH/MAGL inhibitors to suppress acute and anticipatory nausea in rat models. Psychopharmacology (Berl) 233:2265–2275. https://doi.org/10.1007/s00213-016-4277-y

Parker LA, Mechoulam R (2003) Cannabinoid agonists and antagonists modulate lithium-induced conditioned gaping in rats. Integr Physiol Behav Sci 38:133–145. https://doi.org/10.1007/BF02688831

Parker LA, Mechoulam R, Schlievert C (2002) Cannabidiol, a non-psychoactive component of cannabis and its synthetic dimethylheptyl homolog suppress nausea in an experimental model with rats. Neuroreport 13:567–570. https://doi.org/10.1097/00001756-200204160-00006

Parker LA, Mechoulam R, Schlievert C, Abbott L, Fudge ML, Burton P (2003) Effects of cannabinoids on lithium-induced conditioned rejection reactions in a rat model of nausea. Psychopharmacology (Berl) 166:156–162. https://doi.org/10.1007/s00213-002-1329-2

Parker LA, Niphakis MJ, Downey R, Limebeer CL, Rock EM, Sticht MA, Morris H, Abdullah RA, Lichtman AH, Cravatt BF (2015) Effect of selective inhibition of monoacylglycerol lipase (MAGL) on acute nausea, anticipatory nausea, and vomiting in rats and Suncus murinus. Psychopharmacology (Berl) 232:583–593. https://doi.org/10.1007/s00213-014-3696-x

Patel N, Kommineni N, Surapaneni SK, Kalvala A, Yaun X, Gebeyehu A, Arthur P, Duke LC, York SB, Bagde A, Meckes DG Jr, Singh M (2021) Cannabidiol loaded extracellular vesicles sensitize triple-negative breast cancer to doxorubicin in both in-vitro and in vivo models. Int J Pharm 607:120943. https://doi.org/10.1016/j.ijpharm.2021.120943

Percie du Sert N, Ho WS, Rudd JA, Andrews PL (2010) Cannabinoid-induced reduction in antral pacemaker frequency: a telemetric study in the ferret. Neurogastroenterol Motil 22:1257–1266., e324. https://doi.org/10.1111/j.1365-2982.2010.01581.x

Percie du Sert N, Rudd JA, Apfel CC, Andrews PL (2011) Cisplatin-induced emesis: systematic review and meta-analysis of the ferret model and the effects of 5-HT receptor antagonists. Cancer Chemother Pharmacol 67:667–686. https://doi.org/10.1007/s00280-010-1339-4

Pertwee RG (2008) The diverse CB1 and CB2 receptor pharmacology of three plant cannabinoids: delta9-tetrahydrocannabinol, cannabidiol and delta9-tetrahydrocannabivarin. Br J Pharmacol 153:199–215. https://doi.org/10.1038/sj.bjp.0707442

Pertwee RG, Rock EM, Guenther K, Limebeer CL, Stevenson LA, Haj C, Smoum R, Parker LA, Mechoulam R (2018) Cannabidiolic acid methyl ester, a stable synthetic analogue of cannabidiolic acid, can produce 5-HT(1A) receptor-mediated suppression of nausea and anxiety in rats. Br J Pharmacol 175:100–112. https://doi.org/10.1111/bph.14073

Polito S, MacDonald T, Romanick M, Jupp J, Wiernikowski J, Vennettilli A, Khanna M, Patel P, Ning W, Sung L, Dupuis LL (2018) Safety and efficacy of nabilone for acute chemotherapy-induced vomiting prophylaxis in pediatric patients: a multicenter, retrospective review. Pediatr Blood Cancer 65:e27374. https://doi.org/10.1002/pbc.27374

Pomeroy M, Fennelly JJ, Towers M (1986) Prospective randomized double-blind trial of nabilone versus domperidone in the treatment of cytotoxic-induced emesis. Cancer Chemother Pharmacol 17:285–288. https://doi.org/10.1007/BF00256701

Prieto González JM, Vila Silván C (2021) Safety and tolerability of nabiximols oromucosal spray: a review of more than 15 years' accumulated evidence from clinical trials. Expert Rev Neurother 21:755–778. https://doi.org/10.1080/14737175.2021.1935879

Ray AP, Griggs L, Darmani NA (2009) Delta 9-tetrahydrocannabinol suppresses vomiting behavior and Fos expression in both acute and delayed phases of cisplatin-induced emesis in the least shrew. Behav Brain Res 196:30–36. https://doi.org/10.1016/j.bbr.2008.07.028

Razavi D, Delvaux N, Farvacques C, De Brier F, Van Heer C, Kaufman L, Derde MP, Beauduin M, Piccart M (1993) Prevention of adjustment disorders and anticipatory nausea secondary to adjuvant chemotherapy: a double-blind, placebo-controlled study assessing the usefulness of alprazolam. J Clin Oncol 11:1384–1390

Richards JR (2017) Cannabinoid hyperemesis syndrome: a disorder of the HPA axis and sympathetic nervous system? Med Hypotheses 103:90–95. https://doi.org/10.1016/j.mehy.2017.04.018

Riedel G, Fadda P, McKillop-Smith S, Pertwee RG, Platt B, Robinson L (2009) Synthetic and plant-derived cannabinoid receptor antagonists show hypophagic properties in fasted and non-fasted mice. Br J Pharmacol 156:1154–1166. https://doi.org/10.1111/j.1476-5381.2008.00107.x

Rocha JF, Santos A, Gama H, Moser P, Falcão A, Pressman P, Wallace Hayes A, Soares-da-Silva P (2021) Safety, tolerability, and pharmacokinetics of FAAH inhibitor BIA 10-2474: a double-blind, randomized, Placebo-controlled study in healthy volunteers. Clin Pharmacol Ther. https://doi.org/10.1002/cpt.2290

Rock EM, Ayoub SM, Limebeer CL, Gene A, Wills KL, DeVuono MV, Smoum R, Di Marzo V, Lichtman AH, Mechoulam R, Parker LA (2020a) Acute naloxone-precipitated morphine withdrawal elicits nausea-like somatic behaviors in rats in a manner suppressed by N-oleoylglycine. Psychopharmacology (Berl) 237:375–384. https://doi.org/10.1007/s00213-019-05373-2

Rock EM, Bolognini D, Limebeer CL, Cascio MG, Anavi-Goffer S, Fletcher PJ, Mechoulam R, Pertwee RG, Parker LA (2012) Cannabidiol, a non-psychotropic component of cannabis, attenuates vomiting and nausea-like behaviour via indirect agonism of 5-HT(1A) somatodendritic autoreceptors in the dorsal raphe nucleus. Br J Pharmacol 165:2620–2634. https://doi.org/10.1111/j.1476-5381.2011.01621.x

Rock EM, Boulet N, Limebeer CL, Mechoulam R, Parker LA (2016a) Cannabinoid 2 (CB2) receptor agonism reduces lithium chloride-induced vomiting in Suncus murinus and nausea-induced conditioned gaping in rats. Eur J Pharmacol 786:94–99. https://doi.org/10.1016/j.ejphar.2016.06.001

Rock EM, Connolly C, Limebeer CL, Parker LA (2016b) Effect of combined oral doses of Δ(9)-tetrahydrocannabinol (THC) and cannabidiolic acid (CBDA) on acute and anticipatory nausea in rat models. Psychopharmacology (Berl) 233:3353–3360. https://doi.org/10.1007/s00213-016-4378-7

Rock EM, Goodwin JM, Limebeer CL, Breuer A, Pertwee RG, Mechoulam R, Parker LA (2011) Interaction between non-psychotropic cannabinoids in marihuana: effect of cannabigerol (CBG) on the anti-nausea or anti-emetic effects of cannabidiol (CBD) in rats and shrews. Psychopharmacology (Berl) 215:505–512. https://doi.org/10.1007/s00213-010-2157-4

Rock EM, Kopstick RL, Limebeer CL, Parker LA (2013a) Tetrahydrocannabinolic acid reduces nausea-induced conditioned gaping in rats and vomiting in Suncus murinus. Br J Pharmacol 170:641–648. https://doi.org/10.1111/bph.12316

Rock EM, Limebeer CL, Aliasi-Sinai L, Parker LA (2019) The ventral pallidum as a critical region for fatty acid amide hydrolase inhibition of nausea-induced conditioned gaping in male

Sprague-Dawley rats. Neuropharmacology 155:142–149. https://doi.org/10.1016/j.neuropharm.2019.05.031
Rock EM, Limebeer CL, Mechoulam R, Piomelli D, Parker LA (2008) The effect of cannabidiol and URB597 on conditioned gaping (a model of nausea) elicited by a lithium-paired context in the rat. Psychopharmacology (Berl) 196:389–395. https://doi.org/10.1007/s00213-007-0970-1
Rock EM, Limebeer CL, Navaratnam R, Sticht MA, Bonner N, Engeland K, Downey R, Morris H, Jackson M, Parker LA (2014) A comparison of cannabidiolic acid with other treatments for anticipatory nausea using a rat model of contextually elicited conditioned gaping. Psychopharmacology (Berl). https://doi.org/10.1007/s00213-014-3498-1
Rock EM, Limebeer CL, Parker LA (2015a) Effect of combined doses of Δ(9)-tetrahydrocannabinol (THC) and cannabidiolic acid (CBDA) on acute and anticipatory nausea using rat (Sprague-Dawley) models of conditioned gaping. Psychopharmacology (Berl) 232:4445–4454. https://doi.org/10.1007/s00213-015-4080-1
Rock EM, Limebeer CL, Smoum R, Mechoulam R, Parker LA (2021) Effect of oleoyl glycine and oleoyl alanine on lithium chloride induced nausea in rats and vomiting in shrews. Psychopharmacology (Berl). https://doi.org/10.1007/s00213-021-06005-4
Rock EM, Limebeer CL, Smoum R, Mechoulam R, Parker LA (2022) Evaluation of sex differences in the potential of Δ(9)-tetrahydrocannabinol, cannabidiol, cannabidiolic acid, and oleoyl alanine to reduce nausea-induced conditioned gaping reactions in Sprague-Dawley rats. Cannabis Cannabinoid Res. https://doi.org/10.1089/can.2022.0158
Rock EM, Limebeer CL, Ward JM, Cohen A, Grove K, Niphakis MJ, Cravatt BF, Parker LA (2015b) Interference with acute nausea and anticipatory nausea in rats by fatty acid amide hydrolase (FAAH) inhibition through a PPARα and CB1 receptor mechanism, respectively: a double dissociation. Psychopharmacology (Berl) 232:3841–3848. https://doi.org/10.1007/s00213-015-4050-7
Rock EM, Moreno-Sanz G, Limebeer CL, Petrie GN, Angelini R, Piomelli D, Parker LA (2017) Suppression of acute and anticipatory nausea by peripherally restricted fatty acid amide hydrolase inhibitor in animal models: role of PPARα and CB(1) receptors. Br J Pharmacol 174:3837–3847. https://doi.org/10.1111/bph.13980
Rock EM, Parker LA (2013a) Effect of low doses of cannabidiolic acid and ondansetron on LiCl-induced conditioned gaping (a model of nausea-induced behaviour) in rats. Br J Pharmacol 169:685–692. https://doi.org/10.1111/bph.12162
Rock EM, Parker LA (2013b) Suppression of lithium chloride-induced conditioned gaping (a model of nausea-induced behaviour) in rats (using the taste reactivity test) with metoclopramide is enhanced by cannabidiolic acid. Pharmacol Biochem Behav 111:84–89. https://doi.org/10.1016/j.pbb.2013.08.012
Rock EM, Parker LA (2015) Synergy between cannabidiol, cannabidiolic acid, and Delta(9)-tetrahydrocannabinol in the regulation of emesis in the Suncus murinus (house musk shrew). Behav Neurosci 129:368–370. https://doi.org/10.1037/bne0000057
Rock EM, Sticht MA, Duncan M, Stott C, Parker LA (2013b) Evaluation of the potential of the phytocannabinoids, cannabidivarin (CBDV) and Delta(9) -tetrahydrocannabivarin (THCV), to produce CB1 receptor inverse agonism symptoms of nausea in rats. Br J Pharmacol 170:671–678. https://doi.org/10.1111/bph.12322
Rock EM, Sullivan MT, Collins SA, Goodman H, Limebeer CL, Mechoulam R, Parker LA (2020b) Evaluation of repeated or acute treatment with cannabidiol (CBD), cannabidiolic acid (CBDA) or CBDA methyl ester (HU-580) on nausea and/or vomiting in rats and shrews. Psychopharmacology (Berl) 237:2621–2631. https://doi.org/10.1007/s00213-020-05559-z
Rock EM, Sullivan MT, Pravato S, Pratt M, Limebeer CL, Parker LA (2020c) Effect of combined doses of Δ(9)-tetrahydrocannabinol and cannabidiol or tetrahydrocannabinolic acid and cannabidiolic acid on acute nausea in male Sprague-Dawley rats. Psychopharmacology (Berl) 237:901–914. https://doi.org/10.1007/s00213-019-05428-4
Roscoe JA, Morrow GR, Aapro MS, Molassiotis A, Olver I (2011) Anticipatory nausea and vomiting. Support Care Cancer 19:1533–1538. https://doi.org/10.1007/s00520-010-0980-0

Royet JP, Meunier D, Torquet N, Mouly AM, Jiang T (2016) The neural bases of disgust for cheese: an fMRI study. Front Hum Neurosci 10:511. https://doi.org/10.3389/fnhum.2016.00511
Russo EB, Burnett A, Hall B, Parker KK (2005) Agonistic properties of cannabidiol at 5-HT1a receptors. Neurochem Res 30:1037–1043. https://doi.org/10.1007/s11064-005-6978-1
Russo EB, Cuttler C, Cooper ZD, Stueber A, Whiteley VL, Sexton M (2022a) Survey of patients employing cannabigerol-predominant cannabis preparations: perceived medical effects, adverse events, and withdrawal symptoms. Cannabis Cannabinoid Res 7:706–716. https://doi.org/10.1089/can.2021.0058
Russo EB, Spooner C, May L, Leslie R, Whiteley VL (2022b) Cannabinoid hyperemesis syndrome survey and genomic investigation. Cannabis Cannabinoid Res 7:336–344. https://doi.org/10.1089/can.2021.0046
Sadhasivam S, Zhang X, Chidambaran V, Mavi J, Pilipenko V, Mersha TB, Meller J, Kaufman KM, Martin LJ, McAuliffe J (2015) Novel associations between FAAH genetic variants and postoperative central opioid-related adverse effects. Pharmacogenomics J 15:436–442. https://doi.org/10.1038/tpj.2014.79
Sallan SE, Zinberg NE, Frei E 3rd (1975) Antiemetic effect of delta-9-tetrahydrocannabinol in patients receiving cancer chemotherapy. N Engl J Med 293:795–797. https://doi.org/10.1056/NEJM197510162931603
Sam TS, Cheng JT, Johnston KD, Kan KK, Ngan MP, Rudd JA, Wai MK, Yeung JH (2003) Action of 5-HT3 receptor antagonists and dexamethasone to modify cisplatin-induced emesis in Suncus murinus (house musk shrew). Eur J Pharmacol 472:135–145. https://doi.org/10.1016/s0014-2999(03)01863-6
Sanger GJ, Andrews PL (2006) Treatment of nausea and vomiting: gaps in our knowledge. Auton Neurosci 129:3–16. https://doi.org/10.1016/j.autneu.2006.07.009
Schelling G, Hauer D, Azad SC, Schmoelz M, Chouker A, Schmidt M, Hornuss C, Rippberger M, Briegel J, Thiel M, Vogeser M (2006) Effects of general anesthesia on anandamide blood levels in humans. Anesthesiology 104:273–277. https://doi.org/10.1097/00000542-200602000-00012
Schmidt ME, Liebowitz MR, Stein MB, Grunfeld J, Van Hove I, Simmons WK, Van Der Ark P, Palmer JA, Saad ZS, Pemberton DJ, Van Nueten L, Drevets WC (2021) The effects of inhibition of fatty acid amide hydrolase (FAAH) by JNJ-42165279 in social anxiety disorder: a double-blind, randomized, placebo-controlled proof-of-concept study. Neuropsychopharmacology 46:1004–1010. https://doi.org/10.1038/s41386-020-00888-1
Sharkey KA, Cristino L, Oland LD, Van Sickle MD, Starowicz K, Pittman QJ, Guglielmotti V, Davison JS, Di Marzo V (2007) Arvanil, anandamide and N-arachidonoyl-dopamine (NADA) inhibit emesis through cannabinoid CB1 and vanilloid TRPV1 receptors in the ferret. Eur J Neurosci 25:2773–2782. https://doi.org/10.1111/j.1460-9568.2007.05521.x
Sink KS, McLaughlin PJ, Wood JA, Brown C, Fan P, Vemuri VK, Peng Y, Olszewska T, Thakur GA, Makriyannis A, Parker LA, Salamone JD (2008) The novel cannabinoid CB1 receptor neutral antagonist AM4113 suppresses food intake and food-reinforced behavior but does not induce signs of nausea in rats. Neuropsychopharmacology 33:946–955. https://doi.org/10.1038/sj.npp.1301476
Steele N, Gralla RJ, Braun DW Jr, Young CW (1980) Double-blind comparison of the antiemetic effects of nabilone and prochlorperazine on chemotherapy-induced emesis. Cancer Treat Rep 64:219–224
Sticht MA, Limebeer CL, Rafla BR, Abdullah RA, Poklis JL, Ho W, Niphakis MJ, Cravatt BF, Sharkey KA, Lichtman AH, Parker LA (2016) Endocannabinoid regulation of nausea is mediated by 2-arachidonoylglycerol (2-AG) in the rat visceral insular cortex. Neuropharmacology 102:92–102. https://doi.org/10.1016/j.neuropharm.2015.10.039
Sticht MA, Rock EM, Parker LA (2013) 2-Arachidonoylglycerol interferes with lithium-induced vomiting in the house musk shrew, Suncus murinus. Physiol Behav 120:228–232. https://doi.org/10.1016/j.physbeh.2013.08.015

Strewe C, Feuerecker M, Nichiporuk I, Kaufmann I, Hauer D, Morukov B, Schelling G, Chouker A (2012) Effects of parabolic flight and spaceflight on the endocannabinoid system in humans. Rev Neurosci 23:673–680. https://doi.org/10.1515/revneuro-2012-0057

Sukpiriyagul A, Chartchaiyarerk R, Tabtipwon P, Smanchat B, Prommas S, Bhamarapravatana K, Suwannarurk K (2023) Oral tetrahydrocannabinol (THC):cannabinoid (CBD) cannabis extract adjuvant for reducing chemotherapy-induced nausea and vomiting (CINV): a randomized, double-blinded, placebo-controlled, crossover trial. Int J Womens Health 15:1345–1352. https://doi.org/10.2147/ijwh.s401938

Sun Y, Alexander SP, Garle MJ, Gibson CL, Hewitt K, Murphy SP, Kendall DA, Bennett AJ (2007) Cannabinoid activation of PPAR alpha; a novel neuroprotective mechanism. Br J Pharmacol 152:734–743. https://doi.org/10.1038/sj.bjp.0707478

Suárez J, Romero-Zerbo SY, Rivera P, Bermúdez-Silva FJ, Pérez J, De Fonseca FR, Fernández-Llebrez P (2010) Endocannabinoid system in the adult rat circumventricular areas: an immunohistochemical study. J Comp Neurol 518:3065–3085. https://doi.org/10.1002/cne.22382

Sweet DL, Miller NJ, Weddington W, Senay E, Sushelsky L (1981) delta 9-Tetrahydrocannabinol as an antiemetic for patients receiving cancer chemotherapy. A pilot study. J Clin Pharmacol 21: 70S–75S. https://doi.org/10.1002/j.1552-4604.1981.tb02577.x

Szaflarski JP, Bebin EM, Comi AM, Patel AD, Joshi C, Checketts D, Beal JC, Laux LC, De Boer LM, Wong MH, Lopez M, Devinsky O, Lyons PD, Zentil PP, Wechsler R (2018) Long-term safety and treatment effects of cannabidiol in children and adults with treatment-resistant epilepsies: Expanded access program results. Epilepsia 59:1540–1548. https://doi.org/10.1111/epi.14477

Tait RJ, Caldicott D, Mountain D, Hill SL, Lenton S (2016) A systematic review of adverse events arising from the use of synthetic cannabinoids and their associated treatment. Clin Toxicol (Phila) 54:1–13. https://doi.org/10.3109/15563650.2015.1110590

Takeda S, Okajima S, Miyoshi H, Yoshida K, Okamoto Y, Okada T, Amamoto T, Watanabe K, Omiecinski CJ, Aramaki H (2012) Cannabidiolic acid, a major cannabinoid in fiber-type cannabis, is an inhibitor of MDA-MB-231 breast cancer cell migration. Toxicol Lett 214: 314–319. https://doi.org/10.1016/j.toxlet.2012.08.029

Takizawa M, Hatta T, Iitsuka H, Katashima M, Sato Y, Kuroishi K, Nagashima H (2020) Safety, tolerability, pharmacokinetics, and pharmacodynamics of ASP3652, a reversible fatty acid amide hydrolase inhibitor, in healthy, nonelderly, Japanese men and elderly, Japanese men and women: a randomized, double-blind, placebo-controlled, single and multiple oral dose, phase I study. Clin Ther 42:906–923. https://doi.org/10.1016/j.clinthera.2020.03.021

Thomas A, Baillie GL, Phillips AM, Razdan RK, Ross RA, Pertwee RG (2007) Cannabidiol displays unexpectedly high potency as an antagonist of CB1 and CB2 receptor agonists in vitro. Br J Pharmacol 150:613–623. https://doi.org/10.1038/sj.bjp.0707133

Thomas A, Stevenson LA, Wease KN, Price MR, Baillie G, Ross RA, Pertwee RG (2005) Evidence that the plant cannabinoid Delta9-tetrahydrocannabivarin is a cannabinoid CB1 and CB2 receptor antagonist. Br J Pharmacol 146:917–926. https://doi.org/10.1038/sj.bjp.0706414

Torii Y, Saito H, Matsuki N (1991) Selective blockade of cytotoxic drug-induced emesis by 5-HT3 receptor antagonists in Suncus murinus. Jpn J Pharmacol 55:107–113. https://doi.org/10.1254/jjp.55.107

Tramer MR, Carroll D, Campbell FA, Reynolds DJ, Moore RA, McQuay HJ (2001) Cannabinoids for control of chemotherapy induced nausea and vomiting: quantitative systematic review. BMJ 323:16–21. https://doi.org/10.1136/bmj.323.7303.16

Tuerke KJ, Leri F, Parker LA (2009) Antidepressant-like effects of paroxetine are produced by lower doses than those which produce nausea. Pharmacol Biochem Behav 93:190–195. https://doi.org/10.1016/j.pbb.2009.05.011

Tuerke KJ, Limebeer CL, Fletcher PJ, Parker LA (2012) Double dissociation between regulation of conditioned disgust and taste avoidance by serotonin availability at the 5-HT3 receptor in the posterior and anterior insular cortex. J Neurosci 32:13709–13717. https://doi.org/10.1523/JNEUROSCI.2042-12.2012

Ueno S, Matsuki N, Saito H (1988) Suncus murinus as a new experimental model for motion sickness. Life Sci 43:413–420. https://doi.org/10.1016/0024-3205(88)90520-6

van Esbroeck ACM, Janssen APA, Cognetta AB 3rd, Ogasawara D, Shpak G, van der Kroeg M, Kantae V, Baggelaar MP, de Vrij FMS, Deng H, Allarà M, Fezza F, Lin Z, van der Wel T, Soethoudt M, Mock ED, den Dulk H, Baak IL, Florea BI, Hendriks G, De Petrocellis L, Overkleeft HS, Hankemeier T, De Zeeuw CI, Di Marzo V, Maccarrone M, Cravatt BF, Kushner SA, van der Stelt M (2017) Activity-based protein profiling reveals off-target proteins of the FAAH inhibitor BIA 10-2474. Science 356:1084–1087. https://doi.org/10.1126/science.aaf7497

Van Sickle MD, Duncan M, Kingsley PJ, Mouihate A, Urbani P, Mackie K, Stella N, Makriyannis A, Piomelli D, Davison JS, Marnett LJ, Di Marzo V, Pittman QJ, Patel KD, Sharkey KA (2005) Identification and functional characterization of brainstem cannabinoid CB2 receptors. Science 310:329–332. https://doi.org/10.1126/science.1115740

Van Sickle MD, Oland LD, Ho W, Hillard CJ, Mackie K, Davison JS, Sharkey KA (2001) Cannabinoids inhibit emesis through CB1 receptors in the brainstem of the ferret. Gastroenterology 121:767–774. https://doi.org/10.1053/gast.2001.28466

Van Sickle MD, Oland LD, Mackie K, Davison JS, Sharkey KA (2003) Delta9-tetrahydrocannabinol selectively acts on CB1 receptors in specific regions of dorsal vagal complex to inhibit emesis in ferrets. Am J Physiol Gastrointest Liver Physiol 285:G566–G576. https://doi.org/10.1152/ajpgi.00113.2003

Vaziri ND, Thomas R, Sterling M, Seiff K, Pahl MV, Davila J, Wilson A (1981) Toxicity with intravenous injection of crude marijuana extract. Clin Toxicol 18:353–366. https://doi.org/10.3109/15563658108990042

Venkatesan T, Zadvornova Y, Raff H, Hillard CJ (2016) Endocannabinoid-related lipids are increased during an episode of cyclic vomiting syndrome. Neurogastroenterol Motil 28:1409–1418. https://doi.org/10.1111/nmo.12843

Vinciguerra V, Moore T, Brennan E (1988) Inhalation marijuana as an antiemetic for cancer chemotherapy. N Y State J Med 88:525–527

Wada JK, Bogdon DL, Gunnell JC, Hum GJ, Gota CH, Rieth TE (1982) Double-blind, randomized, crossover trial of nabilone vs. placebo in cancer chemotherapy. Cancer Treat Rev 9 Suppl B:39–44. https://doi.org/10.1016/s0305-7372(82)80034-0

Wallace D, Martin A-L, Park B (2007) Cannabinoid hyperemesis: marijuana puts patients in hot water. Australas Psychiatry 15:156–158. https://doi.org/10.1080/10398560701196778

Wang Y, Ray AP, McClanahan BA, Darmani NA (2009) The antiemetic interaction of Delta9-tetrahydrocannabinol when combined with tropisetron or dexamethasone in the least shrew. Pharmacol Biochem Behav 91:367–373. https://doi.org/10.1016/j.pbb.2008.08.008

Weber K, Häcker R, Hardisty JF, Harris SB, Hayes AW (2020) Oral repeated-dose toxicity studies of BIA 10-2474 in cynomolgus monkeys. Regul Toxicol Pharmacol 111:104547. https://doi.org/10.1016/j.yrtph.2019.104547

Wooldridge LM, Ji L, Liu Y, Nikas SP, Makriyannis A, Bergman J, Kangas BD (2020) Antiemetic effects of cannabinoid agonists in nonhuman primates. J Pharmacol Exp Ther 374:462–468. https://doi.org/10.1124/jpet.120.265710

Zagzoog A, Mohamed KA, Kim HJJ, Kim ED, Frank CS, Black T, Jadhav PD, Holbrook LA, Laprairie RB (2020) In vitro and in vivo pharmacological activity of minor cannabinoids isolated from Cannabis sativa. Sci Rep 10:20405. https://doi.org/10.1038/s41598-020-77175-y

Zheng Y, Wang XL, Mo FF, Li M (2014) Dexamethasone alleviates motion sickness in rats in part by enhancing the endocannabinoid system. Eur J Pharmacol 727:99–105. https://doi.org/10.1016/j.ejphar.2014.01.047

Zhong W, Shahbaz O, Teskey G, Beever A, Kachour N, Venketaraman V, Darmani NA (2021) Mechanisms of nausea and vomiting: current knowledge and recent advances in intracellular emetic signaling systems. Int J Mol Sci 22. https://doi.org/10.3390/ijms22115797

Delta-9-tetrahydrocannabinol and Cannabidiol for Pain: Preclinical and Clinical Models

H. M. Harris, C. F. Moore, B. W. Jenkins, M. F. Bedillion, E. M. Weerts, and C. A. Arout

Contents

1 Introduction 390
1.1 Neurophysiology of Pain and Inflammation 391
1.2 The Role of the Endocannabinoid System in Pain and Inflammation 392
2 Preclinical Studies of Cannabinoids' Effects on Pain 394
2.1 Preclinical Pain Assessments 395
2.2 Preclinical Models of Pain States 397
2.3 Cannabinoid Efficacy on Nociception (Acute Pain) 399
2.4 Cannabinoid Efficacy on Nociceptive Pain and Inflammation 401
2.5 Cannabinoid Efficacy on Neuropathic Pain 403
3 Clinical Studies of Cannabinoids' Effects on Pain 405
3.1 Human Experimental Pain Assessments 405
3.2 Human Laboratory Studies of Cannabinoid Analgesia 409
3.3 Randomized Controlled Trials in Pain Populations 413
4 Discussion/Additional Considerations 416
4.1 Intoxication and Abuse Liability 416
4.2 Cannabinoids in Conjunction with Opioids 417
4.3 Minor Cannabinoids and Terpenes 417
4.4 Tolerance and Withdrawal 418
4.5 Peri-Operative Issues 418
5 Concluding Remarks 419
References 419

Abstract Cannabinoids are increasingly being used to manage pain resulting from a variety of conditions. Both preclinical animal models and human studies have played

Author Moore C. F. shared first-author

H. M. Harris, M. F. Bedillion, and C. A. Arout (✉)
Department of Psychiatry, Division on Substance Use Disorders, Columbia University Irving Medical Center and New York State Psychiatric Institute, New York, NY, USA
e-mail: caroline.arout@nyspi.columbia.edu

C. F. Moore, B. W. Jenkins, and E. M. Weerts
Division of Behavioral Biology, Department of Psychiatry and Behavioral Sciences, Johns Hopkins University School of Medicine, Baltimore, MD, USA

Curr Topics Behav Neurosci (2026) 76: 389–431
https://doi.org/10.1007/7854_2025_604
Published Online: 29 August 2025

a crucial role in advancing our knowledge of cannabinoids, their involvement in pain mechanisms, and their potential utility as novel analgesics. This chapter first reviews basic pain neurobiology and the most common experimental pain paradigms, which provide a basis for our discussion of preclinical, human laboratory, and clinical research characterizing the effectiveness of cannabinoids for managing pain. While a substantial body of literature exists describing these effects, findings are complex and largely mixed, dependent on the cannabinoid administered, route of administration, and pain modality/syndrome tested. Herein, we highlight the need for more rigorous, placebo-controlled research defining the therapeutic efficacy of cannabinoids. The chapter concludes by emphasizing the need for further investigation of other cannabis constituents (e.g., minor cannabinoids and terpenes), potential interactions between cannabinoids and other analgesic medications, as well as other emerging issues in the intersection between cannabinoids and pain management.

Keywords Cannabidiol · Cannabis · Delta-9-tetrahydrocannabinol · Pain

1 Introduction

Nearly 21% of the population (approximately 52 million people) report suffering from chronic pain, with an approximate 7% subset reporting high-impact, debilitating chronic pain (Rikard 2023). Current FDA-approved standard analgesic drugs include mu-opioid agonists (e.g., morphine), anti-inflammatory steroids (e.g., cortisone), and nonsteroidal anti-inflammatory drugs (NSAIDs; e.g., aspirin, ibuprofen), which are not effective for all patients and can lead to other serious health issues. For example, NSAIDs can cause severe gastrointestinal bleeding and ulcers (Tai and McAlindon 2021), and opioids have significant risks for dependence and overdose and generally are not effective for chronic pain (Phillips et al. 2017). Thus, patients with pain syndromes are increasingly seeking alternative analgesic medications.

With widespread legalization of cannabis for recreational and medical purposes and decreasing stigma surrounding its use (Farrelly et al. 2023), cannabis use continues to rise globally (Wang et al. 2024). The prevalence of medical cannabis use has become increasingly ubiquitous, particularly in countries and territories/US states with legalized medical cannabis (Rhee and Rosenheck 2023), where 42–64% of patients currently report pain management as their primary indication for use (Azcarate et al. 2020; Kosiba et al. 2019; Sexton et al. 2016). Further, the majority (62%) of medical indications approved across states are pain related (Boehnke et al. 2019a). Notable subsets of pain patients (36%) also report substituting cannabis for prescription opioids (Corroon Jr. et al. 2017), and individuals who recreationally use cannabis also report self-medicating for pain with cannabis (Bonn-Miller et al. 2014; Sexton et al. 2016; Walsh et al. 2013). However, the use of cannabis as a medicine is challenging. Cannabis that contains >0.3% delta-9-tetrahydrocannabinol (THC) content remains classified as a Schedule I substance, creating considerable ambiguity surrounding the role of cannabis in medical practice. While preclinical and clinical research has played a critical role in characterizing the putative pain-relieving effects of cannabis, healthcare providers often report a lack of knowledge on the subject. In fact, only about 9% of medical schools address cannabis in their curriculum (Holman et al. 2022). The absence of empirically based guidelines preclude healthcare providers from making clear, evidence-based recommendations regarding medical cannabis use, particularly for the treatment of pain. Thus, understanding the role of cannabinoids in pain management is of significant public health importance.

1.1 Neurophysiology of Pain and Inflammation

Pain is a complex and multidimensional experience that includes both unpleasant sensory and negative emotional components (e.g., anhedonia, pain aversion, and anxiety) associated with actual or potential tissue damage (Raja et al. 2020). Neurotransmitter and neuromodulatory systems, including the endocannabinoid system (ECS), modulate both the sensory and affective dimensions of pain (Fine and Rosenfeld 2013). Acute pain serves an adaptive role by alerting the body to noxious stimuli, thereby facilitating avoidance of further tissue damage (Becker et al. 2018). Chronic pain, defined as pain lasting for at least 3 months, can be psychologically and functionally debilitating (Rikard 2023). Pain processing and perception occur via ascending pain pathways in three stages: (1) peripheral nociception: initial detection of noxious stimuli by nociceptors in the periphery (skin, muscles, viscera), (2) spinal transmission: conduction of sensory information via primary afferent neurons synapsing in the dorsal horn of the spinal cord, and (3) central processing: integration and perception of pain information by higher-order brain regions in the central nervous system (CNS) (Yam et al. 2018). Pain perception begins when nociceptors (primary afferent sensory neurons) are activated by noxious chemical, thermal, or mechanical stimulation (Schmidt et al. 1995) as these neurons express a variety of stimulus-sensitive receptors, such as transient receptor potential (TRP channels), voltage-gated sodium channels (Nav), bradykinin receptors, and prostaglandin receptors (Caterina and Julius 1999). This activation can occur directly as a result of tissue damage or indirectly via inflammatory mediators such as cytokines, prostaglandins, or growth factors, which enhance nociceptor sensitivity and contribute to pain signaling (Cook et al. 2018; Sorkin et al. 2018). Once activated, these peripheral afferents synapse onto second-order neurons in the dorsal horn of the spinal cord, where excitatory and inhibitory interneurons shape this signal before it is relayed to supraspinal brain regions via the spinothalamic and other ascending tracts (Yam et al. 2018). Pain is processed through a network of brain regions including the somatosensory cortex and limbic system which together encode the sensory, emotional, and cognitive aspects of pain. The thalamus is a key relay station that sends information to these higher cortical areas (Kenshalo Jr and Willis Jr 1991). Several key brain regions also send projections down to the dorsal horn of the spinal cord, including the *periaqueductal gray* (PAG) and *rostral ventromedial medulla (brainstem)*. This top-down system is responsible for the descending modulation of pain and is often disrupted in chronic pain conditions (Willis and Westlund 1997).

Inflammatory mediators are key contributors to both the initiation and maintenance of pain. Among them, cytokines play a central role. Classically, cytokines are subdivided as pro- and anti-inflammatory based on the immune response, although it is recognized that inflammation is rarely purely binary (Liu et al. 2021). Key classic pro-inflammatory cytokines, including tumor necrosis factor alpha (TNF-α), interleukin (IL)-6, IL-1β, interferon gamma (IFN-γ), and chemokines (CXCL1, CCL2), are produced by activated microphages and other immune cells following tissue injury or in chronic inflammation. Conversely, anti-inflammatory cytokines such as IL-10, IL-4, and IL-13 mediate anti-inflammatory signaling by either directly modulating nociceptors or indirectly by inducing release of further inflammatory mediators (Kidd and Urban 2001; Henshaw et al. 2021). For example, upon cytokine-mediated activation, the *cyclooxygenase enzymes (COX-1 and COX-2)* convert arachidonic acid into prostaglandins, which are lipid mediators involved in pain and inflammation. Other inflammatory biochemical mediators include bradykinin, 5-hydroxytryptamine (5-HT; serotonin), and histamines (Abdulkhaleq et al. 2018). Endocannabinoids, which are endogenous bioactive lipids, contribute to the complex interplay between these mediators, influencing the immune response and our perception of pain (Donvito et al. 2018).

Following repeated tissue damage, sensory nerve fibers at the site of an injury become sensitized, such that they have a reduced threshold for activation and an increased magnitude of response to stimuli. This phenomenon of peripheral sensitization includes hyperalgesia (i.e., increased pain sensitivity to an already-painful stimulus) and allodynia (i.e., pain in response to a non-noxious stimulus) (Raja et al. 2020; Jensen and Finnerup 2014). Cytokines released from immune cells during inflammation play a key role in pain sensitization and sustained inflammation. For example, TNF-α and IL-1 increase sensitization of sensory neurons and produce both mechanical and thermal hyperalgesia (Kidd and Urban 2001; Szallasi and Blumberg 1999). COX-2 is upregulated during injury and promotes the production of *PGE2,* which increases nociceptor sensitivity (Jang et al. 2020). Prolonged or repeated activation of pain pathways can lead to central sensitization, a phenomenon in which the CNS becomes hypersensitive to both noxious and non-noxious stimuli, contributing to chronic pain conditions (Latremoliere and Woolf 2009). Central sensitization provides an explanation for the presence of ongoing pain in the absence of ongoing primary nociception, tissue damage, or inflammation. Endocannabinoid signaling can influence both peripheral and central sensitization, making it a compelling target for chronic pain.

1.2 The Role of the Endocannabinoid System in Pain and Inflammation

The ECS is widely distributed throughout the nervous system and plays a crucial role in the regulation of pain and inflammation (Manzanares et al. 2006). The ECS is comprised of two G-protein coupled receptors, cannabinoid receptor type 1 (CB1R) and cannabinoid receptor type 2 (CB2R), with endogenous ligands anandamide (AEA) and 2-arachidonylglycerol (2-AG) and their metabolic enzymes. The endocannabinoids AEA and 2-AG are synthesized on demand and modulate neuronal excitability and synaptic transmission (Di Marzo 2009). Their actions are primarily mediated through CB1R, expressed predominantly in neurons within the central and peripheral nervous systems, and CB2R, expressed predominantly on immune cells (Turcotte et al. 2016).

Peripherally, CB1R and CB2R are expressed on nociceptors and immune cells, where their activation can dampen nociceptive signaling. On primary afferent neurons responsible for detection and transmission of pain signals, CB1R activation reduces excitability and inhibits the release of pro-nociceptive neurotransmitters (Pernía-Andrade et al. 2009). CB2R activation on immune cells reduces the release of pro-inflammatory cytokines and chemokines (Beltramo et al. 2006, Malfait et al. 2000). TRP channels, such as transient receptor potential vanilloid 1 (TRPV1) and transient receptor potential ankyrin 1 (TRPA1), which detect noxious thermal and chemical stimuli, are also activated by cannabinoids/endocannabinoids, leading to a desensitization of nociceptors and contributing to peripheral antinociceptive effects (Benítez-Angeles et al. 2020; Petrocellis et al. 2001; Akopian et al. 2008). In the spinal cord, CB1Rs are densely expressed in the dorsal horn (Parnell et al. 2023), and their activation inhibits substance P and glutamatergic transmission, reducing nociceptive input at the spinal level (Zhang et al. 2010; Kelly and Chapman 2001). Notably, spinal endocannabinoid tone is increased under conditions of inflammation and injury, suggesting a compensatory role in limiting pain hypersensitivity (Lim et al. 2003; Garcia-Ovejero et al. 2009). The ECS also modulates pain via supraspinal mechanisms, including actions along the spinothalamic tract and within the thalamus (Turcotte et al. 2016; Martin et al. 1996). Through these pathways, the ECS influences both the sensory and emotional components of pain.

The ECS is also a critical regulator of inflammation. Activation of the ECS is associated with attenuation of pro-inflammatory cytokine release (i.e., TNF-α, IL-1β, and IL-6), enhanced anti-inflammatory cytokine production (i.e., IL-10), and decreased inflammatory cell recruitment (Turcotte et al. 2015). CB2R are upregulated in CNS and the periphery following injury, including during acute and chronic inflammation (Hsieh et al. 2011; Zhang et al. 2003; Romero-Sandoval and Eisenach 2007). Interestingly, the biosynthetic precursor of endocannabinoids (arachidonic acid) is a substrate for COX enzymes; COX-2 can oxidize endocannabinoids into pro- or anti-inflammatory mediators (e.g., prostaglandin-ethanolamides and prostaglandin-glycerol esters) (Fowler 2007). Further, CB1R activation may inhibit COX-2 expression or reduce prostaglandin release, thereby contributing to anti-inflammatory and analgesic effects (Zhang and Chen 2008). Together, these findings highlight the ECS as an intricate signaling system that is activated under conditions of chronic pain or inflammation to restore homeostasis and limit excessive nociceptive signaling (Di Marzo 2018).

Notably, preclinical and clinical studies have supported the analgesic effects of phytocannabinoids, showing their potential to alleviate pain through various mechanisms, including the modulation of CB1R and CB2R and reduction of inflammation (Henshaw et al. 2021). Cannabis is a cultivated plant that includes many chemically diverse varieties or "chemovars" (McPartland and Small 2020; Small 2015). Typically, cannabis is categorized as THC-dominant, cannabidiol (CBD)-dominant, or balanced, but varieties of cannabis contain different concentrations of other phytocannabinoids and chemical compounds (e.g., terpenes, flavonoids). THC is a CB1R and CB2R partial agonist; THC produces psychoactive, motor impairing, and dependence producing effects via CB1R (Cooper and Haney 2008) that may limit its utility as an analgesic. CBD is a nonintoxicating cannabinoid with a complex pharmacological profile (Reimann-Philipp et al. 2020; Spindle et al. 2019). CBD has minimal activity at CB2R, but has very low binding affinity for CB1R (Pertwee 2012). It may act via negative allosteric modulation of CB1R and potential signaling mechanisms include serotonin-1A (5-HT_{1A}), mu and delta opioid receptors (Kathmann et al. 2006), and transient receptor potential V1 (TRPV1) and V2 (TRPV2), GPR55, GPR18 from the TRP superfamily of nonselective, ligand-gated cation channels, and other TRP channels (Russo and Marcu 2017; McPartland et al. 2015). CBD has been purported to augment the therapeutic effects and/or mitigate the adverse effects of THC when co-administered (see review by (Boggs et al. 2018)). Ongoing efforts in cannabis-based drug development are focused on identifying novel formulations that may have therapeutic efficacy for different types of pain (Finn et al. 2021).

Although the cannabis plant contains other minor cannabinoids and terpenes, THC and CBD are the primary focus of this chapter given their dominant role in preclinical and clinical pain research, and the plant components that likely play the most meaningful role in pain modulation. In the following sections, we briefly review the evidence for these two primary cannabinoids in the modification of experimental pain as a function of dose and route of administration. For a more detailed review of minor cannabinoids and terpenes, see Baron (2018) (Baron 2018).

2 Preclinical Studies of Cannabinoids' Effects on Pain

Preclinical animal models of nociception are essential for investigating pain mechanisms and developing new analgesics. A vast literature of animal studies support the utility of ECS agonists in alleviating acute and chronic pain (Finn et al. 2021; Soliman et al. 2021). These models serve as a reliable tool for investigating preliminary efficacy of novel analgesics and for measuring behaviors in response to noxious stimulation, as described below. To best illustrate the complex multidimensional state of pain, studies typically use a combination of behavioral tests that measure sensory thresholds, evoked pain, spontaneous pain, as well as pain comorbidities (e.g., anhedonic-, anxiety-like behavior) (Huerta et al. 2024). Table 1 lists commonly used assessments for measuring pain in preclinical and clinical studies.

2.1 Preclinical Pain Assessments

2.1.1 Thermal Pain Tests

Commonly used thermal pain tests in rodents include the tail-flick test, hot-plate test, and Hargreaves test. These tests involve the application of a noxious heat stimulus (i.e., a plate, probe, beam, water) to the paw or tail. This elicits a reflexive behavioral response from the animal, such as paw or tail withdrawal, which is used as a measure of pain latency (Deuis et al. 2017; Allen and Yaksh 2004). Decreased pain sensitivity (i.e., analgesia), is evident by an increased latency of the withdrawal response. An increase in pain sensitivity (i.e., hyperalgesia), is reflected as a shorter withdrawal latency.

2.1.2 Mechanical Pain Tests

Responses to mechanical stimuli, such as pressure or touch, are also used to assess both allodynia and hyperalgesia. The mainstay in mechanical pain sensitivity testing is the von Frey test (VF), which uses a series of small, flexible monofilaments of different strengths (0.008–300 g) that are typically applied to the hind paw of the animal (Deuis et al. 2017; Chaplan et al. 1994). This test is initiated by starting with a low monofilament strength and applying increasingly larger monofilaments until a withdrawal response is observed. The dependent measure of this test is typically the 50% withdrawal threshold, an estimation of the force that would produce a response 50% of the time. This is calculated with a formula that incorporates the response pattern of the animal, the value of the final filament tested, and the average increment between filaments (in log units) (Chaplan et al. 1994; Kruger 2001; Yuan et al. 2016). An alternative version of this test is the electronic VF (eVF). This test uses a single filament whereby the maximum pressure applied to the hind paw to produce a withdrawal response is processed by a force transducer. The benefit of the eVF is that the withdrawal response is determined within a single test, thereby removing the confound of repeated testing which can lead to unreliable measurements (Yuan et al. 2016). A dynamic plantar aesthesiometer is another instrument used to assess mechanical sensitivity using the VF method. With this device, the same rigid filament is used and the force applied increases linearly over time. Latency and grams of force at the time of paw withdrawal reflex are automatically detected and recorded (Campana and Rimondini 2015). The VF test of mechanical pain sensitivity has clinical relevance, as it models human pain conditions including neuropathy, postoperative pain, and osteoarthritis, as well as others that are characterized by increased sensitivity (Flatters 2008; Gregory et al. 2013; Jali et al. 2024; O'Brien et al. 2014).

The paw pressure test is another commonly used method to assess mechanical pain sensitivity in rodents (Möller et al. 1998). In this test, a gradually increasing force is applied to the animal's hind paw or tail using a blunt probe or specialized

Table 1 Preclinical and clinical modeling techniques for different types of pain

Model name	Description of procedure	Type of pain	Representative citations
Preclinical experimental pain models			
Acute pain sensitivity	No manipulation	None (Nociception)	
Acute injury, postoperative	Surgical incision of the hindpaw, skin, or muscle	Nociceptive	Brennan et al. (1996)
Migraine	Microinjection of 10% allyl isothio-cyanate (AITC) onto the dura	Nociceptive	Edelmayer et al. (2012)
Capsaicin	Injection of capsaicin, a selective TRPV1 agonist, into the hindpaw or joint	Nociceptive/ acute inflammatory	Gilchrist et al. (1996)
Formalin	Injection of formalin into the hindpaw; evokes an initial acute phase (phase I: 0–15) followed by an inflammatory phase (phase II:15-60 m)	Nociceptive/ acute inflammatory	Tjølsen et al. (1992)
Carrageenan	Injection of λ-carrageenan into the hindpaw or joint	Acute inflammatory	Ea and Nuss (1962)
Acetic acid	Injection of dilute acetic acid into the intraperitoneal cavity	Acute vis-ceral / inflammatory	Collier et al. (1968)
Complete Freund's Adjuvant (CFA)	Injection of CFA (heat-killed myobacterium) into the hindpaw or joint	Chronic inflammatory	Noh et al. (2021)
Chemotherapy-induced peripheral neuropathy (CIPN)	Repeated injections of anti-neoplastic agents (e.g., paclitaxel, cisplatin)	Neuropathic	Aley et al. (1996) Authier et al. (2009)
Spared nerve injury (SNI)	Surgical ligation of two of the three branches of the sciatic nerve	Neuropathic	Decosterd and Woolf (2000)
Chronic constriction injury (CCI)	Loose ligation of the common sciatic nerve with sutures	Neuropathic	Bennett and Xie (1988)
Clinical experimental pain models			
Heat/capsaicin sensitization	Heat stimulation applied to skin sur-face area using thermode, and/or cap-saicin cream or intradermal capsaicin applied	Nociceptive	Petersen and Rowbotham (1999) Dirks and Petersen (2003)
Thermal temporal summation (TTS)	Heat stimulation applied repeatedly to same skin surface area	Nociceptive/ neuropathic	Kong et al. (2019)
Conditioned pain modulation (CPM)	Conditioned pain stimulus applied before or after test pain stimulus	Nociceptive	Pud et al. (2009)

Models listed are the ones most common in practice. This list is not exhaustive. Citations include seminal studies defining the methodology and/or recent updates to the technique

device, such as an analgesy-meter, until the animal exhibits a withdrawal response, vocalization, or other behavioral sign of discomfort (Vivancos et al. 2004). The force at which this response occurs is used as a measure of the mechanical pain threshold.

2.1.3 Spontaneous Pain Behaviors

Spontaneous (i.e., non-reflexive) pain behaviors can be assessed in rodents to capture pain-related experiences that occur in the absence of an external stimulus (Tappe-Theodor and Kuner 2014). Unlike evoked responses (e.g., withdrawal to thermal or mechanical stimuli), spontaneous behaviors reflect the animal's internal state and may be useful for understanding diverse components of pain. Common spontaneous pain behaviors that can be measured using home-cage monitoring and/or video observations include guarding, flinching, paw lifting, licking or biting of the affected area, facial expressions of pain (e.g., grimacing), and reductions in general activity, grooming, eating, or drinking. Other spontaneous pain tests include assessments of weight-bearing and gait, voluntary wheel running behavior, burrowing, and ultrasonic vocalizations. More complex assessments of non-evoked pain include conditioned place preference (King et al. 2009) or passive avoidance (LaBuda and Fuchs 2000). Generally, tests of spontaneous pain behaviors are less utilized than evoked-pain tests, though these measures are critical for evaluating ongoing pain states and for testing analgesics that may preferentially target tonic or affective dimensions of pain.

2.2 Preclinical Models of Pain States

Pain is an extremely heterogeneous condition and, therefore, there are various animal models that have been generated to study the different etiologies, phases, and aspects of pain. A list of commonly used animal models for the study of pain are listed in Table 2.

2.2.1 Nociception and Nociceptive Pain

Nociception refers to the transduction of noxious environmental stimuli by primary sensory neurons (nociceptors) via TRP channels, which activate *Aδ and C fibers* and send pain signals to the central nervous system. Nociceptive pain is the unpleasant sensory and emotional experience associated with, or resembling that associated with, actual or potential tissue damage (Loeser and Treede 2008). As nociceptive pain involves tissue injury, this may evolve into inflammatory pain as immune cells infiltrate the site and release pro-nociceptive mediators (e.g., prostaglandins, cytokines), leading to peripheral sensitization. Inflammatory pain is therefore considered a mechanism-based subtype of nociceptive pain. Importantly, in animal models, we

Table 2 Preclinical and clinical pain assessments used in studies, defined by stimulus type and outcomes measured

Experimental pain assessments	Stimulus type	Outcomes	Unit of measurement
Preclinical			
Evoked behaviors			
Hargreaves	Thermal	Paw withdrawal	Seconds (s)
Hot plate	Thermal	Paw/tail withdrawal	Seconds (s)
Tail flick (beam)	Thermal	Tail withdrawal	Seconds (s)
Tail flick (water)	Thermal	Tail withdrawal	Seconds (s)
Von Frey (fibers)	Mechanical	Paw withdrawal	Gram-force (gf)
Electronic von Frey	Mechanical	Paw withdrawal	Gram-force (gf)
Dynamic plantar aesthesiometer	Mechanical	Paw withdrawal	Gram-force (gf), latency (s)
Paw pressure	Mechanical	Struggling/vocalizing	Grams (g), Newtons (N)
Acetic acid	Chemical	Stretching/Writhing/Abdominal constrictions	Frequency (n)
Spontaneous behaviors			
Pain suppressed behaviors	No manipulation	E.g., Burrowing, locomotor activity, wheel running	Duration (s), frequency (n), distance (cm)
Grimace scale	No manipulation	Orbital tightening, nose bulge, cheek bulge, ear position, whisker change	Each outcome assigned a severity score of 0, 1, or 2
Clinical			
Cold Pressor test	Thermal	Pain threshold (time of first feeling of pain), pain tolerance (total time submerged minus threshold)	Seconds (s)
Contact heat	Thermal	Heat pain threshold	Degrees (Celsius)
Thermal temporal summation (TTS)	Thermal	Pain intensity on numerical scale, response magnitude, change rate, peak effects, time to peak effects	Scale of 0–10 or 0–100

cannot directly measure an animal's subjective pain experience. Instead, we use behavioral readouts such as thermal/mechanical nociceptive thresholds, described above, as proxies for pain sensitivities and nociceptive processing.

Inflammatory pain arises in the context of tissue injury and immune activation, manifesting as spontaneous pain and heightened sensitivity to both noxious (hyperalgesia) and normally innocuous (allodynia) stimuli as well as local swelling (edema) reflects the underlying pro-inflammatory response (Kidd and Urban 2001).

It is characteristic of clinical conditions such as postoperative pain, burns, and traumatic injuries. Despite its prevalence, inflammatory pain remains a significant clinical challenge, as current therapies are often limited by inadequate efficacy or poor tolerability. Inflammatory pain models in rodents typically involve the local injection of an irritating substance that produces a robust inflammatory response and are used to mirror the physiological response to tissue injury (Burma et al. 2017; Rafiyan et al. 2023). These models have been used extensively to evaluate opioids, NSAIDS, and other analgesics for the treatment of acute and chronic inflammatory pain. Different models focus on specific aspects of inflammation and pain, from the initial acute pain response (e.g., abdominal constriction, capsaicin, and carrageenan) to more persistent, chronic pain (Complete Freund's Adjuvant, CFA) and central sensitization (Rafiyan et al. 2023).

2.2.2 Neuropathic Pain Models

Neuropathic pain refers to pain caused by pathology of the peripheral or central somatosensory system (Treede et al. 2008). This contrasts with nociceptive pain, which is generated from injury of peripheral nociceptors and includes inflammatory pain. While the etiologies of the different types of chronic pain are fundamentally different, there are often common mechanisms underlying the generation of each pain state. For example, post-traumatic inflammation in the spinal cord contributes to neuropathic pain formation (Schomberg et al. 2012). The two most common neuropathic pain models in animals are controlled nerve injury models (e.g., spared nerve injury (SNI), chronic constriction injury (CCI)) and neuropathy induced by chemotherapy (e.g., chemotherapy-induced peripheral neuropathy, CIPN) or induction of diabetes. See Table 2 and (Jaggi et al. 2011; Abboud et al. 2021) for a review of preclinical neuropathic pain models.

2.3 *Cannabinoid Efficacy on Nociception (Acute Pain)*

2.3.1 THC

One of the most documented and known effects of THC is that it dose dependently produces antinociception to an acute painful stimulus in otherwise pain-free animals. The antinociceptive effects of THC have been shown following oral (p.o.), intraperitoneal (i.p.), subcutaneous (s.c.), intrathecal (i.t.), intracerebroventricular (i.c.v.), intravenous (i.v.), and vapor administration of THC in rats and mice (Bicher and Mechoulam 1968; Chesher et al. 1973; Moore and Weerts 2022; Devsi et al. 2020; Lichtman and Martin 1991; Martin 1985; Welch et al. 1995). As evidenced by analgesia produced via i.t. or i.c.v. routes of administration, as well as systemic administration, THC elicits antinociception by both spinal and supraspinal mechanisms (Lichtman and Martin 1991).

The antinociceptive effects of THC are blocked by CB1R antagonists (Nguyen et al. 2016; Welch et al. 1998; Tseng and Craft 2004; Lichtman and Martin 1997). There has been much work demonstrating that there are cannabinoid-opioid system interactions in antinociception/analgesia. THC's antinociceptive effects have been linked with the release of dynorphin A in the spinal cord, and the kappa-opioid receptor antagonist nor-BNI specifically blocks the antinociceptive effects of THC, sparing other THC-induced behaviors (e.g., catalepsy, hypolocomotion) (Mason Jr et al. 1999). Pre-proenkephalin-knockout mice show reduced levels of antinociception from THC (Valverde et al. 2000). THC also enhances the antinociceptive effects of the mu-opioid receptor (MOR) agonist morphine, and these additive effects rely on both the CB1R and mu-opioid receptor (Cichewicz 2004). These studies indicate an intimate connection between cannabinoid and opioid signaling pathways in the modulation of pain perception (Carta et al. 1999).

2.3.2 CBD

Studies investigating CBD effects on nociceptive pain have shown that CBD alone is largely ineffective at producing antinociception in naive animals. Oral CBD (30–400 mg/kg) had no effect in rodents on thermal and mechanical pain tests (Chesher et al. 1973; Sofia et al. 1975a). Moore and Weerts (2022) reported similar findings; no antinociceptive effects of oral CBD were detected in the tail flick test, and, in fact, increased pain sensitivity was observed after the highest dose of CBD (30 mg/kg) in the VF test 5-h after administration. Similarly, no antinociceptive effects were observed after vaporized CBD in rats (Javadi-Paydar et al. 2019) or CBD smoke exposure in mice (Varvel et al. 2006). Acute antinociceptive effects were also not observed following i.p. CBD administration in pain-naive rats (1.25–50 mg/kg) (Britch et al. 2017; Booker et al. 2009) or mice (30 mg/kg) (Abraham et al. 2020); however, in one study, 30 mg/kg CBD administered i.p. produced antinociceptive effects in male rats and female rats in the late diestrous phase of the estrous cycle only (Arantes et al. 2024).

2.3.3 THC and CBD Interactions

THC and CBD may functionally interact on several levels to influence antinociception, including through synergistic or antagonistic actions via different receptor systems, altering receptor binding kinetics (e.g., negative allosteric modulation of CB1R), or pharmacokinetic interactions (e.g., metabolism). Several studies have shown that pre-treatment with CBD can potentiate the antinociceptive effects of THC. An early study in mice showed that the 10 and 20 mg/kg of CBD combined with 5 mg/kg THC significantly potentiated the THC's antinociceptive effects, with latencies being 4–5 times higher than THC alone (Karniol and Carlini 1973). A more recent study in mice also saw potentiation of THC effects by CBD, where pretreatment with 30 mg/kg CBD (i.v.) potentiated the antinociceptive effects of a subthreshold dose of THC (0.3 mg/kg), but had no effect on a higher dose of THC (3 mg/kg) that did have antinociceptive effects on its own (Varvel et al. 2006). However, pretreatment with CBD did not alter the antinociceptive effects of cannabis smoke exposure in the same study (Varvel et al. 2006). A study in rats testing CBD pretreatment (10, 30 mg/kg, i.p.) prior to THC (1.8, 3.2, 5.6, or 10 mg/kg, i.p.) found that the effects on thermal and mechanical nociception were weak and inconsistent, and suggesting that if anything, CBD may prolong THC's duration of action on antinociception in the paw pressure test (Britch et al. 2017).

2.4 *Cannabinoid Efficacy on Nociceptive Pain and Inflammation*

2.4.1 THC

THC produces analgesic effects in a variety of models of experimentally induced nociceptive and inflammatory pain. For example, in a model of acute migraine pain, 0.32 mg/kg, i.p. THC blocked pain-related depression in wheel running while having no effect on controls: a higher dose (1 mg/kg) had sedative effects (Kandasamy et al. 2018). In the acetic acid model of acute inflammatory pain, THC (2.5–10 mg/kg, i.p., and 5–10 mg/kg, p.o.) reduced abdominal writhing and stretching in several studies (Sofia et al. 1975a; Harris et al. 2019). Several studies using the carrageenan-induced acute inflammation model have shown that THC reduces hyperalgesia, allodynia, and edema (swelling) (Jenkins et al. 2025; Rock et al. 2018; Conti et al. 2002). Specifically, pretreatment with THC (1 mg/kg, p.o.) or nabilone (2.5 mg/kg, p.o.), a synthetic THC analog, reduced both carrageenan-induced hyperalgesia and edema (Rock et al. 2018; Conti et al. 2002) in male rats. Another study showed that THC reduced carrageenan-induced hyperalgesia and allodynia in both male and female rats, but had no effect on edema (Jenkins et al. 2025). In a study using the CFA model of chronic inflammatory pain, twice daily THC (0.5–4.0 mg/kg, i.p.) across 3 days dose-dependently reduced hyperalgesia (Britch et al. 2020). However, THC showed minimal effect on serum cytokine levels or CFA-induced paw edema (Britch et al. 2020). Thus, while the antinociceptive effects of THC extend to models of nociceptive and inflammatory pain, the direct impact of THC inflammatory processes is still under investigation.

2.4.2 CBD

CBD has produced mixed results in nociceptive and inflammatory pain models (Finn et al. 2021; Soliman et al. 2021). A study testing post-surgical allodynia in the VF assay found that CBD (0.3–30 mg/kg, i.p.) reduced allodynia with greater effects in males (Arantes et al. 2024). A study using the formalin test found that CBD (10 mg/kg, i.p.) effectively produced antinociception during phase I (acute/nociceptive pain phase), but not phase II (inflammatory pain phase) in male mice (Sepulveda et al. 2022a). Studies with the acetic acid-induced abdominal writhing assay have shown either antinociceptive effects or no effects of CBD. Studies reporting positive results found that CBD at doses of 5.6, 30, 56, and 90 mg/kg (i.p.) reduced writhing behavior when 0.4% acetic acid was administered (Foss et al. 2021; Silva et al. 2017). In contrast, doses of up to 400 mg/kg i.p. CBD did not affect writhing elicited by higher concentrations of acetic acid (0.6–0.7%) (Booker et al. 2009; Harris et al. 2022; Sofia et al. 1975b). While all of these studies used male mice, some methodological differences may explain the discrepant results (e.g., timing of evaluations, use of 0.4% acetic acid in (Foss et al. 2021; Silva et al. 2017) vs. 0.6 in (Booker et al. 2009), 0.7% in (Harris et al. 2022)).

CBD has shown mixed results in models of acute inflammation. In the carrageenan model of acute inflammatory pain, when given sub-chronically across 3 days following the carrageenan injection, oral CBD (5–40 mg/kg, p.o.) effectively decreased edema and thermal hyperalgesia in male rats (Costa et al. 2004a, b). Further, when given acutely following the carrageenan injection, CBD (7.5–40 mg/kg, p.o.) reduced edema (Costa et al. 2004a). In contrast, when given prophylactically (i.e., before the carrageenan injection), CBD (10–30 mg/kg, p.o.) had no effect on hyperalgesia nor allodynia and only minimal effects on edema in male rats only (Jenkins et al. 2025). In a separate study, prophylactic CBD (10 mg/kg, p.o.) did reduce hyperalgesia but had no effect on edema in male rats (Rock et al. 2018).

CBD has also had mixed results in the CFA model of chronic inflammation. Following CFA administration, acute administration of CBD (20 mg/kg, p.o.), reduced hyperalgesia and allodynia (Costa et al. 2007). CBD also reduced biochemical inflammatory mediators such as prostaglandin E2 (PGE2), lipid peroxides, nitric oxide (NO), and glutathione enzymes (Costa et al. 2007). It is important to note that this dose of CBD only reduced endothelial NO synthase (eNOS) expression and did not reduce inducible NOS (iNOS) in the inflamed/damaged CFA tissue. CBD did not reduce the levels of other inflammatory signaling molecules such as NF-κB or TNFα, indicating CBD's anti-inflammatory actions are mediated through different mechanisms (Costa et al. 2007). Britch and colleagues (2020) also tested CBD's effects on the CFA model, but with repeated administration of CBD (1.25–10 mg/kg, i.p.). In this study, treatment with CBD showed minimal efficacy on mechanical allodynia, such that significant effects were only observed following repeated i.p. administration of the 2.5 mg/kg dose of CBD (Britch et al. 2020). While largely ineffective on measure of pain, CBD treatment decreased serum levels of IL-1β, IL-10, and IFN-γ and increased IL-6 in rats with chronic inflammation (Britch et al. 2020). Overall, CBD shows limited efficacy in alleviating pain in most models once inflammation is established. However, many studies have demonstrated effects on modulating inflammatory mediators and signaling pathways (Henshaw et al. 2021). Therefore, CBD may be effective when given through alternative dosing strategies (e.g., repeated dosing, preventatively) as compared to being given as an acute analgesic.

2.4.3 THC and CBD Interactions

Despite hypotheses that THC and CBD have the potential to produce synergistic effects on pain and inflammation, this has been observed only in a subset of studies, with others reporting no interactions or antagonistic (i.e., sub-additive) effects in acute and chronic pain models (Booker et al. 2009; Abraham et al. 2020; Varvel et al. 2006; Welburn et al. 1976; Sanders et al. 1979; Sepulveda et al. 2022b; Mitchell et al. 2021). In the acetic acid test, THC alone decreased abdominal writhes, but CBD appears to antagonize this effect: CBD did not produce anti-inflammatory effect on its own, and CBD (5 and 10 mg/kg, p.o.) counteracted the antinociceptive effects of THC in a dose-dependent manner (Welburn et al. 1976). Finn et al. (2004) found that CBD (5 mg/kg) did not alter effects of 1–2.5 mg/kg THC on a formalin-evoked acute pain test. However, the authors noted that formalin-induced pain behaviors usually subside within approximately 60 min after injection and any interactions between the cannabinoids at later stages may have been overlooked (Finn et al. 2004).

An antagonistic effect of THC:CBD combinations was also observed in a rat carrageenan model. THC (1, 3, and 10 mg/kg, p.o.) and CBD (10, 30, 100 mg/kg, p. o.) were tested individually and in combination on inflammatory pain and edema (Jenkins et al. 2025). Results from this study showed that THC decreased pain sensitivity, while CBD had no effect on pain sensitivity but had a modest anti-inflammatory effect on reducing edema in males. Interestingly, the combination of THC and CBD produced sub-additive effects on pain sensitivity, suggesting that CBD interfered with the pain-relieving effects of THC (Jenkins et al. 2025).

2.5 *Cannabinoid Efficacy on Neuropathic Pain*

2.5.1 THC

A meta-analysis of preclinical studies on cannabinoids and pain indicates that cannabinoids, and particularly THC, may be the most effective in treating neuropathic pain as compared to other pain types (Finn et al. 2021; Soliman et al. 2021). THC has shown efficacy at reversing or alleviating pain hypersensitivity in multiple models of neuropathic pain. THC (2 mg/kg, i.p.) decreased mechanical allodynia in male mice with chemotherapy-induced peripheral neuropathy (CIPN) (Harris et al. 2016) and in male rats with spared nerve injury (SNI) at doses of 2.5 and 5 mg/kg, i.p. (Pearl-Dowler et al. 2023). In a model of volitional THC consumption, male and female mice with partial sciatic nerve ligation (pSNL) that were allowed to consume THC in gelatin (~1 mg/15 ml) daily, beginning 1 day post-ligation, showed decreased mechanical allodynia when tested from 5 to 22 days post-pSNL (Abraham et al. 2020). Importantly, mice did not show signs of developing tolerance to THC's effects after 7 days (relative to acute THC effects in drug-naïve controls) (Abraham et al. 2020). Intrathecal THC (3–300 nmol) also dose-dependently decreased mechanical and cold allodynia in male mice with chronic constriction injury (CCI) (Casey et al. 2022).

2.5.2 CBD

CBD has also shown efficacy in preclinical models of neuropathic pain when given acutely or repeatedly. In the CIPN model, CBD alleviated mechanical allodynia in male (Harris et al. 2016) and female mice (Ward et al. 2014). These effects were blocked by a 5HT1A receptor antagonist (Ward et al. 2014). When given prophylactically, CBD (2.5 mg/kg, i.p. and 2.5, 25 mg/kg, p.o.) prevented the development of CIPN in male mice (Foss et al. 2021). In the CCI model, acute CBD (3–300 nmol, i.t.) dose dependently decreased mechanical and cold allodynia in male mice (Casey et al. 2022). In another study, chronic dosing of CBD (2.5–20 mg/kg, p.o.) daily for 7 days post-surgery decreased thermal hyperalgesia and mechanical allodynia in male mice with CCI (Costa et al. 2007). A separate study using the CCI model also showed effectiveness of CBD (0.3, 3, 10, or 30 mg/kg, i.p.) on reducing hyperalgesia, and CBD also blocked the emergence of anxiety-like behaviors in the open field test induced by CCI (Silva-Cardoso et al. 2021). Beneficial effects in the CCI models may be mediated via TRPV1 channels (Costa et al. 2007; Silva-Cardoso et al. 2021). In male and female mice with pSNL, volitional consumption of CBD in gelatin (1 mg/15 ml) decreased mechanical allodynia (Abraham et al. 2020). One study using the CCI model in male mice demonstrated that an acute dose of CBD (5 mg/kg, i.p.) combined with transcutaneous electrical nerve stimulation (TENS) had an additive effect on mechanical allodynia (Malta et al. 2023).

2.5.3 THC and CBD Interactions

Several studies have shown potential advantages in use of THC:CBD combinations in neuropathic pain models. When administered in a 1:1 ratio, THC and CBD combinations dose-dependently decreased allodynia in male mice with CCI (Casey et al. 2017). This study tested a range of doses (0.01–56 mg/kg, s.c.) administered at a 1:1 ratio and observed a biphasic effect where lower dose combinations had increased potency, but lower efficacy, and higher dose combinations had decreased potency but increased efficacy (Casey et al. 2017). Notably, the higher combination doses also produced unwanted THC-related side effects (e.g., motor disruption, catalepsy, and sedation) (Casey et al. 2017). A subsequent study investigated whether an orally administered THC:CBD combination (17.8 mg/kg THC, 100 mg/kg CBD; a 1:5.6 ratio) would be effective in the CCI model. This THC:CBD combination reduced allodynia in an additive to synergistic manner but had a poor side-effect window (Mitchell et al. 2021). Similarly, a 1:1 ratio of THC:CBD produced synergy in the CIPN model (paclitaxel) (King et al. 2017). Low doses of CBD and THC combined in a 1:1 ratio (0.16 mg/kg Δ9-THC + 0.16 mg/kg CBD), but not higher doses, prevented the development of mechanical allodynia, and were more effective than either THC and CBD alone at these doses in male mice (King et al. 2017). This effect was further demonstrated in the oxaliplatin CIPN model but not the vincristine CIPN model (King et al. 2017). Overall, these data suggest that select doses of THC:CBD combinations may be effective at decreasing neuropathic pain and minimizing potential side effects associated with THC administration.

3 Clinical Studies of Cannabinoids' Effects on Pain

Below, we will briefly review the pain paradigms most often used in human studies investigating cannabinoid-induced changes in pain perception, followed by a narrative review of the current state of cannabinoids' (THC and CBD) therapeutic utility for pain.

3.1 Human Experimental Pain Assessments

The use of experimental pain paradigms in human laboratory research to determine the analgesic efficacy of pharmacological agents has been controversial. While these models provide a unique and rigorous way to control and apply noxious stimuli, they are often questioned over their subjectivity (i.e., reliance on self-reported pain threshold and tolerance, which varies from individual to individual) and clinical relevance (i.e., experimentally induced pain is dissimilar from clinical pain) (Kumar Reddy et al. 2012). Yet, the perception of pain (as well as other drug effects) is a subjective experience, so the expectation that any experimental measurement can and should eliminate that characteristic is counterintuitive. Thus, experimental pain paradigms are an important and necessary tool to better understand the human laboratory science of pain.

These paradigms present the largest benefit in their utility to characterize potential alterations in pain perception that may mechanistically translate to clinical pain syndromes (e.g., neuropathic pain). In other words, they are useful in determining if there is a signal that a drug may be effective in altering neurobiological signals that present clinically as pain (i.e., hyperalgesia) or the reduction of such (i.e., analgesia). These paradigms typically result in a latency measurement (in seconds) to pain perception, and/or a rating on a visual analogue or numeric rating scale of pain intensity. In human laboratory studies, these paradigms can be quickly and easily assessed, some of which in relatively quick succession that allows researchers to capture a time course of a drug's effect on pain perception (Olesen et al. 2012).

In humans, pain is experienced as two sequential and qualitatively distinct perceptions, mediated by different afferent free nerve ending fibers. "First pain" is the initial brief, localized sharp pain primarily mediated by myelinated Aδ afferent nerve fibers. "Second pain" is mediated by unmyelinated C-fibers, has a slower onset and longer duration, and is perceived as diffuse, dull pain (Jarvis 1990; Ploner et al.

2002). Experimental pain paradigms are useful in modeling the activity of these afferent nerve fibers in clinical syndromes (Olesen et al. 2012) and can be utilized in such a way that the contributions of these distinct patterns of afferent nerve fiber activity can be understood in the context of drug administration. These paradigms are characterized based on the stimulus they utilize and subsequent pain pathways they interact with, including thermal, mechanical, and chemical/inflammatory pathways. Herein, we discuss the primary experimental pain paradigms used in human studies of cannabis effects, their methodologies, and findings in studies where cannabis is either administered experimentally (human laboratory studies) or clinically (clinical case studies and trials). Tables 1 and 2 also list commonly used human models and tests for the study of pain.

3.1.1 Thermal Pain Tests

Thermal paradigms induce pain by way of temperature, and in human laboratory studies, most often include variations on the cold pressor test (CPT) and noxious heat.

The Cold Pressor Test

The CPT is a validated pain induction technique in which volunteers immerse their hand into cold water while blood pressure and heart rate are monitored. Volunteers indicate when the cold stimulus first becomes painful (pain detection/threshold) and when the pain becomes intolerable (pain tolerance), and the latency (typically in seconds) to these responses is recorded as a quantitative outcome. A safety cut-off of 3–5 min is implemented to prevent tissue damage to the skin (Olesen et al. 2012; Compton et al. 2012).

The CPT recruits primarily Aδ afferent nerve fibers and can be utilized to model acute (e.g., injuries, post-surgical pain, etc.) and chronic pain conditions (e.g., drug-induced hyperalgesia, fibromyalgia, and other forms of neuropathic pain) (Cruz-Almeida and Fillingim 2014; Hubli et al. 2018). This paradigm is often utilized in human studies due to its reliability and validity in detecting changes in pain perception (e.g., allodynia, hyperalgesia, and analgesia) both as a function of various clinical factors (e.g., long-term opioid use and psychiatric comorbidities), and in response to acute drug administration, such as in human laboratory studies of either healthy volunteers or drug-using populations (Compton et al. 2010; Conley et al. 1997; Kowalczyk et al. 2006; Webb and Kamali 1998; Koenig et al. 2014; Cooper et al. 2013; Cooper et al. 2018). Perhaps the greatest strength of this paradigm is that it is easily repeatable, leaving no lasting tissue damage; assessments are conducted with as little as 10 min in between, which allows the ability to track the effects of cannabis administration on changes in pain perception in real time.

Heat Pain Tests

Heat pain is typically applied using a controlled, fixed intensity heat stimulus via a device (e.g., Medoc's TSA-II Neurosensory Analyzer). A typical paradigm includes placing a thermode on the body, often to the thenar eminence of the palm, and heat stimulation is applied in a ramp-and-hold method. In one common paradigm, the baseline temperature is set to 32.0 °C, increases at a rate of 1 °C/s up to 46.5 °C, and remains constant; this process is conducted once to yield a latency (in seconds) to pain detection and pain tolerance. Alternatively, to achieve an experimental measure of thermal temporal summation (TTS), this process is repeated numerous times in quick succession; each pulse of heat is held constant for approximately 5 s and repeated numerous times with 5 s in between each stimulus (a typical protocol includes four pulses of heat). Throughout the duration of stimulation, participants continuously rate the magnitude of perceived pain using a computerized visual analog scale (VAS). Pain ratings from the first and last stimulation are compared; the last stimulus being rated as more painful than the first is indicative of temporal summation and sensitization of afferent fibers (i.e., central sensitization) (Cruz-Almeida and Fillingim 2014).

Through this quick, repetitive stimulus, TTS induces central sensitization of C-fibers in the spinal cord (Cruz-Almeida and Fillingim 2014), which becomes an experimental correlate of the "wind-up" phenomenon in dorsal horn neurons. C-fibers have shown decreased activation thresholds and increased firing with repeated stimulation (i.e., temporal summation; (Jarvis 1990)), reflecting the neurobiological changes that underlie chronic pain (Edwards et al. 2005; Mendell 1966; Pfau et al. 2012). Opioid-dependent chronic pain patients have demonstrated exacerbated TTS responses relative to healthy controls and chronic pain patients on non-opioid medications (Chen et al. 2009; Zhang et al. 2015). To our knowledge, no published studies have looked at changes in TTS responses as a function of cannabis administration. As cannabinoid receptor activation in the dorsal horn has been linked to suppression of "wind-up" and central sensitization (a potential mechanism of cannabis' efficacy for neuropathic pain (Lee et al. 2018)), it is possible that cannabis administration, and/or abstinence, can impact TTS responses.

This paradigm can also be modified to assess conditioned pain modulation (CPM), where heat pain can be applied to one surface using a thermode, while an additional noxious stimulus (typically submersion of the opposite hand or foot in cold water) can be leveraged to determine the role of the central nervous system (e.g., the spine) in regulating and perceiving pain on the contralateral side. In other words, analgesic medications should potentiate the ability of a noxious stimulus on one side of the body to inhibit/reduce central pain signaling resulting from a noxious stimulus applied to the contralateral side (Yarnitsky 2010; Leone and Truini 2019).

3.1.2 Mechanical Pain Tests

Mechanical pain paradigms utilize controlled mechanical stimuli to induce pain, most commonly by way of applied pressure or stretching of tissues. In humans, mechanical pain is most often measured using behavioral responses (e.g., withdrawal, pain rating). The common techniques include pressure pain and VF and are useful in determining analgesic potential, hyperalgesia, allodynia, and extent of pain pathologies. Applying pressure activates mechanoreceptors, which primarily include Pacinian corpuscles, Merkel's disks, and Ruffini's endings. These receptors then link to Aδ and C-fibers (discussed above) to convey pain to the central nervous system (Olesen et al. 2012).

Pressure Pain Test

Inducing pain with pressure involves the use of a device, such as an algometer, to apply controlled amounts of force to a specific part of the body. Pressure is increased at a constant rate, typically at 0.5–1 kgf/cm^2 per second, and is manipulated to assess latency (in seconds) to pain threshold and tolerance, or the maximum amount of pressure tolerated in units of force (e.g., kgf/cm^2 or N/cm^2). Trials are often repeated and averaged for a final measurement of pain detection and/or threshold (Chesterton et al. 2007; Arendt-Nielsen and Yarnitsky 2009). If a measure of temporal summation is desired, this test can also be easily repeated similarly to the heat pain paradigm described above.

Von Frey Test

Assessing pain with VF monofilaments has experimental and clinical relevance; this paradigm is used as a screening tool to ensure normal nerve functioning prior to enrollment in a human laboratory study, or it can be utilized to determine the extent/progression of a clinical neuropathic pain diagnosis. In the human laboratory, the former paradigm is used to determine change in baseline monofilament sensitivity after drug administration. These monofilaments are thin nylon fibers that ascend in calibrated amounts of force (typically ranging from 0.008 to 300 g) that can be applied to the skin. In humans, ascending fibers are applied with force to the skin perpendicularly until they bend. If an individual does not detect the stimulation, the next size monofilament is applied until sensation and/or pain is noted. The first monofilament to elicit a detectable sensation is noted as the mechanical threshold. That which elicits pain is noted as pain threshold, and the monofilament that results in intolerable pain is the outcome for pain tolerance (Rolke et al. 2006). A limitation of this paradigm is the subjective force each filament is applied with by the experimenter; however, electronic, automatic applicators are available to remove this bias.

3.1.3 Other Pain-Induction Paradigms

Capsaicin

Capsaicin-induced pain models are clinically relevant in that they model neuropathic pain conditions (Olesen et al. 2012). This paradigm produces long-lasting areas of increased pain sensitivity and is conducted using an intradermal injection of capsaicin or by applying a capsaicin patch or cream to the skin. Following capsaicin administration, a map of secondary hyperalgesia is assessed (in cm^2) at baseline and again after drug infusion using a VF monofilament. Typically, a digital photograph is then taken and analyzed using a computer software package to calculate the hyperalgesic surface area (Dirks and Petersen 2003).

Capsaicin activates the TRPV1 receptor system, leading to activation of Aδ and C-fibers that communicate pain to the brain (discussed above; (LaMotte et al. 1992)). This paradigm is less often used in human laboratory studies due to several limitations: (1) a physician must administer the capsaicin injection; (2) it is not repeatable within sessions and thus is not ideal for determining changes in drug effects over time (i.e., a time course of analgesic effects).

Electrical Stimulation

Electrical stimulation is a final paradigm that is often used to determine the analgesic potential of drugs, including cannabis. In this paradigm, a device similar to a TENS unit (e.g., Innervator Model) provides stimulation to a skin surface area using electrodes. Electrical current (mAmps) is typically started at 0 mAmps and progressively increased at specified intervals until pain is noted at a pre-determined significant rating on a numeric rating scale (NRS; ranging from no pain to the worst pain imaginable). The lowest current that produces a painful sensation is noted as the pain threshold, while the current eliciting a predetermined significant score on the VNS is noted as pain tolerance. Multiple applications of stimulation are typically performed at each time point and averaged as the outcome (Rolke et al. 2006). Neurobiologically, the noxious electrical stimulus bypasses specialized sensory nerve endings and mediates pain via peripheral nerve fiber activation (Olesen et al. 2012).

3.2 *Human Laboratory Studies of Cannabinoid Analgesia*

Human laboratory studies present a unique opportunity to rigorously investigate pharmacological parameters of cannabinoid analgesia including dose, route of administration, and time course of action (Arout et al. 2018). While true analgesic efficacy is tethered to study population and pain diagnosis, patients with chronic pain syndromes present with numerous sources of variance that can impact the ability to interpret efficacy of a compound in altering pain, including a) the use of other analgesic medications, b) intrapersonal differences in pain perception, c) daily, within-person fluctuations in pain, and d) comorbid medical and psychiatric conditions that can also impact pain (e.g., anxiety and/or depression, which can manifest as distress intolerance and/or pain catastrophizing). For these reasons, the effects of compounds must first be characterized in healthy populations using human laboratory models of pain, within which inter- and intra-person differences can be accounted for, and experimental pain paradigms where the nuances of the noxious stimuli are stable and can be controlled and manipulated (i.e., intensity, duration, location, etc.) are used (Olesen et al. 2012). It is also important to note that drug effects differ based on experience with the drug, which can be manipulated using human laboratory models (i.e., individuals' baseline tolerance, comfort with subjective effects, etc.). As such, we will review human laboratory findings in the context of those not currently using cannabis vs. individuals currently using cannabis.

3.2.1 THC

Studies among *individuals who are either cannabis-naïve or not currently using cannabis* have largely revealed no analgesic potential following various routes of acute THC administration and, in fact, several have reported increased pain sensitivity (hyperalgesia). In a small sample ($n = 12$) of a single dose of oral THC (20 mg) in cannabis-naïve participants, thermal (CPT and heat pain), pressure, and electrical pain were assessed every hour for 8 h post-THC administration. THC did not produce analgesia on any paradigm, and in fact increased thermal pain sensitivity (i.e., hyperalgesia) in the CPT and heat pain models, though this effect was not significant (Naef et al. 2003). Another study in six cannabis-experienced (no use in the past month) participants assessed thermal (cold and heat), VF, electrical, or capsaicin-induced pain following IV-administered THC (0.01, 0.03 mg/kg) relative to placebo over 4 h across three testing days. While all active doses of THC produced euphoria, none produced analgesia on any measures of experimental pain at any timepoint (Schindler et al. 2020). In another study of 24 volunteers who were drug-abstinent, the analgesic potential of a single dose of oral THC (10 mg) was tested against a battery of pain tests, including electrical stimulation, pressure, and thermal pain (CPT and heat) over 10 h post-administration. Under these conditions, oral THC did not produce any analgesia compared to placebo and, in fact, induced hyperalgesia on the electrical stimulation and pressure pain paradigms (van Amerongen et al. 2018a). Another study utilizing several doses of nabilone (1, 2, or 3 mg) using topical and intradermal capsaicin models in men not currently using cannabis ($n = 24$), significant subjective effects were observed (e.g., increases in "sedation," etc.), supporting that these doses were indeed psychoactive. However, no changes in pain perception were evident (Kalliomäki et al. 2012). Other findings have been similar; oral THC (2.5 or 5 mg dronabinol vs. placebo) did not alter

experimental pain outcomes (pressure, CPT, heat) in 10 otherwise healthy volunteers that were not currently using cannabis (Babalonis et al. 2019).

In slight contrast, the THC analogue nabilone (0.5, 1 mg) has been found to reduce temporal summation at a 1 mg dose in cannabis-abstinent women, but not in men (total $n = 20$). Importantly, reduction in temporal summation is not one in the same with analgesia, and rather these findings should be interpreted as preventing hyperalgesia (Redmond et al. 2008). In 15 cannabis abstinent individuals, there was variable change in intradermal capsaicin-induced hyperalgesia following smoked active cannabis (2%, 4%, 8% THC) vs. placebo. In this study, cannabis' effects on pain were dose-dependent; the 4% THC strength produced modest analgesia, while 8% induced hyperalgesia 45 min post-cannabis administration (Wallace et al. 2007). Finally, an oral THC extract (20 mg) did not produce analgesia in 18 non-cannabis using women undergoing intradermal capsaicin and electrical stimulation pain models (Kraft et al. 2008). Altogether, these human laboratory studies suggest that a range of THC doses across varying routes of administration have limited analgesic potential and may in fact increase pain sensitivity in those who are not regularly using cannabis.

Data from *individuals who currently use cannabis* has yielded more positive findings of THC-induced analgesia. For instance, dose-dependent analgesia following both smoked cannabis (0, 1.98, 3.56% THC) and oral THC (0, 10, 20 mg) administration was observed using the CPT (Cooper et al. 2013). Compared to placebo, both smoked cannabis (3.56% THC) and oral THC (20 mg) increased latency to pain detection (i.e., analgesia), while the 1.98% smoked cannabis dose and 20 mg oral THC conditions increased pain tolerance. All active cannabis conditions also decreased ratings of pain intensity and bothersomeness. Importantly, while peak analgesic effects did not differ between smoked and oral conditions, oral THC produced longer-lasting analgesia (Cooper et al. 2013); this is significant for medicinal cannabis patients who often report smoking cannabis to quickly relieve pain, while using oral formulations to carry their pain relief through longer periods of time (e.g., a work shift, social obligations, sleep, etc.). These findings were later extended to include sex differences; cannabis-induced (3.56–5.60% THC smoked cannabis vs. placebo) analgesia was more pronounced in men vs. women. Specifically, active cannabis decreased pain sensitivity and increased pain tolerance on the CPT in men, but effects in women were limited to increased pain tolerance (Cooper and Haney 2016). Dose-dependent analgesia was also reported in a small sample of individuals who used cannabis regularly ($n = 5$), where more puffs (18) of active smoked cannabis (3.56% THC) yielded significant increases in heat pain tolerance vs. less puffs (0, 3, or 9) (Greenwald and Stitzer 2000). Taken together with findings from individuals who do not regularly use cannabis largely showing no analgesic potential, these studies suggest that experience with THC's effects may be an important factor in its potential to alter pain perception.

3.2.2 CBD

Very few studies exist examining the analgesic potential of CBD alone in otherwise healthy volunteers. Acute oral CBD (0, 200, 400, 800 mg) administration in individuals who do not use cannabis (n = 17) did not significantly alter CPT pain detection or tolerance latencies, and rather increased subjective ratings of painfulness (Arout et al. 2022). Another study reported similar negative findings; an acute oral dose of CBD (800 mg vs. placebo) was unsuccessful in altering intradermal capsaicin pain outcomes in 20 non-cannabis using individuals (Schneider et al. 2022). In a follow-up study by the same group, 1,600 mg of acutely administered oral CBD was also unsuccessful in reversing experimentally induced remifentanil hyperalgesia in healthy volunteers (Dieterle et al. 2022). Finally, no effect of an acutely administered dose of CBD oil (50 mg, sublingual administration) was found in healthy volunteers; there were no changes in pain using a noxious heat paradigm (De Vita et al. 2022).

Overall, these studies do not support that *acute* oral CBD pretreatment in isolation alters experimental pain. Studies have reached similar conclusions for the lack of CBD effect on pain, postulating that higher doses of CBD may be needed to exert analgesic or anti-hyperalgesic effects. It is also possible that prolonged dosing paradigms (vs. the acute administration paradigms utilized in these studies) are needed for CBD to exert effects, particularly considering CBD's anti-inflammatory mechanisms. Finally, it is possible that experimentally induced pain states are not sensitive to CBD's effects; preclinical studies (reviewed above) have shown CBD to exert analgesic effects primarily in neuropathic pain models, and many of the clinical studies that have reported analgesic effects of CBD are in patients with neuropathic pain syndromes where THC was also administered. As such, it is possible that outcomes in a clinical chronic pain population with neurobiological dysfunction may be more sensitive to CBD's potential effects. Findings also suggest that CBD may be ineffective in isolation but may have synergistic effects when combined with THC.

3.2.3 THC and CBD Interactions

Data suggest that individuals who use cannabis medically are increasingly seeking out products that contain a combination of THC and CBD, with most medicinal cannabis patients (>70%) preferring CBD-dominant chemovars (Boehnke et al. 2019b). Further, there is evidence that as individuals become more comfortable with cannabis, they seek out products with higher THC content (Pernía-Andrade et al. 2009; Kim et al. 2019). Pain diagnosis also putatively influences chemovar preference, as patients with neuropathic pain typically opt for lower THC:CBD products, while those battling cancer pain prefer products that are higher in THC: CBD varietals (Kim et al. 2019). Thus, characterizing the impact of THC and CBD combinations on pain is critically important. Yet, there are few rigorously controlled human laboratory studies that have touched on this.

3.3 *Randomized Controlled Trials in Pain Populations*

While randomized controlled clinical trials (RCTs) have occasionally incorporated some of the experimental pain paradigms discussed above, by far, the most commonly used tools to measure changes in pain are VAS, numeric rating scales (NRS), and the Brief Pain Inventory (BPI). VAS and NRS use slightly different techniques to arrive at the same goal: a pain rating. VAS often utilizes the Wong-Baker FACES Pain Rating Scale (Wong 1988), which provides photos of cartoon faces with expressions ranging from smiling (indicating no pain) to frowning (worst pain imaginable). This scale is particularly useful for children and/or non-verbal patients to indicate their current level of pain. A VAS includes a horizontal or vertical line with anchors at either end to denote value ranges (e.g., 0–100 mm, where higher ratings are indicative of increased pain), and patients are asked to bisect the line where it represents their current pain level (Huskisson 1974). These are both similar to NRS, which typically request patients to verbally report, using a scale of 0–10 that represents their current pain level (0 = no pain, 10 = worst pain imaginable) (Jensen et al. 1986). The subjective nature of these scales presents certain limitations; however, the field is actively establishing a more reliable, easily accessible and objective biomarker of pain [e.g., a laboratory blood test for inflammatory markers (Farrell et al. 2024), changes in pupil diameter (Krafthöfer et al. 2024), etc.].

RCTs are a gold standard for determining medication efficacy as compared to placebo or an active control in clinical populations. There are a number of RCTs reporting on various formulations of cannabinoids for a range of chronic pain conditions, although results are mixed (Stockings et al. 2018) and largely support the use of cannabinoids for neuropathic pain conditions (Aviram and Samuelly-Leichtag 2017). Many RCTs have investigated nabiximols (1:1 ratio of oromucosal THC:CBD) for neuropathic pain conditions, indicating potential utility in isolation or as an adjunct; a few have studied THC or CBD products in isolation. Note, many of these studies acknowledged allowing patients to continue their established analgesic regimen, with cannabinoid products added adjunctively. Studies that conclusively indicated cannabinoids as an adjunct to opioid therapy are briefly summarized separately, with additional considerations discussed below.

3.3.1 Neuropathic Pain

Although findings are mixed, the most promising evidence supporting the use of cannabinoids for managing chronic pain comes from studies of neuropathic pain conditions, which can range in origin and are often clinically consolidated when discussed in studies. Leveraging a flexible dosing paradigm of oral THC, CBD, or a combination (dose ranges: THC-only condition, 2.5–25 mg; CBD-only condition, 5–50 mg; THC:CBD combination condition, 2.5 mg THC:5 mg CBD–25 mg THC:50 mg CBD) in patients with peripheral neuropathy (n = 115), no condition significantly reduced NRS pain scores (Zubcevic et al. 2023); these negative findings are supported by a similar RCT (Serpell et al. 2014).

In contrast, analgesia was demonstrated following a THC, CBD, or nabiximols administration vs. placebo over a 2-week period in 24 patients with various types of chronic pain, including multiple sclerosis-associated pain, spinal cord injury, brachial plexus damage, and limb amputation (Wade et al. 2003). Other similar studies in patients with various neuropathic pain conditions have reported courses of nabiximols to be effective in reducing pain as measured by a NRS (Langford et al. 2013; Nurmikko et al. 2007). In another study utilizing flexible dosing of a THC (2.5 mg), CBD (2.5 mg), or a matched THC:CBD ratio (nabiximols) vs. placebo for neuropathic pain over a 12-week period, patients ($n = 34$) receiving THC or nabiximols reported significant reductions in pain ratings, while CBD-only did not differ from placebo (Notcutt et al. 2004).

A handful of studies have tested the increasingly ecologically relevant non-combustible vaporized/inhaled route of administration. For instance, a single, inhaled dose of either 0.5 mg or 1 mg of THC vs. placebo reduced neuropathic pain on a VAS, which persisted for 2.5 h post-administration (Almog et al. 2020). A single average dose of 15.3 ± 2.1 mg THC oil has also reportedly improved pain inhibition via a CPM paradigm in a population of men with chronic neuropathic pain (Weizman et al. 2024). A low dose of vaporized cannabis (1.29%, 3.53% THC) also produced significant reductions in neuropathic pain intensity in patients not currently using cannabis; importantly, these doses of THC did not induce abuse-related effects (Wilsey et al. 2013). Another study investigated active vaporized cannabis vs. placebo in patients with neuropathic pain; across doses of 1%, 4%, or 7% THC, and analyzed analgesic efficacy as a function of plasma THC level. They reported a U-shaped effect, where lower and higher plasma levels did not correlate with any reduction in pain, but levels ranging between 16 ng/mL and 31 ng/mL were associated with analgesia (Wallace et al. 2020). In another study of patients with neuropathic pain, a 3×/day single inhalation of 9.4% THC cannabis (vs. placebo) resulted in a significant reduction of pain on a NRS (Ware et al. 2010). Similarly, vaporized cannabis (15.1 ± 0.1 mg THC) also reduced neuropathic pain by 45% in patients using cannabis regularly (Eisenberg et al. 2014).

Several RCTs have also investigated smoked cannabis; Wilsey et al. (2016) reported efficacy of cannabis (2.9% and 6.7% THC) in reducing neuropathic pain using a NRS (Wilsey et al. 2016). In a study of patients with diabetic neuropathy, higher doses of smoked cannabis (7%) reduced pain as measured by VF stimulation as compared to lower active cannabis doses (1%, 4%) and placebo (Wallace et al. 2015). In a study of patients with HIV-related neuropathy, active cannabis (potency range of 1–8% THC administered 4×/day for 5 consecutive days) resulted in 46% of patients reporting at least 30% reduction in pain intensity vs. placebo (Ellis et al. 2008).

Chemotherapy-induced neuropathic pain (CIPN) is a particularly treatment-resistant form of neuropathic pain and can be particularly problematic because this adverse effect drastically limits retention in cancer treatment. Recent studies have investigated cannabinoids' potential to alleviate CIPN; a course of nabiximols yielded a trend toward NRS pain improvement in a small sample of patients with CIPN, warranting further study (Lynch et al. 2014). Other recent studies have been less encouraging. In an 8-week, placebo-controlled study of oral CBD (100 mg CBD:5 mg THC, TID) in CIPN patients ($n = 12$), a subset of patients in both active and placebo groups requested dose reductions, and findings indicated no difference in pain in the previous 24 h for the active condition vs. placebo. Active cannabis also worsened ratings of pain interference (Haney et al. 2025). More placebo-controlled studies are needed to better understand the potential efficacy of cannabinoids for CIPN.

Other open-label studies using cannabinoid products in patients with neuropathic pain have reported mixed effects. One study reported no clinically significant effects following 6 weeks of escalating nabilone administration in patients with neuropathic pain (Frank et al. 2008), while another found THC:CBD oromucosal administration to induce successful analgesia in patients with various neuropathic pain syndromes (Hoggart et al. 2015). However, these two studies were not placebo controlled.

3.3.2 Fibromyalgia

Cannabinoids are of interest in patients with fibromyalgia, particularly due to this condition's similarity to neuropathic pain; yet, reported efficacy is mixed. In a study of women with fibromyalgia ($n = 17$), patients received 8 weeks of a THC-dominant cannabis oil (24.44 mg/mL THC, 0.51 mg/mL CBD) vs. placebo. Patients in the active condition reported an impressive reduction of 4.5 points on a 0–10 NRS (Chaves et al. 2020). In a separate study testing three active, inhaled cannabis chemovars (22.4 mg THC:<1 mg CBD; 13.4 mg THC:17.8 mg CBD; or 18.4 mg CBD:<1 mg THC) vs. placebo ($n = 20$), no effect of any condition was found using an electrical stimulation pain paradigm, but both high-THC conditions reduced experimental pressure pain. However, clinical relevance of these doses is unclear given the finding that none of the doses reduced NRS ratings of fibromyalgia pain (Van De Donk et al. 2019). In a third study, titrated daily doses of nabilone (up to 1 mg over 4 weeks) were reported to be efficacious for reducing fibromyalgia pain when compared to placebo (Skrabek et al. 2008).

3.3.3 Multiple Sclerosis-Related Pain

There are a handful of studies investigating various routes of cannabinoid administration in patients with multiple sclerosis (MS)-induced pain, and most are promising. In one placebo-controlled, crossover study, daily nabilone (0.5 mg per day for 1 week, followed by 1 mg/day for 3 weeks vs. placebo) reduced pain on a NRS

(Wissel et al. 2006). Similarly, another study utilized either 1.5 or 5 mg/3x day of oral THC tablets vs. placebo for 4 weeks and reported modest improvements in pain ratings (van Amerongen et al. 2018b). Additionally, patients with MS-related chronic pain were given a maximum daily dronabinol dose of 10 mg per day for 3 weeks and reported modest analgesic effects compared to placebo (Svendsen et al. 2004). Finally, flexible nabiximols dosing (up to 48 sprays/day for 4 weeks) resulted in significant NRS reductions compared to placebo (Rog et al. 2005).

3.3.4 Other Types of Pain

A smaller number of studies have investigated cannabinoids' effects in other clinical pain populations. For instance, some have investigated patients with chronic abdominal pain; neither an acute (a single 8 mg dose of oral THC; (de Vries et al. 2016, 2017)) or longer-term administration (7 weeks of adjunctive oral THC doses up to 24 mg; (de Vries et al. 2017)) altered NRS pain scores when compared to placebo. In a sample of patients with sickle cell disease ($n = 23$) using a range of other analgesics, a vaporized ratio of THC:CBD (4.4%:4.9%) vs. placebo 3×/day for 5 days; cannabis did not reduce NRS pain (Abrams et al. 2020).

In a study comparing oromucosal THC (2.7 mg) and THC:CBD (2.7 mg THC and 2.5 mg CBD; nabiximols) extracts to placebo in patients with intractable cancer pain ($n = 177$), patients self-titrated to a dose they were comfortable with. Those receiving nabiximols reported a significant reduction in NRS pain compared to the THC-only and placebo extracts (Johnson et al. 2010). Finally, Blake et al. (2006) noted a small improvement in pain following nabiximols vs. placebo over 5 weeks in a population of patients with rheumatoid arthritis (Blake et al. 2006).

4 Discussion/Additional Considerations

4.1 Intoxication and Abuse Liability

Based largely on findings from human laboratory studies, cannabis has subjective intoxicating effects (e.g., euphoria), which underlies its abuse liability (e.g., self-administration, subjective drug effects, and reinforcing effects; (Haney et al. 1997, 2003, 2004, 2008, 2010, 2013, 2016, 2019), and increases the likelihood that individuals will develop a cannabis use disorder (CUD). In the USA, an estimated 19.4 million individuals reported experiencing past-year CUD (Substance Abuse and Mental Health Services Administration 2019). Repeated use of cannabis throughout the day to manage pain symptoms could increase the risk of escalating use patterns (Hasin 2017; Hasin et al. 2017); thus, patients using cannabis to manage pain are at increased risk for developing a CUD. As both the rates of CUD and cannabis use for pain continue to rise, it highlights the public health importance of monitoring and understanding cannabis-related risks, particularly in the context of pain management.

4.2 *Cannabinoids in Conjunction with Opioids*

Amidst a vast preclinical literature studying cannabinoids in conjunction with opioids for pain (Tucci 2010; Nielsen et al. 2017; Le Foll 2021), there is a growing human literature in this area (Babalonis and Walsh 2020; Guy et al. 2024). Cannabinoids are being increasingly studied in combination with opioids to investigate opioid-sparing/potential synergistic effects on pain-related outcomes, using human laboratory models of healthy adults (e.g., (Naef et al. 2003; Babalonis et al. 2019; Roberts et al. 2006; Dunn et al. 2021)) or cannabis-using volunteers undergoing experimental pain paradigms (e.g., (Cooper et al. 2018)), patients with opioid use disorder (De Aquino et al. 2023), or those with various chronic pain conditions (e.g., (Narang et al. 2008; Abrams et al. 2011; Portenoy et al. 2012; Lichtman et al. 2018; Campbell et al. 2023). Further, some have theorized that cannabis legalization is correlated with decreased rates of opioid misuse and use disorders, though recent data suggests a muddied relationship (Martins et al. 2025). Findings across studies have been mixed; thus, more controlled research is needed to better define the relationship between cannabinoids and opioids.

4.3 *Minor Cannabinoids and Terpenes*

While we have focused on the primary cannabinoids in this chapter, there is a host of preclinical data and an evolving focus in human studies on the potential analgesic efficacy and therapeutic utility of minor cannabinoids [e.g., cannabichromene (CBC), cannabinol (CBN), and cannabigerol (CBG)] and terpenes (e.g., D-limonene, β-myrcene, α-pinene, and β-caryophyllene). CBC, CBN, and CBG interact with TRPV1 and TRPA1 receptors; it is postulated that this mechanism underlies the analgesic efficacy displayed in animal models (Janero and Makriyannis 2014; Walsh et al. 2021; Muller et al. 2019; Liktor-Busa et al. 2021). Further, data suggests that combinations of minor cannabinoids and THC could elicit an entourage effect in reducing pain.

Preclinical studies have shown that terpenes may also possess a range of therapeutic properties (Rock and Parker 2020). Cannabis-derived terpenes act as nonselective modulators at various G-protein–coupled receptors (GPCRs), including opioid, CB1, and CB2 receptors, as well as TRPV channels. Additionally, they can activate anti- or pro-inflammatory pathways to alter nociceptive transmission (Liktor-Busa et al., 2021). Considering their molecular targets and promising preclinical studies, terpenes warrant further research as potential analgesics.

4.4 Tolerance and Withdrawal

A distinct cannabis withdrawal syndrome has been described and includes increased irritability, anxiety, cannabis craving, decreased quality and quantity of sleep, and decreased caloric intake (Budney et al. 1999, 2003, 2004; Haney et al. 1999a, b; Liktor-Busa et al. 2021). However, there is no rigorously controlled data characterizing how cannabis withdrawal may impact pain sensitivity; thus, studies are needed.

While tolerance to cannabis' effects is anecdotally reported, and tolerance breaks ("T-breaks") are a widely used behavioral self-management strategy to reduce cannabis use (Walukevich-Dienst et al. 2025), tolerance to cannabis' analgesic effects has also not been systemically characterized. One study of neuropathic pain patients indirectly suggested that analgesic tolerance does not develop because pain relief from an initial dose of nabiximols remained efficacious for 1 year (Nurmikko et al. 2007). In another 3-month inpatient study (Clark et al. 1981), repeatedly administered low cannabis dose (2% THC) was tested against thermal pain in cannabis-using men. Prior to enrollment, volunteers were variably abstinent for the first 3–4 weeks, followed by a month of daily ad libitum cannabis administration (ranging from 3 to 12 cannabis cigarettes/day), ending with 3–4 weeks of abstinence. While cannabis was initially successful in significantly reducing pain, analgesic efficacy decreased over time relative to baseline measurements. Specifically, those who used higher amounts of cannabis during the study became more sensitive to experimental pain compared to those who used less (Clark et al. 1981). While this study suggests that analgesic tolerance does develop, and corroborates the dose-dependent hyperalgesia found in other studies (reviewed above), it did not control for cannabis use across the exclusively male volunteers. Thus, no well-controlled studies have directly examined whether tolerance develops to the analgesic effects of cannabis either in healthy volunteers or in pain patients. These outcomes hold significant public health relevance; as shown, cannabis use is increasing, particularly for pain management. It is imperative that we understand the full range of cannabis' impact on pain, so that patients can make informed decisions regarding use for pain and are appropriately guided by their treatment providers.

4.5 Peri-Operative Issues

An emerging issue for individuals who use cannabis, which parallels what has been established with other drugs of abuse including opioids (Lee et al. 2014) and nicotine (Khalil et al. 2021; Karsalia et al. 2021), is changes in pain perception following long-term use and upon subsequent abrupt cessation. Anesthesiologists are increasingly reporting a unique profile for individuals who use cannabis, including debilitating anxiety leading up to surgery that requires anxiolytics and difficulties in anesthesia induction necessitating increased sedatives for successful induction (Imasogie et al. 2021; Kosirog et al. 2024). These patients also experience poor postoperative outcomes, including much higher postoperative opioid requirements, and worsened pain and other functional outcomes (e.g., sleep) in the months following surgery when compared to non-cannabis using individuals (Bicket et al. 2024; Ekrami et al. 2024). These reports are extremely concerning for medical patients using cannabis for pain, who may reduce or stop their use entirely leading up to surgery and will subsequently experience cannabis withdrawal-induced effects, including increased anxiety and changes in pain perception. While there are several retrospective studies supporting these reports, rigorously controlled investigations that elucidate how cannabis use may impact perioperative outcomes are nonexistent yet are essential for guiding treatment providers in the medical advice offered to their cannabis-using patients.

5 Concluding Remarks

In conclusion, preclinical studies are largely supportive of cannabinoid efficacy for pain using both experimental pain paradigms and models of chronic pain syndromes. Human laboratory studies and RCTs are more complex and less conclusive; some are supportive while others are negative. Interpretation is obfuscated by a number of methodological inconsistencies, including varying (1) samples of individuals (both healthy and with a clinical pain diagnosis), (2) cannabinoid doses, (3) time frames (acute vs. repeated), (4) routes and methods of administration, (5) assessment methodologies, and (6) in some cases, a lack of placebo control. More rigorous, placebo-controlled studies are needed to firmly characterize the therapeutic potential of cannabinoids, both primary and minor, alone and in combination with other medications for chronic pain management. These studies are crucial in order to provide evidence-based information to help guide physicians when advising patients about the use of cannabinoids for pain.

References

Abboud C et al (2021) Animal models of pain: diversity and benefits. J Neurosci Methods 348: 108997

Abdulkhaleq L et al (2018) The crucial roles of inflammatory mediators in inflammation: a review Vet World 11(5):627

Abraham AD et al (2020) Orally consumed cannabinoids provide long-lasting relief of allodynia in a mouse model of chronic neuropathic pain. Neuropsychopharmacology 45(7):1105–1114

Abrams D et al (2011) Cannabinoid–opioid interaction in chronic pain. Clin Pharmacol Ther 90(6): 844–851

Abrams DI et al (2020) Effect of inhaled cannabis for pain in adults with sickle cell disease: a randomized clinical trial. JAMA Netw Open 3(7):e2010874

Abuse and Mental Health Services Administration (2019) Key substance use and mental health indicators in the United States: Results from the 2018 National Survey on Drug Use and Health, vol 170. HHS Publication No. PEP19-5068, NSDUH Series H-54, pp 51–58

Akopian AN et al (2008) Cannabinoids desensitize capsaicin and mustard oil responses in sensory neurons via TRPA1 activation. J Neurosci 28(5):1064–1075
Aley K, Reichling D, Levine J (1996) Vincristine hyperalgesia in the rat: a model of painful vincristine neuropathy in humans. Neuroscience 73(1):259–265
Allen JW, Yaksh TL (2004) Assessment of acute thermal nociception in laboratory animals. In: Luo ZD (ed) Pain research: methods and protocols. Humana Press, Totowa, pp 11–23
Almog S et al (2020) The pharmacokinetics, efficacy, and safety of a novel selective-dose cannabis inhaler in patients with chronic pain: a randomized, double-blinded, placebo-controlled trial. Eur J Pain 24(8):1505–1516
Arantes ALF et al (2024) Antinociceptive action of cannabidiol on thermal sensitivity and post-operative pain in male and female rats. Behav Brain Res 459:114793
Arendt-Nielsen L, Yarnitsky D (2009) Experimental and clinical applications of quantitative sensory testing applied to skin, muscles and viscera. J Pain 10(6):556–572
Arout CA, Herrmann E, Haney M (2018) Human laboratory models of cannabis use disorder. In: Cannabis use disorders. Springer, pp 75–84
Arout CA et al (2022) A placebo-controlled investigation of the analgesic effects, abuse liability, safety and tolerability of a range of oral cannabidiol doses in healthy humans. Br J Clin Pharmacol 88(1):347–355
Authier N et al (2009) Animal models of chemotherapy-evoked painful peripheral neuropathies. Neurotherapeutics 6(4):620–629
Aviram J, Samuelly-Leichtag G (2017) Efficacy of cannabis-based medicines for pain management: a systematic review and meta-analysis of randomized controlled trials. Pain Physician 20(6): E755–e796
Azcarate PM et al (2020) Medical reasons for marijuana use, forms of use, and patient perception of physician attitudes among the US population. J Gen Intern Med 35:1979–1986
Babalonis S, Walsh SL (2020) Therapeutic potential of opioid/cannabinoid combinations in humans: review of the evidence. Eur Neuropsychopharmacol 36:206–216
Babalonis S et al (2019) Cannabinoid modulation of opioid analgesia and subjective drug effects in healthy humans. Psychopharmacology 236:3341–3352
Baron EP (2018) Medicinal properties of cannabinoids, terpenes, and flavonoids in cannabis, and benefits in migraine, headache, and pain: an update on current evidence and cannabis science. Headache 58(7):1139–1186
Becker S et al (2018) Emotional and motivational pain processing: current state of knowledge and perspectives in translational research. Pain Res Manag 2018(1):5457870
Benítez-Angeles M et al (2020) TRPV1: structure, endogenous agonists, and mechanisms. Int J Mol Sci 21(10):3421
Bennett GJ, Xie Y-K (1988) A peripheral mononeuropathy in rat that produces disorders of pain sensation like those seen in man. Pain 33(1):87–107
Bicher HI, Mechoulam R (1968) Pharmacological effects of two active constituents of marihuana. Arch Int Pharm Ther 172(1):24–31
Bicket MC et al (2024) The association of cannabis use after discharge from surgery with opioid consumption and patient-reported outcomes. Ann Surg 279(3):437–442
Blake DR et al (2006) Preliminary assessment of the efficacy, tolerability and safety of a cannabis-based medicine (Sativex) in the treatment of pain caused by rheumatoid arthritis. Rheumatology 45(1):50–52
Boehnke KF, Gangopadhyay S, Clauw DJ, Haffajee RL (2019a) Qualifying conditions of medical cannabis license holders in the United States. Health Affairs 38(2):295–302
Boehnke KF et al (2019b) Cannabis use preferences and decision-making among a cross-sectional cohort of medical cannabis patients with chronic pain. J Pain 20(11):1362–1372
Boggs DL et al (2018) Clinical and preclinical evidence for functional interactions of Cannabidiol and Delta(9)-tetrahydrocannabinol. Neuropsychopharmacology 43(1):142–154
Bonn-Miller MO et al (2014) Self-reported cannabis use characteristics, patterns and helpfulness among medical cannabis users. Am J Drug Alcohol Abuse 40(1):23–30

Booker L et al (2009) Evaluation of prevalent phytocannabinoids in the acetic acid model of visceral nociception. Drug Alcohol Depend 105(1–2):42–47
Brennan TJ, Vandermeulen EP, Gebhart G (1996) Characterization of a rat model of incisional pain. Pain 64(3):493–502
Britch SC et al (2017) Cannabidiol-Delta(9)-tetrahydrocannabinol interactions on acute pain and locomotor activity. Drug Alcohol Depend 175:187–197
Britch SC et al (2020) Antinociceptive and immune effects of Delta-9-tetrahydrocannabinol or Cannabidiol in male versus female rats with persistent inflammatory pain. J Pharmacol Exp Ther 373(3):416–428
Budney AJ, Novy PL, Hughes JR (1999) Marijuana withdrawal among adults seeking treatment for marijuana dependence. Addiction 94(9):1311–1322
Budney AJ et al (2003) The time course and significance of cannabis withdrawal. J Abnorm Psychol 112(3):393
Budney AJ et al (2004) Review of the validity and significance of cannabis withdrawal syndrome. Am J Psychiatry 161(11):1967–1977
Burma NE et al (2017) Animal models of chronic pain: advances and challenges for clinical translation. J Neurosci Res 95(6):1242–1256
Campana G, Rimondini R (2015) Mechanical nociception measurement in mice and rats with automated Von Frey equipment. Methods Mol Biol 1230:229–231
Campbell CM et al (2023) Within-subject, double-blind, randomized, placebo-controlled evaluation of combining the cannabinoid dronabinol and the opioid hydromorphone in adults with chronic pain. Neuropsychopharmacology 48(11):1630–1638
Carta G, Gessa GL, Nava F (1999) Dopamine D2 receptor antagonists prevent Δ9-tetrahydrocannabinol-induced antinociception in rats. Eur J Pharmacol 384(2–3):153–156
Casey SL, Atwal N, Vaughan CW (2017) Cannabis constituent synergy in a mouse neuropathic pain model. Pain 158(12):2452–2460
Casey SL et al (2022) Intrathecal actions of the cannabis constituents Δ(9)-tetrahydrocannabinol and Cannabidiol in a mouse neuropathic pain model. Int J Mol Sci 23(15):8649
Caterina MJ, Julius D (1999) Sense and specificity: a molecular identity for nociceptors. Curr Opin Neurobiol 9(5):525–530
Chaplan SR et al (1994) Quantitative assessment of tactile allodynia in the rat paw. J Neurosci Methods 53(1):55–63
Chaves C, Bittencourt PCT, Pelegrini A (2020) Ingestion of a THC-rich cannabis oil in people with fibromyalgia: a randomized, double-blind, placebo-controlled clinical trial. Pain Med 21(10): 2212–2218
Chen L et al (2009) Altered quantitative sensory testing outcome in subjects with opioid therapy. Pain 143(1):65–70
Chesher GB et al (1973) The effect of cannabinoids on intestinal motility and their antinociceptive effect in mice. Br J Pharmacol 49(4):588–594
Chesterton LS et al (2007) Interrater reliability of algometry in measuring pressure pain thresholds in healthy humans, using multiple raters. Clin J Pain 23(9):760–766
Cichewicz DL (2004) Synergistic interactions between cannabinoid and opioid analgesics. Life Sci 74(11):1317–1324
Clark WC et al (1981) Effects of moderate and high doses of marihuana on thermal pain: a sensory decision theory analysis. J Clin Pharmacol 21(S1):299S–310S
Collier HO et al (1968) The abdominal constriction response and its suppression by analgesic drugs in the mouse. Br J Pharmacol Chemother 32(2):295–310
Compton P et al (2010) Gabapentin improves cold-pressor pain responses in methadone-maintained patients. Drug Alcohol Depend 109(1):213–219
Compton P et al (2012) Hyperalgesia in heroin dependent patients and the effects of opioid substitution therapy. J Pain 13(4):401–409
Conley KM et al (1997) Modulating effects of a cold water stimulus on opioid effects in volunteers. Psychopharmacology 131(4):313–320

Conti S et al (2002) Antiinflammatory action of endocannabinoid palmitoylethanolamide and the synthetic cannabinoid nabilone in a model of acute inflammation in the rat. Br J Pharmacol 135(1):181–187
Cook AD et al (2018) Immune cytokines and their receptors in inflammatory pain. Trends Immunol 39(3):240–255
Cooper ZD, Haney M (2008) Cannabis reinforcement and dependence: role of the cannabinoid CB1 receptor. Addict Biol 13(2):188–195
Cooper ZD, Haney M (2016) Sex-dependent effects of cannabis-induced analgesia. Drug Alcohol Depend 167:112–120
Cooper ZD, Comer SD, Haney M (2013) Comparison of the analgesic effects of dronabinol and smoked marijuana in daily marijuana smokers. Neuropsychopharmacology 38(10):1984
Cooper ZD et al (2018) Impact of co-administration of oxycodone and smoked cannabis on analgesia and abuse liability. Neuropsychopharmacology 43:2046
Corroon JM Jr, Mischley LK, Sexton M (2017) Cannabis as a substitute for prescription drugs – a cross-sectional study. J Pain Res 10:989–998
Costa B et al (2004a) Oral anti-inflammatory activity of cannabidiol, a non-psychoactive constituent of cannabis, in acute carrageenan-induced inflammation in the rat paw. Naunyn Schmiedeberg's Arch Pharmacol 369(3):294–299
Costa B et al (2004b) Vanilloid TRPV1 receptor mediates the antihyperalgesic effect of the nonpsychoactive cannabinoid, cannabidiol, in a rat model of acute inflammation. Br J Pharmacol 143(2):247–250
Costa B et al (2007) The non-psychoactive cannabis constituent cannabidiol is an orally effective therapeutic agent in rat chronic inflammatory and neuropathic pain. Eur J Pharmacol 556(1): 75–83
Cruz-Almeida Y, Fillingim RB (2014) Can quantitative sensory testing move us closer to mechanism-based pain management? Pain Med 15(1):61–72
De Aquino JP et al (2023) THC modulates pain sensitivity among persons receiving opioid agonist therapy for opioid use disorder: a within-subject, randomized, placebo-controlled laboratory study. Addict Biol 28(9):e13317
De Vita MJ et al (2022) The effects of cannabidiol and analgesic expectancies on experimental pain reactivity in healthy adults: a balanced placebo design trial. Exp Clin Psychopharmacol 30(5): 536
de Vries M et al (2016) Single dose delta-9-tetrahydrocannabinol in chronic pancreatitis patients: analgesic efficacy, pharmacokinetics and tolerability. Br J Clin Pharmacol 81(3):525–537
de Vries M et al (2017) Tetrahydrocannabinol does not reduce pain in patients with chronic abdominal pain in a phase 2 placebo-controlled study. Clin Gastroenterol Hepatol 15(7): 1079–1086.e4
Decosterd I, Woolf CJ (2000) Spared nerve injury: an animal model of persistent peripheral neuropathic pain. Pain 87(2):149–158
Deuis JR, Dvorakova LS, Vetter I (2017) Methods used to evaluate pain behaviors in rodents. Front Mol Neurosci 10:284
Devsi A et al (2020) A pharmacological characterization of Cannabis sativa chemovar extracts. J Cannabis Res 2:17
Di Marzo V (2009) The endocannabinoid system: its general strategy of action, tools for its pharmacological manipulation and potential therapeutic exploitation. Pharmacol Res 60(2): 77–84
Di Marzo V (2018) New approaches and challenges to targeting the endocannabinoid system. Nat Rev Drug Discov 17(9):623–639
Dieterle M et al (2022) Pain response to cannabidiol in opioid-induced hyperalgesia, acute nociceptive pain, and allodynia using a model mimicking acute pain in healthy adults in a randomized trial (CANAB II). Pain 163(10):1919–1928
Dirks J, Petersen KL (2003) The heat/capsaicin sensitization model: a methodologic study. J Pain 4(3):122–128

Donvito G et al (2018) The endogenous cannabinoid system: a budding source of targets for treating inflammatory and neuropathic pain. Neuropsychopharmacology 43(1):52–79
Dunn KE et al (2021) Within-subject, double-blinded, randomized, and placebo-controlled evaluation of the combined effects of the cannabinoid dronabinol and the opioid hydromorphone in a human laboratory pain model. Neuropsychopharmacology 46(8):1451–1459
Ea WCR, Nuss G (1962) Carrageenin-induced edema in hind paw of the rats as an assay for antiinflammatory drugs. Proc Soc Exp Biol Med 111:544–547
Edelmayer RM et al (2012) Activation of TRPA1 on dural afferents: a potential mechanism of headache pain. Pain 153(9):1949–1958
Edwards RR et al (2005) Quantitative assessment of experimental pain perception: multiple domains of clinical relevance. Pain 114(3):315–319
Eisenberg E, Ogintz M, Almog S (2014) The pharmacokinetics, efficacy, safety, and ease of use of a novel portable metered-dose cannabis inhaler in patients with chronic neuropathic pain: a phase 1a study. J Pain Palliat Care Pharmacother 28(3):216–225
Ekrami E et al (2024) Association between cannabis use and opioid consumption, pain, and respiratory complications after surgery: a retrospective cohort analysis. Anesth Analg 139(4): 724–733
Ellis RJ et al (2008) Smoked medicinal cannabis for neuropathic pain in HIV: a randomized, crossover clinical trial. Neuropsychopharmacology 34(3):672–680
Farrell SF et al (2024) C-reactive protein (CRP) is associated with chronic pain independently of biopsychosocial factors. J Pain 25(2):476–496
Farrelly KN et al (2023) The impact of recreational cannabis legalization on cannabis use and associated outcomes: a systematic review. Subst Abuse 17:11782218231172054
Fine PG, Rosenfeld MJ (2013) The endocannabinoid system, cannabinoids, and pain. Rambam Maimonides Med J 4(4):e0022
Finn DP et al (2004) Effects of coadministration of cannabinoids and morphine on nociceptive behaviour, brain monoamines and HPA axis activity in a rat model of persistent pain. Eur J Neurosci 19(3):678–686
Finn DP et al (2021) Cannabinoids, the endocannabinoid system, and pain: a review of preclinical studies. Pain 162(Suppl 1):S5–S25
Flatters SJL (2008) Characterization of a model of persistent postoperative pain evoked by skin/muscle incision and retraction (SMIR). Pain 135(1–2):119–130
Foss JD et al (2021) Behavioural and pharmacological effects of cannabidiol (CBD) and the cannabidiol analogue KLS-13019 in mouse models of pain and reinforcement. Br J Pharmacol 178(15):3067–3078
Fowler C (2007) The contribution of cyclooxygenase-2 to endocannabinoid metabolism and action. Br J Pharmacol 152(5):594–601
Frank B et al (2008) Comparison of analgesic effects and patient tolerability of nabilone and dihydrocodeine for chronic neuropathic pain: randomised, crossover, double blind study. BMJ 336(7637):199–201
Garcia-Ovejero D et al (2009) The endocannabinoid system is modulated in response to spinal cord injury in rats. Neurobiol Dis 33(1):57–71
Gilchrist HD, Allard BL, Simone DA (1996) Enhanced withdrawal responses to heat and mechanical stimuli following intraplantar injection of capsaicin in rats. Pain 67(1):179–188
Greenwald MK, Stitzer ML (2000) Antinociceptive, subjective and behavioral effects of smoked marijuana in humans. Drug Alcohol Depend 59(3):261–275
Gregory N et al (2013) An overview of animal models of pain: disease models and outcome measures. J Pain 14(11). https://doi.org/10.1016/j.jpain.2013.06.008
Guy D et al (2024) Pharmacodynamic effects following co-administration of cannabinoids and opioids: a scoping review of human experimental studies. Pain Med 25(7):423–434
Haney M et al (1997) Factors influencing marijuana self-administration by humans. Behav Pharmacol 8(2–3):101–112

Haney M et al (1999a) Abstinence symptoms following oral THC administration to humans. Psychopharmacology 141(4):385–394
Haney M et al (1999b) Abstinence symptoms following smoked marijuana in humans. Psychopharmacology 141(4):395–404
Haney M et al (2003) Nefazodone decreases anxiety during marijuana withdrawal in humans. Psychopharmacology 165(2):157–165
Haney M et al (2004) Marijuana withdrawal in humans: effects of oral THC or divalproex. Neuropsychopharmacology 29(1):158
Haney M et al (2008) Effects of THC and lofexidine in a human laboratory model of marijuana withdrawal and relapse. Psychopharmacology 197(1):157–168
Haney M et al (2010) Effects of baclofen and mirtazapine on a laboratory model of marijuana withdrawal and relapse. Psychopharmacology 211(2):233–244
Haney M et al (2013) Nabilone decreases marijuana withdrawal and a laboratory measure of marijuana relapse. Neuropsychopharmacology 38(8):1557–1565
Haney M et al (2016) Oral cannabidiol does not alter the subjective, reinforcing or cardiovascular effects of smoked cannabis. Neuropsychopharmacology 41(8):1974–1982
Haney M et al (2019) Guanfacine decreases symptoms of cannabis withdrawal in daily cannabis smokers. Addict Biol 24(4):707–716
Haney M et al (2025) Oral cannabis for Taxane-induced neuropathy: a pilot randomized placebo-controlled study. Cannabis Cannabinoid Res. https://doi.org/10.1089/can.2025.0028
Harris H et al (2016) Effects of Delta-9-tetrahydrocannabinol and Cannabidiol on cisplatin-induced neuropathy in mice. Planta Med 82(13):1169–1172
Harris HM et al (2019) Role of cannabinoids and terpenes in cannabis-mediated analgesia in rats. Cannabis Cannabinoid Res 4(3):177–182
Harris HM et al (2022) Differential effects of Cannabidiol and a novel Cannabidiol analog on oxycodone place preference and analgesia in mice: an opioid Abuse deterrent with analgesic properties. Cannabis Cannabinoid Res 7(6):804–813
Hasin DS (2017) US epidemiology of cannabis use and associated problems. Neuropsychopharmacology 43:195
Hasin DS et al (2017) US adult illicit cannabis use, cannabis use disorder, and medical marijuana laws: 1991-1992 to 2012-2013. JAMA Psychiatry 74(6):579–588
Henshaw FR et al (2021) The effects of cannabinoids on pro- and anti-inflammatory cytokines: a systematic review of in vivo studies. Cannabis Cannabinoid Res 6(3):177–195
Hoggart B et al (2015) A multicentre, open-label, follow-on study to assess the long-term maintenance of effect, tolerance and safety of THC/CBD oromucosal spray in the management of neuropathic pain. J Neurol 262:27–40
Holman A et al (2022) Healthcare provider and medical cannabis patient communication regarding referral and medication substitution: the Canadian context. J Cannabis Res 4(1):32
Hsieh GC et al (2011) Central and peripheral sites of action for CB(2) receptor mediated analgesic activity in chronic inflammatory and neuropathic pain models in rats. Br J Pharmacol 162(2): 428–440
Hubli M, Bolt D, Krassioukov AV (2018) Cold pressor test in spinal cord injury—revisited. Spinal Cord 56(6):528
Huerta MÁ et al (2024) Strategies for measuring non-evoked pain in preclinical models of neuropathic pain: systematic review. Neurosci Biobehav Rev 163:105761
Huskisson EC (1974) Measurement of pain. Lancet 304(7889):1127–1131
Imasogie N, Rose RV, Wilson A (2021) High quantities: evaluating the association between cannabis use and propofol anesthesia during endoscopy. PLoS One 16(3):e0248062
Jaggi AS, Jain V, Singh N (2011) Animal models of neuropathic pain. Fundam Clin Pharmacol 25(1):1–28
Jali AM et al (2024) Navigating preclinical models and medications for peripheral neuropathy: a review. Pharmaceuticals 17(8):1010

Janero DR, Makriyannis A (2014) Terpenes and lipids of the endocannabinoid and transient-receptor-potential-channel biosignaling systems. ACS Chem Neurosci 5(11):1097–1106
Jang Y, Kim M, Hwang SW (2020) Molecular mechanisms underlying the actions of arachidonic acid-derived prostaglandins on peripheral nociception. J Neuroinflammation 17(1):30
Jarvis RG (1990) Pain and sensory perversions, in clinical methods: the history, physical, and laboratory examinations, 3rd edn. Butterworths
Javadi-Paydar M et al (2019) Vapor inhalation of cannabidiol (CBD) in rats. Pharmacol Biochem Behav 184:172741
Jenkins BW, Moore CF, Weerts EM (2025) Cannabidiol interactions with Δ-9-tetrahydrocannabinol on antinociception after carrageenan-induced inflammatory pain in male and female rats. J Pharmacol Exp Ther 392(7):103625
Jensen TS, Finnerup NB (2014) Allodynia and hyperalgesia in neuropathic pain: clinical manifestations and mechanisms. Lancet Neurol 13(9):924–935
Jensen MP, Karoly P, Braver S (1986) The measurement of clinical pain intensity: a comparison of six methods. Pain 27(1):117–126
Johnson JR et al (2010) Multicenter, double-blind, randomized, placebo-controlled, parallel-group study of the efficacy, safety, and tolerability of THC: CBD extract and THC extract in patients with intractable cancer-related pain. J Pain Symptom Manag 39(2):167–179
Kalliomäki J et al (2012) Lack of effect of central nervous system-active doses of nabilone on capsaicin-induced pain and hyperalgesia. Clin Exp Pharmacol Physiol 39(4):336–342
Kandasamy R et al (2018) Anti-migraine effect of Δ(9)-tetrahydrocannabinol in the female rat. Eur J Pharmacol 818:271–277
Karniol IG, Carlini EA (1973) Pharmacological interaction between cannabidiol and delta 9-tetrahydrocannabinol. Psychopharmacologia 33(1):53–70
Karsalia M et al (2021) Predictors of postoperative pain using PROMIS pain interference two-years following knee surgery. J Orthop 25:271–277
Kathmann M et al (2006) Cannabidiol is an allosteric modulator at mu- and delta-opioid receptors. Naunyn Schmiedeberg's Arch Pharmacol 372(5):354–361
Kelly S, Chapman V (2001) Selective cannabinoid CB1 receptor activation inhibits spinal nociceptive transmission in vivo. J Neurophysiol 86(6):3061–3064
Kenshalo DR Jr, Willis WD Jr (1991) The role of the cerebral cortex in pain sensation. In: Normal and altered states of function. Springer, pp 153–212
Khalil H, Shajrawi A, Henker R (2021) Predictors of severe postoperative pain after orthopedic surgery in the immediate postoperative period. Int J Orthop Trauma Nurs 43:100864
Kidd B, Urban L (2001) Mechanisms of inflammatory pain. Br J Anaesth 87(1):3–11
Kim A et al (2019) Patterns of medical cannabis use among cancer patients from a medical cannabis dispensary in New York State. J Palliat Med 22(10):1196–1201
King T et al (2009) Unmasking the tonic-aversive state in neuropathic pain. Nat Neurosci 12(11): 1364–1366
King KM et al (2017) Single and combined effects of Δ9 -tetrahydrocannabinol and cannabidiol in a mouse model of chemotherapy-induced neuropathic pain. Br J Pharmacol 174(17):2832–2841
Koenig J et al (2014) Two-week test–retest stability of the cold pressor task procedure at two different temperatures as a measure of pain threshold and tolerance. Pain Pract 14(3):E126
Kong J-T et al (2019) Novel characterization of thermal temporal summation response by analysis of continuous pain vs time curves and exploratory modeling. J Pain Res Volume 12:3231–3244
Kosiba JD, Maisto SA, Ditre JW (2019) Patient-reported use of medical cannabis for pain, anxiety, and depression symptoms: systematic review and meta-analysis. Soc Sci Med 233:181–192
Kosirog J et al (2024) Marijuana and endoscopy: the effects of marijuana on sedation. Gastrointest Endosc 100(2):177–182
Kowalczyk WJ et al (2006) Sex differences and hormonal influences on response to cold pressor pain in humans. J Pain 7(3):151–160
Kraft B et al (2008) Lack of analgesia by oral standardized cannabis extract on acute inflammatory pain and hyperalgesia in volunteers. Anesthesiology 109(1):101–110

Krafthöfer J et al (2024) Pupillometry as a potential objective measurement of pain assessment in healthy volunteers. J Pain Res 17:2037–2042

Kruger L (2001) Methods in pain research. CRC Press

Kumar Reddy KS et al (2012) Human experimental pain models: a review of standardized methods in drug development. Journal Res Med Sci 17(6):587

LaBuda CJ, Fuchs PN (2000) A behavioral test paradigm to measure the aversive quality of inflammatory and neuropathic pain in rats. Exp Neurol 163(2):490–494

LaMotte RH, Lundberg L, Torebjörk H (1992) Pain, hyperalgesia and activity in nociceptive C units in humans after intradermal injection of capsaicin. J Physiol 448(1):749–764

Langford R et al (2013) A double-blind, randomized, placebo-controlled, parallel-group study of THC/CBD oromucosal spray in combination with the existing treatment regimen, in the relief of central neuropathic pain in patients with multiple sclerosis. J Neurol 260:984–997

Latremoliere A, Woolf CJ (2009) Central sensitization: a generator of pain hypersensitivity by central neural plasticity. J Pain 10(9):895–926

Le Foll B (2021) Opioid-sparing effects of cannabinoids: myth or reality? Prog Neuro-Psychopharmacol Biol Psychiatry 106:110065

Lee D et al (2014) Preoperative opioid use as a predictor of adverse postoperative self-reported outcomes in patients undergoing spine surgery. JBJS 96(11):e89

Lee G et al (2018) Medical cannabis for neuropathic pain. Curr Pain Headache Rep 22(1):8

Leone C, Truini A (2019) The CPM effect: functional assessment of the diffuse noxious inhibitory control in humans. J Clin Neurophysiol 36(6):430–436

Lichtman AH, Martin BR (1991) Spinal and supraspinal components of cannabinoid-induced antinociception. J Pharmacol Exp Ther 258(2):517–523

Lichtman AH, Martin BR (1997) The selective cannabinoid antagonist SR 141716A blocks cannabinoid-induced antinociception in rats. Pharmacol Biochem Behav 57(1–2):7–12

Lichtman AH et al (2018) Results of a double-blind, randomized, placebo-controlled study of nabiximols oromucosal spray as an adjunctive therapy in advanced cancer patients with chronic uncontrolled pain. J Pain Symptom Manag 55(2):179–188. e1

Liktor-Busa E et al (2021) Analgesic potential of terpenes derived from Cannabis sativa. Pharmacol Rev 73(4):1269–1297

Lim G et al (2003) Upregulation of spinal cannabinoid-1-receptors following nerve injury enhances the effects of Win 55,212-2 on neuropathic pain behaviors in rats. Pain 105(1–2):275–283

Liu C et al (2021) Cytokines: from clinical significance to quantification. Adv Sci 8(15):2004433

Loeser JD, Treede R-D (2008) The Kyoto protocol of IASP basic pain terminology. Pain 137(3): 473–477

Lynch ME, Cesar-Rittenberg P, Hohmann AG (2014) A double-blind, placebo-controlled, cross-over pilot trial with extension using an Oral mucosal cannabinoid extract for treatment of chemotherapy-induced neuropathic pain. J Pain Symptom Manag 47(1):166–173

Malta I et al (2023) Synergistic effect of cannabidiol and transcutaneous electrical nerve stimulation on neuropathic and inflammatory pain in mice. Neuroreport 34(3):165–169

Manzanares J, Julian M, Carrascosa A (2006) Role of the cannabinoid system in pain control and therapeutic implications for the management of acute and chronic pain episodes. Curr Neuropharmacol 4(3):239–257

Martin BR (1985) Structural requirements for cannabinoid-induced antinociceptive activity in mice. Life Sci 36(16):1523–1530

Martin WJ, Hohmann AG, Walker JM (1996) Suppression of noxious stimulus-evoked activity in the ventral posterolateral nucleus of the thalamus by a cannabinoid agonist: correlation between electrophysiological and antinociceptive effects. J Neurosci 16(20):6601–6611

Martins SS et al (2025) The relationship of medical and recreational cannabis laws with opioid misuse and opioid use disorder in the USA: does it depend on prior history of cannabis use? Int J Drug Policy 136:104687

Mason DJ Jr, Lowe J, Welch SP (1999) Cannabinoid modulation of dynorphin a: correlation to cannabinoid-induced antinociception. Eur J Pharmacol 378(3):237–248

McPartland JM, Small E (2020) A classification of endangered high-THC cannabis (Cannabis sativa subsp. indica) domesticates and their wild relatives. PhytoKeys 144:81–112
McPartland JM et al (2015) Are cannabidiol and Delta(9) -tetrahydrocannabivarin negative modulators of the endocannabinoid system? A systematic review. Br J Pharmacol 172(3):737–753
Mendell LM (1966) Physiological properties of unmyelinated fiber projection to the spinal cord. Exp Neurol 16(3):316–332
Mitchell VA et al (2021) Oral efficacy of Δ(9)-tetrahydrocannabinol and cannabidiol in a mouse neuropathic pain model. Neuropharmacology 189:108529
Möller KÄ, Johansson B, Berge O-G (1998) Assessing mechanical allodynia in the rat paw with a new electronic algometer. J Neurosci Methods 84(1–2):41–47
Moore CF, Weerts EM (2022) Cannabinoid tetrad effects of oral Δ9-tetrahydrocannabinol (THC) and cannabidiol (CBD) in male and female rats: sex, dose-effects and time course evaluations. Psychopharmacology 239(5):1397–1408
Muller C, Morales P, Reggio PH (2019) Cannabinoid ligands targeting TRP channels. Front Mol Neurosci 11:487
Naef M et al (2003) The analgesic effect of oral delta-9-tetrahydrocannabinol (THC), morphine, and a THC-morphine combination in healthy subjects under experimental pain conditions. Pain 105(1–2):79–88
Narang S et al (2008) Efficacy of dronabinol as an adjuvant treatment for chronic pain patients on opioid therapy. J Pain 9(3):254–264
Nguyen JD et al (2016) Inhaled delivery of Δ9-tetrahydrocannabinol (THC) to rats by e-cigarette vapor technology. Neuropharmacology 109:112–120
Nielsen S et al (2017) Opioid-sparing effect of cannabinoids: a systematic review and meta-analysis. Neuropsychopharmacology 42:1752
Noh ASM et al (2021) Effects of different doses of complete Freund's adjuvant on nociceptive behaviour and inflammatory parameters in polyarthritic rat model mimicking rheumatoid arthritis. PLoS One 16(12):e0260423
Notcutt W et al (2004) Initial experiences with medicinal extracts of cannabis for chronic pain: results from 34 'N of 1'studies. Anaesthesia 59(5):440–452
Nurmikko TJ et al (2007) Sativex successfully treats neuropathic pain characterised by allodynia: a randomised, double-blind, placebo-controlled clinical trial. Pain 133(1):210–220
O'Brien PD, Sakowski SA, Feldman EL (2014) Mouse models of diabetic neuropathy. ILAR J 54(3):259–272
Olesen AE et al (2012) Human experimental pain models for assessing the therapeutic efficacy of analgesic drugs. Pharmacol Rev 64(3):722–779
Parnell J et al (2023) Cannabinoid CB1 receptor expression and localization in the dorsal horn of male and female rat and human spinal cord. Canadian J Pain 7(2):2264895
Pearl-Dowler L, Posa L, Lopez-Canul M, Teggin A, Gobbi G (2023) Anti-allodynic and medullary modulatory effects of a single dose of delta-9-tetrahydrocannabinol (THC) in neuropathic rats tolerant to morphine. Progress in Neuro-Psychopharmacology and Biological Psychiatry, 127, 110805.
Pernía-Andrade AJ et al (2009) Spinal endocannabinoids and CB1 receptors mediate C-fiber–induced heterosynaptic pain sensitization. Science 325(5941):760–764
Pertwee RG (2012) Targeting the endocannabinoid system with cannabinoid receptor agonists: pharmacological strategies and therapeutic possibilities. Philos Trans R Soc Lond Ser B Biol Sci 367(1607):3353–3363
Petersen KL, Rowbotham MC (1999) A new human experimental pain model: the heat/capsaicin sensitization model. Neuroreport 10(7):1511–1516
Petrocellis LD et al (2001) The activity of anandamide at Vanilloid VR1 receptors requires facilitated transport across the cell membrane and is limited by intracellular metabolism *. J Biol Chem 276(16):12856–12863
Pfau DB et al (2012) Quantitative sensory testing of neuropathic pain patients: potential mechanistic and therapeutic implications. Curr Pain Headache Rep 16:199–206

Phillips JK et al (2017) Pain management and the intersection of pain and opioid use disorder. In: Pain management and the opioid epidemic: balancing societal and individual benefits and risks of prescription opioid use. National Academies Press (US)
Ploner M et al (2002) Cortical representation of first and second pain sensation in humans. Proc Natl Acad Sci 99(19):12444–12448
Portenoy RK et al (2012) Nabiximols for opioid-treated cancer patients with poorly-controlled chronic pain: a randomized, placebo-controlled, graded-dose trial. J Pain 13(5):438–449
Pud D, Granovsky Y, Yarnitsky D (2009) The methodology of experimentally induced diffuse noxious inhibitory control (DNIC)-like effect in humans. Pain 144(1–2):16–19
Rafiyan M et al (2023) Experimental animal models of chronic inflammation. Curr Res Immunol 4: 100063
Raja SN et al (2020) The revised International Association for the Study of Pain definition of pain: concepts, challenges, and compromises. Pain 161(9):1976–1982
Redmond WJ et al (2008) Analgesic and antihyperalgesic effects of nabilone on experimental heat pain. Curr Med Res Opin 24(4):1017–1024
Reimann-Philipp U et al (2020) Cannabis Chemovar nomenclature misrepresents chemical and genetic diversity; survey of variations in chemical profiles and genetic markers in Nevada medical cannabis samples. Cannabis Cannabinoid Res 5(3):215–230
Rhee TG, Rosenheck RA (2023) Increasing use of cannabis for medical purposes among US residents, 2013–2020. Am J Prev Med 65(3):528–533
Rikard SM (2023) Chronic pain among adults—United States, 2019–2021. MMWR Morb Mortal Wkly Rep 72:379
Roberts JD, Gennings C, Shih M (2006) Synergistic affective analgesic interaction between delta-9-tetrahydrocannabinol and morphine. Eur J Pharmacol 530(1):54–58
Rock EM, Parker LA (2020) Constituents of Cannabis sativa. In: Cannabinoids and neuropsychiatric disorders, pp 1–13
Rock EM, Limebeer CL, Parker LA (2018) Effect of cannabidiolic acid and Δ9-tetrahydrocannabinol on carrageenan-induced hyperalgesia and edema in a rodent model of inflammatory pain. Psychopharmacology 235(11):3259–3271
Rog DJ et al (2005) Randomized, controlled trial of cannabis-based medicine in central pain in multiple sclerosis. Neurology 65(6):812–819
Rolke R et al (2006) Quantitative sensory testing in the German Research Network on Neuropathic Pain (DFNS): standardized protocol and reference values. Pain 123(3):231–243
Romero-Sandoval A, Eisenach JC (2007) Spinal cannabinoid receptor type 2 activation reduces hypersensitivity and spinal cord glial activation after paw incision. Anesthesiology 106(4): 787–794
Russo EB, Marcu J (2017) Cannabis pharmacology: the usual suspects and a few promising leads. Adv Pharmacol 80:67–134
Sanders J, Jackson DM, Starmer GA (1979) Interactions among the cannabinoids in the antagonism of the abdominal constriction response in the mouse. Psychopharmacology 61(3):281–285
Schindler EA et al (2020) In an exploratory randomized, double-blind, placebo-controlled, cross-over study, psychoactive doses of intravenous delta-9-tetrahydrocannabinol fail to produce antinociceptive effects in healthy human volunteers. Psychopharmacology 237:3097–3107
Schmidt R et al (1995) Novel classes of responsive and unresponsive C nociceptors in human skin. J Neurosci 15(1):333–341
Schneider T et al (2022) Pain response to cannabidiol in induced acute nociceptive pain, allodynia, and hyperalgesia by using a model mimicking acute pain in healthy adults in a randomized trial (CANAB I). Pain 163(1):e62–e71
Schomberg D et al (2012) Neuropathic pain: role of inflammation, immune response, and ion channel activity in central injury mechanisms. Ann Neurosci 19(3):125–132
Sepulveda DE et al (2022a) Evaluating the antinociceptive efficacy of cannabidiol alone or in combination with morphine using the formalin test in male and female mice. Cannabis Cannabinoid Res 7(5):648–657

Sepulveda DE et al (2022b) Combinations of Cannabidiol and Δ(9)-tetrahydrocannabinol in reducing chemotherapeutic induced neuropathic pain. Biomedicine 10(10)
Serpell M et al (2014) A double-blind, randomized, placebo-controlled, parallel group study of THC/CBD spray in peripheral neuropathic pain treatment. Eur J Pain 18(7):999–1012
Sexton M et al (2016) A cross-sectional survey of medical cannabis users: patterns of use and perceived efficacy. Cannabis Cannabinoid Res 1(1):131–138
Silva NR et al (2017) Antinociceptive effects of HUF-101, a fluorinated cannabidiol derivative. Prog Neuro-Psychopharmacol Biol Psychiatry 79:369–377
Silva-Cardoso GK et al (2021) Cannabidiol effectively reverses mechanical and thermal allodynia, hyperalgesia, and anxious behaviors in a neuropathic pain model: possible role of CB1 and TRPV1 receptors. Neuropharmacology 197:108712
Skrabek RQ et al (2008) Nabilone for the treatment of pain in fibromyalgia. J Pain 9(2):164–173
Small E (2015) Evolution and classification of Cannabis sativa (marijuana, hemp) in relation to human utilization. Bot Rev 81:189–294
Sofia RD, Vassar HB, Knobloch LC (1975a) Comparative analgesic activity of various naturally occurring cannabinoids in mice and rats. Psychopharmacologia 40(4):285–295
Sofia RD, Diamantis W, Ludwig BJ (1975b) Comparative anti-inflammatory, analgesic, and antipyretic activities of 7-chloro-3,3a-dihydro-2-methyl-2H,9H-isoxazolo-(3,2-b)(1,3)-benzoxazin-9-one and 5-chlorosalicylic acid in rats. J Pharm Sci 64(8):1321–1324
Soliman N et al (2021) Systematic review and meta-analysis of cannabinoids, cannabis-based medicines, and endocannabinoid system modulators tested for antinociceptive effects in animal models of injury-related or pathological persistent pain. Pain 162:S26–S44
Sorkin LS et al (2018) Origins of antidromic activity in sensory afferent fibers and neurogenic inflammation. In: Seminars in immunopathology, vol 40. Springer
Spindle TR, Bonn-Miller MO, Vandrey R (2019) Changing landscape of cannabis: novel products, formulations, and methods of administration. Curr Opin Psychol 30:98–102
Stockings E et al (2018) Cannabis and cannabinoids for the treatment of people with chronic noncancer pain conditions: a systematic review and meta-analysis of controlled and observational studies. Pain 159(10):1932–1954
Strand N et al (2023) Medical cannabis: a review from the American Society of Pain and Neuroscience. J Pain Res 16:4217–4228
Svendsen KB, Jensen TS, Bach FW (2004) Does the cannabinoid dronabinol reduce central pain in multiple sclerosis? Randomised double blind placebo controlled crossover trial. BMJ 329(7460):253
Szallasi A, Blumberg PM (1999) Vanilloid (capsaicin) receptors and mechanisms. Pharmacol Rev 51(2):159–211
Tai FWD, McAlindon ME (2021) Non-steroidal anti-inflammatory drugs and the gastrointestinal tract. Clin Med 21(2):131–134
Tappe-Theodor A, Kuner R (2014) Studying ongoing and spontaneous pain in rodents–challenges and opportunities. Eur J Neurosci 39(11):1881–1890
Tjølsen A et al (1992) The formalin test: an evaluation of the method. Pain 51(1):5–17
Treede R-D et al (2008) Neuropathic pain: redefinition and a grading system for clinical and research purposes. Neurology 70(18):1630–1635
Tseng AH, Craft RM (2004) CB1 receptor mediation of cannabinoid behavioral effects in male and female rats. Psychopharmacology 172(1):25–30
Tucci S (2010) Editorial [Hot topic: addiction and pain: cannabinoid and opioid interactions (Guest Editor: Sonia Tucci)]. Curr Drug Targets 11(4):392–392
Turcotte C et al (2015) Regulation of inflammation by cannabinoids, the endocannabinoids 2-arachidonoyl-glycerol and arachidonoyl-ethanolamide, and their metabolites. J Leukoc Biol 97(6):1049–1070
Turcotte C et al (2016) The CB 2 receptor and its role as a regulator of inflammation. Cell Mol Life Sci 73:4449–4470

Valverde O et al (2000) Cannabinoid withdrawal syndrome is reduced in pre-proenkephalin knock-out mice. J Neurosci 20(24):9284–9289
van Amerongen G et al (2018a) Effect profile of paracetamol, Δ9-THC and promethazine using an evoked pain test battery in healthy subjects. Eur J Pain 22(7):1331–1342
van Amerongen G et al (2018b) Effects on spasticity and neuropathic pain of an oral formulation of Δ9-tetrahydrocannabinol in patients with progressive multiple sclerosis. Clin Ther 40(9): 1467–1482
Van De Donk T et al (2019) An experimental randomized study on the analgesic effects of pharmaceutical-grade cannabis in chronic pain patients with fibromyalgia. Pain 160(4):860–869
Varvel SA et al (2006) Interactions between THC and cannabidiol in mouse models of cannabinoid activity. Psychopharmacology 186(2):226–234
Vivancos G et al (2004) An electronic pressure-meter nociception paw test for rats. Braz J Med Biol Res 37:391–399
Wade DT et al (2003) A preliminary controlled study to determine whether whole-plant cannabis extracts can improve intractable neurogenic symptoms. Clin Rehabil 17(1):21–29
Wallace M et al (2007) Dose-dependent effects of smoked cannabis on capsaicin-induced pain and hyperalgesia in healthy volunteers. Anesthesiology 107(5):785–796
Wallace MS et al (2015) Efficacy of inhaled cannabis on painful diabetic neuropathy. J Pain 16(7): 616–627
Wallace MS et al (2020) A secondary analysis from a randomized trial on the effect of plasma tetrahydrocannabinol levels on pain reduction in painful diabetic peripheral neuropathy. J Pain 21(11–12):1175–1186
Walsh Z et al (2013) Cannabis for therapeutic purposes: patient characteristics, access, and reasons for use. Int J Drug Policy 24(6):511–516
Walsh KB, McKinney AE, Holmes AE (2021) Minor cannabinoids: biosynthesis, molecular pharmacology and potential therapeutic uses. Front Pharmacol 12:777804
Walukevich-Dienst K et al (2025) What are tolerance breaks and why do young adults take them? A qualitative investigation of temporary abstinence breaks from cannabis. Cannabis 8(1):150
Wang Q et al (2024) Prevalence of cannabis use around the world: a systematic review and meta-analysis, 2000–2024. China CDC Weekly 6(25):597
Ward SJ et al (2014) Cannabidiol inhibits paclitaxel-induced neuropathic pain through 5-HT1A receptors without diminishing nervous system function or chemotherapy efficacy. Br J Pharmacol 171(3):636–645
Ware MA et al (2010) Smoked cannabis for chronic neuropathic pain: a randomized controlled trial. CMAJ 182(14):E694–E701
Webb J, Kamali F (1998) Analgesic effects of lamotrigine and phenytoin on cold-induced pain: a crossover placebo-controlled study in healthy volunteers. Pain 76(3):357–363
Weizman L et al (2024) Oral delta-9-tetrahydrocannabinol (THC) increases parasympathetic activity and supraspinal conditioned pain modulation in chronic neuropathic pain male patients: a crossover, double-blind, placebo-controlled trial. CNS Drugs 38(5):375–385
Welburn PJ et al (1976) Effect of cannabinoids on the abdominal constriction response in mice: within cannabinoid interactions. Psychopharmacologia 46(1):83–85
Welch SP, Thomas C, Patrick GS (1995) Modulation of cannabinoid-induced antinociception after intracerebroventricular versus intrathecal administration to mice: possible mechanisms for interaction with morphine. J Pharmacol Exp Ther 272(1):310–321
Welch SP, Huffman JW, Lowe J (1998) Differential blockade of the antinociceptive effects of centrally administered cannabinoids by SR141716A. J Pharmacol Exp Ther 286(3):1301–1308
Willis WD, Westlund K (1997) Neuroanatomy of the pain system and of the pathways that modulate pain. J Clin Neurophysiol 14(1):2–31
Wilsey B et al (2013) Low-dose vaporized cannabis significantly improves neuropathic pain. J Pain 14(2):136–148
Wilsey B et al (2016) An exploratory human laboratory experiment evaluating vaporized cannabis in the treatment of neuropathic pain from spinal cord injury and disease. J Pain 17(9):982–1000

Wissel J et al (2006) Low dose treatment with the synthetic cannabinoid Nabilone significantly reduces spasticity-related pain: a double-blind placebo-controlled cross-over trial. J Neurol 253: 1337–1341
Wong D (1988) Pain in children: comparison of assessment scales. Pediatr Nurs 14(1):9–17
Yam MF et al (2018) General pathways of pain sensation and the major neurotransmitters involved in pain regulation. Int J Mol Sci 19(8):2164
Yarnitsky D (2010) Conditioned pain modulation (the diffuse noxious inhibitory control-like effect): its relevance for acute and chronic pain states. Curr Opin Anesthesiol 23(5):611–615
Yuan T et al (2016) Assessment of itch and pain in animal models and human subjects. In: Ma C, Huang Y (eds) Translational research in pain and itch. Springer, Dordrecht, pp 1–22
Zhang J, Chen C (2008) Endocannabinoid 2-arachidonoylglycerol protects neurons by limiting COX-2 elevation. J Biol Chem 283(33):22601–22611
Zhang J et al (2003) Induction of CB2 receptor expression in the rat spinal cord of neuropathic but not inflammatory chronic pain models. Eur J Neurosci 17(12):2750–2754
Zhang G et al (2010) Cannabinoid CB1 receptor facilitation of substance P release in the rat spinal cord, measured as neurokinin 1 receptor internalization. Eur J Neurosci 31(2):225–237
Zhang Y et al (2015) Increased pain sensitivity in chronic pain subjects on opioid therapy: a cross-sectional study using quantitative sensory testing. Pain Med 16(5):911–922
Zubcevic K et al (2023) Oral capsules of tetra-hydro-cannabinol (THC), cannabidiol (CBD) and their combination in peripheral neuropathic pain treatment. Eur J Pain 27(4):492–506

The Use of Cannabis-Based Medicine in Selected Neurological Disorders

Natalia Szejko, Kamila Saramak, and Kirsten R. Müller-Vahl

Contents

1 Introduction 434
2 The Use of Cannabis-Based Medicine in Movement Disorders 434
2.1 Tic Disorders 434
2.2 Parkinson's Disease 439
2.3 Other Movement Disorders: Dystonia, Huntington's Disease (HD), and Restless Legs Syndrome (RLS) 443
2.4 Aging, Cognition, and Dementia 446
3 Conclusions 448
References 448

Abstract Cannabis-based medicine (CBM) is used in a wide variety of different neurological disorders. While the use of CBM in the treatment of pain, AIDS wasting, loss of appetite, and spasticity is well established, CBM application in movement disorders and neurodegenerative disorders is still an emerging topic. The purpose of this chapter is to summarize current evidence behind the use of CBM in selected neurological diseases, mainly movement and neurodegenerative disorders. The best evidence for efficacy of CBM is for Tourette syndrome resulting in an improvement of tics and psychiatric comorbidities. In this indication, delta-9-tetra-hydrocannabinol (THC)-containing CBMs are recommended. There is limited evidence that CBMs are also effective in Parkinson's disease in which they may improve tremor, but also non-motor symptoms such as pain and sleeping problems. With respect to other neurodegenerative diseases, there is limited evidence that CBMs may improve behavioral symptoms in Huntington's disease. In addition, it

N. Szejko and K. R. Müller-Vahl (✉)
Clinic of Psychiatry, Social Psychiatry and Psychotherapy, Hannover Medical School, Hannover, Germany
e-mail: mueller-vahl.kirsten@mh-hannover.de

K. Saramak
Department of Neurology, Medical University of Innsbruck, Innsbruck, Austria

Curr Topics Behav Neurosci (2026) 76: 433–454
https://doi.org/10.1007/7854_2024_564
Published Online: 31 December 2024

has been speculated that CBMs may have neuroprotective effects, but this has not yet been confirmed in the clinical setting.

Keywords Cannabis-based medicine · Movement disorders · Neurology · Tourette syndrome

1 Introduction

Cannabis-based medicine (CBM) is used in a wide variety of different neurological disorders. However, the degree of evidence for use of CBMs differs depending on indication (Hidding et al. 2024) with clear benefit in some types of treatment resistant epilepsy (Devinsky et al. 2024), chronic (neuropathic) pain (Pantoja-Ruiz et al. 2022), and spasticity in multiple sclerosis (Fragoso et al. 2020). In addition, CBMs have also been tested in a variety of other neurological disorders including different movement disorders and neurodegenerative diseases (Bosnjak Kuharic et al. 2021; Figura et al. 2022; Kluger et al. 2022; Broers and Bianchi 2023). The purpose of this chapter is to summarize current degree of evidence behind the use of CBMs in selected neurological disorders with a special attention on movement disorders and neurodegenerative diseases.

2 The Use of Cannabis-Based Medicine in Movement Disorders

2.1 *Tic Disorders*

2.1.1 Alterations in the ECS

Previous studies have demonstrated that CBMs could have beneficial effects on a variety of symptoms accompanying different kinds of movement disorders (Oikonomou and Jost 2022). However, until today the best investigated indications are primary tic disorders including Gilles de la Tourette syndrome (GTS) (Szejko and Müller-Vahl 2022). The pathophysiology of tic disorders is complex and multifactorial with diverse neurotransmitters involved (Szejko 2022), mainly dopamine (Maia 2018), but also serotonin (Müller-Vahl et al. 2019), glutamatergic acid (Kanaan et al. 2017), histamine (Hartmann et al. 2012), GABA (Draper et al. 2014), and endocannabinoids (Szejko et al. 2018). In a small number of studies, it has been shown that changes in the endocannabinoid system (ECS) may contribute to the occurrence of tics. Recently, Müller-Vahl et al. (2020) measured levels of various endocannabinoids in cerebrospinal fluid (CSF) in adults with GTS compared to a control group. CSF levels of different endocannabinoids ("N"--arachidonoylethanolamine (AEA, anandamide), 2-arachidonoylglycerol (2-AG),

the endocannabinoid-like molecule palmitoyl ethanolamide (PEA), and the lipid arachidonic acid (AA)) were significantly elevated in GTS compared to controls. Interestingly, levels of 2-AG correlated with the severity of comorbid attention deficit/hyperactivity disorder (ADHD). The authors concluded that elevated level of endocannabinoids may either represent an epiphenomenon or may be secondary or compensatory to changes in other neurotransmitter systems, but alternatively may also represent a direct involvement of the ECS in the pathology of GTS.

Moreover, in a small number of neuroimaging studies, alterations in the ECS have been investigated in GTS. Berding et al. (2004) conducted a single photon emission computed tomography (SPECT) study using a marker of the ECS ([123I] AM281) in six patients with GTS. CB1 receptor binding was reduced after treatment with delta-9-tetrahydrocannabinol (THC, dronabinol) (Berding et al. 2004). However, as no control group was included, it was not possible to establish whether CB1 receptor binding is altered in patients with GTS.

Finally, two studies have been published investigating genetic variations in the cannabinoid receptor gene (CNR1) in GTS. One study showed a significant association of the GTS clinical phenotype with the rs2023239 variant of the CNR1 (Szejko et al. 2020), while the other did not show any association (Gadzicki et al. 1999).

2.1.2 Data from Retrospective Reports on Self-Medication and Use of Different CBMs

First reports of self-medication in GTS with inhaled cannabis flowers have been published by Sandyk and Awerbuch (1988) in 1988. The authors reported the effect of cannabis use in three male patients aged between 15 and 39 who benefited in terms of their tics as well with respect to different psychiatric comorbidities after having smoked 0.5–2 cannabis cigarettes per day. Subsequently, Hemming and Yellowlees (1993) published a case study of a 36-year-old male with tics whose symptoms improved significantly after smoking one cannabis cigarette per day. Following these reports, Muller-Vahl (1998) reported the results of a retrospective study on self-medication with inhaled cannabis flowers in GTS in a sample of 64 patients, of whom 17 reported having used cannabis. However, patients differed significantly in terms of use of cannabis. While some participants had used only two cigarettes over lifetime, others smoked regularly three to four cannabis cigarettes per day. Of the 17 participants reporting prior use of cannabis, 14 patients reported an improvement or even a remission of vocal and/or motor tics. In some cases, an improvement of other symptoms including premonitory urges and psychiatric comorbidities had been reported. No serious side effects were described.

More recently, Abi-Jaoude et al. (2017) analyzed retrospectively the effectiveness and safety of different kinds of CBM treatment (inhalation, ingestion with food, and smoking) in 19 adults with GTS. The authors were unable to obtain reliable information with regard to respective cannabis chemovars or THC/cannabidiol (CBD) content and the estimated average total daily dose varied substantially from less than 0.1 g to 10 g cannabis flowers per da (median = 1 g daily). All patients reported a symptom relief. As measured with the Yale Global Tic Severity Scale (YGTSS), the average tic reduction was 60%. In addition, all participants also experienced improvements of psychiatric comorbidities. Most commonly reported side effects were sensations of being "high," problems with concentration, aggravation of anxiety, and changes in appetite. However, no serious adverse events (AEs) were reported.

Subsequently, Thaler et al. (2019) published the results of an interview with 42 patients with GTS who were using inhaled medicinal cannabis (MC) for at least 1 year. In this study, a Likert score (0–5) was used to assess effectiveness. The overall benefit was 3.85/5 confirming a clinically relevant impact of CBMs on tics. However, six participants stopped using cannabis due to different reasons. One participant reported an acute psychotic episode. However, this side effect was not confirmed by a physician. Other AEs included self-reported hallucinations, irritability or confusion, and cognitive decline.

In 2019, Milosev et al. (2019) analyzed retrospectively a large sample of patients with GTS ($n = 98$) who used different CBMs. In line with the aforementioned reports, the vast majority of patients (85%) reported benefitting from CBMs treatment with a self-rated tic improvement of about 60%. More than half of patients experienced also an improvement in terms of psychiatric comorbidities and quality of life. While side effects were relatively frequent and were reported by 50% of patients, all of them were rated as mild or moderate. When comparing different CBMs, most patients (66%) selected cannabis flowers over other other CBMs such as dronabinol or oral cannabis extracts due to subjectively reported better efficacy and tolerance. Most patients preferred THC-rich strains.

More recently, Anis et al. (2022) described subjects' satisfaction and consumption habits regarding MC amongst 25 patients with GTS in a retrospective longitudinal study. The most common mode of administration was smoking (80%). The mean duration of MC use was 4.0 ± 2.3 years (range, 0.5–10 years). Patients reported a continuous increase in MC dose of 0.6–0.7 g/year. The average reduction of tics was rated as 75%. Treatment was well tolerated, although the majority of participants (88%) reported mild side effects.

2.1.3 Results from Prospective Observational and Open-Label Studies

To date, two studies have evaluated effects of CBMs in patients with chronic tic disorders in a prospective, but uncontrolled fashion. A group from Israel led by Barchel et al. investigated long-term effects of treatment with different CBMs (including different inhaled MC chemovars as well as oral THC and CBD oils) in 70 patients with tics at baseline and at six-months follow-up (Barchel et al. 2024). The majority of patients (about 70%) used THC-dominant CBMs and more than one CBMs. The mean daily dose of THC was 123 mg, while for CBD it was 50.5 mg. At follow-up assessment, a significant improvement in quality of life and a reduction in the number of medications were detected. Moreover, 67% of patients reported an improvement of obsessive-compulsive symptoms and 89% an improvement of anxiety. In addition, a not statistically improvement in motor and vocal tics as well as in general mood was reported. However, no information was given on clinical symptoms depending on the kind of CBMs.

An open-label, Phase 2 study was published by Bloch et al. investigating the effects of oral SCI 110, formerly THX 110, a combination of THC and PEA, in 16 adults with GTS (Bloch et al. 2021). It was hypothesized that the combination of THC (2.5–10 mg/day) with PEA (800 mg/day) may result in a so-called entourage effect with increased effectiveness, since PEA potentiates responses to the endocannabinoid anandamide, but unchanged AEs (Gabrielsson et al. 2016).

During this 12-week trial, tics significantly improved within week one. The average tic improvement was rated as 20%.

2.1.4 Data from Randomized Controlled Trials

The first two randomized controlled trials (RCTs) investigating efficacy and safety of THC in GTS were published in the early 2000s (Müller-Vahl et al. 2002, 2003). In the first study, 12 adults with GTS (Müller-Vahl et al. 2002) were treated once with diverse doses of oral THC (2.5–10 mg) that were adjusted to sex and weight. Treatment with THC resulted in a statistically significant improvement of tics ($p = 0.015$) and obsessive-compulsive symptoms ($p = 0.041$). From additional analyses, it is suggested that the best improvement was achieved in a group treated with higher doses of THC (7.5–10 mg per day). Treatment was well tolerated and no serious AEs were detected. These results were complemented by additional analyses aiming to investigate the relationship between plasma concentrations of THC and its metabolites and tic severity. The authors demonstrated that improvement of tics was positively correlated with plasma levels of 11-hydroxy-delta-(9)-tetrahydrocannabinol (11-OH-THC), the active metabolite of THC. In a follow-up study, the same group included 24 adults with GTS (Müller-Vahl et al. 2003). Patients were treated for 6 weeks with oral THC (dose adjusted to tolerance with a maximum dose of 10 mg THC per day). Again, treatment resulted in a statistically significant reduction of tics at different time points and was well tolerated.

Recently, the same group published the results of the largest RCT using CBM in GTS: in the CANNA-TICS trial efficacy and safety of the oromucosal spray nabiximols was tested (Müller-Vahl et al. 2023). In this study, 97 participants were included and randomized to nabiximols (average dose = 19.5 mg THC, 19.5 mg CBD) or placebo at a 2:1 ratio. The primary endpoint defined as a tic reduction of ≥25% according to the Total Tic Score (TTS) of the YGTSS after 13 weeks of treatment, was narrowly missed. However, a much larger number of patients in the nabiximols group (14/64, 21.9%) compared to the placebo group (3/33, 9.1%) met the responder criterion. In line with this, secondary endpoints also showed a tic reduction. Subgroup analyses suggested a better tic reduction in males, in patients with more severe tics as well as in those with comorbid ADHD. In a substudy of the CANNA-TICS trial, the authors in addition explored participants' ability to drive under the influence of nabiximols (Müller-Vahl et al. 2024). In this study, 64 patients were enrolled. Nabiximols had no negative impact on driving skills. Remarkably, nabiximols even improved fitness to drive in a subset of patients who were unfit to drive before start of treatment, but fit to drive after treatment with nabiximols.

Another RCT investigating the effects of acute CBMs administration in GTS was published in 2023 by Abi-Jaoude et al. (2023). In this double-blind, crossover study, each participant received a single 0.25 g dose of vaporized MC of 10% THC, 9%/9% THC/CBD, 13% CBD, and placebo at 2-week intervals. Altogether, nine adults were included. There was no statistically significant impact of MC on tic severity, but significant effects especially of 10% THC and to a lesser extent of 9%9% THC/CBD – but not of CBD – on premonitory urges. Plasma THC and its metabolites correlated with improvements on all measures.

In 2023, another RCT was published by Mosley et al. (2023). This was a double-blind, crossover trial in which adults with severe GTS received escalating doses of an oral oil (1 mL) containing 5 mg/mL of THC and 5 mg/mL of CBD followed by a 6-week of placebo or vice versa followed by a 4-week washout period. Apart from the assessment of tics, psychiatric comorbidities were also evaluated and a cognitive battery was used aiming to check the impact of CBMs. Altogether, 22 patients were enrolled in this study. Treatment with CBMs resulted in a significant tic reduction according to YGTSS. Moreover, the authors found a significant association between plasma 11-OH-THC and the primary outcome. The most common side effect in the active group were memory problems.

Finally, in two RCTs efficacy and safety of an endocannabinoid modulator was tested on tics, premonitory urges, and psychiatric comorbidities (Müller-Vahl et al. 2021, 2022a). In both studies, Lu AG06466 (previously called ABX-1431), a highly selective inhibitor of monoacylglycerol lipase (MAGL), the most important enzyme involved in the degradation of the endocannabinoid 2-arachidonoylglycerol (2-AG), was administered. The pilot study, which included a single-site and 20 adults with GTS, found that a single dose of Lu AG06466 resulted in a significant reduction of tics and premonitory urges (Müller-Vahl et al. 2022a). In the follow-up multi-center study including 49 patients, treatment with Lu AG06466 over 12 weeks had no beneficial effects (Müller-Vahl et al. 2021). However, in a small subgroup of patients with comorbid ADHD, Lu AG06466 had a positive effect on ADHD symptoms although this effect was not statistically significant. Overall, in both studies, Lu AG06466 was well tolerated and no serious AEs were detected.

2.1.5 Summary

There is moderate to good evidence suggesting that CBMs are effective in the treatment of tics in patients with GTS. Based on current data, it can be assumed that both oral and inhaled THC-containing CBMs including pure THC, cannabis extracts such as nabiximols, and MC are similarly effective, while pure CBD seem to be ineffective in reducing tics. Overall, CBM are well tolerated in this group of patients, but side effects may occur. Although up to now no studies have been performed investigating long-term effects on tics, based on clinical experience it can be assumed that treatment effects last over years. Accordingly, CBMs are recommended in the guidelines of the European Society for the Study of Tourette Syndrome ESSTS (Müller-Vahl et al. 2022b) and the guidelines of the American Academy of Neurology (Pringsheim et al. 2019) in otherwise treatment refractory patients.

2.2 *Parkinson's Disease*

The second most investigated movement disorder with respect to the efficacy of CBMs is Parkinson's disease (PD) (Figura et al. 2022). It is widely known that PD symptoms are not solely restricted to motor abnormalities; they also manifest with a variety of non-motor symptoms such as pain, urinary problems, constipation, orthostatic hypotension, and psychiatric problems. Psychiatric symptoms include depression, cognitive problems, and hallucinations and often require administration of therapies with high abuse liability in conjunction with levodopa.

2.2.1 Data from Retrospective Reports on Self-Medication and Use of Different CBMs

The vast majority of previously published studies in PD using CBMs are based on surveys in which patients report their own experiences (Lotan et al. 2014; Yenilmez et al. 2021; Erga et al. 2022). The largest of these survey-based studies was published in 2022 and included 1,881 respondents with PD (Holden et al. 2022). Of these, 73% reported that they use CBMs due to medical reasons. Remarkably, 31% of patients did not inform their doctor about this kind of (self-)treatment. Interestingly, 87% of participants were aware of the type of CBMs they were taking. While the majority (55%) used CBD-dominant CBMs, 30% used THC-dominant products and another 15% preferred balanced products containing equal amounts of THC and CBD. In this patient group, the most common mode of administration was oral. The most frequently reported beneficial effects were improvements in non-motor symptoms including pain, anxiety, agitation, and disrupted sleep. The profile of side effects was quite typical and similar to other patient groups and included dry mouth, dizziness, and cognitive problems.

2.2.2 Data from Randomized Controlled Trials

Effects on Motor Symptoms

In the following, we want to describe the most important RCTs on CBMs in PD in more detail, specifically as they relate to motor symptoms associated with PD. In 2014, Chagas et al. (2014b) investigated the effects of oral CBD on motor symptoms in PD. In this study, 21 patients with PD without dementia and comorbid psychiatric disorders were included. Patients were randomized to placebo or two different doses of CBD (75 mg/day or 300 mg/day). Apart from the motor symptoms as measured per Unified Parkinson's Disease Rating Scale (UPDRS), quality of life and possible neuroprotective effects were also assessed through the levels of brain-derived neurotrophic factor (BDNF) and magnetic resonance spectroscopy (MRS). While there were no differences between groups in the UPDRS, plasma BDNF levels, and MRS measures, there was a statistically significant difference between the placebo and the high dose (300 mg/day) CBD group regarding quality of life.

In 2024, the largest study was published by Liu et al. (2024) including 61 patients. In this RCT, effectiveness of an orally administered CBD-dominant cannabis extract was investigated. Participants were randomized to CBD/THC ($n = 31$) or placebo ($n = 30$). Mean final dose (CBD/THC group) was 191.8 ± 48.9 mg CBD and 6.4 ± 1.6 mg THC daily. However, the cannabis extract was not superior compared to placebo in the treatment of PD motor symptoms.

In another double-blind, placebo-controlled phase 2/3 study with 18 patients, oral CBD had no positive effects on restless legs syndrome (RLS) in patients with PD (De Almeida et al. 2023). Six patients were randomized to the CBD group in doses of 75–300 mg, and 12 received placebo capsules. They were followed up for 14 weeks.

In an RCT from Thailand (Kanjanarangsichai et al. 2022), the effectiveness of an oral CBD-enriched cannabis extraction product (CBDEP) on motor and non-motor symptoms as well liver and kidney metabolism was assessed. Forty patients were assigned either to the CBDEP or placebo group. All patients were prescribed a low initial dose with gradual titration within 2 weeks based on individual response – including side effects – followed by 6 weeks of stable dosing. The CBDEP group received mean CBD and THC dosages of 15.59 ± 5.04 mg/day and 0.61 ± 0.19 mg/day, respectively. No effect on disease progression, motor and non-motor symptoms was demonstrated. However, CBDEP was safe and improved the levels of blood urea nitrogen, serum albumin, serum globulin, and albumin/globulin ratio in patients with PD.

Effects on Non-motor Symptoms

Several studies have also investigated the effects of CBMs on non-motor symptoms in PD and showed some promising results (Lotan et al. 2014; Shohet et al. 2017; Peball et al. 2020; Yenilmez et al. 2021). The first RCT was released in 2020 by the group from the Medical University of Innsbruck (Peball et al. 2020). In this phase 2 placebo-controlled, double-blind, parallel-group, enriched enrollment randomized withdrawal trial, the synthetic THC analog nabilone was orally administered to 47 patients with stable motor PD, but presented with challenging non-motor symptoms including anxiety and disrupted sleep. Patients underwent an open-label oral nabilone titration. They started with a dose of 0.25 mg in the evening and concluded with a twice daily dosing. Nabilone was titrated in 0.25 mg-increments every one to four days. Dose adjustments were performed until patients met the responder criterion defined as a patient-based rating of their non-motor symptoms (NMS) as "much improved" or "very much improved" on the 7-point Clinical Global Impression of Improvement Scale (CGI-I). Subsequently those patients who did respond to the treatment ($n = 19$) were randomized into a placebo or intervention group (median nabilone dose = 0.75 mg) at a 1:1 ratio of 4 weeks. The primary efficacy criterion was the change of the UPDRS. The authors found that nabilone is effective for the treatment of some non-motor symptoms, mainly anxiety and sleeping problems.

The acute effects of oral CBD administration (300 mg) on anxiety and tremor in PD patients were investigated in Brazil in 2020. Anxiety and tremor were induced by the Simulated Public Speaking Test (SPST) in 24 individuals (de Faria et al. 2020). Acute CBD administration resulted not only in decreased anxiety, but also in an improvement in the tremor amplitude in the anxiogenic situation. Similar results were found in several observational studies (Chagas et al. 2014a; Lotan et al. 2014; Shohet et al. 2017; Balash et al. 2017; Figura et al. 2022; Holden et al. 2022).

In 2023, Di Luca et al. (2023) performed a phase 1 RCT investigating the effect of three oral formulations of THC/CBD (18:0, 10:10, and 1:20) in eight patients with PD specifically on pain. Study participants were assessed at days 14 (mid-titration), 28 (end of titration/start of maintenance), 35 (end of maintenance), and by phone call every 3 days to assess drug compliance and occurrence of AE. The mean maximal tolerated doses (MTD) were numerically similar across the three groups: 0.8 mL/day with 10:10 formulation (7.84 mg of THC and 7.92 mg of CBD), 0.9 mL/day for 18:0 (17.01 mg of THC and 0.18 mg of CBD), and 0.9 mL/day for 1:20 (0.9 mg of THC and 18 mg of CBD).

Another intriguing premise was tested in the recent trial by Weber et al. (2024). This was a trial of oral CBD used for the treatment of a common comorbid dermatological condition in PD, seborrheic dermatitis (SD). The study was based on recent reports on sebum reduction after CBD administration. Study participants were assessed in the clinic at days 14, 28, 35, and by phone call every 3 days. Medication compliance was also assessed by reconciling/weighing remaining cannabis oil at each visit. CBD blood levels were not evaluated. The initial dose was 0.1 mL/day followed by an increase by 0.1 mL/day every 3 days up to maximum of 1.0 mL/day while the oil formulation was 100 mg/mL CBD and 3.3 mg/mL THC. During the titration phase, if participants did not tolerate a dose increase, then they entered the maintenance phase at the corresponding MTD dose and remained on the MTD for a minimum of 1 week. A total of 27 patients received placebo while 16 were randomized to the CBD group. The within-group pre-post change was not statistically significant for either group, but they differed from each other because there was an estimated improvement for the CBD group and an estimated worsening for the placebo group.

Effects on Levodopa-Induced Dyskinesia

Two studies have investigated the effects of CBMs on levodopa-induced dyskinesia (LID). In 2001, Sieradzan et al. (2001) performed a double-blind, randomized, placebo-controlled, crossover pilot study including nine patients with idiopathic PD and found that the THC analog nabilone reduces LID significantly according to the Rush Dyskinesia Disability Scale (RDRS). Nabilone (0.03 mg/kg to nearest whole milligram) or placebo was administered in two split doses 12 h and 1 h before levodopa administration. In addition, the authors assessed in an in vitro assay whether nabilone reduces GABA reuptake in the globus pallidus internus (GPi) and through this mechanism may improve LID. Nabilone has been shown to decrease reuptake of [3 H]-GABA. Interestingly, nabilone did not have impact on antiparkinsonian effect of levodopa in the patient sample.

Three years later, in a 4-week dose-escalation trial including 17 patients with PD, Carroll et al. (2004) investigated the effect of Cannador®, an ethanolic extract of *Cannabis sativa* standardized to 2.5 mg of THC and 1.25 mg of CBD per capsule. The primary outcome was improvement of dyskinesias according to the UPDRS dyskinesia score, while secondary outcome measures included the RDRS as well as several other examiner and self-assessments on dyskinesia and activities of daily living following a levodopa challenge. While Cannador® was well tolerated, it had no pro- or antiparkinsonian effect. There was also no effect on LID.

Effects on Cognition and Behavior

Regarding the safety profile, one of the main considerations is the impact of CBMs on cognition and psychotic symptoms that quite frequently occur in PD, especially in advanced stages. This matter was investigated by Domen et al. (2023). The authors investigated cognitive safety data from the RCT of oral CBD/THC administered for 16.3 days with dosage escalation to twice per day. Participants started at 1.25 mg/kg/day daily for 4 $\pm$ 1 days, increased to 1.25 mg/kg/day twice daily (minimum 6 h apart) for 14 $\pm$ 4 days, and adjusted as needed for tolerability. Results suggested that CBMs had a small detrimental impact on cognitive performance. In a sub-analysis of the trial by Ellmerer et al. (2022) the authors compared saccadic paradigms, a marker of cognitive performance, between patients randomized to placebo (0 mg BID) and nabilone (1 mg BID); no significant differences between the groups were found indicating that nabilone does not worsen performance providing further support of its safety.

Neuroprotective Effects

Most recent developments focused on the possibility of neuroprotective effects that CBMs may have on the disease progression in PD. However, these studies are limited to preclinical work (Jiang et al. 2024).

2.2.3 Results Based on Systematic Reviews and Meta-Analyses

In the first meta-analysis on CBMs in PD published by Thanabalasingam et al. (2021) in 2021 altogether 15 studies were included, six of which were I RCTs (Sieradzan et al. 2001; Mesnage et al. 2004; Carroll et al. 2004; Chagas et al. 2014b; de Faria et al. 2020; Peball et al. 2020) while the rest were observational studies (Venderová et al. 2004; Zuardi et al. 2009; Lotan et al. 2014; Finseth et al. 2015; Shohet et al. 2017; Kindred et al. 2017; Balash et al. 2017; Micheli et al. 2020; Yenilmez et al. 2021). In 12/15 (80%) trials, CBM was used alongside with antiparkinsonian medication, most often levodopa. According to this review, observational studies that did not include any comparator showed efficacy of CBMs. In their meta-analysis, the authors included three RCTs that used an active comparator (Mesnage et al. 2004; Chagas et al. 2014b; de Faria et al. 2020; Peball et al. 2020), However, in this meta-analysis, the authors combined three RCTs that used completely different CBMs that should not be compared with each other due to their different mechanisms of action. While Mesnage et al. (2004) used the cannabinoid receptor *antagonist* SR 141716, Chagas et al. (2014b) performed a study using pure CBD, and finally, Peball et al. (2020) investigated the effectiveness of the THC analog nabilone. Thus, any conclusions drawn from this meta-analysis are not meaningful.

In another systematic review and meta-analysis of clinical trials published in 2022 by Urbi et al. (2022), 18 non-randomized trials and five RCTs were identified. However, all RCTs included small sample sizes ranging from nine to 38 patients. Of note, different CBMs had been used such as CBD, nabilone, and a cannabis extract. In summary, the authors report potential beneficial effects of both THC and CBD in patients with PD, particularly with regard to the treatment of tremor and non-motor symptoms such as anxiety, pain, sleep quality, and improvement in quality of life.

2.2.4 Summary

There is low to moderate evidence suggesting that CBMs may improve tremor and non-motor symptoms such as anxiety, pain, sleep quality, and quality of life in patients with PD. CBMs may also have a positive impact on LID in PD. However, until today it is not clear which CBMs are most effective, since a variety of CBMs have been used in studies to date including CBD, nabilone, and cannabis extracts with different THC:CBD ratios. Overall, CBMs are well tolerated with no evidence of clinically relevant cognitive decline in this group of patients.

2.3 Other Movement Disorders: Dystonia, Huntington's Disease (HD), and Restless Legs Syndrome (RLS)

There are a small number of studies investigating the use of CBMs in the treatment of other movement disorders such as tremor (Fox et al. 2004), atypical parkinsonian syndromes (Hounie and Almeida Vasques 2019), parkinsonism secondary to toxic injury of the central nervous system (Szejko et al. 2021), dystonia (Muller-Vahl et al. 1999), Huntington's disease (HD) (Saramak et al. 2022; Oikonomou and Jost 2022), ataxias (Meinck et al. 1989), RLS (Megelin and Ghorayeb 2017; Ghorayeb 2020), and myoclonus (Zöllner et al. 2019), but these studies are mainly limited to single case reports or case series, and studies with a better degree of evidence are limited to small sample sizes (Oikonomou and Jost 2022). In this section, we discuss the use of CBMs in patients with dystonia, HD, and RLS, since studies in these clinical populations are the most robust.

2.3.1 Dystonia

The current database supporting the use of CBMs for the treatment of dystonia is quite limited. The majority of available studies is limited to anecdotal cases of generalized and focal dystonia such as blepharospasm, cervical dystonia, Meige syndrome, and task-specific dystonia treated with different CBMs, mainly CBD and dronabinol (for an overview, see (Mascia et al. 2020)). In 2002, Fox et al. (2002) investigated the effect of acute nabilone administration (0.03 mg/kg) in 15 patients with generalized and segmental primary dystonia in a placebo-controlled crossover trial over a period of 2 weeks. However, no significant reduction in dystonia could be detected after treatment with nabilone. Zloto et al. (2022) investigated the effectiveness of a medical cannabis oil (3.2% THC and 0.1% CBD) in the treatment of blepharospasm in a prospective, double-blind, placebo-controlled study in six patients. The study lasted overall for a period of 12 weeks. During the last 6 weeks, the treatment group used an average of 6.27 drops and the placebo group used an average of 5.36 drops (one drop was equivalent to approximately 0.04 mL and contained approximately 1.2 mg THC.). Only patients who did not respond to treatment with botulinum toxin injection for blepharospasm were eligible. CBM treatment resulted in a significant reduction of spasm events and the mean duration of spasm attack. Similar results were found in a subsequent study from the USA. In this RCT, 12 patients were enrolled and several beneficial kinematic effects for blepharospasm were found after low dose oral CBD treatment (Silkiss et al. 2023). Participating patients remained on a 3-month BTX injection cycle and were randomized to group A (adjunctive CBD + Botox in cycle 1 followed by placebo + Botox in cycle 2) or group B (placebo + Botox in cycle 1 followed by adjunctive CBD + Botox in cycle 2). Medication was provided by Greenwich Biosciences, Inc. (Carlsbad, CA) in the form of sublingual CBD at 200 mg/day (100 mg BID) under the brand name of Epidiolex.

One large cross-sectional survey study explored the experience with CBMs in 158 patients with laryngeal dystonia and voice tremor (Millman et al. 2023). Overall, 53.8% of participants had tried CBMs at some point for the treatment of dystonia, while 52.9% were actively using CBMs as part of their regular treatment. Most patients used CBMs as additional treatment and not as standalone treatment. Overall effectiveness was rated as somewhat effective by 42.4% of patients, but 45.9% rated use of CBMs as ineffective. Effectiveness was associated with an improvement of voice strain and anxiety. In another observational study, Radke et al. (2017) reported an improvement in three of five patients with blepharospasm after treatment with CBMs. In all cases, the mode of administration was oral and consisted of capsule or tincture. The daily dose was individual in every case (2.5 mg THC, 47.5 mg CBD capsule; 5 mg THC 5 mg CBD /1 mL tincture; 5 mg THC 5 mg CBD /1 mL tincture).

In addition, it has been speculated that beneficial effects of CBMs in dystonia may be related to a reduction in pain. In a recently published study from Israel (Anis et al. 2023), the authors interviewed 23 subjects with dystonia and asked about the use of MC for dystonia and associated pain. The most commonly used mode of administration was smoking, followed by oil and a combination of both. On average, participants were using CBMs for 2.5 ± 2.9 years and reported mean symptom improvement of 63% in the group with more widespread dystonia (generalized, hemi, and multifocal), while benefit in a group with more focal symptoms was lower and was rated to be 32% on average. There was a positive impact on pain and quality of life.

2.3.2 Huntington's Disease

Another neurological disease that is on the verge between neurodegeneration and movement disorder in which CBMs have been tested is HD. HD is a genetically determined chronic neurodegenerative disorder in which apart from motor symptoms, mainly chorea, also cognitive and psychiatric symptoms co-occur. It is widely known that even in the early stages of HD there is a reduction of CB1 receptors in grey matter which has an obvious impact on the neurotransmission in the ECS suggesting a dysbalance in the ECS in patients with HD (Glass et al. 2000; Van Laere et al. 2010). In humans, a reduction in CB1 receptor binding has been shown in the prefrontal cortex using positron emission tomography (PET) (Ceccarini et al. 2019). In postmortem studies on brain tissues of individuals with HD – but also in PET imaging studies – in addition a reduction of CB1 receptors in the basal ganglia has been demonstrated (Glass et al. 2000; Pazos et al. 2008; Van Laere et al. 2010).

First reports on preclinical uses of CBMs in HD showed conflicting results (Lastres-Becker et al. 2003a, b; Sagredo et al. 2009; Valdeolivas et al. 2012). A first RCT was performed in 1991 by Consroe et al. (1991) who investigated efficacy and safety of oral CBD in a small group of 15 neuroleptic-free patients with HD. The effects of oral CBD (10 mg/kg/day for 6 weeks) and placebo (sesame oil for 6 weeks) were ascertained weekly under a double-blind, randomized crossover design. The primary outcome was motor improvement as per Marsden and Quinn Chorea Severity Scale (Mestre 2017). There was no statistically significant improvement of motor symptoms, but also no difference between the control and intervention groups in terms of the profile of side effects. Several years later, Curtis et al. (2009) published the results of an RCT using nabilone. In this trial, 37 patients were included and treated with 1 or 2 mg nabilone daily for 5 weeks. The treatment effect of nabilone on motor symptoms was small according to the Unified Huntington's

Disease Rating Scale (UHDRS). However, there was an improvement in two neuropsychiatric outcome measures (UHDRS behavior and neuropsychiatric inventory) after treatment with nabilone. No statistically significant differences were found between lower (1 mg) and higher (2 mg) doses of nabilone. Side effects were reported at similar rates for the placebo and nabilone groups. One patient discontinued treatment due to severe sedation caused by nabilone. In 2016, the results of another RCT conducted by Lopez-Sendon Moreno et al. (2016) were released using nabiximols in the treatment of HD. Nabiximols and placebo were randomly assigned to 25 participants who were administered up to 12 sprays/day (corresponding to 32.4 mg THC and 30 mg CBD) for 12 weeks. No differences in motor and non-motor symptoms between the placebo and treatment groups were detected.

In addition, a case series ($n = 7$) of patients with early-onset HD has been published (Saft et al. 2018). In this study, different types of CBMs were used (nabiximols, dronabinol, and nabilone). In all patients, there was an improvement of motor symptoms, mainly dystonia which is a typical clinical feature in early-onset HD. Finally, Müller-Vahl et al. (1999) published a case of significant worsening of chorea after treatment with a single dose of nabilone, while Curtis et al. (2006) described a case of a 43-year-old women whose choreatic movements, in contrast to the previous case, improved after treatment with nabilone.

2.3.3 Restless Legs Syndrome

From different surveys it is known that patients with RLS are using CBMs to treat their symptoms (Hazekamp et al. 2013; Szejko et al. 2024). However, up until now, only two small case series with six (Megelin and Ghorayeb 2017) and 12 patients (Ghorayeb 2020), respectively, have been published reporting effects of CBMs. In the first study by Megelin and Ghorayeb (2017), patients used smoked cannabis or sublingual administration of CBD. In another study by Ghorayeb, experience with smoked cannabis flowers as a self-medication was described. In both studies, no information about THC and/or CBD amount was given. Remarkably, in both of these studies, 6/6 and 8/12 patients, respectively, reported 100% improvement of symptoms while the other 3/12 patients described an improvement of 60–90%.

In a survey in patients with terminal kidney insufficiency ($n = 192$) (Samaha et al. 2020), the authors described 15 cases who used cannabis for the treatment of RLS. Of these, eight patients reported a symptom improvement. Finally, in one double-blind placebo-controlled trial ($n = 18$) (De Almeida et al. 2023), oral CBD was tested in patients with RLS in PD. The dose of CBD ranged between 75 and 300 mg. No differences between the placebo and CBD groups were detected.

2.4 *Aging, Cognition, and Dementia*

Another neurological indication from the spectrum of neurodegeneration in which CBMs have been tested are cognitive problems (Mueller and Fixen 2021; Khoury

et al. 2022). This issue is especially important since the population is getting older and the prevalence of individuals who are using CBMs in the geriatric population is increasing (Solomon et al. 2021). As misuse of cannabis may have detrimental effects on cognitive performance, the first issue that was explored in previous studies was the effect of CBMs on cognitive profile in elderly adults. So far there is only limited evidence on this topic but preliminary data suggest that CBMs do not have such detrimental effect in the older age (Yoo et al. 2020; Weinstein and Sznitman 2020; Pocuca et al. 2021). On the contrary, it seems that cannabis use in elderly could even be associated with improved brain health (Weston-Green et al. 2021). In addition, evidence from both clinical and preclinical studies demonstrated that phytocannabinoids have immunoregulatory effects and can reduce neuroinflammation (Hodges and Ashpole 2019; Baban et al. 2021; Watson et al. 2022).

Furthermore, several lines of evidence suggest an involvement of the ECS in Alzheimer's disease (AD) pathology, since it is highly expressed in the hippocampus and cortex and is involved in synaptic responsiveness and plasticity and thus, plays an essential role in brain memory and cognitive function. Postmortem analyses revealed alterations in several components of the ECS in human AD brains such as decreased levels of cannabinoid (CB) 1 and increased levels of CB2 receptors in different cortices, selective abundance of increased expression of CB2 receptors in plaques, and increased expression of endocannabinoid degradation enzymes (e.g., anandamide (AEA) and 2-AG) (Xiong and Lim 2021).

Accordingly, CBD has been suggested as a new promising compound in AD, since it operates as a negative allosteric modulator at CB1 receptors, as an inverse agonist of CB2 receptors, inhibits the G protein-coupled receptor 55, acts as an inverse agonist for other G protein-coupled orphan receptors, but also activates a wide spectrum of other non-CB receptors (such as the Transient Receptor Potential Vanilloid (TRPV) channels, serotonin receptor 5-hydroxytryptamine, peroxisome proliferator-activated (PPARγ) receptors, N-methyl-D-aspartate receptor, adenosine (A2A) receptors, α-amino-3-hydroxy-5-methyl-4-isoxazolepropionic acid receptors, opioid receptors, and dopamine receptors), is an allosteric modulator of μ- and δ-opioid receptors, and inhibits enzymatic hydrolysis and AEA uptake through its action on CB1, CB2, and TRPV1 receptors, and, thus, indirectly enhances endogenous AEA signaling by inhibiting its intercellular degradation (Peres et al. 2018; Xiong and Lim 2021). In preclinical studies, CBD exhibited complex neuroprotective, antioxidant, and anti-inflammatory properties. In AD animal models, CBD improved social recognition memory, reduced amyloid plaque deposition, and stimulated hippocampal neurogenesis (Abate et al. 2021). It has been speculated that possible neuroprotective function of CBD not only refers to various consequences of modulating the ECS, but also its influence on the regulation of multiple CB receptor-independent signaling pathways (including reduction of oxidative stress, exotoxicity, amyloid plaques, and neurofibrillary tangles formation) as well as CBD's very complex anti-inflammatory properties.

In humans, so far, there is only one small RCT ($n = 48$) available demonstrating positive effects of acute oral CBD on episodic memory in healthy individuals

(Englund et al. 2013). In addition, a limited number of clinical studies haves been performed testing the effects of CBMs other than CBD in the treatment of patients with dementia. In the recently published Cochrane systematic review (Bosnjak Kuharic et al. 2021), all articles about therapeutic use of CBMs in cognitive problems were analyzed. A sample of 126 patients from four RCTs was included in the analysis. The majority of patients had AD, vascular dementia, or multifactorial cognitive problems. The included trials tested the following CBMs: the oral THC formulation Namisol®, dronabinol, and the synthetic THC analog nabilone. Improvement of cognitive performance was not evident in any of these trials.

3 Conclusions

In summary, there is a growing interest for the use of CBMs in the treatment of movement disorders and neurodegenerative diseases. Previous studies have shown that THC-containing CBMs may be useful for the treatment of tics in patients with chronic tic disorders. Importantly, CBMs may improve not only motor and vocal tics, but also comorbid psychiatric disorders such as ADHD. In addition, CBMs may be beneficial in other movement disorders, mainly PD where they could provide improvement mainly in non-motor symptoms such as pain, anxiety, and sleeping problems. However, there is only very limited evidence for use of CBMs to treat tremor, but not for other PD motor symptoms. In patients with dystonia, especially generalized and selected types of dystonia (blepharospasm), CBMs may be used as an adjunctive treatment with botulinum toxin injections. There could also be some amelioration of pain. So far, it is unclear, whether CBMs may have beneficial effects on non-motor symptoms in HD. From limited uncontrolled case studies, it is suggested that CBMs may improve RLS. On the contrary, while preclinical studies have demonstrated that cannabinoids – and in particular CBD – may have neuroprotective effects, this has not been confirmed in the clinical research. Based on current data, it can be assumed that CBMs are safe in the elderly patients and do not have detrimental effects on cognitive functioning.

References

Abate G, Uberti D, Tambaro S (2021) Potential and limits of cannabinoids in Alzheimer's disease therapy. Biology (Basel) 10. https://doi.org/10.3390/BIOLOGY10060542

Abi-Jaoude E, Chen L, Cheung P et al (2017) Preliminary evidence on cannabis effectiveness and tolerability for adults with Tourette syndrome. J Neuropsychiatry Clin Neurosci. https://doi.org/10.1176/appi.neuropsych.16110310

Abi-Jaoude E, Bhikram T, Parveen F et al (2023) A double-blind, randomized, controlled crossover trial of Cannabis in adults with Tourette syndrome. Cannabis Cannabinoid Res. https://home.liebertpub.com/can 8:835–845. https://doi.org/10.1089/CAN.2022.0091

Anis S, Zalomek C, Korczyn AD et al (2022) Medical Cannabis for Gilles de la Tourette syndrome: an open-label prospective study. Behav Neurol. https://doi.org/10.1155/2022/5141773

Anis S, Faust-Socher A, Sverdlov D et al (2023) A single-center real-life study on the use of medical cannabis in patients with dystonia. Front Neurol 14. https://doi.org/10.3389/FNEUR.2023.1218300

Baban B, Khodadadi H, Salles ÉL et al (2021) Inflammaging and Cannabinoids. Ageing Res Rev 72. https://doi.org/10.1016/J.ARR.2021.101487

Balash Y, Bar-Lev Schleider L, Korczyn AD et al (2017) Medical Cannabis in Parkinson disease: real-life patients' experience. Clin Neuropharmacol 40:268–272. https://doi.org/10.1097/WNF.0000000000000246

Barchel D, Stolar O, Ziv-Baran T et al (2024) Use of medical cannabis in patients with Gilles de la Tourette's syndrome in a real-world setting. Cannabis Cannabinoid Res 9:293–299. https://doi.org/10.1089/CAN.2022.0112

Berding G, Müller-Vahl K, Schneider U et al (2004) [123I]AM281 single-photon emission computed tomography imaging of central cannabinoid CB1 receptors before and after Δ9-tetrahydrocannabinol therapy and whole-body scanning for assessment of radiation dose in Tourette patients. Biol Psychiatry 55:904–915. https://doi.org/10.1016/j.biopsych.2004.01.005

Bloch MH, Landeros-Weisenberger A, Johnson JA, Leckman JF (2021) A Phase-2 pilot study of a therapeutic combination of Δ9-tetrahydracannabinol and palmitoylethanolamide for adults with Tourette's syndrome. J Neuropsychiatry Clin Neurosci 33:328–336. https://doi.org/10.1176/APPI.NEUROPSYCH.19080178

Bosnjak Kuharic D, Markovic D, Brkovic T et al (2021) Cannabinoids for the treatment of dementia. Cochrane Database Syst Rev 9. https://doi.org/10.1002/14651858.CD012820.PUB2

Broers B, Bianchi F (2023) Cannabinoids for behavioral symptoms in dementia: an overview. Pharmacopsychiatry. https://doi.org/10.1055/A-2262-7837/BIB

Carroll CB, Bain PO, Teare L et al (2004) Cannabis for dyskinesia in Parkinson disease: a randomized double-blind crossover study. Neurology 63:1245–1250. https://doi.org/10.1212/01.WNL.0000140288.48796.8E

Ceccarini J, Ahmad R, Van De Vliet L et al (2019) Behavioral symptoms in Premanifest Huntington disease correlate with reduced frontal CB1R levels. J Nucl Med 60:115–121. https://doi.org/10.2967/JNUMED.118.210393

Chagas MHN, Eckeli AL, Zuardi AW et al (2014a) Cannabidiol can improve complex sleep-related behaviours associated with rapid eye movement sleep behaviour disorder in Parkinson's disease patients: a case series. J Clin Pharm Ther 39:564–566. https://doi.org/10.1111/JCPT.12179

Chagas MHN, Zuardi AW, Tumas V et al (2014b) Effects of cannabidiol in the treatment of patients with Parkinson's disease: an exploratory double-blind trial. J Psychopharmacol 28:1088–1092. https://doi.org/10.1177/0269881114550355

Consroe P, Laguna J, Allender J et al (1991) Controlled clinical trial of cannabidiol in Huntington's disease. Pharmacol Biochem Behav 40:701–708. https://doi.org/10.1016/0091-3057(91)90386-G

Curtis A, Rickards H (2006) Nabilone could treat chorea and irritability in Huntington's disease. J Neuropsychiatry Clin Neurosci 18:553–554. https://doi.org/10.1176/JNP.2006.18.4.553

Curtis A, Mitchell I, Patel S et al (2009) A pilot study using nabilone for symptomatic treatment in Huntington's disease. Mov Disord 24:2254–2259. https://doi.org/10.1002/MDS.22809

De Almeida CMO, Brito MMC, Bosaipo NB et al (2023) The effect of Cannabidiol for restless legs syndrome/Willis-Ekbom disease in Parkinson's disease patients with REM sleep behavior disorder: a post hoc exploratory analysis of phase 2/3 clinical trial. Cannabis Cannabinoid Res 8:374–378. https://doi.org/10.1089/CAN.2021.0158

de Faria SM, de Morais FD, Tumas V et al (2020) Effects of acute cannabidiol administration on anxiety and tremors induced by a simulated public speaking test in patients with Parkinson's disease. J Psychopharmacol 34:189–196. https://doi.org/10.1177/0269881119895536

Devinsky O, Jones NA, Cunningham MO et al (2024) Cannabinoid treatments in epilepsy and seizure disorders. Physiol Rev 104:591–649. https://doi.org/10.1152/PHYSREV.00049.2021

Di Luca DG, Gilmour GS, Fearon C et al (2023) A phase Ib, double blind, randomized study of cannabis oil for pain in Parkinson's disease. Mov Disord Clin Pract 10:1114–1119. https://doi.org/10.1002/MDC3.13754

Domen CH, Sillau S, Liu Y et al (2023) Cognitive safety data from a randomized, double-blind, parallel-group, placebo-controlled phase IIb study of the effects of a Cannabidiol and Δ9-tetrahydrocannabinol drug on Parkinson's disease-related motor symptoms. Mov Disord 38: 1341–1346. https://doi.org/10.1002/MDS.29447

Draper A, Stephenson MC, Jackson GM et al (2014) Increased GABA contributes to enhanced control over motor excitability in Tourette syndrome. Curr Biol 24:2343–2347. https://doi.org/10.1016/j.cub.2014.08.038

Ellmerer P, Peball M, Carbone F et al (2022) Eye tracking in patients with Parkinson's disease treated with Nabilone-results of a phase II, placebo-controlled, double-blind. Parallel-Group Pilot Study Brain Sci:12. https://doi.org/10.3390/BRAINSCI12050661

Englund A, Morrison PD, Nottage J et al (2013) Cannabidiol inhibits THC-elicited paranoid symptoms and hippocampal-dependent memory impairment. J Psychopharmacol 27:19–27. https://doi.org/10.1177/0269881112460109

Erga AH, Maple-Grødem J, Alves G (2022) Cannabis use in Parkinson's disease – a nationwide online survey study. Acta Neurol Scand 145:692–697. https://doi.org/10.1111/ANE.13602

Figura M, Koziorowski D, Sławek J (2022) Cannabis in Parkinson's disease – the patient's perspective versus clinical trials: a systematic literature review. Neurol Neurochir Pol 56:21–27. https://doi.org/10.5603/PJNNS.A2022.0004

Finseth TA, Hedeman JL, Brown RP et al (2015) Self-reported efficacy of cannabis and other complementary medicine modalities by Parkinson's disease patients in Colorado. Evid Based Complement Alternat Med 2015. https://doi.org/10.1155/2015/874849

Fox SH, Kellett M, Moore AP et al (2002) Randomised, double-blind, placebo-controlled trial to assess the potential of cannabinoid receptor stimulation in the treatment of dystonia. Mov Disord 17:145–149. https://doi.org/10.1002/MDS.1280

Fox P, Bain PG, Glickman S et al (2004) The effect of cannabis on tremor in patients with multiple sclerosis. Neurology 62:1105–1109. https://doi.org/10.1212/01.WNL.0000118203.67138.3E

Fragoso YD, Carra A, Macias MA (2020) Cannabis and multiple sclerosis. Expert Rev Neurother 20:849–854. https://doi.org/10.1080/14737175.2020.1776610

Gabrielsson L, Mattsson S, Fowler CJ (2016) Palmitoylethanolamide for the treatment of pain: pharmacokinetics, safety and efficacy. Br J Clin Pharmacol 82:932–942. https://doi.org/10.1111/BCP.13020

Gadzicki D, Müller-Vahl K, Stuhrmann M (1999) A frequent polymorphism in the coding exon of the human cannabinoid receptor (CNR1) gene. Mol Cell Probes 13:321–323. https://doi.org/10.1006/mcpr.1999.0249

Ghorayeb I (2020) More evidence of cannabis efficacy in restless legs syndrome. Sleep Breath 24: 277–279. https://doi.org/10.1007/S11325-019-01978-1

Glass M, Dragunow M, Faull RLM (2000) The pattern of neurodegeneration in Huntington's disease: a comparative study of cannabinoid, dopamine, adenosine and GABA(A) receptor alterations in the human basal ganglia in Huntington's disease. Neuroscience 97:505–519. https://doi.org/10.1016/S0306-4522(00)00008-7

Hartmann A, Worbe Y, Arnulf I (2012) Increasing histamine neurotransmission in Gilles de la Tourette syndrome. J Neurol 259:375–376

Hazekamp A, Ware MA, Muller-Vahl KR, et al (2013) The medicinal use of cannabis and cannabinoids--an international cross-sectional survey on administration forms. J Psychoactive Drugs 45:199–210. https://doi.org/10.1080/02791072.2013.805976

Hemming M, Yellowlees PM (1993) Effective treatment of Tourette's syndrome with marijuana. J Psychopharmacol 7:389–391. https://doi.org/10.1177/026988119300700411

Hidding U, Mainka T, Buhmann C (2024) Therapeutic use of medical Cannabis in neurological diseases: a clinical update. J Neural Transm (Vienna) 131:117–126. https://doi.org/10.1007/S00702-023-02719-1

Hodges EL, Ashpole NM (2019) Aging circadian rhythms and cannabinoids. Neurobiol Aging 79: 110–118. https://doi.org/10.1016/J.NEUROBIOLAGING.2019.03.008

Holden SK, Domen CH, Sillau S et al (2022) Higher risk, higher reward? Self-reported effects of real-world Cannabis use in Parkinson's disease. Mov Disord Clin Pract 9:340–350. https://doi.org/10.1002/MDC3.13414

Hounie AG, Almeida Vasques MA (2019) Neurological improvement with medical Cannabis in a progressive Supranuclear Palsy patient: a case report. Med Cannabis Cannabinoids 2:65–68. https://doi.org/10.1159/000503864

Jiang Y-K, Mmeng L, Syi W et al (2024) Protective effect of phenylpropionamides in the seed of Cannabis Sativa L. on Parkinson's disease through autophagy. Fitoterapia 175:10.1016/J.FITOTE.2024.105883

Kanaan AS, Gerasch S, García-García I et al (2017) Pathological glutamatergic neurotransmission in Gilles de la Tourette syndrome. Brain 140:218–234. https://doi.org/10.1093/BRAIN/AWW285

Kanjanarangsichai A, Mitarnun W, Mitarnun W et al (2022) Cannabidiol-enriched cannabis extraction product in Parkinson's disease: a randomized, double-blind, and placebo-controlled trial in Buriram Hospital. J Neurosci Rural Pract 13:663–668. https://doi.org/10.25259/JNRP-2022-6-19

Khoury R, Maliha P, Ibrahim R (2022) Cannabis use and misuse in older adults. Clin Geriatr Med 38:67–83. https://doi.org/10.1016/J.CGER.2021.07.003

Kindred JH, Li K, Ketelhut NB et al (2017) Cannabis use in people with Parkinson's disease and multiple sclerosis: a web-based investigation. Complement Ther Med 33:99–104. https://doi.org/10.1016/J.CTIM.2017.07.002

Kluger BM, Huang AP, Miyasaki JM (2022) Cannabinoids in movement disorders. Parkinsonism Relat Disord 102:124–130. https://doi.org/10.1016/J.PARKRELDIS.2022.08.014

Lastres-Becker I, Bizat N, Boyer F et al (2003a) Effects of cannabinoids in the rat model of Huntington's disease generated by an intrastriatal injection of malonate. Neuroreport 14:813–816. https://doi.org/10.1097/00001756-200305060-00007

Lastres-Becker I, De Miguel R, De Petrocellis L et al (2003b) Compounds acting at the endocannabinoid and/or endovanilloid systems reduce hyperkinesia in a rat model of Huntington's disease. J Neurochem 84:1097–1109. https://doi.org/10.1046/j.1471-4159.2003.01595.x

Liu Y, Bainbridge J, Sillau S et al (2024) Short-term Cannabidiol with Δ-9-tetrahydrocannabinol in Parkinson's disease: a randomized trial. Mov Disord. https://doi.org/10.1002/MDS.29768

López-Sendón Moreno JL, García Caldentey J, Trigo Cubillo P et al (2016) A double-blind, randomized, cross-over, placebo-controlled, pilot trial with Sativex in Huntington's disease. J Neurol 263:1390–1400. https://doi.org/10.1007/S00415-016-8145-9

Lotan I, Treves TA, Roditi Y, Djaldetti R (2014) Cannabis (medical marijuana) treatment for motor and non-motor symptoms of Parkinson disease: an open-label observational study. Clin Neuropharmacol 37:41–44. https://doi.org/10.1097/WNF.0000000000000016

Maia T v., Conceição VA (2018) Dopaminergic disturbances in Tourette syndrome: an integrative account. Biol Psychiatry 84:332–344. https://doi.org/10.1016/J.BIOPSYCH.2018.02.1172

Mascia MM, Carmagnini D, Defazio G (2020) Cannabinoids and dystonia: an issue yet to be defined. Neurol Sci 41:783–787. https://doi.org/10.1007/S10072-019-04196-5

Megelin T, Ghorayeb I (2017) Cannabis for restless legs syndrome: a report of six patients. Sleep Med 36:182–183. https://doi.org/10.1016/J.SLEEP.2017.04.019

Meinck HM, Schönle PW, Conrad B (1989) Effect of cannabinoids on spasticity and ataxia in multiple sclerosis. J Neurol 236:120–122. https://doi.org/10.1007/BF00314410

Mesnage V, Houeto JL, Bonnet AM et al (2004) Neurokinin B, neurotensin, and cannabinoid receptor antagonists and Parkinson disease. Clin Neuropharmacol 27:108–110. https://doi.org/10.1097/00002826-200405000-00003

Mestre TA, Joao Forjaz M, Mahlknecht P et al (2017) Rating scales for motor symptoms and signs in Huntington's disease: critique and recommendations. https://doi.org/10.1002/mdc3.12571

Micheli FE, Groppo J, Contartese ML et al (2020) Cannabis in patients with Parkinson's disease in Argentina. A cross sectional study. Parkinsonism Relat Disord 78:66–67. https://doi.org/10.1016/J.PARKRELDIS.2020.06.033

Millman N, van der Woerd B, Sund LT, Johns M (2023) Cannabinoid use in the treatment of laryngeal dystonia and vocal tremor: a pilot investigation. J Voice 0. https://doi.org/10.1016/j.jvoice.2023.05.006

Milosev LM, Psathakis N, Szejko N et al (2019) Treatment of Gilles de la Tourette syndrome with cannabis-based medicine: results from a retrospective analysis and online survey. Cannabis Cannabinoid Res 4. https://doi.org/10.1089/can.2018.0050

Mosley PE, Webb L, Suraev A et al (2023) Tetrahydrocannabinol and Cannabidiol in Tourette syndrome. NEJM Evidence 2. https://doi.org/10.1056/EVIDOA2300012

Mueller A, Fixen DR (2021) Use of cannabis for agitation in patients with dementia. Sr Care Pharm 35:312–317. https://doi.org/10.4140/TCP.N.2020.312

Muller-Vahl KRK (1998) Cannabinoids: possible role in patho-physiology and therapy of Gilles de la Tourette syndrome. Acta Psychiatr Scand 98:502–506

Muller-Vahl KR, Kolbe H, Schneider U, Emrich HM (1999) Cannabis in movement disorders. Forsch Komplementmed 6(Suppl 3):23–27. https://doi.org/10.1159/000057153

Müller-Vahl KR, Schneider U, Emrich HM (1999) Nabilone increases choreatic movements in Huntington's disease. Mov Disord 14:1038–1040. https://doi.org/10.1002/1531-8257(199911)14:6<1038::aid-mds1024>3.0.co;2-7

Müller-Vahl KR, Schneider U, Koblenz A et al (2002) Treatment of Tourette's syndrome with Δ9-tetrahydrocannabinol (THC): a randomized crossover trial. Pharmacopsychiatry 35:57–61. https://doi.org/10.1055/S-2002-25028/ID/24/BIB

Müller-Vahl KR, Schneider U, Prevedel H et al (2003) Delta 9-tetrahydrocannabinol (THC) is effective in the treatment of tics in Tourette syndrome: a 6-week randomized trial. J Clin Psychiatry 64:459–465. https://doi.org/10.4088/JCP.V64N0417

Müller-Vahl KR, Szejko N, Wilke F et al (2019) Serotonin transporter binding is increased in Tourette syndrome with obsessive compulsive disorder. Sci Rep 9. https://doi.org/10.1038/S41598-018-37710-4

Müller-Vahl KR, Bindila L, Lutz B et al (2020) Cerebrospinal fluid endocannabinoid levels in Gilles de la Tourette syndrome. Neuropsychopharmacology 45:1323–1329. https://doi.org/10.1038/s41386-020-0671-6

Müller-Vahl KR, Fremer C, Beals C et al (2021) Monoacylglycerol lipase inhibition in Tourette syndrome: a 12-week, randomized, controlled study. Mov Disord 36:2413–2418. https://doi.org/10.1002/MDS.28681

Müller-Vahl KR, Fremer C, Beals C et al (2022a) Endocannabinoid modulation using Monoacylglycerol lipase inhibition in Tourette syndrome: a phase 1 randomized, placebo-controlled study. Pharmacopsychiatry 55:148–156. https://doi.org/10.1055/A-1675-3494

Müller-Vahl KR, Szejko N, Verdellen C et al (2022b) European clinical guidelines for Tourette syndrome and other tic disorders: summary statement. Eur Child Adolesc Psychiatry 31:377–382. https://doi.org/10.1007/S00787-021-01832-4

Müller-Vahl KR, Pisarenko A, Szejko N et al (2023) CANNA-TICS: efficacy and safety of oral treatment with nabiximols in adults with chronic tic disorders – results of a prospective, multicenter, randomized, double-blind, placebo controlled, phase IIIb superiority study. Psychiatry Res:323. https://doi.org/10.1016/J.PSYCHRES.2023.115135

Müller-Vahl KR, Pisarenko A, Ringlstetter R et al (2024) The effect of Nabiximols on driving ability in adults with chronic tic disorders: results of a substudy analysis of the double-blind, randomized, placebo-controlled CANNA-TICS trial. https://home.liebertpub.com/can. https://doi.org/10.1089/CAN.2023.0114

Oikonomou P, Jost WH (2022) Randomized controlled trials on the use of cannabis-based medicines in movement disorders: a systematic review. J Neural Transm (Vienna) 129:1247–1256. https://doi.org/10.1007/S00702-022-02529-X

Pantoja-Ruiz C, Restrepo-Jimenez P, Castañeda-Cardona C et al (2022) Cannabis and pain: a scoping review. Braz J Anesthesiol 72:142–151. https://doi.org/10.1016/J.BJANE.2021.06.018
Pazos M, Sagredo O, Fernandez-Ruiz J (2008) The endocannabinoid system in Huntingtons disease. Curr Pharm Des 14:2317–2325. https://doi.org/10.2174/138161208785740108
Peball M, Krismer F, Knaus HG et al (2020) Non-motor symptoms in Parkinson's disease are reduced by Nabilone. Ann Neurol 88:712–722. https://doi.org/10.1002/ANA.25864
Peres FF, Lima AC, Hallak JEC et al (2018) Cannabidiol as a promising strategy to treat and prevent movement disorders? Front Pharmacol 9. https://doi.org/10.3389/FPHAR.2018.00482
Pocuca N, Walter TJ, Minassian A et al (2021) The effects of cannabis use on cognitive function in healthy aging: a systematic scoping review. Arch Clin Neuropsychol 36:673–685. https://doi.org/10.1093/ARCLIN/ACAA105
Pringsheim T, Okun MS, Müller-Vahl K et al (2019) Practice guideline recommendations summary: treatment of tics in people with Tourette syndrome and chronic tic disorders. Neurology. https://doi.org/10.1212/WNL.0000000000007466
Radke PM, Mokhtarzadeh A, Lee MS, Harrison AR (2017) Medical cannabis, a beneficial high in treatment of Blepharospasm? An early observation. Neuroophthalmology 41:253–258. https://doi.org/10.1080/01658107.2017.1318150
Saft C, Von Hein SM, Lucke T et al (2018) Cannabinoids for treatment of dystonia in Huntington's disease. J Huntingtons Dis 7:167–173. https://doi.org/10.3233/JHD-170283
Sagredo O, González S, Aroyo I et al (2009) Cannabinoid CB2 receptor agonists protect the striatum against malonate toxicity: relevance for Huntington's disease. Glia 57:1154–1167. https://doi.org/10.1002/glia.20838
Samaha D, Kandiah T, Zimmerman D (2020) Cannabis use for restless legs syndrome and uremic pruritus in in patients treated with maintenance dialysis: a survey. Can J Kidney Health Dis 7. https://doi.org/10.1177/2054358120954944
Sandyk R, Awerbuch G (1988) Marijuana and Tourette's syndrome. J Clin Psychopharmacol 8: 444–445. https://doi.org/10.1097/00004714-198812000-00021
Saramak K, Szejko N, Saramak K, Szejko N (2022) Endocannabinoid system as a new therapeutic avenue for the treatment of Huntington's disease. From pathophysiology to treatment of Huntington's disease. https://doi.org/10.5772/INTECHOPEN.104566
Shohet A, Khlebtovsky A, Roizen N et al (2017) Effect of medical cannabis on thermal quantitative measurements of pain in patients with Parkinson's disease. Eur J Pain 21:486–493. https://doi.org/10.1002/EJP.942
Sieradzan KA, Fox SH, Hill M et al (2001) Cannabinoids reduce levodopa-induced dyskinesia in Parkinson's disease: a pilot study. Neurology 57:2108–2111. https://doi.org/10.1212/WNL.57.11.2108
Silkiss RZ, Koppinger J, Truong T et al (2023) Cannabidiol as an adjunct to botulinum toxin in Blepharospasm – a randomized pilot study. Transl Vis Sci Technol 12:17–17. https://doi.org/10.1167/TVST.12.8.17
Solomon HV, Greenstein AP, Delisi LE (2021) Cannabis use in older adults: a perspective. Harv Rev Psychiatry 29:225–233. https://doi.org/10.1097/HRP.0000000000000289
Szejko N (2022) Update and recent progress in the neurobiology of Tourette syndrome. Int Rev Mov Disord 3:131–158. https://doi.org/10.1016/BS.IRMVD.2021.11.003
Szejko N, Müller-Vahl K (2022) Cannabinoids: possible role in the pathophysiology and therapy of Gilles de la Tourette syndrome. Int Rev Mov Disord 4:269–294. https://doi.org/10.1016/BS.IRMVD.2022.04.002
Szejko N, Jakubovski E, Müller-Vahl K et al (2018) Possible role of the endocannabinoid system in Tourette syndrome. Recent Adv Cannabinoid Res. https://doi.org/10.5772/INTECHOPEN.79895
Szejko N, Fichna JP, Safranow K et al (2020) Association of a variant of CNR1 gene encoding cannabinoid receptor 1 with Gilles de la Tourette syndrome. Front Genet 11. https://doi.org/10.3389/FGENE.2020.00125

Szejko N, Burger F, Sidoroff V, Wenning GK (2021) Case report: secondary bilateral parkinsonism and dystonia treated with dronabinol. F1000Res 9. https://doi.org/10.12688/F1000RESEARCH.26476.3

Szejko N, Becher E, Heimann F et al (2024) Medicinal use of different cannabis strains: results from a large prospective survey in Germany. Pharmacopsychiatry. https://doi.org/10.1055/A-2261-2269

Thaler A, Arad S, Schleider LBL et al (2019) Single center experience with medical cannabis in Gilles de la Tourette syndrome. Parkinsonism Relat Disord 61:211–213. https://doi.org/10.1016/J.PARKRELDIS.2018.10.004

Thanabalasingam SJ, Ranjith B, Jackson R, Wijeratne DT (2021) Cannabis and its derivatives for the use of motor symptoms in Parkinson's disease: a systematic review and meta-analysis. Ther Adv Neurol Disord 14. https://doi.org/10.1177/17562864211018561

Urbi B, Corbett J, Hughes I et al (2022) Effects of cannabis in Parkinson's disease: a systematic review and meta-analysis. J Parkinsons Dis 12:495–508. https://doi.org/10.3233/JPD-212923

Valdeolivas S, Satta V, Pertwee RG et al (2012) Sativex-like combination of phytocannabinoids is neuroprotective in malonate-lesioned rats, an inflammatory model of Huntington's disease: role of CB1 and CB2 receptors. ACS Chem Nerosci 3:400–406. https://doi.org/10.1021/cn200114w

Van Laere K, Casteels C, Dhollander I et al (2010) Widespread decrease of type 1 cannabinoid receptor availability in Huntington disease in vivo. J Nucl Med 51:1413–1417. https://doi.org/10.2967/JNUMED.110.077156

Venderová K, Ružička E, Vořísšek V, Višňovský P (2004) Survey on cannabis use in Parkinson's disease: subjective improvement of motor symptoms. Mov Disord 19:1102–1106. https://doi.org/10.1002/MDS.20111

Watson KK, Bryan AD, Thayer RE et al (2022) Cannabis use and resting state functional connectivity in the aging brain. Front Aging Neurosci 14. https://doi.org/10.3389/FNAGI.2022.804890

Weber I, Zagona-Prizio C, Sivesind TE et al (2024) Oral Cannabidiol for seborrheic dermatitis in patients with Parkinson disease: randomized clinical trial. JMIR Dermatol 7:e49965. https://doi.org/10.2196/49965

Weinstein G, Sznitman SR (2020) The implications of late-life cannabis use on brain health: a mapping review and implications for future research. Ageing Res Rev 59. https://doi.org/10.1016/J.ARR.2020.101041

Weston-Green K, Clunas H, Jimenez Naranjo C (2021) A review of the potential use of Pinene and linalool as terpene-based medicines for brain health: discovering novel therapeutics in the flavours and fragrances of cannabis. Front Psych 12. https://doi.org/10.3389/FPSYT.2021.583211

Xiong Y, Lim CS (2021) Understanding the modulatory effects of Cannabidiol on Alzheimer's disease. Brain Sci 11. https://doi.org/10.3390/BRAINSCI11091211

Yenilmez F, Fründt O, Hidding U, Buhmann C (2021) Cannabis in Parkinson's disease: the patients' view. J Parkinsons Dis 11:309–321. https://doi.org/10.3233/JPD-202260

Yoo HB, DiMuzio J, Filbey FM (2020) Interaction of cannabis use and aging: from molecule to mind. J Dual Diagn 16:140–176. https://doi.org/10.1080/15504263.2019.1665218

Zloto O, Weisman A, Avisar I et al (2022) Medical cannabis oil for benign essential blepharospasm: a prospective, randomized controlled pilot study. Graefes Arch Clin Exp Ophthalmol 260:1707–1712. https://doi.org/10.1007/S00417-021-05533-1

Zöllner JP, Noda AH, Rosenow F, Strzelczyk A (2019) Improving post-hypoxic myoclonus using cannabidiol. Seizure 67:38–39. https://doi.org/10.1016/J.SEIZURE.2019.03.003

Zuardi AW, Crippa JAS, Hallak JEC et al (2009) Cannabidiol for the treatment of psychosis in Parkinson's disease. J Psychopharmacol 23:979–983. https://doi.org/10.1177/0269881108096519

Correction to: Genetic Tools in Rodents to Study Cannabinoid Functions

Krisztina Monory, Inigo Ruiz de Azua, and Beat Lutz

Correction to:
Chapter 550 in: Curr Topics Behav Neurosci
https://doi.org/10.1007/7854_2024_550

The original version of the chapter was inadvertently published without the following corrections. The chapter has now been corrected.

Correction:

The last name of the author: Inigo Ruiz de Azua is "Ruiz de Azua" instead of "de Azua". This has now been corrected in the xml files as requested by the author.

The updated version of this chapter can be found at
https://doi.org/10.1007/7854_2024_550

Curr Topics Behav Neurosci (2026) 76: 455
https://doi.org/10.1007/7854_2025_584
Published Online: 08 August 2025

GPSR Compliance
The European Union's (EU) General Product Safety Regulation (GPSR) is a set of rules that requires consumer products to be safe and our obligations to ensure this.

If you have any concerns about our products, you can contact us on

ProductSafety@springernature.com

In case Publisher is established outside the EU, the EU authorized representative is:

Springer Nature Customer Service Center GmbH
Europaplatz 3
69115 Heidelberg, Germany

www.ingramcontent.com/pod-product-compliance
Lightning Source LLC
Chambersburg PA
CBHW071923160426
43267CB00023B/22
* 9 7 8 3 0 3 2 0 9 1 3 2 1 *